S. Morand, B.R. Krasnov, R. Poulin (Eds.)

Micromammals and Macroparasites

From Evolutionary Ecology to Management

S. Morand, B.R. Krasnov, R. Poulin (Eds.)

Micromammals and Macroparasites

From Evolutionary Ecology to Management

With 79 Figures

Springer

Serge Morand, Ph.D.
Center for Biology and Management of Populations
Campus International de Baillarguet
CS 30016 34988 Montferrier-sur-Lez cedex, France

Boris R. Krasnov, Ph.D.
Mitrani Department of Desert Ecology, Jacob Blaustein Institutes for Desert Research
Ben-Gurion University of the Negev, Sede-Boqer Campus
84990 Midreshet Ben-Gurion, Israel and
Ramon Science Center, P.O. Box 194, 80600 Mizpe Ramon, Israel

Robert Poulin, Ph.D.
Professor, Department of Zoology
University of Otago
P.O. Box 56, Dunedin 9054, New Zealand

Cover illustrations: Micromammals: a lagomorph, *Ochotona daurica* (middle left); a rodent, *Gerbillus dasyurus* (middle right); an insectivore, *Suncus murinus* (bottom right; photographs taken by Georgy I. Shenbrot, Ben-Gurion University of the Negev, Israel). Macroparasites: a cestode, *Meggittina cricetomydis* (top); a nematode, *Heligmosomoides glareoli* (bottom left); photographs taken by Boyko Georgiev, Natural History Museum, UK, and Central Laboratory of General Ecology, Bulgarian Academy of Sciences, Bulgaria; capitulum of a nymph of a tick, *Haemaphysalis leporispalustris* (middle center); scanning electron micrograph taken by Lance A. Durden, Georgia Southern University, USA.

ISBN10 4-431-36024-7 Springer-Verlag Tokyo Berlin Heidelberg New York
ISBN13 978-4-431-36024-7 Springer-Verlag Tokyo Berlin Heidelberg New York

Library of Congress Control Number: 2006930402

Springer is a part of Springer Science+Business Media
springer.com

Printed in Japan

Typesetting: Camera-ready by the editors and authors
Printing and binding: Kato Bunmeisha, Japan

Printed on acid-free paper

Preface

The idea for a book reviewing current knowledge on mammals and their parasites emerged during a visit by one of us (SM) to the laboratory of another of us (BK) at the Jacob Blaustein Institutes for Desert Research, Ben-Gurion University of the Negev (Israel) in December 2004, with RP becoming associated with the project from its very beginning.

Frankly, we decided to restrict our focus to macroparasites, i.e., metazoan parasites such as helminths and arthropods. A second volume at least would be necessary to cover microparasites, i.e., viruses, bacteria, and protozoans. We also decided to restrict our scope to small (=micro) mammals, because they are the most abundant and diversified species in the order Mammalia. Moreover, most of our knowledge on the interactions between mammals and their macroparasites concerns small mammalian species, mainly rodents, but also insectivores, lagomorphs, and bats.

Our idea was to associate disciplinary fields (taxonomy, phylogenetics, physiology, genetics, ecology, evolution, conservation biology, mathematical epidemiology) that may not have enough opportunities to exchange and debate ideas. What better opportunity can there be than a book on the evolution and ecology of host–parasite interactions, and moreover, a book that focuses on and emphasizes a particular group of hosts and their parasites? A symposium on "Parasites and mammals: A macroecological perspective" (organized by BK and SM) at the 9th International Mammalogical Congress in Sapporo (Japan), held in August 2005, allowed us to finalize the project with Springer Japan.

The book is conceived for a broad audience. Students will find up-to-date reviews and state-of-the-art syntheses in several domains. We hope that they will find ideas and opportunities for new research and new applications. Senior researchers, who try to maintain themselves at the forefront of their discipline, will also be interested readers. They are forced to specialize, leaving them little time for exploring other fields, even those closely related to their interest. This volume is organized in order that they will easily find reviews, summaries, data and references. Environmental managers, veterinarians, and conservationists have to use the results of fundamental science for their daily tasks: evaluating different options to manage natural populations and habitats. They have to deal with and/or know that parasitism and diseases are important emerging problems. They need to have a clear picture of current knowledge, and the contributions in this book will prove invaluable.

The volume is divided into six parts, including a brief opening introduction explaining what micromammals and macroparasites are. The second part presents the major taxa that parasitize small mammals: helminths (trematodes, cestodes, nematodes, acanthocephalans) and arthropods (ticks, mites, lice, fleas and bat flies). We did not include dipterans that are not normally considered as parasites but as blood feeders. Besides, the main victims of dipterans are large rather than small mammals. In addition, we did not consider the chewing lice (Ischnocera, Amblycera) because they are generally understood to be commensals rather than parasites. A review of the diversity of species, life traits and life cycles, and also of the known effects of these parasites on their hosts, is provided for each of these taxa. The third part deals with some ecological and evolutionary patterns of parasite associations: parasite species diversity, host specificity, co-speciation and co-phylogeography. The fourth part explores the processes that operate in parasite associations at both higher (populations and communities) and lower (individuals) levels of biological organization. Mathematical epidemiology, community ecology, physiology (with endocrinology, metabolism and immunology) and genetics are explored. The fifth part provides practical examples or applications of ecological concepts to management purposes: conservation biology, and the ecology of human and animal health. The volume ends with a conclusion that explores the future of host–parasite interactions in the face of global change.

Serge Morand
Center for Biology and Management of Populations
Campus International de Baillarguet
France

Boris R. Krasnov
Mitrani Department of Desert Ecology
Jacob Blaustein Institutes for Desert Research
Ben-Gurion University of the Negev, Sede-Boqer Campus
and Ramon Science Center
Israel

Robert Poulin
Department of Zoology
University of Otago
New Zealand

Contents

Part VI. Conclusion

List of authors

BARNARD, CHRISTOPHER J. (CHAPTER 22)
School of Biology, University of Nottingham, Biology Building, University Park, Nottingham, NG7 2RD, UK

BEHNKE, JERZY M. (CHAPTER 22)
School of Biology, University of Nottingham, Biology Building, University Park, Nottingham, NG7 2RD, UK

BOUAMER, SALAH (CHAPTER 4)
Plant-Pathogen Interactions Division, Rothamsted International, Harpenden, Hertfordshire AL5 2JQ, UK

BRAY, RODNEY A. (CHAPTER 3)
Department of Zoology, Natural History Museum, Cromwell Road, London SW7 5BD, UK

CASANOVA, JUAN-CARLOS (CHAPTERS 5, 23)
Laboratory of Parasitology, Department of Microbiology and Parasitology, Faculty of Pharmacy, University of Barcelona, Avda. Diagonal, 08028 Barcelona, Spain

CATTADORI, ISABELLA M. (CHAPTER 18)
Center for Infectious Disease Dynamics, Department of Biology, 208 Mueller Lab, Pennsylvania State University, University Park, PA 16802, USA

CHARBONNEL, NATHALIE (CHAPTER 20)
Center for Biology and Management of Populations, Campus International de Baillarguet, CS 30016 34988 Montferrier-sur-Lez, France

CHRISTE, PHILIPPE (CHAPTER 27)
Department of Ecology and Evolution, Faculty of Biology and Medicine, University of Lausanne, Le Biophore, CH-1015 Lausanne, Switzerland

CRAIG, PHILIP S. (CHAPTER 25)
Biosciences Research Institute, School of Environment and Life Sciences, University of Salford, Salford Crescent, Manchester M5 4WT, UK

DEGEN, ALLAN A. (CHAPTER 19)
Wyler Department of Dryland Agriculture, Jacob Blaustein Institutes for Desert Research, Ben-Gurion University of the Negev, 84990 Midreshet Ben-Gurion, Israel

DICK, CARL W. (CHAPTER 11)
Department of Zoology, Field Museum of Natural History, 1400 South Lake Shore Drive, Chicago, IL 60605-249, USA

DOWLING, ASHLEY P. G. (CHAPTER 7)
Department of Entomology, University of Kentucky, S-225 Agricultural Science Center North, Lexington, Kentucky, 40546-0091, USA

DUPLANTIER, JEAN-MARC (CHAPTER 24)
Center for Biology and Management of Populations, UR 22, IRD, BP 1386, Dakar, Senegal

DURDEN, LANCE A. (CHAPTER 6)
Department of Biology, Institute of Arthropodology, and U.S. National Tick Collection, Georgia Southern University, P.O. Box 8042, Statesboro, Georgia 30460-8042, USA

FELIU, CARLOS (CHAPTER 2)
Laboratory of Parasitology, Department of Microbiology and Parasitology, Faculty of Pharmacy, University of Barcelona, Avda. Diagonal, 08028 Barcelona, Spain

FERRARI, NICOLA (CHAPTER 17)
Department of Animal Pathology, Faculty of Veterinary Medicine, University of Milan, Via Celoria 10, 20133 Milan, Italy

FONS, ROGER (CHAPTER 2)
Laboratoire Arago, University of Pierre and Marie Curie, CNRS URA 2156, F-66650 Banyuls-sur-Mer, France

GEORGIEV, BOYKO B. (CHAPTER 3)
Department of Zoology, Natural History Museum, Cromwell Road, London SW7 5BD, UK, and Central Laboratory of General Ecology, Bulgarian Academy of Sciences, 2 Gagarin Street, 1113 Sofia, Bulgaria

GIRAUDOUX, PATRICK (CHAPTER 25)
Department of Environmental Biology, UsC INRA, University of Franche-Comté, 1 place Leclerc, 25030 Besançon, France

GOÜY DE BELLOCQ, JOELLE (CHAPTER 20)
Center for Biology and Management of Populations, Campus International de Baillarguet, CS 30016 34988 Montferrier-sur-Lez, France

HAUKISALMI, VOITTO (CHAPTER 18)
Finnish Forest Research Institute, Vantaa Research Centre, Jokiniemenkuja 1, FI-01301 Vantaa, Finland

HENTTONEN, HEIKKI (CHAPTER 18)
Finnish Forest Research Institute, Vantaa Research Centre, Jokiniemenkuja 1, FI-01301 Vantaa, Finland

HUDSON, PETER J. (CHAPTER 18)
Center for Infectious Disease Dynamics, Department of Biology, 208 Mueller Lab, Pennsylvania State University, University Park, PA 16802, USA

HUGOT, JEAN-PIERRE (CHAPTERS 4, 14)
Department of Systematics and Evolution, National Museum of Natural History, 45 rue Buffon, 75005 Paris, France

KIM, KE CHUNG (CHAPTER 9)
Frost Entomological Museum, Department of Entomology, Center for Biodiversity Research, Institutes of the Environment, Pennsylvania State University, University Park, PA 16802, USA

KRASNOV, BORIS R. (CHAPTERS 1, 10, 12, 13, 28)
Marco and Louise Mitrani Department for Desert Ecology, Jacob Blaustein Institutes for Desert Research, Ben-Gurion University of the Negev, Sede Boqer Campus, 84990 Midreshet Ben-Gurion, Israel and Ramon Science Center, Ben-Gurion Blvd. 1, 80600 Mizpe Ramon, Israel

KUDRYASHOVA, NAINA I. (CHAPTER 8)
Zoological Museum, Moscow State University, 6 Bolshaya Nikitskaya str., 125009 Moscow, Russia

LEIRS, HERWIG (CHAPTER 26)
Evolutionary Biology Group, University of Antwerp, Groenenborgerlaan 171, B-2020 Antwerp, Belgium and Danish Pest Infestation Laboratory, Danish Institute of Agricultural Sciences, Department of Integrated Pest Management, Skovbrynet 14, DK-2800 Kongens Lyngby, Denmark

LITTLEWOOD, D. TIMOTHY J. (CHAPTER 3)
Department of Zoology, Natural History Museum, Cromwell Road, London SW7 5BD, UK

MAGNANOU, ELODIE (CHAPTER 16)
Laboratoire Arago, University of Pierre and Marie Curie, CNRS URA 2156, F-66650 Banyuls-sur-Mer, France

MARTIN II, LYNN B. (CHAPTER 21)
Department of Psychology and Neuroscience, Institute for Behavioral Medicine Research, Ohio State University, Columbus, OH 43210, USA

MEDVEDEV, SERGEY G. (CHAPTER 10)
Department of Parasitology, Zoological Institute, Russian Academy of Sciences, 1 Universitetskaya embankment, 199034 St Petersburg, Russia

MICHAUX, JOHAN (CHAPTER 27)
Zoogeographical Research Unit, Zoological Institute, Institute of Zoology, University of Liege, Quai Van Beneden 22, 4020 Liege, Belgium

MIQUEL, JORDI (CHAPTER 2)
Laboratory of Parasitology, Department of Microbiology and Parasitology, Faculty of Pharmacy, University of Barcelona, Avda. Diagonal, 08028 Barcelona, Spain

MORAND, SERGE (CHAPTERS 1, 4, 12, 13, 15, 16, 20, 27, 28)
Center for Biology and Management of Populations, Campus International de Baillarguet, CS 30016 34988 Montferrier-sur-Lez, France

NELSON, RANDY J. (CHAPTER 21)
Department of Psychology and Neuroscience, Institute for Behavioral Medicine Research, Ohio State University, OH 43210 Columbus, Ohio, USA

NIEBERDING, CAROLINE (CHAPTER 15)
Zoogeographical Research Unit, Zoological Institute, Institute of Zoology, University of Liege, Quai Van Beneden 22, 4020 Liege, Belgium

PATTERSON, BRUCE D. (CHAPTER 11)
Department of Zoology, Field Museum of Natural History, 1400 South Lake Shore Drive, Chicago, IL 60605-249, USA

PLEYDELL, DAVID R. J. (CHAPTER 25)
Department of Environmental Biology, UsC INRA, University of Franche-Comté, 1 place Leclerc, 25030 Besançon, France

POULIN, ROBERT (CHAPTERS 1, 12, 13, 28)
Department of Zoology, University of Otago, 340 Great King Street, P. O. Box 56, Dunedin, New Zealand

PUGLIESE, ANDREA (CHAPTER 17)
Department of Mathematics, University of Trento, Via Sommarive 14, 38050 Povo, Italy

RAOUL, FRANCIS (CHAPTER 25)
Department of Environmental Biology, UsC INRA, University of Franche-Comté, 1 place Leclerc, 25030 Besançon, France

RIBAS, ALEXSIS (CHAPTERS 5, 23)
Laboratory of Parasitology, Department of Microbiology and Parasitology, Faculty of Pharmacy, University of Barcelona, Avda. Diagonal, 08028 Barcelona, Spain

RIZZOLI, ANNAPAOLA (CHAPTER 17)
Center of Alpine Ecology, 38040 Viote del Monte Bondone, Trento, Italy

ROSÀ, ROBERTO (CHAPTER 17)
Center of Alpine Ecology, 38040 Viote del Monte Bondone, Trento, Italy

SEGOVIA, JUAN M. (CHAPTERS 2, 23)
Laboratory of Parasitology, Department of Microbiology and Parasitology, Faculty of Pharmacy, University of Barcelona, Avda. Diagonal, 08028 Barcelona, Spain

SENE, MARIAMA (CHAPTER 24)
Center for Biology and Management of Populations, UR 22, IRD, BP 1386, Dakar, Senegal

SHATROV, ANDREY B. (CHAPTER 8)
Department of Parasitology, Zoological Institute, Russian Academy of Sciences, 1 Universitetskaya embankment, 199034 St Petersburg, Russia

SINGLETON, GRANT R. (CHAPTER 26)
International Rice Research Institute, DAPO Box 7777 Metro Manila, Philippines

TORRES, JORDI (CHAPTER 2)
Laboratory of Parasitology, Department of Microbiology and Parasitology, Faculty of Pharmacy, University of Barcelona, Avda. Diagonal, 08028 Barcelona, Spain

VANISCOTTE, AMÉLIE (CHAPTER 17)
Department of Environmental Biology, UsC INRA, University of Franche-Comté, 1 place Leclerc, 25030 Besançon, France

WEIL, ZACHARY M. (CHAPTER 21)
Department of Psychology and Neuroscience, Institute for Behavioral Medicine Research, Ohio State University, Columbus, OH 43210, USA

Part I. Introduction and definitions

1 Micromammals and macroparasites: Who is who and how do they interact?

Serge Morand, Boris R. Krasnov, Robert Poulin and A. Allan Degen

1 Introductory remarks

Parasites are associated with their hosts, by definition, over both a long evolutionary term and in an ecologically transient time. Mammals and their parasites have co-interacted in a historical framework, which is revealed by cophylogenetic studies (Page 2003; Hugot et al. 2003). The interactions between micromammals and their macroparasites can be investigated in the light of history, i.e. within a phylogenetic framework. Although macroparasites are easy to define as metazoan parasites corresponding to well-defined clades, micromammals are more problematical and they necessitate a more thorough full definition (see below).

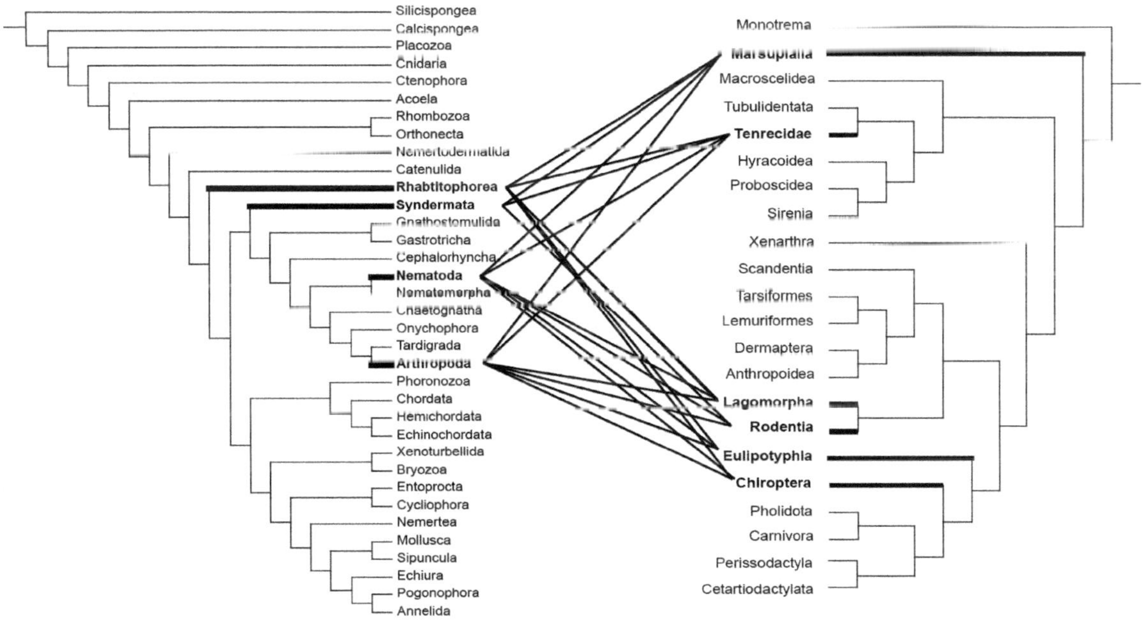

Fig. 1. Tangled trees of Metazoa (with groups including parasites in black) and the Mammalia (with groups including small-bodied forms in black)

Four major phyla of metazoans include members that parasitize micromammals (Fig. 1): the Rhabdtitophorea (cestodes and trematodes), the Syndermata (acanthocephalans), the Nematoda and the Arthropoda (fleas, lice, ticks, mites and flies). They are found as parasites of practically all micromammals. They have direct or indirect life-cycle, and mammals can be either intermediate host (such as for larval cestodes) or definitive host. Arthropod parasites are mostly ectoparasites, whereas helminths (cestodes, nematodes, trematodes and acanthocephalans) are internal. Some of these internal parasites show complex migrations within the host.

Parasites have evolved specialized adaptations to find and exploit their hosts, and these have in turn evolved mechanisms to avoid or to eliminate infections (Hart 1990; Moore 2002). The first line of defense involves behavioral activities such as grooming or avoiding potentially infected habitats or congeners. The second line of defense involves non- and specific immune responses, which can be costly in terms of energetic requirements. Costly defenses are at the basis of several physiological trade-offs.

All these reciprocal interactions between mammals and their parasites occur in within a complex network of other ecological interactions, giving them opportunities for new adaptation and even for new evolutionary outcomes.

2 Micromammals

Terrestrial mammals range in body mass from less than 2 g for the Etruscan (*Suncus etruscus*) and pygmy (*Sorex tscherskii*) shrews to more than 5 tons for the African elephant (*Loxodonta africana*). However, the frequency distribution of mammalian body masses is highly skewed, with the great majority of mammals weighing between a few grams and several kilograms (Fig. 2). In addition, there are a large number of mammals above 20 kg, but a paucity of species between 5 and 20 kg. The definition of a micromammal (=small mammal) is rather arbitrary. In their article on the energetics of small mammals, Grodzinski and Wunder (1975) restricted body mass to the range between 3 and 300 g and Happold (1984), in his article on small mammals of the Sahara, used an upper body mass of 3 kg. Heusner (1991) designated 20 kg in dividing mammals into small and large sizes. The International Biological Programme (IBP) Small Mammals Working Group decided that mammals weighing up to 5 kg are to be classified as small (Boulière 1975).

This definition will be partly adopted in the present text. This is because, using this guideline, Artiodactyla such as the 1.6 kg lesser mouse deer (*Tragulus javanicus*), 4 kg dik-diks (*Madoqua phillipsi* and *Madoqua guentheri*), duiker (*Cephalus dorsalis*) and suni (*Neotragus pygmaeus*) would be considered as small mammals, but Rodentia such as the 9 kg agouti (*Agouti paca*) and 15 kg Indian crested porcupine (*Hystrix indica*) would not.

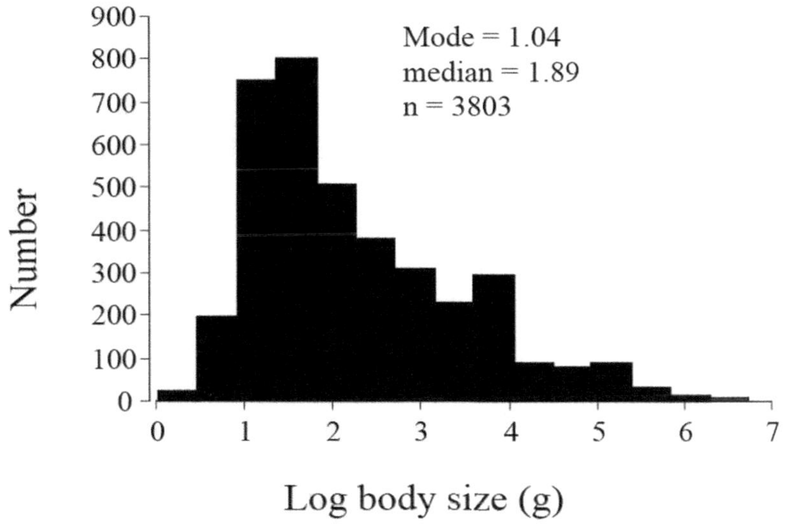

Fig. 2. Body size distribution of terrestrial mammals including Chiroptera (data from Walker's Mammals of the World, Nowak 2003). Note that half of the described mammals weight less than 100 g.

Consequently, we decided to adopt not only a purely size-related but also a taxonomic approach. Therefore, we included in our consideration mammals of the orders Rodentia, Insectivora and Chiroptera as well as most Lagomorpha and some marsupials. Together, these taxa contain more genera and species than all other orders combined. It should be noted, however, that bats (Chiroptera) differ from other micromammals in that they are "metabolically" more similar to large mammals. This, for example, is reflected in their relatively long lifespan and gestation period. Indeed, in general, bat lifespan is about 3.5 times longer than that of other mammals of comparable body sizes (Jurgens and Protero 1987; Wilkinson and South 2002). Nevertheless, bats share with other small mammals many other ecological characteristics. They are conspicuous and important components of any biota. Their populations are large and many of them inhabit large territories. As such, they represent an important element of biodiversity all over the world.

Micromammals are a major component of predator diets and perform vital ecosystem services, particularly in seed and spore dispersal and germination. Many of them are also keystone species (e. g., ecological engineers). Consequently, the existence of countless other animals and plants depends on small mammals. As a result, micromammals have to be one of the primary targets of conservation effort. On the other hand, many micromammals are aggressive agricultural pests that are responsible for huge harvest losses in many countries. They are also hosts for numerous parasite species and reservoirs for many diseases dangerous for both humans and livestock. For example, huge plague epidemics that struck Europe, Asia and Africa in the 6th, 14th, 17th and early 20th centuries with a total death toll of about 137 million victims were related to small mammals and their flea parasites.

This duality (being an important positive component of biodiversity on one hand and an important negative factor of human well-being on the other hand) is the driving force behind the intense study effort devoted to small mammals worldwide. Small mammals offer the most spectacular and explosive examples of evolutionary radiations in modern mammals and are also of interest in that light. In addition, the ubiquity of small mammals and the large sizes of their populations made them one of the favourite models for studies aimed at elucidating fundamental rules and patterns of various physiological, behavioural, ecological and evolutionary processes.

Conservation of biodiversity as well as control of animal populations is impossible without understanding the factors that govern the dynamics of populations and communities of target organisms. Parasites are one of these factors. They strongly affect the abundance and composition of populations and communities of their hosts. Understanding the relationships between micromammals and their parasites is, therefore, crucially important for our attempts to manage small mammal populations from both conservation and control points of view.

3 Macroparasites

Parasites are traditionally divided into two main groups: microparasites and macroparasites. Microparasites are primarily single-celled organisms, including viruses, bacteria and protozoans, as well as some multicellular organisms of small size such as myxozoans, that typically reproduce directly within the cells of the host. They are generally associated with disease in which transmission is direct, but can also be indirectly transmitted via alternate hosts or vectors. Macroparasites are "large" metazoan parasites, including several major taxa of endoparasitic helminths (worms) and ectoparasitic arthropods. In contrast to microparasites, macroparasites are characterized by longer generation times, and (except for some trematodes and cestodes in their intermediate hosts) by the absence of direct multiplication within the host. Thus, eggs are produced while the parasites are in or on the host, or off the host in the case of many arthropod ectoparasites, with each offspring then infecting a host different from that on which its parents lived. Immune responses elicited by macroparasites generally depend on the number of parasites present in a given host, and tend to be of relatively short duration, i.e. there is usually no long-lasting acquired immunity following an initial infection. Macroparasite infections therefore tend to be of a persistent nature, with hosts being continually reinfected (Anderson and May 1979).

All the above issues have led to a sharp increase in empirical, comparative and theoretical studies of small mammal-parasite relationships during the last two decades. Patterns and processes in small mammalian host- macroparasite systems have been documented and studied at a variety of scales, across various habitats, in different biogeographic regions and for various parasite taxa. All these efforts call for regular syntheses of original data and generalizations. The present book is an attempt to compile and generalize such data on the relationships between small mammals and their metazoan parasites.

We intentionally restrict ourselves to consideration of macroparasites only, and put aside the role of small mammals in the transmission of viral, fungal, bacterial and rickettsial infections as well as the regulating role of microparasites in populations and communities of small mammals. The reason for this is that including microparasites into our synthesis would require a great deal more space. In addition, patterns of macro- and microparasite relationships with their hosts are often strikingly different; furthermore, these two groups of parasites seem to develop along quite different evolutionary pathways.

4 A complex of dynamic interactions...

Hosts are unequal with respect to parasite infections, at the individual level, among populations, or among species. Why is that so? Even if a clear picture emerges from our existing knowledge, the pattern of parasite diversity must be confronted with ecological hypotheses. Numerous hypotheses have been proposed and we review the relative importance of host attributes that explain the large disparity in parasite species richness among and within host species. Similarly, some macroparasite species are very host specific, whereas others are not. We try to analyse the reasons and we explore the consequences of this.

Macroparasites have the potential to regulate their host populations due to their sub-lethal effects that cause reductions in host survival, host fecundity or progeny size. Population modelling is a tool that allows the investigators to better understand the potential roles of parasites in host regulation, but also to predict emergences of the microparasitic diseases they transmit (arthropod vectors such as fleas and ticks).

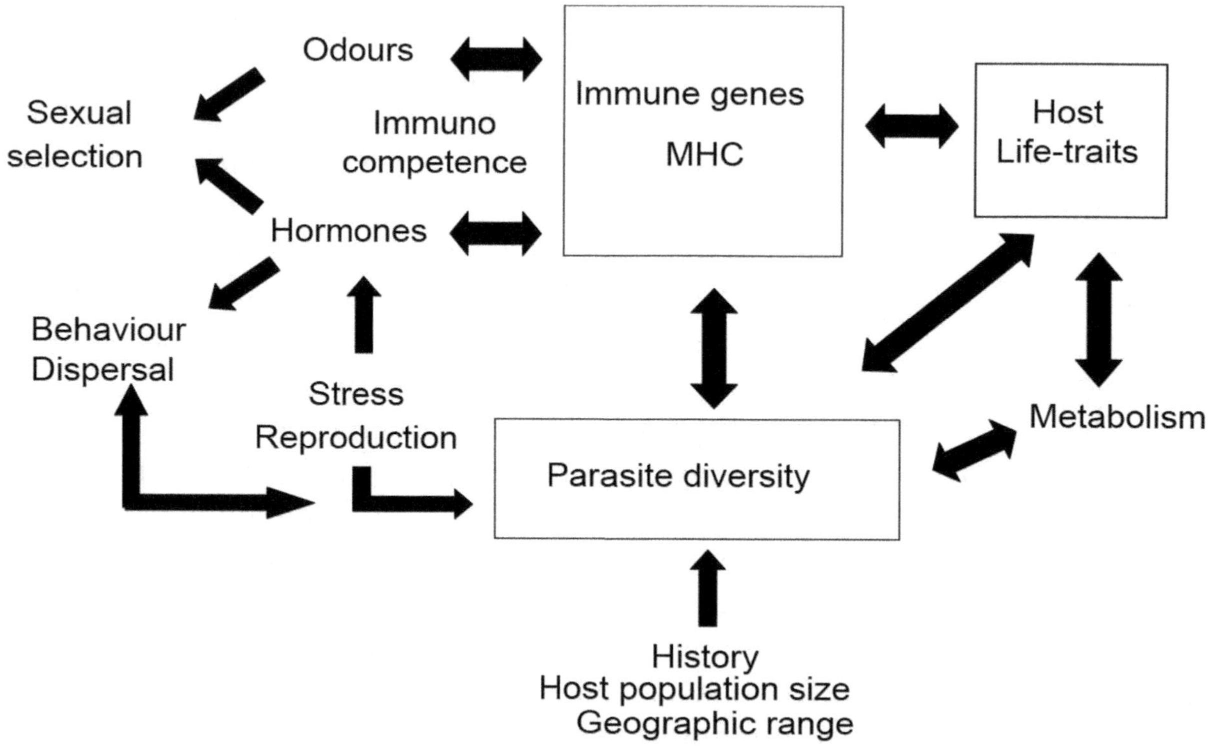

Fig. 3. Parasite diversity (parasite species richness), its determinants, and its interactions with host genetics, physiology and behaviour

Hosts can avoid parasite infection with their first line of defence, i.e. behaviour, or with the second line of defence, i.e. immune systems. Both lines of defences involve genetic background and physiological adaptation, which may be paid at the expense of other physiological functions (Fig. 3).

However, the world is full of worms, fleas and lice, and whatever its choice a host has few chances of escaping infection. The host has then to manage with the parasite, and vice versa. The detrimental effect of the infection and/or the manipulation of the host immune system may impose strong selective pressures, which may compromise many aspects of host life including behaviour or survival.

5 … with a human component

The human footprint on the earth is dramatically modifying the epidemiological environment (Daily and Erhlich 1996). Climate change, biotic invasion and landscape modification are affecting the biology of hosts and their parasites, which are displaced within and outside their geographical ranges. Parasites are becoming greater threats for biological conservation, but we show how parasites have their own roles and values and should be conserved. The alteration of the epidemiological environment increases the potential contacts between humans and parasites and pathogens of wildlife, favouring the risks of emerging zoonoses. Humans, by their outgrowing activities, affect the very nature of the host-parasite coevolutionary dynamics (Thompson 2005). The changes that affect our planet will encourage collaborations between evolutionary ecologists, epidemiologists, conservationists, physicians and veterinarians.

References

Anderson RM, May RM (1979) Population biology of infectious diseases: Part I. Nature 280:361–367

Boulière F (1975) Mammals, small and large: The ecological implications of size. In: Golley FB, Petrusewicz K, Ryszkowski L (eds) Small mammals: their productivity and population dynamics. Cambridge Univ Press, Cambridge, pp 1–8

Daily GC, Ehrlich PR (1996) Global change and human susceptibility to disease. Ann Rev Energy Environ 21:125–144

Grodzinski W, Wunder BA (1975) Ecological energetics of small mammals. In: Golley FB, Petrusewicz K, Ryszkowski L (eds) Small mammals: Their productivity and population dynamics. Cambridge Univ Press, Cambridge, pp 173–204

Happold DCD (1984) Small mammals. In: Cloudsley-Thompson JL (ed) Sahara desert: Key environments. Pergamon Press, London, pp 251–257

Hart BL (1990) Behavioral adaptations to pathogens and parasites: Five strategies. Neurosci Biobehav Rev 14:273–294

Heusner AA (1991) Size and power in mammals. J Exp Biol 160:25–54

Hugot J-P, Demanche C, Barriel V, Dei-Cas E, Guillot J 2003. Phylogenetic systematics and evolution of primate-derived *Pneumocystis* based on mitochondrial or nuclear DNA sequence comparison. Syst Biol 52:735–44

Jurgens KD, Prothero J (1987) Scaling of maximal lifespan in bats. Comp Biochem Physiol A 88:361–367

Moore J (2002) Parasites and the behavior of animals. Oxford Univ Press, Oxford

Nowak RM (2002) Walker's mammals of the world. John Hopkins Univ Press, Baltimore

Page RDM (ed) (2003) Tangled trees: Phylogeny, cospeciation, and coevolution. Chicago Univ Press, Chicago

Thompson JN (2005) The geographic mosaic of coevolution. Univ Chicago Press, Chicago

Wilkinson GS, South JM (2002) Life history, ecology and longevity of bats. Aging Cell 1:124–131

Part II. Major taxa parasitic on micromammals

2 Digenean trematodes

Carlos Feliu, Jordi Torres, Jordi Miquel, Juan Matías Segovia
and Roger Fons

1 Introductory remarks

Trematodes of the class Digenea belong to the phylum Platyhelminthes (see Gibson et al. 2002 for a recent update on taxonomy). They are parasites with complex life cycles and often use micromammals as definitive hosts. Trematodes are distributed all around the world. In Europe, for example, monographs or articles that deal with the helminthofauna of small mammals have always reported the ubiquitous presence of digenean trematodes (López-Neyra 1947; Prokopic and Genov 1974; Merkusheva and Bobkova 1981; Genov 1984; Goüy de Bellocq et al. 2002). In this chapter, we offer general information on the trematode faunas of micromammals, with a focus on parasites of small mammals of the Iberian Peninsula. Recent extensive studies of helminths in this Peninsula provide an opportunity to investigate various aspects of the ecology of these helminths.

2 Distribution of trematodes among small mammals

Available data on digeneans parasitic in small mammals are summarized in Tables 1-4. These tables have been compiled from an extensive bibliographical search (CAB Abstracts 1973-2005) with additional data from Yamaguti (1971), Gibson et al. (2002) and Jones et al. (2005), and show the records of trematode families in mammalian families belonging to four orders (Insectivora, Rodentia, Lagomorpha and Chiroptera). In general, small mammals harbour a great variety of trematodes representing a total of 37 families. Rodentia harbours the most diverse trematode fauna consisting of 30 families, while Lagomorpha have the least diverse spectrum of trematodes with seven families only.

Parasitic digeneans belonging to 19 families have been recorded from four families of Insectivora (Table 1). All of these digenean families have been reported for Soricidae. They were found mainly in the most extensively studied genera such as *Crocidura, Neomys* and *Sorex*. In contrast, among tenrecids (represented by 10 genera), only *Limnogale mergulus* was found to be parasitized by trematodes (omphalometrid *Neoglyphe polylecithos* and plagiorchiid *Plagiorchis limnogale*; Yamaguti 1971).

Table 1. Trematodes parasitic in Insectivora

	Erinaceidae	Soricidae	Talpidae	Tenrecidae
Brachylaimidae	+	+	+	
Cyathocotylidae		+	+	
Dicrocoeliidae	+	+		
Diplostomidae	+	+	+	
Echinostomatidae	+	+	+	
Heterophyidae		+	+	
Lecithodendriidae	+	+		
Microphallidae		+	+	
Nanophyetidae		+	+	
Omphalometridae		+	+	+
Opisthorchiidae		+		
Panopistidae	+	+		
Plagiorchiidae		+	+	+
Pleurogenidae		+		
Prosthogonimidae		+		
Psilostomidae		+		
Schistosomatidae	+	+		
Strigeidae		+		
Troglotrematidae		+	+	

Among Rodentia, data on trematodes are available for 13 of 29 recent families (Table 2). Of these, Muridae show the most diverse trematode fauna with 29 of the 30 digenean families recorded for Rodentia. The least diverse trematode faunas occur in four rodent families that each harbour a single digenean family. These rodent families are Chinchillidae with *Lagidium viscacia boxi* parasitized by *Fasciola hepatica* (Fasciolidae) (Led et al. 1979); Dasyproctidae with *Dasyprocta agouti* parasitized by two species of Cladorchiidae (*Cladorchis pyriformis* and *Stichorchis giganteus*) (Yamaguti 1971); Erethizontidae with *Erethizon dorsatum* parasitized by *Schistosoma douthitti* (Schistosomatidae) (Choquette et al. 1973); and Heteromyidae with *Liomys pictus* parasitized by *Brachylaima bravoae* (Brachylaimidae) (Yamaguti 1971).

Table 2. Digenean trematodes parasitic in Rodentia

	Cas	Cav	Chi	Das	Dip	Ech	Ere	Het	Hyd	Mur	Myc	Myx	Sci
Achillurbaniidae										+			
Brachylaimidae								+		+		+	+
Cladorchiidac	+	+		+					+	+	+		
Collyriclidae										+			
Dicrocoeliidae	+									+	+	+	+
Diplostomidae										+		+	+
Echinostomatidae	+									+	+		
Fasciolidae	+	+	+							+	+		
Gastrodiscidae	+									+	+		
Gymnophallidae										+			
Leucochloridiidae										+			
Leucochloridio-morphidae										+			
Heterophyidae										+			
Lecithodendriidae										+		+	+
Microphallidae										+			
Nanophyetidae										+			
Notocotylidae		+			+				+	+	+		
Nudacotylidae									+	+			
Omphalometridae										+			+
Opisthorchiidae	+									+			
Panopistidae										+		+	
Paragonimidae										+			
Paramphistomidae						+							
Plagiorchiidae	+				+					+		+	+
Prosthogonimidae										+			
Psilostomidae	+									+			
Schistosomatidae	+	+			+	+	+			+			
Strigeidae										+			
Troglotrematidae										+		+	
Zygocotylidae										+			

Cas Castoridae, *Cav* Caviidae, *Chi* Chinchillidae, *Das* Dasyproctidae, *Dip* Dipodidae, *Ech* Echimyidae, *Ere* Erethizontidae, *Het* Heteromyidae, *Hyd* Hydrochaeridac, *Mur* Muridac, *Myc* Myocastoridae, *Myx* Myoxidae, *Sci* Sciuridae.

All seven trematode families characteristic of Lagomorpha were found in Leporidae (Table 3). In contrast, only two trematode species from two families have been reported for Ochotonidae: *Dicrocoelium dendriticum* (Dicrocoeliidae) in *Ochotona alpina* (Gvozdev and Orlov 1985) and *Ochotona hyperborea* (Sakamoto et al. 1982) and *Hasstilesia ochotonae* (Hasstilesiidae) in *Ochotona rutila* (Zdarska and Soboleva 1990).

Table 3. Digenean trematodes parasitic in Lagomorpha

	Ochotonidae	Leporidae
Dicrocoeliidae	+	+
Echinostomatidae		+
Fasciolidae		+
Hasstilesiidae	+	+
Heterophyidae		+
Nudacotylidae		+
Schistosomatidae		+

Table 4. Digenean trematodes parasitic in Chiroptera

	Emb	Hip	Meg	Mol	Mor	Nat	Noc	Nyc	Phy	Pte	Rhl	Rhp	Ves
Anenterotrematidae	+			+	+	+			+				+
Brachylaimidae												+	+
Cyathocotylidae								+					
Dicrocoeliidae	+	+	+	+				+	+	+	+	+	+
Diplostomidae									+				
Echinostomatidae			+										
Heterophyidae							+						+
Lecithodendriidae	+	+	+	+	+	+	+	+	+	+	+	+	+
Mesotretidae											+		+
Microphallidae									+		+		+
Notocotylidae					+								+
Nudacotylidae									+				
Plagiorchiidae	+	+	+	+	+	+	+		+		+	+	+
Pleurogenidae													+
Rhopaliidae									+				

Emb Emballonuridae, *Hip* Hipposideridae, *Meg* Megadermatidae, *Mol* Molossidae, *Mor* Mormoopidae, *Nat* Natalidae, *Noc* Noctilionidae, *Nyc* Nycteridae, *Phy* Phyllostomidae, *Pte* Pteropodidae, *Rhl* Rhinolophidae, *Rhp* Rhinopomatidae, *Ves* Vespertilionidae.

Finally, 15 digenean families were found in 13 families of Chiroptera (Table 4). Vespertilionidae and Phyllostomidae harbour the richest digenean faunas (10 and eight parasite families, respectively). Lecithodendriidae parasitize all of the 13 chiropteran families. Plagiorchiidae and Dicrocoeliidae are also well represented (recorded in 11 and 10 chiropteran families, respectively). Representation of the remaining trematode families among chiropteran families is more restricted, including Cyathocotylidae with *Prohemistomum azimi* parasitizing *Nycteris thebaica* (Nycteridae) (Saoud and Ramadan 1977); Diplostomidae with *Neodiplostomum vaucheri* parasitizing *Chrotopterus auritus* (Phyllostomidae) (Dubois

1983); Echinostomatidae with *Echinochasmus megadermi* and *E. perfoliatus* found in *Megaderma lyra* (Megadermatidae) (Salem 1975); Nudacotylidae with *Nudacotyle* species reported from the phyllostomid genera *Carollia*, *Artibeus* and *Phyllonycteris* (Zdzitowiecki 1980; Zdzitowiecki and Rutkowska 1980; Vélez and Thatcher 1990); Pleurogenidae with *Maxbraunium baeri* parasitic on *Myotis siligorensis* (Vespertilionidae) (Yamaguti 1971); and Rhopaliidae with *Rhopalias coronatus* found in *Carollia perspicillata* (Phyllostomidae) (Marshall and Miller 1979).

3 Small mammals as experimental models for studies of Digenea

Several small mammals have been used in experiments to investigate the life cycles of parasites and to serve as models of human infections. Although the ideal scenario would be to monitor in the laboratory the entire trematode life cycle, it has not always been possible. Often, only the adult stage has been studied. In both situations, the availability of adult trematodes required experimental infection of the definitive hosts, usually rodents. However, some experimental models involved also lagomorphs and, on rarer occasions, bats and insectivores (see Yamaguti 1971 for details).

Among rodents, most experimental infections with digeneans have been carried out on Muridae, though guinea pigs (Caviidae: *Cavia porcellus*) has also been frequently used in experiments. Among murids, Murinae (*Mus* and *Rattus*) have been used for experimental infections mainly by Brachylaimidae, Echinostomatidae, Heterophyidae, Paragonimidae, Plagiorchiidae and Schistosomatidae. Other Murinae, such as *Apodemus*, *Mastomys*, *Praomys*, *Thallomys*, *Arvicantis* and *Bandicota* have been less frequently used, although they also have been experimentally infected with Fasciolidae, Brachylaimidae, Schistosomatidae and Paragonimidae. Gerbillinae of the genus *Meriones*, especially *M. unguiculatus*, have been extensively used in experiments with digeneans. Another gerbilline, *Tatera indica*, has also been experimentally infected with Fasciolidae (Sahba et al. 1972) and Schistosomatidae (Massoud 1973). Among Cricetinae, *Mesocricetus auratus* has served as experimental model mainly for investigations with Schistosomatidae, Echinostomatidae and Heterophyidae. Other cricetines (*Phodopus*, *Tscherskia* and *Cricetulus*) have also been used in experiments with digeneans of particular public health importance such as *Fasciola hepatica*, *Dicrocoelium dendriticum*, *Opistorchis felineus*, *Clonorchis sinensis* and *Schistosoma mansoni* (Gitsu and Kovalenko 1983; Zelya and Sergeeva 1987; Terasaki et al. 2003; etc.). Additional models include schistosomatids in *Cricetomys* sp. and *Saccostomus campestres* (Crycetominae); *Schistosomatium*, *Notocotylus* and *Nudacotyle* in *Microtus pennsylvanicus* and *M. montanus*; and *Euryhelmis* in *Ondatra zibethica* (Arvicolinae); as well as some schistosomatids, brachylaimids, heterophyids and opisthorchiids in Sigmodontinae (*Sigmodon*, *Holochilus*, *Holochistus*, *Nectomys*, *Zygodontomys* and *Peromyscus*) (Pitchford 1975; Zajac and Williams 1981; Gitsu and Kovalenko 1983; Kawazoe and Pinto 1983; McKown et al. 2000; Terasaki et al. 2003).

Representatives of other rodent families (Echimyidae, Dasyproctidae, Hystricidae, Myoxidae, Sciuridae and Myocastoridae) have rarely been used for experimental infections with digeneans. Nevertheless, *Myocastor coypus* (Myocastoridae) has been described as a useful experimental model for *Fasciola*, *Schistosoma*, *Paragonimus* and *Clonorchis* (Kuntz et al. 1975; Hatsushika et al. 1979). *Paragonimus skrjabini* and *Brachylaima ruminae* have been maintained in *Hystrix hodgnosi* (Hystricidae) and *Eliomys quercinus* (Myoxidae) (Yamaguti 1971; Mas-Coma and Montoliu 1986). Finally, *Proechimys* (Echimyidae), *Dasyprocta* (Dasyproctidae), *Marmota* and *Callosciurus* (Sciuridae) have been infected with some *Schistosoma* species (Chiu and Kao 1973; Anderson et al. 1991).

4 Digeneans and Iberian small mammals

The mammal fauna of the Iberian Peninsula comprises 65 species of small mammals (14 Insectivora, 21 Rodentia, four Lagomorpha and 26 Chiroptera; Palomo and Gisbert 2002). Continuous extensive studies of the helminth fauna of Iberian small mammals started in the mid 70s (beginning with insectivores and rodents, then involving bats and, finally, hares and rabbits). At present, detailed information on the helminth faunas and, in particular, on trematodes, of a substantial number of the Iberian micromammalian species is available. The information provided here has been extracted from our own studies that included detailed examination of around 7000 individual hosts (500 Insectivora, 5500 Rodentia and 1000 Lagomorpha). Data on trematodes parasitic in Chiroptera were taken from the literature.

4.1 Faunistic aspects

Forty-six species of digeneans have been found in Iberian micromammals. In fact, this number may be greater, because of some unclear and unresolved taxonomical issues (Mas-Coma and Montoliu 1986; Botella et al. 1993; Esteban et al. 1999; Gracenea and González-Moreno 2002). These 46 trematodes were recorded in 36 host species (10 Insectivora, 13 Rodentia, two Lagomorpha and 11 Chiroptera).

Table 5 shows the number of families, genera and species of trematodes that infect small mammals belonging to different orders and families, whereas numbers of digenean families and species recorded in each host species are shown in Fig. 1.

Table 5. Number of families, genera and species of trematodes recorded in Iberian small mammals

Hosts	Trematodes		
	Families	Genera	Species
INSECTIVORA			
Talpidae	4	4	4
Erinaceidae	2	2	3
Soricidae	8	9	11
RODENTIA			
Myoxidae	6	7	8
Muridae	10	12	15
LAGOMORPHA			
Leporidae	2	2	2
CHIROPTERA			
Rhinolophidae	3	5	8
Vespertilionidae	3	6	12
Molossidae	2	3	3

A high diversity of trematode species is the characteristic for Muridae and Soricidae, whereas the lowest number of trematode species has been found in Leporidae and Erinaceidae. One of the reasons for this is the high number of host species within both Muridae (17) and Soricidae (nine), and their broad distribution across the Peninsula. Among dormice (Myoxidae), trematodes have been found only in *Eliomys quercinus*, whereas no trematode has been recorded in *Glis glis*. Leporidae is the only family that does not have specific trematodes, but its members are hosts of euryxenous trematodes (*Fasciola hepatica* and *Dicrocoelium dendriticum*) that are frequently found in other mammals. Rhinolophidae and Vespertilionidae are parasitized by a relatively high number of genera and species of trematodes, although these belong to only few families. Bats are the most frequently infected by Plagiorchiidae and Lecithodendriidae (Botella et al. 1993; Esteban et al. 1990, 1992, 1999).

Each trematode family has been recorded in either one (Collyriclidae, Echinostomatidae, Fasciolidae, Mesotretidae, Nanophyetidae, Notocotylidae, Omphalometridae and Prosthogonimidae), two (Brachylaimidae, Microphallidae, Panopistidae, Plagiorchiidae, Psilostomidae, Troglotrematidae) or three orders of hosts (Dicrocoeliidae, Lecithodendriidae). The most speciose genera of digeneans found in Iberian small mammals are *Brachylaima*, *Plagiorchis* and *Prosthodendrium*, each represented by four species.

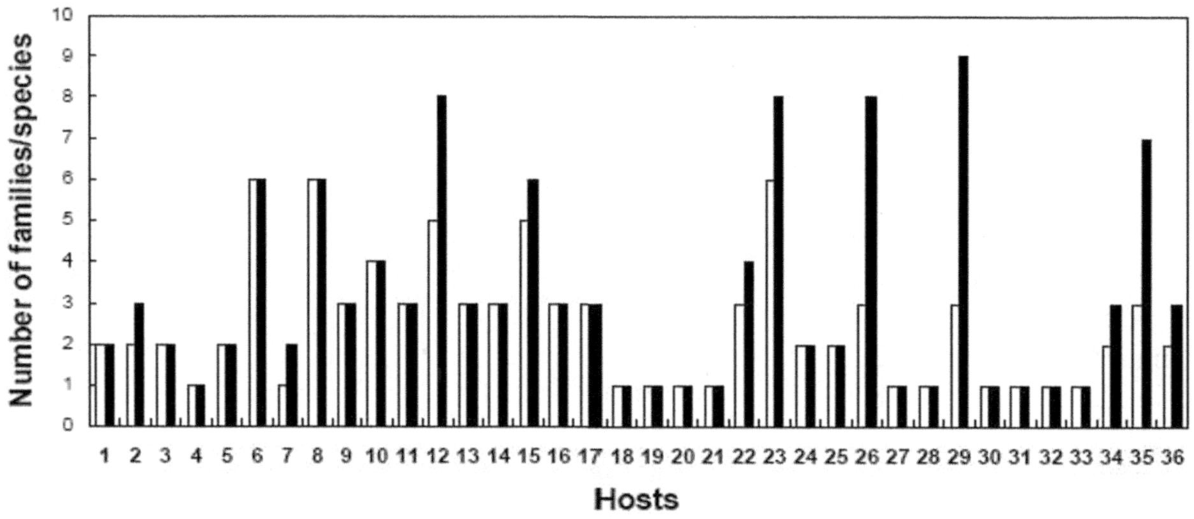

Fig. 1. Number of families and species of digenean trematodes recorded in different species of Iberian small mammals. 1 *Talpa europaea,* 2 *Talpa occidentalis,* 3 *Galemys pyrenaicus,* 4 *Erinaceus europaeus,* 5 *Atelerix algirus,* 6 *Crocidura russula,* 7 *Crocidura suaveolens,* 8 *Neomys fodiens,* 9 *Sorex minutus,* 10 *Sorex araneus,* 11 *Rattus rattus,* 12 *Rattus norvegicus,* 13 *Mus domesticus,* 14 *Mus spretus,* 15 *Apodemus sylvaticus,* 16 *Clethrionomys glareolus,* 17 *Microtus agrestis,* 18 *Microtus lusitanicus,* 19 *Microtus arvalis,* 20 *Microtus cabrerae,* 21 *Microtus duodecimcostatus,* 22 *Arvicola sapidus,* 23 *Eliomys quercinus,* 24 *Oryctolagus cuniculus,* 25 *Lepus granatensis,* 26 *Rhinolophus ferrumequinum,* 27 *Rhinolophus hipposideros,* 28 *Rhinolophus mehelyi,* 29 *Pipistrellus pipistrellus,* 30 *Myotis daubentonii,* 31 *Myotys nattereri,* 32 *Myotis myotis,* 33 *Eptesicus serotinus,* 34 *Plecotus auritus,* 35 *Miniopterus schreibersii,* 36 *Tadarida teniotis*

Among host species, the highest diversity of trematodes is characteristic of *Pipistrellus pipistrellus* (nine species), *Rattus norvegicus*, *Eliomys quercinus*, *Rhinolophus ferrumequinum* (eight species each) and *Miniopterus schreibersii* (seven species). All these hosts are broadly distributed across the Iberian Peninsula, and some of them (*P. pipistrellus*, *R. norvegicus*) live in peridomestic habitats (Palomo and Gisbert 2002). The lowest number of trematode species (one) is characteristic of herbivorous hosts that have a patchy distribution in Iberia (*Microtus* spp.) as well as Rhinolophidae and Vespertilionidae, which have been only occasionally studied for parasites.

The majority of Iberian digeneans show relatively low host specificity in terms of the number of host species they exploit. However, host specificity of some trematodes appears to be relatively high in terms of the taxonomic composition of the host spectrum. For example, all seven hosts of *Plagiorchis vespertilionis* and six hosts of *Mesotretes peregrinus* belong to Chiroptera; all six hosts of *Notocotylus neyrai* belong to Rodentia, whereas *Nephrotrema truncatum* has been recorded in five insectivores and one rodent. High host specificity is characteristic of Nanophyetidae, Collyriclidae, some Dicrocoeliidae and some Lecithodendriidae. Other trematode species have low host specificity in terms of both the number of host species used and their phylogenetic affinities. For example, some Psilostomidae can infect insectivores, rodents and birds (Montoliu et al. 1987; Cordero del Campillo et al. 1994).

The taxonomy of digenean trematodes in general, and those inhabiting the Iberian Peninsula in particular, is far from being clear. Many new species have been described during the last decades. Synonymy issues also need to be resolved (see Cordero del Campillo et al. 1994). Nevertheless, recent studies on spermiogenesis and spermatozoon of these digeneans may be helpful in clarifying the taxonomic issues (Ndiaye et al. 2004; Miquel et al. 2006). The use of molecular tools to determine phylogenetic relationships of digeneans is also growing (Littlewood et al. 1998; Cribb et al. 2001).

4.2 Ecological and zoogeographical aspects

The role of host factors such as host sex or age on the composition and structure of trematode assemblages is poorly known. Nevertheless, host sex has been shown to have no effect on trematode assemblages in *Talpa occidentalis* and *Erinaceus europaeus* (Casanova et al. 1996; Feliu et al. 2001). Prevalence of *Dicrocoelium dendriticum* did not differ between male and female *Oryctolagus cuniculus* (Blasco et al. 1996). No strong effect of host age and reproductive status on infection parameters of digenean trematodes has been found, although the highest prevalence of infection has been reported in adult rabbits, especially in pregnant females and in males with no scrotal testicles (Molina 1999). The variation in trematode infection among host individuals within host species can, however, be masked by seasonal effects. For example, intensity of infection of *Erinaceus europaeus* by *Brachylaima erinacei* increases substantially during summer and autumn (Feliu et al. 2001), whereas *Dicrocoelium dendriticum* is absent in its rabbit hosts in autumn (Molina et al. 1998).

On the other hand, the structure of trematode assemblages is strongly affected by the dietary habits of the host. For example, among rodents, trematode assemblages are richer in omnivorous and insectivorous (murines and *Eliomys quercinus*) than in herbivorous species (arvicolines) (Fig. 1). In some hosts, diet varies among geographic localities, and so does the composition of trematode assemblages, being poorer in individuals from localities where the diet is strictly herbivorous (see Torres and Feliu 1990 for an example with *Arvicola sapidus*). The life style of a host species also plays an important role in determining the composition of trematode assemblages. For example, among rodents, the richest trematode assemblages are found in terrestrial and/or semi-aquatic species, whereas the poorest assemblages are characteristic of subterranean or arboreal hosts (Feliu et al. 1992, 1997).

Different trematode species demonstrate various patterns of geographic distribution across Iberia. Some species are distributed across the entire Peninsula (e.g., *Plagiorchis vespertilionis*; Cordero del Campillo et al. 1994), whereas the distribution of other species is much more restricted. For example, *Dollfusinus frontalis*, a relatively common species in the Balearic Islands, has been recorded in the Eastern Peninsula only once (Galán-Puchades et al. 1994). Some species are endemic to the Pyrenees (e.g., *Brachylecithum eliomydis, Macyella apodemi*) or are detected in very specific habitats (deltas, albuferas; e.g., *Euparyphium albuferensis*, *Echinostoma friedi*). Finally, the distribution of some oioxenous or stenoxenous species mimics that of their hosts (e.g., *Ityogonimus* spp., *Matovius galemydis*; see Cordero del Campillo et al. 1994; Ribas 2005 for details).

4.3 Biological aspects

Although details of the life cycles of many trematodes inhabiting the Iberian Peninsula are still unknown, the available data allow to distinguish at least seven types of trematode life cycles (Table 6).

While the first intermediate hosts are either pulmonate gastropods or prosobranchs, the taxonomic affinity of the second intermediate hosts (in species with triheteroxenous life cycles) varies greatly (insects, annelids, crustaceans, pulmonate gastropods). Unfortunately, studies of the biological cycles of digeneans in Iberia are scarce and almost always focused on three families, namely Brachylaimidae, Echinostomatidae and Microphallidae. Nevertheless, a study on *Plagiorchis* (Plagiorchiidae), parasitic in *Apodemus sylvaticus* has been completed recently (Esteban et al. unpublished data). In addition, some information on a brachylaimid (*Brachylaima ruminate*) and on two panopistids (*Pseudoleucochloridium pericardicum* and *Dollfusinus frontalis*) parasitic in insectivores and rodents from regions geographically close to Iberia (eastern French Pyrenees and the Balearic Islands) is available (Mas-Coma and Montoliu 1986, 1987, 1995). The study of trematode life cycles may help to clarify some taxonomic issues (Gracenea et al. 1993; Esteban et al. 1997; Toledo et al. 2000; González-Moreno 2002; Gracenea and González-Moreno 2002).

Table 6. Types of life cycles for the Iberian digenean parasites of small mammals

Life cycle	Digenean species	First intermediate host	Second intermediate host
Terrestrial			
Triheteroxenous	*Brachylaima mascomai*	Pulmonate	Pulmonate
Triheteroxenous	*Dicrocoelium dendriticum*	Pulmonate	Insect
Aquatic			
Diheteroxenous	*Psilotrema spiculigerum*	Prosobranch	Metacercarie encysted on vegetation
Triheteroxenous	*Hypoderaeum conoideum*	Pulmonate	Pulmonate
Triheteroxenous	*Plagiorchis vespertilionis*	Pulmonate	Insect
Triheteroxenous	*Nephrotrema truncatum*	Prosobranch	Annelid
Triheteroxenous	*Maritrema feliui*	Prosobranch	Crustacean

5 Concluding remarks

Digenean trematodes are characteristic parasites of small mammals all over the world. In spite of the availability of faunistic data, it seems that still many species remain to be described. In particular, the use of molecular techniques and ultrastructural data of the spermatozoon and the spermiogenesis will undoubtedly help toward this end. The details of the life cycles of many species of trematode parasites of micromammals are still unclear, although some species have been thoroughly studied using experimental models. The relationships between a variety of host-related and environmental-related factors and the populations and communities of digenean trematodes also require further investigations.

References

Anderson WI, King JM, Uhl EM, Hornbuckle WE, Tennant BC (1991) Pathology of experimental *Schistosoma mansoni* infection in the Eastern woodchuck (*Marmota monax*). Vet Pathol 28:245–247

Blasco S (1996) Sobre las helmintofaunas de *Erinaceus europaeus* Linnaeus, 1758 (Insectivora: Erinaceidae), *Arvicola terrestris* (Linnaeus, 1758) (Rodentia: Arvicolidae) y *Oryctolagus cuniculus* (Linnaeus, 1758) (Lagomorpha: Leporidae) en la España peninsular. Ph. D. thesis, University of Barcelona

Botella P, Sánchez L, Esteban JG (1993) Helminthfauna of bats in Spain. III. Parasites of *Pipistrellus pipistrellus* (Schreber, 1774) (Chiroptera: Vespertilionidae). Res Rev Parasitol 53:63–70

Casanova JC, Feliu C, Torres J (1996) Helminthic community of the dwarf mole, *Talpa occidentalis* Cabrera 1907 (Insectivora: Talpidae), in Iberian Peninsula. Riv Parassitol 13:439–449

Chiu JK, Kao CT (1973) Studies on host-parasite relationships of Ilan strain of *Schistosoma japonicum* in animals. I. Prepatent period. Chin J Microbiol 6:72-79

Choquette LPE, Broughton E, Gibson GG (1973) *Schistosomatium douthitti* (Cort, 1914) Price, 1929 in a porcupine (*Erethizon dorsatum*) in eastern Ontario, Canada. Can J Zool 51:1317

Cordero del Campillo M, Castañón Ordóñez L, Reguera Feo A (1994) Indice-Catálogo de zooparásitos Ibéricos. Universidad de León, Secretariado de Publicaciones, León

Cribb TH, Bray RA, Littlewood DTJ, Pichelin SP, Herniou EA (2001) The Digenea. In: Littlewood DTJ, Bray RA (eds) Interrelationships of the Platyhelminthes. Taylor and Francis, London New York, pp 168-185

Dubois S (1983) Un neodiplostome peruvien, *Neodiplostomum* (*N.*) *vaucheri* n. sp. (Trematoda: Strigeoidea: Diplostomidae), parasite d'une Chauve-souris. Rev Suisse Zool 90:179–182

Esteban JG, Oltra-Ferrero JL, Mas-Coma S (1990) Helmintofauna de los murciélagos de España. II. Parásitos de *Miniopterus schreibersii* (Kuhl, 1819) (Chiroptera: Vespertilionidae). Rev Ibér Parasitol 50:199–209

Esteban JG, Oltra-Ferrero JL, Mas-Coma S (1992) Helmintofauna de los murciélagos de España. I. Interés sanitario de los Quiropteros y estado actual de conocimientos sobre sus helmintos parásitos. Circ Farm 314:107–122

Esteban JG, Toledo R, Sánchez L, Muñoz-Antoli C (1997) Life-cycle of *Euparyphium albuferensis* n. sp. (Trematoda: Echinostomatidae) from rats in Spain. Syst Parasitol 38:211–219

Esteban JG, Botella P, Toledo R, Oltra-Ferrero JL (1999) Helminthfauna of bats in Spain. IV. Parasites of *Rhinolophus ferrumequinum* (Schreber, 1774) (Chiroptera: Rhinolophidae). Res Rev Parasitol 59:57–68

Feliu C, Torres J, Casanova JC, Miquel J (1992) Consideraciones sobre el binomio helmintofauna-modo de vida en los Roedores ibéricos. In: Hernández Rodríguez S (ed) In memoriam al Profesor Doctor D. Francisco de Paula

Martínez Gómez. Universidad de Córdoba, Servicio de Publicaciones, Córdoba, pp 353–366

Feliu C, Renaud F, Catzeflis F, Hugot J-P, Durand P, Morand S (1997) A comparative analysis of parasite species richness of Iberian rodents. Parasitology 115:453–466

Feliu C, Blasco S, Torres J, Miquel J, Casanova JC (2001) On the helminthfauna of *Erinaceus europaeus* Linnaeus, 1758 (Insectivora, Erinaceidae) in the Iberian Peninsula. Res Rev Parasitol 61:31–37

Galán-Puchades MT, Fuentes MV, Cerezuela AM, Mas-Coma S (1994) New record of *Dollfusinus frontalis* Biocca et Ferretti, 1958 (Trematoda: Brachylaimidae) in continental Europe. Res Rev Parasitol 54:249–254

Genov T (1984) Helminths of insectivorous mammals and rodents in Bulgaria. Publ House Bulgarian Acad Sci, Sofia

Gibson DI, Jones A, Bray RA (2002) Keys to the Trematoda, vol 1. CABI and Nat Hist Mus, London

Gitsu GA, Kovalenko FP (1983) Susceptibility of *Phodopus sungarus*, golden hamsters, cotton rats and white rats to experimental opisthorchiasis (in Russian). Meditzinskaya Parazitologiya i Parazitarnye Bolezni [Medical Parasitology and Parasitic Diseases] 5:63–64

González-Moreno O (2002) Contribución al conocimiento de la bionomía de digénidos de la Familia Brachylaimidae Joyeux et Foley, 1930, en el delta del Llobregat (Barcelona). Ph. D. thesis, University of Barcelona

Goüy de Bellocq J, Morand S, Feliu C (2002) Patterns of parasite species richness of western Palearctic micromammals: Island effects. Ecography 25:173–183

Gracenea M, González-Moreno O (2002) Life cycle of *Brachylaima mascomai* n. sp. (Trematoda: Brachylaimidae), a parasite of rats in the Llobregat Delta (Spain). J Parasitol 88:124–133

Gracenea M, Montoliu I, Deblock S (1993) Contribution a l'étude des Microphallidae Travassos, 1920 (Trematoda). XLV. Description de *Maritrema feliui* n. sp., parasite de musaraignes (mammifères) en Espagne. Ann Parasitol Hum Comp 68:76–81

Gvozdev EV, Orlov GI (1985) The helminth fauna of *Ochotona alpina* in the western Altai (in Russian). In: Anonymous (ed) Helminths of animals in the Kazakhstan ecosystems. Nauka, Alma-Ata, pp 109–113

Hatsushika R, Shimizu M, Kawakami S (1979) Experimental infection of nutrias with the metacercariae of *Paragonimus miyazakii* Kamo et al., 1961 (in Japanese). Japan J Parasitol 28:335–338

Jones A, Bray RA, Gibson DI (2005) Keys to the Trematoda, vol 2. CABI and Nat Hist Mus, London

Kawazoe U, Pinto ACM (1983) Epidemiological importance of some wild animals in schistosomiasis *mansoni*. Rev Saude Publ 17:345–366

Kuntz RE, Moore JA, Huang TC (1975) Miscellaneous mammals (*Dasypus novemcinctus*, *Myocastor coypus*, *Procyon lotor*, and *Bassariscus astutus*) as hosts for *Schistosoma haematobium* (Iran). Proc Oklahoma Acad Sci 55:143–146

Led JE, Yannarella FG, Scasso DA, Denegri GM (1979) *Lagidium viscaccia boxi*, a new reservoir host of *Fasciola hepatica* in Argentina. Vet Argentina 2:31–37

Littlewood DTJ, Bray RA, Clough KA (1998) A phylogeny of the Platyhelminthes: Towards a total-evidence solution. Hydrobiologia 383:155-160

López-Neyra CR (1947) Helmintos de los vertebrados Ibéricos, tom 1. Consejo Superior de Investigaciones Científicas, Granada

Marshall ME, Miller GC (1979) Some digenetic trematodes from Ecuadorian bats including five new species and one new genus. J Parasitol 65:909–917

Mas-Coma S, Montoliu I (1986) The life cycle of *Brachylaima ruminae* n. sp. (Trematoda: Brachylaimidae), a parasite of rodents. Z Parasitienkd 72:739–753

Mas-Coma S, Montoliu I (1987) The life cycle of *Dollfusinus frontalis*, a brachylaimid trematode of small mammals (Insectivora and Rodentia). Int J Parasitol 17:1063–1079

Mas-Coma S, Montoliu I (1995) Life cycle of *Pseudoleucochloridium pericardicum* n.sp. (Trematoda: Brachylaimidae), a parasite of shrews (Insectivora: Soricidae) in the Oriental Pyrenees. Res Rev Parasitol 55:155–171

Massoud J (1973) Studies on the schistosomes of domestic animals in Iran. I. Observations on *Ornithobilharzia turkestanicum* (Skrjabin, 1913) in Khuzestan. J Helminthol 47:165–180

McKown RD, Ridley RK, Kennedy GA (2000) The hispid cotton rat (*Sigmodon hispidus*) as an experimental host for the bovine liver fluke (*Fasciola hepatica*). Vet Parasitol 87:125–132

Merkusheva IV, Bobkova AF (1981) Helminths of domestic and wild animals in Belorussia: A catalogue (in Russian). Nauka i Tekhnika, Minsk

Miquel J, Fournier-Chambrillon C, Fournier P, Torres J (2006) Spermiogenesis and spermatozoon ultrastructure of the cranial digenean *Troglotrema acutum* (Leuckart, 1842). J Parasitol (in press)

Molina X (1999) Estudio parasitoecológico de las helmintofaunas de Lepóridos (Lagomorpha) en diferentes ecosistemas (peninsulares–insulares). Ph. D. thesis, University of Barcelona

Molina X, Casanova JC, Feliu C, Castién E (1998) Studies of wild rabbit (*Oryctolagus cuniculus*) populations in Navarra (Spain). III. Qualitative and quantitative composition of the rabbit parasitic helminthfauna. Game Wildl 15:123–133

Montoliu I, Gracenea M, Feliu C, Torres J (1987) *Psilotrema spiculigerum* (Mühling, 1898) (Trematoda: Psilostomidae) parásito intestinal de *Arvicola sapidus* (Miller, 1908) (Rodentia: Arvicolidae) en el Delta del Ebro (NE de la Península Ibérica). Rev Ibér Parasitol 47:341–345

Ndiaye PI, Miquel J, Bâ CT, Marchand B (2004) Spermiogenesis and ultrastructure of the spermatozoon of the liver fluke *Fasciola gigantica* Cobbold, 1856 (Digenea, Fasciolidae), a parasite of cattle in Senegal. J Parasitol 90:30–40

Palomo LJ, Gisbert J (2002) Atlas de los mamíferos terrestres de España. Organismo Autónomo de Parques y Jardines, Madrid

Pitchford RJ (1975) Introduction of *Schistosoma leiperi* Le Roux 1955 and *Schistosoma margrebowiei* Le Roux 1933 to the laboratory. Trans R Soc Trop Med Hyg 69:362

Prokopic J, Genov T (1974) Distribution of helminths in micromammals (Insectivora and Rodentia) under different ecological and geographical conditions. Studie CSAV, Prague

Ribas A (2005) Systematic, faunistic and ecologic study of the parasitic helminthfaunas of talpids in the south eastern Europe. Ph. D. thesis, University of Barcelona

Sahba GH, Arfaa F, Farahmandian I, Jalali H (1972) Animal fascioliasis in Khuzestan, southwestern Iran. J Parasitol 58:712–716

Sakamoto T, Abe S, Suga M, Kono I, Yasuda N (1982) *Dicrocoelium dendriticum* (Rudolphi, 1819) Looss, 1899 from pika *Ochotona hyperborea yesoensis* Kishida in Hokkaido, Japan. Bull Fac Agric Kagosh Univ 32:119–123

Salem JB (1975) A new species of *Echinochasmus* from the bats found in Hyderabad, Andhra Pradesh (India). Riv Parassitol 36:33–36

Saoud MFA, Ramadan MM (1977) On a new trematode, *Prohemistomum azimi* n.sp. (Trematoda: Cyathocotylidae) from the Egyptian slit-faced bat. Z Parasitenkd 53:281–285

Terasaki K, Noda Y, Shibahara T, Itagaki T, Fukuda K, Tsuchiya K (2003) Experimental fascioliasis in the rat-like hamster, *Tscherskia triton*, and other rodent hosts. Parasitol Int 52:147–154

Toledo R, Muñoz-Antoli C, Esteban JG (2000) The life cycle of *Echinostoma friedi* n. sp. (Trematoda: Echinostomatidae) in Spain and a discussion on the relationships within the "*revolutum*" group based on cercarial chaetotaxy. Syst Parasitol 45:199–217

Torres J, Feliu C (1990) El Delta del Ebro un ecosistema especial en el NE de la Península Ibérica: Consideraciones helmintoecológicas en relación a las especies de micromamíferos. Circ Farm 48:25–34

Vélez I, Thatcher V (1990) *Nudacotyle carollia* nov. sp. (Trematoda, Nudacotylidae) intestinal parasite of *Carollia perspicillata* L. (Chiroptera) in Colombia. Rev Brasil Zool 7:69–72

Yamaguti S (1971) Synopsis of digenetic trematodes of vertebrates, vol 1. Keigaku Publ, Tokyo

Zajac AM, Williams JF (1981) The pathology of infection with *Schistosomatium douthitti* in the laboratory mouse and the meadow vole, *Microtus pennsylvanicus*. J Comp Pathol 91:1–10

Zdarska Z, Soboleva TN (1990) Scanning electron microscopic investigation of the trematode *Hasstilesia ochotonae* Gvosdev, 1962. Folia Parasitol 37:347–348

Zdzitowiecki K (1980) *Nudacotyle quartus* sp. n. (Trematoda, Nudacotylidae), the parasite of insectivorous bats from Cuba. Acta Parasitol Pol 26:215–218

Zdzitowiecki K, Rutkowska MA (1980) The helminthfauna of bats (Chiroptera) from Cuba. III. A review of trematodes. Acta Parasitol Pol 26:201–214

Zelya OP, Sergeeva MN (1987) Peculiarities in the shedding of *Schistosoma mansoni* eggs by experimentally infected golden hamsters and *Phodopus sun-*

gorus. Meditzinskaya Parazitologiya i Parazitarnye Bolezni [Medical Parasitology and Parasitic Diseases] 3:67–71

3 Cestodes of small mammals: Taxonomy and life cycles[1]

Boyko B. Georgiev, Rodney A. Bray and D. Timothy J. Littlewood

1 Introductory remarks

Cestodes are diverse, ubiquitous parasites of vertebrates. They are frequent parasites of small mammals. It is not possible to provide a concise description of their taxonomy and life cycles without omitting many of the complexities, controversies and fascinating novelties of such a diverse group. Therefore, this chapter is introductory and not exhaustive. Its aim is to introduce the reader to the biology of cestodes and to present the basic components of their diversity in rodents, insectivores, lagomorphs and chiropterans as well as the major patterns of cestode life cycles, in which these mammalian groups participate. The appearance of cestodes in their contemporary vertebrate host range comes as a result of multiple independent evolutionary invasions, host shifts and coevolutionary history with their hosts. Appearing well before the origin of mammals, cestodes have not failed to follow their vertebrate hosts wherever they have gone. Not least, whether as intermediate or definitive hosts, small mammals have provided significant opportunities for cestodes to diverge in space and time. Likewise, cestodes have played an integral role in the evolutionary ecology of small mammals.

[1] We are grateful to C. Vaucher (Natural History Museum, Geneva), V.V. Tkach (University of North Dakota, Grand Forks), T. Genov and G.P. Vasileva (Central Laboratory of General Ecology, Bulgarian Academy of Sciences, Sofia) for critical comments on the manuscript. This study was undertaken as part of a NERC award (NER/A/S/2003/00313) to DTJL and RAB. Additional funding was provided by the Natural History Museum, London.

2 General information on the class Cestoda

The tapeworms (cestodes) are considered a class within the phylum Platyhelminthes (flatworms). All are internal parasites, with only the egg (and, in some groups, a short-living larva hatching from the egg) existing outside a host. As a rule, adult cestodes are parasitic in vertebrates. The larvae occur in invertebrate and/or vertebrate hosts. Cestodes are distributed in almost all terrestrial, marine, brackish and freshwater habitats where vertebrate animals live.

2.1 Body organisation and life cycles

Typically, the body of the tapeworms is dorso-ventrally flattened and highly elongate, resembling tape. Its length ranges between 0.6 mm (some parasites of shrews) to 30 m (some parasites of cetaceans). Usually, the body of tapeworms consists of three distinct regions (Fig. 1): scolex (plural scoleces), neck and strobila (plural strobila).

The scolex is the most anterior part of the body. Its main function is the attachment of the parasite to the intestinal wall. It may bear spines, hooks, glands releasing adhesive secretions, grooves (bothria), suckers or tentacles, or various combinations of these depending on the systematic position of the species. The scoleces of the cestodes of several orders are characterised by the presence of an apical organ consisting mostly of muscular and/or glandular tissue. In the order Cyclophyllidea (which includes almost all cestodes parasitic in small mammal), the apical organ is typically a rostellum, which is characterised by an immense variability of structure (Fig. 2). It is protrusible and most commonly provided with one or two rows of hooks. In some families (e.g., Hymenolepididae, Dilepididae), the rostellum can be withdrawn in a special muscular pouch (rostellar sac). The protruded rostellum penetrates into the intestinal wall of the host, anchoring there by the hooks. Some cyclophyllideans (e.g. Anoplocephalidae) lack rostellar apparatus (Fig. 1).

The neck is the region just posterior to the scolex. It is a zone of proliferation giving rise to the strobila. The strobila consists of proglottides (singular proglottis) arranged in a linear series. Each proglottis contains a complete set of reproductive organs. The strobila may consist of few (2-3), several dozens (in the majority of species) or numerous (hundreds to thousands) proglottides. The development of each proglottis starts at the neck, resulting from the division of stem cells. Most commonly, the formation of proglottides at the neck is an enduring process lasting the entire life of the

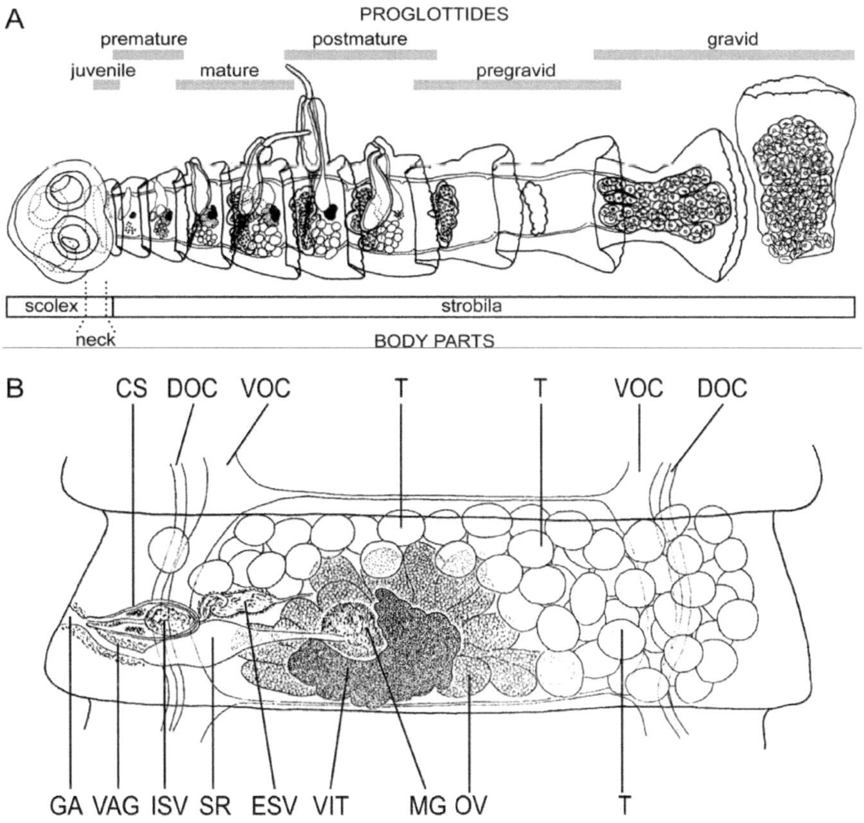

Fig. 1. A. Parts of the body of *Leporidotaenia pseudowimerosa* (Anoplocephalidae), a parasite of rabbits, *Oryctolagus cuniculus*, in Spain (the general view redrawn from Genov et al. 1990, reproduced with kind permission of Springer Science and Business Media). **B.** Mature proglottis of *Paranoplocephala aquatica* (Anoplocephalidae) (redrawn from Genov et al. 1996, reproduced with kind permission of Springer Science and Business Media), a parasite of aquatic voles *Arvicola terrestris* and *Ondatra zibethica*, demonstrating the structure of genital systems. Abbreviations: CS, cirrus sac; DOC, dorsal osmoregulatory canal; ESV, external seminal vesicle; GA, genital atrium, ISV, internal seminal vesicle; MG, Mehlis' gland; OV, ovary; SR, seminal receptacle; T, testes; VAG, vagina; VIT, vitellarium; VOC, ventral osmoregulatory canal.

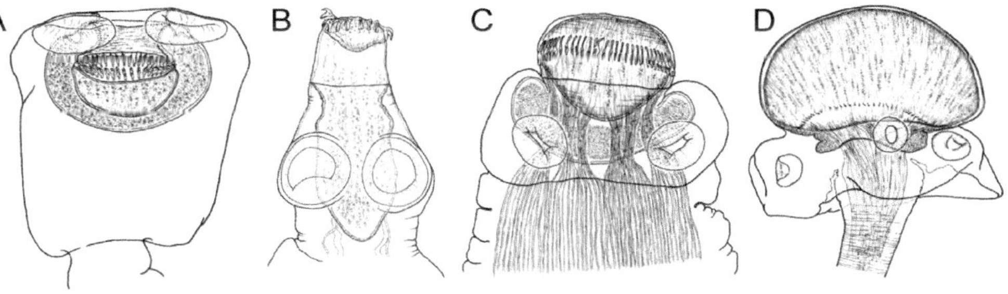

Fig. 2. Scoleces of cestodes of the family Hymenolepididae demonstrating the variability of the structure of the rostellar apparatus. *Coronacanthus magnihamatus* (A) and *Triodontolepis boyanensis* (B) from the water shrew *Neomys fodiens*; *Hylmilepis prokopici* (C) and *H. nagatyi* (D) from *Crocidura* spp. (A, B, modified fromr Vasileva et al. 2005; C, D, modified from Vasileva et al. 2004)

cestode in the definitive host. Just posterior to the neck, the proglottides are short, containing undifferentiated cells (juvenile proglottides). With the appearance of a new proglottis at the neck, already formed proglottides move posteriorly. This coincides with growth and the gradual development of the reproductive organs. Posterior to the juvenile proglottides, each strobila typically contains premature (with primordia of genital organs), mature (with developed and functioning male and female genital systems), postmature (in which the uteri are filled with developing eggs and gonads gradually degenerate), pregravid (uterus well-developed but eggs not entirely formed) and gravid (containing uteri with ripe eggs) proglottides. As a rule, the gravid proglottides in the terminal position on the strobila detach. They pass to the environment with the host's excrement or disintegrate along their route and only eggs are released. Some cestodes (e.g., Pseudophyllidea) have uterine pores and eggs can be released one by one.

Out of 15 cestode orders, formation of proglottides occurs in 11. The representatives of three orders (Gyrocotylidea, Amphilinidea and Caryophyllidea) have only a single set of genital organs per body (i.e., no proglottides); they are often referred to as "monozoic" cestodes. The order Spathebothriidea exhibits an intermediate pattern of body organisation: an internal multiplication of reproductive organs down the strobila occurs but no externally distinct units are formed.

Cestodes lack a gut during all developmental stages. They feed through the body surface, i.e. the tegument. The latter is a syncytial tissue peculiar to parasitic flatworms. Its main functions are protective (against immune reactions and enzymes of the host) and digestive (as a major site of absorption, metabolic transformations and transport of nutrients). The tegument consists of a surface syncytial layer (distal cytoplasm) linked by cytoplasmic bridges with cell bodies (cytons) situated deep beneath the superficial muscle layers. The secretions of cytons continuously renovate the distal cytoplasm, which acts as a contact zone between the parasite body and the tissues and fluids of the host. Usually there are three layers of superficial musculature surrounding parenchyma. Strong longitudinal muscular bundles pass along the entire strobila and are responsible for the movements of the body. The nervous system is represented by paired ganglia situated in the scolex and arising from them major anterior and posterior longitudinal nerves; the latter run through the strobila. There are also numerous transverse commissures connecting longitudinal nerves and smaller nerves emanating from them and reaching to the musculature and the receptors. The osmoregulatory system comprises flame cells scattered in the parenchyma. Narrow ducts connect these cells with the major longitudinal canals of the system passing along the strobila (typically, two dorso-lateral and two ventro-lateral, see Fig. 1). This system is responsible for the excretion of metabolic products and eliminates excess water from the body.

The majority of tapeworms, including all the species occurring in small mammals, are hermaphroditic. As a rule, each mature proglottis contains one male reproductive system and one female reproductive system, but two of each per proglottis is not rare.

The male reproductive system includes from one (some Hymenolepididae) to several hundreds (some Taeniidae) of testes. Each testis is provided with an outgoing duct (vas efferens). These ducts unite into a common wider duct (vas deferens), which transports the sperm to the male copulatory organ (cirrus). The latter is situated within a muscular pouch (cirrus-sac). Along its course, in order to have greater sperm storage capacity, vas deferens may form seminal vesicles before entering the cirrus sac (external seminal vesicle) and (or) within it (internal seminal vesicle), e.g. in the Hymenolepididae and Anoplocephalidae; in other cases, the same effect is achieved by a highly convoluted vas deferens (Dilepididae). The cirrus is a muscular (smooth or spinous) organ, which is able to invaginate (to be withdrawn) in the cirrus-sac or to evaginate (project) through its pore.

The female reproductive system includes the ovary, vitellarium, ootype, uterus, vagina, seminal receptacle and the ducts connecting them (Fig. 1). During copulation, sperm passes into the female system through the vagina and is stored in the seminal receptacle. As oocytes mature in the ovary, they pass from it into the oviduct. A duct coming from the seminal receptacle joins to the oviduct. The junction of these ducts forms a fertilization chamber. The vitellarium may be a compact organ (e.g., in Cyclophyllidea) or may consist of follicles scattered in the parenchyma, with outgoing ducts uniting into a common vitelline duct (in the majority of orders). The vitelline duct is linked with the oviduct where one or more vitelline cells become associated with each zygote. Together they pass into the ootype, which is normally surrounded by glandular tissue (Mehlis' gland) producing a secretion forming a thin envelope around the zygote and associated vitelline cells. The young eggs pass from the ootype through the uterine duct into the uterus where they complete their development.

The life cycle of cestodes involves at least two hosts, a final or definitive (harbouring the sexually reproducing, adult cestode) and an intermediate host (in which larvae, known also as metacestodes, develop). The two hosts are in ecological association making parasite transmission possible: the intermediate host occurs in habitats where the definitive host feeds and defecates. The intermediate host is a component of the diet of the definitive host. The principal scheme of the cestode life cycle is as follows. Parasite eggs are released with host's faeces into the environment. Each egg contains an embryo (oncosphere) provided with six embryonic hooks and several glandular cells and is surrounded by several protective envelopes. The egg is eaten by the intermediate host. The oncosphere hatches in the gut of the latter and, using its hooks and glands, penetrates through the wall of the gut. It locates in the body cavity or in any internal organ. There, it metamorphoses into an infective metacestode possessing a fully- or an almost fully developed scolex. The definitive host is infested by eating infected intermediate hosts. The scolex of the metacestode attaches to the intestinal wall of the definitive host and the strobila is formed. Passage of cestodes from one host to another occurs universally through food chains.

There are some cestode species, which have life cycles differing from the above summary. For example, some species have two intermediate hosts or mobile embryos able to swim. Also, the embryos of the Amphilinidea and the Gyrocotylidea are not oncospheres but lycophores, which possess 10 embryonic hooks.

Metacestodes of various orders and families exhibit great morphological variability (Fig. 3). This resulted in a complex terminology associated with metacestode forms (Chervy 2002) and can make it difficult to compare and contrast developmental and ecological elements of life cycles. The procercoid is the metacestode in the first intermediate host, having an elongate body and a cercomer (e.g., in Pseudophyllidea). Entering into the second intermediate host, it develops into the plerocercoid. The latter possesses a differentiated scolex and is able to infect the definitive host. The most widespread type of metacestodes in the Cyclophyllidea is the cysticercoid, a metacestode with a developed scolex retracted (encysted) into the body. Among the Taeniidae, the most widespread metacestode is the cysticercus, which has a scolex introverted into the bladder-like posterior body part.

In terms of known intermediate host species and (or) described morphology of larval stages, the life cycles of only about 5% of cestode species are known. Having in view that each cestode occurs in at least two microhabitats during its life, i.e. in one definitive and in at least one intermediate host, this percentage exposes an enormous gap in the knowledge: for 95% of the cestode species, we know only one of at least two microhabitats (hosts) utilised by them.

Two major approaches have been used in cestode life-cycle studies, each characterised by a number of advantages and restrictions. The first approach is the experimental laboratory infection of potential intermediate hosts with eggs collected from an adult cestode. The main advantages are the reliable identification of the metacestodes studied (based on the adult specimens used as a source of infective eggs) and the possibility of describing the subsequent stages of metacestode development in an exact temporal scale. The main disadvantage of this approach is the lack of certainty that an animal species being successfully infected in experimental conditions is really the intermediate host in natural conditions. The second approach is the examination of potential intermediate hosts in order to find naturally infected animals. These studies provide informative results for cyclophyllidean cestodes only because in this order the morphogenesis of the scolex mostly occurs in the intermediate host and the majority of the species possess rostellar hooks identical in adults and metacestodes. However, it is not applicable in numerous cases when congeneric species have identical rostellar hooks and have been differentiated on the basis of the strobilar morphology only. For the further development of these studies, the application of molecular markers can be of great value.

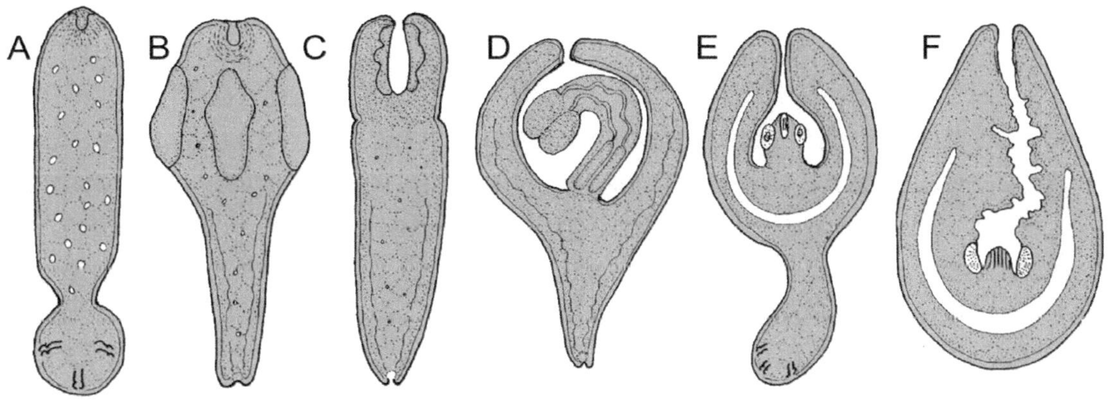

Fig. 3. Morphological types of metacestodes (redrawn from Chervy 2002, reproduced with kind permission of Springer Science and Business Media). A. Procercoid. B. Plerocercoid. C. Merocercoid. D. Plerocercus. E. Cysticercoid. F. Cysticercus

The present account contains basic information about the body organisation and the life cycles of tapeworms. The reader can obtain more detailed data from zoology textbooks (e.g., Ruppert and Barnes 1994), reference readings (Caira and Littlewood 2000; Georgiev 2003) or specialised monographs (Wardle and McLeod 1952; Joyeux and Baer 1962; Arme and Pappas 1984a, b; Smyth and McManus 1989; Coil 1991).

2.2 Classification, taxonomic diversity and phylogenetic relationships

According to a recent estimate (Georgiev 2003), the class Cestoda encompasses about 5100-5200 species, 680 genera, 72 families and 15 orders. Out of them, 3100 species, 380 genera and 18 families belong to the order Cyclophyllidea. A contemporary source on cestode classification at the ordinal, familial and generic level is the book by Khalil et al. (1994). The last comprehensive source on the species diversity of tapeworms was by Schmidt (1986).

In the past, two subclasses have been recognised within this class: Cestodaria, including the monozoic orders Gyrocotylidea and Amphilinidea, and Eucestoda, comprising remaining orders (mostly polyzoic but also the monozoic Caryophyllidea). Some authorities consider Amphilinidea and Gyrocotylidea as distinct classes within the phylum Platyhelminthes (as Amphilinida and Gyrocotylida, respectively). Sometimes the order Caryophyllidea is placed out of the Eucestoda and believed to be close to the amphilinideans and gyrocotylideans. However, mostly as a result of recent extensive phylogenetic studies based on gene sequences and morphology, a wide consensus has been achieved on several points (e.g., Hoberg et al. 2001; Olson et al. 2001). The Cestoda, comprising Gyrocotylidea, Amphilinidea and Eucestoda are believed to form a monophyletic and highly derived flatworm group. The tapeworms, together with the monogeneans and the trematodes, belong to the monophyletic taxon Neodermata. The monogeneans are believed to be the closest relatives of the tapeworms; the two groups are included in the higher taxon Cercomeromorphae. Within the Cestoda, the Gyrocotylidea have a position basal to the branch containing the remaining taxa, i.e. Amphilinidea plus eucestode orders (Fig. 4). Among the Eucestoda, the monozoic Caryophyllidea are considered basal to the remaining groups. Among the polyzoic orders, these having as a rule four suckers or bothridia on the scolex (known as tetrafossate, e.g., Tetraphyllidea, Proteocephalidea and Cyclophyllidea) are considered more derived than those having two bothria or bothridia (difossate). For more detailed information on cestode phylogeny, see Hoberg et al. (2001) and Olson et al. (2001).

The representatives of 10 orders (Gyrocotylidea, Caryophyllidea, Spathebothriidea, Haplobothriidea, Diphyllidea, Trypanorhyncha, Tetraphyllidea, Litobothriidea, Lecanicephalidea and Nippotaeniidea) are entirely associated with fishes as definitive hosts. The majority of the species belonging to three other orders (Amphilinidea, Pseudophyllidea and Proteocephalidea) are also parasites of fishes; however, some taxa of these orders have colonised tetrapods. Among them, the Pseudophyllidea is relevant in the context of this volume because some species occur as larvae in small mammals. The order Tetrabothriidea includes parasites of marine birds and marine mammals only.

The most species-rich order is the Cyclophyllidea. Its members occur as adults in tetrapods, mostly in birds and mammals (a few species are parasites of reptiles and amphibians). All the species occurring as adults in small mammals belong to this order (Fig. 4). Intermediate hosts of cyclophyllideans include arthropods, annelids, molluscs or mammals (only intermediate hosts or first intermediate hosts), fishes, amphibians, reptiles, birds or mammals (second intermediate hosts).

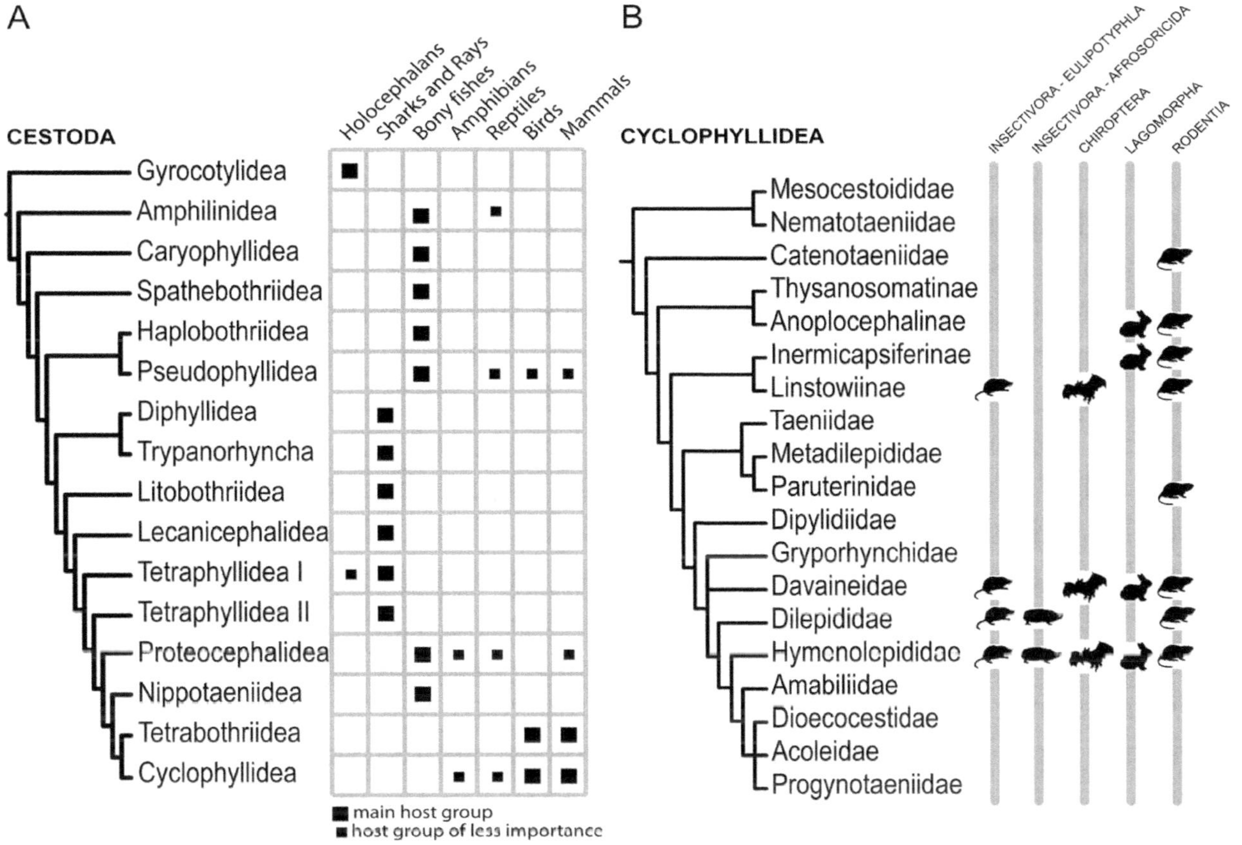

Fig. 4. A. Phylogeny of the class Cestoda drawn from multiple sources including morphological and molecular estimates both published (e.g. Hoberg et al. 2001; Olson et al. 2001) and unpublished. Range of definitive hosts is indicated. **B.** Phylogeny of the order Cyclophyllidea. Cladogram modified from Hoberg et al. (1999). The family-group taxa, for which small mammals are known as definitive hosts, are indicated. Their distribution across the tree suggests numerous events of colonisation and minor role of the co-diversification for the formation of the cestode fauna of the small mammal orders

3 Small mammals as definitive hosts of cestodes

The fauna of the cestodes of small mammals is relatively well studied in Europe, northern Asia and North America. These studies showed that adult cestodes are widespread and abundant in small mammals. In many areas, their species diversity is greater than that of their host groups. For example, Vaucher (1971) examined eight species of Soricidae from Europe and recorded 28 cestode species in them. Genov (1984) summarised data from the study of 33 species of insectivores and rodents in Bulgaria and reported

41 species of adult cestodes. Feliu et al. (1997) recorded 17 species of adult cestodes from 16 species of rodents from the Iberian Peninsula.

Unfortunately, even for the well-studied areas in the Northern Hemisphere, few reviews are available and those that are need updating (e.g., Stiles 1896; Baer 1927; Spasskii 1951; Gvozdev et al. 1970; Baer and Tenora 1970; Vaucher 1971; Rausch 1975, 1976; Beveridge 1978; Ryzhikov et al. 1978; Tenora and Murai 1978, 1980; Genov 1984; Tenora et al. 1985). Most primary information is scattered in numerous sources, mostly journal articles. Outside temperate latitudes, data are even more fragmentary, although detailed and comprehensive studies have been carried out in a few tropical areas (Hunkeler 1974; Quentin 1964, 1971).

No current estimate of the species diversity of cestodes from lagomorphs, rodents and insectivores has been made. According to Sawada (1997), 117 nominal cestode species have been described from bats. Even in well-studied areas in Europe and North America, new cestode species have been discovered from rodents and insectivores during the last few years (Haukisalmi et al. 2002; Tkach et al. 2003; Vasileva et al. 2005; Haukisalmi and Henttonen 2005). Considering that the diversity of small mammals in tropical areas is much greater than that in temperate latitudes, and that most of the cestodes described are genus- or species-specific to their definitive hosts, we predict that currently no more than half of the cestode species occurring in small mammals are described.

All cestodes occurring in small mammals as adults belong to the order Cyclophyllidea (Table 1, Appendix I). Currently, the validity of 15 families of this order is widely recognised (e.g., Khalil et al. 1994). Members of six families have been recorded in small mammals as definitive hosts (Fig. 4). One of them, the Catenotaeniidae, has a host range entirely restricted to rodents, which is indicative for the co-diversification of the two groups (Quentin 1971, 1994).

The families Hymenolepididae and Anoplocephalidae are represented by considerable diversity in small mammals, containing ca. 45 and ca. 25 genera, respectively, with host ranges restricted to small mammal orders; however, these families exhibit higher diversity in other groups of tetrapods, i.e. hymenolepidids in birds (Czaplinski and Vaucher 1994) and anoplocephalids in other mammals (ruminants, macropodid marsupials), birds and reptiles (Beveridge 1994). Nevertheless, the considerable diversity of anoplocephalines in the lagomorphs and in murid rodents (Appendix I) and the restricted host specificity of the majority of the genera suggest the important role of these host groups for the diversification of the Anoplocephalinae (see also Wickström et al. 2005). Similarly, the co-diversification with the Soricidae seems to be the major event for the formation of the current diversity of mammalian hymenolepidids.

Table 1. Cestode genera occurring as adults in the main groups of small mammals (Insectivora, Chiroptera, Lagomorpha and Rodentia). The cestode genera marked by an asterisk (*) occur in this group of hosts only

Host group	Family	Genera	Source
Insectivora	Anoplocephalidae (Linstowiinae)	*Mathevotaenia*	Schmidt (1986)
	Davaineidae	*Raillietina*	Sawada (1999)
	Dilepididae	*Dilepis*, *Hepatocestus**, *Monocercus** (=*Molluscotaenia*), *Multitesticulata**, *Polycercus*	Bona (1994), Gulyaev and Kornienko (1998), Sawada (1999)
	Hymenolepididae	*Blarinolepis**, *Brachylepis**, *Coronacanthus**, *Cryptocotylepis**, *Ditestolepis** (=*Sinuterilepis*), *Ecrinolepis**, *Hilmylepis**, *Hymenolepis*, *Karpenolepis**, *Lineolepis**, *Lockerrauschia**, *Mathevolepis**, *Neomylepis**, *Neoskrjabinolepis**, *Protogynella**, *Pseudhymenolepis**, *Pseudobothrialepis**, *Pseudodiorchis**, *Skrjabinacanthus**, *Soricinia**, *Spalania**, *Spasskylepis**, *Staphylocystis**, *Staphylocystoides** (=*Zarnowskiella*), *Talpolepis**, *Triodontolepis**, *Urocystis**, *Vaucherilepis**, *Vigisolepis**, *Vogelepis**	Czaplinski and Vaucher (1994), Tkach and Kornyushin (1997), Tkach (1998), Gulyaev and Kornienko (1999), Karpenko and Gulyaev (1999), Karpenko and Chechulin (2000), Tkach et al. (2003), Mel'nikova et al. (2004), Gulyaev et al. (2004), Gulyaev and Mel'nikova (2005)
Lagomorpha	Anoplocephalidae (Anoplocephalinae)	*Andrya*, *Cittotaenia**, *Diuterinotaenia**, *Ectopocephalium**, *Leporidotaenia**, *Mosgovoyia**, *Neandrya**, *Schizorchis**	Beveridge (1994), Haukisalmi and Wickström (2005)
	Anoplocephalidae (Inermicapsiferinae)	*Inermicapsifer*	Beveridge (1994)
	Davaineidae	*Fuhrmannetta*, *Paroniella*, *Raillietina*, *Vadifresia*	Jones and Bray (1994), Movsesyan (2003a, b)
	Hymenolepididae	*Gvosdevilepis**	Czaplinski and Vaucher (1994)
Chiroptera	Anoplocephalidae (Linstowiinae)	*Cycloskrjabinia**, *Mathevotaenia*, *Oochoristica*	Beveridge (1994), Schmidt (1986)
	Davaineidae	*Raillietina*	
	Hymenolepididae	*Milina** (=*Myotolepis*), *Gopalaia* (*genus inquirendum*), *Pseudoligorchis**, *Rodentolepis*, *Vampirolepis**	Czaplinski and Vaucher (1994)

Table 1. Continued.

Host group	Family	Genera	Source
Rodentia	Catenotaeniidae	*Catenotaenia**, *Hemicatenotaenia**, *Meggittina**, *Pseudocataenotaenia**, *Quentinia**, *Skrjabinotaenia**	Quentin (1994)
	Anoplocephalidae (Anoplocephalinae)	*Andrya*, *Anoplocephaloides* (= *Paranoplocephaloides*), *Bertiella* (=*Indotaenia*), *Ctenotaenia**, *Diandrya**, *Gallegoides**, *Hokkaidocephala**, *Moniezia*, *Monoecocestus**, *Parandrya**, *Paranoplocephala**, *Pseudocittotaenia**, *Sdarikovina**, *Viscachataenia**	Beveridge (1994), Gulyaev (1996), Gulyaev and Chechulin (1996), Tenora et al. (1999), Denegri et al. (2003)
	Anoplocephalidae (Linstowiinae)	*Mathevotaenia* (=*Schizorchodes*), *Sinaiotaenia**, *Witenbergitaenia**	Beveridge (1994)
	Anoplocephalidae (Inermicapsiferinae)	*Inermicapsifer*, *Metacapsifer*	Beveridge (1994)
	Paruterinidae	*Orthoskrjabinia*	Georgiev and Kornyushin (1994)
	Davaineidae	*Delamuretta**, *Dollfusoquenta**, *Fuhrmannetta*, *Paroniella*, *Raillietina* (=*Erchanella*, *Tenoretta*), *Skrjabinia*, *Vadifresia*	Jones and Bray (1994), Movsesyan (2003a, 2003b), Spasskii (1994)
	Dilepididae	*Alproma**, *Dilepis*, *Hunkeleria**	Bona (1994)

The families Dilepididae, Davaineidae and Paruterinidae include mostly parasites of birds, and individual taxa are associated with small mammals as definitive hosts, or occur accidentally in them (Bona 1994; Jones and Bray 1994; Georgiev and Kornyushin 1994; Movsesyan 2003a, 2003b).

3.1 Catenotaeniidae

This family contains about 35 species (Schmidt 1986). All are intestinal parasites of rodents: Sciuridae, Muridae, Heteromyidae, Geomyidae and Caviidae (Quentin 1994). The geographical range of the family includes all continents except Australia and Antarctica. Morphologically, catenotaeniids are diagnosed by a uterus consisting of a longitudinal stem and lateral branches (similar to that of the family Taeniidae). Their scolex is provided with suckers only. A rostellar apparatus is lacking, but sometimes adults have a vestigial "apical sucker" (metacestodes have apical organ).

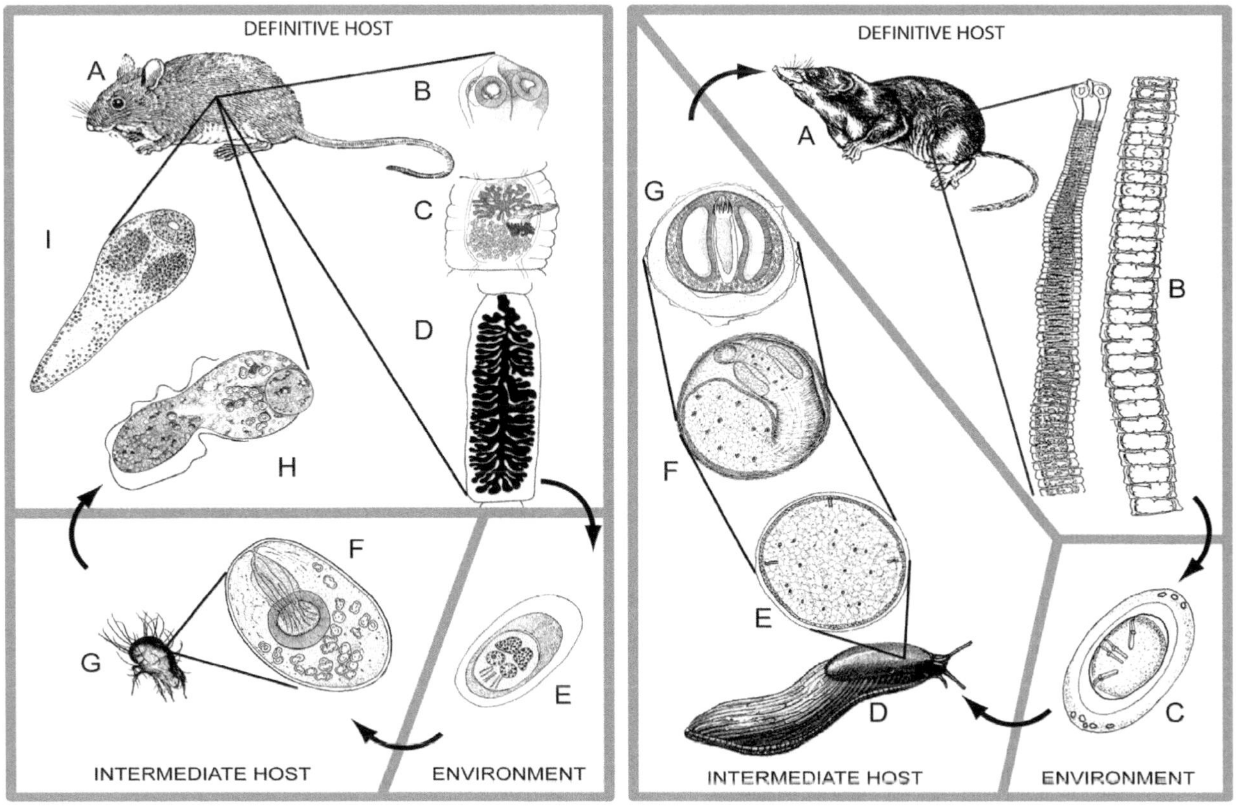

Fig. 5. Life cycles of *Catenotaenia pusilla* (Catenotaeniidae) and *Monocercus arionis* (Dilepididae). **Left.** *C. pusilla.* A. House mouse, *Mus musculus* (definitive host). B, C, D. Body parts of adult cestode (B, scolex; C, mature proglottis; D, gravid proglottis). E. Egg. F. Infective metacestode (merocercoid) in the body cavity of the intermediate host. G. Tyroglyphid mite, *Glycyphagus domesticus* (intermediate host). H. Merocercoid excysting under the action of digestive enzymes in the gut of the definitive host. I. Young cestode with developing scolex in the intestine of the definitive host. (B-F, H, I – modified from Joyeux and Baer 1945). **Right.** *M. arionis* (=*Molluscotaenia crassiscolex*) (Dilepididae). A. Common shrew, *Sorex araneus* (definitive host). B. Adult cestode. C. Egg. D. Slug, *Arion lusitanicus* (intermediate host). E-G. Gradual stages of metacestode development in the intermediate host. G. Infective monocysticercoid. (B – modified from Mel'nikova and Gulyaev 2004; C, E, F – modified from Jourdane 1972; G – modified from Kisielewska 1958a)

The life cycle of only one species has been described (Joyeux and Baer 1945). *Catenotaenia pusilla* is a common intestinal parasite of the house mouse and has been reported also from other rodents. Tyroglyphid mites have been demonstrated as its intermediate host (Fig. 5). The metacestode (merocercoid) develops to the infective stage within some 15 days. It has a large apical organ ("apical sucker") but the suckers are not developed. The final stage of the scolex morphogenesis is in the intestine of the definitive host, where degeneration of the apical organ and differentiation of suckers occur.

3.2 Dilepididae

This family is characterised by great taxonomic diversity: the number of the valid genera exceeds 100 (Bona 1994) and the number of species is not less than 500 (Matevosyan 1963; Schmidt 1986). However, the dilepidids are widespread in birds and only few of them occur in small mammals (Table 1). The family has a cosmopolitan distribution.

The members of this family are usually characterised by a complex rostellar apparatus consisting of a rostellum, rostellar pouch and two (rarely one) rows of rostellar hooks. Usually, the testes in the mature proglottis are numerous, situated posteriorly to the female gonads or around them. They lack seminal vesicles and the function of sperm storage is carried out by a highly convoluted vas deferens.

The life cycles of about 30 species of Dilepididae are known in terms of recorded intermediate hosts and described metacestodes. However, almost all these are parasites of birds. The range of intermediate hosts includes annelids, molluscs and arthropods.

Monocercus arionis is an intestinal parasite of shrews of the genus *Sorex* throughout northern Eurasia (Fig. 5). Its metacestode has been recorded in more than 20 species of terrestrial gastropods (Kisielewska 1958a; Jourdane 1972; Genov 1984). *Monocercus estavarensis* has a similar life cycle, parasitizing terrestrial gastropods as intermediate hosts and shrews as definitive hosts (Jourdane 1972). Another species, *Multitesticulata filamentosa*, is a parasite of moles (*Talpa europaea*) in Europe; its intermediate hosts are myriapods (Genov 1984).

Sometimes immature specimens of the genera *Dilepis* (e.g. *Dilepis undula*) and *Polycercus* (e.g. *P. paradoxa*) have been recorded from insectivores. These are parasites of birds, typically occurring in thrushes and woodcocks, respectively. Their intermediate hosts are earthworms.

3.3 Hymenolepididae

With the number of species exceeding 900 (McLaughlin 2003), this is the most species-rich cestode family. The majority of hymenolepidids are parasites of birds (aquatic or terrestrial), but at least a quarter of the species have been described from mammals (Schmidt 1986; McLaughlin 2003). The family has a cosmopolitan distribution reflecting the broad geographical distribution of its hosts.

Morphologically, the hymenolepidids are characterised by a complex rostellar apparatus similar to that of dilepidids; however, generally they have only a single row of rostellar hooks. There are also some genera, which have no rostellar hooks and the rostellum is lacking or rudimentary. The male reproductive system contains a small number of testes: usually three, sometimes two or even one, and only exceptionally more than three. The role of sperm storage is performed by seminal vesicles. The proglottides are typically much wider than long, which results in a higher number of proglottides relative to strobila size than in most other families (e.g., Dilepididae). It is possible that this strobilar organisation results in more frequent and more widespread dissemination of eggs by hymenolepidids.

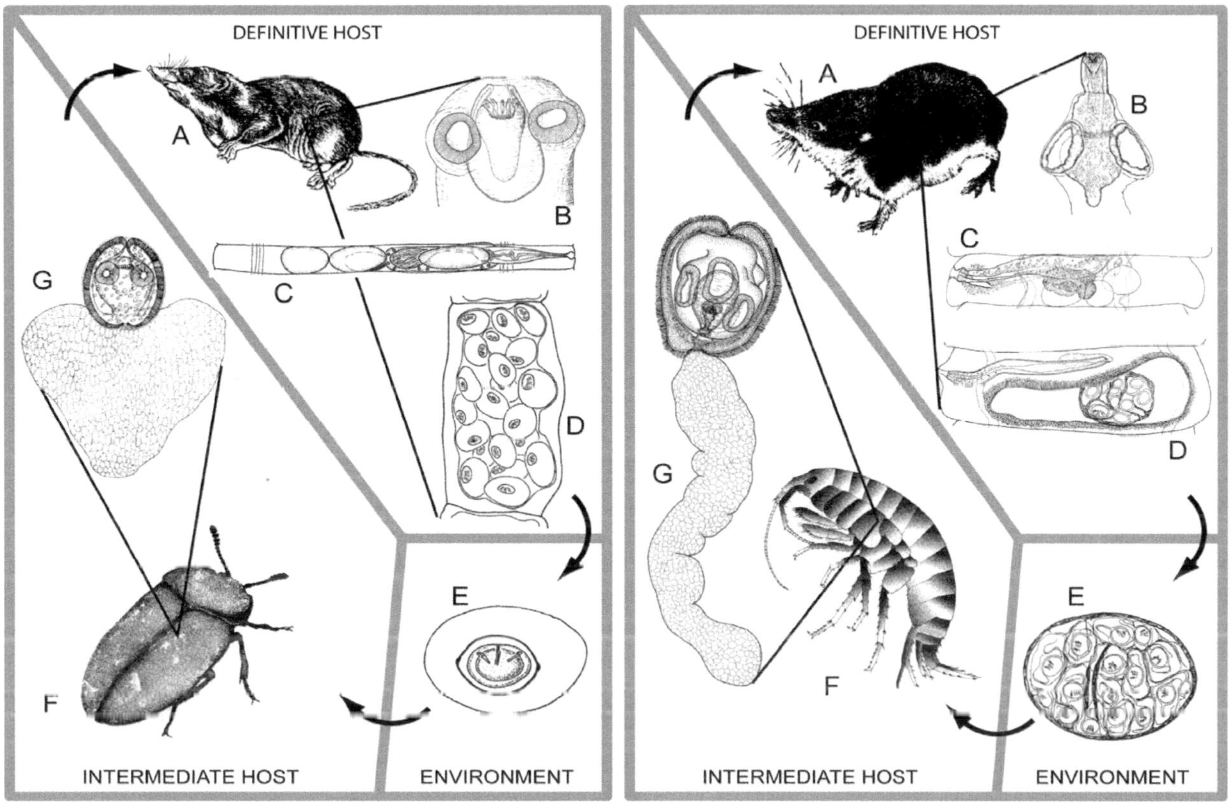

Fig. 6. Life cycles of hymenolepidids from soricids. **Left.** *Neoskrjabinolepis schaldybini*. A. Common shrew, *Sorex araneus* (definitive host). B-D. Adult cestode from the intestine of the definitive host (modified from Vaucher 1971). B. Scolex. C. Mature proglottis. D. Gravid proglottis. E. Egg (modified from Jourdane 1971). F. Red-breasted carrion beetle, *Oiceoptoma thoracica*, (intermediate host). G. Cysticercoid from the body cavity of the intermediate host (modified after Prokopič 1968b). **Right.** *Vaucherilepis trichophorus*. A. Water shrew, *Neomys fodiens* (definitive host). B-D. Adult cestode from the intestine of the definitive host. B. Scolex. C. Mature proglottis. D. Gravid proglottis. E. Eggs are distributed in the environment embedded in uterine capsule, adaptation for group transmission and possibly for attraction of the intermediate host. F. *Gammarus* (*Rivulogammarus*) *balcanius* (intermediate host). G. Cysticercoid from the body cavity of the intermediate host (B-D, E, G, modified from Tkach et al. 2003)

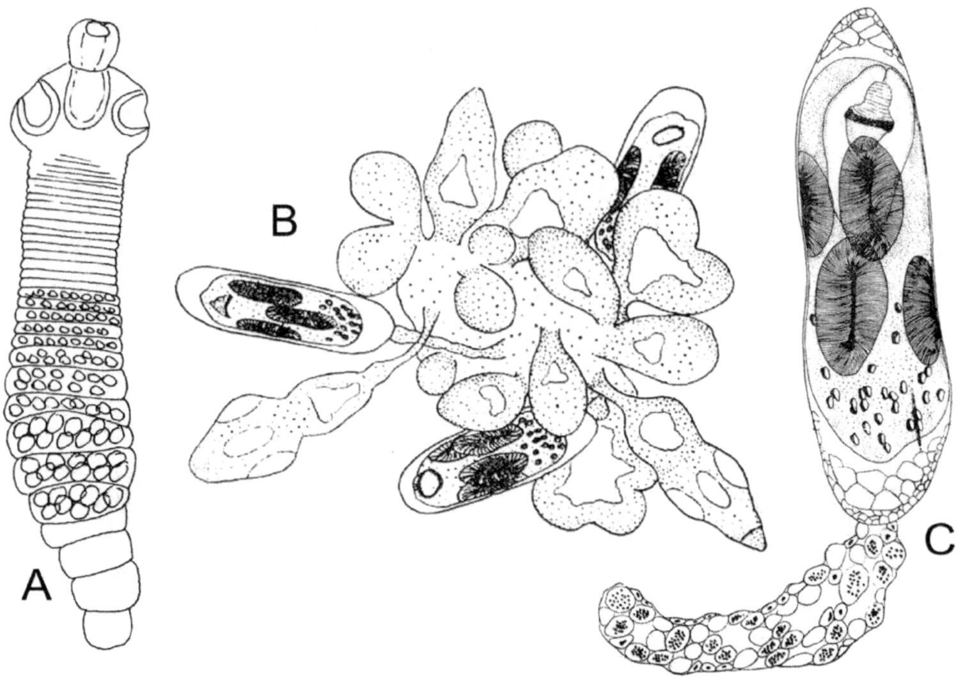

Fig. 7. *Urocystis prolifer* (Hymenolepididae). A. Adult worm from the intestine of a shrew (redrawn from Baer and Della Santa 1960). Adults (up to 20,000 specimens per host) are 0.25-0.55 mm long (Baer and Della Santa 1960) and their gravid proglottides contain few eggs (7-14, according to Kisielewska 1960). B. Early stage of metacestode development in the body cavity of the myriapod *Glomeris connexa*. The development of the oncosphere gives the beginning of cellular mass of irregular shape producing cysticercoids by budding. C. Fully developed cysticercoid. (B and C modified from Kisielewska 1960)

The life cycles of about 200 hymenolepidid species are known. Arthropods and annelids are their intermediate hosts in both aquatic and terrestrial environments.

The number of the hymenolepidid genera occurring in insectivores exceeds 30; almost all are specific parasites to the families or the genera. A number of taxa occurring in *Sorex* spp. as definitive hosts use insects as intermediate hosts. These include, e.g., *Vigisolepis spinulosa* recorded in the collembolans *Tomocerus flavesvens* (see Prokopič 1968a), *Lineolepis scutigera* in fleas of the genus *Ctenophthalmus* (see Quentin and Beaucournu 1966), *Neoskrjabinolepis schaldybini* in coleopterans of the genera *Catops* (Leiodidae) and *Oiceoptoma* (Silphidae) (see Kisielewska 1958b; Prokopič 1968b; Vaucher 1971) (Fig. 6), *Staphylocystis furcata* in a wide range of coleopterans (Geotrupidae, Carabidae and Silphidae) and orthopterans (Ryšavý and Prokopič 1965; Ryšavý 1989). *Urocystis prolifer* (Fig. 7) is another widespread intestinal parasite of *Sorex* spp. in Europe (Baer and Della Santa 1960; Genov 1984). Its intermediate hosts are glomerid diplopods in the body cavities of which metacestodes reproduce asexually by budding (Joyeux 1922; Kisielewska 1960).

Similarly, hymenolepidids parasitizing *Crocidura* spp. also use insects and myriapods as intermediate hosts. For example, *Staphylocystis brusatae* has been recorded in sand flies *Phlebotomus* spp. (Quentin et al. 1972), *S. uncinata* in coleopterans *Silpha* spp. (Vaucher 1971; Genov 1984), *S. scalaris* and *S. pistillum* in glomerids (Joyeux and Baer 1936). Asexual reproduction of metacestodes has been recorded in the latter species (Joyeux and Baer 1936). *Pseudhymenolepis redonica* is hyperapolytic (Fig. 7), i.e. its proglottides are detached before reaching maturation and live separately in the intestine (Joyeux and Baer 1936). Fleas (Quentin and Beaucournu 1966) and opiliones (Gabrion 1977) serve as its intermediate hosts.

About 20 hymenolepidid species of the genera *Coronacanthus*, *Triodontolepis*, *Neomylepis, Pseudobothrialepis* and *Vaucherilepis* occur in water shrews of the genus *Neomys* (Vaucher 1971; Tkach et al. 2003; Vasileva et al. 2004). For all these genera, use of freshwater amphipods as intermediate hosts in the life cycles has been demonstrated (Fig. 6).

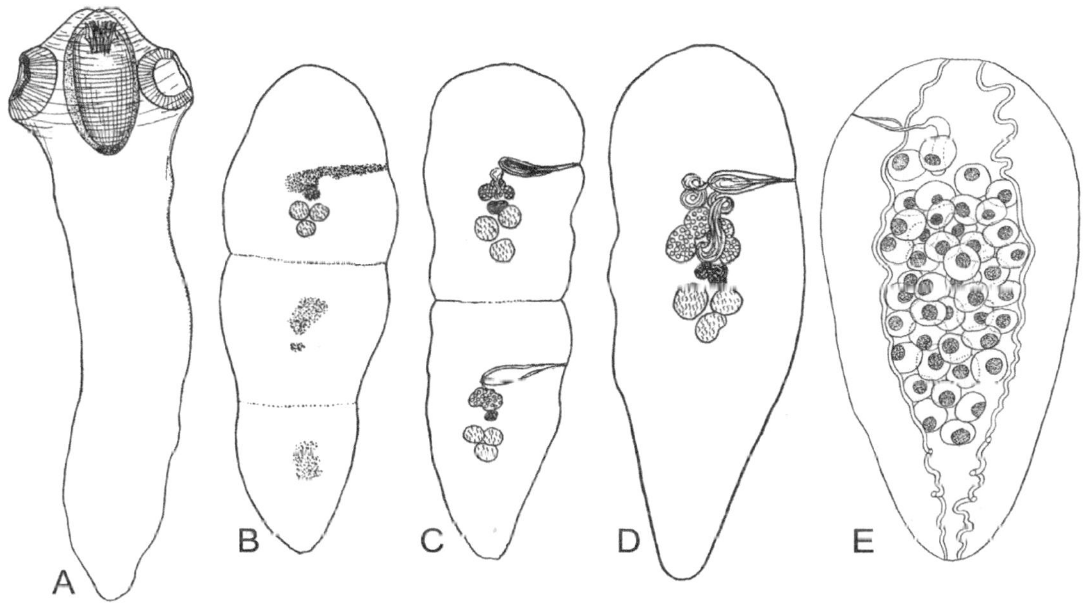

Fig. 8. *Pseudhymenolepis redonica* (Hymenolepididae), a hyperapolytic intestinal parasite of *Crocidura* spp. The scolex and the neck (A) are attached at the intestinal wall. The neck forms strobilar segments (B-E), which detach from it before their maturation and continue their development in intestine. Modified after Joyeux and Baer (1936)

The hymenolepidid cestodes of rodents are also characterised by considerable taxonomic diversity (Table 1). *Hymenolepis diminuta* has a cosmopolitan distribution, frequently recorded in rats. Burt (1980) presented a list of 99 definitive host species (93 rodent species) and 66 species of intermediate hosts (29 coleopterans, 2 dermapterans, 2 embiopterans, 11 lepidopterans, 9 orthopterans, 11 siphonapterans and 2 diplopods).

Rodentolepis fraterna occurs mostly in the domestic mouse, sometimes also in other rodents; its metacestodes have been shown to develop in fleas and tenebrionid beetles. However, this species is unique among tapeworms in that eggs swallowed by the definitive host may give rise to cysticercoids located in the intestinal villi. Therefore, the life cycle can be completed in the same host, without participation of invertebrate intermediate host.

The metacestodes of *Rodentolepis asymmetrica* (a parasite mostly of voles) have been recorded in acarines *Archipteria coleoptera* (see Prokopič and Mauer 1969) and *Ceratozetella sellnicki* (see Pavlichenko et al. 1992).

Hymenolepidids are the main cestode group occurring in bats (Table 1). However, no life cycle is known for any species.

3.4 Anoplocephalidae

The Anoplocephalidae are characterised by the lack of a rostellum, i.e. the scolex has four suckers only. Currently, four subfamilies are recognised in it, i.e. Anoplocephalinae, Linstowiinae, Inermicapsiferinae and Thysanosomatinae. The former three subfamilies occur in small mammals. However, in this concept the Anoplocephalidae seems unlikely to be monophyletic (Beveridge 1994; Hoberg et al. 1999). The representatives of the subfamily Anoplocephalinae are characterised by the presence of eggs containing a conically-elongate envelope (modification of the embryophore) termed the "pyriform apparatus" (Fig. 9).

Typically, as with hymenolepidids, they have well-expressed seminal vesicles (external and internal). The intermediate hosts are mostly mites and definitive hosts (mostly herbivorous mammals) accidentally ingest them while grazing. The Anoplocephalinae has a cosmopolitan distribution occurring in equids, ruminants, rodents, lagomorphs, marsupials, dermopterans, primates and some birds (Spasskii 1951; Beveridge 1994).

Most of the life cycle information about anoplocephaline cestodes comes from studies on the genera and species occurring in ruminants and equids, where oribatean mites are implicated intermediate hosts. Data on life cycles of anoplocephalids from small mammals are scarce. However, the life cycles of *Cittotaenia denticulata* and *Mosgovoyia ctenoides*, occurring in rabbits, have been studied in detail (Stunkard 1941). Their intermediate hosts are oribatean mites of various families (Fig. 9). The life cycle of the species of the genus *Monoecocestus*, parasitic in porcupines, is similar (Freeman 1952). There are data that collembolans participate as intermediate hosts in the life cycle of *Paranoplocephala omphalodes*, a parasite of voles, in the Russian tundra (Smirnova and Shalayeva 1986).

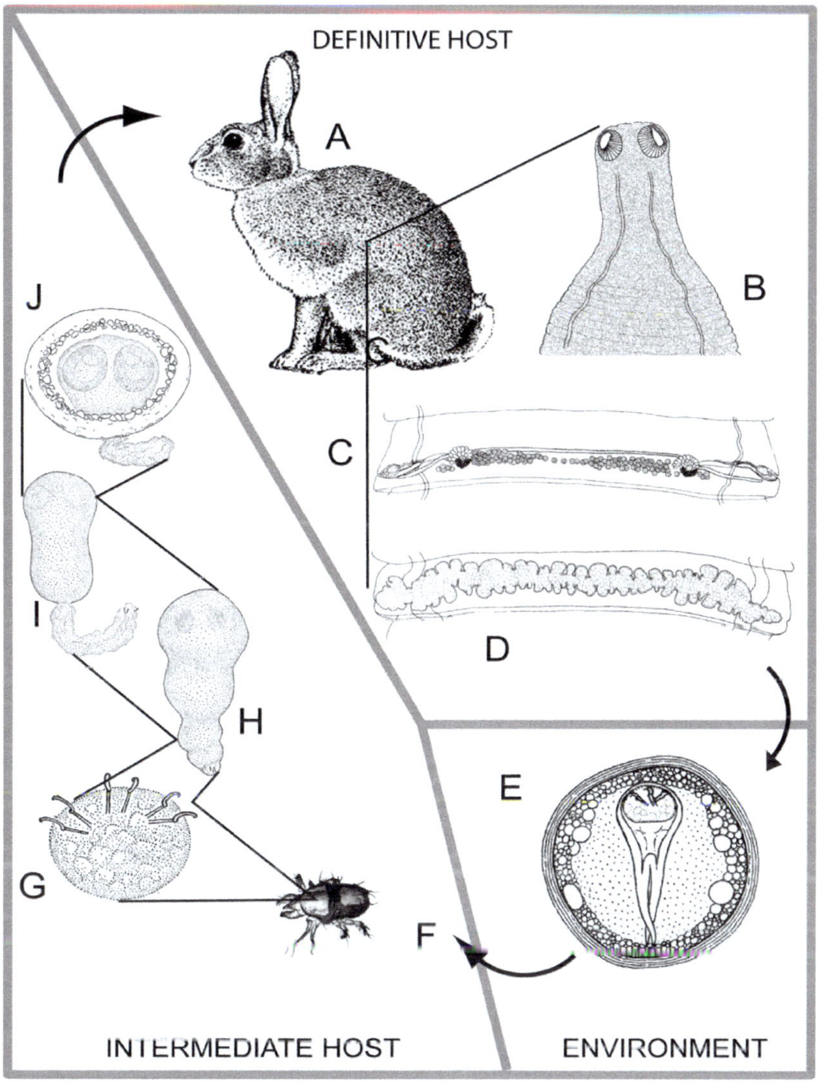

Fig. 9. Life cycle of *Mosgovoyia ctenoides*. A. Rabbit, *Oryctolagus cuniculus* (definitive host). B-D. Parts of the adult worm from the intestine of the definitive hosts (modified from Beveridge 1978). B. Scolex. C. Mature proglottis. D. Gravid proglottis. E. Egg. Note the pyriform apparatus (modified from Stunkard 1941). F. Mite *Scutovertex minutus* (intermediate host). G-J. Gradual stages of the metacestode development from the body cavity of the intermediate host (modified from Stunkard 1941). G. Oncosphere. H, I. Gradual stages of body differentiation. J. Infective cysticercoid. According to Stunkard (1941), about 70 days are needed for the metacestode to reach the infective encysted stage

4 Small mammals as intermediate and paratenic hosts of cestodes

Small mammals participate as intermediate and paratenic hosts in the life cycles of pseudophyllidean cestodes (Diphyllobothriidae), as the only intermediate hosts in the life cycles of cyclophyllidean cestodes of the families Taeniidae and Paruterinidae, as intermediate and paratenic hosts in the life cycles of the Mesocestoididae (*Mesocestoides*) and as host of metacestodes (whether as intermediate or paratenic disputed) of the Dipylidiidae (*Joyeuxiella*). Definitive hosts of all these cestodes are carnivorous mammals and/or birds of prey.

4.1 Diphyllobothriidae

According to the taxonomic concept adopted here (Bray et al. 1994), the family includes pseudophyllideans occurring as adults in reptiles, birds and mammals; their life cycles typically include copepods as hosts of procercoids (first intermediate host) and fishes as hosts of plerocercoids (second intermediate host). However, the species of the cosmopolitan genus *Spirometra*, which includes intestinal parasites of carnivorous mammals, use tetrapods as second intermediate and paratenic hosts, i.e. amphibians, reptiles and mammals (Dubinina 1951; Mueller 1974; Uchida 2003). Their life cycles have been studied in detail, mostly because plerocercoids (termed spargana, singular sparganum) are agents of the disease human sparganosis and because they were found to secrete a hormone-like factor stimulating an overgrowth of mammals (Mueller 1974; Hirai 2003).

S. erinaceieuropaei is widespread in the Old World, in many parts of it geographical range (e.g., in Europe) exhibiting a focal pattern of distribution. Its life cycle was studied in the Volga Delta (Dubinina 1951), where mostly frogs (*Rana* spp.) and grass snakes (*Natrix* spp.) were infected with plerocercoids. However, rodents also participated, supposedly being infested either by ingesting crustaceans with procercoids (i.e. as a second intermediate host) or by swallowing developed plerocercoids with the flesh of intermediate hosts (i.e. as a paratenic host). Dubinina (1951) believed that the transmission route using rodents had secondary importance in the Volga Delta region. In contrast, a study carried out in the Srebarna Reserve, a wetland associated with the lower Danube in Bulgaria (Genov 1969), showed that two species of frogs and nine species of mammals (rodents, insectivores and mustelid carnivores) were infected with plerocercoids of *S. erinaceieuropaei*. Two carnivore species, stray domestic cats and wild cats (*Felis silvestris*), were recorded as definitive hosts. While the most abundant frog species, *Rana ridibunda*, had a prevalence of plerocercoid infection of 8.6%, the prevalence in the most abundant insectivores *Crocidura leucodon* and *C. suaveolens* reached 29.6 and 25.0%, respectively. Therefore, the importance of the small mammal transmission route in that ecosystem was comparable with, or perhaps of more importance, than that of the amphibian route. Use of small mammals (rodents) was also demonstrated for the life cycle of the North American *Spirometra mansonoides* (see Mueller 1938).

4.2 Taeniidae

This family includes about 50 species, which are intestinal parasites of carnivorous mammals or humans as adults and their metacestodes occur in the internal organs, musculature, body cavity, connective tissue or other parenteral sites ("external to gut") of mammals (Abuladze 1964; Verster 1969; Loos-Frank 2000). The family includes five species of primary medical importance and about 20 species of basic veterinary importance. The main intermediate hosts of taeniids are two groups of mammals: the ruminants and small mammals (mostly rodents). They all become infected by eating food contaminated with taeniid eggs. Abuladze (1964) listed 112 species of rodents, 16 species of lagomorphs and two species of insectivores as intermediate hosts of taeniids.

Currently, the validity of two genera, *Taenia* and *Echinococcus*, is widely accepted (see Rausch 1994). The former genus exhibits a morphological uniformity of the adult cestodes but diverse morphology of metacestodes; their structural peculiarities have sometimes been used as characters for splitting *Taenia* into several genera, e.g. *Hydatigera*, *Multiceps*, *Taeniarhynchus* and *Tetratirotaenia* (see Abuladze 1964). We adopt here the concept of Verster (1969) and Rausch (1994) recognising the latter generic names as synonyms of *Taenia*, a course of action in line with a recent phylogenetic study (Hoberg et al. 2000).

According to a recent review (Loos-Frank 2000), out of 44 species of *Taenia*, 20 use small mammals as intermediate hosts: metacestodes of 16 occur in rodents, of eight in lagomorphs and of two in insectivores. The following morphological modifications of metacestodes of *Taenia* spp. are found in small mammals (terminology follows Hoberg et al. 2000 and Chervy 2002, see Figs. 3 and 10):

- Cysticercus. This is the basic modification of metacestode in the Taeniidae, and is characterised by a scolex introverted (invaginated) during the development into the posterior bladder-like body part. This modification is widespread among *Taenia* spp. Examples for species having cysticerci in their life cycles are *T. crassiceps* with a wide range of carnivores as definitive hosts (frequent in European foxes) and a wide range of rodents and lagomorphs as intermediate hosts, in which it occurs in connective tissue and in the musculature; *T. pisiformis* with range of definitive hosts similar to that of the previous species and metacestodes occurring mostly in rabbits and hares. Hoberg et al. (2000) considered the cysticercus as plesiomorphic and the other metacestode modifications in taeniids as its derivatives.

- Strobilocercus. This is similar to the cysticercus but has, in addition to a scolex and neck, a segmented region ("metacestode strobila"). This metacestode occurs in several species previously referred to *Hydatigera*, e.g. *T. taeniaeformis*, which as an adult is found mostly in cats and as a metacestode in the liver of a wide range of rodents.
- Fimbriocercus (armatetrathyridium). This type of cysticercus is characterised by having an elongated unsegmented body. It occurs in the life cycle of several species, e.g. *Taenia martis*, which as an adult parasitizes martens (*Martes* spp.) and as a metacestode occurs in the thoracal cavity of bank voles (has been referred to the genus *Fimbriotaenia*).
- Coenurus. This metacestode forms a large bladder, which is full of liquid and is internally lined by a special layer (termed germinative membrane), which buds off multiple scoleces. It occurs in the species previously recognised as members of *Multiceps*, e.g. *T. serialis,* a tapeworm widespread in canids and having a metacestode occurring mostly in the intermuscular and subcutaneous connective tissue of rabbits and hares.
- Polycephalic. In this metacestode, several scoleces are situated on elongate stalks arising by exogenous budding from a central bladder, which later regresses. It is known for several species, e.g. for *T. twitchelli* from the wolverine in North America, with the metacestode found in the pleural and abdominal cavity of various rodents, mostly voles.

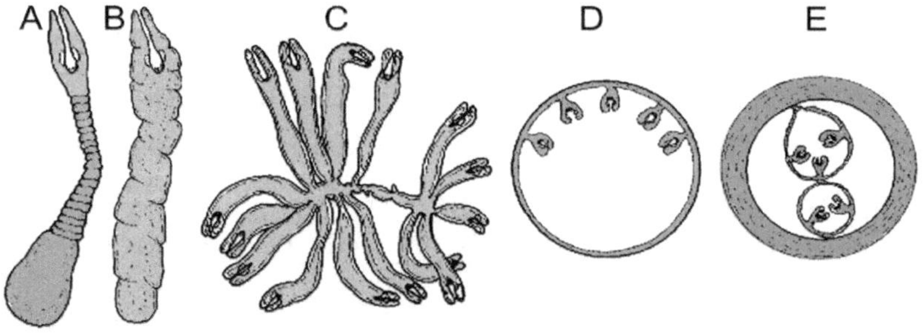

Fig. 10. Schematic presentation of the types of taeniid larvae (from Chervy 2002, reproduced with kind permission of Springer Science and Business Media). A. Strobilocercus. B. Fimbriocercus. C. Polycephalic. D. Coenurus. E. Echinococcus or hydatid (for cysticercus, see Fig. 3).

As seen from the above descriptions, the metacestodes belonging to the modifications "coenurus" and "polycephalic" are able to reproduce asexually in the intermediate host.

The second genus of the Taeniidae, *Echinococcus*, includes cestodes, which as adults are minute, only few millimetres long and consist of a few proglottides. They also occur in intestines of carnivores. Their metacestode is known as an echinococcus (= hydatid) and is located in the internal organs of herbivorous animals. Structurally, it resembles a coenurus but its germinative membrane is capable of producing daughter bladders, which, on their turn, can also form both daughter bladders and scoleces. Thus, the rate of the asexual reproduction of metacestodes is greatly increased, resulting in the proliferation of thousands of scoleces from a single oncosphere. Humans can also become infected as intermediate hosts. This disease is one of the major current problems of the medical parasitology.

The validity of four species is recognised (Rausch 1995; Thompson 1995). These are (after Rausch 1995): (1) *E. granulosus*, a cosmopolitan parasite occurring as an adult mostly in canids and as a metacestode in a wide range of ungulates. This species exhibits an immense genetic diversity associated with various life-cycle patterns (including the range of intermediate hosts) and, to a lesser extent, with the geographical distribution (for a survey, see McManus and Thompson 2003); the 'strains' perhaps deserve to be given subspecies or species status. (2) *E. multilocularis*, a species occurring in the Northern Hemisphere only, with a circumpolar distribution in the tundra and focal distribution in temperate latitudes, mostly in highland areas. Its definitive hosts are canids and felids, and its intermediate hosts are rodents, insectivores and lagomorphs. (3) *E. oligarthus*, a Neotropical species occurring from Central America (Costa Rica) to the subantarctic areas of Argentina, with wild felids as definitive hosts and rodents (mostly dasyproctids) as intermediate hosts. (4) *E. vogeli*, having a restricted geographical range in Central America and northern South America, with the bush dog (*Speothos venaticus*, Canidae) as definitive host and the paca (*Cuniculus paca*, Dasyproctidae) and few other rodent species as intermediate hosts; the participation of hunters' dogs in its life cycle as definitive hosts is possible. Recently, a fifth species (*E. shiquicus*) was proposed, occurring in Tibet and having a life cycle associated with the Tibetan fox *Vulpes ferrilata* and the plateau pika *Ochotona curzoniae*, distinguished mostly on the basis of genetic differences (see Xiao et al. 2005); however, its validity needs further confirmation, especially because its morphology is poorly described.

4.3 Paruterinidae

The number of the genera included in this family exceeds 20. Most of them are parasites of insectivorous birds. For two genera only, Paruterina (parasites of owls) and *Cladotaenia* (parasites of birds of prey), small mammals (rodents and insectivores) are known as intermediate hosts (Georgiev and Kornyushin 1994). Life cycles of several species of these two genera have been described in detail (Freeman 1957, 1959). The metacestodes were found in the liver and mesenteric lymph of various experimentally and naturally infected rodents. They have elongated body, their scolex is invaginated within the body and a cercomer and bladder are lacking. The terms "plerocercoids" (Freeman 1957, 1959), "cladothyridia" (Abuladze 1964) and "merocercoids" (Chervy 2002) have been used for them.

5 Concluding remarks

Cestodes are widespread in small mammals. Their taxonomic diversity is rather well studied in the Holarctic but poorly known in tropical areas. All cestodes using small mammals as definitive hosts belong to the order Cyclophyllidea. The numbers of the families and the genera occurring in each major group of small mammals (i.e., insectivores, bats, lagomorphs and rodents) are 4 and 37, 3 and 9, 3 and 14, and 6 and 47, respectively. Studies on the life cycles of the cestodes from small mammals are relatively few. No life cycle is known for any cestode species occurring in bats. As a rule, cestode species occurring in small mammals as definitive hosts are characterised by two-host life cycles, involving invertebrates as intermediate hosts. Most of the invertebrate hosts are terrestrial arthropods (insects, acari, myriapods, in some cases arachnids), rarely molluscs; only the species parasitizing water shrews are known to have life cycles involving crustaceans as intermediate hosts. Some life cycles demonstrate adaptations for group infestation, asexual reproduction within the intermediate host or even a possibility of infestation without the participation of an intermediate host. Small mammals participate as the only intermediate host in the life cycles of taeniid species parasitic in carnivore mammals (*Taenia*, *Echinococcus*) and paruterinid species parasitic in birds of prey (*Paruterina*, *Cladotaenia*). They can also provide alternative transmission routes as second intermediate hosts of some pseudophyllidean cestodes (*Spirometra*) and as (second?) intermediate hosts or paratenic hosts of some cyclophyllidean cestodes (*Mesocestoides*, *Joyeuxiella*).

References

Abuladze KI (1964) Taeniata – tapeworms of animals and man, and diseases caused by them (in Russian). In: Skrjabin KI (ed) Osnovy tsestodologii [Essentials of cestodology], vol 4. Nauka, Moscow

Arme C, Pappas PW (eds) (1984a) Biology of the Eucestoda, vol 1. Acad Press, London

Arme C, Pappas PW (eds) (1984b) Biology of the Eucestoda, vol 2. Acad Press, London

Baer JG (1927) Monographie des cestodes de la famille des Anoplocephalidae. Bulletin Biologique de la France et de la Belgique, Supplément 10:1–241

Baer JG, Della Santa E (1960) Matériaux pour servir à une révision des espèces du genre *Hymenolepis* Weinland (Cestoda), parasites de Musaraignes. I. *Hymenolepis prolifer* (Villot, 1880) Stammer, 1955. Bulletin de la Société Neuchâteloise des Sciences Naturelles 83:71–77

Baer JG, Tenora F (1970) Some species of *Hymenolepis* from rodents and from primates. Acta Scientarum Naturalium Academiae Scientarum Bohemoslovacae, Brno 4:1–32

Beveridge I (1978) A taxonomic revision of the genera *Cittotaenia* Riehm, 1881, *Ctenotaenia* Railliet, 1893, *Mosgovoyia* Spasskii, 1951 and *Pseudocittotaenia* Tenora, 1976 (Cestoda: Anoplocephalidae). Mémoires du Muséum National d'Histoire Naturelle, Nouvelle Série, Série A, Zoologie 107:1–64

Beveridge I (1994) Family Anoplocephalidae Cholodkovsky, 1902. In: Khalil LF, Jones A, Bray RA (eds) Keys to the cestode parasites of vertebrates. CABI, Wallingford, pp 315–366

Bona F (1994) Family Dilepididae Railliet & Henry, 1909. In: Khalil LF, Jones A, Bray RA (eds) Keys to the cestode parasites of vertebrates. CABI, Wallingford, pp 443–554

Bray RA, Jones A, Andersen KI (1994) Order Pseudophyllidea Carus, 1863. In: Khalil LF, Jones A, Bray RA (eds) Keys to the cestode parasites of vertebrates. CABI, Wallingford, pp 205–247

Burt MDB (1980) Aspects of life history and systematics of *Hymenolepis diminuta*. In: Arai HP (ed) Biology of the tapeworm *Hymenolepis diminuta*. Acad Press, New York, pp 1–57

Caira JN, Littlewood DTJ (2000) Worms, Platyhelminthes. In: Levin SA (ed) Encyclopedia of biodiversity, vol 5. Acad Press, London, pp 863–899

Chervy L (2002) The terminology of larval cestodes or metacestodes. Syst Parasitol 52:1–33

Coil WH (1991) Platyhelminthes: Cestoidea. In: Harrison FW, Bogitsh BJ (eds) Microscopic anatomy of invertebrates, vol 3. Platyhelminthes and Nemertinea. Wiley-Liss, London, pp 211–283

Czaplinski B, Vaucher C (1994) Family Hymenolepididae Ariola, 1899. In: Khalil LF, Jones A, Bray RA (eds) Keys to the cestode parasites of vertebrates. CABI, Wallingford, pp 595–663

Denegri G, Dopchiz MC, Elissondo MC, Beveridge I (2003) *Viscachataenia* n. g., with the redescription of *V. quadrata* (von Linstow, 1904) n. comb. (Cestoda: Anoplocephalidae) in *Lagidium viscacia* (Rodentia: Chinchillidae) from Argentina. Syst Parasitol 54:81–88

Dubinina MN (1951) On the biology and distribution of *Diphyllobothrium erinacei-europaei* (Rud., 1819) Iwata, 1933 (in Russian). Zool Zh 30:421–429

Feliu C, Renaud F, Catzeflis F, Hugot JP, Durand P, Morand S (1997) A comparative analysis of parasite species richness of Iberian rodents. Parasitology 115:453–466

Freeman RS (1952) The biology and life history of *Monoecocestus* Beddard, 1914 (Cestoda: Anoplocephalidae) from the porcupine. J Parasitol 38:111–129

Freeman RS (1957) Life cycle and morphology of *Paruterina rauschi* n. sp. and *P. candelabraria* (Goeze, 1782) (Cestoda) from owls, and significance of plerocercoids in the order Cyclophyllidea. Can J Zool 35:349–370

Freeman RS (1959) On the taxonomy of the genus *Cladotaenia*, the life histories of *C. globifera* (Batsch, 1786) and *C. circi* Yamaguti, 1935, and a note on distinguishing between the plerocercoids of the genera *Paruterina* and *Cladotaenia*. Can J Zool 37:317–340

Gabrion C (1977) Présence de cysticercoïdes d'un cestode cyclophyllide chez un arachnide, *Phalangium opilio*. Ann Parasitol Hum Comp 52:229–234

Genov T (1969) Natural infection of vertebrates in the Srebarna Reserve with plerocercoids of *Spirometra erinaceieuropaei* (Rudolphi, 1819) – Cestoidea, Diphyllobothriidae (in Bulgarian). Izvestiya na Tsentralnata Khelmintologichna Laboratoriya 13:197–209

Genov T (1984) Helminths of insectivores and rodents in Bulgaria (in Bulgarian). Izdatelstvo na Balgarskata Akademiya na Naukite, Sofia

Genov T, Murai É, Georgiev BB, Harris EA (1990) The erection of *Leporidotaenia* n. g. (Cestoda: Anoplocephalidae) for *Anoplocephaloides* spp. parasitizing Leporidae (Lagomorpha). Syst Parasitol 16:107–126

Genov T, Vasileva GP, Georgiev BB (1996) *Paranoplocephala aquatica* n. sp. (Cestoda, Anoplocephalidae) from *Arvicola terrestris* and *Ondatra zibethica* (Rodentia), with redescriptions and comments on related species. Syst Parasitol 34:135–152

Georgiev BB (2003) Cestoda (tapeworms). In: Thoney DA, Schlager N (eds) Grzimek's animal life encyclopedia (second edition), vol 1. Lower Metazoans and lesser Deuterostomes. Gale, Detroit, pp 225–243

Georgiev BB, Kornyushin VV (1994) Family Paruterinidae Fuhrmann, 1908 (*sensu lato*). In: Khalil LF, Jones A, Bray RA (eds) Keys to the cestode parasites of vertebrates. CABI, Wallingford, pp 559–584

Gulyaev VD (1996) On the taxonomic independence of *Anoplocephaloides* spp. (Cestoda: Anoplocephalidae) with serial alternation of genital atria (in Russian). Parazitologiya 30:263–269

Gulyaev VD, Chechulin AI (1996) *Parandrya feodorovi* gen. et sp. nov., a new cestode (Cyclophyllidea: Anoplocephalidae) from Siberian voles (in Russian). Parazitologiya 30:132–140

Gulyaev VD, Kornienko SA (1998) On morphological peculiarities of cysticercoids of *Monocercus* (Cestoda: Cyclophyllidea: Dilepididae) (in Russian). Parazitologiya 32:141–145

Gulyaev VD, Kornienko SA (1999) On the morphology of *Cryptocotylepis globosoides* (Cestoda: Hymenolepididae) – a cestode from water shrews from the Palaearctic region (in Russian). Parazitologiya 33:49–54

Gulyaev VD, Mel'nikova YA (2005) A new cestode genus from moles, *Talpolepis* gen. n., and redescription of *T. peipingensis* (Hsü, 1935) comb. n. (Cyclophyllidea: Hymenolepididae) (in Russian). In: Anonymous (ed) Problemy tsestodologii [Problems of cestodology]. Russian Acad Sci, Zool Inst, St. Petersburg, pp 130–139

Gulyaev VD, Lykova KA, Mel'nikova YA., Bibik EV (2004) On independence of the species *Spasskylepis ovaluteri* (Cestoda, Cyclophyllidea, Hymenolepididae), a parasite of shrews in Palaearctic Region (in Russian). Zool Zh 83:387–394

Gvozdev EV, Kontrimavichus VL, Ryzhikov KM, Shaldybin LS (1970) Key to the helminth parasites of Lagomorpha in the USSR (in Russian). Nauka, Moscow

Haukisalmi V, Henttonen H (2005) Description of *Paranoplocephala yoccozi* n. sp. (Cestoda: Anoplocephalidae) from the snow vole *Chionomys nivalis* in France, with a review of anoplocephalid cestodes of snow voles in Europe. Parasite 12:203–211

Haukisalmi V, Wickström LM (2005) Morphological characterisation of *Andrya* Railliet, 1893, *Neandrya* n. g. and *Paranoplocephala* Lühe, 1910 (Cestoda: Anoplocephalidae) in rodents and lagomorphs. Syst Parasitol 62:209–219

Haukisalmi V, Henttonen H, Niemimaa J, Rausch RL (2002) Description of *Paranoplocephala etholeni* n. sp. (Cestoda: Anoplocephalidae) in the meadow vole *Microtus pennsylvanicus*, with a synopsis of *Paranoplocephala* s. l. in Holarctic rodents. Parasite 9:305–314

Hirai K (2003) *Spirometra erinaceieuropaei* (2) Growth hormone-like factor produced by plerocercoids of *Spirometra erinaceieuropaei*. In: Otsuru M, Kamegai S, Hayashi S (eds) Progress of medical parasitology in Japan, vol 8. Meguro Parasitological Museum, Tokyo, pp 273–280

Hoberg EP, Jones A, Bray RA (1999) Phylogenetic analysis among the families of the Cyclophyllidea (Eucestoda) based on comparative morphology, with new hypotheses for co-evolution in vertebrates. Syst Parasitol 42:51–73

Hoberg EP, Jones A, Rausch RL, Eom KS, Gardner SL (2000) A phylogenetic hypothesis for the species of the genus *Taenia* (Eucestoda: Taeniidae). J Parasitol 86:89–98

Hoberg EP, Mariaux J, Brooks DR (2001) Pylogeny among the orders of the Eucestoda (Cercomeromorphae): Integrating morphology, molecules and total evidence. In: Littlewood DTJ, Bray RA (eds) Interrelationships of the Platyhelminthes. Taylor & Francis, London, pp 112–126

Hunkeler P (1974) Les Cestodes parasites de petits mammifères (Rongeurs et Insectivores) de Côte-d'Ivoire et de Haute-Volta. Rev Suisse Zool 80:809–930

Jones A, Bray RA (1994) Family Davaineidae Braun, 1900. In: Khalil LF, Jones A, Bray RA (eds) Keys to the cestode parasites of vertebrates. CABI, Wallingford, pp 407–441

Jourdane J (1971) Helminthes parasites des micromammifères de Pyrénées-Orientales. II. Les Plathelminthes de Soricinae. Ann Parasitol Hum Comp 46:553–574

Jourdane J (1972) Etude expérimentale du cycle biologique de deux espèces de *Choanotaenia* intestinaux des Soricidae. Z Parasitenkd 38:333–343

Joyeux C (1922) Recherches sur l'*Urocystis prolifer* Villot. Note préliminaire. Bull Soc Zool France 47:53–58

Joyeux C, Baer JG (1936) Quelques helminthes nouveaux et peu connus de la musaraigne, *Crocidura russula* Herm. Rev Suisse Zool 43:25–50

Joyeux C, Baer JG (1945) Morphologie, évolution et position systématique de *Catenotaenia pusilla* (Goeze, 1782), cestode parasite de Rongeurs. Rev Suisse Zool 52:13–51

Joyeux C, Baer JG (1962) Classe des cestodes. Cestoidea Rudolphi. In: Grassé PP (ed) Traité de Zoologie. Anatomie, systématique, biologie. Masson, Paris, pp 347–560

Karpenko SV, Chechulin AI (2000) Redescription of *Spalania crassisaccata* (Cestoda, Cyclophyllidea, Hymenolepididae) from shrews of Kunashir Island (in Russian). Zool Zh 79:1147–1154

Karpenko SV, Gulyaev VD (1999) *Brachylepis* gen. n. – a new cestode genus (Cyclophyllidea: Hymenolepididae) from shrews of Siberia and Russian Far East (in Russian). Parazitologiya 33:410–419

Khalil LF, Jones A, Bray RA (eds) (1994) Keys to the cestode parasites of vertebrates. CABI, Wallingford

Kisielewska K (1958a) The life cycle of *Choanotaenia crassiscolex* (Linstow, 1890) (Dilepididae) and some data relating to the formation of its cysticercoids. Bulletin de L'Academie Polonaise des Sciences 6:79–84

Kisielewska, K (1958b) Cysticercoid of the tapeworm *Neoskrjabinolepis singularis* (Cholodkovsky, 1912) Spassky, 1954 in a beetle of the family Catopidae. Bulletin de l'Academie Polonaise des Sciences 6:205–208

Kisielewska K (1960) Life cycle of the tapeworm *Pseudodiorchis prolifer* (Villot, 1890) comb. nova (= *Pseudodiorchis multispinosa* Żarnowski, 1955). Acta Parasitol Pol 8:197–204

Loos-Frank B (2000) An up-date of Verster's "Taxonomic revision of the genus *Taenia* Linnaeus" in table format. Syst Parasitol 45:155–183

Makarikov AA, Gulyaev VD, Chechulin AI (2005) *Arvicolepis* gen. n., a new cestode genus from rodents and redescription of *Arvicolepis transfuga* (Spassky et Merkusheva, 1967) comb. nov. (Cyclophyllidea, Hymenolepididae) (in Russian). In: Anonymous (ed) Problemy tsestodologii [Problems of cestodology]. Russian Acad Sci, Zool Inst, St. Petersburg, pp 178–186

Mas-Coma S, Tenora F (1997) Proposal of *Arostrilepis* n. gen. (Cestoda: Hymenolepididae). Res Rev Parasitol 57:93–101

McManus DP, Thompson RCA (2003) Molecular epidemiology of cystic echinococcosis. Parasitology 127:S37–S51

Matevosyan EM (1963) Dilepidoidea – tapeworms of domesticated and life animals (in Russian). In: Skrjabin KI (ed) Osnovy tsestodologii [Essentials of cestodology], vol 3. Publ House Akad Sci USSR, Moscow, pp 1–688

McLaughlin JD (2003) An annotated checklist of hymenolepidid cestodes described from birds: 1983-2002. Parassitologia 45:33–45

Mel'nikova YA, Gulyaev VD (2004) A new cestode species, *Monocercus dokuchaevi* sp. nov. (Cyclophyllidea: Dilepididae) from shrews in North-Eastern Asia (in Russian). Parazitologiya 38:68–73

Mel'nikova YA, Lykova KA, Gulyaev VD (2004) *Ecrinolepis kontrimavichusi* sp. nov. (Cyclophyllidea, Hymenolepididae), a new cestode in shrews from the Far East Southern Primorye (in Russian). Zool Zh 83:1275–1279

Movsesyan SO (2003a) Davaineata – tapeworms of animals and man, part I (in Russian). In: Beer SA (ed) Osnovy tsestodologii [Essentials of cestodology], vol 13(1). Nauka, Moscow, pp 1–395

Movsesyan SO (2003b) Davaineata – tapeworms of animals and man, part II (in Russian). In: Beer SA (ed) Osnovy tsestodologii [Essentials of cestodology], vol 13(2). Nauka, Moscow, pp 1–263

Mueller JF (1938) The life history of *Diphyllobothrium mansonoides* Mueller, 1935, and some considerations with regards to sparganosis in the United States. Am J Trop Med 18:41–68

Mueller JF (1974) The biology of *Spirometra*. J Parasitol 60:3–14

NHM 1988-2003, Host-parasite database. Natural History Museum, London, UK. http://www.nhm.ac.uk/research-curation/projects/host-parasites/

Olson PD, Littlewood DTJ, Bray RA, Mariaux J (2001) Interrelationships and evolution of the tapeworms (Platyhelminthes: Cestoda). Mol Phyl Evol 19:443–467

Pavlichenko PG, Tkach VV, Kornyushin VV (1992) *Ceratozetella sellnicki* (Acari, Oribatida) – a new intermediate host of the cestode *Rodentolepis asymmetrica* (in Russian). Vestnik Zoologii 5:76–79

Prokopič J (1968a) A description of the cysticercoid of the cestode *Vigisolepis spinulosa* (Cholodkowsky, 1906) found in Collembola. Folia Parasitol 15:266

Prokopič J (1968b) *Oecetoma thoracica* L., a new intermediate host of the cestode *Neoskrjabinolepis singularis* (Cholodkowsky, 1912). Folia Parasitol 15:105

Prokopič J, Mauer Z (1969) Některé poznatky o biologii tasemnic z drobnych savkù. Zprávy Československé Společnosti Parasitologické 9:22–30

Quentin JC (1964) Cestodes de rongeurs de République Centrafricaine. Cahiers de La Maboké 2:117–140

Quentin JC (1971) Cestodes *Skrjabinotaenia* de Rongeurs Muridés et Dendromuridés de Centrafrique. Hypothèse sur l'évolution des Cestodes Catenotaeniinae. Cahiers de La Maboké 9:57–79

Quentin JC (1994) Family Catenotaeniidae Spasskii, 1950. In: Khalil LF, Jones A, Bray RA (eds) Keys to the cestode parasites of vertebrates. CABI, Wallingford, pp 367–374

Quentin JC, Beaucournu JC (1966) Cysticercoïdes d'Hymenolepididae parasites d'Insectivores chez des Siphonaptères. C R Acad Sci Paris Sér D 262:2059–2062

Quentin JC, Jourdane J, Rioux JA, Leger N, Houin R, Croset H (1972) Présence du cysticercoïde d'*Hymenolepis brusatae* Vaucher, 1971, chez *Phlebotomus perniciosus* Newstead, 1911 et *Phletotomus mascittii* Grassi, 1908. Ann Parasitol Hum Comp 46:589–594

Rausch RL (1975) Cestodes of the genus *Hymenolepis* Weinland, 1858 (*sensu lato*) from bats in North America and Hawaii. Can J Zool 53:1537–1551

Rausch RL (1976) The genera *Paranoplocephala* Lühe, 1910 and *Anoplocephaloides* Baer, 1923 (Cestoda: Anoplocephalidae), with particular reference to species in rodents. Ann Parasitol Hum Comp 51:513–562

Rausch RL (1994) Family Taeniidae Ludwig, 1886. In: Khalil LF, Jones A, Bray RA (eds) Keys to the cestode parasites of vertebrates. CABI, Wallingford, pp 665-672

Rausch RL (1995) Life cycle patterns and geographic distribution of *Echinococcus* species. In: Thompson RCA, Lymbery AJ (eds) *Echinococcus* and hydatid disease. CABI, Wallingford, pp 89–134

Ruppert EE, Barnes RD (1994) Invertebrate zoology, 6th edition. Saunders College Publ, Forth Worth

Ryšavý B (1989) Orthoptera as intermediate hosts of *Staphylocystis furcata* (Stieda, 1862) (Cestoda: Hymenolepididae). Folia Parasitol 36:43–47

Ryšavý B, Prokopič J (1965) Neue Zwischenwirte des Bandwurmes *Staphylocystis furcata* Stieda, 1862 (Hymenolepididae). Z Parasitenk 25:371–374

Ryzhikov KM, Gvozdev EV, Tokobaev MM, Shaldybin LS, Matsaberidze GV, Merkusheva IV, Nadtochiy EV, Khokhlova IG, Sharpilo LD (1978) Key to the helminth parasites of rodents from the fauna of the USSR. Cestodes and trematodes (in Russian). Nauka, Moscow

Sawada I (1997) A world checklist of cestode species from Chiroptera. Nara (Japan), 18 pp (CABI Abstracts Record No: 19970803813)

Sawada I (1999) A checklist of cestode species from Asian Insectivora. Nara Sangyo University Journal of Industry and Economics 13:109–157

Schmidt GD (1986) CRC handbook of tapeworm identification. CRC Press, Boca Raton, Florida

Smirnova L, Shalajeva N (1986) The role of Collembola in the circulation of micromammalia cestodes in zonal tundra. Wiadomości Parazytologiczne 32:489–491

Smyth JD, McManus DP (1989) The physiology and biochemistry of cestodes. Cambridge Univ Press, Cambridge

Spasskii AA (1951) Anoplocephalata – tapeworms of domestic and wild animals (in Russian). In: Skryabin KI (ed) Osnovy tsestodologii [Essentials of cestodology], vol 1. Publ House Acad Sci USSR, Moscow, pp 1–736

Spasskii AA (1994) Three new genera of Davaineidae (Cestoda: Cyclophyllidea) - parasites of mammals (in Russian). Buletinul Academiei de Stiinte a Republicii Moldova. Stiinte Biologice si Chimice 6:66

Springer MS, Stanhope MJ, Madsen O, de Jong WW (2004) Molecules consolidate the placental mammal tree. Trends Ecol Evol 19:430–438

Stiles CW (1896) A revision of the adult tapeworms of hares and rabbits. Proc US Natl Mus 19:145–235

Stunkard HW (1941) Studies on the life history of the anoplocephaline cestodes of hares and rabbits. J Parasitol 27:299–325

Tenora F, Murai É (1978) Anoplocephalidae (Cestoda) parasites of Leporidae and Sciuridae in Europe. Acta Zoologica Academiae Scientiarum Hungaricae 24:415–429

Tenora F, Murai É (1980) The genera *Anoplocephaloides* and *Paranoplocephala* (Cestoda), parasites of Rodentia in Europe. Acta Zoologica Academiae Scientiarum Hungaricae 26:263–284

Tenora F, Murai É, Vaucher C (1985) On some *Paranoplocephala* species (Cestoda: Anoplocephalidae) parasitizing rodents (Rodentia) in Europe. Parasitologia Hungarica 18:29–48

Tenora F, Ganzorig S, Koubková B, Kamiya M (1999) Gynandry in *Hokkaidocephala* Tenora, Gulyaev, Kamiya, 1999 (Anoplocephalidae, Cestoda) parasitizing Rodentia in Japan. Helminthologia 36:269–273

Thompson RCA (1995) Biology and systematics of *Echinococcus*. In: Thompson RCA, Lymbery AJ (eds) *Echinococcus* and hydatid disease. CABI, Wallingford, pp 1–50

Tkach VV (1998) *Neomylepis* gen. n. – a new genus of hymenolepidid tapeworms (Cestoda, Cyclophyllidea), parasites of water shrews (in Russian). Vestnik Zoologii 32:90–93

Tkach VV, Kornyushin VV (1997) On the systematic position of *Hymenolepis blarinae* Rausch & Kuns, 1950 (Cestoda, Hymenolepididae), with a redescription of the holotype. Syst Parasitol 37:21–25

Tkach VV, Vasileva GP, Genov T (2003) Description of *Vaucherilepis trichophorus* sp. nov., gen. nov. (Cyclophyllidea, Hymenolepididae) from water shrews and gammarid crustaceans in Bulgaria and Ukraine. Acta Parasitol 48:87–97

Uchida A (2003) *Spirometra erinaceieuropaei* (1) Biology. In: Otsuru M, Kamegai S, Hayashi S (eds) Progress of medical parasitology in Japan, vol 8. Meguro Parasitological Museum, Tokyo, pp 261–272

Vasileva GP, Vaucher C, Tkach VV, Genov T (2004) Taxonomic revision of *Hilmylepis* Skryabin & Matevosyan, 1942 (Cyclophyllidea: Hymenolepididae). Syst Parasitol 59:45–63

Vasileva GP, Tkach VV, Genov T (2005) Two new hymenolepidid species (Cestoda, Hymenolepididae) from water shrews *Neomys fodiens* Pennant (Insectivora, Soricidae) in Bulgaria. Acta Parasitol 50:56–64

Vaucher C (1971) Les Cestodes parasites des Soricidae d'Europe. Etude anatomique, révision taxonomique et biologie. Rev Suisse Zool 78:1–113

Verster A (1969) A taxonomic revision of the genus *Taenia* Linnaeus, 1758 *s. str.* Onderstepoort J Vet Res 36:3–58

Wardle RA, McLeod JA (1952) The zoology of tapeworms. Univ Minnesota Press, Minneapolis

Wickström LM, Haukisalmi V, Varis S, Hantula J, Henttonen H (2005) Molecular phylogeny and systematics of anoplocephaline cestodes in rodents and lagomorphs. Syst Parasitol 62:83–99

Xiao N, Qiu J, Nakao M, Li T, Yang W, Chen X, Schantz PM, Craig PS, Ito A (2005) *Echinococcus shiquicus* n. sp., a taeniid cestode from Tibetan fox and plateau pika in China. Int J Parasitol 35:693–701

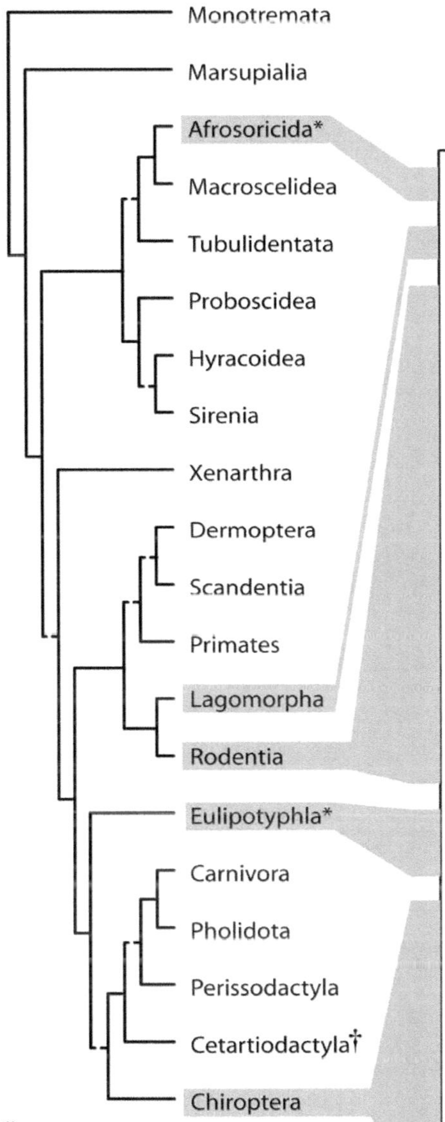

* = Insectivora

† = Artiodactyla+Cetacea

	CESTODA																																
	Anoplocephalidae																											Catenotaen.					
	Anoplocephalinae																					In.		Linstow.										
	Andrya	Anoplocephaloides	Bertiella	Cittotaenia	Ctenotaenia	Diandrya	Diuterinotaenia	Ectopocephalium	Gallegoides	Hokkaidocephala	Leporidotaenia	Moniezia	Monoecocestus	Mosgovoyia	Neandrya	Parandrya	Paranoplocephala	Pseudocittotaenia	Schizorchis	Sudarikovina	Viscachataenia	Inermicapsifer	Metacapsifer	Cycloskrjabinia	Mathevotaenia	Sinaiotaenia	Witenbergitaenia	Catenotaenia	Hemicatenotaenia	Meggittina	Pseudocatenotaenia	Quentinia	Skrjabinotaenia
Afrosoricida																																
Chrysochloridae																																
Tenrecidae																																
Lagomorpha																																
Leporidae	x			x							x			x	x							x											
Ochotonidae						x	x											x													
Rodentia																																
Anomaluridae																																
Aplodontidae																																
Castoridae																																
Ctenodactylidae																																
Dipodidae																								x								
Geomyidae		x															x	x										x					
Heteromyidae														x															x					
Muridae	x	x	x						x	x			x			x	x			x		x	x		x	x	x	x	x	x	x		x
Myoxidae																																
Pedetidae																																
Sciuridae		x		x	x							x				x								x			x	x				
Abrocomidae																																
Agoutidae																																
Bathyergidae																																
Capromyidae											x	x																				
Caviidae													x																			x	
Chinchillidae		x																	x												
Ctenomyidae													x																				
Dasyproctidae																																
Dinomyidae																																
Echimyidae													x																				
Erethizontidae													x																				
Heptaxodontidae																																
Hydrochaeridae													x																				
Hystricidae																																
Myocastoridae													x																				
Octodontidae	x																															
Petromuridae																																
Thryonomyidae																																
Eulipotyphla																																
Erinaceidae																								x								
Solenodontidae																																
Soricidae																								x								
Talpidae																																
Chiroptera																																
Craseonycteridae																																
Emballonuridae																																
Furipteridae																																
Megadermatidae																																
Molossidae																																
Mormoopidae																																
Mystacinidae																																
Myzopodidae																																
Natalidae																																
Noctilionidae																																
Nycteridae																																
Phyllostomidae																								x								
Pteropodidae																																
Rhinolophidae																																
Rhinopomatidae																																
Thyropteridae																																
Vespertilionidae																							x	x								

Appendix I. Distribution of cestode genera in orders and families of small mammals (as definitive hosts). The generic concepts are those presented in Khalil et al. (1994), with some subsequent developments added (see Table 1). The information is completed on the basis of the Host-Parasite Database maintained at the Division of Parasitology, Natural History Museum, London (NHM, 1988-2003). The phylogeny of the mammals is drawn from a molecular estimate adapted from Springer et al. (2004). Abbreviations: In – Inermicapsiferinae; Linstow. – Linstowiinae; P – Paruterinidae; Catenotaen. – Catenotaeniidae.

CESTODA

	Hymenolepididae																																										
	Arostrilepis	Arvicolepis	Biarinolepis	Brachylepis	Chitinolepis	Coronacanthus	Cryptocotylepis	Ditestolepis	Ecrinolepis	Gvosdevilepis	Hilmylepis	Hunkelepis	Hymenandrya	Hymenolepis	Lineolepis	Lockerrauschia	Lophurolepis	Mathevolepis	Neomylepis	Milina	Neoskrjabinolepis	Paraoligorchis	Protogynella	Pseudandrya	Pseudanoplocephala	Pseudhymenolepis	Pseudobothrialepis	Pseudodiorchis	Pseudoligorchis	Rodentolepis	Skrjabinacanthus	Soricinia	Spalania	Spasskylepis	Staphylocystis	Staphylocystoides	Talpolepis	Triodontolepis	Urocystis	Vampirolepis	Vaucherilepis	Vigisolepis	Vogelepis
Afrosoricida																																											
Chrysochloridae																																											
Tenrecidae																																											
Lagomorpha																																											
Leporidae										x																																	
Ochotonidae																																											
Rodentia																																											
Anomaluridae																																											
Aplodontidae																																											
Castoridae																																											
Ctenodactylidae																																											
Dipodidae																														x													
Geomyidae													x	x																													
Heteromyidae																																											
Muridae	x	x			x							x		x			x					x		x	x					x													
Myoxidae														x																x													
Pedetidae																																											
Sciuridae														x																													
Abrocomidae																																											
Agoutidae																																											
Bathyergidae																																											
Capromyidae																																											
Caviidae																																											
Chinchillidae																																											
Ctenomyidae																																											
Dasyproctidae																																											
Dinomyidae																																											
Echimyidae																																											
Erethizontidae																																											
Heptaxodontidae																																											
Hydrochaeridae																																											
Hystricidae																																											
Myocastoridae																																											
Octodontidae																																											
Petromuridae																																											
Thryonomyidae																																											
Eulipotyphla																																											
Erinaceidae														x																													
Solenodontidae																																												
Soricidae			x			x	x	x	x		x				x	x		x	x		x		x			x	x	x			x	x	x	x	x	x		x	x		x	x	x
Talpidae																																			x		x						
Chiroptera																																											
Craseonycteridae																																											
Emballonuridae																				x																				x			
Furipteridae																																											
Megadermatidae																																											
Molossidae																																								x			
Mormoopidae																																								x			
Mystacinidae																																											
Myzopodidae																																											
Natalidae																																											
Noctilionidae																																											
Nycteridae																																											
Phyllostomidae																																								x			
Pteropodidae																																								x			
Rhinolophidae																				x										x										x			
Rhinopomatidae																				x																							
Thyropteridae																																											
Vespertilionidae																				x																				x			

	Dilepididae							Davaineidae							P
	Alproma	Dilepis	Hepatocestus	Hunkeleria	Monocercus	Multitesticulata	Polycercus	Delamuretta	Dollfusoquenta	Fuhrmannetta	Paroniella	Raillietina	Skrjabinia	Vadifresia	Orthoskrjabinia
Afrosoricida															
Chrysochloridae		x													
Tenrecidae															
Lagomorpha															
Leporidae										x	x	x		x	
Ochotonidae															
Rodentia															
Anomaluridae															
Aplodontidae															
Castoridae															
Ctenodactylidae															
Dipodidae															x
Geomyidae															
Heteromyidae															
Muridae	x	x		x				x	x			x	x	x	x
Myoxidae															
Pedetidae															
Sciuridae												x			
Abrocomidae															
Agoutidae															
Bathyergidae												x			
Capromyidae															
Caviidae															
Chinchillidae															
Ctenomyidae															
Dasyproctidae												x			
Dinomyidae															
Echimyidae												x			
Erethizontidae															
Heptaxodontidae															
Hydrochaeridae															
Hystricidae															
Myocastoridae															
Octodontidae															
Petromuridae															
Thryonomyidae												x			
Eulipotyphla															
Erinaceidae												x			
Solenodontidae															
Soricidae		x	x		x		x								x
Talpidae		x				x	x								
Chiroptera															
Craseonycteridae															
Emballonuridae															
Furipteridae															
Megadermatidae															
Molossidae															
Mormoopidae															
Mystacinidae															
Myzopodidae															
Natalidae															
Noctilionidae															
Nycteridae															
Phyllostomidae															
Pteropodidae												x			
Rhinolophidae															
Rhinopomatidae															
Thyropteridae															
Vespertilionidae												x			

Appendix I. Continued

4 Nematodes

Serge Morand, Salah Bouamer and Jean-Pierre Hugot

1 Introductory remarks

More than 25,000 species of nematodes have been described including some 10,000 free-living marine and terrestrial nematodes, 3,500 parasites of invertebrates, and 12,000 parasitic nematodes of vertebrates (Poulin and Morand 2000; Hugot et al. 2001). These are assigned to more than 2270 genera and 256 families (Anderson 1992). When considering the estimated number of living species, which May (1988) evaluated at 1,000,000 and Hammond (1992) at 500,000, Nematoda are the most speciose phylum after Arthropoda. If we consider the lower estimate, only 5.3 % of the living species in Nematoda have been described. Knowing that, on average, 364 new nematode species were listed per year in the Zoological Records between 1979 and 1988 (Hammond 1992), it could take 1,300 years to achieve an extensive record of the living species of Nematoda. Although our knowledge of nematode species diversity is poor, we know a lot about the life-history of nematodes and in particular those parasitizing small mammals.

Nematodes have a uniform development and simple anatomical structure, but they display a great diversity of life-cycle with direct (monoxeny) and indirect (heteroxeny) transmission. Nematodes rarely kill their mammal hosts; they are sub-lethal parasites. Due to their impact on host-reproduction and survival, they have the potential to regulate their host populations.

2 Taxonomy and phylogeny

On the basis of complete 18S ribosomal RNA (rRNA) sequences, Aguinaldo et al. (1997) proposed that nematodes were related to arthropods in a clade of moulting animals they called Ecdysozoa.

According to recent molecular phylogenies (Blaxter et al. 1998; de Ley and Blaxter 2002), five major clades in the phylum Nematoda are recognized, with seven independent transitions from free-living towards parasitism on vertebrates.

Here, we follow Anderson (2000), the "CIH Keys to the nematode parasites of vertebrates" (Anderson et al. 1974-1983) and the recent phylogenetic framework based on molecular methods (Blaxter et al. 1998; de Ley and Blaxter 2002) (Fig. 1).

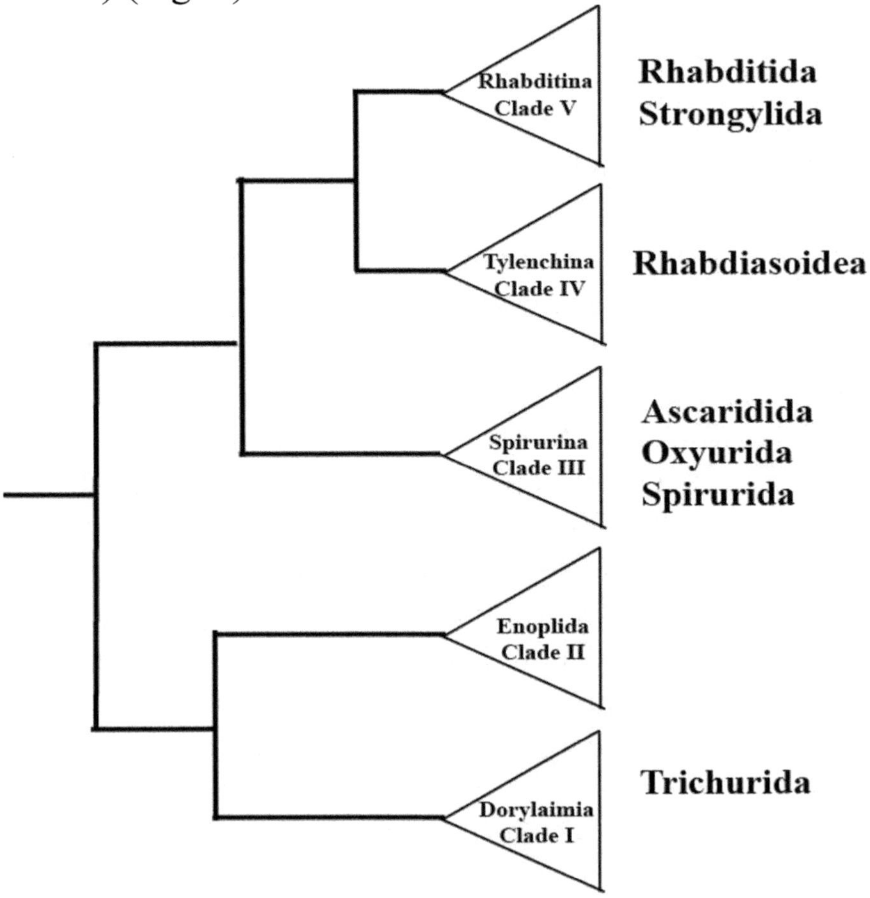

Fig. 1. Phylogenetic relationships among the five major nematode clades, with groups known to parasitize small mammals indicated in bold. The phylogeny was adapted from De Ley and Blaxter (2002).

Seven nematode orders have members that are found in small mammals (Fig. 1): Rhabditida, Strongylida, Panagrolaimida, Ascaridida, Oxyurida, Spirurida and Trichurida. Nematode superfamilies with some member species (as examples) that parasitize small mammals (small marsupials, insectivores, rodents, lagomorphs and elephant shrews) are as follows:

- **Clade I**
    - Order Trichurida
        - Superfamily Trichinelloidea

        - Trichuridae (whipworm): *Trichuris leporis* (caecum of rabbits), *Trichuris muris* (caecum of rats).
        - Capillariidae: *Calodium* (*Capillaria*) *hepaticum* (liver of various small mammals).
        - Trichinellidae: *Trichinella muris* (muscle cells of mice and rats), *Trichinella nativa* (rodents), *Trichinella pseudospiralis* (rodents).
    - Superfamily Muspiceoidea
        - *Muspicea borreli* (house mice; protandrous hermaphrodite).

- **Clade III**
  - Order Ascaridida
    - Superfamily Ascaridoidea (roundworms)
        - Ascarididae: *Baylisascaris laevis* (intestine of marmots and ground squirrels in North America); *Toxocara pteropodis* (intestine of flying foxes).
    - Superfamily Heterakoidea
        - Aspidoderidae, *Paraspidodera indica* (intestine of squirrels from India).
    - Superfamily Seuratoidea
        - Seuratidae: *Seuratum cadarachense* (intestine of Gliridae).
    - Superfamily Subuluroidea
        - *Maupasina weissi* (caecum of elephant shrews).
  - Order Oxyurida (pinworms)
    - Superfamily Oxyuroidea
        - Heteroxynematidae: *Aspiculuris tetraptera* (intestine of the Old World mice and rats).
        - Oxyuridae: *Passalurus ambiguus* (large intestine of lagomorphs), *Syphacia obvelata* (large intestine of rodents).
  - Order Spirurida
    - Superfamily Gnathostomatoidea

- *Gnathostoma spp.* (intestine of various small mammals).
  - Superfamily Physalopteroidea
    - *Physaloptera hispida* (stomach of cotton rats in North America).
  - Superfamily Rictularioidea
    - *Rictularia proni* (intestine of mice).
  - Superfamily Spiruroidea
    - *Protospirura numidica* (stomach and oesophagus of rodents), *Mastophorus muris* (stomach of Muridae).
  - Superfamily Filarioidea
    - Onchocercidae: *Achantocheilonema mansonbabri* (subcutaneous tissues and fascia of rodents; fleas are vectors), *Litomosoides carinii* (pleural and body cavity of cotton rats; microfilariae in blood; mites are vectors).
  - Superfamily Acuarioidea
    - Acuariidae: *Stammerinema suffodiax* (intestine of a dasyurid marsupial from Australia); *Antechiniella suffodiax* (intestine of a water rat from Australia).
  - Superfamily Thelazioidea
    - Pneumospiruridae; *Metathelazia caballeroi* (lungs of a Chiroptera in Malaysia); *Thelazia iheringi* (ocular cavity of a rat from Brazil).

- **Clade IV**
  - Superfamily Rhabdiasoidea
    - Rhabdiasidae: *Strongyloides ratti* (rodents), *Strongyloides venezuelensis* (Old World rats), *Parastrongyloides spp.* (insectivores and marsupials; dioecious parasitic forms).

- **Clade V**
  - Order Rhabditida
    - Superfamily Rhabditoidea
      - Rhabditidae: *Pelodera strongyloides* (free-living, occasionally invade dermis),

Rhabditis orbitalis (L3 can invade conjunctival sac of rodents).

- Order Strongylida (bursate nematodes)
  - Superfamily Ancylostomatoidea
    - *Ancylostoma* spp. (intestine of rodents).
  - Superfamily Metastrongyloidea
    - Chabertiidae: *Oesophagostomum selfi* (intestine of rats from Taiwan).
  - Superfamily Ancylostomatoidea
    - Ancylostomidae: *Cyclodontostomum purvisi* (intestine of rats in Indonesia).
  - Superfamily Strongyloidea
    - Strongyloidae: *Lamothiella romerolagi* (intestine of a rat and lagomorphs from Mexico); *Characostomum howelli* (intestine of a rat from Tanzania).
  - Superfamily Molineoidea
    - Molineidae: *Molineus neotetraci* (intestine of Insectivora and Chiroptera from China).
  - Superfamily Trichostrongyloidea
    - *Heligmosomoides polygyrus* (intestine of mice), *Nippostrongylus braziliensis* (*Rattus* spp.), *Obeliscoides cuniculi* (intestine of rabbits).
  - Superfamily Metastrongyloidea (lungworms)
    - *Parastrongylus cantonensis* (pulmonary arteries of rats) and *P. costaricensis* (mesenteric arteries of rodents), *Protostrongylus boughtoni* (lungs of snowshoe hares in Canada; gastropods are intermediate hosts).

The numbers of nematode genera and families differ greatly among the major groups of small mammals, with rodents harbouring the highest diversity in both genera and families (Table 1).

Table 1. Diversity of nematodes (numbers of genera and families) reported in four groups of small mammals.

Nematode diversity	Rodents	Lagomorphs	Chiropterans	Insectivores
Number of genera	141	32	34	6
Number of families	36	10	14	6

3 Life-cycle and biology

Nematodes display a wide variety of life cycles, some simple and direct (monoxenous) and some complex and indirect (heteroxenous) with one or more intermediate hosts (Anderson 1988), with or without host tissue migration (Read and Skorping 1995; Morand 1996). Despite the diversity and sometimes complexity of nematode life cycles, all of them can be related to the same basic pattern with two phases. The first phase takes place inside the definitive host where maturation and reproduction usually occur, and the pre-parasitic phase occurs either as a free-living larva in the external environment or inside an intermediate host.

Maupas (1900) noted that free-living rhabditoids passed through 5 stages separated by 4 moults and that the third stage (L3) initiated new populations when all other stages died due to depletion of environment resources. This developmental rule applies to the great majority of nematodes including parasites with five stages, plus the egg, constituting the basic pattern: the four larval stages (L1, L2, L3, L4) and one adult stage. Sometimes, the sexually immature adult stage is called L5. In all nematodes of clades III, IV and V, the L3 is the infective stage, whether the nematode requires an intermediate host or not, has free-living stages, or develops in the egg (Chabaud 1955).

3.1 Modes of infection

The routes of nematode infection are various (Adamson 1986; Anderson 1988) (Table 1) and include:

- Skin penetration by the infective third-stage larvae in Rhabditida (*Strongyloides* spp.) and in some Trichostrongylidae of rodents (*Nippostrongylus brasiliensis*);

- Oral ingestion of eggs containing the infective stage in Trichuridae (*Trichuris leporis*, *Trichuris muris*) and in Oxyuroidea (*Syphacia spp.*);

- Ingestion of the infective L3 contained in the tissues of the intermediate hosts in the heteroxenous groups Metastrongyloidea and Spirurida;
- Injection of the infective L3 by blood-sucking arthropod vectors in Filaroidea;
- Ingestion of eggs by coprophagy in rodents and lagomorphs in Oxyuroidae;
- Ingestion of eggs by allo- or auto-grooming in Oxyuroidae, Trichostrongylidae and Muspiceoidea;
- Ingestion of the infected flesh though cannibalism and scavenging in some Capillariidae and Trichinellidae, with the same mammal species serving both as definitive and intermediate host;
- Autoinfection with internal cycles in Strongyloididae (*Strongyloides* spp.);
- Transplacental and transmammary transmission in *Strongyloides* spp. and *Toxocara pteropodis*.

Table 2. Diversity of the modes of infection in the parasitic nematodes of small mammals.

Example	Infective stage	Route of infection	Internal migration	Intermediate host	Paratenic host
Trichuridae	Egg (L3)	Oral	No (larvae in intestinal cell)	No	No
Trichinellidae	L1	Oral	Yes (larvae in skeletal muscle fiber)	No	Yes
Trichonstrongylidae	L3	Oral/skin	No/Yes	No	No
Metastrongylidae	L3	Oral	Yes (L3)	Yes	Yes
Ascarididae	Egg (L2)	Oral	No/Yes	No/Yes	No/Yes
Oxyuridae	Egg (L3)	Oral	No	No	No
Physalopteridae	L3	Oral	No	Yes	No
Onchocercidae	L3	Skin	Yes	Yes	No

3.1.1 Grooming

Hernandez and Sukhdeo (1995) stressed the significance of self- and allogrooming in the transmission of *Heligmosomoides polygyrus*. They showed experimentally that higher numbers of nematodes were recovered from mice that were allowed to self-groom compared to infection levels in mice that had been prevented from self-grooming.

3.1.2 Cannibalism and necrophagy

Capillaria hepatica is the only known nematode of mammals that depends on the death of the definitive host for transmission. The adults live in the liver parenchyma and the females deposit eggs into the sinuous tracts, in which they are encapsulated. Eggs exit when liberated by cannibalism or necrophagy. Eggs then pass through faeces and are released into the environment where they can remain viable for over one year. Embryonated eggs, that are eaten, hatch in the intestine and L1 larvae thus liberated migrate to the liver where maturation takes place.

3.1.3. Paratenesis and paratenic host

Many nematodes use a paratenic host in their transmission. The paratenic host is "an organism which serves to transfer a larval stage or stages from one host to another but in which little or no development takes place" (Anderson 1999). Such a transmission process is called paratenesis.

For example, Mustelidae do not acquire mestastrongyloid nematodes directly by eating infected gastropods (the first obligatory intermediate host), but by preying upon paratenic hosts - shrews or rodents. When an infected gastropod is eaten by a shrew or a rodent, the infective larval nematode invades the tissues and usually encapsulates in the liver of these hosts. As there is no development of the encapsulated larvae, these hosts are purely paratenic. Modes of transmission that involve paratenic small mammal hosts is also widespread in the spiruroid parasites of carnivores.

3.1.4 Transplacental transmission

Toxocara pteropodis is a nematode parasite of flying-foxes (Megachiroptera) of the genus *Pteropus* in Australia, Oceania and southeastern Asia. Prociv (1989) reported that adult nematodes mature in the intestines of suckling bats. The eggs pass in faeces until the end of infection, which ends at about the time of weaning. Third-stage larvae, hatched from infective eggs, are ingested by adult bats and migrate to the liver. In females, at the end of parturition, some hepatic larvae are mobilized and pass through the mammary glands into the intestine of the neonate bat.

3.1.5 Arrested development

Arrested development or hypobiosis is well known in Trichostrongylidea parasitic on lagomorphs and rodents. This phenomenon occurs when a large percentage of L3 and L4 remain in the intestine mucosae long after the end of the normal prepatent period. Extrinsic factors such as fluctuating temperatures and intrinsic factors such as immunity influence the induction and the length of the arrested development, as well as the number of arresting larvae.

Arrested development plays an important role because it allows species with limited adult life spans to survive when external conditions are unsuitable for the development and survival of larval stages. In temperate regions where transmission cannot occur in winter, larvae acquired in autumn may remain arrested in the host gut mucosa until spring, to await favourable external conditions for transmission. Host immunity may also cause arrest. The number of arrested larvae in a host may be related to the number of adult nematodes already present (Behnke and Parish 1979). When adults die, there is a relaxation of the immunity, which allows some arrested larvae to leave the gut to mature and replace adult nematodes.

3.2 Sex determination and sexual patterns

The great majority of vertebrate parasitic nematodes are gonochorous with sex determined chromosomally. XX/XO sex determination is very common across Nematoda. It occurs in both the free-living (*Caenorhabditis elegans*) and numerous parasitic (*Trichinella spiralis*, *Haemonchus contortus*, *Strongyloides ratti*) nematodes, suggesting that the ancestral phylogenetic state is possibly the XX/XO sex determination. Only a handful of nematodes are known to have Y chromosomes. These are *Brugia malayi*, *Onchocerca volvulus*, *Baylisascaris transfuga*, *Contracaecum incurvum* and *Trichuris muris*. Since Y chromosomes are only known in these few distantly related nematodes, White (1973) suggested that they probably emerged recently.

Nematode species display a great variety of reproduction patterns. In *Strongyloides ratti*, parasitic females are XX and produce male (XO) and female (XX) larvae. Male larvae are only able to develop into free-living adult males, whereas female larvae can either develop into free-living adult females or develop directly into new parasitic females (through infective L3). When the free-living adult males mate, all the progeny inherit the paternal X chromosome and thereby all the progeny are XX females. Host conditions and, in particular, host immune response influence the sex ratio of the progeny of parasitic females, with the sex ratio becoming more male-biased in hosts mounting an anti-*S. ratti* immune response (Harvey et al. 2000).

Oxyuroid nematodes exhibit haplodiploidy, where males are haploid and arise by partial parthenogenesis via unfertilized eggs, while females are diploid and develop from fertilized eggs (Adamson 1989). This may result in a high degree of inbreeding and a decrease in genetic diversity compared to gonochoric species (Mueller-Graf et al. 1999).

Traumatic insemination occurs in oxyuroid nematodes such as *Passalurus ambiguus* in rabbits with males injecting sperm through the female larvae's cuticle (Fig. 2). Females have developed specialized internal structures for sperm storage and transfer to the reproductive system when mature (Chabaud et al. 1983; Hugot et al. 1982) (Fig. 2).

Protandrous hermaphrodite forms are recognized within the Muspiceoidea (*Muspicea, Riouxgolvania, Lukonema*).

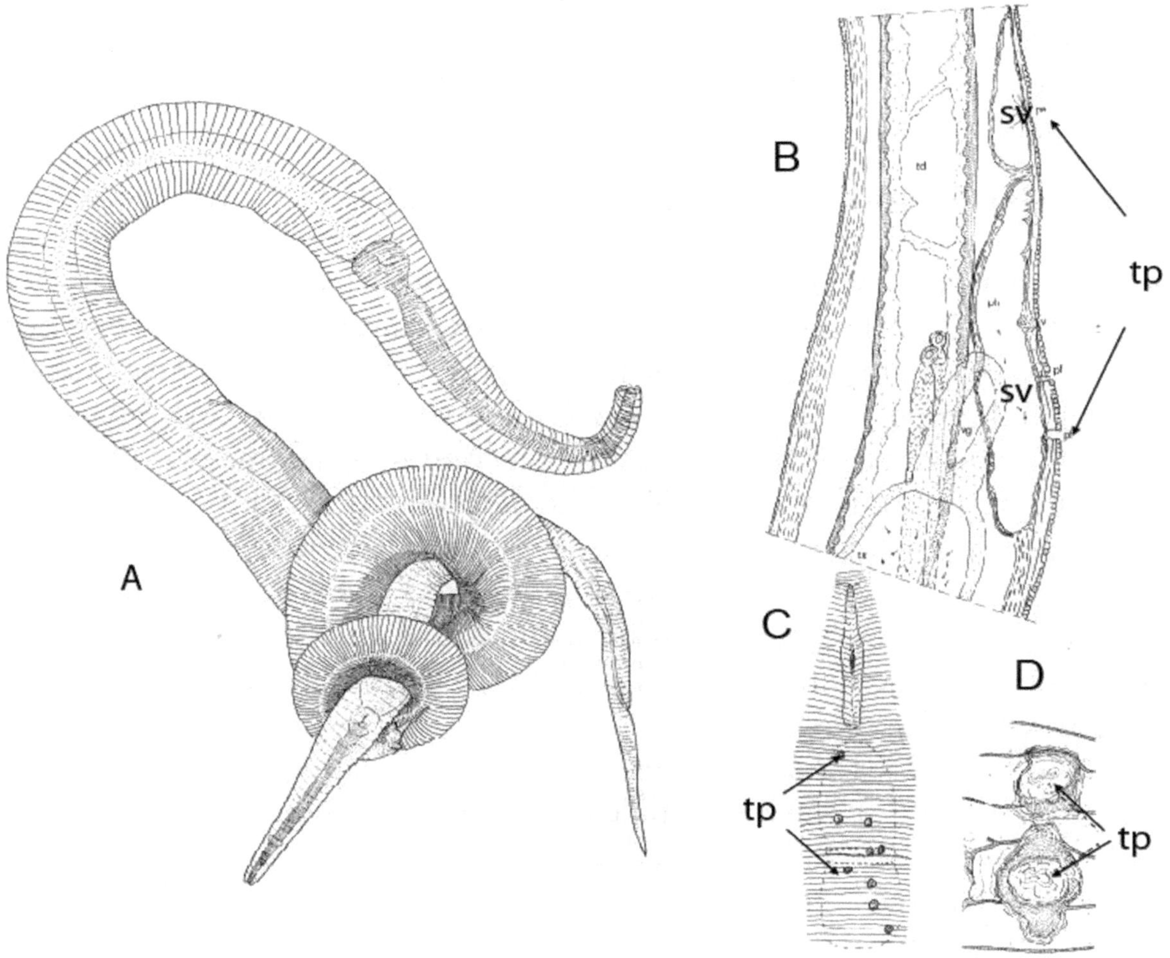

Fig. 2. A. Mating between an adult male (large worm) and immature female (stage 4 larvae) of *Hilgertia seurati*; B. Transversal section showing traumatic pore (tp) and sperm vesicule (sv) in larval female of *Passalurus ambiguous*; C-D. Traumatic insemination pores in larval female of *Passalurus ambiguous* (modified after Hugot 1982; Hugot et al. 1992)

4 Nematodes as laboratory models

Very few species of nematodes have served as laboratory models to investigate host-parasite interactions and diseases that affect humans or domestic animals. Most of our knowledge comes from these nematodes and among them are the following:

- *Trichinella spiralis*, which is responsible for the human disease trichinellosis, is maintained in laboratory rodents. This unique macroparasite infects epithelial cells of the small intestine.

- *Trichuris muris* is a mouse model for the human parasite *T. trichiura*, highly prevalent in many parts of the developing world. This parasite lives within host gut epithelial cells just as *Trichinella spiralis* does. The adult nematode manages to penetrate the epithelial cells without immediately inducing necrosis and death and to suppress immune responses.

- *Heligmosomoides polygyrus* is used as a model for nematode diseases in domestic animals (strongyloids of livestock), and also serves to investigate host immune responses.

- *Nippostrongylus brasiliensis* is a parasite of rodents and used as a model for hookworm disease.

- *Strongyloides ssp.* (used as a model for human strongyloidiasis) and *Parastrongyloides spp.* have the ability to live as free-living or as parasitic worms within some small mammals (rodents, marsupials). Manipulation of culture conditions and host physiology may influence the nematode's life span and whether it is free-living or as parasitic. These nematode models are useful to study genes that are required for parasitism or that contribute to longevity (Mitreva et al. 2004).

- *Litomosoides carinii* is a rodent filarioid nematode used as a model for human filariosis (*Wuchereria brancrofti*, *Brugia malayi*, *Onchocerca volvulus*).

5 Evolution of life history traits

Adaptation to parasitism imposes some constraints on the parasites. For example, co-variation of host and parasite body sizes has been observed in oxyuroids of vertebrates, with large mammals harbouring large-sized nematodes, even after controlling for confounding phylogenetic effects (Morand et al. 1996; Morand and Poulin 2002). The studies of Morand (1996) and Morand and Sorci (1998) have emphasized that the evolution of life history traits of nematode parasites are similar to those of free-living nematodes with the habitat-specific mortality driving the evolution of life traits in both free-living and parasitic forms. Morand (1996) compiled life history traits information (body size, life expectancy of adult stage and free-living stage, maturation time or time to patency, total reproductive output) of free-living, as well as plant, insect and vertebrate parasitic nematodes. Most of the correlations between life traits, although confirming the previous study of Skorping et al (1991), enabled him to propose a causal chain of life history evolution in parasitic nematodes with the key parameter being the adult age-dependent mortality.

Nematode maturation within the host presumably occurs at the time that maximizes reproductive success and depends on age-dependent mortality. Time until maturation in the host, i.e. time to patency, is a determinant of body size, which positively correlates with reproductive output in parasitic nematodes (Morand 1996; Morand and Poulin 2000) and also in parasitic platyhelminths (Trouvé and Morand 1998; Trouvé et al. 1998).

Adult nematode mortality rate, which results from host mortality and host-induced mortality, should favour advanced or delayed time to patency. High mortality rate should favour advanced time to patency in order to reach sexual maturity earlier, whereas low mortality rate should favour delayed time to patency in order to achieve a large body size at sexual maturity (and hence large total reproductive output).

Read and Skorping (1995) showed that tissue migration is a trait selectively advantageous for some nematodes. Species that undertake migration in host tissues during their larval development delay their maturation and tend to grow larger than those that develop directly in the gut.

The first comparative test was provided by Sorci et al. (1997), who showed a significant relationship between host longevity and adult parasite body size, using the primate-oxyurid system, supporting the hypothesis that long-living hosts select large body-sized nematodes. The second test was based on an optimality modelling approach (Morand and Poulin 2000). The optimality model derived a relationship between the time to patency and the inverse of the sum of parasite mortality and host mortality. A comparative test was then performed and the slope of this relationship based on collected data was found to be consistent with the slope expected by the optimality model. Hence, high levels of parasite mortality select for a reduction in time to patency, whereas greater host longevity favours delayed parasite maturity (Morand and Poulin 2000). Although these results contrast with those of Gemmill et al. (1999), they give strong support to the comparative study of Sorci et al. (1997) on the co-variation of parasite and host life-history traits.

6 Effect of nematodes on host dynamics

Macroparasites have been reported to influence behaviour, energetic demands, reproduction, survival and mate choice of their hosts. Several studies have demonstrated experimentally the detrimental effects that parasites can have on fecundity and survival in wild small mammal populations. For example, *Obeliscoides cuniculi* reduced survival of the snowshoe hare *Lepus californicus*. If these parasites reduce host fecundity or survival, it is then theoretically possible that they might regulate host numbers (Tompkins and Begon 1999).

The best evidence that the impact of nematodes can reduce host populations comes from laboratory and enclosure experiments. However, the possibility that these nematodes may regulate host populations has been investigated in very few studies. Scott (1987) conducted the first experimental study that demonstrated the ability of a nematode to regulate the abundance of a mammal population. She performed a free-running experimental investigation using mice and the directly transmitted nematode *H. polygyrus*. In the absence of the parasite, the uninfected-mouse population achieved an equilibrium size 20-times larger than that of the infected mouse population. Removal of the parasite from infected mouse populations allowed them to recover similar equilibrium densities as non-infected mouse populations.

Singleton and Spratt (1986) showed that *C. hepatica* induces reduction of natality and survival in laboratory mice. Singleton and McCallum (1990) reported that *C. hepatica* has the potential to control house mouse plagues in Australia. This nematode has the distinctive feature of dependence on the death of its host for transmission (see above). Using a mathematical model, McCallum and Singleton (1989) found that the necessity of host death to ensure the parasite transmission has a destabilizing influence on the host dynamics, leading to oscillations in the mouse population. The model suggested that the parasite could maintain host density far below the carrying capacity. However, manipulative field experiments did not corroborate the efficiency of this parasite in controlling the mouse populations (Singleton and Chambers 1996).

Sub-lethal nematodes have the potential to induce changes in the life history optima of their hosts. Recently, Kristan (2004) investigated the effects of *H. polygyrus* on the house mouse *Mus musculus* and found that reproduction investment was increased when female mice were infected, with parasitized females having larger offspring than unparasitized females. Moreover, she showed the existence of maternal effects. Only offspring from parasitized females were able to eliminate their own infection.

Consequently, offspring susceptibility to *H. polygyrus* could be modified by maternal parasite infection. One possible mechanism for this might include interactions of the maternal antibodies with the offspring's immune system.

7 Effect of nematodes on host behaviour

Nematodes can also alter the behaviour of their hosts. Kavaliers and Colwell (1995) showed that female mice could discriminate between the odours of parasitized and non-parasitized male mice. In particular, female mice discriminated between the chemical signals of uninfected males and males infected with *H. polygyrus*. Females preferentially choose uninfected males as they find the odours of uninfected males more attractive than those of infected males.

8 Concluding remarks

Nematodes are greatly diverse in terms of species number but also in various kinds of life-cycle, modes of infection and life-history traits. This diversity is far for being completely known. Hence, recently, Clarke et al. (2004) presented evidence that an un-described nematode parasite of the very well investigated wood mouse *Apodemus sylvaticus* may be sexually transmitted. These authors found larval nematodes in the epididymides of males, which suggests that they would be transmitted to females during ejaculation.

In order to better understand the origins of all these facets of nematode biodiversity, we need a more detailed phylogenetic framework of this phylum. By mapping parasite traits onto the nematode phylogeny, we will better estimate the phylogenetical constraints and the ecological adaptation that may have shaped the diversity of nematodes, and particularly their interactions with small mammals.

Although rarely causing the death of their hosts, nematodes have the capacity to alter both the physiology and the behaviour of their hosts. Nematodes can regulate the population dynamics of their hosts, and this may contribute to the extinction spiral of small host populations (see Christe et al. in this volume), although much more experimental and theoretical work is needed. Finally, if nematodes are a threat to biodiversity, they are also of great concern with regards to human health (see Casanova and Ribas; Leirs and Singleton in this volume).

References

Adamson ML (1986) Modes of transmission and evolution of life histories in zooparasitic nematodes. Can J Zool 64:1375–1384

Adamson ML (1989) Evolutionary biology of the Oxyurida (Nematoda): Biofacies of a haplodiploid taxon. Adv Parasitol 28:175–228

Anderson RC (1988) Nematode transmission patterns. J Parasitol 74:30–45

Anderson RC (1999) Nematode parasites of vertebrates. Their development and transmission. CABI, Wallingford

Anderson RC, Chabaud AG, Willmott S (eds) (1974-1983) CIH keys to the nematode parasites of vertebrates. Commonwealth Agricultural Bureaux, Farnham Royal

Aguinaldo AM, Turbeville JM, Linford LS, Rivera MC, Garey JR, Raff RA, Lake JA (1997) Evidence for a clade of nematodes, arthropods and other molting animals. Nature 387:489–493

Behnke JM, Parish HA (1979) *Nematospiroides dubius*: Arrested development of larvae in immune mice. Exp Parasitol 47:116–127

Blaxter ML, de Ley P, Garey JR, Liu LX, Scheldeman P, Vierstraete A, Vanfleteren JR, Mackey LY, Dorris M, Frisse LM, Vida JT, Thomas WK (1998) A molecular evolutionary framework for the phylum Nematoda. Nature 392:71–75

Chabaud AG (1955) Essai d'interprétation phylétique des cycles évolutifs chez nematodes parasites des vertebras. Ann Parasitol Hum Comp 30:83–126

Chabaud A, Bain O, Hugot J-P, Rausch RL, Rausch VR (1983) Organe de Man et insémination traumatique. Rev Nematol 6:127–131

Clarke JR, Anderson TJC, Bandi C (2004) Sexual transmission of a nematode parasite of wood mice (*Apodemus sylvaticus*)? Parasitology 128: 561–568

De Ley P, Blaxter ML (2002) Systematic position and phylogeny. In: Lee D (ed) The biology of nematodes. Harwood Acad Publ, Reading, pp 1–30

Gemmill AW, Skørping A, Read AF (1999) Optimal timing of first reproduction in parasitic nematodes. J Evol Biol 12:1148–1156

Hammond PM (1992) Species inventory. In: Groombridge B (ed) Global diversity, status of the earth's living resources. Chapman and Hall, London, pp 17–39

Harvey SC, Gemmill AW, Read AF, Viney ME (2000) The control of morph development in the parasitic nematode *Strongyloides ratti*. Proc R Soc Lond B 267:2057–2063

Hernandez AD, Sukhdeo MVK (1995) Host grooming and the transmission strategy of *Heligmosomoides polygyrus*. J Parasitol 81:856–869

Hugot J-P (1982) Sur le genre *Hilgertia* (Oxyuridae, Nematoda) parasite de rongeurs Ctenodactylidae. Bull Mus Hist Nat Paris 4: 419–431

Hugot J-P, Bain O, Cassone J (1982). Insémination traumatique et tube de ponte chez l'oxyure parasite du lapin domestique. C R Acad Sci Paris 294:707–710

Hugot J-P, Baujard P, Morand S (2001) Biodiversity in helminths and nematodes as a field of study: An overview. Nematology 3:1–10

Kavaliers M, Colwell DD (1995) Odours of parasitized males induce adverse responses in female mice. Anim Behav 50:1161–1169
Kavaliers M, Choleris E, Pfaff DW (2005) Genes, odours and the recognition of parasitized individuals by rodents. Trends Parasitol 21:423–429
Kristan DM (2004) Intestinal nematode infection affects host life history and offspring susceptibility to parasitism. J Anim Ecol 73:227–238
Maupas EF (1900) Modes et formes de reproduction des nématodes. Arch Zool Exp Gen 7:563–628
May RM (1988) How many species are there on the Earth? Science 241:1441–1449
McCallum HI, Singleton GR (1989) Models to assess the potential of *Capillaria hepatica* to control outbreaks of house mice. Parasitology 98:425–437
Mitreva M, McCarter JP, Martin J (2004) Comparative genomics of gene expression in the parasitic and free-living nematodes *Strongyloides stercoralis* and *Caenorhabditis elegans*. Genome Res 14:209–220
Morand S (1996) Life-history traits in parasitic nematodes: A comparative approach for the search of invariants. Funct Ecol 10:210–218
Morand S, Poulin R (2000) Optimal time to patency in parasitic nematodes: Host mortality matters. Ecol Let 3:186–190
Morand S, Poulin R (2002) Body size-density relationships and species diversity in parasitic nematodes: Patterns and likely processes. Evol Ecol Res 4:951–961.
Morand S, Sorci G (1998) Determinants of life-history evolution in nematodes. Parasitol Today 14:193–196
Morand S, Legendre P, Gardner SL, Hugot J-P (1996) Body size evolution of oxyurid parasites: The role of hosts. Oecologia 107:274–282
Mueller-Graf C, Durand P, Feliu C, Hugot J-P, Renaud F, Santalla F, Morand S (1999) Genetic variability and epidemiology of two species of nematodes (*Heligmosomoides polygyrus* and *Syphacia stroma*) of the same host (*Apodemus* sp.) Parasitology 118:425–432
Murray DL, Cary JR, Keith LB (1997) Interactive effects of sublethal nematodes and nutritional status on snowshoe hare vulnerability to predation J Anim Ecol 66:250–264
Poulin R, Morand S (2000) The diversity of parasites. Q Rev Biol 75:277–293
Prociv, P (1989) *Toxocara pteropodis* and visceral larva migrans. Parasitol Today 5:106–109
Read AF, Skørping A (1995) The evolution of tissue migration by parasitic nematode larvae. Parasitology 111:359–371
Scott ME (1987) Regulation of mouse colony abundance by *Heligmosomoides polygyrus*. Parasitology 95:111–124
Singleton GR, Chambers LK (1996) A manipulative field experiment to examine the effect of *Capillaria hepatica* (Nematoda) on wild mouse populations in southern Australia. Int J Parasit 26:383–398
Singleton GR, McCallum HI (1990) The potential of *Capillaria hepatica* to control mouse plagues. Parasitol Today 6:190–193

Singleton GR, Spratt DM (1986) The effects of *Capillaria hepatica* (Nematoda) on natality and survival to weaning in BALB/c mice. Aust J Zool 34:687–681
Skørping A, Read AF, Keymer AE (1991) Life history covariation in intestinal nematodes of mammals. Oikos 60:365–372
Sorci G, Morand S, Hugot J-P (1997) Host-parasite coevolution: Comparative evidence for covariation of life history traits in primates and oxyurid parasites. Proc R Soc Lond B 264:285–289
Tompkins DM, Begon M (1999) Parasites can regulate wildlife populations. Parasitol Today 15:311–313
Trouvé S, Morand S (1998) The evolution of parasites' fecundity. Int J Parasitol 28:1817–1819
Trouvé S, Sasal P, Jourdane J, Renaud F, Morand S (1998) Revisited views on the evolution of life-history traits in parasitic and free-living platyhelminthes. Oecologia 115:370–378
White MJD (1973) Animal cytology and evolution. Cambridge Univ Press, New York

5 Acanthocephalans

Alexis Ribas and Juan Carlos Casanova

1 Introductory remarks

Acanthocephala (thorny-headed worms) is a parasitic phylum that was described as recently as in the 18th century by I.T. Koelreuther during his work at the Russian Academy of Science. At their adult stage, the thorny-headed worms parasitize terrestrial and aquatic host species from all vertebrate classes including humans. They attach to the intestinal wall of the vertebrate host by an invaginable proboscis armed with hooks. Some species (e.g., *Moniliformis moniliformis*) cause diseases (acanthocephaliasis) in humans (Goldsmith et al. 1974; Ikeh et al. 1992; Anosike et al. 2000). The intermediate hosts of acanthocephalans are usually various arthropods.

Some features of Acanthocephalans are similar to those of Plathelminthes and Nematoda. For example, similarly to cestodes, they lack an alimentary tract. The phylogenetic position of the phylum among Metazoa is not resolved. Some authors consider it to be a close relative of (or even belonging to) Rotifera (Garey et al. 1998). The mitochondrial genome of only a single species is known to date (*Leptothynchoides thecatus*; Steinauer et al. 2005) and, thus, additional studies are required to elucidate the phylogenetic position of acanthocephalans. In this chapter we will briefly review the main biological features of Acanthocephala and list species parasitizing small mammals.

2 Life cycle

Complete life cycles are known for only about 25 species of acanthocephalans (approximately 5% of known species). Laboratory studies of acanthocephalan life cycles typically used *Tenebrio molitor* as intermediate hosts (Crompton and Nickol 1985; Carmichael and Moore 1991).

As mentioned above, the adult acanthocephalans occur in the intestine of the definitive host. They lack an alimentary tract and absorb nutrients through their entire body surface. The proboscis used for attachment is armed with a large number of chitinized hooks, which damage the host tissues causing inflammation in the intestinal wall (Martin et al. 1983).

Acanthocephalans are characterized by female-biased sexual size dimorphism (Poulin and Morand 2000; Sasal et al 2000). Fertilization is internal. During mating, the male bursa is wrapped around the posterior end of a female and the male's cirrus is introduced into the female's gonopore. Sperm migrates up the vagina to the uterus. After insemination, the male cement glands secrete a cement plug sealing up the gonopore and thus preventing sperm escape and potentially further insemination. Eggs are passed to the environment with the host's faeces and do not hatch until ingested by a suitable intermediate host. Intermediate hosts are usually insects in species parasitic in terrestrial animals and crustaceans (mainly amphipods, isopods and crabs) in species parasitic in aquatic hosts (Annuar and Paran 1976; Allely et al. 1992; Abe et al. 1997).

In the intermediate host, the parasite goes through several developmental larval stages until it attains the infective stage (cystacanth) (see Ravindranath and Anantaram 1977; Sato et al. 2005 for details). When infected intermediate hosts are eaten by the definitive host, the cystacanth develops into the adult worm in the latter. Thus, acanthocephalans exploit existing food chains and circulate through predator-prey interactions.

The parasitized intermediate hosts can be affected by a decrease in male mating success, as shown by Bollache et al. (2001) with male gammarids infected by *Pomphorhynchus laevis* and *Polymorphus minutus* having lower reproductive success than uninfected males. More importantly, practically all acanthocephalans appear capable of altering the behaviour or coloration of their intermediate hosts, in ways that make them more susceptible to predation by the definitive host (Moore 1984).

Sexual competition between adult males for access to females has been demonstraded in *Corynosoma magdaleni*, parasite of the Saimaa ringed seal (*Phoca hispida saimensis*) where larger males of this acanthocephalan are positively selected (Sinisalo et al. 2004).

Vertebrates can also act as paratenic hosts, in which the parasites are usually encapsulated in the body cavity or the muscles. The larvae of *Macracanthorhynchus catulinus*, a parasite of carnivores is also found in the muscles and body cavity of several species of rodents, being transferred to the definitive host by ingestion.

3 Host specificity

Some acanthocephalan species are characterized by a high degree of opportunism in relation to the host taxonomic affinity. For example, *M. moniliformis* can be found in many species of rodents. Acanthocephalans are more host-specific at the level of host class. However, exchanges of acanthocephalans among warm-blooded animals are possible although very rare. Species of the genus *Centrorhynchus* are usually found in birds, but *C. ninni* is found in *Mustela vison* (Torres et al. 2003). Nevertheless, exchanges between cold and warm blooded animals seem impossible due to physiological conditions.

4 Classification, number of species and associations with small mammals

Although acanthocephalans are not encountered as commonly as parasitic flatworms (trematodes and tapeworms) or nematodes, they are found in many species of fishes, amphibians, birds, and mammals. However, in many surveys of small mammal parasites, acanthocephalans can be identified to the generic level only. This is because of the difficulty in identification of the larvae as they lack reproductive organs, which are the key characters for species identification.

Three classes with about 500 species belonging to 141 genera are recognized in the phylum (Golvan 1994). Thus, Acanthocephala is much less speciose than other helminth phyla such as Platyhelminthes and Nematoda. Class Archiacanthocephala parasitizes terrestrial hosts, mainly birds and mammals, class Palaeacanthocephala is represented by parasites of most classes of aquatic vertebrates (fish, birds and mammals), whereas members of the class Eoacanthocephala exploit aquatic and terrestrial lower vertebrates, namely fish and reptiles.

The following list of Acanthocephala that use small mammals as their hosts is based on studies of Ward and Nelson (1967), Kamiya et al. (1968), Petrochenko (1971), Schmidt (1975), Khairul (1977), Leong (1979), Uga et al. (1983), Pfaffenberger et al. (1985), Deveaux et al. (1988), Schmidt and Edmons (1989), Asakawa et al. (1992), Cordero del Campillo et al. (1994), El Shazy et al. (1994), Golvan (1994), Yen et al. (1996), Coady and Nickoll (2000), Feliu et al. (2000), Ribas (2005), Dimitrova and Gibson (2005) and Tantelán (2005). It should be noted also that birds of prey can share acanthocephalan species with their small mammalian prey.

- AFROTROPICAL REGION
  - In tenrecs
    - *Promoniliformis ovocristatus* (Madagascar).
- AUSTRALASIAN REGION
  - In marsupials (Paramelidae and Dasyuridae)
    - *Australiformis semoni*.
- NEARCTIC REGION
  - In marsupials
    - *Hamanniella tortuosa*.
  - In insectivores
    - *Prosthorhynchus cylindraceus* (shrews), *Moniliformis clarki* (moles).
  - In rodents
    - *Moniliformis clarki* (various host species).
- NEOTROPICAL REGION
  - In marsupials
    - *Hamanniella microcephala*.
- PALAEARCTIC REGION
  - In insectivores
    - *Oligocanthorhynchus circumflexus* (moles), *Macracanthorhynchus erinacei* (hedgehogs), *M. aegypticus* (hedgehogs), *Centrorhynchus alauconis* (shrews), *C. appendiculatum* (shrews), *Prosthorhynchus cylindraceus* (shrews), *Polymorphus minutus* (water shrews), *Nephridiacanthus major* (hedgehogs).
  - In rodents
    - *Oligocanthorhynchus citilli*, *O. thumbi*, *Macracanthorhynchus catulinus*, *Moniliformis acomysi* (spiny mice and gerbils), *M. siciliensis*, *Mediorhynchus conirostis* (spiny mice), *Oncicola travassosi* (spiny mice).
- COSMOPOLITAN
  - In various hosts
    - *Macracanthorhynchus hirudinaceus*, *Moniliformis moniliformis*.

5 Host-acanthocephalan interactions

Microhabitat preferences of acanthocephalans inside a definitive host's intestine seem to be determined by the intestinal sugar gradients (Starling 1985). Fitness of parasites appears to be dependent on how close they come to the optimal attachment site. For example, in rats experimentally infected with *M. moniliformis,* the relative position of parasites in the intestine influenced their reproductive success (Lawlor et al. 1990).

Knowledge on the physiological and pathological aspects of the relationships between acanthocephalans and small mammals is scarce, most concerning *M. moniliformis* (see review in Taraschewski 2000). The effects of adults of this species on a small mammalian host have been studied mainly in experimentally infected laboratory rats. Host tissue are damaged by mechanic actions of the acanthocephalan hooks and by the substances excreted by the parasite. Furthermore, worms damage the mucosa and partly the tunica properia with their proboscis, but never reach the muscular layers of the host's intestinal wall. The secretion-excretion products of acanthocephalans, which are probably involved in both host immune reaction and immunomodulation, are unknown. The inflammatory responses, with the presence of immune cells around the site of fixation of *M. moniliformis* in the gut, are accompanied by increasing levels of specific immununoglobulins.

The behaviour of an insect intermediate host is altered by larval acanthocephalans, making a host more susceptible to predation by small mammals (Moore 1984; Libersat and Moore 2000; Moore and Freehling 2002). For example, *M. moniliformis* alters the behaviour of the cockroach *Periplaneta americana* and several other cockroach species. The alterations include a retarded evasion strategy of the infected cockroach when suddenly exposed to light and disturbance, which may then favour the predation by rats and consequently their infection. There is no indication that parasitized small mammals may also suffer from increased predation when parasitized by acanthocephalans.

6 Concluding remarks

The number of acanthocephalan species in small mammals is very low compared with other metazoan parasites. An example can be given from the Iberian Peninsula, where helminthological surveys of small mammals have been very intensive with nearly all species of small mammals surveyed (16 species of rodents). Hence, only *M. moniliformis* in *Rattus norvegicus* was recovered (Cordero del Campillo et al. 1994; Feliu et al. 1997).

Small rodents are scarcely infected by acanthocephalans in comparison with other groups of parasites. This could be related to the aquatic origin of the group with fish being the original hosts (Herlyn et al. 2003). Mammals were later colonized by these parasites. Few phylogenetic studies are available (Herlyn et al. 2003; Steinauer et al. 2005), but they are crucial for providing scenarios of colonization of terrestrial hosts and the specific adaptation for a very different environment.

References

Abe N, Nonaka N, Iwaki T, Ganzirig JA, Alexander Y, Oku Y, Kamiya M (1997) Description of *Moniliformis* worms from grey red-backed vole, *Clethrionomys rufocanus bedfordiae,* in Hokkaido, Japan. Parasit Int 46:41–44

Allely Z, Moore J, Gotelli NJ (1992) *Moniliformis moniliformis* infection has no effect on some behaviors of the cockroach *Diploptera punctata.* J Parasitol 78:524–526

Anosike JC, Njoku AJ, Nwoke BEB, Okoro OU, Okere AN, Ukaga CN, Adimonye RN (2000) Human infections with *Moniliformis moniliformis* (Bremser 1811) Travassos 1915 in south-eastern Nigeria. Ann Trop Med Parasitol 94:837–838

Anuar AK, Paran TP (1976) *Periplaneta americana* (L.) as intermediate host of *Moniliformis moniliformis* (Bremser) in Penang, Malaysia. Southeast Asian J Trop Med Public Health 7:415–416

Asakawa M, Tamura T, Fukumoto S, Ohbayashi M (1992) Parasitic helminths of small mammals in a sandbar of Lake Saroma, Hokkaido, Japan (in Japanese). J Rakuno Gakuen Univ 17:9–16

Bollache L, Gambade G, Cezilly F (2001) The effects of two acanthocephalan parasites, *Pomphorhynchus laevis* and *Polymorphus minutus*, on pairing success in male *Gammarus pulex* (Crustacea : Amphipoda). Behav Ecol Sociobiol 49:296–303

Carmichael LM, Moore J (1991) A comparison of behavioral alterations in the brown cockroach, *Periplaneta brunnea*, and the American cockroach, *Periplaneta americana*, infected with the acanthocephalan, *Moniliformis moniliformis*. J Parasitol 77:931–936

Coady NR, Nickol BB (2000) Assessment of parenteral *Plagiorhynchus cylindraceus* (Acantocephala) infections in shrews. Comp Parasitol 67:32–39

Cordero del Campillo M, Castañón Ordóñez L, Reguera Feo A (1994) Indice-catálogo de zooparásitos Ibéricos. Universidad de León, Secretariado de Publicaciones, León

Crompton DWT, Nickol BB (1985) Biology of Acanthocephala. Cambridge Univ Press, Cambridge

Deveaux TP, Schmidt GD, Krishnasamy M (1988) Two new species of *Moniliformis* (Acanthocephala: Moniliformidae) from Malaysia. J Parasitol 74:322–325

Dimitrova ZM, Gibson DI (2005) Some species of *Centrorhynchus* Lühe, 1911 (Acantocephala: Centrorhynchidae) from the collection of the Natural History Museum, London. Syst Parasitol 62:117–134

El Shazly AM, Morsy TA, El Kady GA, Ragheb DA, Handousa AE, Ahmed MM, Younis TA, Habib KS (1994) The helminthic parasites of rodents in Dakahlia Governorate, with reference to their Egyptian helminth fauna. J Egypt Soc Parasitol 24:413–428

Feliu C, Renaud F, Catzeflis F, Hugot JP, Durand P, Morand S (1997) Comparative analysis of parasite species richness of Iberian rodents Parasitology 115:453–466

Feliu C, Blasco S, Torres J, Miquel J, Casanova J-C (2001) On the helminthfauna of *Erinaceus europaeus* Linnaeus, 1758 (Insectivora, Erinaceidae) in the Iberian peninsula. Rev Iber Parasitol 61:31–37

Garey JR, Schmidt-Rhaesa A, Near TJ, Nadler SA (1998) The evolutionary relationships of rotifers and acantocephalans. Hydrobiologia 387/388:83–91

Goldsmith JM, Smith ME, Fleming F (1974) Human infection with *Moniliformis* sp. in Rhodesia. Ann Trop Med Parasitol 68:363–364

Golvan YJ (1994) Nomenclatura of the Acantocephala. Res Rev Parasitol 54:135–205

Guinnee MA, Moore J (2004) The effect of parasitism on host fecundity is dependent on temperature in a cockroach-acanthocephalan system. J Parasitol 90:673–677

Herlyn H, Piskurek O, Schmitz J, Ehlers U, Zischler H (2003) The syndermatan phylogeny and the evolution of acanthocephalan endoparasitism as inferred from 18S rDNA sequences. Mol Phylogenet Evol 26:155–164

Ikeh EI, Anosike JC, Okon E (1992) Acanthocephalan infection in man in northern Nigeria. J Helminthol 66:241–242

Kamiya M, Chinzei H, Sasa M (1968) A survey on helminth parasites of rats in southern Amami, Japan (in Japanese). Japan J Parasitol 17:436–444

Khairul AA (1977) *Moniliformis moniliformis* Bremser (*Moniliformis dubius*, Mayer) and its prevalence in *Rattus* spp. in Penang, Malaysia. Med J Malaysia 32:111–113

Lawlor BJ, Read AF, Keymer AE, Parveen G, Crompton DWT (1990) Non-random mating in a parasitic worm: Mate choice my males? Anim Behav 40:870–876

Leong TS, Lim BL, Yap LF, Krishnasamy M. (1979) Parasite fauna of the house rat *Rattus rattus diardii* in Kuala Lumpur and nearby villages. Southeast Asian J Trop Med Public Health 10:122–126

Libersat F, Moore J (2000) The parasite *Moniliformis moniliformis* alters the escape response of its cockroach host *Periplaneta americana*. J Insect Behav 13:103–110

Martin J, Crompton DW, Singhvi A (1983) Mucosal lesions in the rat associated with the attachment of *Moniliformis* (Acanthocephala). J Helminthol 57:85–90

Moore J (1984) Altered behavioral responses in intermediate hosts: an acanthocephalan parasite strategy. Amer Nat 123:572–577

Moore J, Freehling M (2002) Cockroach hosts in thermal gradients suppress parasite development. Oecologia 133:261–266

Petrochenko VI (1971) Acanthocephala of domestic and wild mammals (in Russian). Nauka, Moscow

Pfaffenberger GS, Kemether K, de Bruin D (1985) Helminths of sympatric populations of kangaroo rats (*Dipodomys ordii*) and grasshopper mice (*Onychomys leucogaster*) from the high plains of eastern New Mexico. J Parasitol 71:592–595

Poulin R, Morand S (2000) Testis size, body size and male-male competition in acanthocephalan parasites. J Zool Lond 250:551–558

Ravindranath MH, Anantaraman S (1977) The cystacanth of *Moniliformis moniliformis* (Bremser, 1811) and its relationship with the haemocytes of the intermediate host (*Periplaneta americana*). Z Parasitenkd 53:225–37

Ribas A (2005) Systematic, faunistic and ecologic study of the parasitic helminthfaunas of talpids in the south western Europe. Ph. D. thesis, University of Barcelona

Sasal P, Jobet E, Faliex E, Morand S (2000) Evidence of sexual competition in an acanthocephalan parasite of fish. Parasitology 120:65–69

Sato H, Suzuki K, Uni S, Kamiya H (2005) Recovery of the everted cystacanth of seven acanthocephalan species of birds from feral raccoons (*Procyon lotor*) in Japan. Japan Vet Med Sci 67:1203–1206

Schmidt GD (1975) *Sphaerirostris wertheimae* sp. n. and other acantocephala from vertebrates of Israel. J Parasitol 61:298–300

Schmidt GD, Edmonds SJ (1989) *Australiformis semoni* (Linstow, 1898) n. gen., n. comb. (Acanthocephala: Moniliformidae) from marsupials of Australia and New Guinea. J Parasitol 75:215–217

Sinisalo T, Poulin R, Hogmander H, Juuti T, Valtonen ET (2004) The impact of sexual selection on *Corynosoma magdaleni* (Acanthocephala) infrapopulations in Saimaa ringed seals (*Phoca hispida saimensis*) Parasitology 128:179–185

Starling JA (1985) Feeding, nutrition and metabolism. In: Crompton DWT, Nickol BB (eds) Biology of the Acanthocephala. Cambridge Univ Press, Cambridge, pp 125–212

Steinauer ML, Nickol, BB, Broughton R, Ortí G (2005) First sequenced mitochondrial genome from the phylum Acanthocephala (*Leptorhynchoides thecatus*) and its phylogenetic position within Metazoa. J Mol Evol 60:706–715

Tantaleán M, Sánchez L, Gómez L, Huiza A (2005) Acantocéfalos del Perú. Rev Peru Biol 12:83–92

Taraschewski H (2000) Host-parasite interactions in Acanthocephala: A morphological approach. Adv Parasitol 46:1–161

Torres J, Mañas S, Palazon S, Ceña JC, Miquel J, Feliu C (2003) Helminth parasites of *Mustela lutreola* (Linnaeus, 1761) and *M. vison* Schreber, 1777 in Spain. Acta Parasitol 48:55–59
Uga S, Matsumura T, Araki K (1983) A survey of the parasitic helminths of wild rats at a zoo in Hyogo Prefecture (in Japanese). Japan J Parasitol 32:597–600
Yen CM, Wang JJ, Lee JD, Chen YP, Chen ER (1996) Parasitic infections among wild rats from two areas of Kaohsiung. Kaohsiung J Med Sci 12:145–149
Ward HL, Nelson DR (1967) Acantocephala of the genus *Moniliformis* from rodents of Egypt with the description of a new species from the Egyptian spiny mouse (*Acomys cahirinus*). J Parasitol 53:150–156

6 Taxonomy, host associations, life cycles and vectorial importance of ticks parasitizing small mammals

Lance A. Durden

1 Introductory remarks

Ticks transmit more kinds of pathogens than any other group of blood-feeding arthropods and, among arthropods, are second in importance only to mosquitoes in their public health impact worldwide (Hoogstraal 1985; Sonenshine et al. 2002; Goodman et al. 2005). In this chapter, small mammals are interpreted as being the smaller marsupials (Marsupialia) and bats (Chiroptera), most insectivores (Insectivora), most rodents (Rodentia), and all tree shrews (Scandentia), elephant shrews (Macroscelidea), and lagomorphs (Lagomorpha). Taxonomy, host-associations, life cycles and vectorial trends for ticks parasitizing these groups of mammals are considered from a worldwide perspective. Because of the widespread and abundant nature of small mammals and their associated ticks, these two groups of ecological partners have immense importance in many ecosytems (Durden and Keirans 1996). From a human perspective, these associations have most relevance with respect to tick-borne zoonotic pathogens that utilize small mammals as reservoir or amplifying hosts.

2 Tick taxonomy

The recently published and widely accepted taxonomic treatment of the world's ~825 species of ticks by Horak et al. (2002) is followed in this chapter. However, the state of tick taxonomy is currently in a state of partial flux as some researchers place more importance in molecular versus morphological tick phylogenies and taxon recognitions (Barker and Murrell 2004; Horak et al. 2002). Future works should start to reach a concensus on interpretations from these two approaches (Beati and Keirans 2001; Klompen et al. 2000). Significantly, the classifications of both Horak et al. (2002) and Barker and Murrell (2004) treat some formerly recognized ixodid (hard tick) taxa as follows: 1) *Boophilus* as a junior synonym (or subgenus) of *Rhipicephalus*; 2) *Aponomma* as a junior synonym of *Amblyomma*; 3) *Bothriocroton* as a full genus (originally described as a subgenus) for Australasian members of the former *Aponomma*. Further, the bat-associated argasid (soft tick) genera *Antricola* and *Nothoaspis* are treated as junior synonyms of *Carios*.

3 Overview of small mammal-tick associations worldwide

Underlining the importance of small mammals as hosts of ticks, representatives of 13 of the 17 genera of ticks (76%) recognized by Horak et al. (2002) are known to be ectoparasites of small mammals in at least one active stage (larvae, nymphs and/or adults) of their life cycle. These include all 4 currently recognized argasid genera (*Argas, Carios, Ornithodoros* and *Otobius*) and 8 of the 12 currently recognized ixodid genera (*Amblyomma, Anomalohimalaya, Dermacentor, Haemaphysalis, Hyalomma, Ixodes, Rhipicentor* and *Rhipicephalus*). The true hosts of the rarely collected *Nuttalliella namaqua*, the sole representative of the third tick family, the Afrotropical Nuttalliellidae, are unknown but 1 specimen has been recovered from the skin of a rodent so the host repertoire of this tick may also include small mammals (Keirans et al. 1976). Therefore, the only 4 tick genera (all belonging to the family Ixodidae) recognized by Horak et al. (2002) without members that are known to feed on small mammals in any of their active stages are *Bothriocroton, Cosmiomma, Nosomma* and *Margaropus*. However, it is likely that immatures (or possibly even adults in the case of *Bothriocroton*) of some species within these genera also occasionally feed on small mammals; future research on the host associations for these ticks will provide more definitive data. With respect to host-tick associations, Klompen et al. (1996) advocated that ecological/habitat specificity is often more important than host specificity for several tick species – this seems to be especially true for many argasid species.

Smaller marsupials (opossums, etc.) are parasitized by immature and adult stages of several species of ixodids in Australiasia and the Americas, and sometimes also by certain argasids (e.g., *Ornithodoros hermsi, O. parkeri* and *O. turicata* in North America, and *O. macmillani* in Australia).

Nevertheless, there are few host-specific ticks of smaller marsupials, particularly for species that parasitize this host group as adult ticks. Rather, and especially in North America, opossums are often parasitized by immature and adult ticks that belong to species with wide host ranges. In eastern North America, these include all active stages of at least three tick species (*Amblyomma americanum, Dermacentor variabilis* and *Ixodes scapularis*) that are important vectors of zoonotic pathogens.

Tick species with adult stages that parasitize small mammals (small marsupials, insectivores, rodents, lagomorphs and elephant shrews) in different zoogeographical regions are as follows:

- AFROTROPICAL REGION
    - On insectivores
        - Argasidae: *Argas brumpti, A. echinops, A. foleyi.*
        - Ixodidae: *Haemaphysalis elongata, H. erinacei, H. simplex, H. simplicima, H. subelongata, H. theilerae, H. tiptoni, Ixodes alluaudi, I. bedfordi, I. dawsi, I. lunatus.*
    - On rodents
        - Argasidae: *Argas brumpti, A. eboris, A. foleyi, A. zumpti, Ornithodoros arenicolous, O. erraticus, O. graingeri, O. grenieri, O. sonrai.*
        - Ixodidae: *Haemaphysalis anoplos, H. calcarata, H. houyi, H. nesomys, H. taufliebi, Ixodes albignaci, I. bedfordi, I. elongatus, I. minutae, I. myotomys, I. nesomys, I. randrianasoloi, I. rhabdomysae, I. transvaalensis, Rhipicephalis simpsoni.*
    - On lagomorphs
        - Ixodidae: *Rhipicephalus arnoldi, R. deltoideus.*
    - On elephant shrews
        - Ixodidae: *Ixodes nchisiensis, I. rasus* group, *I. vanidicus, Rhipicephalus oculatus, R. pravus.*
- AUSTRALASIAN REGION
    - On marsupials
        - Argasidae: *Ornithodoros macmillani*
        - Ixodidae: *Ixodes antechini, I. fecialis.*
- NEARCTIC REGION
    - On marsupials
        - Ixodidae: *Amblyomma americanum, Dermacentor variabilis, Ixodes scapularis.*
    - On insectivores

    - Ixodidae: *Ixodes eastoni, I. soricis.*
  - On rodents
    - Argasidae: *Carios talaje, Ornithodoros eremicus, O. hermsi, O. nicollei, O. parkeri, O. turicata, Otobius sparnus*? (adults of *Otobius* spp. do not typically feed but *O. sparnus* was recently transferred from *Ornithodoros* – Horak et al., 2002).
    - Ixodidae: *Ixodes angustus, I. eadsi, I. eastoni, I. hearlei, I. jellisoni, I. marmotae, I. marxi, I. minor, I. muris, I. peromysci, I. sculptus, I. spinipalpis, I. woodi.*
  - On lagomorphs
    - Argasidae: *Ornithodoros hermsi, O. parkeri, O. turicata.*
    - Ixodidae: *Amblyomma inornatum, Dermacentor parumapertus, Haemaphysalis leporispalustris, Ixodes dentatus, I. ochotonae, I. spinipalpis.*
- NEOTROPICAL REGION
  - On marsupials
    - Argasidae: *Carios chironectes, C. marmosae.*
    - Ixodidae: *Ixodes loricatus, I. luciae.*
  - On rodents
    - Argasidae: *Carios aragaoi, C. casebeeri, C. chironectes, C. daviesi, C. echimys, C. puertoricensis, C. rudis, C. talaje, C. tuttlei, Ornithodoros rostratus.*
    - Ixodidae: *Amblyomma pacae, Ixodes andinus, I. capromydis, I. dampfi, I. galapagoensis, I. guatemalensis, I. jonesae, I. lasallei, I. nectomys, I. nuttalli, I. sigelos, I. sinaloa, I. tamaulipas, I. tancitarius, I. tecpanensis, I. tiptoni, I. tropicalis, I. uruguayensis.*
  - On lagomorphs
    - Ixodidae: *Amblyomma inornatum, Haemaphysalis leporispalustris, Ixodes dicei, I. pomerantzi.*
- ORIENTAL REGION
  - On insectivores
    - Ixodidae: *Ixodes granulatus.*
  - On tree shrews
    - Ixodidae: *Ixodes granulatus, I. malayensis.*
  - On rodents

        - Ixodidae: *Haemaphysalis atherurus, H. bandicota, H. bartelsi, H. kadarsani, H. kutchensis, H. kyasanurensis, H. sciuri, H. verticalis, Ixodes audyi, I. granulatus, I. himalayensis, I. kuntzi, I. petauristae, I. radfordi, I. werneri, Rhipicephalus ramachandrai.*
    - On lagomorphs
        - Ixodidae: *Haemaphysalis kutchensis.*
- PALAEARCTIC REGION
    - On insectivores
        - Argasidae: *Argas brumpti, Ornithodoros arenicolous, O. marocanus, O. tartakovskyi, O. tholozani.*
        - Ixodidae: *Anomalohimalaya lama, Ixodes hexagonus.*
    - On rodents
        - Argasidae: *Argas brumpti, A. bureschi, A. delanoi, Ornithodoros alactagalis, O. arenicolous, O. erraticus, O. normandi, O. tartakovskyi, O. tholozani.*
        - Ixodidae: *Anomalohimalaya cricetuli, A. lama, A. lotozskyi, Haemaphysalis verticalis, Ixodes angustus, I. apronophorus, I. crenulatus, I. laguri, I. nipponensis, I. occultus, I. pomerantzevi, I. redikorzevi, I. trianguliceps, Rhipicephalus fulvus.*
    - On lagomorphs
        - Ixodidae: *Anomalohimalaya lama, Haemaphysalis hispanica, H. pentalagi, Ixodes festai, I. hyatti, I. sachalinensis, I. shahi, Rhipicephalus leporis, R. pumilio.*

Although some insectivores (especially aquatic species) are infrequently parasitized by ticks, there is a wide variety of ixodids and a few argasids that parasitize insectivores (shrews, hedgehogs, tenrecs, moles, etc.) in various parts of the world. Perhaps the most diverse group of insectivore-associated ticks is the array of species of *Haemaphysalis* that parasitize various tenrecs (family Tenrecidae) in Madagascar with many of these tick species exhibiting relatively specialized morphologies (Hoogstraal and Kim 1985). Most adult-stage ixodids that parasitize insectivores belong to the genera *Haemaphysalis* or *Ixodes* whereas insectivore-associated argasids belong to the genera *Argas* or *Ornithodoros*. Some insectivore-associated ticks also sometimes parasitize other mammals; one example is the hedgehog-associated western Palaearctic species *Ixodes hexagonus*, immature stages of which also feed on rodents and adults on carnivores, etc.

Tree shrews are parasitized by immature stages of various species of *Amblyomma, Dermacentor, Haemaphysalis* and *Ixodes* in their native southeast Asia (Audy et al. 1960; Durden, unpublished) and also by most or all active stages (including adults) of at least two species of *Ixodes*. The latter group includes one species (*I. malayensis*) that may be host specific to tree shrews, and another (*I. granulatus*) that also parasitizes rodents and, occasionally, some other mammals in the Oriental zoogeographical region.

Bats are parasitized by many species of argasid ticks in different parts of the world. However, bat-associated ticks are not included in the above list because, in many cases, only (or mainly) immature stages of argasids are typically collected from bats with adults being rare or unknown for some species. The reason for this is because larval argasids typically attach to their hosts for extended periods (days) whereas nymphs and adults feed rapidly (minutes or hours) and are typically only found on the host when they are feeding. The importance of bats as hosts for members of the argasid genus *Carios* is underlined by calculations of the number of bat-associated species in each of the four argasid genera recognized by Horak et al. (2002). For example, 2 of 57 (4%) recognized species of *Argas* feed on bats, whereas 61 of 87 (70%) species of *Carios*, 4 of 38 (11%) *Ornithodoros*, and 0 of 3 (0%) of *Otobius* are known to parasitize bats. Moreover, as highlighted by Klompen et al. (1996), the host-specificity of several argasids is low so that, for example, many species of bat-associated *Carios* spp. may feed on 2 or more (sometimes several) different bat species. A few species even feed on bats in addition to other non-mammalian vertebrates; an example is the nearctic *Carios concanensis* which readily feeds on both bats and birds. In addition to argasids, a few species of prostriate ixodids (members of the genus *Ixodes*) also parasitize bats. Examples include *I. vespertilionis* mainly in the Palaearctic region, and *I. kopsteini, I. paradoxus* and *I. simplex* mainly in the Oriental region. Because bats are volant mammals, some species have large geographical ranges that are often mirrored by several of their ectoparasite species, including ticks.

Perhaps related to their taxonomic and zoogeographical diversity, rodents are parasitized by the widest variety of ticks, with immature and adult stages of many tick species being dependent on rodents for their survival. In fact, members of all 13 of the tick genera known to parasitize small mammals, have been found on rodents. This includes some highly specialized or geographically restricted ticks, such as the 3 members of the

Himalayan (Palaearctic) ixodid genus *Anomalohimalaya*. Typically, immature stages of ixodids are recorded as ectoparasites of rodents especially in temperate regions. However, adults of many ixodid species, especially some members of the genus *Ixodes*, also parasitize rodents, especially in tropical and subtropical regions. Moreover, a few abundant species of rodent-associated ixodids in the northern hemisphere are vectors of important zoonotic pathogens. Rodents are also important hosts of certain argasids, especially members of the genus *Ornithodoros*; following the tick list of Horak et al. (2002), 5 of 57 (9%) species of *Argas* are known to parasitize rodents, compared to 8 of 87 (9%) species of *Carios*, 16 of 38 (42%) *Ornithodoros* and 1 of 3 (33%) *Otobius*. Some species of rodent-associated *Ornithodoros* in both the Old and New worlds are vectors of zoonotic relapsing fever spirochetes.

Lagomorphs are parasitized by host specific (or nearly host specific) ixodid species in each of the major zoogeographical regions where they occur as part of the native mammal fauna. In addition to larvae and nymphs, adult stages of many of these ixodids are typical lagomorph ectoparasites. Lagomorph-associated ixodids mostly belong to the genera *Haemaphysalis* and *Ixodes* especially in the northern hemisphere, but adults of at least 4 species of *Rhipicephalus* and 1 species of *Anomalohimalaya* parasitize lagomorphs in the Old World, as well as at least 1 species of *Amblyomma* in the New World, and 1 species of *Dermacentor* in North America. Also in North America, at least 4 species of argasids frequently parasitize lagomorphs, including 3 species of *Ornithodoros* and immature stages of *Otobius lagophilus* (adults of this tick do not feed). Other opportunistically feeding species of burrow-dwelling *Ornithodoros* in the Afrotropical, Neotropical and Palaearctic regions will also feed on lagomorphs (Hoogstraal 1985).

Elephant shrews are known to be parasitized by a fairly wide variety of ixodid species in their native sub-Saharan Africa including some species with medical and veterinary importance (Fourie et al. 1995). Most of these ticks are represented by larval and nymphal stages but adults of a few tick species also parasitize elephant shrews.

4 Life cycles of ticks parasitizing small mammals

Argasid and ixodid ticks have very different life cycle strategies (Oliver 1989; Sonenshine et al. 2002) both of which are well adapted for exploiting small mammals as hosts. Argasid ticks have 6-legged larvae that typically feed and attach (once) to the host for an extended period (3-10 days).

After moulting, the first 8-legged nymphal stage of argasids then feeds relatively rapidly (~30 minutes to a few hours) on a host. Additional nymphal stages (2-8 depending on the species), each separated by a moult and feeding rapidly on the host, typically follow with the exact number of nymphal stages usually being dictated by the species and sex of the tick (females may have 1 more nymphal instar than conspecific males). Nymphal argasid instars typically feed once before moulting to the next stage but occasionally they feed twice especially if the first blood meal was small (Oliver 1989). Final instar argasid nymphs molt into adults but the morphological difference between nymphs and adults is subtle, with nymphs lacking a genital aperture. Similarly, male and female argasids are morphologically similar with the appearance of the genital aperture being the easiest method of distinguishing the 2 sexes. Like nymphs, adult argasids feed relatively rapidly on the host but they take multiple blood meals during their lifespans and females typically have multiple gonotrophic cycles and lay several separate, small egg batches each consisting of ~100 eggs. There are exceptions to this generalized argasid life cycle. For example, adults of at least 2 of the 3 species of *Otobius* and some species of bat-associated *Carios* spp. (those formerly assigned to *Antricola*) do not feed as adults. Further, larvae and first-instar nymphs of a small number of argasids do not feed before moulting to the next stage (Oliver 1989), and larvae of 1 bird-parasitizing species, *Argas cucumerinus*, are known to feed rapidly (in 7-25 minutes).

Because larval argasids typically attach to their hosts for extended periods, dispersal to new locations and host sites is typically achieved during this life stage. Also, the multiple feeding strategies of most argasid nymphs and adults often necessitates that these stages sequester themselves in host nests or burrows where the host is readily available for frequent blood meals. As Klompen et al. (1996) have discussed, this means that host nests and burrows, rather than specific hosts, become the focus of many argasid species and habitat rather than host specificity may develop. Different vertebrate species may share the same burrow or nest system and the argasid ticks residing in that burrow may feed on all of its vertebrate inhabitants and visitors. For example, in North America, rodent burrows occupied by *O. hermsi, O. parkeri* or *O. turicata*, may also be occupied at various times by snakes, lizards, tortoises, burrowing owls, shrews, lagomorphs and small carnivores, all of which may be fed on by these ticks. If the ticks in these burrow systems are infected with relapsing fever spirochetes, then these burrows also become foci for zoonotic pathogen transmission.

Ixodid ticks parasitizing small mammals almost invariably follow a 3-host (multi-host) life cycle with each of the active feeding stages (larva, nymph and adult) feeding on a different host individual. One- and 2-host ticks (which remain on the same host individual after at least 1 of the 2 moults) appear to be rare on small mammals although larvae of 1-host ticks, such as *Rhipicephalus* (*Boophilus)* spp., are occasionally recovered from small mammals. Many species of 3-host ixodid ticks feed on a variety of progressively larger hosts as these ticks moult into larger stages (larva→nymph→adult), especially in temperate regions. This means that a large number of small mammal species in any given region may be parasitized by immature stages of one or more ixodid species such as members of the genera *Amblyomma*, *Dermacentor*, *Haemaphysalis*, *Hyalomma, Ixodes* and/or *Rhipicephalus*. In fact, immature stages of some of the most important vector ticks in the northern hemisphere feed on several small mammal species in this way. Examples of these ticks include *Ixodes scapularis, I. pacificus* and *Dermacentor variabilis* in North America, and *I. ricinus* and *I. persulcatus* in Eurasia. However, there are many species of ixodids in which the adult stages parasitize small mammals, especially in subtropical and tropical regions and a large number of these ticks appear to be host-specific (or nearly so) to various mammals.

Unlike argasids, each active feeding stage of ixodid ticks feeds just once and each stage attaches to a host for an extended period (2-12 days) which aids in the dispersal of all active stages of these ticks. After each stage has engorged with blood, it detaches from the host, drops to the ground (leaf litter, etc.), moults to the next stage and, after a certain period, then searches (quests) for another host. Adult ixodids mate either on or off the host; members of the genus *Ixodes* can do either, but members of all other ixodid genera must mate on the host after the female has attached and started to feed (Oliver 1989). Adults of several species of small mammal-associated *Ixodes* not only mate off the host but also do so in the host nest. Because males of some of these *Ixodes* spp. rarely (or never) feed on the host, males of these "nidicolous" tick species are typically found only in the host nest or burrow.

Morphologically, ixodid larvae have 6-legs, nymphs and adults have 8-legs, and males have the hardened dorsal plate (scutum) virtually covering the entire dorsal surface. This "entire scutum" of male ixodids prevents significant enlargement (engorgement) of their bodies and adult male ixodids therefore take small blood meals (or none at all) whereas larvae, nymphs and females all engorge significantly. After mating and engorgement, female ixodids detach from the host and, over several days, lay one large egg mass (~1,000-10,000 eggs) before dying.

5 Ticks and small mammals as vectors and reservoirs of zoonotic pathogens

It could be argued that a wide variety of vertebrate pathogens (especially certain viruses, bacteria and protozoa) have evolved strategies to utilize the close associations between certain ticks and their small mammals hosts in order to perpetuate and amplify themselves and the infections they cause (Sonenshine et al. 2002, Goodman et al. 2005) (Table 1). For example, intimate associations between *Ixodes* ticks and small mammals in North America, Europe and central Asia promote the maintenance and spread of the Lyme disease. Similarly, in Eurasia, intimate associations between *Ixodes ricinus, I. persulcatus* and tick-borne encephalitis (TBE) virus promote the maintenance of this zoonotic disease.

Table 1. Important zoonotic tick-borne diseases for which small mammals are reservoir hosts

Disease	Agent	Main vector(s)	Distrubution
Tick-borne encephalitis	*Flavivirus*	*Ixodes ricinus, I. persulcatus*	Eurasia
Louping ill	*Flavivirus*	*I. ricinus*	United Kingdom
Kyasanur forest disease	*Flavivirus*	*Haemaphysalis spinigera*	Indian subcontinent
Powassan encephalitis	*Flavivirus*	Various ticks	N. America
Colorado tick fever	*Colitvirus*	*Dermacentor andersoni*	N. America
Crimean-Congo hemorrhagic fever	*Nairovirus*	*Hyalomma* spp.	Africa, Asia, Europe
Tick-borne relapsing fever	*Borrelia* spp.	*Ornithodoros* spp.	Old and New Worlds
Lyme disease	*Borrelia* spp	*Ixodes* spp.	N. Hemisphere
Rocky Mountain spotted fever	*Rickettsia rickettsii*	*D. andersoni, D. variabilis*	N. America
Tick typhus	*Rickettsia* spp.	Various ticks	Almost worldwide
Human granulocytic anaplasmosis	*Anaplasma phagocytophilum*	*Ixodes* spp.	N. Hemisphere
Human monocytic ehrlichiosis	*Ehrlichia chaffeensis*	*Amblyomma americanum*, other ticks in Eurasia	N. Hemisphere
Tularemia	*Francisella tularensis*	Various ticks	Almost worldwide
Human babesiosis	*Babesia* spp.	Ixodes spp.	N. Hemisphere

With respect to argasids, various species of *Borrelia* spirochetes are transmitted by characteristic species of *Ornithodoros* (often exhibiting tick specificity). Rodent burrows occupied by these *Ornithodoros* spp. ticks act as foci for tick-borne relapsing fever spirochetes and persons camping, hiking or resting near these burrows can be infected via relatively rapid infectious soft tick bites (Hoogtraal 1985). These and other zoonotic diseases with pathogens that exploit small mammal-tick associations in various parts of the world are listed in Table 1. Some other tick-borne infections of small mammals, such as certain members of the protozoan genus *Babesia*, are confined to particular mammals where the pathology to their hosts ranges from inapparent to severe. Clearly, intimate associations between small mammals and ticks not only have major ecological implications but also significant epidemiological consequences on a worldwide basis.

References

Audy JR, Nadchatram M, Lim B-L (1960) Host distribution of Malayan ticks (Ixodoidea). Malaysian Parasites 49:225–246

Barker SC, Murrell A (2004) Systematics and evolution of ticks with a list of valid genus and species names. Parasitology 129S:515–S536

Durden LA, Keirans JE (1996) Host-parasite co-extinction and the plight of tick conservation. Amer Entomol 42:87–91

Fourie LJ, Du Toit JS, Kok DJ, Horak IG (1995) Arthropod parasites of elephant-shrews with particular reference to ticks. Mammal Rev 25:31–37

Goodman JL, Dennis DT, Sonenshine DE (eds) (2005) Tick borne diseases of humans. Am Soc Microbiol Press, Washington

Hoogstraal H (1985) Argasid and nuttalliellid ticks as parasites and vectors. Adv Parasitol 24:135–238

Hoogstraal H, Aeschlimann A (1982) Tick-host specificity. Bull Soc Entomol Suisse 55:5–32

Hoogstraal H, Kim KC (1985) Tick and mammal coevolution, with emphasis on *Haemaphysalis*. In: Kim KC (ed) Coevolution of parasitic arthropods and mammals. John Wiley, New York, pp 505–568

Horak I, Camicas J-L, Keirans JE (2002) The Argasidae, Ixodidae and Nutalliellidae (Acari: Ixodida): A world list of valid tick names. Exp App Acarology 28:27–54

Keirans JE, Clifford CM, Hoogstraal H, Easton ER (1976) Discovery of *Nuttalliella namaqua* Bedford (Acarina: Ixodoidea: Nuttalliellidae) in Tanzania and redescription of the female based on scanning electron microscopy. Ann Entomol Soc Amer 69:926–932

Klompen JSH, Black IV WC, Keirans JE, Oliver Jr JH (1996) Evolution of ticks. Ann Rev Entomol 41:141–161

Klompen JSH, Black IV WC, Keirans JE, Norris DE (2000) Systematics and biogeography of hard ticks, a total evidence approach. Cladistics 16:79–102
Oliver Jr JH (1989) Biology and systematics of ticks (Acari: Ixodoidea). Ann Rev Ecol Syst 20:397–430
Sonenshine DE, Lane RS, Nicholson WL (2002) Ticks (Ixodida). In: Mullen G, Durden L (eds) Medical and veterinary entomology. Acad Press, San Diego, pp 517–558

7 Mesostigmatid mites as parasites of small mammals: Systematics, ecology, and the evolution of parasitic associations

Ashley P.G. Dowling

1 Introductory remarks

The Acari represent the most diverse assemblage within the Arachnida and are currently subdivided into three major lineages: Acariformes, Opilioacariformes, and Parasitiformes. This chapter focuses on the Mesostigmata, a morphologically and ecologically diverse group of parasitiform mites, many of which are parasitic on small mammals.

The majority of mesostigmatid species are found as free-living predators of other arthropods and nematodes in a wide variety of habitats. Only a small proportion of total species diversity is parasitic and all confirmed vertebrates parasites are currently placed within the cohort Dermanyssina. More specifically, most of the parasite diversity is found in the superfamily Dermanyssoidea, which will be the primary focus of this chapter. The diversity of ecological strategies found in the Dermanyssoidea is great and includes everything from predatory to endoparasitic mites. This chapter aims to introduce the reader to the systematics, biology, and ecological strategies of dermanyssines associated with small mammals and will examine the evolution of parasitism within this amazingly diverse group of mites.

2 Taxonomy, phylogeny, and distribution

Over the past few decades, mesostigmatid classification has been fairly stable. For a summary of mesostigmatid classification see Krantz (1978; note that Krantz uses the term Gamasida in place of Mesostigmata). The Mesostigmata is a suborder within the Parasitiformes, one of three currently recognized acarine orders. Within the Mesostigmata almost all vertebrate parasites are found in the superfamily Dermanyssoidea with a couple exceptions.

Unlike Mesostigmata, the taxonomy of the Dermanyssoidea has been in a perpetual state of flux since the late 19th century. Many classification schemes (Berlese 1892, 1913; Vitzthum 1943; Zumpt and Patterson 1951; Baker and Wharton 1952; Evans and Till 1966, 1979; Radovsky 1967; Karg 1971; Krantz 1978; Johnston 1982), often contradictory, have been proposed by different researchers, typically based more upon ecology and host associations than any character evidence. Because dermanyssoids exhibit such wide ecological amplitude and display high levels of morphological variability, it has been difficult to determine phylogeny based upon morphological characters alone. The remaining discussion follows the classification scheme of Johnston (1982) with modifications based on Radovsky (1985) and recent molecular evidence by Dowling (2005).

Most classifications have placed all parasitic mesostigmatid groups in the Dermanyssoidea, primarily based on their parasitic lifestyles, which has often led to highly modified morphological features. Until recently, no modern phylogenetic methods had been used to test the monophyly of the Dermanyssoidea or the phylogenetic relationships within the entire subfamily. Systematics of the group has been based entirely on morphological characteristics and ecological associations. Casanueva (1993) and Strong (1995) conducted phylogenetic studies, but both focused on arthropod associated Laelapidae, effectively ignoring the evolution of vertebrate parasitism. Morphology presents a problem to phylogenetic reconstruction because of the amount of convergence due to multiple independent evolutions of a parasitic lifestyle. Dowling (2005) was the first to use molecular sequence data to test dermanyssoid relationships across the superfamily and within the Dermanyssina. Results suggest that the two obligate bat parasite families, Spinturnicidae and Spelaeorhynchidae, do not belong within the Dermanyssoidea, but do remain within the Dermanyssina. Further discussions of dermanyssoid systematics and evolution will exclude these two families, but parasitic adaptations in each family will be discussed later as they are both important bat parasites. The families Laelapidae, Haemogamasidae, Hirstionyssidae, Macronyssidae, Dermanyssidae, Halarachnidae, Dasyponyssidae, Manitherionyssidae, and Hystrichonyssidae all contain species that are parasitic on small mammals. Specific adaptations to parasitism in each group will be discussed later.

3 Morphology and development

The primitive dermanyssoid varies little from the typical Mesostigmata form. These mites are predatory in nature and found in a wide range of habitats, primarily soil and decomposing leaf litter. They are typically 0.2-2.0 mm in length with long legs for active locomotion. The first pair of legs is usually elongated, slender, and used as sensory structures and all four tarsi contain paired claws and an ambulacral sucker. The gnathosoma contains a mouth, palps, and chelicerae. The palps are paired leg-like structures used primarily as sensory apparatus and sometimes to manipulate food items. The chelicerae are a pair of retractable structures equipped with an anterior claw comprised of a fixed dorsal chela and a moveable ventral chela. Both chelae are often equipped with one or more teeth and the chelicerae are used for capturing, tearing, and puncturing prey. Like most arachnids, mesostigmatid mites are typically fluid feeders and the chelicerae macerate the food item in preparation for feeding. The gnathosoma and legs are attached to the opithosoma, which is unsegmented and contains a number of sclerotized shields for protection and as rigid points for muscle attachment. The remainder of the opithosoma is typically comprised of soft, expandable cuticle. Genital openings in adults are positioned ventrally, in a mid to anterior position and the anus is positioned ventral and posterior. Respiration occurs through a pair of spiracular openings (stigmata) located ventrally, and usually between coxae III and IV, hence the group name Mesostigmata.

Development in the Mesostigmata includes an egg, larva, protonymph, deutonymph, and adult. The larva is hexapod, lacking the fourth pair of legs, while the nymphal and adult stages are all octopod. Each stage has distinct setal and idiosomal shielding patterns. The larva typically is non-feeding and possesses very little, if any shielding, and reduced setation. Each successive molt increases both shielding and setation until the full complement of each is reached at the adult stage.

Morphological features and development are fairly uniform among the free-living predatory Mesostigmata and primitive Dermanyssoidea, but a high degree of variability in many features is found in taxa with parasitic and commensal associations formed with vertebrates and arthropods. Evans (1963) noted extreme variability in palpal chaetotaxy among species that have developed close associations with other animals. The chelicerae have undergone amazing modifications across the parasitic groups. Multiple groups have independently converged on slender, edentate chelicerae that function similar to stylets of blood-sucking insects. Paedomorphosis also seems to be a common occurrence among parasitic dermanyssoids, especially the endoparasitic lineages, as reflected by reductions in setal count, sclerotization, and dorsal shield shape and size. Variations in the life cycle also exist among the parasitic groups. While all stages are present, many may be suppressed and completed inside the female or passed inside the ecdysed cuticle of the preceding stage. The Macronyssidae and Halarachnidae are two groups with independent modifications to the life cycle. The Macronyssidae have a highly suppressed, non-feeding deutonymph and halarachnids are characterized by suppression of both nymphal stages. This variability in morphology and development among vertebrate associates has been one of the main confounding factors in attempts to accurately classify the Dermanyssoidea.

4 Evolution of parasitism

The ecological amplitude of dermanyssoid mites is phenomenal, and life-histories across the superfamily include free-living, soil dwelling predators, arthropod predators in vertebrate and invertebrate nests or colonies, facultative and obligatory vertebrate parasites, and respiratory and auditory endoparasites of birds, mammals, and lepidosaurs. The Dermanyssoidea provide a unique opportunity to study the evolution of parasitism because of the many intermediate forms between predators and parasites represented among extant lineages. The full range of ecological associations is even represented within single genera such as *Androlaelaps* and *Haemogamasus*, which will be discussed later. The remainder of this chapter will focus on the evolutionary transition to parasitism and the adaptations that have evolved in each of the parasitic dermanyssine families.

4.1 Transitions to parasitism: Pre-adaptation or opportunistic exposure

Two competing hypotheses prevail in describing the evolutionary transition from a predatory to parasitic lifestyle. These have been described as type A and type F routes or pathways to parasitism by Waage (1979). In Type A routes, associations with hosts preceded adaptations for parasitic feeding. It is widely agreed on that the diversity of feather mites followed this pathway, where they went from feeding on bird debris in the nest to permanently inhabiting the body of the bird (Proctor 2003). For example, Fain and Hyland (1985) suggested that parasitic psoroptoids and analgids in the acariform acarine lineage Astigmata probably arose from nidicolous Pyroglyphidae. Type F routes involved adaptations to feeding on a host that preceded the actual association, such as the stylet mouthparts of nectar feeding mosquitoes and plant feeding Hemiptera that were easily adaptable to blood-feeding (Radovsky 1985). The evolution of parasitism in dermanyssoid mites was likely a combination of the two pathways.

Mesostigmatid mites in general are very well pre-adapted to parasitism. The chelicerae, even in the most primitive free-living predatory forms, can effectively feed on secretions, scales, scabs, and even tear into the skin of young vertebrates to reach a blood meal. The chelicerae of many free-living dermanyssoids are more generalized than those of other predatory mesostigmatids, which may have provided the necessary advantage to invade the nidicolous niche. In fact, the morphological change from the general dermanyssoid type in some parasites is so subtle that without the context of a host, it would be difficult to tell that the mite was an obligate parasite (Evans 1955; Radovsky 1985). Even though this generalized cheliceral form is suitable for parasitism, slender edentate chelae specialized for piercing skin are widely found throughout the Dermanyssoidea.

A second key feature that helped lead to the radiation of parasitic types within Dermanyssoidea and not in the other mesostigmatid groups may be the utilization of an exceptional number of niches by primitive *Hypoaspis* types (Radovsky 1985). Members of the *Hypoaspis*-complex are found in soil, litter, decaying substrates, the nests of social insects, burrows and galleries of beetles, and in the nests and on the bodies of mammals and birds. Most *Hypoaspis* species studied are predators (Karg 1961; Nelzina et al. 1967) and have not been shown to have any predilection to feeding on the host, but the association as a predator in vertebrate nests has been hypothesized to be the origin of vertebrate parasitism in dermanyssoid mites (Radovsky 1969). While most Mesostigmata have characteristics suitable for parasitism, it may be that hypoaspidines were the first predators to colonize and utilize vertebrate nests and they competitively limited other predatory mites. Mesostigmatid groups, such as the Parasitidae and Ologamasidae, contain species that are obligate nest predators, but the Laelapidae are typically the most abundant and commonly encountered. Constant exposure to a potential host while occupying the nest niche may have possibly opened the door for dermanyssoid radiation into a parasitic lifestyle.

4.1.1 Androlaelaps and Haemogamasus

The transition from predator to obligate parasite is most readily observed within two genera, *Androlaelaps* and *Haemogamasus*. The laelapine lineage from *Hypoaspis* is thought to have begun with *Androlaelaps*, due to many shared morphological characters between the two genera (Radovsky 1969). *Androlaelaps* are found worldwide and exhibit varying degrees of dependence on a diversity of vertebrate hosts. Reytblat (1965) compared the feeding behavior of four species of *Androlaelaps, A. fahrenholzi, A. longipes, A. casalis,* and *A. semidesertus*. The degree of adaptation for parasitism was based upon reproductive success of laboratory-reared mites fed on a blood diet versus an arthropod diet as well as their ability to feed from a host. *A. longipes* and *A. casalis* had comparable numbers of offspring on blood or arthropods, but both had highest reproductive output when fed a mixed diet. *A. fahrenholzi* and *A. semidesertus* were unable to reproduce on a diet of arthropods alone, showing dependence on a host. All four possess typical *Hypoaspis*-type chelate-dentate chelicerae and readily inflicted wounds to start blood flow from suckling mice. *A. fahrenholzi* frequently feeds from preexisting wounds, dried blood, scabs, as well as on other small arthropods (Reytblat 1965; Radovsky 1985).

The genus *Haemogamasus* exhibits a range of feeding ecologies as well, from non-parasitic predators to polyphagous and facultative nidicoles to obligatory hematophages (Radovsky 1985). *Haemogamasus pontiger* is the only *Haemogamasus* often found free-living and is common in detritus on warehouse floors (Hughes 1961) and in granaries (Evans and Till 1966). Furman (1959a) found that 63% of *H. pontiger* in laboratory studies would feed from flowing blood on mice hosts, but unlike *A. fahrenholzi*, none would feed on dried blood. Hughes (1961) found that *H. pontiger* can complete its lifecycle if only supplied with wheat germ as food indicating that *H. pontiger* is a predator and saprophage with zero dependence on a host.

The *Haemogamasus reidi* group (Williams et al. 1978) includes all *Haemogamasus* that do not possess special cheliceral adaptations for skin penetration as found in the *Haemogamasus liponyssoides* group, but are obligatory nidicoles, thus excluding *H. pontiger*. Members of this group are thought to typically feed from preexisting wounds, rather than puncturing the host's skin (Furman 1959a, b, 1968; Goncharova and Buyakova 1960). *H. reidi* can complete development and reproduce on a diet entirely of blood or arthropods (Furman 1959b). *Haemogamasus clitelli* had reduced rates of reproduction when restricted to either blood or arthropod meals, and if restricted entirely to blood showed an increasing tendency to cannibalize their young (Nelzina and Danilova 1956). *Haemogamasus nidi* was able to sustain reproduction on blood, but was not able to reproduce on an all-arthropod diet. Reproduction was highest on a mixed diet. Finally, *H. citelli*, associated with ground squirrels, was found to actively engorge on newborn gophers in the nest, potentially puncturing the skin, though more likely frequently feeding from preexisting wounds (Nelzina and Danilova 1956). Feeding from the adult ground squirrels was rarely witnessed, presumably due to the thicker and tougher skin of the adult host. The *H. reidi* group is very reminiscent of the *Androlaelaps* group previously discussed in its lack of modification specific to parasitism and the varying levels of blood feeding and dependence on a host found across species.

The *liponyssoides* group consists of several species of *Haemogamasus*, all of which are obligatory hematophages. The defining characteristic of the group is highly modified, slender and edentate chelicerae used to pierce the host skin rather than tear at it (Radovsky 1985). In laboratory conditions, *H. liponyssoides* would not feed on arthropods and only reluctantly and poorly on free-flowing blood (Radovsky 1960). *H. liponyssoides* frequently fed from both adults and suckling rodents by penetrating the skin to cause blood flow. Species in the *liponyssoides* group are also able to engorge (i.e. taking in more than their weight at a single meal) more so than other *Haemogamasus*, a trait commonly found in obligate blood-feeding parasites (Radovsky 1985).

The fact that two independent genera show such a graded transition from predator to obligate parasite among species is a strong indication that the transition began with opportunistic feeding in the vertebrate nest environment. The morphological components were available, as exhibited by active blood feeding in *Androlaelaps*, and the active predation of a *Hypoaspis*-like ancestor on microarthropods in vertebrate nests, provided the opportunity for host interaction. The ability to utilize a variety of nutrients for development and reproduction, as exhibited by species of *Androlaelaps* and *Haemogamasus*, also likely played a major role in the successful colonization of parasitic niches by dermanyssoid mites.

4.2 Host associations and specializations for parasitism

The remaining families of parasites are comprised of members that are entirely parasitic and in turn show many specialized adaptations to parasitism. The remainder of this chapter will focus on the unique modifications and diversity of host associations found in each group. A table summarizing the host associations of parasitic Mesostigmata will follow this section (Table 1).

4.2.1 Laelapidae, Haemogamasidae, and Hirstionyssidae

These three families are the most commonly found dermanyssoids associated with rodents and insectivores and the Laelapidae are the most speciose group. As discussed earlier, the Laelapidae and Haemogamasidae show a graded transition from predatory to parasitic mites. For most laelapids associated with mammals it is unknown whether they are nidicolous, facultative, or obligate parasites and one suspects that parasitism has arisen in the Laelapidae multiple times. The most common genera found are *Laelaps* and *Echinolaelaps*, both laelapids and restricted to muroid rodents. *Haemogamasus* is the most common haemogamasid genus and is found on various rodent groups and insectivores. Neither group exhibits extreme modifications to parasitism, but do posses a range of cheliceral changes towards the edentate type found in many other parasitic dermanyssoid groups. The hirstionyssids however, have slender edentate chelicerae designed for piercing and are clearly parasitic. The genus *Echinonyssus* is cosmopolitan and can be found on most small mammal groups except Chiroptera.

4.2.2 Macronyssidae

Members of the Macronyssidae are obligate parasites of a wide range of hosts including bats, rodents, lizards, and birds and are cosmopolitan in distribution. Most members of this group still retain the dermanyssoid behavior of inhabiting the nest or roost and only making contact with the host when a blood meal is necessary. The Macronyssidae are characterized by an actively feeding protonymph, an inactive, non-feeding, and highly regressed deutonymph, and an actively feeding adult. This unique life-cycle modification is shared by all macronyssids and found nowhere else among the Mesostigmata, except in the Rhinonyssidae, a family of dermanyssoid avian endoparasites, clearly derived from the Macronyssidae (Dowling 2005).

Macronyssid mites are divided into two groups, the Macronyssinae and the Ornithonyssinae (Radovsky 1967, 1969), both of which have unique parasitic qualities. The Macronyssinae are primarily bat parasites, except for two genera, *Acanthonyssus* and *Argitis*, found on Neotropical rodents. Ornithonyssines have a much broader host range, including bats, rodents, lizards, snakes, and birds. Interestingly enough, although the widespread and apparent long standing association with bats, no macronyssids are found in association with any Megachiroptera.

Morphologically the two groups are easy to differentiate based upon characteristics associated with feeding. Ornithonyssines, unlike macronyssines, engorge heavily during prolonged feeding periods, resulting in bodies that are much less sclerotized and more able to greatly expand than macronyssines. Overall, dorsal and ventral shielding is reduced in ornithonyssines and although chelicerae in all macronyssids are edentate, ornithonyssine chelicerae are much further specialized for piercing skin than macronyssines.

4.2.3 Dermanyssidae

The family Dermanyssidae is primarily parasitic on birds, but contains a few species readily found on rodents. The group is recognizable by their highly specialized chelicerae. The second segment of each chelicera is elongated and the chela is small and edentate. The chelicerae apparently function as a stylet for piercing host tissue and sucking blood, much the way that blood sucking insects do (Radovsky 1969, 1985). Rodent associated dermanyssids are in the genus *Liponyssoides* and are nest inhabiting parasites that only contact the host when feeding.

4.2.4 Halarachnidae

Halarachnids are endoparasites of a wide range of mammal orders. The family can be divided into two subfamilies, the Halarachninae, which are respiratory endoparasites, and the Raillietiinae, which are auditory endoparasites. Both subfamilies share a unique developmental modification where the larva is an active, non-feeding stage responsible for transmission between hosts, followed by two suppressed, non-feeding nymphal stages. These nymphal stages are passed very quickly and oftentimes the deutonymph is passed inside the ecdysed protonymphal cuticle (Radovsky 1985). The adult is an actively feeding stage.

Only a few species of raillietines are known, all in the genus *Raillietia,* and infect the auditory passages of cattle, goats, antelope, and wombats. They are not known to typically damage the ear or negatively affect the host except in cases of very heavy infestations.

Halarachnines parasitize the respiratory tracts, mostly the nasal passages, of a wide range of mammals (Furman 1979) across a large geographic scale. The breadth of the host associations and distribution is suggestive of an ancient association with mammals and is worthy of a closer look. Based on host associations, four major groups of halarachnines exist. The first involves an infestation of pinnipeds, with the genus *Halarachne*, on phocid seals and *Orthohalarachne* on otariid seals and walruses. The second group includes two genera, *Pneumonyssus* and *Rhinophaga* that are primarily associated with Old World cercopithecid and pongid primates, but also parasitize African procaviids (Elephant shrews) and hystricids (porcupines). Domrow (1974) described *P. capricornii* from a New Guinea phalangerid that is very similar to *P. simicola* from Asian macaques, suggesting a host switch from primates to marsupials. The genus *Pneumonyssus* is the only halarachnid group found primarily in the lungs, all others typically reside in the nasal passages (except adults of *Orthohalarhne*, which are also found in the lungs of their hosts). The third group involves the genus *Zumptiella* parasitic in rodents, specifically Holarctic Sciuridae and African Pedetidae. One species has been found in an African mongoose, but whether or not it represents a natural association or a predator to prey transfer has not been determined (Furman 1979). The final genus, *Pneumonyssoides*, has a bizarre host range. Species are known only from African bush pigs and wart hogs, the domestic dog, and Neotropical primates in the family Cebidae.

The most commonly studied halarachnids are *Pneumonyssoides caninum*, found in the domestic dog, and species of *Pneumonyssus* (primarily *P. simicola*) because of their common presence in primates used for research. *Pneumonyssoides caninum* receives attention because of its presence in domesticated dogs, but rarely causes symptoms any more severe than extra mucous production (Yunker 1973) and antibody response to antigens (Gunnarsson and Zakrisson 2000). One case of *P. caninum* infection has been reported in a fox (*Vulpes vulpes*), but no other reports exist to suggest this is a typical association (Bredal et al 1997).

Unlike the rather asymptomatic *P. caninum*, the monkey lung mite, *Pneumonyssus simicola* (and other species) can cause severe pathogenecity and because the hosts are often laboratory research primates (Yunker 1973; Kim 1977, 1980; Leathers 1978), infestations garner great attention. Infestation, or pulmonary acariasis causes nodules or tubercules, each containing numerous adult mites, to form in the lungs of the host (Sundararaj et al. 1992; Hiraoka et al. 2001). Infestations can severely affect the health of the host, leading to death in the case of heavy infestations.

4.2.5 Spinturnicidae

Spinturnicid mites are more morphologically modified and adapted to an ectoparasitic lifestyle than any other mesostigmatid group. Spinturnicids are obligate blood-feeding ectoparasites of bats and are almost exclusively found on the wing and tail membranes. In some genera, such as *Periglischrus*, *Ancystropus*, and *Meristaspis*, adult females develop a greatly enlarged opithosoma and are found attached to the face, ears, and along the arm bones. Adult females of *Paraspinturnix* have only been found in the anal orifices of bats (Rudnick 1960), but all other stages of all these genera are typically found on the membranes. To facilitate a life on the wing membrane of bats, spinturnicids posses long, thick, and robust legs and large hooked tarsal claws on a comparatively small diamond shaped body. These mites can cling to the wings very effectively, but at the same time can release from the membrane and run rapidly across it. Spinturnicid mites are also nymphiparous, with the egg and larval stages passed intrauterine, which releases the most vulnerable stages (egg and larva) from pressures associated with living permanently on the wing membranes. Only the actively feeding stages are independently active on the host.

Spinturnicidae are found in association with all bat groups worldwide and Rudnick (1960) has produced the most extensive review of spinturnicid host associations and taxonomy. Based on the currently known geographic distributions and host groups, spinturnicids appear to have had a long coevolutionary history with the Chiroptera. Two genera, *Meristaspis* and *Ancystropus*, are exclusively found on Megachiroptera, while all other species are found only on Microchiroptera. Most genera are specific to the family of bats they parasitize. For example, *Eyndhovenia* and *Paraperiglischrus* are only associated with rhinolophid bats, *Periglischrus* only on phyllostomids, *Spinturnix* on natalids and vespertilionids, and *Paraspinturnix* only with vespertilionid bats.

4.2.6 Spelaeorhynchidae

Spelaeorhynchids superficially resemble ticks with a soft expandable body and small dorsal and ventral plates, but the resemblance stops there. The dorsal and ventral plates on the body are small, but heavily sclerotized and form part of a strongly sclerotized gnathosomal ring. The chelicerae are chelate dentate (unlike ticks) and are short and stout with very large teeth. Spelaeorhynchids use their chelicerae for attachment to host tissue and often the leading edge of the gnathosomal ring and first pair of legs are embedded in the tissue. Because the anterior end of the adult female is typically embedded in the host, the location of the genital opening in spelaeorhynchids is uniquely positioned close to the anus, rather than in the more standard anterior position of most mesostigmatid mites. Like spinturnicids, they are also nymphiparous, passing the egg and larval stages in the female. Based on morphological features, it is impossible to suggest phylogenetic relationships to any specific group of Mesostigmata. Molecular evidence does suggest a relationship to the Spinturnicidae, which is circumstantially supported by the common host group and the intrauterine passing of egg and larva. Due to problems associated with long branch attraction in molecular systematics, and the fact that these two families represent the longest branches in current hypotheses, the suggested relationship is questioned.

Table 1. Summary of commonly found genera of Mesostigmata parasitic on mammals

Mite Family and Genera	Primary Host Associations
Laelapidae	
Laelaps, Echinolaelaps	Muridae
Andreacarus	Nesomyinae, Tenrecidae
Tur, Gigantolaelaps	New World Rodentia
Neolaelaps, Notolaelaps	Megachiroptera
Haemogamasidae	
Haemogamasus	Rodentia, Insectivora
Hirstionyssidae	
Echinonyssus	Rodentia, Insectivora, Carnivora
Dermanyssidae	
Liponyssoides	Muridae
Macronyssidae	
Macronyssus, Ichoronyssus	Vespertilionidae, Rhinolophoidea
Bewsiella, Megistonyssus	Rhinolophidae
Parichoronyssus	Phyllostomidae, Emballonuridae
Radfordiella, Macronyssoides	Phyllostomidae
Chiroptonyssus	Molossidae
Parasteatonyssus	Molossidae
Steatonyssus	Microchiroptera
Ornithonyssus, Acanthonyssus	Neotropical Rodentia
Argitis	Neotropical Rodentia
Halarachnidae	
Halarachne	Phocidae
Orthohalarachne	Otariidae, Odobenidae
Pneumonyssus, Rhinophaga	Cercopithecidae, Pongidae, Procaviidae, Hystricidae
Zumptiella	Sciuridae, Pedetidae
Pneumonyssoides	Suidae, Canidae, Cebidae
Dasyponyssidae	
Dasyponyssus, Xenarthronyssus	Dasypodidae
Manitherionyssidae	
Manitherionyssus	Manidae
Hystrichonyssidae	
Hystrichonyssus	Hystricidae
Spinturnicidae	
Meristaspis, Ancystropus	Pteropodidae
Eyndhovenia, Paraperiglischrus	Rhinolophidae
Periglischrus, Cameronieta	Phyllostomidae
Spinturnix, Paraspinturnix	Vespertilionidae
Spelaeorhynchidae	
Spelaeorhynchus	Phyllostomidae

4.2.7 Dasyponyssidae, Hystrichonyssidae, and Manitherionyssidae

The remaining three mesostigmatid mites parasitic on small mammals are all known from only one or two species. All are highly modified for an ectoparasitic lifestyle and their phylogenetic relationships to each other as well as to other dermanyssines is unknown. They are not presumed to be related to one another and are only treated together here because of the lack of overall knowledge regarding the families. The Dasyponyssidae are split into two monotypic genera (*Dasyponyssus* and *Xenarthronyssus*) and are restricted to armadillos in Central and South America. Manitherionyssidae and Hystrichonyssidae are both monotypic and restricted to the African pangolin and Asian porcupine, respectively. Characteristics in the legs, claws, and body of manitherionyssids are similar to that of dasyponyssids, but these similarities may simply be due to their evolution on scaled mammals. Hystrichonyssids are the most highly modified of the three families, with extremely thin, elongated chelicerae designed for piercing. The chelicerae resemble those of most Dermanyssidae, discussed earlier, except that hystrichonyssid chelicerae are elongated at the basal digit, rather than at the second digit.

References

Baker EW, Wharton GW (1952) An introduction to acarology. Macmillan, New York

Berlese A (1882–1892) Acari, Myriapoda et Scorpiones hucusque in Italia reperta. Ordo Mesostigmata (Gamasidae). Fusc 52, No 2

Berlese A (1913) Acarotheca Italica. Rizzi, Firenze

Bredel WP, Gjerde BK, Kippenes H (1997) *Pneumonyssoides caninum*, the canine nasal mite, reported for the first time in a fox (*Vulpes vulpes*). Vet Parasitol 73:291–297

Casanueva ME (1993) Phylogenetic studies of the free living and arthropod associated Laelapidae (Acari: Mesostigmata). Gayana (Zoologia) 57:21–46

Domrow R (1974) Notes on halarachnine larval morphology and new species of *Pneumonyssus* Banks (Acari: Dermanyssidae). J Aust Entomol Soc 13:17–26

Dowling APG (2005) Molecular systematics and the evolution of parasitic associations of dermanyssoid mites (Acari: Dermanyssoidea). Ph. D. thesis, University of Michigan

Evans GO (1955) A review of the laelaptid paraphages of the Myriapoda with descriptions of three new species (Acarina: Laelaptidae). Parasitology 45:352–368

Evans GO (1963) Some observations on the chaetotaxy of the pedipalps in the Mesostigmata (Acari). Ann Mag Nat Hist 13:513–527

Evans GO (1992) Principles of acarology. CABI, Wallingford

Evans GO, Till WM (1966) Studies on the British Dermanyssidae (Acari: Mesostigmata). Part II. Classification. Bull British Mus (Nat Hist Zool) 14:109–370

Evans GO, Till WM (1979) Mesostigmatic mites of Britain and Ireland (Acari: Mesostigmata). An introduction to their external morphology and classification. Trans Zool Soc Lond 35:139–270

Fain A, Hyland Jr KW (1985) Evolution of astigmatid mites on mammals. In: KC Kim (ed) Coevolution of parasitic arthropods and mammals. John Wiley, New York, pp 641–658

Furman DP (1959a) Feeding habits of symbiotic mesostigmatid mites of mammals in relation to pathogen-vector potentials. Am J Trop Med Hyg 8:5–12

Furman DP (1959b) Observations on the biology and morphology of *Haemogamasus ambulans* (Thorell) (Acarina: Haemogamasidae). J Parasitol 45:274–280

Furman DP (1968) Effects of the microclimate on parasitic nest mites of the dusky footed wood rat *Neotoma fuscipes* Baird. J Med Entomol 5:160–168

Furman DP (1979) Specificity adaptation and parallel evolution in the endoparasitic Mesostigmata of mammals In: Rodriguez JG (ed) Recent advances in acarology, vol. 2. Acad Press, New York, pp 329–337

Goncharova AA, Buyakova TG (1960) Biology of *Haemogamasus mandschuricus* Vtzth in Transbaikalia (Far Eastern area) (in Russian). Parasitologicheskyi Sbornik [Parazitological Collection] 19:155–163

Gunnarsson L, Zakrisson G, Christensson D, Uggla A (2003) Transmission of the canine nasal mite, *Pneumonyssoides caninum*. Acta Parasitol 48:64–67

Hiraoka E, Sato T, Shirai W, Kimura J, Nogami S, Itou M, Shimizu K (2001) A case of pulmonary acariasis in lung of Japanese macaque. J Vet Med Sci 63:87–89

Hughes AM (1961) The mites of stored food. Min Agr Fish London Tech Bull 9:1–287

Johnston DE (1982) Acari. In: Parker SP (ed) Synopsis and classification of living organisms, vol 2. McGraw Hill, New York, pp 111–117

Karg W (1961) Ökologische Untersuchungen von edaphischen Gamasiden (Acarina, Parasitiformes). Pedobiol 1:53–98

Karg W (1971) Die freilebenden Gamasina (Gamasides). Raubmilben Die Tierwelt Deutschlands 59. Fisher Verlag, Jena

Kim JCS (1977) Pulmonary acariasis in Old World monkeys. Vet Bull 47:249–255

Kim KC, Haas VL, Keyes MC (1980) Populations, microhabitat preference, and effects of infestation of two species of *Orthohalarachne* (Halarachnidae: Acarina) in the northern fur seal. J Wildl Dis 16:45–51

Krantz GW (1978) A manual of acarology, 2nd edn. Oregon State Univ Book Stores, Corvallis

Leathers CW (1978) Pulmonary acariasis in an infant colony born rhesus monkey *Macaca mulatta*. Lab Animal Sci 28:102–103

Nelzina EN, Danilova GM (1956) Materials on the biology of mites of the family Haemogamasidae (Gamasoidea, Parasitiformes) 1. Nutrition of *Haemogama-*

sus citelli Breg and Nelz and *H. nidi* Mich (in Russian). Meditsinskaya Parasitologiya i Parasitarnye Bolezni [Medical Parasitology and Parasitic Diseases] 25:352–358

Nelzina EN, Danilova GM, Klinova ZI (1967) Gamasoid mites (Gamasoidea, Parasitiformes) – one of the basic components of burrow biocenoses (in Russian). Parazitologiya 1:412–421

Proctor HC (2003) Feather mites (Acari: Astigmata): Ecology, behavior and evolution. Ann Rev Entomol 48:185–209

Radovsky FJ (1960) Biological studies on *Haemogamasus liponyssoides* Ewing (Acarina: Haemogamasidae). J Parasitol 46:410–417

Radovsky FJ (1967) The Macronyssidae and Laelapidae (Acarina: Mesostigmata) parasitic on bats. Univ Calif Publ Entomol 46:1–288

Radovsky FJ (1969) Adaptive radiation in parasitic Mesostigmata. Acarologia 11:450–483

Radovsky FJ (1985) Evolution of mammalian mesostigmatid mites. In: Kim KC (ed) Coevolution of parasitic arthropods and mammals. John Wiley, New York, pp 441–504

Radovsky FJ (1994) The evolution of parasitism and the distribution of some dermanyssoid (Mesostigmata) mites on vertebrate hosts. In: Houck MA (ed) Mites: Ecological and evolutionary analysis of life-history patterns. Chapman and Hall, New York, pp 186–217

Reytblat AG (1965) Biology of the gamasid mite *Haemolaelaps semidesertus* Breg (Gamasoidea: Parasitiformes) (in Russian). Zool Zh 44:863–870

Rudnick A (1960) A revision of the mites of the family Spinturnicidae (Acarina). Univ Calif Publ Entomol 17:157–284

Strong KL (1995) A phylogeny of laelapid mites associated with arthropods with special reference to *Hypoaspis* Canestrini (Mesostigmata: Laelapidae) associated with Australian burrowing cockroaches (Blattodea: Blaberidae). Ph. D. thesis, Australian National University

Sundararaj A, Thilagarajan N, Thanikachalam M, Viswanathan S, Manohar BM (1992) Pulmonary akariasis in lion tailed macaque monkeys (*Macaca silenus*). Indian Vet J 69:1085–1087

Vitzthum HG (1943) Acarina. In: Bronn HG (ed) Klassen und Ordnungen des Tierreichs, Bd 5, Abt 4, Buch 5. Akad Verlag, Leipzig, pp 1–1011

Waage J K (1979) The evolution of insect/vertebrate associations. Biol J Linn Soc 12:187–224

Williams GL, Smiley RL, Redington BC (1978) A taxonomic study of the genus *Haemogamasus* in North America with descriptions of two new species (Acari: Mesostigmata: Laelaptidae). Int J Acarol 4:235–273

Yunker CE (1973) Mites. In: Flynn RJ (ed) Parasites of laboratory animals. Iowa State Univ Press, Ames, pp 425–492

Zumpt F, Patterson PM (1951) Further notes on laelaptid mites parasitic on vertebrates. A preliminary study to the Ethiopian fauna. J Ent Soc S Afr, Pretoria 14:63–93

8 Taxonomy, life cycles and the origin of parasitism in trombiculid mites[1]

Andrey B. Shatrov and Naina I. Kudryashova

1 Introductory remarks

Acariform mites of the family Trombiculidae are the only larval parasites from the cohort Parasitengona that attack vertebrate hosts. Trombiculid larvae, or chiggers, cause skin irritations and itching in their hosts and serve as vectors of the causative agents of tsutsugamushi disease (Ewing 1944a; Audy 1961; Traub and Wisseman 1974; Kawamura et al. 1995). Deutonymphs and adults of these mites are soil dwellers that prey on various arthropods and, in particular, on their eggs. Adults and deutonymphs of the majority of trombiculid species have never been observed on the soil surface (Wharton 1946; Daniel 1961, 1965) and, therefore, the taxonomy of trombiculid mites is based solely on their larvae.

2 Taxonomy and phylogeny

2.1 Taxonomy

The first mention of trombiculid mites dates back to 1758 when *Acarus batatas* was first described in "Systema Naturae" by Linnaeus. The separate genus *Trombicula*, with six species, was established only in 1905, and was included initially in the family Trombidiidae (Oudemans 1912). Soon thereafter, detailed examination of the genus established a separate subfamily Trombiculinae, which has been raised to the rank of a family (Ewing 1944b). This family was composed originally of 26 species belonging to two subfamilies, Trombiculinae and Hemitrombiculinae. Concomi-

[1]Partly supported by a grant N 03-04-49081 from the Russian Foundation for Fundamental Research.

tantly, the monotypic subfamily Leeuwenhoekiinae was established, which also had been included in Trombiculidae (Womersley 1944). However, a year later, this group was given the status of family (Womersley 1945). Since then, the question about the rank of this taxon has been debatable. Nevertheless, Ewing (1949) included four subfamilies in Trombiculidae, namely Hemitrombiculinae, Walchiinae, Leeuwenhoekiinae, and Trombiculinae. He proposed terminology for the taxonomic characters based on the external morphology of larvae for the first time. In the following years, a large number of new taxa were described, and the necessity for new generalizations on taxonomy and classification of the family emerged. Wharton et al. (1951) presented a new classification of larvae in which Trombiculidae was divided into the four subfamilies, Leeuwenhoekiinae, Walchiinae, Apoloniinae, and Trombiculinae. Later, the subfamily Walchiinae was renamed Gahrliepiinae (Womersley 1952).

Some taxonomists (Audy 1954; Radford 1946b; Vercammen-Grandjean 1968; Vercammen-Grandjean and Langston 1976; Kolebinova 1992) supported the establishment of the independent family Leeuwenhoekiidae with two subfamilies (nominative and Apoloniinae). However, other taxonomists continued to consider leeuwenhoekiids as a subfamily within Trombiculidae (Ewing 1949; Wharton et al. 1951; Wharton and Fuller 1952; Lakshana 1973; Nadchatram and Dohany 1974; Goff et al. 1982; Domrow and Lester 1985). Consequently, there are three main above-genus classifications of the trombiculids, all equally accepted.

I.

- Family Trombiculidae Ewing, 1944
    - Subfamily Gahrliepiinae Womersley, 1952
    - Subfamily Trombiculinae Ewing, 1929
- Family Leeuwenhoekiidae Womersley, 1945
    - Subfamily Leeuwenhoekiinae Womersley, 1944
        - Tribe Leeuwenhoekiini Vercammen-Grandjean, 1968
        - Tribe Whartoniini Vercammen-Grandjean, 1968
    - Subfamily Apoloniinae Wharton, 1947
        - Tribe Apoloniini Vercammen-Grandjean, 1968
        - Tribe SauracarelliniVercammen-Grandjean, 1968

II.

- Family Trombiculidae
    - Subfamily Trombiculinae
        - Tribe Trombiculini Vercammen-Grandjean, 1960

- Tribe Schoengastiini Vercammen-Grandjean, 1960
- Tribe Gahrliepiini Nadchatram et Dohany, 1974

o Subfamily Leeuwenhoekiinae
- Tribe Leeuwenhoekiini
- Tribe Whartoniini

o Subfamily Apoloniinae
- Tribe Apoloniini
- Tribe Sauracarellini

III.

- Family Trombiculidae
  - o Subfamily Walchiinae Ewing, 1946
    - Tribe Walchiini Wen, 1984
    - Tribe Schoengastiellini Wen, 1984
    - Tribe Gahrliepiini Nadchatram et Dohany, 1974
    - Tribe Intermedialiini Wen, 1984
  - o Subfamily Trombiculinae
- Family Leeuwenhoekiidae
  - o Subfamily Apoloniinae
  - o Subfamily Leeuwenhoekiinae
    - Tribe Leeuwenhoekiini
    - Tribe Whartoniini

2.2 Phylogeny

There is a great difficulty in studying phylogenetic relationships and the pathways of evolution in trombiculid mites since there is no paleontologic material on this group, although representatives of the closely related recent families Erythraeidae and Trombidiidae *sensu lato* were found in Baltic amber (Oligocene) (Dubinin 1962). In addition, the spectrum of potential hosts of trombiculids is extraordinarily wide, and their host specificity is extremely low. Therefore, the use of trombiculid-host associations in the construction of the phylogeny of this taxon is highly problematic.

Ewing (1949) made the first attempt to depict the phylogeny of trombiculids. He assumed parasitism on vertebrates to be basal compared with parasitism on invertebrates, since trombiculid mites are more archaic morphologically than mites parasitic on invertebrates. He suggested that hypothetical ancestors of trombiculids were larval parasites of vertebrates that gave rise to two taxa – Trombiculidae and Trombidiidae (parasites of arthropods). According to this scheme, subfamilies within Trombiculidae form the Gahrliepiinae (=Walchiinae) – Leeuwenhoekiinae – Trombiculinae phyletic lineage.

Other taxonomists (e.g., Vercammen-Grandjean et al. 1973) suggested that hypothetical ancestors of trombiculids were closely related to representatives of the genus *Schoutedenichia* (Trombiculinae), and the evolutionary tendency was thought to be expressed in an increase in the number of setae on scutum, palps and legs. This could lead to the following evolutionary sequence: Gahrliepiinae – Trombiculinae – Apoloniinae – Leeuwenhoekiinae.

In contrast, Robaux (1973) considered the reduction of setae and leg segments in evolution of Parasitengona as derived characters. He demonstrated that basal and derived characters in various families of Parasitengona may be present in various proportions in larval and subsequent developmental stages. This could lead to differential interpretations of the evolutionary events. Robaux compared morphological characters of mites at different developmental stages with proposed evolution of the larval parasitism which, in his opinion, has arisen from free-living carnivorous larvae via larvae parasitic on arthropods to larvae parasitic on vertebrates (Robaux 1973). As a result, in the phylogeny proposed by Robaux (1973), Trombiculidae appeared to be more derived than other families of Trombidioidea. Feider (1959) studied the classification of trombidiids and the newly established phalanx Trombidia. Based on (a) morphology of both larvae and adults and (b) geographic distribution of mites, he suggested equal phylogenetic rank of Leeuwenhoekiidae, Trombiculidae and Trombellidae.

Kudryashova (1998) studied relationships within Trombiculidae based mostly on larval morphology. According to her study, trombiculid mites represent a clearly delimited taxon whose position is isolated from other closely related families of terrestrial Parasitengona (Trombellidae and Trombidiidae *sensu lato*). Characteristic larval features of trombiculids include the presence of only one dorsal shield (scutum), which is located somewhat posterior to the frontal margin of the idiosoma and bears one pair of trichobotria and 4-7 tactile mechanoreceptive branched setae; trombidioid (instead of styletoid) type of chelicerae (that is also characteristic for the later stages of development); united coxae I and II; presence and position of urstigma in the postero-lateral margin of the coxa I; specific number of setae on the segments of I-III leg pairs; presence of 2 claws and empodium on leg tarsae; presence of solenidium, various number of branched setae as well as sometimes smooth subterminal seta on palp tarsus; a constant number 1-1-3 of setae on the palp trochanter, genu and tibia respectively; and the presence of oligotrichic setae on idiosoma, often organized in regular rows. Analysis of these characteristics demonstrated that Leeuwenhoekiinae should be considered as the most basal subfamily, whereas Trombiculinae and Gahrliepiinae are derived (Kudryashova 1998). Contradicting earlier ideas (e.g., Vercammen-Grandjean et al. 1973), this study suggests that Leeuwenhoekiinae should be placed at the base and Gahrliepiinae at the top of phylogenetic tree. Thus, the new version of classification of Trombiculidae presented by Kydryashova (1998) is:

- Family Trombiculidae
    - Subfamily Leeuwenhoekiinae
        - Tribe Leeuwenhoekiini
        - Tribe Whartoniini
    - Subfamily Apoloniinae
        - Tribe Apoloniini
        - Tribe Sauracarellini
    - Subfamily Trombiculinae
        - Tribe Trombiculini
        - Tribe Schoengastiini
    - Subfamily Gahrliepiinae

Welbourn (1991) proposed the relationships within Parasitengona based on a cladistic analysis of morphological characters of larvae. Results of this analysis demonstrated that families Trombiculidae and Trombidiidae (formerly thought as closely related) belong to two different superfamilies (Trombiculoidea and Trombidioidea) and Leeuwenhockiidae is a separate family within Trombiculoidea. However, Wen (1984, 2004) suggested that all trombiculids parasitic on vertebrates should be incorporated into a single superfamily Trombiculoidea, with three families (Leeuwenhoekiidae, Trombiculidae and Walchiidae).

3 Life cycles and development

Biology of trombiculids has been studied mainly in the laboratory (Elton and Keay 1936; Michener 1946a, b; Wharton 1946; Williams 1946; Radford 1946a, b; Jenkins 1947, 1948; Cockings 1948; Mehta 1948; Richards 1948; Schluger 1949; Neal and Barnett 1961; Sasa 1961; Shoshina 1964, 1965; Audy and Lavoipierre 1966; Kaufmann and Traub 1966; Ito 1967; Jameson 1967, 1968; Shirasaka and Sasa 1967; Nadchatram 1968; Kudryashova 1972; Everett et al. 1973; Kulkarni and Mahadev 1973; Cunningham et al. 1975; Simonová 1977, 1983; Vasilyeva 1977; Kulkarni

1988; Southcott and Frances, 1991; Takahashi et al. 1993, 1995; Shatrov 1996, 2000, 2003). The life cycle of trombiculids is characterized by alternating active and inactive (regressive, quiescent) instars (Fig. 1). When trombiculids are reared in the laboratory, their larvae feed on the hosts from several hours to several days. Engorged larvae drop off hosts and may serve as "a starting point" for a laboratory colony. Mites are reared in darkness in tightly closed containers with plaster-charcoal medium, under various air temperatures and high relative humidity (about 90-100%). Active deutonymphs and adult mites can feed successfully on the eggs of springtails *Sinella curviseta*. Starving (unfed) larvae can feed on various small mammals (mostly rodents) either using special capsules stuck to the host's skin or directly on-host in the isolated containers over a water layer. These methods achieve an output of 80-90% successfully fed larvae (Nadchatram 1968; Kulkarni 1988; Southcott and Frances 1991). Although it is very difficult to find active postlarval instars in the soil or on the soil surface in northern countries (Elton and Keay 1936; Daniel 1961), this is not the case for some tropical species (Michener 1946a, b; Radford 1946a, b; Wharton 1946; Cockings 1948; Jenkins 1948; Nadchatram 1968).

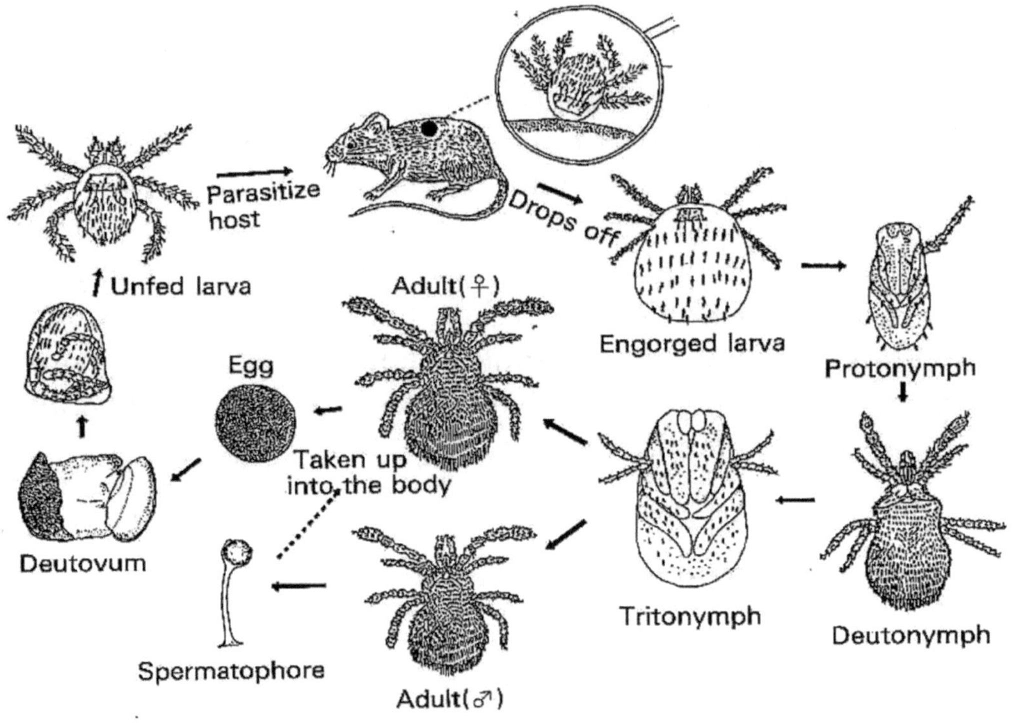

Fig. 1. Life cycle of *Leptotrombidium pallidum* (after Takahashi 1988, reprinted with permission from the author). Sizes of instars are not scaled. See text for explanations

The life cycle of trombiculids is characterized by seven stages (Figs. 1, 2), namely egg, regressive quiescent prelarva, active parasitic heteromorph larva, regressive quiescent protonymph, active soil-dwelling predatory deutonymph, regressive quiescent tritonymph, and, finally, active soil dwelling predatory adult mite. Active deutonymphs look almost identical to adult mites, but are slightly smaller. Sexual dimorphism is not apparently evident.

The most conspicuous feature of the life cycle is the constant duration of quiescent periods (deutovum=prelarva, nympho-chrysalis=deutonymph, imago-chrysalis=tritonymph) and the variable duration of active stages (larva, deutonymph and adult mite). Duration of quiescent stages does not typically exceed 25-30 days, whereas active deutonymphs could live more than 500 days, and lifespan of adult mites in the laboratory could exceed 900-1000 days (e.g., *Hirsutiella zachvatkini*; Shatrov 1996, 2000, 2003). Eggs of this boreal species may remain "dormant" (diapaused) for more than 300-400 days without losing the ability to develop. Unfed larvae survive without food for 200 and more days and can feed successfully thereafter. In boreal species, an egg-to-egg cycle ranges from 150 to 400 days but it is shorter in tropical species (see Shatrov 2000 for details).

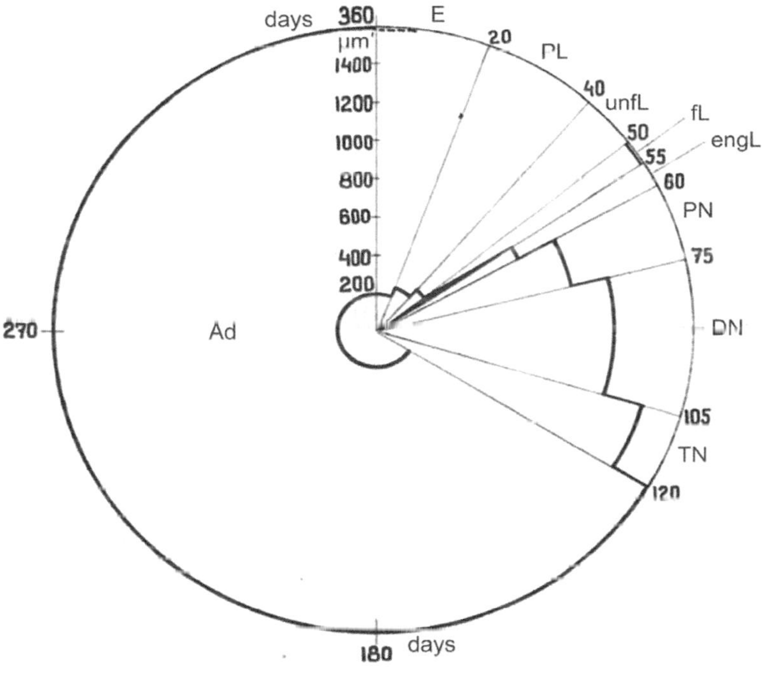

Fig. 2. Cyclogram of the life cycle of a trombiculid mite (duration of stages and axial measurements of instars are shown). The circle is a timeline approximately equal to one year; radii denote the axial sizes of the instars. *Ad* – adult mite, *DN* – deutonymph, *E* – egg, *engL* – engorged larva; *fL* – feeding larva (the time of parasitism), *PL* – prelarva, *PN* – protonymph, *TN* – tritonymph, *unfL* – unfed larva (after Shatrov 2000, reprinted with permission from Zoological Institute of Russian Academy of Sciences)

Male mites produce spermatophores either during their entire life or for certain period (Simonová 1983). Females usually have one to three oviposition cycles during their life with intervals between them. These cycles are frequently weekly outlined temporal periods (Shatrov 1996, 2003). Egg production varies greatly among individual females. Usually, females lay separate eggs, although many tropical and sub-tropical species may lay eggs in clutches. Egg production per female during two oviposition cycles varies from several tens to several hundreds.

The seasonal pattern of mite development and stages at which they overwinter (hibernate) are still unclear. Although various speculations were proposed (see Williams 1946; Mehta 1948; Shoshina 1965; Jameson 1967, 1968; Vasilyeva 1977), only a few experimental studies have been carried out (Takahashi et al. 1993; Takahashi et al. 1995). Results of these experiments demonstrated that (a) mites could retard their development at any stage in the cold period (when the temperature decreases below 10°C), and (b) all larvae feeding in late autumn did not undergo further development until spring (Takahashi et al. 1993). However, larvae, which fed in October, developed rapidly into active deutonymphs, which became dormant in winter, whereas adults appeared as late as in the next May (Takahashi et al. 1995). Mites that started to be reared in spring were transformed into adults by August, whereas the next generation gave rise to adults that again became dormant in winter. At constant temperature in the laboratory, mites continue their development without any conspicuous diapause, except, sometimes, for diapaused eggs. However, a short temperature increase may emulate spring conditions and trigger egg development. Boreal mites that dwell in soil where temperature rarely increases above 15–17°C, even in summer, demonstrate relatively slow activity and have an univoltine or even semi-univoltine life cycle, when adult mites emerging in summer produce a single larval generation or, in the case of the late emergence, do not produce larvae and become dormant in winter.

Thus, the main life strategy of trombiculid mites can be characterized, on the one hand, by rapid nymphal development and, on the other hand, by high viability of adult mites, which are apparently able to produce eggs during at least two summer seasons. Increase of air temperature in spring stimulates egg development and synchronizes the entire life cycle (Jameson 1967). Consequently, it is likely that room temperature, commonly used in the laboratory rearing of trombiculids, cannot be considered as an optimum (see Neal and Barnett 1961; Jameson 1967, 1968; Everett et al. 1973; Cunningham et al. 1975; Simonová 1977, 1983) but rather it shortens and intensifies the life cycle of mites. In other words, the natural life of mites appears to be mostly "cold and hungry", although we still do not know much about it (see also Daniel 1961).

4 Trombiculid-host relationships

4.1 Hosts of trombiculids

As mentioned above, the host spectrum of trombiculid larvae is very broad. They parasitize all groups of terrestrial vertebrates. However, the question of the host specificity of trombiculids is still debatable. Most likely, trombiculids demonstrate preferences to a particular local habitat, within which they attack and feed on all or the majority of vertebrate species that occupy this habitat (Kudryashova 1998). Nevertheless, in a given area, a given trombiculid species can prefer a specific host among the vertebrates. This can be related to co-occurrence of a trombiculid species and a host species in a local habitat or patch, density of host population, characteristic host behavior as well as ecological characteristics of a trombiculid.

An example of among-host distribution of mites from the largest trombiculid genus, *Leptotrombidium*, is presented in Table 1. More than half of the 178 species are associated with Rodentia. Furthermore, 62 mite species were found only on rodents, whereas the remaining 78 species also occurred on other animals. Although many trombiculid species that exploit rodents occur also on insectivores, some trombiculids characteristic of rodents attack also bats, birds or reptiles. Chiggers exploiting bats appear to be most host-specific. Eleven of 16 trombiculid species found on Chiroptera are highly specific to this taxon, whereas the other five species were also found on hosts from other taxa. In general, the most preferable hosts for trombiculid larvae in various biomes seem to be rodents. Humans and other primates are occasional hosts for trombiculid larvae. Some species and genera of trombiculids parasitize hosts belonging to a particular order or a particular class. For example, larvae of *Hannemania*, *Vercammenia*, and *Endotrombicula* feed on amphibians only, whereas *Siseca*, *Vatacarus*, *Iguanacarus* and *Fonsecia* are parasitic on reptiles. *Toritrombicula* exploit small forest birds, whereas *Blankaartia*, *Neacariscus*, *Neoschoengastia*, *Hypogastia*, *Rhabdotella*, and *Mackiena* attack waterbirds (except for *B. acuscutellaris* that also parasitize small mammals associated with aquatic habitats). *Tecomatlana*, *Pentagonaspis*, *Audytrombicula*, *Myotrombicula*, *Chiroptella*, *Oudemansidium*, *Willmannium*, *Sasatrombicula*, *Rudnicula*, *Speleocola*, *Riedlinia*, *Trombigastia*, *Perissopalla*, *Trisetica*, and *Whartonia* are characteristic chiggers of bats.

Nevertheless, the strong preference of a trombiculid species for a particular host taxon can be disrupted frequently. For example, among amphibian parasites, some *Endotrombicula* species were found also on mollusks of the family Urocyclidae (Vercammen-Grandjean and Benoit 1971), whereas some *Vercammenia* species were found on lizards *Mabuya striata* (Vercammen-Grandjean and Langston 1976). The frog parasite *Eutrombicula goeldii* has been collected also from lizards, birds, marsupials, insectivores, bats, rodents and even tapirs (Brennan and Reed 1975). Despite high host specificity of bat chiggers, some *Whartonia* species have been recorded on rodents (Loomis and Crossley 1963). Therefore, host switching is a common phenomenon in trombiculids (Loomis 1956). Furthermore, it appears that trombiculid larvae are tightly associated with particular habitats in which they are able to feed on almost every vertebrate species. In other words, the trombiculid-host association can be established whenever spatial niches of a mite and a host coincide (Kudryashova 1998).

Table 1. Distribution of 178 *Leptotrombidium* species among different host groups

	1	2	3	4	5	6	7	8	9
1. Marsupials	**2**	0	0	1.4	1.3	0	0	9.1	0
2. Man	-	**6**	13.3	18.1	6.4	0	16.5	16.6	0
3. Other primates	-	2	**7**	4.1	6.3	0	18.7	11.5	0
4. Rodents	2	6	6	**140**	42.2	3.3	20.4	11.7	2.1
5. Insectivores	1	5	5	65	**77**	3.3	1.9	16.5	2.6
6. Chiropterans	-	-	-	5	3	**15**	4.4	2.7	5.6
7. Other mammals	-	5	6	29	2	2	**31**	15.2	6.3
8. Birds	2	4	3	17	14	1	7	**22**	0
9. Reptiles	-	-	-	3	2	1	2	-	**3**

Numbers in the upper row correspond to the numbers of host groups indicated in the left column. Bold numbers in the diagonal row are the numbers of mite species; above diagonal row – pair-wise Jaccard's indices of similarity of (%); below diagonal row – pair-wise numbers of common species between host groups (after Kudryashova 1998, reprinted with permission from KMK Scientific Press)

4.2 Trombiculid feeding

Trombiculid larvae are characterized by extra-oral digestion. Most trombiculids usually feed on the surface of the host epidermis and cut the stratum corneum with their chelicerae. Preferable areas of the host body that these parasites attack are the abdomen, head, ears, armpits, genitalia and the zone around the tail and anus. In some cases, trombiculid larvae burrow themselves subdermally (partly or totally), and a connective tissue capsule is formed within the host's skin around the parasite. This is characteristic for trombiculids parasitic on amphibians (Ewing 1926; Hyland 1961; Grover et al. 1975) as well as for species parasitic on mammals (*Acomata-*

carus, Apollonia, Cheladonta, Euschoengastia, Gahrliepia, Intercutestrix, Neoschoengastia, Schoutedenichia) (Audy and Vercammen-Grandjean 1955; Vercammen-Grandjean 1958, 1959; Brennan and Yunker 1966, 1969; Easton and Krantz 1973, Schramlová 1978). In such cases, the feeding time is prolonged. In lizards, specific adaptive structures of skin, known as "mite pockets", may evolve to decrease the possible damage from mite feeding (Arnold 1986; Goldberg and Holshuh 1992).

In most cases, however, when feeding on their natural hosts, trombiculid larvae develop a characteristic feeding tube (stylostome) in the host's skin (Jones 1950; Hoeppli and Schumacher 1962; Schumacher and Hoeppli 1963; Voigt 1970; Hase et al. 1978; Shatrov 2000). The walls of the stylostome consist of a complex glycoprotein structure that originated mainly from the solidifying saliva of a larva and do not include cellular elements of the host tissues (Fig. 3). The length and width of the stylostome range among trombiculid species between 100 and 200 μm and 30 and 70 μm, respectively, whereas the diameter of the axial channel is about 5 μm. The host's tissues around the stylostome are destroyed and necrotized. Beneath the distal end of the stylostome, an interstitial food cavity is formed (Fig. 3). It contains lymphoid and epithelioid cellular elements and, apparently, serves as a reservoir of the nutritional medium for the larva. Nevertheless, the nutritional medium itself is represented by liquid components and contains neither cellular elements nor their fragments. The larval feeding is accompanied by an immunological response of the host's connective tissue such as hyperemia of the superficial capillaries and cellular infiltration of the affected area (Wright et al. 1988; Shatrov 2000).

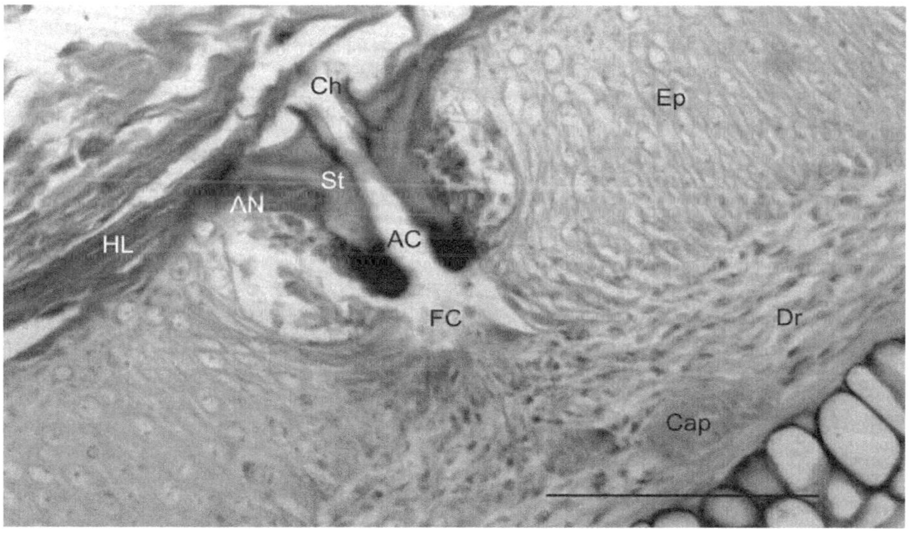

Fig. 3. Stylostome of *Neotrombicula pomeranzevi* in the ear skin of *Clethrionomys rufocanus*. *AC* – axial channel, *AN* – area of necrosis, *Cap* – capillary, *Ch* – chelicerae, *Dr* – derma, *Ep* – epidermis, *FC* – food cavity, *HL* – stratum corneum. Scale – 100 μm

Larvae of *Euschoengastia rotundata* represent another example of trombiculid feeding. This parasite occurs frequently on the abdominal skin of voles, *Clethrionomys rufocanus*, where it forms very visible whitish capsules (Fig. 4). A capsule *per se* is represented by a solid saliva secretion that is situated on the concave and suppressed epidermis of a host (Shatrov 2000). A stylostome is located at the bottom of the capsule. The epidermis below the stylosome is dissolved locally by lytic agents of the chigger saliva. The capsule is surrounded by the epidermis undergoing strong hyperplasia as well as hyper- and parakeratosis. Invagination of the epidermis is primarily due to intensive edema of the skin connective tissue around the site of the mite attachment and is accompanied by a slow exudative inflammatory response with polymorphic leucocytes and macrophages predominant in the infiltrate. Feeding *E. rotundata* larvae do not submerge below the epidermal layer into the dermis, and, therefore, can be considered as true ectoparasites.

The structure of the trombiculid stylostome is species-specific and does not depend on the host species (Hase et al. 1978). In addition, morphology of a stylostome and character of the host's skin response may be an indirect indicator of the capability of a given trombiculid species to serve as an effective vector of rickettsiae – causative agents of tsutsugamushi disease (Boese 1972).

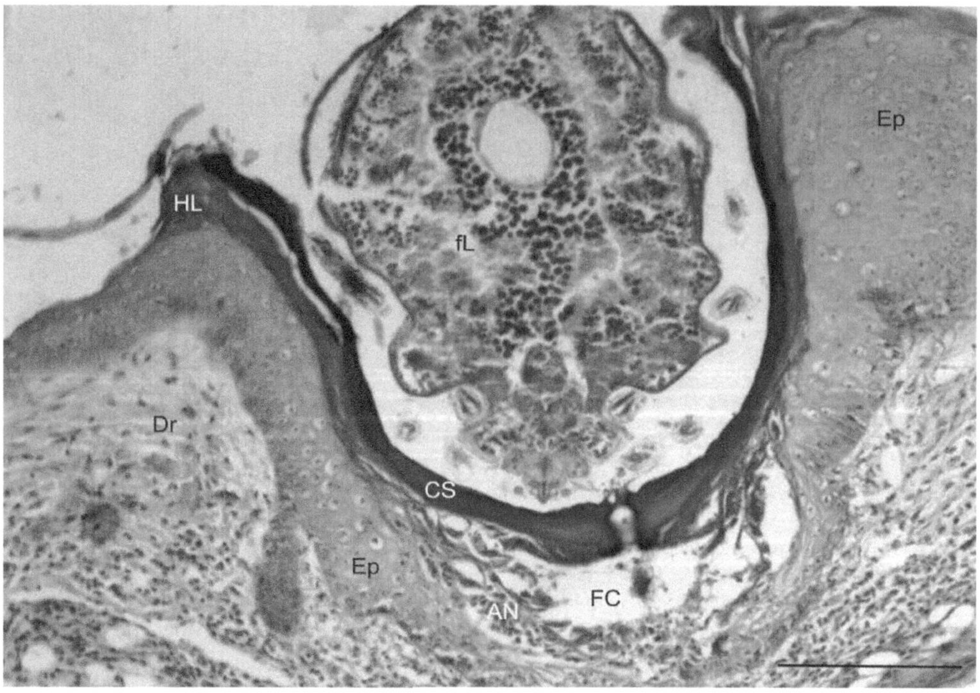

Fig. 4. Capsule of *Euschoengastia rotundata* in the abdominal skin of *C. rufocanus*. *AN* – area of necrosis, *CS* – capsule, *Dr* – derma, *Ep* – epidermis, *FC* – food cavity, *fL* – feeding larva, *HL* – stratum corneum. Scale – 100 μm

5 Origin of parasitism in trombiculid mites

The origin of parasitism in any animal taxon, including trombiculid mites, is associated with both the pattern of ontogenesis and the possession of morphological pre-adaptations that favour and facilitate transition to parasitism of particular developmental stages at a particular historical period (Shatrov 2000).

The parasitic life of a trombiculid individual is extremely short (approximately 3-5 days) in comparison with the duration of its entire life cycle (which may last up to 1000 and more days) (Fig. 2). Furthermore, parasitism of trombiculid larvae does not seem to be accompanied by strong morphological adaptations (Shatrov 2000, 2001). However, the parasitic strategy of trombiculids is characterized by a very effective mode of concentrated feeding on semi-digested (by extra-oral digestion) organic (mainly protein) substrate. Indeed, the duration of uninterrupted larval feeding on a host (3-5 days) equals approximately the total time spent by adult mites during their lifetime for feeding on arthropod eggs (3.5 days) (assuming that during 500 days of life, a mite consumes an average of 2 eggs per day, whereas it requires about 5 min to suck an egg dry; Shatrov 2000, 2001). Such a mode of larval feeding could have originated only on the basis of a general pre-adaptation to utilize natural protein diet. For the remaining and the greatest part of the life cycle, including the reproductive stage, a trombiculid lives as a free-living animal.

Small size (about 200 μm) and a ridged cuticle of a heteromorph trombiculid larva (Fig. 5) are thought to be a result of somatic conditions of embryogenesis of small, yolk-rich eggs. Indeed, a large amount of yolk leads (a) to fast embryonization of the first larval instar, which then transforms into a quiescent non-feeding and non-motile prelarva with a lecitotrophic aphagy (Grandjean 1938a, b; Robaux 1970, 1974; Coineau 1976) and (b) to small-sized heteromorphic second larva. To continue development, this small larva has to grow to a relatively large (900-1200 μm) size of postlarval instars (Fig. 2). This can be considered as a kind of essential ontogenic threshold, which must be quickly attained. In addition, the ridged cuticle of a larva guarantees a rapid increase of the body surface during transition to the next developmental stage (Shatrov 2000). A rapid increase in size can be attained either by repeated consumption of numerous invertebrates or, at later evolutionary stages (Fig. 6), by parasitism on vertebrate animals as sources of large food quantities that can be consumed during a relatively short period. The latter seems to be more effective than the former mode. This pattern seemed to provoke the appearance of non-feeding and regressive nymphal stages (calyptostases) – protonymph and tritonymph - in the trombiculid ontogeny, thus accelerating the process of individual development (Grandjean 1947, 1957).

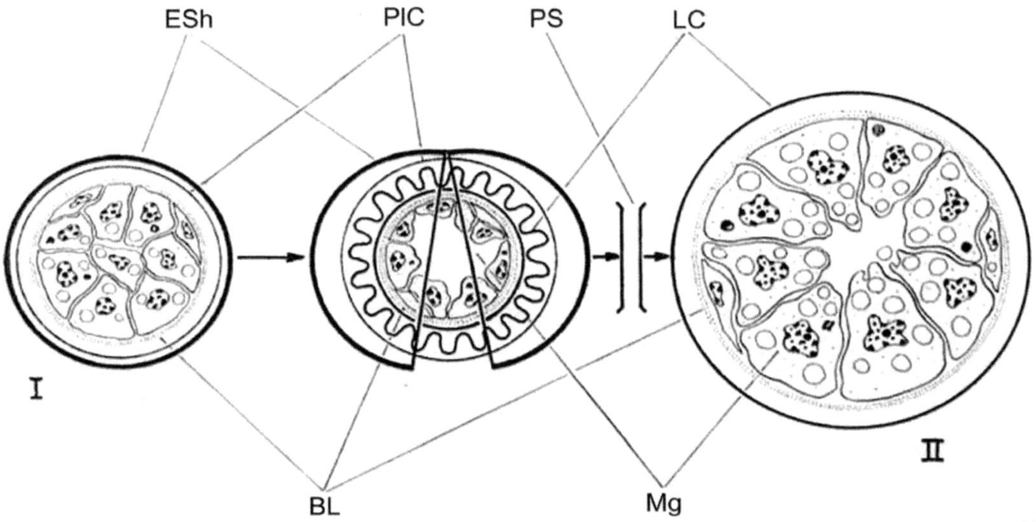

Fig. 5. Changes in shape, size and midgut epithelium of mites during transformation from egg (*I*) to fed larva (*II*). *BL* – basal lamina, *ESh* – eggshell, *LC* – larval cuticle, *Mg* – midgut, *PlC* – prelarva cuticle, *PS* – parasitism (after Shatrov 2000, reprinted with permission from Zoological Institute of Russian Academy of Sciences)

If we accept the hypotheses that larval parasitism in arthropods never spreads to the imaginal stage, and that the original transition to parasitism can be realized only in actively feeding stages of the arthropod life cycle (Beklemishev 1970), it can be suggested that the evolutionary appearance of larval parasitism was possible only when the quiescent instars already existed in the ontogeny (Fig. 6). The presence of the quiescent and, hence, non-parasitic prelarva in the ontogeny of trombiculids supports this assumption, indicating that parasitism originated after the evolutionary establishment of regressive prelarva. Consequently, parasitism in trombiculid mites should be considered as a relatively young evolutionary phenomenon, totally dependent on their life history and specific habitat adaptations.

Considering ecological reasons for the origin of parasitism in trombiculid mites, two opposing views have been expressed: 1) the primary parasitism of larvae was on vertebrates (Ewing 1949); and 2) it was on arthropods (Audy 1961; Beklemishev 1970). Because there is no paleontological datum on trombiculids (but not on other mite taxa, see below), evolutionary history of this group of mites can be presumptive only, based on general principles of paleogeography (Wharton and Fuller 1952) and comparative morphology. If the hypothesis of Beklemishev (1970) is accepted, it remains unclear how trombiculid larvae shifted from parasitism on insects to parasitism on vertebrates.

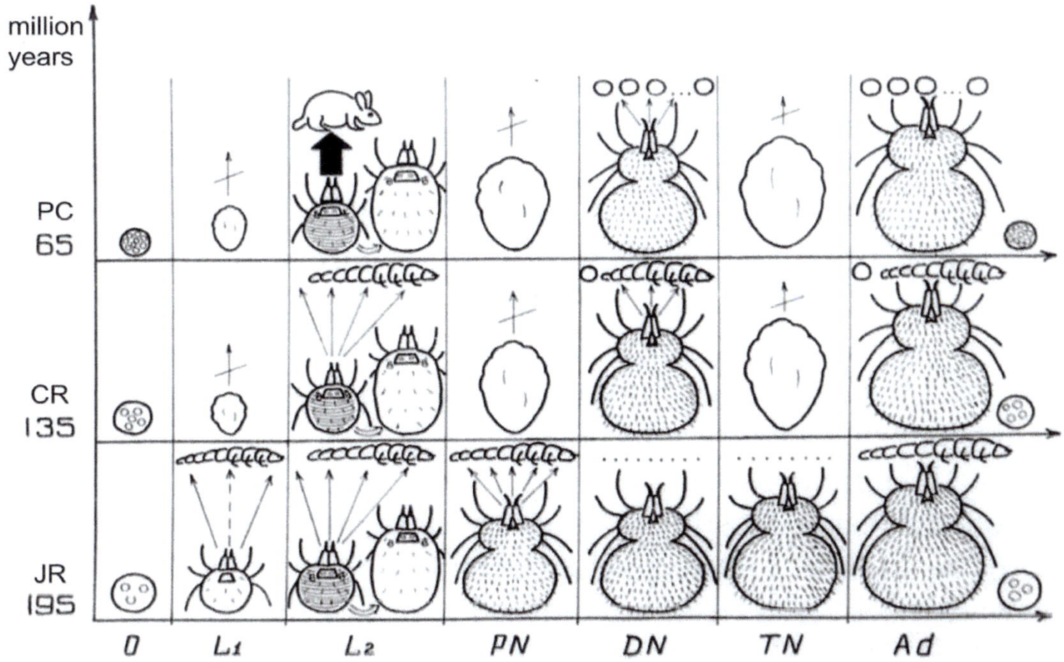

Fig. 6. Hypothetical evolution of ontogeny in trombiculid mites. Abscissa – successive stages of ontogeny: *O* – egg (ovum), L_1 – first larva, L_2 – second larva, larva proper, *PN* – protonymph, *DN* – deutonymph, *TN* – tritonymph, *Ad* – adult mite; ordinate – geologic time: *PC* – Paleocene, *CR* – Cretaceous, *JR* – Jurassic. Solid arrows – repeated carnivory on soil arthropods and their eggs; strikethrough arrows –aphagy of quiescent regressive instars; bold arrow – larval parasitism on vertebrates (after Shatrov 2000, reprinted with permission from Zoological Institute of Russian Academy of Sciences)

It is most likely that, originally, vertebrate animals (presumably mammals) simply "collected" free-living predatory larvae, which, in turn, attempted to attach and feed. Prolonged feeding is an apparent result of the adaptation of larvae with the short mouthparts to penetrate the epidermis of a potential vertebrate using a newly developed peculiar feeding tube – stylostome. Homogeneous, protein-rich food acquired using this feeding strategy could facilitate the transition to obligate parasitism. Beklemishev (1970) argued that, in arthropods with originally bite-sucking mouthparts, haematophagy arose at once, and thus the nature of the food extracted from a host did not change during their evolution. Therefore, further morphological adaptations, such as mouthparts changes, were minor. Trombiculids feed mainly on lymph and tissue fluids, which can be no less satisfying than blood, especially given the primitively organized alimentary system of these mites. Consequently, trombiculid larvae, originally being entomophagous predators with bite-sucking mouthparts in some conditions (e.g., pastures; see Beklemishev 1970) could easily become parasitic lymphophages of vertebrate animals, especially mammals. Furthermore, due to their wide polyphagy, trombiculids seem to be ecologically close to free-living blood-sucking insects, such as dipterans. In other words, a potential victim or a host of trombiculid larvae can be any vertebrate animal that comes into contact with them (Wharton and Fuller 1952; Traub and Wisseman 1974). The pattern of food acquisition by a trombiculid larva from a host resulted in the establishment of tight and continuous larva-host association, and trombiculids transformed from a free carnivorous life to an obligate larval parasite. The stylostome that is characteristic for trombiculid larvae and some other closely related Parasitengona can be considered as an universal tool to acquire large amounts of liquid food during a relatively short time from a wide spectrum of hosts.

Concerning historical aspects of trombiculid parasitism, it should be noted that, on one hand, fossilized trombidiform mites (water species) are known from the Middle Jurassic, whereas many recent families of terrestrial mites were discovered from the mineral resins of the upper Cretaceous (Vainstein 1978). On the other hand, the radiation of trombiculid mites and the evolutionary rise of larval parasitism seemed to occur after differentiation of their ontogeny into quiescent and active stages, i.e. not earlier than Paleocene (Fig. 6).

Warton (1946) suggested that habitat distribution of trombiculid mites is determined by environmental requirements of free-living stages, whereas their geographical distribution depends also on spreading of larvae by their hosts. Parasitism of trombiculid larvae on birds, reptiles and amphibians has likely arisen after parasitism on mammals. Furthermore, in these cases, trombiculid parasitism can differ from the usual trombiculid pattern (Ewing 1926; Hyland 1961; Audy et al. 1972). Larvae of some species are able to attack and successfully feed not only on vertebrates, but also on arthropods (Audy 1951). This is a characteristic mainly for the basal subfamily Leeuwenhoekiinae (Ewing 1949; Warton and Fuller 1952).

6 Concluding remarks

The ontogenetic threshold related to the large size difference between heteromorphic larva and post-larva has inevitably led to larval parasitism and regression of nymphal instars in trombiculid mites. It is most likely that mammals are primary hosts of trombiculid larvae, whereas the use of other terrestrial vertebrates as hosts arose later. An association with a host takes only a short time from the entire life cycle of a trombiculid mite; a mite spends the rest of its life as a free-living soil-dweller. As a result, host selection by a trombiculid larva is determined mainly by habitat distribution of a host rather than by its taxonomic affinity.

References

Arnold EN (1986) Mite pockets of lizards, a possible means of reducing damage by ectoparasites. Biol J Linn Soc 29:1–21

Audy JR (1950) Occurrence of trombiculid mites on arthropods. Nature 165:193

Audy JR (1954) Malaysian parasites. IX. Notes on the taxonomy of trombiculid mites with description of a new subgenus. Stud Inst Med Res Malaya 26:123–170

Audy JR (1961) The ecology of scrub typhus. In: May JM (ed) Studies in disease ecology. Hafner, New York, pp 389–432

Audy JR, Lavoipierre MMJ (1966) Parasitic mites. In: Smith CN (ed) Insect colonization and mass production. Acad Press, New York, London, pp 25–48

Audy JR, Vercammen-Grandjean PH (1955) Endoparasitism in trombiculid mites. Nature 175:263–264

Audy JR, Radovsky FJ, Vercammen-Grandjean PH (1972) Neosomy: Radical intrastadial metamorphosis associated with arthropod symbioses. J Med Entomol 9:487–494

Beklemischev VN (1970) Biocenological base of comparative parasitology (in Russian). Nauka, Moscow

Boese JL (1972) Tissue reactions at the site of attachment of chiggers. J Med Entomol 9:591

Brennan JM, Reed JT (1975) A list of Venezuela chiggers, particularly of small mammalian hosts (Acarina: Trombiculidae). B Young Univ Sci Bull Biol Ser 20:45–75

Brennan JM, Yunker CE (1966) Endoparasitic chiggers. III. *Neoschongastia vellata* n.sp. (Acarina) an intradermal parasite of boreal rodents in Montana. J Med Entomol 3:338–339

Brennan JM, Yunker CE (1969) Endoparasitic chiggers: V. New genera, species and records from Venezuela and Brazil (Acarina: Trombiculidae). J Med Entomol 6:299–304

Cockings KL (1948) Successful methods of trapping *Trombicula* (Acarina) with notes on rearing *T. deliensis* Walch. Bull Entomol Res 39:281–296

Coineau Y (1976) La premiere prelarve connue du genre Eupodes: *Eupodes strandtmanni*, n. sp. Acarologia 18:56–64

Cunningham JR, Price MA, Kunz SE (1975) *Neoschongastia americana*: Life cycle in the laboratory. J Econ Entomol 68:495–496

Daniel M (1961) The bionomics and developmental cycle of some chiggers (Acariformes, Trombiculidae) in the Slovak Carpathians. Československá Parasitol 8:31–118

Daniel M (1965) Some questions of the dispersal of adult trombiculid mites in the soil. Acarologia 7:527–531

Domrow R, Lester LN (1985) Chiggers of Australia (Acari: Trombiculidae): An annotated checklist, keys and bibliography. Austral J Zool, Suppl Ser 114:1–111

Dubinin VB (1962) Class Acaromorpha. Acari (in Russian). In: The bases of paleontology. Arthropods – tracheal and cheliceral. Nauka, Moscow, pp 450–473

Easton ER, Krantz GW (1973) A *Euschoengastia* species (Acari: Trombiculidae) of possible medical and veterinary importance in Oregon. J Med Entomol 10:225–226

Elton C, Keay G (1936) The seasonal occurrence of harvest mites (*T. autumnalis* Shaw) on voles and mice near Oxford. Parasitology 28:110–115

Everett RE, Price MA, Kunz SE (1973) Biology of the chigger *Neoschongastia americana* (Acarina: Trombiculidae). Ann Ent Soc Amer 66:429–435

Ewing HE (1926) The life history and biology of the tree toad chigger, *Trombicula hylae* Ewing. Ann Ent Soc Amer 19:261–267

Ewing HE (1929) A manual of external parasites. C. C. Thomas, Springfield, Illinois

Ewing HE (1944a) The trombiculid mites (chigger mites) and their relation to disease. J Parasitol 30:339–365

Ewing HE (1944b) Notes on the taxonomy of the trombiculid mites. Proc Biol Soc Wash 57: 101–104

Ewing HE (1949) The origin and classification of the trombiculid mites, or Trombiculidae. J Washington Acad Sci 39:229–237

Feider Z (1959) Proposal on a new classification of trombidiid mites (in Russian). Zool Zh 38:537–549

Goff ML, Loomis RB, Wilbourn WC, Wrenn WJ (1982) A glossary of chigger terminology (Acari: Trombiculidae). J Med Entomol 19:221–238

Goldberg SR, Holshuh HJ (1992) Ectoparasite-induced lesions in mite pockets of the yarrow's spiny lizard, *Sceloporus jarrovii* (Phrynosomatidae). J Wildlife Dis 28:537–541

Grandjean F (1938a) Description d'une nouvelle prelarve et remarques sur la bouche des Acariens. Bull Soc Zool France 63:58–68

Grandjean F (1938b) Sur l'ontogenia des Acariens. CR Acad Sci Paris 206:146–150

Grandjean F (1947) L'harmonie et la dysharmonie chronologiques dans l'evolution des stases. CR Acad Sci Paris 225:1047–1050

Grandjean F (1957) L'evolution selon l'age. Arch Sci Geneve 10:477–526

Grover JJ, Duszynski DW, Bogan BC (1975) Histochemistry of the tissue capsule surrounding intradermal mites, *Hannemania* spp. (Acarina: Trombiculidae) in New Mexico Amphibians. J Parasitol 61:382–384

Hase T, Roberts LW, Hildebrandt PK, Cavanaugh DC (1978) Stilostome formation by *Leptotrombidium* mites (Acarina: Trombiculidae). J Parasitol 64:712–718

Hoeppli R, Schumacher HH (1962) Histological reactions to trombiculid mites, with special reference to "natural" and "unnatural" hosts. Z Tropenmed Parasitol 13:419–428

Hyland KE (1961) Parasitic phase of chigger mite, *Hannemania hegeneri* on experimantally infested amphibians. Exp Parasitol 11:212–225

Ito T (1967) An outline of research history of the tsutsugamushi disease in Niigata prefecture of Japan. Acta Med Biol 15:1–7

Jameson EW (1967) Preliminary experimental studies on the life cycle of *Euschöngastia radfordi*. Acta Med Biol 15:21–25

Jameson EW (1968) Experimental rearing of some Japanese trombiculid mites. Japan J Exp Med 38:279–288

Jenkins DW (1947) A laboratory method of rearing chiggers affecting man (Acarina, Trombiculidae). Ann Ent Soc Amer 40:56–68

Jenkins DW (1948) Trombiculid mites affecting man. I. Bionomics in relation to epidemiology in the United States. Am J Hyg 48:22–35

Jones BM (1950) The penetration of the host tissue by the harvest mite, *Trombicula autumnalis* Shaw. Parasitology 40:247–260

Kaufmann T, Traub R (1966) Preliminary notes on parthenogenesis in a species of Trombiculid mites, *Leptotrombidium arenicola* Traub, 1960 (Acarina; Trombiculidae). J Med Entomol 3:256–258

Kawamura AJ, Tanaka H, Tamura A (eds) (1995) Tsutsugamushi disease. Univ Tokyo Press, Tokyo

Kolebinova MG (1992) Acariformes, Trombidioidea: Trombiculidae, Leeuwenhoekiidae (in Bulgarian). Fauna of Bulgaria 21:1–172

Kudryashova NI (1972) Laboratory breeding of trombiculid mites (Acariformes, Trombiculidae) (in Russian). Meditsinskaya Parasitologiya i Parasitarnye Bolezni [Medical Parasitology and Parasitic Diseases] 41:40–44

Kudryashova NI (1998) Chigger mites (Acariformes, Trombiculidae) of East Palaearctics (in Russian). KMK Sci Press, Moscow

Kulkarni SM (1988) Some observations on the life history and behavior of three species of trombiculids in India. In: Channabasavanna GP, Viraktamath CA (eds) Progress in acarology, vol 1. Oxford and IBH, New Delhi, pp 133–138

Kulkarni SM, Mahadev PVM (1973) Successful colonization of Trombiculid mites with notes on life histories of three species. Ind J Med Res 61:352–360

Lakshana P (1973) A checklist of the trombiculid mites of Thailand (Prostigmata: Trombiculidae). US Army Medical Component South East Asia Treaty Organization, Bangkok, pp 1–45

Loomis RB (1956) The chigger mites of Kansas (Acarina, Trombiculidae). Univ Kansas Sci Bull 37:1159–1443

Loomis RB, Crossley DA (1963) New species and new records of chiggers (Acarina: Trombiculidae) from Texas. Acarologia 5:371–383

Mehta DR (1948) Studies on typhus in the Simla Hills. Pt. IX. On the life-history of *Trombicula deliensis* Walch, a suspected vector of typhus in the Simla Hills. Ind J Med Res 36:159–171

Michener CD (1946a) Observations on the habits and life history of a chigger mite, *Eutrombicula batatas* (Acarina: Trombiculinae). Ann Ent Soc Amer 39:101–118
Michener CD (1946b) A method of rearing chigger mites (Acarina, Trombiculinae). Am J Trop Med Hyg 26:251–256
Mitchell RD, Nadchatram M (1969) Schizeckenosy: The substitute for defecation in chigger mites. J Nat Hist 3:121–124
Nadchatram M (1968) A technique for rearing trombiculid mites (Acarina) developed in a tropical laboratory. J Med Entomol 5:465–469
Nadchatram M, Dohany AL (1974) A pictorial key to the subfamilies, genera and subgenera of Southeast Asian chiggers (Acari, Prostigmata, Trombiculidae). Bull Inst Med Res Malaysia 16:1–67
Neal TG, Barnett HC (1961) The life cycle of the scrub typhus chigger mite, *Trombicula akamushi*. Ann Ent Soc Amer 54:196–203
Oudemans AC (1912) Die bis jetzt bekannten Larven von Thrombidiidae und Erythraeidae. Zool Jahrb Jena Suppl 14:223
Radford CD (1946a) Notes on *Trombicula deliensis* Walch, 1923 (Acarina; Trombiculidae), with description of the adult. Parasitology 37:42–45
Radford CD (1946b) Larval and nymphal mites (Acarina; Trombiculidae) from Ceylon and the Maldive islands. Parasitology 37:46–54
Richards WS (1948) The life history of the British harvest mite, *Trombicula autumnalis*, with a description of sampling methods for larvae. Trans R Soc Trop Med Hyg 42:6–7
Robaux P (1970) La prelarve de *Campylothrombium barbarum* Lucas (Acari – Thrombidiidae). Acarologia 12:131–135
Robaux P (1973) Importance de l'etude des caracteres morphologiques, de la biologie et de l'ecologie a toutes les stases, pour etablir la phylogenese des Acariens voisins des Thrombidions. Acarologia 15:121–128
Robaux P (1974) Recherches sur le developpment et la biologie des acariens "Thrombidiidae". Mem Mus Natl Hist Nat (N.S.) Ser A Zool 85:1–186
Sasa M (1961) Biology of chiggers. Ann Rev Entomol 6:221–244
Schluger EG (1949) Some data on biology of *Trombicula pavlovskyi* (in Russian). Voprosy Krayevoi, Obschey i Eksperimentalnoy Parasotologii [Problems of the Local, General and Experimental Parasitology] 4:65–66
Schramlová J (1978) Skin lesion produced by the larvae of *Cheladonta costulata* (Willmann, 1952) (Acarina: Trombiculidae) and the feeding mechanism of this parasite. Folia Parasitol 25:261–270
Schumacher HH, Hoeppli R (1963) Histochemical reactions to trombiculid mites, with special reference to the structure and function of the "stylostome". Z Tropenmed Parasitol 14:192–208
Shatrov AB (1996) Some peculiarities of the life cycle and biology of chiggers in laboratory. Entomol Rev 76:1197–1208
Shatrov AB (2000) Trombiculid mites and their parasitism on vertebrate hosts (in Russian). St Petersburg Univ Publ, St Petersburg
Shatrov AB (2001) On the origin of parasitism in trombiculid mites (Acariformes: Trombiculidae). Acarologia 41:205–214

Shatrov AB (2003) Further observations on the life cycle and life strategy of a trombiculid mite *Hirsutiella zachavtkini* (Schluger, 1948) (Acariformes: Trombiculidae) in the laboratory. Acarina 11:73–90

Shirasaka A, Sasa M (1967) Observations on the life history in the laboratory of some Japanese and Thai trombiculid mites. Japan J Exp Med 37:129–140

Shoshina MA (1964) Laboratory culturing of chiggers *Trombicula zachvatkini* Schlug. 1948 (in Russian). Meditsinskaya Parasitologiya i Parasitarnye Bolezni [Medical Parasitology and Parasitic Diseases] 3:741–742

Shoshina MA (1965) Survival of larvae of trombiculid mite *Trombicula zachvatkini* Schlug. 1948 under low temperature (in Russian). Meditsinskaya Parasitologiya i Parasitarnye Bolezni [Medical Parasitology and Parasitic Diseases] 34:355–356

Simonová V (1977) Laboratory culturing of chiggers (Acarina: Trombiculidae). Folia Parasitol 24:285–288

Simonová V (1983) Developmental cycle of chiggers under laboratory conditions. Folia Parasitol 30:79–87

Southcott RV, Frances SP (1991) The life history stages of *Odontacarus (Leogonius) barrinensis* (Womersley) (Acari: Trombiculidae: Leeuwenhoekiinae). Int J Acarol 17:275–287

Takahashi M, Machida K, Hori E (1988) Ecological study of *Leptotrombidium (Leptotrombidium) pallidum*. 7. Observation of development under natural conditions. J Saitama Med School 16:179–189

Takahashi M, Machida K, Murata M, Misumi H, Hori E, Kawamura AJr, Tanaka H (1993) Seasonal development of *Leptotrombidium pallidum* (Acari: Trombiculidae) observed by experimental rearing in the natural environment. J Med Entomol 30:320–325

Takahashi M, Morita K, Tsuji O, Misumi H, Otsuji J, Hori E, Kawamura A Jr, Tanaka H (1995) Seasonal development of *Leptotyrombidium akamushi* (Acari: Trombiculidae) under field temperatures. J Med Entomol 32:843–846

Traub R, Wisseman CL (1974) The ecology of chigger-born rickettsiosis (scrub typhus). J Med Entomol 11:237–303

Vainstein BA (1978) System, evolution and phylogeny of trombidiform mites. In: Identification keys of the soil-dwelling mites (Trombidiformes) (in Russian). Nauka, Moscow, pp 228–244

Vasilyeva IS (1977) Data on biology of the free-living phases of *Hirsutiella zachvatkini* and *Leptotrombidium orientalis* (Acariformes, Trombiculidae) (in Russian). Zool Zh 56:162–163

Vercammen-Grandjean PH (1958) Revision du genre *Schoutedenichia* Jad. et Verc. Ann Mus Roy Congo Belge Sci Zool 65:1–101

Vercammen-Grandjean PH (1959) *Acomatacarus (Austracarus) wittebolsi* n.sp. un Trombiculidae larvaire parasite intradermicole de la taupe. Acarologia 1:253–256

Vercammen-Grandjean PH (1968) The chigger mites of the Far East (Acarina: Trombiculidae et Leeuwenhoekiidae). U.S. Army Medical Research and development command, Washington

Vercammen-Grandjean PH, Benoit PLG (1971) Un aoutat Africain parasite d'une limace. Rev Zool Bot Afr 83:180–187

Vercammen-Grandjean PH, Langston RL (1976) The chigger mites of the World. (Acarina: Trombiculidae et Leeuwenhoekiidae). Volume III. *Leptotrombidium* complex. Section A, B, C. G.W. Hooper Foundation Univ California, San Francisco

Vercammen-Grandjeam PH, Langston RL, Audy JR (1973) Tentative nepophylogeny of trombiculids. Folia Parasitol 20:49–66

Voigt B (1970) Histologische Untersuchungen am Stylostom der Trombiculudae (Acari). Z Parasitenkd 34:180–197

Wen T (1984) Revision of the family-group taxa of sand mites with a proposal system of Walchiinae (Acariformes: Trombiculidae et Leeuwenhoekiidae). Sand mites of China (Acariformes: Trombiculidae et Leeuwenhoekiidae). Xuelin Publ, Shanghai

Wen T (2004) Classification of the sand-mite family Walchiidae (Acariformes: Trombiculoidea). Chin J Parasitol Parasit Dis 22:113–118

Welbourn WC (1991) Phylogenetic studies of the terrestrial Parasitengona. In: Dusbabek F, Bukva V (eds) Modern Acarology, vol 2. Academia, Prague and SPB Acad Publ, Hague, pp 163–170

Wharton GW (1946) Observations on *Ascoschöngastia indica* (Hirst 1915) (Acarinida: Trombiculidae). Ecol Monogr 16:151–184

Wharton GW, Fuller HS (1952) A manual of the chiggers. The biology, classification, distribution and importance to man of the larvae of the family Trombiculidae (Acarina). Mem Ent Soc Wash 4:1–185

Wharton GW, Jenkins DW, Brennan JM, Fuller HS, Kohls IM, Philip CB (1951) The terminology and classification of trombiculid mites (Acarina: Trombiculidae). J Parasitol 37:13–31

Williams RW (1946) A contribution to our knowledge of the bionomics of the common North American chigger, *Eutrombicula alfreddugesi* (Oudemans) with a description of a rapid collecting method. Amer J Trop Med Hyg 26:243–250

Womersley H (1944) Notes on and additions to the Trombiculinae and Leeuwenhoekiinae (Acarina) of Australia and New Guinea. Trans R Soc South Australia 68:82–112

Womersley H (1945) Acarina of Australia and New Guinea. Family Leeuwenhoekiidae. Trans R Soc South Australia 69:96–113

Womersley H (1952) The scrub-typhus and scrub-itch mites (Trombiculidae, Acarina) of the Asiatic-Pacific region. Rec South Austr Mus 10:1–673

Womersley H (1954) New genera and species, apparently of Apoloniinae (Acarina, Leeuwenhoekiidae) from the Asiatic-Pacific region. Malaysian Parasites I–XV. Studies for the Institute for Medical Research, Malaya 26:108–119

Wright JD, Hastriter MW, Robinson DM (1984) Observations on the ultrastructure and distribution of *Rickettsia tsutsugamushi* in naturally infected *Leptotrombidium (Leptotrombidium) arenicola* (Acari: Trombiculidae). J Med Entomol 21:17–27

9 Blood-sucking lice (Anoplura) of small mammals: True parasites[1]

Ke Chung Kim

1 Introductory remarks

The sucking lice (Anoplura) are true parasites of the eutherian mammals. True to their mode of parasitic life, the sucking lice are obligate, permanent parasites of specific mammalian hosts, inhabiting their hosts' fur habitats (except human body lice). They are equipped with unique piercing-sucking mouthparts with which these blood-sucking insects directly feed from small blood vessels of host mammals (Snodgrass 1944; Lavoipierre 1967). Their lives are completely dependent on the fate of their mammalian hosts and their close associations with specific taxa of mammals epitomize coevolution of the parasite-host lineages (Kim 1985a). All these factors together also make the sucking lice efficient vectors of the typhus and related bacterial pathogens. e.g., louse-borne/epidemic typhus (Harwood and James 1979). Accordingly, the sucking lice have been subjects of extensive research and they are often used as models for the ecology of ectoparasitic insects (e.g., Wenzel and Tipton 1966; Marshall 1981) and epidemiology of vector-borne diseases (e.g., Zinsser 1935; Busvine 1976).

In 2003, anthropologists determined the age of human clothing by the human louse, *Pediculus humanus*, which consist of two distinct taxa (currently considered as "subspecies"), *P. h. humanus* (body louse) and *P. h. capitis* (head louse) that inhabit and feed in the hair environment on the human scalp. Unlike all other sucking lice, however, the human body louse is adapted to inhabiting habitats of natural fibers used for human clothing. Using molecular clock techniques, Kittler et al. (2003) estimated that body lice diverged from ancestral head lice as early as 72000±42000 years ago. This information certainly stirred up interest in news media and thus scien-

[1] Blair Hedges has kindly assisted in tracking the estimated age of several higher taxa of mammals in preparation of this chapter

tific, cultural and religious circles (Lorenzi 2003; Wade 2004; Berenbaum 2006). In a recent forensic investigation of an unusual rape case, the blood in the gut of a pubic louse (*Pthirus pubis*) collected from the victim was analyzed for human DNA to confirm the suspect (Goff 2000). These are some of many examples which draw scientific and societal interests in the sucking lice which are known rather negatively as "vermin."

The sucking lice certainly are interesting insects to study, as they took over 30 years of my scientific career. In recent years many articles on human lice and their applications have been published in various scientific and technical journals which were also carried by news media worldwide. By being obligate, permanent parasites and largely host-species specific, the sucking lice not only are unique models for studying the biology and evolution of parasites but also offer diverse applications to scientific research and even in forensic investigations. Since a synthesis of taxonomic studies on sucking lice of the world by Ferris (1951), much of which were his own contributions, considerable advances were made on the biodiversity and classification of Anoplura by dedicated scientists like P.T. Johnson (USA), D.I. Blagoveshtechensky (Russia), K.C. Kim (USA), K. Kaneko (Japan), H.W. Ludwig (Germany), H.-J. Kuhn (Germany), T.-H. Chin (China), L.A. Durden (USA), and D.C. Castro (Durden and Musser 1994). Recently, anopluran research has focused on phylogenetic analysis by molecular taxonomy, all the different aspects of human lice, chemical control, and epidemiology of louse-borne diseases.

In this chapter I review the state of Anoplura biodiversity and the association of sucking lice with diverse small mammals, and discuss the dynamics of disruption in distribution and association of sucking lice in diverse lineages of mammals, adaptation in life history strategies in Anoplura, and future perspectives.

2 The sucking lice, true parasites

The sucking lice (Anoplura) are dorsoventrally flattened, wingless insects with elaborate piercing-sucking mouthparts that are similarly developed in all species of diverse lineages and with highly modified legs and claws adapted for grasping host hairs. As obligate, permanent parasites, most of the sucking lice are closely associated with the fur environment of the mammalian hosts in which temperature is relatively constant and optimal, although skin and fur temperatures vary among different areas of the body surface of a host animal (e.g., Marshal 1981; Piotrowski 1992).

In general, the relationship between the sucking lice and mammalian hosts is relatively stable and the impact of sucking lice infestation is hardly lethal to host animals and host populations. This relationship makes the sucking lice "good parasites" and they may even be considered as ecological and evolutionary partners rather than "vermin," coevolving along the parallel phylogenetic processes.

2.1 Life history strategy

Being obligate, permanent parasites, the sucking lice, once established on the host animal, live, feed, reproduce, and die in the fur environment from generation to generation until the host animal's life is ended. They go through a predictable life cycle: egg, three instars of larval stage, and adult, consecutively for most species. The eggs are glued to the hair near the skin in the fur environment. The adult female of the human head louse *P. h. capitis*, for example, lays an average of 57 eggs with 7.5 eggs per day. From oviposition by adult females, the egg requires about 8.5 days of incubation at 30°C and after hatching the larva (=nymph) goes through three relatively simple larval stages and the mature 3rd instar moults to become an adult, the development taking on average 9.7 (from 8.5 to 12.2) days. The female adult life lasts on average 13.3 (from 9 to 22) days. Thus, the human lice complete the entire life cycle in about 18 days. The sucking lice have literally ready-made food resources which are available for feeding at any time, usually twice a day for human lice (Busvine 1976). An exception to the general biology of sucking lice is the human body louse (*P. h. humanus*) which inhabits a highly specialized microhabitat - human clothing rather than animal hair and fur. Females lay eggs on man-made fibers, and the longevity of a female is much longer (20-21 days) than that of the head louse.

2.2 Lice and external environment

The sucking lice of aquatic mammals such as seals and sea lions and river otters are highly adapted to cold and wet environments. Species of Echinophthriidae, such as *Antarctophthirus callorhini* of the Northern fur seal (*Callorhinus ursinus*), *A. ogmorhini* of the Weddell seal (*Leptonychotes weddellii*), *Lepidophthirus macrorhini* of the Southern elephant seal (*Mirounga leonine*), are usually resilient and can survive for many days at low temperature, for many hours in submergence, and for many days of starvation (Busvine 1976; Murray 1976). For example, starving human head lice can survive 55 hours at 23°C, whereas starving *A. ogmorhini* can survive 12 days at 6°C and *L. macrorhini* – 6-8 weeks at 5-10°C. *A. ogmorhini* can even survive supercooling by an exposure to -20°C for 36 hours (Murray et al. 1965).

Life history of the sucking lice of marine carnivores becomes opportunistic due to the life cycle and beaching behaviour of the host mammals and is also limited to the period of the host breeding season out of water (e.g., Murray and Nichols 1965; Kim 1971). The life history of the sucking lice parasitic on *C. ursinus* takes place in two different habitats: *Proechinophthirus fluctus* inhabiting the fur and *A. callorhini* inhabiting the skin at the base of flippers, anal and genital orifice, eyelids and nostrils. They are not only opportunistic but their life cycle is also precarious because the hosts breed on the beach areas in the Pribilof Islands of the Bering Sea but by the end of summer they migrate to the south as far south as Baja California. On land, the adults of these lice on the pregnant female migrate to newborn pups during parturition and almost immediately mate and start the new life cycle on the pups. These species go through 2-3 generations on the growing pups as well as older seals (Kim 1971). The louse populations established on juvenile or adult hosts begin a long, slow life cycle on the migrating seals for many months and then they become mature adults when their host animals return to the breeding ground next spring. At this point the sucking lice again begin the new generation of louse populations on the seals, particularly on newborn pups on land (Kim 1989).

3 Anoplura biodiversity

Ever since the Linnean taxonomy was established and consistently applied in biology, world taxonomists have recorded about 1.75 million species (Heywood and Watson 1995) which barely represents 18% of the extant global biodiversity (if the figure of 10 million species as an average estimate of global biodiversity is accepted). As this labour took approximately 250 years, it will be an enormous task to explore and describe the remainder of global biodiversity, approximately 8.25 million species. We must come to grips very soon with how extant global biodiversity is to be explored and documented, particularly for the backyard biodiversity that is the essence of ecosystem function and the source of sustainable development at the grassroots. Considering the rapidly increasing human population that already passed 6.4 billion, biodiversity and biological resources must be studied and conserved for the sustainable economic development which all of us stride for, requiring ecosystem management for our backyards, whether in rich or poor countries, or for cities or the countryside.

Considering the state of global biodiversity, the Anoplura biodiversity is relatively better known because of the dedicated efforts of taxonomic specialists during the productive period from the late 1800s through the 20th century. Global Anoplura biodiversity is commonly estimated to be around 1500 species (Kim and Ludwig 1978; Kim 1985a, b; Kim et al. 1990), of which 532 species were listed as valid in "The Sucking Lice of the World" by Durden and Musser (1994). Considering the mammalian biodiversity that is far better known than that of most other animal taxa (perhaps 95% or more described), about 64% of the living mammals, equivalent approximately to 2671 species, are suspected to harbour sucking lice, of which the known species of sucking lice were recorded from only about 31% of potential mammalian host species. In other words, there still are over 828 species of living mammals, which could yield new species of sucking lice if they are closely examined for ectoparasites (Kim et al. 1990).

The sucking lice have evolved closely with eutherian mammals through their parallel lineages through a long evolutionary process. As stated earlier, their intimate biological relationships resulted in a high level of monoxeny with some species still being oligoxenous or polyxenous, perhaps since as early as the late Cretaceous (Kim and Ludwig 1982). The associations of Anoplura and Mammalia show that 29 genera of sucking lice are associated with a single mammalian family, six genera with two mammalian families, three genera with three mammalian families, and four genera with four to six mammalian families (Ludwig 1968). Recent studies based on molecular data in relation to fossil records show that the geological timing of placental mammal diversification is closely aligned with Anoplura phylogeny, thus forming close parasite-host associations (e.g., Kim 1982, 1985b; Springer et al. 2003).

Viewed from the perspective of placental mammalian cladogenesis, the emergence of mammalian splits and new clades closely mirrors the phylogeny and distribution pattern of parallel lineages of sucking lice and eutherian mammals (Kim 1982, 1985b; Springer et al. 2003). These phylogenetic patterns strongly suggest that sucking lice and their mammalian hosts had a high level of early associations and coevolution between them which led to close phylogenetic parallelism (Tables 1 and 2). Today's sucking lice are found on the species of diverse eutherian mammals: Artiodactyla, Carnivora, Pinnipedia, Dermoptera, Hyracoidea, Insectivora, Lagomorpha, Macroscelidae, Perissodactyla, Primates, Rodentia, Scandentia, and Tubulidentata (Table 1) (Kim 1985a; Kim et al. 1990; Durden and Musser 1994).

Strangely, some taxa within those orders that harbour large numbers of anopluran lineages are completely devoid of sucking lice and exclusively infested with other ectoparasites such as ischnocerans (Trichodectidae) on pocket gophers (Geomyidae, Rodentia), whereas some mammalian lineages such as Monotremata, Marsupialia, Xenarthra, Pholidota, Chiroptera, Cetacea, Proboscidea, and Sirenia, are not associated with sucking lice but with other parasitic arthropods such as Hemiptera, Diptera, and Acarina (Kim 1985a; Kim et al. 1990).

Table 1. Mammalian hosts and their sucking lice (Kim 1988)

Host mammals	Sucking lice
Artiodactyla	Linognathidae
	Haematopinidae (*Haematopinus*)
Camelidae	Microthoraciidae (*Microthoracius*)
Tayassuidae	Pecaroecidae (*Pecaroecus*)
Perissodactyla	Linognathidae
	Haematopinidae (*Haematopinus*)
Suidae	Ratemiidae (*Ratemia*)
Carnivora	
Fissipedia	Echinophthiriidae (*Latagophthirus*)
Pinnipedia	Echinophthiriidae
Macroscelidea	Neolinognathidae (*Neolinognathus*)
Insectivora	Hoplopleuridae (*Ancistroplax*, *Haematopinoides*)
Rodentia	Hoplopleuridae
	Polyplacidae
Sciuridae	Enderleinellidae
Lagomorpha	Polyplacidae (*Haemodipsus*)
Dermoptera	Hamophthiriidae (*Hamophthius*)
Scandentia	Polyplacidae (*Docophthirus, Sathrax*)
Primates	Polyplacidae
Cercopithecidae	Pedicinidae
Anthropoidea	Pthiridae, Pediculidae
Hyracoidea	Linognathidae (*Prolinognathus*)
Tubulidentata	Hybophthiridae (*Hybophthirus*)

In the cladistic analysis, no direct concordance exists between the family cladograms of Anoplura and their mammalian hosts. When close resolutions were made for specific family clades, however, highly closely parallel phylogeny emerged between anopluran and mammalian lineages (Kim 1988). The Anoplura cladogram recognized three primary lineages, Polyplacoid, Microthracoid, and Pediculoid groups (Table 2). It is interesting to note that a recent analysis of the placental mammal diversification related to the Cretaceous-Tertiary (K-T) boundary shows that Dermoptera and Scandentia are sister taxa that split before the K-T boundary, more than 80 million years ago (Springer et al. 2003), as was the case with *Hamophthirius* and *Sathrax-Docophthirus* of the Polyplacoid lineage of

Anoplura (Table 1 and 2). A closer examination of the cladistic disconcordance reveals that there are considerable similarities between cladistic patterns between the sucking lice and eutherian mammals. It suggests that the initial colonization and primary infestations of diverse mammals by the sucking lice must have taken place erratically at different times and regions before the cladogenesis of mammalian hosts was undertaken. Then, these sucking lice closely co-evolved with the host mammals with sporadic host shifts that established new host associations, in the end resulting in parallel lineages between them throughout the entire history of their associations (Kim 1985b, 1988).

Table 2. Associations between the sucking lice and mammalian hosts (Kim 1988)

Anoplura clades	Major host groups
Polyplacoid group	
Hamophthridae	Dermoptera
Neolinognathidae	Macroscelidea
Hoplopleuridae	Rodentia, Insectivora
Enderleinellidae	Rodentia (Sciuridae)
Polyplacidae	Rodentia, Primates, etc.
Linognathidae	Artiodactyla, Perissodactyla, Hyracoidea
Microthoracoid group	
Ratemidiidae	Perissodactyla (Suidae)
Microthoraciidae	Artiodactyla (Camelidae)
Echinophthiriidae	Carnivora
Pediculoid group	
Hybophthiriidae	Tubulidentata
Haematopinidae	Artiodactyla, Perisodactyla
Pecaroecidae	Artiodactyla (Taayassuidae)
Pedicinidae	Primates (Cercopithecidae)
Pthiridae	Primates (Anthropoidea)
Pediculidae	Primates (Anthropoidea)

4 The sucking lice and eutherian mammals: Coevolutionary partnership

Host associations in the Anoplura and Mammalia system are related to ecological and physiological interactions between parasites and hosts in a short time frame and genetics and coevolution over geological time. Therefore, a scientific approach to determine the origin and age of lineages or understand the phylogenetic processes of coevolutionary relationships of the parasite-host systems demands a synthesis based on multivariate data from comparative studies of morphological, genetic, ecological and geological parameters, where necessary.

The sucking lice have endured well with host mammals, the latter providing rather constant and steady fur environments within each specific mammalian lineage, unless fur habitats were threatened with abrupt environmental changes by host shift that could change habitats or habitat environment. In general, infraspecific genetic variation in Anoplura is relatively small (e.g., Kim et al. 1963) and their genetic variation within a genus-taxon is also relatively small, often with interspecific variation primarily limited to the genital structures. However, there are distinct differences in morphological configuration between specific lineages, such as Hoplopleuridae and Polyplacidae of rodents and Antarctophthriidae of pinnipeds.

Considering information on the biology and behaviour of sucking lice so far available, the sucking lice like many other "pest" species can readily adapt to specific environmental changes. In a laboratory setting, human body lice became resistant to insecticides like DDT within ten generations. For example, the sucking lice associated with the arctoid-fissiped ancestors of modern pinnipeds must have developed behavioural and morphological adaptations to stay on the host and survive in changing environments, as their mammalian hosts frequent the water environment. There are a number of morphological adaptations linked to aquatic habitats such as flattening of body setae to scales and enlargement and elaborate modification of tibia-tarsal segments of mid- and hind legs in the generalist genus, *Antarctophtirus* (Kim 1971, 1975, 1988; Kim et al. 1975).

All species in a given lineage of sucking lice are exclusively parasitic on the specific taxon of their specific parallel lineage of eutherian mammals; e.g., Enderleinellidae associated with Sciuridae; Pediculidae versus Primates; Microthoracidae versus Camelidae. The sucking lice are highly host-specific and over 63% of all known species of Anoplura are monoxenous (one species of parasite on one host species), whereas 24% of species are hetero- or oligoxenous (specific to two or three host species). In other words, most species of sucking lice (87% or more) are associated with one or 2-3 host species. The sucking lice parasitic on rodents show 62% host specificity of which 66% of total known species of the Enderleinellidae are specific to single host species, 62% for Hoplopleuridae, and 58% for Polyplacidiae, while ungulate-infesting taxa (or clades) like Haematopinidae and Linognathidae demonstrate 95% host specificity (Kim 1985b, Kim et al. 1990).

Considering their broad and mostly consistent distribution throughout the diverse lineages of today's global mammal biodiversity, the sucking lice are resilient and persistent parasites, evolved closely along the evolution and radiation of specific mammalian host lineage. It is reasonable to hypothesize that once ancestral sucking lice successfully established on a specific ancestral mammalian species to begin a parasite-host lineage, they must have evolved rapidly along the evolution of their hosts in specific mammalian lineage, as their hosts continued to radiate and evolve along the evolutionary history of mammals (Kim 1985a; Kim 1993). Conversely, anthropogenic stresses that cause today's mass extinction also directly affect the delicate parasite-host relationship, causing the loss of coevolutionary partners, host species and their associated sucking lice.

As a mass extinction by anthropogenic stresses continues, it is likely that as many mammalian species that harbour specific sucking lice become extinct, their parasitic partners will also be lost at the same time (co-extinction). Although we have no specific way to detect and measure this, it is not far-fetched to predict that there are good numbers of host-specific sucking lice already lost by recent extinction of host mammals (Stork and Lyal 1993). Today's anopluran biodiversity is the descent of interactive parasite-host relationships between the sucking lice and their host mammals. In the co-evolutionary process, once established as a clade in a specific parasite-host lineage, the sucking lice appear to have successfully established a phylogenetic base on the host species and its subsequent lineage. In many lineages sucking lice successfully modified their life history strategy to survive in heterogeneous environments; for example, sucking lice on marine carnivores (Murray 1965; Murray and Nicholls 1965; Murray et al. 1965; Kim 1985a).

5 Anoplura biodiversity and micromammals

Micromammals (= small mammals) as defined in this volume include Chiroptera, Insectivora, Rodentia, and Lagomorpha, of which Chiroptera as a specialized taxon are not of concern here because sucking lice are not associated with them at all. The discussion here is limited to the sucking lice of Insectivora, Rodentia and Lagomorpha. The latter order is associated with two anopluran genera: *Haemodipsus* is parasitic on Leporidae and *Hoplopleura ochotonae* is a characteristic louse species of Ochotonidae (pikas). Additional small mammals, used to be considered closely related to Insectivora, harboring polyplacoid sucking lice are Tupaiidae (Scandentia) with *Sathrax* and *Docophthirus* and Macroscelididae (Macroscelidea) with *Neolinognathus*.

5.1 Sucking lice of Insectivora and related mammals

The "Insectivora" used to include all small insectivoran mammals including Monotyphlan families, Macroscelididae and Tupaiidae (their phylogenetic relationships are yet unclear). Today's Insectivora is a monophyletic group that includes six recent Lipotyphlan families: Erinaceidae, Talpidae, Solenodontidae, Tenrectidae, Chrysochloridae, and Soricidae. Their associations with the sucking lice are summarized in Table 3.

Table 3. Anopluran genera parasitic on lipotyphlan and monotyphlan mammals. In parentheses – number of species

Insectivoran families	Anoplura partners
LIPOTYPHLANS	
Talpidae	*Haematopinoides* (1)
Soricidae	*Ancistroplex* (5), *Polyplax* (3)
MONOTYPHLANS	
Macroscelididae (order Macroscelidea)	*Neolinognathus* (2)
Tupaiidae (order Scandentia)	*Sathrax* (1), *Docophthirus* (1)

Among six lipophylan families, only two (Talpidae and Soricidae) are recorded to harbour sucking lice. Of 31 talpid species, only two species, *Parascalopus breweri* and *Scalopus aquaticus*, have lice, with both harbouring a single species of Hoplopleuridae, *Haematopinoides squamosus*. On the other hand, Soricidae, distributed throughout the world except the Polar regions, Australian region, and central and southern South America, are parasitized by specialized *Ancistroplax* (Hoplopleuridae) of which five species are so far recorded from *Soriculus*, *Crocidura*, and *Suncus* (Anderson and Jones 1984). Considering that the sucking lice of Hoplopleuridae are primarily rodent parasites, *Ancistroplax* and *Haematopinoides* must have shifted host from rodents early in the evolution of the Hoplopleuridae lineage and solidly established and evolved along the Insectivore lineage.

The elephant shrews (Macroscelidea) and tree shrews (Scandentia) are two distinct taxa of the menotyphlan insectivores. They are parasitized mainly by two genera of sucking lice, representing two separate phylogenetic lines, *Neolinognathus* and *Sathrax*, respectively.

5.2 Associations of sucking lice and rodents

The species diversity of Anoplura is closely related to the diversity of mammalian hosts within a specific lineage. About 70% of anopluran species are associated with rodents (Kim 1988), which are hosts to three families of sucking lice, namely Enderleinellidae, Polyplacidae, and Hoplopleuridae. Oddly enough, certain mammalian lineages, such as Geomyidae (pocket gophers) are completely devoid of sucking lice, although the Geomyidae is a clade within Rodentia, highly infested by sucking lice. Other rodent clades that are completely devoid of sucking lice include Hystricidae, Aplodontidae, Anomaluridae, Spalacidae as well as several other sciurognath and hystrichognath families (Kim 1985b). The distribution of the genera of the sucking lice among rodents is summarized in Table 4.

Table 4. Anopluran genera parasitic on rodents (Anderson and Jones 1984; Kim 1985b). In parentheses – number of species

Rodent families	Anoplura partners		
	Enderleinellidae	Polyplacidae	Hoplopleuridae
Sciuridae	*Enderleinellus* (45), *Microphthirus* (1), *Werneckia* (5), *Phthiruneulus* (1), *Atopophthirus* (2)	*Neohaematopinus* (22), *Johnsonpthirus* (5), *Linognathoides* (11), *Polyplax* (1)	*Hoplopleura* (10), *Paradoxophthirus* (1)
Heteromyidae		*Fahrenholzia* (12)	
Dipodidae		*Eulinognathus* (16)	*Schizophthirus* (3)
Muridae		*Eulinognathus* (2), *Fahrenholzia* (1), *Neohaematopinus* (2), *Polyplax* (77), *Proenderleinellus* (1), *Mirophthirus* (1), *Typhlomyophthirus* (1)	*Hoplopleura* (122)
Pedetidae		*Eulinognathus* (1)	
Myoxidae			*Schizophthirus* (6)
Bathyergidae		*Eulinognathus* (2)	
Petromuridae		*Scipio* (1)	
Thryonomyidae		*Scipio* (2)	
Chinchillidae		*Cuyana* (1), *Eulinognathus* (1), *Lagidiophthirus* (1)	
Caviidae		*Galeophthirus* (1)	*Trimenopon* (1)
Ctenomyidae		*Eulinognathus* (5)	
Octodontidae			*Hoplopleura* (2)
Abrocomidae		*Polyplax* (1)	
Echimyidae		*Ctenophthirus* (1), *Fahrenholzia* (1)	*Hoplopleura* (3), *Pterophthirus* (3)

Large rodent families are usually infested by relatively large number of louse species. For example, the speciose Sciuridae are colonized by 11 genera and about 100 species of sucking lice, whereas a smaller taxon like the Dipodidae harbours only two genera and 19 species (Kim 1985b; Durden and Musser 1994). The sucking lice of Enderleinellidae are exclusive ectoparasites of Sciuridae which also are infested with four polyplacid and two hoplopleurid genera. Both polyplacid and hoplopleurid lice infest also Muridae, Dipodidae, Echimyidae, and Caviidae. In addition, Polyplacidae are found on Heteromyidae, Pedetidae, Bathyergidae, Petromuridae, Thryonomyidae, Chinchillidae, Ctenomyidae, and Abrocomidae, whereas Hoplopleuridae are found on Octodontidae and Myoxidae. The highest diversity of both polyplacids and hoplopleurids is associated with murid hosts.

6 Coevolution of squirrels and their lice: Unfinished speciation

Squirrels (Sciuridae) constitute one of the largest families of rodents that contain an abundant and diverse group of species well known in many biological and cultural perspectives. They also are primary hosts to the sucking lice of Enderleinellidae at large and polyplacids (*Neohaematopinus* genus-group including *Neohaematopinus*, *Linognathoides*, *Johnsonphthirus*), and straggler species of *Polyplax*. As mentioned above, Enderleinellidae are exclusive parasites of squirrels and their fate and evolution have been closely linked to the evolutionary success of the Sciuridae. There are five genera of Enderleinellidae, namely *Enderleinellus*, *Microphthirus*, *Werneckia*, *Phthirunculus*, and *Atopophthirus*, of which *Enderleinellus* is most diverse (Kim 1966, 1977, 1985b, 1988; Kim and Ludwig 1978; Kim and Adler 1982) (Table 5).

Enderleinellus is a generalist genus which exploits various host species from most tribes within Sciuridae (Table 6). Looking at the host associations of the sucking lice in Sciuridae, it has been observed that the known species of *Enderleinellus* from tree squirrels (*Sciurus*) are monoxenous, one species of lice being associated with a single host species, whereas the species complexes from ground squirrels (*Spermophilus*) are oligoxenous or polyxenous, meaning that one species of parasite is associated with a number of closely related species of ground squirrels (Kim et al. 1963; Kim 1966, 1985c, d; 1988; Kim and Ludwig 1978).

Considering the current state of parasite-host association in Enderleinellidae and Sciuridae (Table 5), most taxa at species and generic level of Enderleinellidae are closely associated with their respective host taxa at similar taxonomic levels. Most sciurids are parasitized by one species of Enderleinellidae except *Spermophilus*. *Atopophthirus, Microphthrius*, and *Phthirunculus* are parasites of flying squirrels (Pteromyinae) and species of both *Atopophthirus* and *Phthirunculus* are parasitic on *Petaurista.* Species of *Enderleinellus* are broadly associated with diverse species of Sciurinae, primarily on tree squirrels (Sciurini, Callosciurini, Protoxerini, Xerini, and Funambulini) and marmots (Marmotini), whereas species of *Werneckia* are parasitic on species of the Funambulini in Africa. Although some of the recorded host associations need to be re-examined and verified for species identity and taxonomic status, most species of *Enderleinellus* are associated with single species of tree squirrels in the New World and of other squirrels in tropical Asia (Calloschiurini, Funambulini) and Africa (Protoxerini, Xerini).

Table 5. Diversity and associations of Enderleinellidae and Sciuridae (Durden and Musser 1994)

Enderleinellidae	Sciuridae
Atopophthirus (2)	Pteromyinae
emersoni	*Petaurista* (Malasia)
setosus	*Petaurista* (Malasia)
Microphthirus (1)	Pteromyinae
uncinatus	*Glaucomys* (Canada, USA.)
Phthirunculus (1)	Pteromyinae
sumatranus	*Petaurista* (Indonesia: Sumatra)
Enderleinellus (45)	Sciurinae
	Tribe Sciurini
arizonensis	*Sciurus alleni, S. arisonensis*(USA: Arizona)
brasiliensis	*Sciurus aestuans* –species complex (Brazil)
deppei	*Sciurus aureogaster, S. granatensis, S. deppei* (Mexico)
extremus	*Sciurus aureogaster, S. deppei* (Guatemala, Mexico)
hondurensis	*Sciurus yucatanensis, S. variegatoides* (Columbia, Honduras, Mexico)
insularis	*Sciurus granatensis* (Venezuela)
kaibabensis	*Sciurus alberti* (USA: Arizona)
kelloggi	*Sciurus giseus*(USA: California)
krochinae	*Sciurus anomalus* (Azerbajian)
longiceps	*Sciurus carolinensis, S. niger* (USA)
mexicanus	*Sciurus aureogastger* (Mexico)
nayaritensis	*Sciurus nayaritensis* (Mexico)
nitzschi	*Sciurus vulgaris* (Eurasia)
oculatus	*Sciurus alleni* (Mexico)
paralongiceps	*Sciurus aberti* (USA)
pratti	*Sciurus colliaei* (Mexico)
urosciuri	*Sciurus igniventris* (Brazil)

venezuelae	*Sciurus granatensis* (Venezuela)
	Tribe Callosciurini
kumadai	*Callosciurus* (Japan)
malaysianus	*Callosciurus* (Borneo, Mayanmar, Malaysia, Thailand)
puvensis	*Callosciurus* (China)
dremomydis	*Dremomys* (China: Sichuan, Thailand)
corrugatus	*Tamiops*, *Callosciurus* (Thailand)
	Tribe Protoxerini
gambiani	*Heliosciurus* (Liberia)
heliosciuri	*Heliosciurus* (Liberia)
	Tribe Xerini
heliosciuri	*Epixerus* (Angola, Kenya, Liberia)
	Tribe Funambulini
nishimarui	*Funambulus* (India)
platyspicatus	*Funambulus* (Ceylon)
euxeri	*Xerus* (Kenya, Dahomey, Liberia, Sudan, Nigeria)
zonatus	*Paraxerus* (Kenya)
larisci	*Lariscus* (Borneo)
menetensis	*Menetes* (Thailand)
nannosciuri	*Nannosciurus* (Indonesia: Java)
	Tribe Marmotini
blagoveshchenskyi	*Marmota* (Kyrgyzstan)
dolichocephalus	*Marmota* (Russia: Yakutia-Sakha)
marmotae	*Marmota* (USA)
tamiasis	*Tamias* (Korea)
disparillus	*Spermophilus* (Russia: Amur)
ferrisi	*Spermophilus* (Bulgaria)
osborni	*Spermophilus* (USA)
propinquus	*Spermophilus* (Kazakhstan, Poland, Romania)
suturalis	*Ammospermophilus*, *Cynomys*, *Spermophilus* (USA)
	Tribe Microsciuini
microsciuri	*Microsciurus* (Columbia, Panama)
	Pteromyinae
replicatus	*Pteromys* (Russia: Tatarstan)
Werneckia (5)	Sciurinae
	Tribe Funambulini
funisciuri	*Funisciurus* (Nigeria)
nigiriensis	*Funisciurus* (Nigeria)
africana	*Funisciurus* (Nigeria)
paraxeri	*Paraxerus* (Kenya)
minuta	*Paraxerus* (Kenya)

Most *Sciurus* species harbour one species of *Enderleinellus*. However, four species (*S. aberti, S. alleni, S. augeogaster,* and *S. granatensis*) are associated with two species of sucking lice (Table 6). In addition, five

Holarctic species of *Enderleinellus* are associated with many species of ground squirrels (*Spermophilus*) (Table 7).

Table 6. Associations between *Sciurus* and *Enderleinellus*

Sciurus	*Enderleinellus*
S. aberti	*E. kaibabensis, E. paralongiceps*
S. aestuans	*E. brasiliensis*
S. alleni	*E. arizonensis, E. oculatus*
S. anomalus	*E. krochinae*
S. arizonensis	?
S. aureogaster	*E. extremus, E. mexicanus*
S. carolinensis	*E. longiceps*
S. colliaei	*E. pratti*
S. deppei	*E. deppei*
S. flammifer	?
S. gilvigularis	?
S. granatensis	*E. insularis, E. venezuelae*
S. griseus	*E. kelloggi*
S. ignitus	*E. urosciui*
S. igniventris	?
S. lis	?
S. nayaritensis	*E. nayaritensis*
S. niger	*E. oculatus*
S. oculatus	*E. oculatus*
S. pucheranii	?
S. pyrrhinus	?
S. richmondi	?
S. sanborni	?
S. spadiceus	?
S. stramineus	?
S. variegatoides	*E. hondurensis*
S. vulgaris	*E. nitschi*
S. yucatanensis	*E. hondurensis*

Table 7. *Enderleinellus* and Sciurid hosts (Tribe Marmotini). * records needs verification of species identity

Sucking lice *(Enderleinellus)*	Squirrels *(*Tribe *Marmotini)*
Enderleinellus blagoveshtchenskyi	*Marmota baibacina*
E. dolichocephalus	*M. camchatica*
E. tamiasis	*Tamias stiatus*
	T. sibiricus
E. disparilis	*Spermophilus undulates*
E. ferrisi	*S. citellus*
E. osborni	*S. (Xerospermophilus)mohavensis*

	S. (X.) tereticaudus
	S. (Otiospermophilus) beecheyi
	S. (O.) variegatus
	S. (O.) atricapillus
	*S. (Spermophilus) beldingi**
E. proqinquus	*S. (Spermophilus) fulvus*
	S. (S.) suslicus
	S. (S.) citellus
E. suturalis	*Ammospermophilus harrisi*
	A. nelsoni
	Cynomys gunnisoni
	C. leucurus
	Spermophilus (S.) beldingi
	S. franklinii
	S. lateralis
	S. mexicanus
	S. richardsonii
	S. spilosoma
	*S. (X.) tereticaudus**
	S. townsendii
	S. tridecemlineatus

Eight species of *Enderleinellus* are parasitic on marmots, prairie dogs, and chipmunks and ground squirrels. Considering the host associations of polyxenous *E. osborni, E. proprinuus,* and *E. suturalis*, it is likely that other species such as *E. disparilis* and *E. ferrisi* are also similarly associated with many other host species beyond those originally recorded (Table 7). The populations of *E. suturalis* from three host species of ground squirrels, *S. tridecemlineatus*, *S. franklini,* and *A. harrisi*, were analyzed to determine if they can be discriminated by morphometric measurements (Kim et al. 1963). The populations of this species were determined to be distinct and could be separated by morphometric characters each of which could be treated as a taxon at the subspecies level. In other words *Enderleinellus* species associated with ground squirrels have not evolved far enough to be recognized as species like those associated with tree squirrels which have distinct characters separating them from other related species.

7 Concluding remarks

The sucking lice (Anoplura) are true parasites and provide an interesting model for speciation, phylogenetic studies, community ecology, and ecosystem function of parasite-host system. As with the global biodiversity of all other organisms, we are a long way from understanding the true extent of Anoplura biodiversity. We should make a determined effort to explore and describe most species of sucking lice from those expected host species of extant mammals because anthropogenic extinction of mammals causes the loss of unknown species of sucking lice, their evolutionary and ecological partners.

Being well established obligate and permanent parasites, the sucking lice could provide better understanding of how parasites-hosts relationships are sustained in balance without serious threats to the survival of host species and how sucking lice evolve as host species split within a specific lineage such as Enderleinellidae and Sciuridae. The more empirical studies on biodiversity, host associations, distribution and community ecology of the sucking lice are pursued and new information established, the better understanding we achieve of parasite-host relationships and evolutionary dynamics of ecological partners. Better understanding of the community structure and the patterns of parasite distribution on host animals should help develop realistic models with sound assumptions and real-term parameters which could provide real-term predictions (e.g., Bittencourt and Rocha 2002; Choe and Kim 1987, 1988, 1989).

We can reach a better understanding of the intricate dynamics of ecosystems involving a community of parasite species interacting with host animals, if we approach the study of parasites-small mammalian host systems with morphological, ecological and molecular parameters. This could provide new means to control and manage parasite infestations of human systems. In the rapidly changing global environment, continued study of the life patterns and harmonious relationship of two ecosystem partners, parasites and host mammals, established through long coevolutionary processes, should offer a better understanding of the dynamics of parasite communities on host animals including humans.

References

Anderson S, Jones Jr J Knox (1984) Orders and families of recent mammals of the World. John Wiley, New York

Berenbaum M (2006) Buzzwords: Lousy designs. Am Entomol 552:4–6

Bittencourt EB, Rocha CFD (2002) Spatial use of rodents (Rodentia: Mammalia) host body surface by ectoparasites. Braz J Biol 62:419–425

Busvine JR (1976) Insects, hygiene and history. Athlone Press, London

Choe JC, Kim KC (1987) Community structure of arthropod ectoparasites on Alaskan seabirds. Can J Zool 65:2992–3005

Choe JC, Kim KC (1988) Microhabitat preference and coexistence of ectoparasitic arthropods on Alaskan seabirds. Can J Zool 66:987–997

Choe JC, Kim KC (1989) Microhabitat selection and coexistence in feather mites (Acari: Analgoidea) on Alaskan seabirds. Oecologia 79:10–14

Durden LA, Musser GG (1994) The sucking lice (Insecta, Anoplura) of the World: A taxonomic checklist with records of mammalian hosts and geographical distributions. Bull Am Mus Nat Hist 218:1–90

Goff ML (2000) A fly for the prosecution: How insect evidence helps solve crimes. Harvard Univ Press, Cambridge

Ferris GH (1951) The sucking lice, vol 1. Mem Pac Coast Entomol Soc 1:1–320

Harwood RF, James MT (1979) Entomology in human and animal health, 6th Ed. Macmillan, New York

Heywood VH, Watson RR (eds) (1995) Global biodiversity assessment. Cambridge Univ Press Cambridge

Kim KC (1966) The species of *Enderleinellus* parasitic on the Sciurini and Tamiasciurini (Anoplura, Hoplopleuridae). J Parasitol 52:988–1024

Kim KC (1971) The sucking lice (Anoplura, Echinophthiriidae) of *Callorhinus ursinus* (Pinnipedia: Otariidae); descriptions and morphological adaptation. Ann Entomol Soc Amer 64:280–292

Kim KC (1975) Ecology and morphological adaptation of the sucking lice (Anoplura, Echinophthiriidae) on the northern fur seal. Rapp P-V Reun Cons Int Explor Mer 169:504–514

Kim KC (1977) *Atopophthirus emersoni*, new genus and new species (Anoplura: Hoplopleuridae) from *Petauristia elegans* (Sciuridae, Rodentia), with a key to the genera of Enderleinellinae. J Med Entomol 14:417–420

Kim KC (1982) Host specificity and phylogeny of Anoplura. Deux Symp Speci. Parasit Parasites Vert 13-17. April 1981. Mem Mus Nat Hist, NS, Zool 123:123–127

Kim KC (ed) (1985a) Coevolution of parasitic arthropods and mammals. John Wiley, New York

Kim KC (1985b) Evolution and host associations of Anoplura. In: Kim KC (ed) Coevolution of parasitic arthropods and mammals. John Wiley, New York, pp 197–231

Kim KC (1985c) Evolutionary aspects of the disjunct distribution of lice on Carnivora. In: Kim KC (ed) Coevolution of parasitic arthropods and mammals. John Wiley, New York, pp 257–294

Kim KC (1985d) Evolutionary interactions and coevolution. In: Kim KC (ed) Coevolution of parasitic arthropods and mammals. John Wiley, New York, pp 601–682

Kim KC (1988) Evolutionary parallelism in Anoplura and eutherian mammals. In: Service MW (ed) Biosystematics of haematophagous insects. Oxford Univ Press, pp 91–114

Kim KC (1993) Insect pests and evolution. In: Kim KC, McPherson BA (eds) Evolution of insect pests: Patterns of variation. John Wiley, New York, pp 3–25

Kim KC, Adler PH (1982) Taxonomic relationships of *Neohaematopinus* to *Johnsonpthirus* and *Linognathoides* (Polyplacidae: Anoplura). J Med Entomo. 12:134–148

Kim KC, Ludwig HW (1978) The family classification of the Anoplura. Syst Entomol 3:249–284
Kim KC, Ludwig HW (1982) Parallel evolution, cladistics and classification of parasitic Psocodea. Ann Entomol Soc Am 75:537–548
Kim KC, Byron Jr BW, Cook EF (1963) A quantitative taxonomic study of the *Enderleinellus suturalis* complex (Hoplopleuridae: Anoplura). Syst Zool 12:134–148
Kim KC, Repenning CA, Moorejohn GV (1975) Specific antiquity of the sucking lice and evolution of otariid seals. Rapp P-V Reun Cons Int Explor Mer 169:544–549
Kim KC, Emerson KC, Traub R (1990) Diversity of parasitic insects: Anoplura, Mallophaga, and Siphonaptera. In: Kosztarab M, Schaefer CW (eds) Systematics of the North American insects and arachnids: Status and needs. Virginia Polytechnic Institute and State University, Virginia Agricultural Experiment Station Information Series 90-1, pp 91–104
Kittler R, Kayser M, Stoneking M (2003) Molecular evolution of *Pediculus humanus* and the origin of clothing. Cur Biol 13:1414–1417
Lavoipierre MMJ (1967) Feeding mechanism of *Haematopinus suis* on the transilluminated mouse ear. Exp Parasitol 20:303–311
Lorenzi R (2003) Human body lice reveal the birthdate of fashion. News in Science, 2 September. http://www.abc.net.au/science/news/stories/s937424.htm
Ludwig HW (1968) Zahl, Vorkommen und Verbreitung der Anopplura. Z Parasitkd 31:254–265
Marshall AG (1981) The ecology of ectoparasitic insects. Acad Press, London
Murray MD (1965) The diversity of the ecology of mammalian lice. In: Freeman P (ed) XII International Congress of Entomology, London, 8-16 July 1964: Proceedings. R Entomol Soc Lond, London, pp 366–367
Murray MD (1976) Insect parasites of marine birds and mammals. In: Cheng L (ed) Marine Insects. North-Holland Publ Co, Amsterdam, pp 79–96
Murray MD, Nicholls DG (1965) Studies on the ectoparasites of seals and penguins. I. Th ecology of the louse *Lepidopthirus macrorhini* Enderlein on the elephant seal, *Mirounga leonina* (L.). Austral J Zool 13:437–454
Murray MD, Smith MSR, Soucek Z (1965) Studies on the ectoparasites of seals and penguins. II. The ecology of the louse *Antarctophthirus ogmorhini* Enderlein on the Weddell seal, *Leptonychotes weddeli* Lesson. Austral J Zool 13:761–771
Piotrowski F (1992) Anoplura (echte Läuse). Handbuch der Zoologie, IV. Arthropoda: Insecta, Teilband 32. Walter de Gruyter, Berlin
Snodgrass RE (1944) The feeding apparatus of biting and sucking insects affecting man and animals. Smithsonian Misc Coll 104:1-113
Springer MS, Murphy WJ, Eizirk E, O'Brien SJ (2003) Placental mammal diversification and the Cretaceous-Tertiary boundary. Proc Natl Acad Sci USA 100:1056–1061
Stork NE, Lyal CHC (1993) Extinction or co-extinction rates. Nature 366:307
Wade N (2004) What a story lice can tell. New York Times, 5 November. http://nytimes.com/2004/10/05/science/05lice.html

Wenzel RL, Tipton VJ (eds) (1966) Ectoparasites of Panama. Field Museum of Natural History, Chicago

Zinsser H (1935) Rats, lice and history: A chronicle of pestilence and plagues. Black Dog and Leventhal Publishers, New York

10 Fleas: Permanent satellites of small mammals[1]

Sergei G. Medvedev and Boris R. Krasnov

1 Siphonaptera at a glance

Fleas (Siphonaptera) represent a relatively small order of secondarily wingless holometabolous insects. All fleas are obligatory haematophagous parasites of higher vertebrates. According to the recent taxonomic scrutiny, the order includes 2005 species and 828 subspecies belonging to 242 genera and 97 subgenera (Medvedev et al. 2005).

The overwhelming majority of fleas parasitize mammals (more than 94% of species; Vatschenok 1988), whereas their association with birds is much less. A total of 214 flea species infest birds, although only 60 species (about 3% of the total number of flea species) can be considered as specific bird parasites (Medvedev 1997a, b). Occurrences of fleas on reptiles are accidental, although they are able to digest blood of these hosts (Vatschenok 1988).

Different flea species vary in the proportion of time they spend on the body of a host or in a host's burrow/nest. Consequently, Ioff (1941) suggested distinguishing between four categories of flea species, namely "nest" fleas, "fur" or "body" fleas, "semistationary" fleas and "stationary" fleas. Nest fleas stay on the host long enough only to take a bloodmeal. Fur fleas spend much time on the host but do not attach to the host by their mouthparts for a long period. They do not lose the ability to move freely and to transfer to other hosts. In semistationary or sticktight fleas (e. g, *Echidnophaga gallinacea*), a female attaches by her mouthparts to the host and spends the rest of her life in that position (but see Vatchenok 1988). Stationary fleas burrow onto the host's skin and become neosomic (Tungidae and Vermipsyllidae). Approximately 70 species can be classified as semistationary and stationary fleas. Most fleas that exploit small mammals are either fur or nest fleas or else take intermediate positions between these

[1] We thank Michael Hastriter for the helpful comments on the earlier version of this chapter.

categories. However, the dichotomy of fur versus nest fleas is problematic (see Krasnov et al. 2004a for details). The idea of this dichotomy led to vigorous discussions in the Russian literature (e. g., Novokrestchenova 1960; Zhovtyi 1963; Nelzina et al. 1963; Vatschenok 1988), although the intensity of these discussions decreased drastically and eventually ceased in the early nineties.

2 Morphological adaptations to parasitism

The anatomy and locomotory patterns of fleas reflect their way of life as obligatory burrow/nest parasites of fur- or feather-covered hosts. Fleas are able to move through the dense host pelage and to withstand the host's anti-parasitic grooming movements. They also are able to jump, to move through the substrate of a host's burrow or nest, and to move on vertical surfaces (e.g., fleas parasitic on bats). Here, we will briefly discuss body size and sexual size dimorphism of fleas and then review some of their morphological features that facilitate the successful exploitation of their hosts.

2.1 Sexual size dimorphism

The length of the body of a flea averages 4-5 mm; whereas the length of a few "giant" flea species is usually no more than 1 cm. Fleas demonstrate sharp female-biased sexual size dimorphism, as is common in arthropods. However, comparative analysis of several flea species showed that size dimorphism in fleas does not decrease with increasing body size (Krasnov et al. 2003), which contradicts Rensch's rule (Rensch 1960; Abouheif and Fairbairn 1997; Colwell 2000). Rensch's rule states that in taxa in which females tend to be larger than males, size dimorphism diminishes in larger species, whereas in taxa in which males tend to be larger than females, size dimorphism increases in larger species. Reviews of the quantitative evidence for Rensch's rule (Reiss 1986; Abouheif and Fairbairn 1997; Fairbairn 1997) indicate that it is a very common allometric trend but that exceptions occur, particularly in taxa in which females are the larger sex as in Siphonaptera.

It is possible that fleas do not conform to assumptions of the various functional hypotheses explaining the evolution of allometry of sexual size dimorphism (Fairbairn 1997; Colwell 2000). For example, natural selection for niche differentiation and resource partitioning are not relevant for fleas, as males and females parasitize the same host individuals and share the same feeding niches. Many of the hypotheses explaining Rensch's rule have invoked either sexual selection on males or stabilizing selection on females or both (Fairbairn 1997). Male fleas have no role in reproduction besides mating and there is no evidence that their mating success depends on their size. Furthermore, fleas do not demonstrate a strong stabilizing selection or strong constraints on female size. For example, egg production and egg size in fleas were reported to be independent of body size (e. g, Vatschenok 1988), but were explained well by the patterns of relationships with their hosts.

2.2 Mouthparts

Flea mouthparts are adapted to extract blood from the host and have been described in details elsewhere (e. g., Snodgrass 1946). Suctorial mouthparts of fleas are composed of epipharinx and two laciniae of the maxillae. The two laciniae and the epipharynx together enclose a food channel for inbound blood. The laciniae form a smaller salivary channel for outbound saliva. These structures have an elongated stylet-like form, and each outer side of the laciniae has two rows of backward-pointed teeth which cut or saw the skin of the host and anchor the mouthparts. The length of the mouthparts and the number and development of the teeth vary among flea species. For example, they are very long in stationary fleas (*Dorcadia* and *Vermipsylla*), shorter and more robust in semi-stationary or stick-tight fleas (*Echidnophaga* and *Hectopsylla*), and more slender and minimally serrated in fur or nest fleas (*Neopsylla*, *Orchopeas*, *Ceratophyllus*, etc.). Among fleas parasitizing small mammals, the longest mouthparts occur in *Brachyctenonotus* and species of *Rhadinopsylla* that parasitize zokors (*Myospalax*).

2.3 Jumping

The jump of fleas is their most conspicuous and fascinating locomotory character that allows these wingless insects to attack their hosts successfully. Classical studies of flea jump were conducted by Bennet-Clark and Lucey (1967), Rothschild et al. (1973, 1975) and Rothschild and Schlein (1975). They found differences in the jumping ability between sexes and among species. For example, male fleas jumped shorter distances than female fleas. This finding is not surprising since males are smaller than females. However, these data were not corrected for body size dimorphism and, thus, it was unclear if size was the only source of sexual difference in locomotory performance or if other factors were involved. A recent study of jump performance of seven flea species parasitic on small mammals demonstrated that the jumping ability of males was less than that of females even when corrected for difference in body size (Krasnov et al. 2003).

Jump length also varies among species. For example, in the study of Rothschild et al. (1975), the longest jump was in the bird flea *Ceratophyllus styx* and the shortest in the bat flea *Ischnopsyllus octactenus*. It was suggested that this variation in jumping ability was a result of morphological differences. Therefore, the size of certain parts of the locomotory apparatus was thought to be indicative of jumping capacity.

The main sources of jumping power are the hind legs of the flea and a rubber-like protein (resilin) located in the pleural arch (Rothschild et al. 1973, 1975; Rothschild and Schlein 1975). The resilin pad is homologous with the wing hinge ligaments in flying insects. The pleural arch and resilin protein were found to be well developed in fleas with high jumping capacity, but were absent or greatly reduced in poor jumping fleas (Rothschild et al. 1975). Tripet et al. (2002) measured pleural height in preserved flea specimens, considered this measurement as an indicator of flea mobility and used it for broad comparisons among species of bird fleas from different geographic and host ranges. They found a negative correlation between flea mobility (expressed via measurements of pleural height) and the degree of host colonialism and a positive correlation between flea mobility and their host range. However, an experimental study with simultaneous measurements of jump performance and morphological measurements on the same flea individuals demonstrated the lack of any correlation between jumping ability and morphometrics of the locomotory system among species as well as between sexes (Krasnov et al. 2003). Moreover, pleural height that has been considered as a reliable indicator of flea jumping ability (Tripet et al. 2002) proved not to be so, for at least jump length. This was also true for other morphological traits such as maximal body length and length of coxa, femur and tibia. Consequently, jumping ability is determined by factors other than linear metrics of the locomotory system. Further studies showed that differences in jumping ability between males and females or among species were well explained by sexual or interspecific differences in resting metabolic rate (Krasnov et al. 2004b).

2.4 Morphological adaptations for locomotion in the host's pelage

Flea locomotion in the host pelage differs from that of other mammal ectoparasites. For example, nycteribiids (bat flies) have a compressed dorsoventrally body and long, spider-like legs (Dick and Patterson, this volume). They are able to do fast sliding movements above the fur of the host. In contrast, a laterally compressed body, high and narrow head capsule and flexible joints of the thorax and abdomen of fleas allow them to move through the host pelage by dividing the hair during forward movement.

The flea thorax consists of three separate modified segments (pro-, meso- and metathorax), whereas an abdomen consists of 10 segments. Thoracic and abdominal segments do not possess posterior walls. Posterior margins of each segment form collars that overlie the anterior margins of the next segment. As a result, these segments are able to "squeeze" in each other. In contrast to most winged insects, the flea mesothorax and metathorax are separated and, thus, fleas lack a pterothorax which is characteristic for other holometabolous insects. It has also been suggested that the lack of a pterotorax was characteristic also for flea ancestors (Medvedev 2003a, 2005) and could be considered as a pre-adaptation to ectoparasitism on fur-covered hosts. Separation of all three thoracic segments, possession of movable between-thoracic sclerites, and highly developed phragmata (especially, a mesothoracic phragma) allow flexibility of the flea body.

In fleas, the prothorax is not reduced as it is in other Holometabola, and it is tightly connected with the head of the flea. The lower part of the prothotax (pleurosternum) is strongly elongated, exceeding at least two times the length of a pronotum. The pleurosternum protrudes anterior to the notum and envelops the posterior part of a head from beneath. As a result, a head and a prothorax together constitute a frontal complex which is movable relatively to other thoracic segments (Medvedev 2003a). Structures of this complex also include maxillar plates and fore coxae. Maxillary plates are the 1st segments of maxilla that possess highly developed collars. Due to an elongated pleurosternum, fore coxae are situated anterior to the notum. The maxillary plates are broadened in the middle, but narrowed at the bases and apices. As a result, the anterior frontal complex of a flea is shaped as a keel which divides the host's hairs or particles of the substrate of its burrow during movement.

2.5 Morphological adaptations to withstand host grooming

Fleas possess various anatomical features that allow the flea to attach themselves to the host's hairs and to resist the host's grooming (Traub 1980; but see Smit 1972). These features are represented by sclerotinized bristles such as helmets, ctenidia, spines and setae. Furthermore, these structures, as well as head shape and modifications of shape and size of spines, have been shown to correlate with particular characteristics of the host's fur (Traub 1985). For example, Egyptian spiny mouse *Acomys cahirinus* is a specific host for *Parapulex chephrenis*. The coat of *A. cahirinus* is characterized by thick but very short hairs and widely spaced long, rigid keratin spines. This coat can be groomed efficiently and fleas can be easily detached. However, the entire body of *P. chephrenis* is covered with sclerotinized bristles which facilitate flea resistance to host grooming. Strongly sclerotized, flattened spines seem to originate from modified bristles and spinelets that were located on reduced parts of the walls of the head, thorax and abdomen (Medvedev 2001a, b). For example, genal ctenidia (combs) could originate from the bristles along the posterior margin of the gena and anterior areas of the antennal fossae. Genal ctenidia are characteristic mainly for fleas of the Southern Hemisphere such as Chimaeropsyllidae (Africa), Macropsyllidae (Australia) and Craneopsyllinae (South and Central America) (Medvedev 2001a). On the contrary, no endemic family or subfamily possessing genal ctenidia is represented in faunas of the Holarctic and Oriental regions.

3 Origin

All flea species share many of their specialized characters such as morphology of head, thorax and genitalia. Consequently, most phylogenetic studies based on morphological characters (e.g., Medvedev 2003b) have suggested that Siphonaptera are monophyletic. Their monophyletic origin is supported by molecular evidence (Whiting 2002a, b).

Although fossil fleas are extremely rare, there are several findings of fleas from the Baltic and Dominican amber. Fleas from Baltic amber (*Palaeopsylla dissimilis*, *P. klebsiana*) are from the Eocene and early Oligocene (Dampf 1911; Peus 1968), whereas *Pulex larimerius* from Dominican amber dates back only to the Miocene (Lewis & Grimaldi 1997). Morphology of the fossil fleas suggests that in the early Cenozoic, when the modern mammalian orders started to appear, fleas already had their characteristic appearance possessing all the main features of morphological specialization to parasitism. Some evidence even suggests that fleas existed as early as in the Mesozoic, although their association with Mesozoic mammals is questionable. Nevertheless, a number of Cretaceous insects were sometimes considered as belonging to Siphonaptera or to their ancestral taxon (Riek 1970; Ponomarenko 1976; Rasnitsyn 1992). These considerations were based on the similarity in some morphological features between fleas and these fossil insects, although the latter did not demonstrate clear morphological evidence of adaptations to jumping. However, Smit (1978) rejected the inclusion of these Cretaceous taxa into Siphonaptera.

Traditionally, the origin of fleas is associated with the appearance of a pelage and fossorial way of life in their hosts. For example, Smit (1972) suggested that flea ancestors were scavengers that lived in hosts' burrows and that they lost their wings during adaptation to parasitism. However, Medvedev (2005) argued that flea ancestors were wingless (otherwise there would be another taxon of flying haematophages such as mosquitos). He theorized that some characteristic flea features (such as winglessness, jumping ability and laterally compressed body) are associated with their life in spatially restricted conditions (e. g, host's burrow) rather than with parasitism *per se*, whereas other flea features (such as suctorial mouthparts, keel-like frontal part of the body) are parasitism-related adaptations.

The highest diversity of fleas on various rodents suggests that the most intensive flea diversification was associated with high diversification of Rodentia, i.e. from the Eocene. Furthermore, according to the hypothesis of the geography of flea origin suggested by Medvedev (1996), they originated in a temperate zone. Geographic distribution of modern flea species supports this hypothesis. Indeed, the majority of flea species are distributed in regions of temperate and subtropical climate and predominate on the mountain landscape.

The most common phylogenetic hypothesis favours the origin of fleas from the Mecoptera-like ancestors. This hypothesis was initially proposed by Tillyard (1935) and supported by Hinton (1958) on the basis of comparison of some larval characters. Close phylogenetic relationships between Siphonaptera and Mecoptera were further supported by analyses of various morphological characters (Kristensen 1981), as well as by molecular data (Whiting 2002a, b). Both morphological and molecular evidence strongly suggests that the closest living relative to Siphonaptera is a mecopteran family Boreidae. Hastriter and Whiting (2003) suggested the following scenario. When the boreid-flea ancestor shifted from free-living in snowy, mossy habitats to living in burrows of its host, it likely lost its wings (but see Medvedev 2005) and acquired its jumping ability. Further adaptations to parasitism included lateral flattening, development of suctorial mouthparts and elaborate ctenidia and setae.

4 Geographic distribution

Fleas are distributed all around the world (including Antarctica where a bird flea *Glaciopsyllus antarcticus* has been recorded), although they undoubtedly were introduced by humans and their pets and livestock to some oceanic islands. Geographic distribution of fleas is characterized by highly unequal numbers of flea taxa among different regions (Medvedev 1996, 1998, 2000a, b). Flea fauna of the Palaearctic region appears to be the most diverse and includes 892 species (approximately 38% of the total number of known species). The number of species in the Nearctic, Afro-Tropical, and Neotropical regions are similar (299, 275 and 289 species, respectively), whereas in the Oriental and Australian regions, the number is considerably less (191 and 68 species, respectively).

In general, flea fauna of the Southern Hemisphere is characterized by species-poor families and subfamilies. These are Malacopsyllidae, Rhopalopsyllidae and Craneopsyllinae in South America, Xiphiopsyllidae and Chimaeropsyllidae in Africa, and Macropsyllidae, Lycopsyllidae and Stephanocircinae in Australia. In contrast, the largest flea families (Hystrichopsyllidae, Ceratophyllidae and Leptopsyllidae) inhabit the Northern Hemisphere. The degree of flea endemism (at least, at the generic level) varies less among regions. The proportion of endemic genera attains as high as 61% in the Afro-Tropical region, being slightly lower in the Neotropical and Australian regions (56 and 58% respectively). Endemic genera compose about 45% of all flea genera in the Palaeoarctic, 37% in the Neoarctic and 42% in the Oriental regions.

5 Life history

5.1 Reproduction

Life history of the majority of fleas is characterized by their periodic burrow/nest parasitism. In most fleas, imagoes periodically attack an endothermic host for a bloodmeal. Female fleas of some species oviposit while on the host and the eggs drop off into the nest or burrow (e.g. *Pectinoctenus pavlovskii*, *Leptopsylla segnis*). Other species mate and oviposit both on-host and off-host (e.g., *Xenopsylla cheopis*). Egg production varies greatly among flea species. For example, female *Xenopsylla dipodilli* usually oviposit after one or two eggs have matured in its oviducts, whereas female *Parapulex chephrenis* oviposit after up to eight eggs have accumulated in the oviducts (Krasnov et al. 2002a). Furthermore, egg production varies within flea species, depending on the host species that the flea exploits (Krasnov et al. 2004c). Two sharpest examples of the influence of a host on flea reproduction are represented by the rabbit fleas *Spilopsyllus cuniculi* and *Cediopsylla simplex*. In these fleas, reproduction is related to the reproductive and, consequently, hormonal cycle of their hosts, *Oryctolagus cuniculus* and *Sylvilagus floridanus*, respectively. Development of the reproductive system, mating, egg maturation and oviposition in these fleas have been shown to be triggered by sex hormones of the pregnant doe rabbits and nestlings (Rothschild and Ford 1966, 1969, 1972, 1973; Sobey et al. 1974).

Under favourable conditions, larvae hatch within 2-21 days. Flea larvae are not obligate parasites on mammalian or avian hosts (except for species *Uropsylla tasmanica*). They usually live in the burrow or nest of a host and feed on organic matter as well as on excrements of imago fleas. Cannibalism in pre-imaginal fleas is also a common occurrence (Lawrence and Foil 2002). In particular, third instar larvae readily cannibalize naked pupae. The third instar larva, on completion of feeding, expels all of its gut content, spins a silken cocoon, and camouflages it by adhering particles of the nest substrate. Thus, in nearly all cases, larval and pupal development is entirely off-host. One exception is that of *Glaciopsyllus antarcticus* (Bell et al. 1988), whose pupae are attached to the down of its avian host *Fulmarus glacialoides*.

5.2 Effect of environmental factors

As a result of the free-living life of flea pre-imago and only periodic contact of imago with its host, fleas are adapted not only to the body of a host, but also to the microclimatic conditions of its burrow or nest. Indeed, environmental factors have been shown to affect the survival of pre-imaginal fleas (e.g., Bacot 1914; Bacot and Martin 1924; Uvarov 1931; Edney 1945, 1947a, b; Krasnov et al. 2001a, b; 2002b). For example, air temperature influences the developmental time and emergence of the Oriental rat flea, *Xenopsylla cheopis*, and the cat flea, *Ctenocephalides felis* (Margalit and Shulov 1972; Silverman and Rust 1983; Metzger and Rust 1997). Relative humidity is another factor affecting fleas. The quiescent adult within the cocoon has a lower respiratory demand than the emerged adult, and its survival is considerably longer under low humidity conditions (Silverman and Rust 1985; Metzger and Rust 1997). It has been speculated that prolonged survival of quiescent adults within the cocoon is due, in part, to a reduction in respiratory water loss because less time is spent with the spiracles open (Silverman and Rust 1985). Larvae, in contrast, cannot close their spiracles, and thus are extremely sensitive to low humidity (Krasnov et al. 2001a). However, larvae and pupae of fleas did not survive at air temperatures >35°C even at optimal relative humidity (Silverman et al. 1981; Silverman and Rust 1983). Adult imagoes spend much time with their spiracles open and, thus, are also strongly affected by relative humidity (Silverman and Rust 1985).

5.3 Seasonality

Survival and reproduction of fleas are dependent on a combination of factors including favourable climatic conditions for development of the immature stages and for adults to survive unpredictable and sometimes lengthy periods without a blood meal. This dependence results in seasonal changes of life history parameters of fleas (abundance, reproduction rate, pattern of parasitizing etc.) (Marshall 1981). Therefore, the annual cycle of a particular flea species in a particular locality is expected to correspond with seasonal climatic fluctuation in this locality. Studies on the annual life cycles of fleas that were conducted in areas with pronounced seasonality in temperature and rainfall demonstrated seasonal changes in the abundance and reproductive patterns of fleas (e.g., Lindsay and Galloway 1998). However, the results of studies in areas with less pronounced climatic seasonality and in urban areas showed much less seasonality in the life history parameters of fleas.

Darskaya (1970) and Vatschenok (1988) proposed a classification of fleas based on the pattern of their annual cycles as follows: (1) adult fleas are active and reproduce all year round; (2) adult fleas are active all year round, but reproduce in the warm season only; (3) adult fleas are active and reproduce in warm season only; (4) adult fleas are active and reproduce most of the year except for the hottest and driest periods when fleas survive in cocoons and (5) adult fleas are active and reproduce in the cold season only. This classification reflects ecological differences between flea species and has little to do with their taxonomic relationships. For example, it appears that the annual cycle of a flea depends primarily on the ecological properties of its host. Indeed, *X. cheopis* has been shown to reproduce all year round (Seal and Bhatacharji 1961) and, thus, its annual cycle corresponds to type 1 of the above classification. Other studied *Xenopsylla* fleas demonstrated seasonal breaks in reproduction and, thus, their cycles are similar to type 2 (Vatschenok 1988 and references therein). This difference can be easily explained in that *X. cheopis* parasitizes mainly commensal rodents and, thus, climatic fluctuations of its environment are much less pronounced than those of congeneric species that parasitize wild rodents. Another example showing that this classification is not related to taxonomy, is that the annual cycle of type 5 is characteristic for fleas of different genera and families (*Coptopsylla*, *Stenoponia*, *Paradoxopsyllus*, *Rhadinopsylla*, *Wagnerina* etc.).

6 Flea-host associations

Medvedev and Lobanov (1999) and Medvedev (2005) compiled a comprehensive database of species and subspecies of fleas on their hosts. Analysis of flea-host associations demonstrated that 70% of all flea-mammal associations involve rodents (Medvedev 2002). In addition, rodents compose 82% of all specific and/or principal hosts for fleas. Primary association of fleas with rodents is observed in all parts of the world (except Australia, where fleas are harboured mainly by marsupials), although fleas are also, albeit weaker, associated with hosts belonging to other mammalian orders. In the Palaearctic region, fleas exploit mainly voles (Arvicolinae), gerbils (Gerbillinae) and hamsters (Cricetinae), whereas in the Nearctic region, principal flea hosts are voles, New World rats and mice (Sigmodontinae), pocket gophers (Geomyidae) and kangaroo rats and pocket mice (Heteromyidae). Mammals from other orders that characteristically harbour fleas are pikas and hares (Lagomorpha) and various Insectivora. Flea hosts in the Neotropics are sigmodontine and caviomorph rodents (Caviidae, Chinchilliade, Capromyidae, Octodontidae etc.) as well as representatives of two marsupial orders, namely American opossums (Didelphimorphia) and shrew opossums (Paucituberculata). In the Afro-tropics, fleas parasitize mainly rats and mice (Murinae), bamboo rats (Rhizomyidae) and African mole-rats (Bathyergidae). Other flea hosts in this region are hyraxes (Hyracoidea) and elephant shrews (Macroscelidea). Fleas from the Oriental region (including Wallacia and Southern Pacific islands) infest mainly murines and squirrels (Sciuridae) as well as a variety of marsupial orders (Dasyuromorphia and some Diprotodontia). Finally, in the Australian region, fleas parasitize murine rodents and various marsupials (Peramelemorphia and Diprotodontia).

Fleas vary greatly in the degree of their host specificity from being highly host specific to being highly host-opportunistic. Traditional classification of ectoparasites by the degree of their host specificity based on the number of host species and their taxonomic positions distinguishes among monoxenous (only one host species), oligoxenous (two or more host species belonging to the same genus), meso- or pleioxenous (two or more host genera belonging to the same family), polyxenous (two or more host families belonging to the same order) and euryxenous (2 or more host orders or even classes) fleas (Marshall 1981; Medvedev 2000a, b). The number of flea species belonging to each of these categories differs among studies. For example, Medvedev (2002) considered 563 flea species (about 34% of all known species) as monoxenous, 78 species as oligoxenous, 234 species as pleioxenous, 259 species as polyxenous and 609 species as euryxenous. In contrast, Traub (1985) defined 233 species as monoxenous (="ultraspecific") and 183 species as oligoxenous (="generispecific"). It should be noted, however, that literature records can be misleading. For example, many fleas are known only from a single finding. Consequently, the number of monoxenous flea species is likely overestimated. However, lists of recorded host-flea associations are often reported without comments, so the relative importance of a particular host for a particular flea cannot be determined. In addition, these lists also include accidental findings and misidentification. Therefore, the number of oligo-, pleio-, poly- and euryxenous species could be also easily overestimated.

Host-specific (monoxenous and oligoxenous) fleas usually exploit ecologically isolated hosts such as fossorial rodents. For example, *Ctenophthalmus spalacis* is a specific parasite of *Spalax*, *Ctenophthalmus inornatus* of *Prometheomys schaposchnikowi*, and *Xenopsylla magdalinae* of *Ellobius*. Host specificity in these fleas may result not only from their adaptations to some features of host body (blood biochemistry, fur structure) but also by their adaptations to specific microclimatic conditions in the host burrow. However, not every species, genus or even family of fossorial rodents harbours host specific fleas (Medvedev 2002).

Host-opportunistic fleas are usually habitat-dependent rather than host-dependent (Krasnov et al. 1997). They occupy a certain habitat, where they exploit a number of ecologically similar host species.

The proportions of host specific and host opportunistic species are unequal among flea families (Medvedev 2002). For example, the highest proportion of monoxenous species has been found among Hystrichopsyllidae (30% of total number of species), whereas the lowest proportion (less than 1% of species) among Coptopsyllidae, Xiphiopsyllidae, Ancistropsyllidae and Lycopsyllidae.

Proportion of oligoxenous species varies among flea families from 3% to 9%, pleioxenous from 11% to 25%, polyxenous from 2% to 15%, and euryxenous from 39% to 46%. In addition, there is no euryxenous species among Ischnopsyllidae (but 31% of the family is composed of polyxenous species) and there are only a few such species among Chimaeropsyllidae.

Finally, specific associations can be distinguished between some flea families, tribes and genera and mammalian orders. For example, fleas of the family Ischnopsyllidae are associated with bats (Chiroptera), Chimaeropsyllidae with elephant shrews (Macroscelidea), and Malacopsyllidae with edentates (Xenarthra). Insectivore hosts are parasitized mainly by fleas from the tribe Doratopsyllini, whereas lagomorphs by fleas from the tribe Spillopsyllini. The genus *Procaviopsylla* is specific for hyraxes (Hyracoidea) and *Neotunga* for pangolins (Pholidota).

7 Concluding remarks

Fleas are strongly specialized to periodic ectoparasitism on small mammals and demonstrate a variety of life histories and associations with their hosts. This makes fleas a very convenient model for testing hypotheses related to evolutionary ecology of parasitism. Some of these studies will be reviewed in further chapters of this book.

References

Abouheif E, Fairbairn DJ (1997) A comparative analysis of allometry for sexual size dimorphism: Assessing Rensch's rule. Amer Nat 149:540–562

Bacot AW (1914) A study of bionomics of the common rat fleas and other species associated with human habitations, with special reference to the influence of temperature and humidity at various periods of the life history of the insect. J Hyg 13:447–654

Bacot AW, Martin CJ (1924) The respective influences of temperature and moisture upon the survival of the rat flea (*Xenopsylla cheopis*) away from its host. J Hyg 23:98–105

Bell PJ, Burton HR, van Franeker JA (1988) Aspects of the biology of *Glaciopsyllus antarcticus* (Siphonaptera: Ceratophyllidae) during the breeding season of a host (*Fulmarus glacialoides*). Polar Biol 8:403–410

Bennet-Clark HC, Lucey EC (1967) The jump of the flea: A study of the energetics and a model of the mechanism. J Exp Biol 47:59–67

Colwell RK (2000) Rensch's rule crosses the line: Convergent allometry of sexual size dimorphism in hummingbirds and flower mites. Amer Nat 156:495–510

Dampf A (1911) *Palaeopsylla klebsiana* n. sp., eine fossiler Floh aus dem baltischen Bernstein. *Schrift.* Physik-Ökonom Gesel Königsberg 51:248–259

Darskaya NF (1970) Ecological comparisons of some fleas of the USSR fauna (in Russian). Zool Zh 49:729–745

Edney EB (1945) Laboratory studies on the bionomics of the rat fleas, *Xenopsylla brasiliensis* Baker and *X. cheopis* Roths. I. Certain effects of light, temperature and humidity on the rate of development and on adult longevity. Bull Entomol Res 35:399–416

Edney EB (1947a) Laboratory studies on the bionomics of the rat fleas, *Xenopsylla brasiliensis* Baker and *X. cheopis* Roths. II. Water relations during the cocoon period. Bull Entomol Res 38:263–280
Edney EB (1947b) Laboratory studies on the bionomics of the rat fleas, *Xenopsylla brasiliensis* Baker and *X. cheopis* Roths. III. Further factors affecting adult longevity. Bull Entomol Res 38:389–404
Fairbairn DJ (1997) Allometry for sexual size dimorphism: Pattern and process in the coevolution of body size in males and females. Ann Rev Ecol Syst 28:659–687
Hastriter MW, Whiting MF (2003) Siphonaptera (fleas). In: Resh VH, Carde R (eds) Encyclopedia of insects. Acad Press/Elsevier, Orlando, pp 1039–1045
Hinton HE (1958) The phylogeny of the Panorpoid orders. Ann Rev Entomol 3:181–206
Ioff IG (1941) Ecology of fleas in relevance to their medical importance (in Russian). Ordzhonikidze Regional Publisher, Pyatygorsk
Krasnov BR, Shenbrot GI, Medvedev SG, Vatschenok VS, Khokhlova IS (1997) Host-habitat relations as an important determinant of spatial distribution of flea assemblages (Siphonaptera) on rodents in the Negev Desert. Parasitology 114:159–173
Krasnov BR, Khokhlova IS, Fielden LJ, Burdelova NV (2001a) Development rates of two *Xenopsylla* flea species in relation to air temperature and humidity. Med Vet Entomol 15:249–258
Krasnov BR, Khokhlova IS, Fielden LJ, Burdelova NV (2001b) The effect of temperature and humidity on the survival of pre-imaginal stages of two flea species (Siphonaptera: Pulicidae). J Med Entomol 38:629–637
Krasnov BR, Khokhlova IS, Oguzoglu I, Burdelova NV (2002a) Host discrimination by two desert fleas using an odour cue. Anim Behav 64:33–40
Krasnov BR, Khokhlova IS, Fielden LJ, Burdelova NV (2002b) Time to survival under starvation in two flea species (Siphonaptera: Pulicidae) at different air temperatures and relative humidities. J Vector Ecol 27:70–81
Krasnov BR, Burdelov SA, Khokhlova IS, Burdelova NV (2003) Sexual size dimorphism, morphological traits and jump performance in seven species of desert fleas (Siphonaptera). J Zool Lond 261:181–189
Krasnov BR, Shenbrot GI, Khokhlova IS (2004a) Sampling fleas: The reliability of host infestation data. Med Vet Entomol 18:232–240
Krasnov BR, Khokhlova IS, Burdelov SA, Fielden LJ (2004b) Metabolic rate and jumping performance in seven species of desert fleas. J Insect Physiol 50:149–156
Krasnov BR, Khokhlova IS, Burdelova NV, Mirzoyan NS, Degen AA (2004c) Fitness consequences of density-dependent host selection in ectoparasites: Testing reproductive patterns predicted by isodar theory in fleas parasitizing rodents. J Anim Ecol 73:815–820
Kristensen NP (1981) Phylogeny of insect orders. Ann Rev Entomol 26:135 –157
Lawrence W, Foil LD (2002) The effect of diet upon pupal development and cocoon formation by the cat flea (Siphonaptera: Pulicidae). J Vector Ecol 27:39–43

Lewis RE, Grimaldi D (1997) A pulicid flea in Miocene amber from the Dominican Republic (Insecta: Siphonaptera: Pulicidae). Amer Mus Nov 3205:1–9

Lindsay LR, Galloway TD (1998) Reproductive status of four species of fleas (Insecta: Siphonaptera) on Richardson's ground squirrel (Rodentia: Sciuridae) in Manitoba, Canada. J Med Entomol 35:423–430

Margalit Y, Shulov AS (1972) Effect of temperature on development of prepupa and pupa of the rat flea, *Xenopsylla cheopis* Rothschild. J Med Entomol 9:117–125

Marshall AG (1981) The ecology of ectoparasitic insects. Acad Press, London

Medvedev SG (1996) Geographical distribution of families of fleas (Siphonaptera). Entomol Rev 76:978–992

Medvedev SG (1997a) Host-parasite relations in fleas (Siphonaptera). I. Entomol Rev 77:318–337

Medvedev SG (1997b) Host-parasite relations in fleas (Siphonaptera). II. Entomol Rev 77:511–521

Medvedev SG (1998) Fauna and host-parasite relations of fleas (Siphonaptera) in the Palaearctic. Entomol Rev 78:292–308

Medvedev SG (2000a) Fauna and host-parasite associations of fleas (Siphonaptera) in different zoogeographical regions of the World. I. Entomol Rev 80:409–435

Medvedev SG (2000b) Fauna and host-parasite associations of fleas (Siphonaptera) in different zoogeographical regions of the World. II. Entomol Rev 80:640–655

Medvedev SG (2001a) On the structure of cephalic ctenidia in fleas (Siphonaptera). Entomol Rev 81:1117–1135

Medvedev SG (2001b) Peculiarities of thoracis and abdominal combs of fleas (Siphonaptera) (in Russian). Parazitologiya 35:291–306

Medvedev SG (2002) Specific features of the distribution and host associations of fleas (Siphonaptera). Entomol Rev 82:1165–1177

Medvedev SG (2003a) Morphological adaptations of fleas (Siphonaptera) to parasitism. I. Entomol Rev 83:1059–1080

Medvedev SG (2003b) Morphological adaptations of fleas (Siphonaptera) to parasitism. II. Entomol Rev 83:1114–1129

Medvedev SG (2005) An attempted system analysis of the evolution of the order of fleas (Siphonaptera) (in Russian). Lectures in Memoriam N. A. Kholodkovsky, No. 57. Russian Entomol Soc and Zool Inst Russian Acad Sci, St Petersburg

Medvedev SG, Lobanov AL (1999) Information-analytical system of the World fauna of fleas (Siphonaptera): Results and prospects. Entomol Rev 79:654–665

Medvedev SG, Lobanov AL, Lyanguzov IA (2005) World database of fleas (Nov 2004 version). In: Bisby FA, Ruggiero MA, Wilson KL, Cachuela-Palacio M, Kimani SW, Roskov YR, Soulier-Perkins A, van Hertum J (eds) Species 2000 and ITIS catalogue of life: 2005 Annual checklist. Species 2000, Reading (CD-ROM)

Metzger ME, Rust MK (1997) Effect of temperature on cat flea (Siphonaptera: Pulicidae) development and overwintering. J Med Entomol 34:173–178
Nelzina EN, Danilova GM, Tchernova GI (1963) Abundance and spatial distribution of micropopulations of hematophagous arthropods in microhabitats of *Citellus pygmaeus* (in Russian). Meditzinskaya Parazitologiya i Parazitarnye Bolezni [Medical Parasitology and Parasitic Diseases] 32:45–54
Novokrestchenova NS (1960) Data on the ecology of fleas of *Citellus pygmaeus* in relation to their epizootological importance (in Russian). Trudy Protivochumnogo Instituta "Microb" [Proceedings of the Anti-Plague "Microb" Institute] 4:444–456
Peus F (1968) Über die beiden Bernstein-Flöhe (Insecta, Siphonaptera). Palaeontol Z 42:62–72
Ponomarenko AG (1976) The new insect from the Cretaceous of Transbaicalia was a probable parasite of pterosaurs (in Russian). Paleontol Zh 3:102–106
Rasnitsyn AP (1992) *Strashila incredibilis*, a new enigmatic mecopteroid insect with possible siphonapteran affinities from upper Jurassic of Siberia. Psyche 99:323–333
Reiss MJ (1986) Sexual dimorphism in body size: Are larger species more dimorphic? J Theor Biol 121:163–72
Rensch B (1960) Evolution above the species level. Columbia Univ Press, New York
Riek EF (1970) Lower Cretaceous fleas. Nature 227:746–747
Rothschild M, Ford R (1966) Hormones of the vertebrate host controlling ovarian regression and copulation of the rabbit flea. Nature 211:261–266
Rothschild M, Ford R (1969) Does a pheromone-like factor from the nestling rabbit stimulate impregnation and maturation in the rabbit flea? Nature 221:1169–1170
Rothschild M, Ford R (1972) Breeding cycle of the flea *Cediopsylla simplex* is controlled by breeding cycle of host. Science 178:625–626
Rothschild M, Ford R (1973) Factors influencing the breeding of the rabbit flea (*Spilopsyllus cuniculi*): A spring-time accelerator and a kairomone in nestling rabbit urine (with notes on *Cediopsylla simplex*, another "hormone bound" species). J Zool Lond 170:87–137
Rothschild M, Schlein J (1975) The jumping mechanism of *Xenospylla cheopis*. I. Exoskeletal structures and musculature. Phil Trans R Soc Lond B 271:457–489
Rothschild M, Schlein J, Parker K, Neville C, Sternberg S (1973) The flying leap of the flea. Sci Amer 229:92–100
Rothschild M, Schlein J, Parker K, Neville C, Sternberg S (1975) The jumping mechanism of *Xenospylla cheopis*. III. Execution of the jump and activity. Phil Trans R Soc Lond B 271:499–515
Seal SC, Bhattacharji LM (1961) Epidemiological studies of plague in Calcutta, Part 1. Bionomics of two species of rat fleas and distribution, densities and resistance of rodents in relation to the epidemiology of plague in Calcutta. Indian J Med Res 49:974–1007

Silverman J, Rust MK (1983) Some abiotic factors affecting the survival of the cat flea, *Ctenocephalides felis* (Siphonaptera: Pulicidae). Environ Entomol 12:490–495

Silverman J, Rust MK (1985) Extended longevity of the pre-emerged adult of the cat flea (Siphonaptera: Pulicidae) and factors stimulating emergence from the pupal cocoon. Ann Entomol Soc Am 78:763–768

Silverman J, Rust MK, Reierson DA (1981) Influence of temperature and humidity on survival and development of the cat flea, *Ctenocephalides felis* (Siphonaptera: Pulicidae). J Med Entomol 18:78–83

Smit FGAM (1972) On some adaptive structures in Siphonaptera. Folia Parasitol 19:5–17

Smit FGAM (1978) Fossil "fleas". Flea News 14:1–2

Snodgrass RE (1946) The skeletal anatomy of fleas (Siphonaptera). Smithsonian Misc Coll 10:1–89

Sobey WR, Menzies W, Conolly D (1974) Myxomatosis: Some observations on breeding the European rabbit flea *Spilopsyllus cuniculi* (Dale) in an animal house. J Hyg 71:453–465

Tillyard RJ (1935) The evolution of the scorpion-flies and their derivatives (order Mecoptera). Ann Entomol Soc Am 28:1–45

Traub R (1980) Some adaptive modifications in fleas. In: Traub R, Starcke H (eds) Fleas. Proceedings of the international conference on fleas. Ashton Wold/Peterborough/UK/21–25 June 1977. Balkema, Rotterdam, pp 33–68

Traub R (1985) Coevolution of fleas and mammals. In: Kim KC (ed) Coevolution of parasitic arthropods and mammals. John Wiley, New York, pp 295–437

Tripet F, Christe P, Møller AP (2002) The importance of host spatial distribution for parasite specialization and speciaton: A comparative study of bird fleas (Siphonaptera: Ceratophyllidae). J Anim Ecol 71:735–748

Uvarov BP (1931) Insects and climate. Trans R Entomol Soc Lond 79:1–247

Vatschenok VS (1988) Fleas – vectors of pathogens causing diseases in humans and animals (in Russian). Nauka, Leningrad

Whiting MF (2002a) Phylogeny of the holometabolous insect orders: Molecular evidence. Zool Scripta 31:3–15

Whiting MF (2002b) Mecoptera is paraphyletic: Multiple genes and phylogeny of Mecoptera and Siphonaptera. Zool Scripta 31:93–104

Zhovtyi IF (1963) Some disputable questions of ecology of rodent fleas in relation to their medical importance (in Russian). Doklady Irkutskogo Protivotchumnogo Institute [Archives of the Irkutsk Anti-Plague Institute of Siberia and Far East] 6:96–104

11 Bat flies: Obligate ectoparasites of bats

Carl W. Dick and Bruce D. Patterson

1 Introductory remarks

Bat flies (Diptera: Hippoboscoidea) are highly specialized ectoparasites and only associate with bats (Mammalia: Chiroptera). They live in the fur and on the wing membranes where they feed on host blood. Bat flies are nominally divided into two cosmopolitan families, Streblidae and Nycteribiidae, but recent phylogenetic studies suggest these are not natural groups (Dittmar et al. 2006). Nycteribiids (275 species) are more speciose in the Eastern Hemisphere, whereas the streblids (227 species) are richer in the Western. Generally, both families are most diverse in the tropics, less diverse in subtropics, and rather impoverished in temperate regions. However, this latitudinal richness gradient is more pronounced in the Western Hemisphere. *Mystacinobia zelandica* (Mystacinobiidae) has been considered a "bat fly" (Holloway 1976). This fly is a roost associate with and phoretic on the endemic New Zealand bat *Mystacina tuberculata*. Unlike members of the Nycteribiidae and Streblidae, *M. zelandica* feeds on guano, not host blood. Molecular analysis places *M. tuberculata* within the Oestroidea (Gleeson et al. 2000). This chapter summarizes current understanding of the taxonomy, life history, and breeding biology of flies allocated to Streblidae and Nycteribiidae, and offers overviews of morphology, behavior, specificity, ecology, and cospeciation in the context of the parasite-host association.

2 Taxonomy

Together with the families Hippoboscidae (bird flies, ked flies) and Glossinidae (tsetse flies), bat flies belong to the dipteran superfamily Hippoboscoidea. This group of "pupiparous" flies represents one of the most derived clades of Diptera (Yeates and Wiegmann 1999), with highly modified development and peculiar life histories. Although phylogenetic analyses of the group are still preliminary, the two bat fly families appear to be monophyletic, sister to either bird flies or tsetse flies. Recognized families of bat flies are small relative to other dipteran families: Streblidae are worldwide in distribution and include 5 subfamilies, 32 genera, and 227 described species. Whereas streblids are found in both Eastern and Western Hemispheres, they largely inhabit tropical and subtropical regions and are more speciose in the West (156 species) than in the East (71 species; Whitaker et al., in press). No species, genus or even subfamily is distributed in both hemispheres. Eastern forms include the Nycteriboscinae (4 genera, 50 species) and the Ascodipterinae (2 genera, 21 species). Western forms include the Nycterophiliinae (2 genera, 6 species), Streblinae (4 genera, 35 species), and Trichobiinae (19 genera, 115 species). All five species of the streblid genus *Megastrebla* (Nycteriboscinae) are associated with pteropodid bats (exclusively Eastern), but the remaining species in the family are associated with the Microchiroptera. In the American tropics, streblids are by far the most diverse upon the bat family Phyllostomidae.

The Nycteribiidae includes 3 subfamilies, 12 genera, and 275 described species. Although species are found in tropical and subtropical regions worldwide, they are richer in the Eastern Hemisphere (222 species versus 53 species in the Western). No nycteribiid species is found in both Eastern and Western Hemispheres. Two subfamilies, Archinycteribiinae (1 genus, 3 species), and Cyclopodiinae (4 genera, 62 species) are exclusively Eastern Hemispheric in distribution, where they associate with Pteropodidae. The Nycteribiinae (7 genera, 210 species) are cosmopolitan in distribution and mainly associated with the families Vespertilionidae and Rhinolophidae. The most species-rich genus, *Basilia*, is also cosmopolitan and comprises 122 nominal species. *Basilia* species are mainly found in association with the Vespertilionidae, but also with the Phyllostomidae. All four species of *Hershkovitzia*, the only genus restricted to the Americas, parasitize bats of the endemic family Thyropteridae.

3 Life history and breeding biology

Little is known regarding life history and reproductive biology of bat flies. Generalizations are based on limited studies of a few species. Generally, all bat flies reproduce via viviparous puparity (Hagan 1951), in which eggs are fertilized internally and all larval stages develop within the female, nourished by intrauterine “milk” glands. Larvae moult twice inside the female, and gravid females deposit a single, terminal (3rd-instar) larva on the roosting substrate. Once deposited, the larva (referred to as a prepupa) immediately forms a puparium. Following a pupal stage that lasts 3-4 weeks (Ching and Marshall 1968), an adult fly emerges and must locate and colonize a host. Prepupal deposition directly on the host's body has been reported, but the pupae were distorted in shape and most failed to develop into an adult fly. Left on the bats, most pupae were groomed off by the host (Ching and Marshall 1968). In general, the life-history strategy of bat flies reflects their obligate association with bats: vulnerable immature stages remain coupled with the host inside the female fly.

Peterson and Wenzel (1987) suggested that the life cycle of nycteribiids is rather uniform. The life cycle of *Basilia hispida* has been described in some detail (Marshall 1970). Flies reached sexual maturity 5-6 days after emergence from the puparium. Males usually copulated with females immediately following prepupal deposition, but sometimes with newly emerged females. A single copulation was sufficient to produce several offspring, suggesting that females store sperm. At 9-day intervals, mature females deposited prepupae on roost substrate, pressed into place with their abdomens. Prepupal deposition occurred between 0900 and 1800 hours, while bats were in their roosts, stimulated largely by elevated temperature. From 25 to 46 days later, with host bats present or absent from the roost, respectively, wingless, teneral adults emerged and randomly walked about until encountering a host. Upon colonizing a host, flies began feeding within 20 min. *Basilia hispida* died within 5-25 hours of being removed from a host, with new mother flies dying sooner and teneral flies dying later. The total life-span of *B. hispida* averaged 136 and 195 days for males and females, respectively, with 5 days pre-partum, 9 days in the larval stages, 25 days in puparium, and 97 days (males) or 156 days (females) in the adult stage (Marshall 1970).

The breeding biology of *Eucampsipoda sundaica* was described by Ching and Marshall (1968). Most features of the life cycle resembled that of *B. hispida*, but during mating, males attached to females for up to 1/2 hour, during which time the female remained fully ambulatory. The interval between successive prepupal depositions was three times faster in *E. sundiaca* than *B. hispida* (3 versus 9 days, respectively). It was noted that some *E. sundiaca* prepupae were deposited directly on the host, but these pupae were distorted in shape and easily removed by the bat (Ching and Marshall 1968).

Life histories of streblid species likewise are poorly known and described for only a few species. The most detailed is Overal's (1980) study of the life cycle of *Megistopoda aranea* from Panama. In this species, prepupae were deposited in the roost, usually near bats. Following about 23 days, the adult emerged from the puparium and located a host for a blood meal before mating. However, females were observed mating only minutes after depositing the prepupa. The time interval between successive depositions of prepupae was about 10 days. Fritz (1983) studied *Trichobius joblingi*, *Speiseria ambigua*, and *Strebla guajiro* in Costa Rica. These species also deposited their prepupae on roost substrates and away from the host, and the pupal stage lasted about 17, 20, and 19 days for these fly species, respectively. A few studies have been conducted on temperate (North American) streblid species. Generally, life histories of temperate species resemble those in tropical regions. However, bat flies remain physically and reproductively active on hibernating bats whose body temperatures ranged from 2-6 degrees C (Ross 1960; Reisen et al. 1976; Caire and Hornuff 1986).

Females of the streblid *Ascodipteron* embed in the skin of their hosts and become endoparasitic (Maa 1965), whereas males (Fig. 1a) are external parasites. Female *Ascodipteron* are the only exception to the ectoparasitic nature of bat flies. Basic life-history details are available for only two species. In *A. namrui*, duration of the pupal stage ranged from 24-35 days, whereas the interval between successive depositions of prepupae ranged from 2-8 days at 26-30° C (Maa 1965). In another species, *A. rhinopomatos*, the pupal duration was nearly equal to *A. namrui* (Theodor and Moscona 1954), but the interval between prepupal depositions was longer, ranging from 6-13 days (Maa 1965).

4 Morphology and behavior

Nycteribiids are completely wingless and spider-like in appearance (Fig. 1b). Although species vary greatly in size (1.5-5.0 mm), their overall aspect is rather similar across the species. Their flight muscles are atrophied, which in turn reduces the overall bulk of the thorax. Their legs and small head all protrude from the dorsal thoracic surface, and the insects are somewhat dorsoventrally flattened. All species possess a head that is folded back against the thorax when at rest. When feeding, the head rotates nearly 180 forward and downward (Peterson and Wenzel 1987). *Basilia hispida* and three species of Central American *Basilia* have been observed only on furred regions of their hosts (Marshall 1971; ter Hofstede et al. 2004). Our observations of other nycteribiids, including *Hershkovitzia* spp., support that observation. Like fleas, nycteribiids possess several ctenidea or combs. The ctenidia are thought to facilitate host attachment, preventing the animal from being brushed backwards from the fur (Traub 1972; Amin 1974), not for protection from abrasive action of host hair

(Marshall 1980a, 1981). Nycteribiids generally move equally well in any direction, and their movements may be very fast when agitated. Such frenetic mobility may allow them to evade host grooming, inferred to be the greatest cause of mortality in adult bat flies (Marshall 1981). When feeding, nycteribiids thrust their bodies downward into the fur. Their mouthparts contact the host's skin and the tip of their abdomen is generally visible at this time.

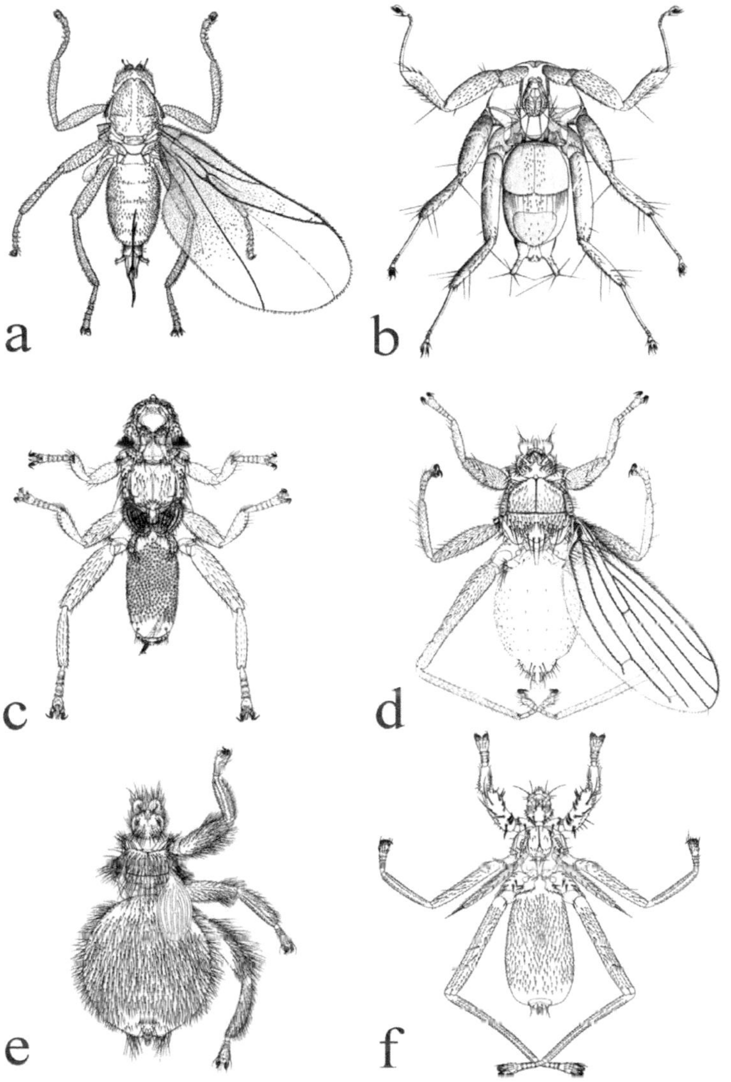

Fig. 1. Representative genera of bat flies. (a) *Ascodipteron africanum* (Streblidae: Ascodipterinae), male, dorsal view, modified from Jobling (1940); (b) *Stylidia biarticulata* (Nycteribiidae: Nycteribiinae), female, dorsal view, modified from Theodor (1967); (c) *Metelasmus pseudopterus* (Streblidae: Streblinae), male, dorsal view, modified from Jobling (1936); (d) *Speiseria ambigua* (Streblidae: Trichobiinae), female, dorsal view, modified from Jobling (1939); (e) *Anatrichobius scorzai* (Streblidae: Trichobiinae), female, dorsal view, from Wenzel et al. (1966); (f) *Neotrichobius stenopterus* (Streblidae: Trichobiinae), female, dorsal view, from Wenzel et al. (1966) (reprinted with permissions from the Field Museum of Natural History, Cambridge University Press and Prof. Yosef Shlein)

The Streblidae also vary greatly in size, with total length generally 1.5-2.5mm, but ranging from 0.73mm (e.g., *Mastoptera minuta*, the smallest bat fly) to 5.50mm (e.g., *Joblingia schmidti*). In contrast to the conservative body plan of nycteribiids (Fig. 1b), streblids possess radically different body plans from strongly laterally compressed (e.g., Nycterophiliinae) to dorsoventrally flattened (e.g., Streblinae; Fig. 1c) to uncompressed (e.g., most Trichobiinae; Figs. 1d-1f) (Wenzel et al. 1966; Dick and Miller, in press). The nycterophiliine species are laterally compressed and resemble fleas in overall morphology and in their movements through hair. The strong and rapid "swimming" movement of these insects makes them especially difficult to capture alive on the host (Dick, personal observations). Other differences among streblid species include extremely elongated legs in some genera of trichobiines (Figs. 1d, f) and a well-developed ctenidium (Fig. 1c) in all species of streblines. Observations of living flies reveal that the long-legged species run across the top of the host's fur. These species are accordingly the most conspicuous parasites when bats are handled. Behaviors of species (e.g. *Strebla* spp.) that possess a ctenidium along the posteroventral margin of the head (Fig. 1c) have been observed in glass vials: when these flies elevate the head, a large gap is formed between the ctenidium and the anteroventral margin of the thorax, which can be clamped closed when the fly pronates the head downward (Dick, personal observations). When on a host, such a motion would secure the fly to the host by grasping host hair within this ctenidiothoracic gap. In all the bat flies we have handled, the most important structures for host attachment appear to be tarsal claws. When streblids are collected alive from the host, nearly always their final and strongest resistance to capture involves grasping hairs or wing membrane with flexed tarsal claws. Attempts to aspirate live flies from hosts failed; flies tightly grasped the host hairs with their tarsal claws (Dick, personal observations).

Although all nycteribiids are wingless (Fig. 1b), most (220 species or 97%) streblids possess wings, but not all these possess functional, macropterous wings (Figs. 1a, d). Of the winged species, 24 species (10.9%) possess non-functional vestigial wings, 7 species the stenopterous form (Fig. 1f) and 17 the brachypterous form (Figs. 1c, e) (Dick and Miller, in press). The remaining 6 species (*Paradyschiria* and *Phalconomus puliciformis*) are apterous, with *Paradyschiria* spp. lacking halters as well. Typically even the fully winged forms are rather weak flyers, but species vary in their proclivity to fly when disturbed. Species of Ascopidterinae are unique among bat flies in that, upon attachment to a host bat, alate females immediately shed their wings, halters, and all leg segments beyond the coxae (Hastriter et al. 2006). The thorax and mouthparts invaginate within the abdomen, and most of the insect is enveloped by host dermal tissue. Only the membranous terminal segments remain exposed. In these flies, attachment occurs on the host's wings, behind the ears, or in the urogenital areas. Thus, females of these species are effectively endoparasites (Hastriter et al., in press).

Adult bat flies typically reside with the host. Females leave the host only to deposit their prepupa, usually on the walls of the roost. Only *Ascodipteron* species are known to deposit their prepupa on the ground, similar to Hippoboscidae and Glossinidae (Maa and Peterson 1987; Jordan 1993). Males were thought not to leave the host at any time. However, in a Puerto Rican cave roost of *Artibeus jamaicensis*, Dick observed both pupae and adults of *Trichobius intermedius* (Streblidae) on the walls of the cave, even after the bats had departed for the night. Adult flies were observed walking among both pupae and pupal exuviae. Of 23 adult flies collected from the roost wall in vicinity of the pupae, 11 were males and 12 females (Dick, unpublished data), so that both males and females leave the host for at least brief periods.

5 Host associations and ecology

As with most obligate parasites, the lives of bat flies are tightly coupled with those of their hosts. Consequently, geographical distributions of bat fly species closely mirror those of their host species (Wenzel et al. 1966; Wenzel 1976). No verified records of bat fly species are known from locations outside the range of their primary host species.

5.1 Autecology

Autecological studies have been undertaken on very few bat fly species. The autecology of *Basilia hispida* (Nycteribiidae) was studied by Marshall (1971). In Malaysia, *B. hispida* is known to associate with two species of vespertilionid bats, *Tylonycteris pachypus* and *T. robustula*. However, prevalence of infestation differed between the two host species; prevalence for 707 *T. pachypus* was 34.8% but only 9.6% for 597 *T. robustula*. In captivity, *B. hispida* lived and reproduced successfully on either sex of either bat species. Flies readily dispersed among bat individuals, but given time, ultimately congregated on *T. pachypus*. Consequently, *T. pachypus* was considered the preferred host and *T. robustula* a secondary host (Marshall 1971). Males of both host species more often harbored parasites than did females. This defies conventional wisdom – females often have higher prevalence and intensity of parasite infestation, purportedly because they offer more dependable vertical transfer of ectoparasites from adult hosts to their offspring.

5.2 Effect of roosting dynamics

Bat species achieve remarkable diversity in the tropics, with local richness in excess of 86 species (e.g., Lim and Engstrom 2004). Such diversified assemblages roost in comparably diversified locations (Kunz and Lumsden 2003), either singly, in social groupings, or in multi-specific associations, some of which are fairly characteristic. Such roosting dynamics of bats appear to be crucial to the ecology of parasitic bat flies. Jobling (1949) believed that bat flies lacked host specificity because multiple bat species often utilize a common roost – the flies effectively use all individuals in the spatially confined roost as one large host population. In fact, he supposed that the polyxenous nature of several bat fly species had evolved due to selective pressures surrounding multiple-species roosts (Jobling 1949). Host population size has been supposed to affect the prevalence and intensity of bat fly infestation, but to date such assertions have only been anecdotal. High densities of host bats would provide a rich substrate for bat flies to feed, and would provide many options for colonization by newly emerged adult flies. Conversely, bats that roost solitarily or in small groups provide limited substrate for bat flies. Over evolutionary time, such pressures should select for dense populations of bat flies on high-density roosters of large colonies, and would discourage robust populations of bat flies on solitary roosting species, or those that roost in small colonies. Bat flies occurring on Belizean bat species were evaluated by ter Hofstede and Fenton (2005). Bats that used cavities as the primary roost structure were hypothesized to support higher densities of bat flies than bats roosting in foliage. Although independent contrasts were impossible because a bat fly phylogeny is lacking, host species known to be cavity roosters had significantly higher parasite loads that those classified as foliage roosters (ter Hofstede and Fenton 2005).

5.3 Multiple species infestation

Bats species often are infested with several bat fly species (Wenzel et al. 1966; Wenzel 1976; Dick and Gettinger 2005). Multiple species parasitization has been documented for both nycteribiids (Whitaker et al., in press) and streblids (Wenzel et al. 1966). However, such patterns have been investigated only among American streblids. Bat species in Venezuela are infested with 0-4 bat fly species; 63% of infested bats hosted 2-4 fly species (Wenzel 1976). Presence-absence analysis revealed that observed combinations of bat fly species occur together more often than expected by chance (Dick 2005). Co-occurring fly species were typically associated positively rather than negatively, so that a high abundance of one species was significantly correlated with a high abundance of the other species (Dick 2005). These results may indicate a mutualistic relationship among co-occurring flies, and contrast with some other parasite co-occurrence studies, in which density compensation seems to be the rule (Gotelli and McCabe 2002; Gotelli and Rohde 2002; Fellis et al. 2003).

Host grooming appears to be the principal cause of insect ectoparasite mortality (Marshall 1981). Grooming pressure may constitute an important selective factor driving the evolution of host-limited parasites. Parasites may become specialized to particular spatial locations or ecological niches due to host grooming behavior (Reiczigel and Rozsa 1998). Plausibly, both the presence and abundance of other parasite species would lessen host grooming pressure on a given species of bat fly. The presence and higher abundance of one parasite species would facilitate the presence and abundance of the other, by reciprocally redirecting grooming pressure from each other. This scenario is similar to that of Reiczigel and Rozsa (1998) who modeled the persistence of two species of parasites on a host over time. Meaningfully, positively associated fly species invariably belonged to different genera (Dick 2005). Different morphologies of co-occurring species may facilitate resource subdivision, including spatial segregation of parasites (ter Hofstede et al. 2004).

5.4 Host specificity

Host specificity gauges the degree to which a parasite species is restricted to a particular host species (Poulin 1998). The degree to which bat flies are host specific has long has been debated (Jobling 1949; Theodor 1957, Wenzel et al. 1966; Marshall 1981; ter Hofstede et al. 2004; Dick and Gettinger 2005). Early studies concluded that bat flies were not highly host specific, presumably because many bat species typically roost together and share a common pool of parasites (Jobling 1949; Theodor 1957). In Panama, 55% of bat fly species were associated with a single host species, while another 15% appeared to be monoxenous but were sometimes recorded on other hosts (Wenzel et al. 1966). In Malaysia, Marshall (1980b) reported that 72% of streblids and 64% of nycteribiids were recorded from a single host, with additional flies restricted to sets of congeneric species. In a study specially designed to eliminate cross-host contamination of parasites, Dick and Gettinger (2005) showed that 99.4% of the 2,467 flies taken in their Paraguayan survey were associated with primary bat host species. All but one of the 15 mismatches resulted from sampling contamination (Dick, unpublished data). Generally, results of modern controlled surveys suggest very high host specificity among bat flies (ter Hofstede et al. 2004; Dick and Gettinger 2005).

The ability to fly presumably is an important factor in the degree of host specificity, with flightless species being more host specific, and winged species being less specific (Jobling 1949). In general, flightless species only able to crawl from host to host should be more host-limited than flying species that may easily move from host to host over great distances. However, nycteribiids appear to be no more host-specific than streblids (Marshall 1980b), despite all being flightless while few streblids are. Recently, ter Hofstede et al. (2004) suggested that mobility has no effect on the degree of host specificity among Belizean bat flies.

Marshall (1976) regarded both families of bat flies as being host specific, with species parasitizing usually one host species, or sometimes two or more species of a given host genus. The degree of host specificity was attributed to many factors, including physical isolation, climate, competition, predation, and morphological and physiological adaptation (Marshall, 1976). Not mentioned were evolutionary responses to host-parasite cospeciation (Clayton et al. 2003; Hugot this volume). Combes (1991) outlined the Filter Concept (FC) in an attempt to explain the evolution of parasite life cycles. The FC encompasses Marshall's (1976) factors listed above. The FC has direct application to the evolution of host specificity. The Encounter Filter excludes potential hosts that the parasite cannot encounter and colonize for behavioral or ecological reasons. The Compatibility Filter excludes host individuals on which the parasite cannot survive due to morphological, physiological, or immunological reasons. The filters together represent selective pressures for the parasite, acting to increase or decrease host specificity (Combes 1991). However, the FC does not account for the parasite's ability to encounter mates and reproduce successfully. Due to cospeciation, lineages of bat flies become reproductively isolated on respective host lineages. Such isolation should also maintain selection for continued specificity on the basis of mate availability; in brief, bat flies are specific to their hosts because that is where they are able to reproduce. This represents a Reproductive Filter for the evolution of host specificity (Dick, unpublished data).

6 Effect of parasites on hosts

While bat roosting dynamics affect the ecology of fly parasites, bat fly parasitism also affects the host mammals. Parasitism is a symbiotic relationship in which one of the participants (the parasite) harms the other participant (the host) or otherwise lives at the expense of the other participant (Roberts and Janovy 2000). Some parasites are quite harmful to their host, whereas the harmful effects of others can scarcely be measured. Although bat fly bites are painful to humans, host bats exhibit no reaction to the nearly constant feeding of bat flies (Wenzel et al. 1966; Dick, personal observations). Bat fly bites do not cause sores or lesions on the bat's skin as they do on humans. Grooming effort by captive bats of one species does not differ depending on the intensity of bat fly infestation. However, bat species with higher parasitism levels groomed more intensely than those host species with lower parasitism rates (ter Hofstede and Fenton 2005). Grooming is a costly behavior for bats (Kunz 1982) and increased grooming may affect the host's time budget for other crucial activities such as foraging for food. Regarding a tangible effect on hosts, parasitism by *Megistopoda proxima* (Streblidae) was correlated with significant weight loss in male *Sturnira lilium* (Linhares and Komeno 2000).

Parasitism may also affect the site fidelity of bats, as has been shown for other host species such as barn swallows (Barclay 1988). Bat flies deposit their prepupae inside the roost, and newly emerged flies depend on the presence of host bats. Moving to a different roost before the adult flies emerge may be an effective means for bats to lower both prevalence and intensity of ectoparasite infestation (Lewis 1995).

7 Phylogeny and cospeciation patterns

Comprehensive, robust phylogenies are lacking for Nycteribiidae and Streblidae, and their relationships to other hippoboscoid flies are poorly understood. There is strong support for monophyly of the Hippoboscoidea (McAlpine 1989; Yeates and Wiegmann 1999; Nirmala et al. 2001; Dittmar et al. 2006). Earlier notions that Streblidae and Nycteribiidae each were monophyletic (McAlpine 1989) were challenged by a molecular analysis of the calypterate Diptera (Nirmala et al. 2001). An expanded analysis based on additional genes argued that bat flies are monophyletic, but that the two principal subdivisions do not follow classical lines. Instead, the Western Hemisphere Streblidae constitute one clade and Eastern Hemisphere Streblidae plus Nycteribiidae comprise another (Dittmar et al.

2006). That conclusion is being reanalyzed using several additional genes and a taxon-dense sampling approach (Dittmar, personal communication). Patterson et al. (1998) provided distributional evidence that bat flies have cospeciated with their bat hosts, finding related groups of flies on host clades. Bat fly families, genera, and species groups often are restricted to particular host families, subfamilies, and genera (Wenzel and Tipton 1966; Wenzel et al. 1966). Additional phylogenetic studies of bat flies are urgently needed to extend current knowledge on many topics.

8 Bat flies as vectors of zoonoses

As blood-feeding parasites, bat flies would appear excellent vectors of zoonoses. Generally high degrees of host specificity (Marshall 1976; Dick and Gettinger 2005) diminish the likelihood of interspecific transfer of bat diseases and pathogens. However, it is likely that bat flies transmit species-specific pathogens within host populations. Both nycteribiid and streblid species are known to infest *Hypsignathus monstrosus, Epomops franqueti*, and *Myonycteris torquata*, three species of Old World fruit bats (Megachiroptera: Pteropodidae). Recent evidence suggests that species of pteropodids may harbor the deadly Ebola virus (Leroy et al. 2005). It is possible that parasitic bat flies not only transfer such viruses among host bats, but given that bat flies occasionally bite humans (Wenzel et al. 1966; Dick, personal observations), it is theoretically possible that bat flies could transmit Ebola to humans.

9 Concluding remarks

Bat flies are highly specialized for a nearly permanent ectoparasitic relationship with their hosts, the Chiroptera. Although known life-histories are rather similar across all taxa, bat flies exhibit a variety of morphological adaptations, most suiting them for the two physical substrates offered to them by their bat hosts, the fur and the flight membranes. These morphological adaptations and the ecological diversity of their hosts make bat flies an excellent group in which to study the parasitic relationship, including morphological accommodations, cospeciation, and coevolution.

References

Amin OM (1974) Comb variations in the rabbit flea *Cediopsylla simplex* (Baker). J Med Entomol 11:227–230

Barclay RMR (1988) Variation in the costs, benefits, and frequency of nest reuse by barn swallows (*Hirundo rustica*). Auk 105:53–60

Caire W, Hornuff L (1986) Overwintering population dynamics of the bat fly *Trichobius major* (Diptera: Streblidae). Southwest Nat 31:126–129

Ching LM, Marshall AG (1968) The breeding biology of the bat-fly *Eucampispoda sundaicum* Theodor, 1955 (Diptera: Nycteribiidae). Malayan Nat J 21:171–180

Clayton DH, Al-Tamimi S, Johnson KP (2003) The ecological basis of coevolutionary history. In: Page RDM (ed) Tangled trees: Phylogeny, cospeciation, and coevolution. Univ Chicago Press, Chicago, pp 310–341

Combes C (1991) Evolution of parasite life cycles. In: Toft CA, Aeschlimann A, Bolis L (eds) Parasite-host associations: Coexistence or conflict? Oxford Univ Press, Oxford, pp 62–82

Dick CW (2005) Ecology and host specificity of bat flies (Diptera: Streblidae) and their chiropteran hosts. Ph. D. Thesis, Texas Tech University

Dick CW, Gettinger D (2005) A faunal survey of streblid bat flies (Diptera: Streblidae) associated with bats in Paraguay. J Parasitol 91:1015–1024

Dick CW, Miller J (in press) Streblidae. In: Brown BV (ed) Diptera of Central America. Los Angeles Co Mus Nat Hist, Los Angeles

Dittmar K, Porter ML, Murray S, Whiting MF (2006) Molecular phylogenetic analysis of nycteribiid and streblid bat flies (Diptera: Brachycera, Calyptratae): Implications for host associations and phylogeographic origins. Mol Phyl Evol 38:155–170

Fellis KJ, Negovetich NJ, Esch GW, Horak IG, Boomker J (2003) Patterns of association, nestedness, and species co-occurrence of helminth parasites in the greater Kudu, *Tragelaphus strepsiceros*, in the Kruger National Park, South Africa, and the Etosha National Park, Namibia. J Parasitol 89:899–907

Fritz GN (1983) Biology and ecology of the bat flies (Diptera: Streblidae) on bats of the genus *Carollia*. J Med Entomol 20:1–10

Gleeson DM, Howitt RLJ, Newcomb RD (2000) The phylogenetic position of the New Zealand batfly, *Mystacinobia zelandica* (Mystacinobiidae: Oestroidea) inferred from Mitochondrial 16S robosomal DNA sequence data. J R Soc New Zeal 30:155–168

Gotelli NJ, McCabe DJ (2002) Species co-occurrence: A meta-analysis of J. M. Diamond's assembly rules model. Ecology 83:2091–2096

Gotelli NJ, Rhode K (2002) Co-occurrence of ectoparasites of marine fishes: A null model analysis. Ecol Lett 5:86–94

Hagan HR (1951) Embryology of the viviparous insects. Ronald Press, New York

Hastriter MW, Dittmar K, Whiting MF (2006) Investigation of taxonomically important morphological features of endoparasitic bat flies of the subfamily As-

codipterinae (Streblidae) by scanning electron microscopy. Zootaxa 1122:57-68
Holloway BA (1976) A new bat-fly family from New Zealand (Diptera: Mystacinobiidae). New Zeal J Zool 3:279–301
Jobling B (1936) A revision of the subfamilies of the Streblidae and the genera of the subfamily Streblinae (Diptera Acalypterae) including a redescription of *Metelasmus pseudopterus* Coquillett and a description of two new species from Africa. Parasitology 28:355–380
Jobling B (1939) On some American genera of the Streblidae and their species, with the description of a new species of *Trichobius* (Diptera: Acalypterae). Parasitology 31:486–497
Jobling B (1940) Description of the young female and of the male of *Ascodipteron africanum* Jobling (Diptera: Streblidae). Parasitology 32:399–400
Jobling B (1949) Host-parasite relationship between the American Streblidae and the bats, with a new key to the American genera and a record of the Streblidae from Trinidad, British West Indies (Diptera). Parasitology 39:315–329
Jordan AM (1993) Tsetse-flies (Glossinidae) In: Lane RP, Crosskey RW (eds) Medical insects and arachnids. SpringerVerlag, Berlin, pp 333–388
Kunz TH (1982) Roosting ecology of bats. In: Kunz TH (ed) Ecology of bats. Plenum Press, New York, pp 1–51
Kunz TH, Lumsden LF (2003) Ecology of cavity and foliage roosting bats. In: Kunz TH, Fenton MB (eds) Bat ecology. Univ Chicago Press, Chicago, pp 3–89
Leroy EM, Kumulungui B, Pourrut X, Rouquet P, Hassanin A, Yaba P, Délicat A, Paweska JT, Gonzalez J-P, Swanepoel R (2005) Fruit bats as reservoirs of Ebola virus. Nature 438:575–576
Lewis SE (1995) Roost fidelity of bats: A review. J Mammal 76:481–496
Lim BK, Engstrom MD (2004) Mammals of Iwokrama Forest. Proc Acad Nat Sci Philadelphia 154:71–108
Linhares AX, Komeno CA (2000) *Trichobius joblingi*, *Aspidoptera falcata*, and *Megistopoda proxima* (Diptera: Streblidae) parasitic on *Carollia perspicillata* and *Sturnira lillium* [sic] (Chiroptera: Phyllostomidae) in Southeastern Brazil: Sex ratios, seasonality, host site preference, and effect of parasitism on the host. J Parasitol 86:167–170
Maa TC (1965) Ascodipterinae of Africa (Diptera: Streblidae). J Med Entomol 4:311–326
Maa TC, Peterson BV (1987) Hippoboscidae. In: McAlpine JF, Peterson BV, Shewell GE, Teskey HJ, Vockeroth JR, Wood DM (eds) Manual of Nearctic Diptera, vol 2. Research Branch, Agriculture Canada, pp 1271–1281
Marshall AG (1970) The life cycle of *Basilia hispida* Theodor 1957 (Diptera: Nycteribiidae) in Malaysia. Parasitology 61:1–18
Marshall AG (1971) The ecology of *Basilia hispida* (Diptera: Nycteribiidae) in Malaysia. J Anim Ecol 40:141–154
Marshall AG (1976) Host-specificity amongst arthropods ectoparasitic upon mammals and birds in the New Hebrides. Ecol Entomol 1:189–199

Marshall AG (1980a) The function of combs in ectoparasitic insects. In: Traub R, Starcke H (eds) Fleas. Proceedings of the international conference on fleas. Ashton Wold/Peterborough/UK/21–25 June 1977. Balkema, Rotterdam, pp 79–87

Marshall AG (1980b) The comparative ecology of insects ectoparasitic upon bats in west Malaysia. In: Wilson DE, Gardner AL (eds) Proceedings. Fifth international bat research conference. Texas Tech Press, Lubbock, pp 135–142

Marshall AG (1981) The ecology of ectoparasitic insects. Acad Press, New York

McAlpine JF (1989) Phylogeny and classification of the Muscomorpha. In: McAlpine JF (ed) Manual of North American Diptera, vol 3. Research Branch, Agriculture Canada, pp 1397–1518

Nirmala X, Hypsa V, Zurovec M (2001) Molecular phylogeny of Calyptratae (Diptera: Brachycera): The evolution of 18S and 16S ribosomal rDNAs in higher dipterans and their use in phylogenetic inference. Insect Mol Biol 10:475–485

Overal WL (1980) Host-relations of the bat fly *Megistopoda aranea* (Diptera: Streblidae) in Panama. Univ Kans Sci Bull 52:1–20

Patterson BD, Ballard JWO, Wenzel RL (1998) Distributional evidence for cospeciation between Neotropical bats and their bat fly ectoparasites. Stud Neotrop Fauna Env 33:76–84

Poulin R (1998) Evolutionary ecology of parasites: From individuals to communities. Chapman and Hall, London

Reisen WK, Kennedy ML, Reisen NT (1976) Winter ecology of ectoparasites collected from hibernating *Myotis velifer* (Allen) in southwestern Oklahoma (Chiroptera: Vespertilionidae). J Parasitol 62:628–635

Reiczigel J, Rozsa L (1998) Host-mediated site segregation of ectoparasites: an individual-based simulation study. J Parasitol 84:491–498

Roberts LS, Janovy Jr J (2000) Foundations of parasitology, 6th ed. McGraw Hill, Chicago

Ross A (1960) Notes on *Trichobius corynorhini* on hibernating bats (Diptera: Streblidae). Wasmann J Biol 18:271–272

Peterson BV, Wenzel RL (1987) Nycteribiidae. In: McAlpine JF, Peterson BV, Shewell GE, Teskey HJ, Vockeroth JR, Wood DM (eds) Manual of Nearctic Diptera, vol 2. Research Branch, Agriculture Canada, pp 1283–1291

ter Hofstede HM, Fenton MB, Whitaker Jr JO (2004) Host and host-site specificity of bat flies (Diptera: Streblidae and Nycteribiidae) on Neotropical bats (Chiroptera). Can J Zool 82:616–626

ter Hofstede HM, Fenton MB (2005) Relationships between roost preferences, ectoparasite density, and grooming behaviour of neotropical bats. J Zool Lond 266:333–340

Theodor O (1957) Parasitic adaptation and host-parasite specificity in the pupiparous Diptera. In: Mayr E (ed) First symposium on host specificity among parasites of vertebrates. Institut de Zoologie, Université de Neuchâtel, Switzerland, pp 50–63

Theodor O (1967) An illustrated catalogue of the Rothschild collection of Nycteribiidae. Trust Brit Nat Hist Mus, London

Theodor O, Moscona A (1954) On bat parasites in Palestine. I. Nycteribiidae, Streblidae, Hempitera, Siphonaptera. Parasitology 44:157–245

Traub R (1972) The relationship between spines, combs and other skeletal features of fleas (Siphonaptera) and the vestiture, affinities and habits of their hosts. J Med Entomol 9:601

Wenzel RL (1976) The streblid batflies of Venezuela (Diptera: Streblidae). B Young Univ Sci Bull 20:1–177

Wenzel RL, Tipton VJ (1966) Some relationships between mammal hosts and their ectoparasites. In: Wenzel RL, Tipton VJ (eds) Ectoparasites of Panama. Field Mus Nat Hist, Chicago, pp 677–723

Wenzel RL, Tipton VJ, Kiewlicz A (1966) The streblid batflies of Panama (Diptera: Calypterae: Streblidae). In: Wenzel, RL, Tipton VJ (eds) Ectoparasites of Panama. Field Mus Nat Hist, Chicago, pp 405–675

Whitaker Jr JO, Ritzi CM, Dick CW (in press) Collecting and preserving ectoparasites for ecological study. In: Kunz TH (eds) Ecological and behavioral methods for the study of bats, 2nd edn. Smithsonian Inst Press, Washington

Yeates DK, Wiegmann BM (1999) Congruence and controversy: Toward a higher-level phylogeny of Diptera. Ann Rev Entomol 44:397–428

Part III. Patterns

12 Patterns of macroparasite diversity in small mammals

Boris R. Krasnov, Robert Poulin and Serge Morand

1 Introductory remarks

One of the recurrent themes in modern ecology is the search for patterns of biodiversity across locations and through time and the explanation of these patterns. In part, this is a consequence of the growing public interest associated with the hot issue of biodiversity conservation. However, the general public usually considers biodiversity as something related to butterflies, birds and large mammals, forgetting usually that parasites too represent an integral part of global biodiversity. Moreover, parasites form a large proportion of the diversity of life, and parasitism is possibly more common than any other feeding strategy (Sukhdeo and Bansemir 1996). Parasites also play important roles in the regulation of populations and communities of their hosts (e. g., Poulin 1998; Combes 2001). Consequently, the number of attempts to explain patterns of parasite species richness within and among host species as well as among geographical areas has increased greatly recently. For example, it is well known that different host species harbour different number of parasite species (e. g., Caro et al. 1997). It is highly improbable that parasite species are distributed randomly among their hosts but rather parasite species richness probably results from multiple interwoven factors (see Combes 2001). In fact, Combes (2001) listed as many as 16 different hypotheses related to correlates of parasite species richness. However, most of these hypotheses have never been tested, whereas tests of others have provided contradictory results. In this chapter, we present a review of the observed patterns of species richness and diversity of parasites associated with parameters of individuals, populations and communities of their small mammal hosts as well as some of the major spatial patterns of parasite diversity in small mammals. Taking the advantage of a series of our recent studies, many of the examples presented in this chapter will be related to fleas (Siphonaptera) parasitic on small mammals.

2 Parasite diversity among mammalian orders

Mammalian orders vary greatly in the number of parasite species they harbour. This is true for both endo- and ectoparasites. For example, total ectoparasite species richness differs drastically among different mammalian clades, achieving its highest values in small-bodied taxa such as Rodentia and Chiroptera (Fig. 1a). This suggests that small-bodied mammals suffer more from parasite diversity than large-bodied mammals. On the one hand, this high diversity of parasites should affect the evolution of the anti-parasite defence systems of small mammals (anti-parasite grooming and immune system). Anti-parasite defence systems are likely to be costly. For example, activation of an immune response and even maintenance of a competent immune system is an energetically demanding process that requires trade-off decisions among competing energy demands for growth, reproduction, thermoregulation, work, and immunity (Sheldon and Verhulst 1996; Lochmiller and Deerenberg 2000). This trade-off is especially important for small-bodied animals with a limited amount of energy resources. However, the mean number of parasite genera or species per mammalian genus or species, respectively, appears to be not particularly high in small-bodied mammals (Fig. 1b). In other words, each small-bodied species harbours relatively few parasite species compared to large-bodied hosts. Nevertheless, the high diversity of small-bodied mammals seems to be in itself an evolutionary driver for the high diversity of parasites (see below). Furthermore, parasite species richness varies greatly within mammalian orders among, for example, families of the same order (Fig. 2). This suggests that even closely-related mammalian hosts harbour different numbers of parasite species. A variety of parameters can be related to these within-order among-species differences in parasite diversity.

3 Parasite diversity and the host's body

The host body is the ultimate habitat for the majority of parasites. Consequently, variation in host body characteristics has often been considered as a primary factor determining among-host variation in parasite diversity (Caro et al. 1997; Feliu et al. 1997; Morand and Poulin 1998; Morand and Harvey 2000; Arneberg 2002; Krasnov et al. 2004a). In particular, host parameters that have been tested as correlates of parasite diversity include body size (Morand and Poulin 1998), metabolic rate (Morand and Harvey 2000) and longevity (Morand and Harvey 2000).

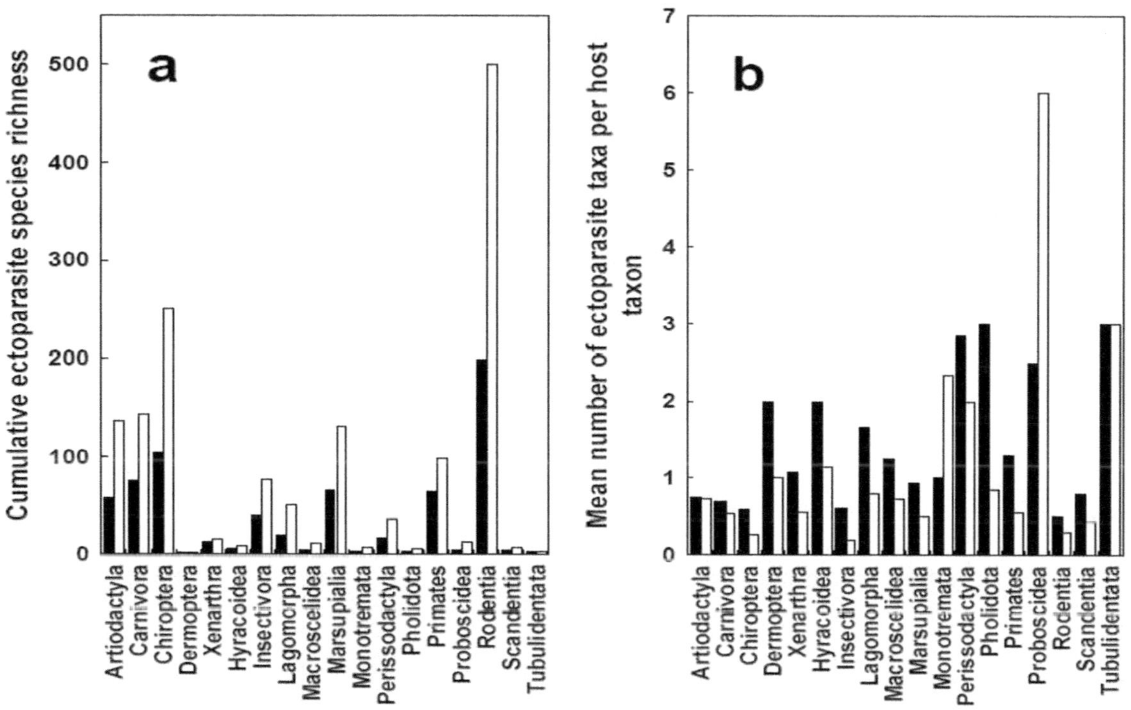

Fig. 1. Number of genera and species of ectoparasitic arthropods in the main mammalian orders (a - cumulative number; b – per mammalian genus or species, respectively). Black columns – genera, white columns - species (data from Kim 1985)

The reasons why a correlation between parasite diversity and host body mass is expected are rather straightforward. Larger hosts likely sustain richer parasite assemblages because they provide more space and a greater variety of niches and, thus, can provide different parasite species with an opportunity for spatial niche diversification. For example, on relatively large hosts, different fleas prefer different body areas (Hsu et al. 2002). Although some studies reported positive correlations between parasite species richness and body size of mammalian hosts (Gregory et al. 1996; Vitone et al. 2004), no relationship between mammalian body size and parasite richness was found in other studies (Poulin 1995; Feliu et al. 1997; Morand and Poulin 1998; Krasnov et al. 2004a). Furthermore, Arneberg (2002) found a positive relationship between strongylid nematode richness and mammalian body mass, but the effect of host population densities had to be controlled for to see this. However, in small mammals, host density can vary greatly on a temporal scale, with ten-fold fluctuations often observed. Consequently, consideration of the mean density of small mammalian populations in the present context is not feasible. The results of Krasnov et al. (2004a) suggest that the conclusions of Poulin (1995) and Morand and Poulin (1998) about the lack of relationship between body size and the species richness of mammalian endoparasites are also valid for ectoparasites. It may differ in fish, however, as a correlation between host body size and parasite richness was reported for fish ectoparasites when the effect of phylogeny was removed (Guégan and Morand 1996). Another but not necessarily alternative explanation for the absence of correlation between body size and ectoparasite richness may be that the main habitat for many ectoparasites (e.g., fleas and some mite taxa) is not the body of a host but rather its burrow or nest. Consequently, ectoparasite richness might be related to the size and the degree of complexity of the host burrow rather than to its body size, although this has never been tested.

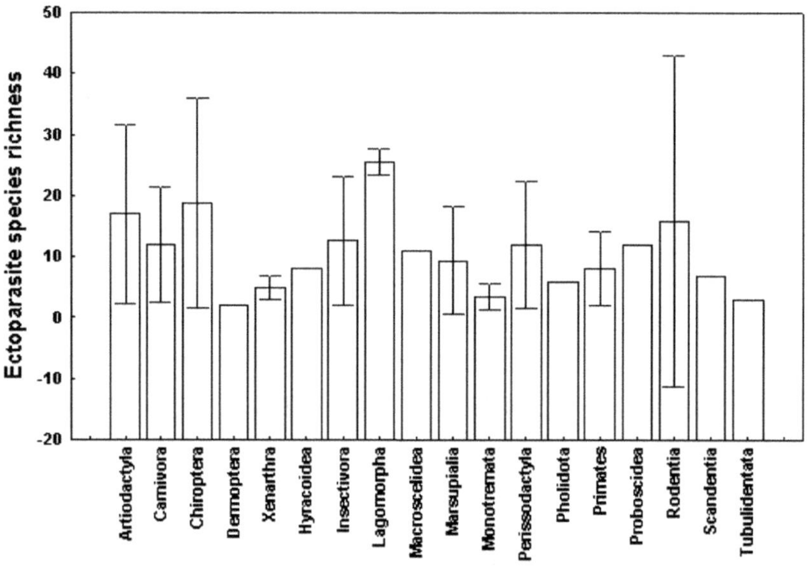

Fig. 2. Mean (±S.D.) parasite species richness per mammalian family among the main mammalian orders (data from Kim 1985)

Metabolic rate is expected to correlate positively with parasite species richness because hosts exposed to diverse infections should invest in a high basal metabolic rate (BMR) in order to compensate for a costly immune response (Morand and Harvey 2000), although some researchers have argued that the cost of the immune response is an energy cost above that of the BMR (Degen 1997). Furthermore, if parasite species richness is expected to be positively correlated with BMR, the correlation of parasite richness with average daily metabolic rate (ADMR) is expected to be even more pronounced. This is because ADMR includes the BMR, the heat increment of feeding for maintenance, locomotory and thermoregulatory costs and, consequently, is thought to be a more appropriate measurement than BMR for evaluating the energy requirements and efficiency of energy utilization of an animal (Degen 1997). A note of caution is necessary here. A positive relationship between BMR or ADMR and parasite species richness does not allow one to determine the direction of causality. For instance, the positive relationship will exist if, as explained above, high parasite species richness drives the evolution of high BMR. However, the positive relationship would also exist if high BMR leads to a higher colonization rate by parasite species, given that hosts with a high BMR will consume more food and thus be exposed to more infective parasite stages, etc. High parasite species richness can be either a *cause* or a *consequence* of high BMR. As with any other correlation involving a host trait and parasite diversity, more than one process can explain an observed pattern.

In any event, BMR in mammals has been shown to correlate positively with helminth species richness (Morand and Harvey 2000; Fig. 3a). In contrast, in rodents, there is no correlation between either BMR or ADMR and flea species richness (Krasnov et al. 2004a). This may suggest that either flea parasitism does not affect negatively a rodent host or does not trigger an immune response, or else the immune response to a particular flea species is equally effective against other flea species (cross-resistance). However, flea parasitism has been shown to have an energetic cost for a host, although the amount of blood consumed by the parasite is extremely small (Khokhlova et al. 2002). This means that the major effects of flea parasitism on the energy expenditure of the host could be through means other than blood depletion, such as stimulation of an immune response to derived molecules from the salivary glands of the fleas. The similarity of salivary components within a parasite taxon can lead to cross-resistance of a host against closely-related parasites (McTier et al. 1981; Njau and Nyiando 1987). The occurrence of cross-reactions of the immune response to different fleas can be responsible for the absence of correlation between host metabolic parameters and flea species richness. In addition, the study of energy requirements for maintenance in a gerbil *Gerbillus dasyurus* under flea parasitism demonstrated that these requirements increased in parasitized individuals in spite of relatively small blood loss, indicating thus that the energetic cost of an immune response was above the ADMR of the rodent (Khokhlova et al. 2002). This suggests that a more relevant parameter relating to flea richness on a host might be the ability of the host to increase its metabolic rate above requirements and not the ADMR itself.

Host longevity can also be an important factor determining the diversity of parasites as a consequence of the continued accumulation of parasites in long-lived species (Bell and Burt 1991; Morand 2000). However, the opposite trend was reported for mammalian hosts and helminth parasites

(Morand and Harvey 2000; Fig. 3b), whereas longevity of small mammalian hosts and flea species richness were found not to be correlated (Stanko et al. 2002). In general, the relationship between host species longevity and parasite diversity is poorly known. At present, there are not enough available data that unequivocally show that this host trait affects parasite diversity. Nevertheless, variation in host longevity has been suggested to be partly responsible for the pattern of variation in parasite diversity, although not in a direct way but rather being mediated via other parameters such as, for example, basal metabolic rate (Morand and Harvey 2000). On the other hand, the longevity of an individual host (=host age) seems to be linked to variation in parasite diversity among host individuals within host species. Energetically costly immunity against parasites is expected to decrease in senescent individuals. This, in turn, can facilitate the co-occurrence of multiple parasite species in long-lived hosts. Krasnov et al. (2006a) found that senescence was accompanied by an increase in mean infrapopulation species richness of fleas in some rodent species, but not in others, and explained this among-host difference by differences in life history parameters such as mobility and shelter structure.

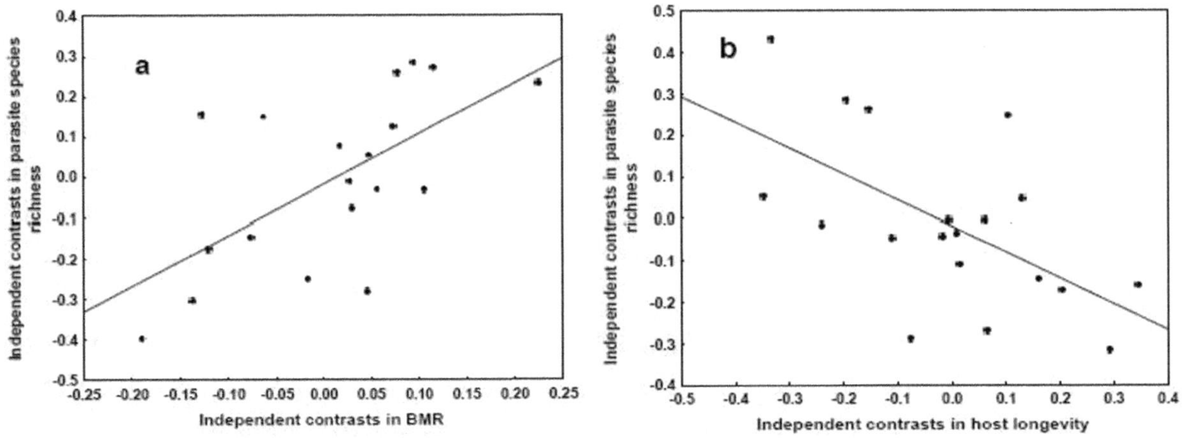

Fig. 3. Partial relationships between (a) BMR (controlled for host body mass and longevity) and parasite species richness (controlled for host sampling effort, body mass and longevity) and (b) host longevity (controlled for host body mass and BMR) and parasite species richness (controlled for host sampling effort, host body mass and BMR) using independent contrasts (modified from Morand and Harvey 2000)

4 Parasite diversity and the host population

Density is the main characteristic of any population of living organisms. It is affected by various intrinsic and extrinsic factors and, in turn, affects a variety of individual and population parameters. In particular, host population density is one of the most important factors influencing the spread and distribution of parasites among host individuals (e.g., Anderson and May 1978). This is because the rate at which host individuals acquire the parasite species may be determined by how many individuals are available for parasite colonization (Morand and Poulin 1998). In addition, high host density can facilitate a process of horizontal parasite transmission both within and among host species and, thus, increase the mean number of parasite species per host individual. Indeed, when the data for 79 species of both small and large mammals were controlled for the confounding effect of phylogeny, there was a positive relationship between helminth species richness and host density (Morand and Poulin 1998). Similar results were reported for 45 species of small and large mammals and directly transmitted nematodes (Arneberg 2002) and for 101 primate species and 231 various parasite taxa (Nunn et al. 2003). However, when the dataset was restricted to rodents or to rodents and insectivores no such relationship among host species was found for the data collected in the Iberian Peninsula (Feliu et al. 1997) and Slovakia (Stanko et al. 2002). Nevertheless, Stanko et al. (2002) found that host density has the greatest influence on the species richness of ectoparasite communities of rodents and insectivores among host populations within host species. The lack of the relationship between host density and parasite species richness in among-host comparisons is not especially surprising because of the huge temporal fluctuations in the density of small mammals for which the idea of "typical" density hardly makes sense (but see Arneberg et al. 1997). However, the existence of this relationship among host populations within host species is easily explained by epidemiological theory (Anderson and May 1978; May and Anderson 1979).

Host social structure is another factor that can be linked with parasite diversity. For example, an increase in social group size increases the rate of contacts between individuals, and, thus, can favour the transmission of contagious parasites (Loehle 1995). Indeed, Côté and Poulin (1995) demonstrated consistent positive correlations between host group size and both the prevalence and intensity of contagious parasites for a variety of host taxa. Tella (2002) found that species richness, both in terms of number of species and number of genera, of blood parasites (Haematozoa) was higher in colonial bird species than in their solitary-breeding sister species. Vitone et al. (2004) reported that sociality played a certain role in accounting for parasite diversity of primates, with group size and the number of females per group explaining a significant amount of variation in the diversity of helminth parasites in non-phylogenetic analyses. However, Poulin (1991), Ranta (1992) and Poulin and Rohde (1997) found no relationship between sociality and parasite species richness in fish hosts. Non-phylogenetic analysis of data on fleas parasitizing 52 rodent hosts demonstrated that social and solitary species did not differ in either species richness of their flea assemblages (controlled for sampling effort) or in the taxonomic distinctness of these assemblages (see Poulin and Mouillot 2003 for details on this measure) (Krasnov et al., unpublished data). The same results were produced by phylogenetically corrected analysis (Krasnov et al., unpublished data). The contradicting results for different host and parasite taxa may be related to the possibility of intense selection in favour of behavioural or immunological barriers to parasite transmission among social hosts (Loehle 1995; Altizer et al. 2004). Another possible reason for the contradicting results is that host sociality can affect differently parasites with different transmission strategies. Indeed, a tighter link with host sociality can be expected for parasites transmitted by direct contact rather than for parasites transmitted by ingestion, or via vectors or the environment (Thrall and Antonovics 1997; Altizer et al. 2004). In addition, it is sometimes difficult to distinguish the role of host social structure from that of host density in their effect on parasite diversity (Morand and Poulin 1998). For example, some rodent species live solitarily at low density while becoming social at high density (e.g., Schradin and Pillay 2005).

5 Parasite diversity and the host community

5.1 Patterns among host species

Factors associated with the structure of host communities can have a profound influence on the parasite diversity of a host species. An example of such factors is the number of sympatric closely-related host species. The richer taxonomic "milieu" of a host increases the probability of lateral transfer of parasites and, thus, can increase the parasite richness of any given host species in that community (Combes 2001). This relationship was supported for ectoparasites of marine fish (Caro et al. 1997; Raibaut et al. 1998) and rodents (Krasnov et al. 2004a), although it was not supported for parasitoids and their insect hosts (Hawkins and Lawton 1987). In the case of rodents, the number of co-existing species belonging to the same subfamily (both across the entire geographic range and locally) is positively correlated with species richness of flea assemblages on a rodent species (Krasnov et al. 2004a). This means that the ability of a flea species to exploit successfully several host species heavily depends on the phylogenetic and/or taxonomic relatedness of these hosts. Indeed, a comparative analysis demonstrated that the abundance of a flea on its auxiliary hosts decreases with increasing taxonomic distance of these hosts from the principal host (Krasnov et al. 2004b). The reasons for this can be similarities among host species in ecological, physiological and/or immunological characters. If, for example, closely-related host species possess similar behavioural or immune defences, a flea could invest less by adapting to a restricted set of host immune defences than it would if its hosts were distantly related and the parasite would be forced to develop multiple adaptations to cope with the array of immune defences of its numerous hosts (Poulin and Mouillot 2004). Another advantage of exploiting closely-related hosts from a parasite perspective is that these hosts often have similar ecological preferences (Brooks and McLennan 1991). Consequently, their habitat distribution can be similar, so a new host encountered by a parasite in the habitat of an original host is possibly a close relative of this original host.

5.2 Patterns among host communities

A positive correlation between species diversity and habitat variety was initially reported by MacArthur (1958, 1964) for birds and then was supported for other organisms, both plants and animals (Rosenzweig 1995 and references therein). However, the question of what is a "habitat" and, consequently, "habitat diversity" arises when diversity is considered in this context, namely whether a habitat is pre-defined and is related to an area of a particular relief, vegetation and soil structure or whether it is a patch with a set of environmental conditions and resources promoting the occupancy, survival and reproduction by individuals of a given species (Morrison et al. 1992; Rosenzweig 1992). Parasites offer a conceptual advantage over free-living animals in this respect because a habitat patch for a parasite is its host, which provides a place for living, foraging and mating. Parasite individuals are distributed across host individuals and, thus, the host population of a particular species can be considered as a habitat for that parasite species. Furthermore, considering a host species as a habitat for a parasite species avoids the disagreement between the two above mentioned habitat concepts. Firstly, a host species is a clearly pre-defined entity. Secondly, parasites clearly distinguish among different host species both in terms of host choice behaviour and fitness reward (e.g., Krasnov et al. 2002a, 2004c). Given a positive correlation between species diversity and habitat diversity in flee-living organisms, a positive correlation between host diversity and parasite diversity can also be expected (Watters 1992, but see "diversity-disease hypothesis" van der Plank 1963).

Krasnov et al. (2004d) studied the relationship between host species richness and parasite species richness using simultaneously collected data on small mammals (Insectivora, Rodentia and Lagomorpha) and their flea parasites in 37 different regions. The data were controlled for the area sampled and sampling effort, and then this relationship was tested using both cross-region conventional analysis and the independent contrasts method (to control for the effects of biogeographic historical relationships among different regions). Both analyses showed a positive correlation between host species richness and flea species richness (Fig. 4).

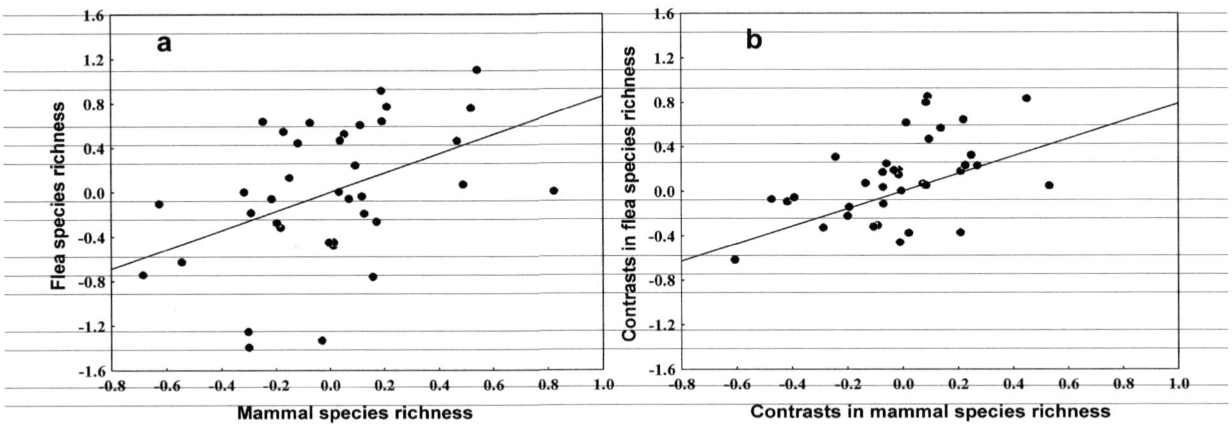

Fig. 4. The relationship between host species richness and flea species richness (both variables controlled for area and host sampling effort) using cross-region comparisons (a) and independent contrasts (b) (modified from Krasnov et al. 2004d; reprinted with permission from Blackwell Publishing)

This demonstrated that the species diversity pattern reported for free-living animals also holds true for parasites. The positive correlation between host species richness and flea species richness suggests that diversification of parasites is a response to diversification of hosts. Diversification of hosts can facilitate an increase in the number of their parasites either by a higher probability of parasite co-diversification (if host diversification stems from host speciation) (Combes 2001; Clayton et al. 2003) or by the introduction of new parasite species (if host diversification stems from host immigration) or both. In any case, the evolutionary reason for the positive host diversity/parasite diversity pattern can be a process of specialization of parasites on different host species, exactly as the specialization of free-living species to a limited range of habitat properties is the reason behind the positive species diversity-habitat diversity pattern (Rosenzweig 1992). This is because "fine habitat subdivision is a coevolved property of the species in a biome" (Rosenzweig 1992, p. 715). The conceptual difference in comparisons between species versus habitat diversity and parasite versus host diversity is mainly in our inability to recognize different habitats in the same manner as animals and plants do, whereas it is much easier to recognize different host species. The absence of a negative relationship between flea species richness and mammal species richness suggests that the relationships between flea parasites and their mammalian hosts reach an equilibrium when neither host defence causes parasite extinction nor parasite pressure leads to host extinction.

6 Parasite diversity and host geographic range

6.1 Patterns within host species

Most studies of parasite diversity have focused on a broad scale (e. g., among different host species), whereas patterns of diversity of parasite communities on a smaller scale (e. g., among different populations of the same host species) have attracted less attention and have been studied in a limited number of host taxa (Kisielewska 1970; Carney and Dick 2000; Poulin and Valtonen 2002; Poulin 2003; Calvete et al. 2004; Proctor and Jones 2004, Krasnov et al. 2005a). Kisielewska (1970) demonstrated that the composition of helminth communities in a vole *Clethrionomys glareolus* was highly variable across locations and seemed to be unpredictable. However, this unpredictability may turn out to be false if the parameters related to the host environment, such as climate and the species composition of sympatric host species, are taken into account. For example, Krasnov et al. (2005a) examined spatial variation in the diversity of fleas on 69 species of small mammals from 24 different regions of the Holarctic. It appeared that flea species richness varied less within than among host species, and was thus a repeatable host species character. This suggests the existence of some threshold of defence against parasites in a host species that limits the host's ability to cope with multiple parasite species (e. g., because of costly defence systems; Schmid-Hempel and Ebert 2003) while maintaining their pressure (expressed as a number of parasite species) at a "tolerable" level (Combes 2001). In contrast with species richness, the taxonomic distinctness of flea assemblages and its variance were not repeatable among populations within a host species. This means that whenever a new exploiter is added to a host's parasite community, this exploiter is a random addition from the regional pool of exploiter species that manages somehow to adapt itself to the new host species.

In almost all host species, similarity in flea assemblages decreased with increases in either or both geographical and "faunal" (dissimilarity in host species composition) distance (Fig. 5), demonstrating thus that the pattern of distance decay of biological similarity found in other organisms is universal (Nekola and White 1999; Poulin 2003). Furthermore, in general, the rate of decrease in the similarity in flea assemblages was lower as a function of geographic distance than against the "faunal" distance, suggesting that, perhaps, differences in the surrounding "milieu" between host populations are a more important determinant of the composition of flea assemblages than mere physical distance.

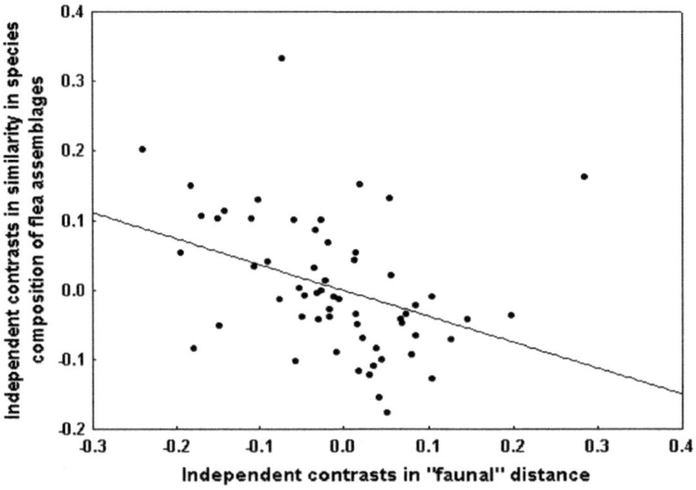

Fig. 5. Relationship between average similarity in flea species composition calculated using the Jaccard index and average "faunal" distance between two host populations. The relationship was computed across 60 mammalian host species using independent contrasts (after Krasnov et al. 2005a; reprinted with permission from Blackwell Publishing)

However, in spite of flea species richness being a true host character, this character varied across the geographic range in many hosts, indicating that diversity of flea assemblages is also influenced by local factors. In most host species, the diversity of flea assemblages correlated with one or more environmental (climatic) variables, in particular mean winter temperature (Fig. 6). This demonstrates that the diversity of flea assemblages on small mammalian hosts is to an important extent mediated by local climatic conditions, highlighting thus how ecological processes interact with co-evolutionary history to determine local parasite biodiversity.

6.2 Parasite diversity in isolated host populations

Isolation and fragmentation of habitats is considered to be one of the most important factors affecting biological diversity (Brown 1995; Rosenzweig

1995). The effect of isolation on species diversity of parasitic organisms has been recently reviewed by Guégan et al. (2005). In particular, they considered those few studies that dealt with the comparison of species richness of parasite communities between island and mainland populations of the same host species. These studies investigated species richness of different parasite and host taxa, namely reptiles and helminths (Dobson et al. 1992), rodents and insectivores and pathogenic leptospires (Collares-Pereira et al. 1997) and rodents and helminths (Goüy de Bellocq et al. 2002, 2003). In all cases, the conclusion was that island host populations harbour poorer parasite communities compared to mainland populations, supporting thus the well-known pattern of depauperation of island fauna and flora in free-living organisms (Rosenzweig 1995). Furthermore, Goüy de Bellocq et al. (2003) suggested that if the evolution of effective immune systems was driven by parasite pressures (Combes and Morand 1999), then host species that inhabit islands and are exposed to low parasite pressures in terms, for example, of parasite species diversity, would invest less in immune systems compared to hosts with higher parasite pressures living on the mainland. Later, Goüy de Bellocq et al. (2005) tested this hypothesis by studying the patterns of polymorphism in the Major Histocompatibility Complex genes in island and mainland populations of *Apodemus sylvaticus* and found that island populations showed a loss (albeit weak) of polymorphism in comparison with their mainland counterparts.

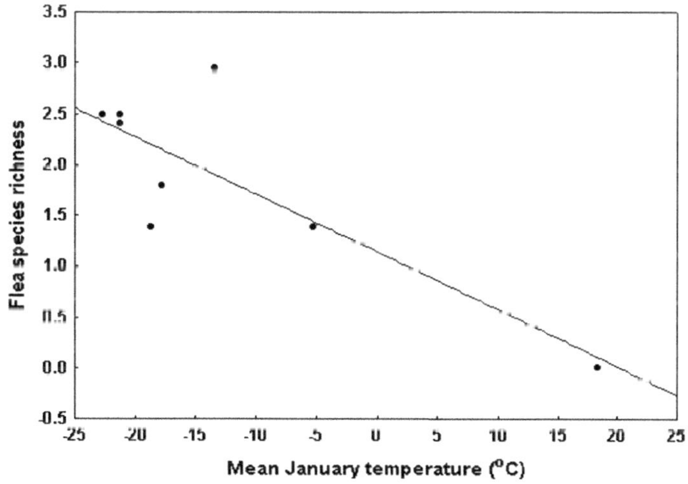

Fig. 6. Relationship between species richness of flea assemblages (log-transformed) and mean January temperature in different populations of *Apodemus uralensis* (data from Krasnov et al. 2005a)

Studies that attempted to link parasite diversity in an island population of a host with the area of that island, at first glance, provide contradictory results. For example, parasite species richness in *Apodemus sylvaticus* on western Mediterranean islands appeared to depend on the island area (Goüy de Bellocq et al. 2002), whereas this was not the case for parasite species richness in *Anolis* lizards (Dobson et al. 1992). However, Guégan et al. (2005) re-analyzed the data of Goüy de Bellocq et al. (2002) and demonstrated that the island area effect seemed to be spurious because the data were flawed by sampling bias. Nevertheless, it is too early to draw a final conclusion about the lack of the effect of island area on parasite diversity, because too few studies of this effect have been carried out.

6.3 Patterns among host species

Hosts that differ in the size and position of their geographic range are expected to differ also in the diversity of their parasite assemblages. A positive correlation between parasite species richness and the size of the host geographic range is expected because hosts with larger geographic ranges would presumably encounter more parasite species. Indeed, this pattern was found in parasite assemblages of rodents (Feliu et al. 1997 for gastrointestinal helminths; Krasnov et al. 2004a for fleas), although it appeared not to be universal (e.g., Clayton and Walther 2001). Combes (2001) suggested that this relationship be interpreted in the framework of the theory of island biogeography (species richness on islands correlates positively with the size of the island; MacArthur and Wilson 1967).

The latitude of the host geographic range is expected to correlate negatively with parasite richness according to the well-known pattern of latitudinal gradient of species richness. The latitudinal gradient is that, in general, the inventory of species declines as one moves further from the equator, either north or south (Rosenzweig 1995), as was repeatedly shown for free-living animals (e. g., Rohde 1992; Rosenzweig 1992). However, studies of this pattern in relation to parasite assemblages provided contradictory results (see Guégan et al. 2005 for recent review). For example, Rohde and Heap (1998) observed this pattern for monogeneans but not for endoparasites of marine fish, whereas Poulin (1995) and Poulin and Mouritsen (2003) failed to find any relationships between latitude and richness of gastrointestinal parasites of birds and mammals or richness of trematodes in intertidal gastropods, respectively. Yet, Krasnov et al. (2004a) found a clear correlation between the latitude of the host geographic range and species richness of flea assemblages, but this trend was the exact opposite of the main latitudinal gradient rule, namely flea species richness increased with the latitude of the centre of the geographic range. The absence of a latitudinal gradient for endoparasites was explained by the relative stability of their environment (inside the host body) (Rohde and Heap 1998). Ectoparasites, in contrast, are exposed to environmental conditions that change with latitude. One of the reasons for the unusual pattern that has been found for fleas may be that only few flea assemblages of both tropical and arctic rodents have been studied. In the data set of Krasnov et al. (2004a), the center of the geographic range was situated at latitudes lower than 20° in only two species and at latitudes higher than 60° in only three species.

Another reason for the positive correlation between flea species richness and latitude might be the deeper burrows in rodents from temperate regions (Kucheruk 1983), as flea assemblages are likely richer in deeper burrows. A pattern opposed to the common and expected latitudinal gradient was also reported by Poulin (2001) for helminth communities in temperate versus tropical fish. Rohde (1996, 1999) questioned the generality of the latitudinal gradient rule and suggested that this rule is a "local" phenomenon that is restricted to the Holarctic above latitudes of 40°-50°N. However, when Krasnov et al. (2004a) limited the data set to 62 Holarctic species with geographic range centers above 40°N, the positive correlation between latitude and flea richness remained. The examples mentioned in this section clearly show that spatial patterns of parasite diversity in general and latitudinal gradient of parasite richness in particular are far from being understood. As Guégan et al. (2005) noted, further studies of the patterns of parasite and microorganism species richness over wide ranges of spatial scales are needed for a deeper comprehension of spatial dynamics of parasite species diversity.

7 Parasite diversity and productivity

Productivity (the rate of energy flow through an ecosystem) is considered to be an important factor influencing distribution of many taxa of animals and plants (Rosenzweig 1992, 1995). Changes in species diversity along productivity gradients have been reported to be linear (Brown and Davidson 1977), concave (Krasnov and Shenbrot 1996) or unimodal (Haedrich et al. 1970; Rex 1981; Abramsky and Rosenzweig 1984; Abramsky 1988; Owen 1988; Shenbrot 1992). However, recent reviews have demonstrated that the very existence of the relationship between productivity and diversity is not often a rule (Mackey and Currie 2001; Mittelbach et al. 2001). Furthermore, some evidence suggests that the relationship between diversity and productivity may differ qualitatively between producer and consumer organisms (Huston 1993).

The only study of the relationship between parasite species diversity and productivity was carried out by Poulin et al. (2003). They used total parasite biovolume per individual host as a measurement for community productivity, and tested the relationship between productivity and species richness among assemblages of endoparasites in 131 vertebrate host species. A linear relationship, but no trace of unimodality, between productivity and parasite species richness was found both across all hosts as well as for each of the five vertebrate groups (fish, amphibians, reptiles, mammals and birds), with no trace of a hump-shaped curve. This study dealt with productivity components that are intrinsic to parasite communities. The effect of productivity components related to extrinsic factors on parasite diversity can be quite different. For example, contrary to endoparasites, ectoparasites are affected not only by host body characteristics but also by the off-host environment. Consequently, some relationship between productivity of the off-host environment and ectoparasites diversity can be expected. Although Krasnov et al.'s (1997) study on the structure of flea assemblages in the desert rodent hosts *Gerbillus dasyurus* and *Meriones crassus* was not specifically designed to test the relationship between flea diversity and habitat productivity, the results of this study demonstrated that the highest species richness of fleas on these hosts was found in habitats with the highest abundance of annual vegetation. The latter is a good estimator of primary production for desert environments. On the other hand, no relationship between either species richness or taxonomic distinctness of flea assemblages and annual precipitation (another good estimator of productivity in dryland ecosystems) was found among populations of the steppe- and desert-dwelling *Cricetulus migratorius* (Krasnov et al. 2005b). In conclusion, the relationship between productivity and species diversity for parasite communities is still poorly known and understood. Further studies of different host-parasite systems carried out on different scales are needed.

8 Parasite diversity and structure of parasite communities

8.1 Co-variance in diversity among parasite taxa within a host

Different parasite taxa exploit different host resources and are often unlikely to interact directly. Indeed, depending on the presence or absence of interspecific interactions, both isolationist and interactive parasite communities can be distinguished (Holmes and Price 1986; Bush et al. 1997). It is commonly accepted that interactive communities are those that comprise parasite species belonging to the same guild, e.g. sharing the same trophic level, whereas parasite species in isolationist communities, though sharing a host, do not exploit the same resources (Poulin 1998). Nevertheless, interactions, although rather indirect than direct, between parasite species belonging to different guilds are also possible. Different parasite taxa exploit different host resources and are often unlikely to interact directly. Some components of host immune defences may operate simultaneously against all kinds of parasites, whereas investment by the host in specific defences against one type of parasites may come at the expense of defence against other parasites. Consequently, both negative and positive relationships among species diversity of parasites belonging to different taxa can be expected. Investigation of the relationships between the species diversity of four higher taxa of ectoparasites (fleas, sucking lice, mesostigmatid mites, and ixodid ticks), and between the species richness of ectoparasites and endoparasitic helminths, across different species of rodent hosts demonstrated positive pairwise correlations between the species richness of fleas, mites and ticks as well as strong positive relationship between the taxonomic distinctness of ecto- and endoparasite assemblages across host species (Krasnov et al. 2005b).

These results, combined with an earlier demonstration that the species richnesses of different groups of endoparasitic helminths covary positively among their vertebrate hosts (Poulin and Morand 2004, pp. 75-79), provide strong evidence of apparent facilitation (*sensu* Levine 1999) among unrelated parasite taxa in the organization of parasite communities. The existence of relationships between the species diversities of different parasite taxa (even those from different guilds) suggests that the host represents an important force shaping parasite communities. For example, the positive relationships among species diversities of the assemblages of different ectoparasites as well as between ecto- and endoparasites could arise from immunodepression in a host subjected to multiple immune challenges from a variety of parasite species. Maintaining several different means of defence is likely more costly than mounting one specific type of defence (Taylor et al. 1998). As a result, the effectiveness of energy allocation to immune defence decreases as the diversity of attack types increases (Jokela et al. 2000). Jokela et al. (2000) argued that in cases when the diversity of attacks is high and, thus, the effectiveness of defence is low, the optimal strategy is to tolerate damage. Consequently, a host subjected to attacks from multiple parasite species is forced to give up its defence and to surrender.

When the diversity of unrelated taxa of free-living organisms covaries positively across localities, the general explanation usually invokes intrinsic differences in rates of colonization and extinction among localities

(Gaston 1996). It is thus possible that intrinsic properties of the various host species could lead to some hosts accumulating parasites of all taxa at a high rate. This can be because of some biochemical, physiological or ecological properties of the host. On the other hand, a host species that is able to resist attacks from many species of one parasite taxon appears to be able also to resist attacks from many species of other parasite taxa. This suggests some level of cross-resistance against distantly related parasites. Indeed, cross-immunity between distantly related parasite taxa has been reported, at least for haematophagous arthropods (see review in Ribeiro 1996). Another, not necessarily alternative, explanation for the observed patterns is that host species can differ in their intrinsic ability to defend themselves against parasites using their immune system. Different, sometimes even closely related, rodent species have been shown to have different abilities to mount both humoral and cell-mediated immune responses (Klein and Nelson 1998a, b). As a result, a rodent with lower intrinsic immunocompetence can be exploited by a higher number of parasite species compared with more immunocompetent species.

8.2 Parasite diversity and parasite abundance

The diversity of species making up the community is likely to affect the abundance of any given species in a community. For example, diffuse competition (when a species competes with a constellation of other species in various combinations and densities; MacArthur 1972) can lead to negative relationships between the abundance of a given parasite species and the species richness or any other measure of diversity of the entire parasite assemblage. This is true if the species in a community interact directly. Indeed, the original model of MacArthur (1972) does not incorporate indirect interactions. Later models of diffuse competition that account for indirect interactions have concluded that a high number of species could reduce the intensity of interactions or even lead to facilitation (Stone and Roberts 1991). Parasites undoubtedly influence one another via their effects on hosts. Suppression of host defence systems resulting from high parasite diversity (that supposedly requires multiple defence responses) could lead to facilitation among parasite species. As a result, the abundance of a given species should be positively correlated with the diversity of co-occurring species.

Krasnov et al. (2005c) studied the relationships between diversity of flea communities on small mammals and abundance of a given flea species. At all scales of analysis, i.e. whether the same flea species on different host species, or different flea species were compared, the consistent result was that the abundance of any given flea species correlates negatively with either the species richness or taxonomic diversity of the flea community (Fig. 7), but correlated positively with the total abundance of other co-occurring flea species.

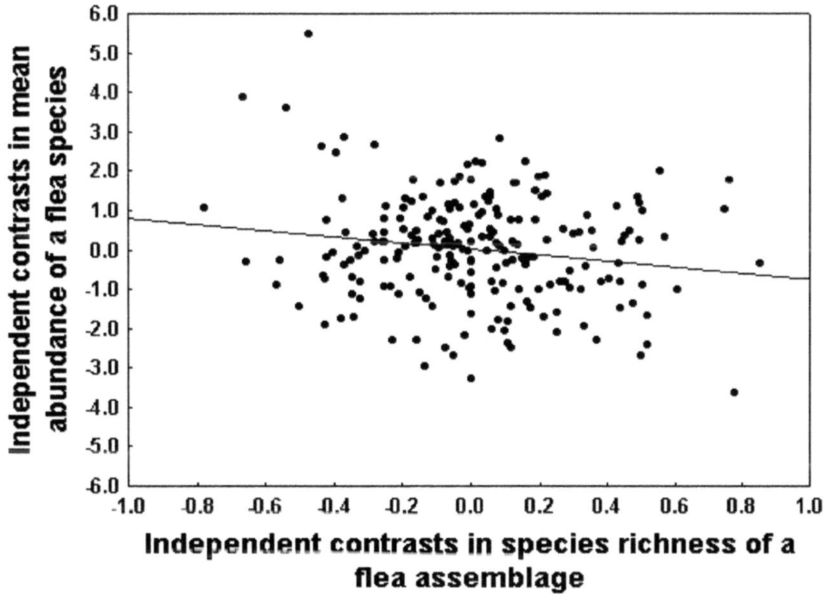

Fig. 7. Relationships between the mean abundance of a flea on a host species and the species richness of the flea assemblage on that host across 230 flea species using independent contrasts (modified from Krasnov et al. 2005c; reprinted with permission from Blackwell Publishing)

This does not support the existence of diffuse competition in flea assemblages, because the more individuals of other flea species are present on a host population, the more individuals of the focal species are there as well. Instead, this suggests important indirect (host-mediated) facilitation among fleas. Furthermore, the abundance of a given flea species was highest in assemblages consisting of few species of limited taxonomic diversity (when co-occurring flea species were closely related with a focal species). The latter can be linked to the higher likelihood of host immunosuppression if the immunogens of the parasites involved are similar which, in turn, is more likely if the parasites are phylogenetically close (a phenomenon opposite to cross-resistance). On the other hand, some form of negative interactions among species seemed to exist also, because the abundance of a given flea species is lower when many other species are also present. The observed pattern supports the idea that both facilitation and competition operate among the same species in a community either simultaneously or with the strength of each process varying in time or space (Levine 1999).

8.3 Parasite diversity and parasite aggregation

Despite the above indications that coexistence of parasites on small mammals is facilitated, the data on parasite abundance and species richness in component and compound communities do not allow testing of the mechanisms of coexistence. These mechanisms can be revealed by studying patterns of parasite abundance and diversity in parasite infracommunities. For example, species coexistence can be facilitated if competing species are distributed such that interspecific aggregation is reduced relative to intraspecific aggregation (aggregation model of coexistence; see Shorrocks and Rosewell 1986; Hartley and Shorrocks 2002). The main assumptions of this model are (a) the fragmented nature of resources and (b) the aggregated distribution of competing species among resource patches, which is exactly the case for parasites. Only three studies of the aggregation model of coexistence in parasites have been carried out to date (Morand et al. 1999; Simkova et al. 2000; Krasnov et al. 2006b). The two former studies dealt with ectoparasites of fish, whereas the third study was carried out on fleas of small mammalian hosts from central Europe. If flea coexistence is facilitated via the aggregation model of coexistence, then the reduction in the level of interspecific aggregation in relation to the level of intraspecific aggregation would be expected to be positively correlated with flea species richness. In this study, the relationships between measures of intraspecific and interspecific aggregations suggested by Ives (1988, 1991) and species richness of flea communities were examined. It appeared that intraspecific aggregation increased compared to interspecific aggregation when infra- and component parasite species richness increased (Fig. 8). However, no relationship between the level of intraspecific aggregation versus interspecific aggregation and taxonomic distinctness of flea assemblages was found. In addition, positive interspecific associations of fleas tended to occur more frequently in species-rich flea assemblages. These results demonstrate that a relatively high diversity of flea assemblages on small mammalian hosts can be attained via mechanisms related to the aggregation model of coexistence.

8.4 Local versus regional parasite diversity

Relationships between local and regional species diversity have often been used to infer the relative importance of local and regional processes in governing local species composition (Srivastava 1999). A linear relationship between local and regional species richness and/or diversity suggests that regional processes strongly control local communities (Cornell and

Lawton 1992), whereas a curvilinear relationship advocates that local richness is independent of regional richness and, thus, local processes play the main role in structuring local communities and impose upper limits on the number of species that are able to coexist (Terborgh and Faaborg 1980; Cornell and Lawton 1992).

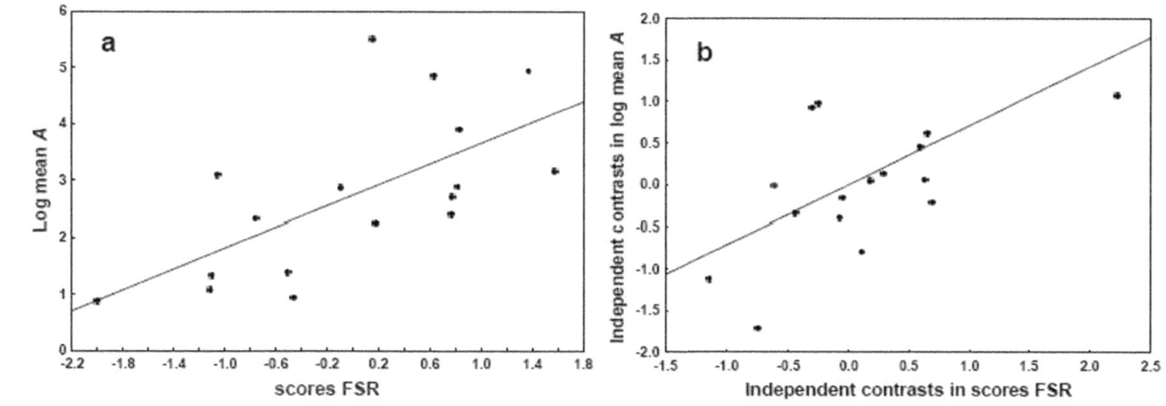

Fig. 8. Relationship between mean relative strength of intraspecific aggregation versus interspecific aggregation (*A*) of a flea assemblage and scores of a principal component of mean and maximum infracommunity species richness and component community species richness (FSR) across 17 small mammal hosts using conventional regression (a) and the independent contrasts method (b) (modified from Krasnov et al. 2006b; reprinted with permission from Blackwell Publishing)

Testing the relationship between local and regional species richness is, at first glance, rather straightforward and can be carried out using regression analysis (e. g., Oberdorff et al. 1998). However, some methodological problems arise (Cresswell et al. 1995; Caley and Schluter 1997; Griffiths 1999; Srivastava 1999; Fox et al. 2000; Loreau 2000; Shurin et al. 2000; Hillebrand 2005), and, thus, the use of local-regional richness plots to test for saturation of diversity has been strongly criticized. Nevertheless, the use of regional to local diversity regressions remains widespread (Valone and Hoffman 2002; Heino et al. 2003; Calvete et al. 2004; Karlson et al. 2004). One of the most important methodological issues is a precise definition of borders for local and regional communities. It is sometimes self-evident for freshwater organisms (e.g., a pond, Shurin et al. 2001), but it is much more difficult for terrestrial or marine organisms. However, for parasites the definition of a community at the lowest hierarchical scale is relatively easy. This is the infracommunity (all parasite individuals of all species within an individual host). The measure of local parasite species richness is, thus, mean (e. g., Morand et al. 1999) or maximum (e. g.,

Calvete et al. 2004) infracommunity parasite species richness. The next hierarchical level is the component community of parasites (all infracommunities within a given host population). Finally, all component parasite communities within a given host species represent either a regional parasite community or a parasite fauna.

Most studies of the relationship between species richness of communities of free-living organisms at different spatial scales have demonstrated that unsaturated communities are the norm, i.e. linear relationships between local and regional species richness predominate (see Srivastava 1999 for review). However, analyses of local versus regional species richness in parasites have revealed that saturated and unsaturated communities are equally common (e. g., Poulin 1996; Morand et al. 1999; Kennedy and Guégan 1996; Calvete et al. 2004; see recent review in Guégan et al. 2005). Guégan et al. (2005) suggested that these contrasting results arise due to the fact that studies reporting linear relationships between infracommunity and component community species richness tested this relationship among host species (e. g., Poulin 1996), whereas studies reporting curvilinear relationships between the two parameters compared different populations of the same host species (e. g., Calvete et al. 2004).

Testing the relationship between parasite infracommunity and parasite component community species richness among host species while simultaneously testing it among different populations within these same host species has rarely been done. Krasnov et al. (2006c) investigated this relationship in communities of fleas on small mammalian hosts, at two different spatial scales: between the richness of flea communities on individual hosts (infracommunities) and that of flea communities on host populations (component communities), and between the richness of component communities and that of the entire regional species pool. At both spatial scales, consistent curvilinear relationships between species richness of the more "local" communities and richness of the more "regional" communities were found.

Demonstrating the existence or absence of saturation in parasite assemblages requires the additional investigation of interspecific interactions such as possible existence of density compensation in species-poor flea infracommunities (Guégan et al. 2005). To test for this, Krasnov et al. (2006c) assessed the relationship between mean flea abundance per host individual and richness of the "local" flea community. There was no strong evidence for density compensation at the infracommunity level (significant linear relationship between mean flea abundance and infracomminity species richness), although its existence at the component community level appeared likely (no relationship between mean flea abundance and component community species richness).

The curvilinear relationship between infracommunity and component community flea richness suggests that the number of species in species-rich flea infracommunities is independent of the species richness of the component community of which they are part. Thus, at first glance, the flea infracommunities are "saturated" and vacant niches seem to be generally unavailable in these communities.

The observed by Krasnov et al. (2006c) pattern may arise because some species can be eliminated or not allowed to invade local communities due to some ecological constraints such as negative interactions among species in an infracommunity (Srivastava 1999; Calvete et al. 2004). However, if negative competitive interactions among flea species in an infacommunity are indeed important, one would expect density compensation in species-poor infracommunities (Cornell 1993). This appears to be not the case for flea infracommunities suggesting that the curvilinear relationship between infracommunity and component community species richness may occur for reasons other than "saturation" due to competitive interspecific interactions.

Indeed, Rohde (1998) demonstrated that curvilinearity in the local versus regional species richness relationship may be caused by processes other than species interactions within a local community. In particular, this curvilinearity may be a consequence of the differential likelihood of parasite species of occurring in an infracommunity because of different transmission rates and lifespans (Rohde 1998). In the case of fleas, these reasons can also be related to differential abiotic preferences of either imago or larval fleas of different species that contribute to the elimination of some flea species from some infracommunities (Krasnov et al. 2001, 2002b). All the above indicate that flea infracommunities are governed by processes acting at higher than "local" levels, and that further species could possibly be added over evolutionary time (Rohde 1998).

The relationship between local and regional flea richness appeared to be the same at the larger scale than at the smaller scale; in other words, the relationship between richness of component communities and that of the regional flea pool seems to be similar to that found for infracommunities versus component communities. However, the absence of a relationship between mean flea abundance and component community species richness suggests the existence of density compensation. Therefore, component communities appeared to be saturated. The causes of this saturation are likely some intrinsic limiting factors that may play an important role in shaping flea component communities. One of the common factors responsible for community saturation is negative interspecific interactions such as competition (Cornell 1993). Although direct interspecific competition between flea imagos within a host population is unlikely (but see Lindsay and Galloway 1997), such competition can occur among larval fleas (Krasnov et al. 2005d). To conclude, similar patterns in the relationships between "local" and "regional" species richness in the same host-parasite system but at different spatial scales may arise because of different mechanisms. This could be one explanation for the contrasting relationships reported between local and regional species richness in earlier studies of different host-parasite systems.

9 Taxonomic diversification of parasite assemblages of small mammal hosts

Diversification of parasite assemblages (increase in species diversity within a host lineage) over evolutionary time can result from at least two different types of evolutionary events (Poulin 1998; Page 2003). First, the parasite taxon can speciate on a host without an accompanying host speciation event and can, thus, produce multiple closely-related parasite lineages on the host's descendants (e.g., Klokenhoff 1980). Second, a new parasite can be acquired via colonization from a different host lineage (host-switching; Hoberg et al. 1997). Mouillot and Poulin (2004) suggested that the relative importance of these two processes in shaping the diversification of parasite assemblages can be indicated by the value of the exponent of the power relationship between the number of higher taxa (e. g., genera) and species richness. In woody plant communities, these exponents appeared to be statistically invariant across and within biogeographical regions, types of plant physiognomy, and geological time (Enquist et al. 2002). However, this appeared not true for assemblages of intestinal helminth parasites, where the value of the exponents varies according to the identity of the vertebrate host taxon, being the highest in fish hosts, lower in bird hosts and the lowest in mammal hosts (Mouillot and Poulin 2004). Mouillot and Poulin (2004) suggested that, for parasites, the value of the exponent of this function might reflect the main pathway of diversification. If an exponent has a value close to one across several comparable parasite assemblages, this would indicate that host-switching has been the main cause of diversification (if each species in an assemblage is taxonomically-independent of the other species, it must therefore have had a separate origin). In contrast, an exponent clearly inferior to one indicates that several species in the same assemblage belong to the same genus or genera, suggesting that they have a common ancestor and that they may have radiated from this common ancestor within the host lineage.

This method was adapted to flea assemblages of small mammalian hosts from 25 Holarctic regions by Krasnov et al. (2005e). They found that the relationships between the number of flea species and the numbers of flea genera on a host species in each region were well described by simple power-functions (Fig. 9).

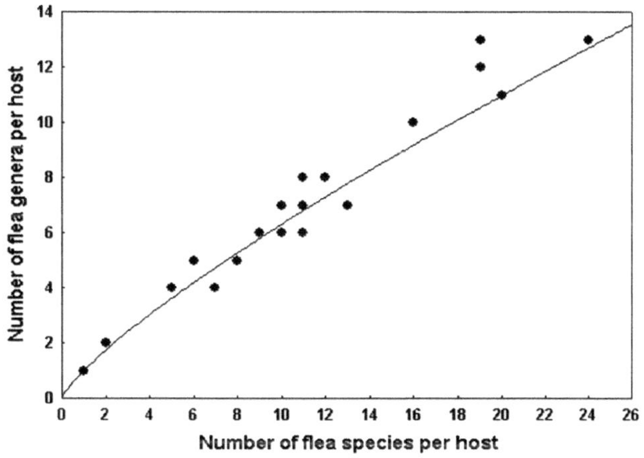

Fig. 9. Relationship between the number of flea species and flea genera per host across 20 small mammal host species in the Volga-Kama region (modified from Krasnov et al. 2005e; reprinted with permission from Blackwell Publishing)

These exponents attained values greater than 0.92 in only 4 of the 25 regions, and were lower than 0.88 in as many as 16 of the 25 regions. Thus, the exponents of species-genera relationships for flea assemblages were, in general, lower that those found by Enquist et al. (2002) for plant communities (0.94) and those found by Mouillot and Poulin (2004) for the communities of helminth parasites in fish and bird hosts (0.97 and 0.92, respectively), but were close to, albeit somewhat higher than, those reported by Mouillot and Poulin (2004) for helminth parasites of mammals (0.83). This suggests that intrahost speciation seems to play a more important role in flea diversification than host-switching (but see Krasnov and Shenbrot 2002; Lu and Wu 1995). This also hints at similar mechanisms influencing the rate of intrahost speciation of ecto- and endoparasites in mammals, and these mechanisms can be related to some, still unknown, host features. However, the lack of invariance of the exponent value of the power function across different regions (0.74-0.95; see Krasnov et al. 2005e), in contrast to that found for plant communities (Enquist et al. 2002), suggests that some local conditions might strongly affect fundamental processes and mechanisms of diversification. These local conditions can be related, for instance, to climate.

10 Parasite diversification in small mammals and climate

The effect of climate on parasite species diversity has been explained mainly by the assumption that higher energy input (e. g., measured as local solar radiation or temperature) determines evolutionary rates (Rohde 1992, 1999). Presumably, a greater input of solar energy leads to faster evolution via increased mutation rates, accelerated physiological processes and shortened generation time (Rohde 1992). If this is so, we might expect that in the relatively colder regions, the main way for a parasite assemblage to diversify is via host switching, and this should lead to roughly only one species per genus on any given host species. In contrast, the number of species per genus can be expected to increase in the relatively warmer regions, where warmer temperatures favour speciation (Rohde 1992). Consequently, a negative relationship between local mean annual temperature and the value of the exponent of the power function between the number of species and the number of higher taxa per host species (described in previous sub-chapter) could be expected. This appeared to be exactly the case for fleas (Krasnov et al. 2005e), suggesting that multiple congeneric species of fleas parasitic on the same host species occurred mainly in warmer regions. This finding supported the hypothesis of Rohde (1992, 1999) about higher evolutionary rates under higher energy input. The increase of the evolutionary rates may be the outcome of an increase in the mutation rate, the acceleration of physiological processes and/or shortened generation time. All these can explain, at least partly, why the flea assemblages in warmer rather than in colder regions diversified more via intrahost speciation, as indicated by the value of the exponent of the power function. It cannot explain, however, why flea assemblages in the colder regions diversified mainly by host switching, especially given that the dispersal abilities of parasite species are restricted at lower temperatures (Rohde 1985, 1992, 1999). Nevertheless, flea transfers from host to host occur mainly when hosts visit each other's burrows (e. g., Ryckman 1971) or via body contact between host individuals (e. g., Krasnov and Khokhlova 2001). Rodent burrows in temperate and colder regions are deeper, more complicated and more frequently visited than those in warmer regions (Kucheruck 1983). These processes can facilitate host switching by fleas, independently of temperature-effects on the mobility of the fleas themselves.

11 Concluding remarks

The diversity of parasite assemblages on small mammalian hosts is affected by a variety of factors. These factors include those related to the host body, those related to the off-host biotic and abiotic environment, as well as those related to parasite community structure. However, studies of patterns of parasite diversity in small mammals within and among host species and across biogeographical areas often provide contradictory results and very few general rules have emerged from these studies (see Poulin and Morand, 2000, 2004 and references therein). Based on all the above, two main directions for future studies can be envisaged, both of which are likely to take us one step closer to uncovering any general rule of parasite diversity. First, we still need further investigations of various host and parasite taxa in clearly understudied regions (e.g., in Africa, South America, South-Eastern Asia). Second, the application to parasite assemblages of modern ecological theories and models that initially have been developed for free-living organisms promises to be a fruitful avenue.

References

Abramsky Z (1988) The role of habitat and productivity in structuring desert rodent communities. Oikos 52:107–114

Abramsky Z, Rosenzweig M (1984) Tilman's predicted productivity-diversity relationship shown by desert rodents. Nature 309:150–151

Altizer S, Nunn CL, Thrall PH, Gittleman JL, Antonovics J, Cunningham AA, Dobson AP, Ezenwa V, Jones KE, Pedersen AB, Poss M, Pulliam JRC (2004) Social organization and parasite risk in mammals: Integrating theory and empirical studies. Ann Rev Ecol Syst 34:517–547

Anderson RM, May RM (1978) Regulation and stability of host-parasite population interactions. I. Regulatory processes. J Anim Ecol 47:219–247

Arneberg P (2002) Host population density and body mass as determinants of species richness in parasite communities: Comparative analyses of directly transmitted nematodes of mammals. Ecography 25:88–94

Arneberg P, Skorping A, Read AF (1997) Is population density a species character? Comparative analyses of the nematode parasites of mammals. Oikos 80:289–300

Bell G, Burt A (1991) The comparative biology of parasite species diversity: Intestinal helminths of freshwater fishes. J Anim Ecol 60:1046–1063

Brooks DR, McLennan DA (1991) Phylogeny, ecology, and behaviour: A research program in comparative biology. Univ Chicago Press, Chicago

Brown JH (1995) Macroecology. Univ Chicago Press, Chicago

Brown JH, Davidson DW (1977) Competition between seed-eating rodents and ants in desert ecosystems. Science 196:880–882

Bush AO, Lafferty KD, Lotz JM, Shostak AW (1997) Parasitology meets ecology on its own terms: Margolis *et al.* revisited. J Parasitol 83:575–583

Caley MJ, Schluter D (1997) The relationship between local and regional diversity. Ecology 78:70–80

Calvete C, Blanco-Aguiar JA, Virgós E, Cabezas-Díaz S, Villafuerte R (2004) Spatial variation in helminth community structure in the red-legged partridge (*Alectoris rufa* L.): Effects of definitive host density. Parasitology 129:101–113

Carney JP, Dick TA (2000) Helminth community in yellow perch (*Perca flavescens* (Mitchill)): Determinants of pattern. Can J Zool 78:538–555

Caro A, Combes C, Euzet L (1997) What makes a fish a suitable host for Monogenea in the Mediterranean? J Helminthol 71:203–210

Clayton DH, Walther BA (2001) Influence of host ecology and morphology on the diversity of Neotropical bird lice. Oikos 94:455–467

Clayton DH, Al-Tamimi S, Johnston KP (2003) The ecological basis of coevolutionary history. In: Page RDM (ed) Tangled trees. Phylogeny, cospeciation and coevolution. Univ Chicago Press, Chicago, pp 310–342

Collares-Pereira M, Korver H, Terpstra WJ, Santos-Reis M, Ramalhinho MG, Mathisa ML, Oom MM, Fons R, Libois R, Petrucci-Fonseca F (1997) First epidemiological data on pathogenic leptospires isolated on the Azorean islands. European J Epidemiol 13:435–441

Combes C (2001) Parasitism. The ecology and evolution of intimate interactions. Univ Chicago Press, Chicago

Combes C, Morand S (1999) Do parasites live in extreme environments? Constructing hostile niches and living in them. Parasitology 119:5107–5110

Cornell HV (1993) Unsaturated patterns in species assemblages: The role of regional processes in setting local species richness. In: Ricklefs RF, Schluter D (eds) Species diversity in ecological communities: Historical and geographical perspectives. Univ Chicago Press, Chicago, pp. 243–252

Cornell HV, Lawton JH (1992) Species interactions, local and regional processes, and limits to richness of ecological communities: A theoretical perspective. J Anim Ecol 61:1–12

Côté IM, Poulin R (1995) Parasitism and group size in social animals: A meta-analysis. Behav Ecol 6:159–165

Cresswell JE, Vidal-Martinez VM, Crichton NJ (1995) The investigation of saturation in the species richness of communities: Some comments on methodology. Oikos 72:301–304

Degen AA (1997) Ecophysiology of small desert mammals. SpringerVerlag, Berlin Heidelberg New York

Dobson AP, Pacala SV, Roughgarden JD, Carper ER, Harris EA (1992) The parasites of *Anolis* lizards in the northern Lesser Antilles. I. Patterns of distribution and abundance. Oecologia 91:110–117

Enquist BJ, Haskell JP, Tiffney BH (2002) General patterns of taxonomic and biomass partitioning in extant and fossil plant communities. Nature 419:610–613

Feliu C, Renaud F, Catzeflis F, Hugot J-P, Durand P, Morand S (1997) Comparative analysis of parasite species richness of Iberian rodents. Parasitology 115:453–466

Fox JW, McGrady-Steed J, Petchey OL (2000) Testing for local species saturation with nonindependent regional species pools. Ecol Lett 3:198–206

Gaston KJ (1996) Spatial covariance in the species richness of higher taxa. In: Hochberg ME, Clobert J, Barbault R (eds) Aspects of the genesis and maintenance of biological diversity. Oxford Univ Press, Oxford, pp 221–242

Goüy de Bellocq J, Feliu C, Morand S (2002) Patterns of parasite species richness of Western Palaeartic micro-mammals: Island effects. Ecography 25:173–183

Goüy de Bellocq J, Sara M, Casanova J-C, Feliu C, Morand S (2003) A comparison of the structure of helminth communities in the woodmouse, *Apodemus sylvaticus*, on islands of the western Mediterranean and continental Europe. Parasitol Res 90:64–70

Goüy de Bellocq J, Delarbre C, Gachelin G, Morand S (2005) Allelic diversity at the MHC-DQA locus of woodmouse populations (*Apodemus sylvaticus*) present in the islands and mainland of the northern Mediterranean. Global Ecol Biogeogr 14:115–122

Gregory RD, Keymer AE, Harvey PH (1996) Helminth parasite richness among vertebrates. Biodivers Conserv 5:985–997

Griffiths D (1999) On investigation local-regional species richness relationships. J Anim Ecol 68:1051–1055

Guégan J-F, Morand S (1996) Polyploid hosts: Strange attractors for parasites. Oikos 77:366–370

Guégan J-F, Morand S, Poulin R (2005) Are there general laws in parasite community ecology? The emergence of spatial parasitology and epidemiology. In: Thomas F, Guégan J-F, Renaud F (eds) Parasitism and ecosystems. Oxford Univ Press, Oxford, pp 22–42

Hawkins BA, Lawton JH (1987) Species richness patterns: Why do some insects have more parasitoids than others? Colloquium INRA 48:131–136

Haedrich R, Rowe G, Polloni P (1980) The megabenthic fauna in the deep sea south of New England, USA. Mar Biol 57:165–179

Hartley S, Shorrocks B (2002) A general framework for the aggregation model of coexistence. J Anim Ecol 71:651–662

Heino J, Muotka T, Paavola R (2003) Determinants of macroinvertebrate diversity in headwater streams: regional and local influences. J Anim Ecol 72:425–434

Hillebrand H (2005) Regressions of local on regional diversity do not reflect the importance of local interactions or saturation of local diversity. Oikos 110:195–198

Hoberg E, Brooks DR, Siegel-Causey D (1997) Host-parasite co-speciation: History, principles, and prospects. In: Clayton DH, Moore J (eds) Host-parasite evolution: General principles and avian models. Oxford Univ Press, Oxford, pp 212–235

Holmes JC, Price PW (1986) Communities of parasites. In: Kikkawa J, Anderson DJ (eds) Community ecology: Patterns and processes. Blackwell Sci, New York, pp 187–213

Hsu M-H, Hsu T-C, Wu W-J (2002) Distribution of cat fleas (Siphonaptera: Pulicidae) on the cat. J Med Entomol 39:685–688

Huston M (1993) Biological diversity, soils, and economics. Science 262:1676–1680

Ives AR (1988) Aggregation and the coexistence of competitors. Ann Zool Fenn 25:75–88

Ives AR (1991) Aggregation and coexistence in a carrion fly community. Ecol Monogr 61:75–94

Jokela J, Schmid-Hempel P, Rigby MC (2000) Dr. Pangloss restrained by the Red Quinn – steps towards a unified defence theory. Oikos 89:267–274

Karlson RH, Cornell HV, Hughes TP (2004) Coral communities are regionally enriched along an oceanic biodiversity gradient. Nature 429:867–870

Kennedy CR, Guégan J-F (1996) The number of niches in intestinal helminth communities of *Anguilla anguilla*: Are there enough spaces for parasites? Parasitology 113:293–302

Khokhlova IS, Krasnov BR, Kam M, Burdelova NV, Degen AA (2002) Energy cost of ectoparasitism: The flea *Xenopsylla ramesis* on the desert gerbil *Gerbillus dasyurus*. J Zool Lond 258:349–354

Kim KC (ed) (1985) Coevolution of parasitic arthropods and mammals. John Wiley, New York

Kisielewska K (1970) Ecological organization of intestinal helminth groupings in *Clethrionomys glareolus* (Schreb.) (Rodentia). III. Structure of helminth groupings in *C. glareolus* populations of various forest biocoenoses in Poland. Acta Parasitol 18:163–176

Klein SL, Nelson RJ (1998a) Sex and species differences in cell-mediated immune responses in voles. Can J Zool 76:1394–1398

Klein SL, Nelson RJ (1998b) Adaptive immune responses are linked to the mating system of arvicoline rodents. Amer Nat 151:59–67

Klockenhoff HF (1980) *Myrsidea karyi* (Mallophaga: Menoponidae), a new species from *Corvus orru* (Passeriformes: Corvidae). Pacific Insects 22:115–122

Krasnov BR, Khokhlova IS (2001) The effect of behavioural interactions on the exchange of flea (Siphonaptera) between two rodent species. J Vector Ecol 26:181–190

Krasnov BR, Shenbrot GI (1996) Structure of darkling beetle communities and factors influencing their spatial distribution in the Negev highlands (Israel). Ecography 19:139–152

Krasnov BR, Shenbrot GI (2002) Coevolutionary events in history of association of jerboas (Rodentia: Dipodidae) and their flea parasites. Israel J Zool 48:331–350

Krasnov BR, Shenbrot GI, Medvedev SG, Vatschenok VS, Khokhlova IS (1997) Host-habitat relations as an important determinant of spatial distribution of flea assemblages (Siphonaptera) on rodents in the Negev Desert. Parasitology 114:159–173

Krasnov BR, Khokhlova IS, Fielden LJ, Burdelova NV (2001) The effect of temperature and humidity on the survival of pre-imaginal stages of two flea species (Siphonaptera: Pulicidae). J Med Entomol 38:629–637
Krasnov BR, Khokhlova IS, Oguzoglu I, Burdelova NV (2002a) Host discrimination by two desert fleas using an odour cue. Anim Behav 64:33–40
Krasnov BR, Khokhlova IS, Fielden LJ, Burdelova NV (2002b) Time to survival under starvation in two flea species (Siphonaptera: Pulicidae) at different air temperatures and relative humidities. J Vector Ecol 27:70–81
Krasnov BR, Shenbrot GI, Khokhlova IS, Degen AA (2004a) Flea species richness and parameters of host body, host geography and host "milieu". J Anim Ecol 73:1121–1128
Krasnov BR, Shenbrot GI, Khokhlova IS, Poulin R (2004b) Relationships between parasite abundance and the taxonomic distance among a parasite's host species: An example with fleas parasitic on small mammals. Int J Parasitol 34:1289–1297
Krasnov BR, Khokhlova IS, Burdelova NV, Mirzoyan NS, Degen AA (2004c) Fitness consequences of density-dependent host selection in ectoparasites: Testing reproductive patterns predicted by isodar theory in fleas parasitizing rodents. J Anim Ecol 73:815–820
Krasnov BR, Shenbrot GI, Khokhlova IS, Degen AA (2004d) Relationship between host diversity and parasite diversity: Flea assemblages on small mammals. J Biogeogr 31:1857–1866
Krasnov BR, Shenbrot GI, Mouillot D, Khokhlova IS, Poulin R (2005a) Spatial variation in species diversity and composition of flea assemblages in small mammalian hosts: Geographic distance or faunal similarity? J Biogeogr 32:633–644
Krasnov BR, Mouillot D, Khokhlova IS, Shenbrot GI, Poulin R (2005b) Covariance in species diversity and facilitation among non-interactive parasite taxa: all against the host. Parasitology 131:557–568
Krasnov BR, Mouillot D, Shenbrot GI, Khokhlova IS, Poulin R (2005c) Abundance patterns and coexistence processes in communities of fleas parasitic on small mammals. Ecography 28:453–464
Krasnov BR, Burdelova NV, Khokhlova IS, Shenbrot GI, Degen AA (2005d) Pre–imaginal interspecific competition in two flea species parasitic on the same rodent host. Ecol Entomol 30:146–155
Krasnov BR., Shenbrot GI, Khokhlova IS, Poulin R (2005e) Diversification of ectoparasite assemblages and climate: An example with fleas parasitic on small mammals. Global Ecol Biogeogr 14:167–175
Krasnov BR, Stanko M, Morand S (2006a) Age-dependent flea (Siphonaptera) parasitism in rodents: A host's life history matters. J Parasitol 92:242–248
Krasnov BR, Stanko M, Khokhlova IS, Mosansky L, Shenbrot GI, Hawlena H, Morand S (2006b) Aggregation and species coexistence in fleas parasitic on small mammals. Ecography 29:159-168
Krasnov BR, Stanko M, Khokhlova IS, Miklisova D, Morand S, Shenbrot GI, Poulin R (2006c) Relationships between local and regional species richness in

flea communities of small mammalian hosts: Saturation and spatial scale. Parasitol Res 98:403–413
Kucheruk VV (1983) Mammal burrows: Their structure, topology and use (in Russian). Fauna and Ecology of Rodents 15:5–54
Levine JM (1999) Indirect facilitation: Evidence and predictions from a riparian community. Ecology 80:1762–1769
Lindsay LR, Galloway TD (1997) Seasonal activity and temporal separation of four species of fleas (Insecta: Siphonaptera) infesting Richardson's ground squirrels, *Spermophilus richardsonii* (Rodentia: Sciuridae), in Manitoba, Canada. Can J Zool 75:1310–1322
Loehle C (1995) Social barriers to pathogen transmission in wild animal populations. Ecology 76:326–335
Lochmiller RI, Deerenberg CB (2000) Trade-offs in evolutionary ecology: just what is the cost of immunity? Oikos 88:87–98
Loreau M (2000) Are communities saturated? On the relationship of ά, β and γ diversity. Ecol Lett 3:73–76
Lu L, Wu H (2005) Morphological phylogeny of *Geusibia* Jordan, 1932 (Siphonaptera: Leptopsyllidae) and the host-parasite relationship with pikas. Syst Parasitol 61:65–78
MacArthur RH (1958) Population ecology of some warblers of Northeastern coniferous forests. Ecology 39:599–619
MacArthur RH (1964) Environmental factors affecting bird species diversity. Amer Nat 98:387–397
MacArthur RH (1972) Geographical ecology. Harper and Row, New York
MacArthur RH, Wilson EO (1967) The theory of island biogeography. Princeton Univ Press, Princeton
Mackey RL, Currie DJ (2001) The diversity-disturbance relationship: is it generally strong and peaked? Ecology 82:3479–3492
May RM, Anderson RM (1979) Population biology of infectious diseases: Part II. Nature 280:455–461
McTier TL, George JE, Bennet SN (1981) Resistance and cross-resistance of guinea pigs to *Dermacentor andersoni* Stiles, *D. variabilis* (Say), *Amblyomma americanum* (Linnaeus) and *Ixodes capularis* Say. J Parasitol 67:813–822
Mittelbach GG, Steiner CF, Scheiner SM, Gross KL, Reynolds HL, Waide RB, Willig MR, Dodson SI, Gough L (2001) What is the observed relationship between species richness and productivity? Ecology 82:2381–2396
Morand S (2000) Wormy world: Comparative tests of theoretical hypotheses on parasite species richness. In: Poulin R, Morand S, Skorping A (eds) Evolutionary biology of host-parasite relationships: Theory meets reality. Elsevier, Amsterdam, pp. 63–79
Morand S, Harvey PH (2000) Mammalian metabolism, longevity and parasite species richness. Proc R Soc Lond B 267:1999–2003
Morand S, Poulin R (1998) Density, body mass and parasite species richness of terrestrial mammals. Evol Ecol 12:717–727
Morand S, Poulin R, Rohde K, Hayward C (1999) Aggregation and species coexistence of ectoparasites of marine fishes. Int J Parasitol 29:663–672

Morrison ML, Marcot BG, Mannan RW (1992) Wildlife-habitat relationships. Concepts and applications. Univ Wisconsin Press, Madison
Mouillot D, Poulin R (2004) Taxonomic partitioning shedding light on the diversification of parasite communities. Oikos 104:205–207
Nekola JC, White PS (1999) The distance decay of similarity in biogeography and ecology. J Biogeogr 26:867–878
Njau BC, Nyindo M (1987) Detection of immune response in rabbits infested with *Rhipicephalus appendiculatus* and *Rhipicephalus evertsi evertsi.* Res Vet Sci 43:217–221
Nunn CL, Altizer S, Jones KE, Sechrest W (2003) Comparative tests of parasite species richness in primates. Amer Nat 162:597–614
Oberdorff T, Hugueny B, Compin A, Belkessam D. (1998) Non-interactive fish communities in the coastal streams of North-western France. J Anim Ecol 67:472–484
Owen JG (1988) On productivity as a predictor of rodent and carnivore diversity. Ecology 69:1161–1165
Page RDM (ed) (2003) Tangled trees: Phylogeny, cospeciation and coevolution. Univ Chicago Press, Chicago
Poulin R (1991) Group-living and the richness of the parasite fauna in Canadian freshwater fishes. Oecologia 86:390–394
Poulin R (1995) Phylogeny, ecology, and the richness of parasite communities in vertebrates. Ecol Monogr 65:283–302
Poulin R (1996) Richness, nestedness, and randomness in parasite infracommunity structure. Oecologia 105:545–551
Poulin R (1998) Evolutionary ecology of parasites: From individuals to communities. Chapman and Hall, London
Poulin R (2001) Another look at the richness of helminth communities in tropical freshwater fish. J Biogeogr 28:737–743
Poulin R (2003) The decay of similarity with geographical distance in parasite communities of vertebrate hosts. J Biogeogr 30:1609–1615
Poulin R, Mauritsen KN (2003) Large-scale determinants of trematode infections in intertidal gastropods. Mar Ecol Progr Ser 254:197–198
Poulin R, Morand S (2000) Diversity of parasites. Q Rev Biol 75:277–293
Poulin R, Morand S (2004) Parasite biodiversity. Smithsonian Inst Press, Washington
Poulin R, Mouillot D (2003) Parasite specialization from a phylogenetic perspective: A new index of host specificity. Parasitology 126:473–480
Poulin R, Mouillot D (2004) The relationship between specialization and local abundance: The case of helminth parasites of birds. Oecologia 140:372–378
Poulin R, Rohde K (1997) Comparing the richness of metazoan ectoparasite communities of marine fishes: Controlling for host phylogeny. Oecologia 110:278–283
Poulin R, Valtonen T (2002) The predictability of helminth community structure in space: A comparison of fish populations from adjacent lakes. Int J Parasitol 32:1235–1243

Poulin R, Mouillot D, George-Nascimento M (2003) The relationship between species richness and productivity in metazoan parasite communities. Oecologia 137:277–285
Proctor HC, Jones DN (2004) Geographical structuring of feather mite assemblages from the Australian brush-turkey (Aves: Megapodiidae). J Parasitol 90:60–66
Ranta E (1992) Gregariousness versus solitude: Another look at parasite faunal richness in Canadian freshwater fishes. Oecologia 89:150–152
Raibaut A, Combes C, Benoit F (1998) Analysis of the parasitic copepod species richness among Mediterranean fish. J Mar Syst 15:185–206
Rex MA (1981) Community structure in the deep sea benthos. Ann Rev Ecol Syst 12:331–353
Ribeiro JMC (1996) Common problems of arthropod vectors of disease. In: Beaty BJ, Marquardt WC (eds) The biology of disease vectors. Univ Press Colorado, Niwot, pp 25–33
Rohde K (1985) Increased viviparity of marine parasites at high latitudes. Hydrobiologia 137:197–201
Rohde K (1992) Latitudinal gradients in species diversity: The search for the primary cause. Oikos 65:514–527
Rohde K (1996) Rapoport's rule is a local phenomenon and cannot explain latitudinal gradients in species diversity. Biodivers Lett 3:10–13
Rohde K (1998) Is there a fixed number of niches for endoparasites of fish? Int J Parasitol 28:1861–1865
Rohde K (1999) Latitudinal gradients in species diversity and Rapoport's rule revisited: A review of recent work and what can parasites teach us about the causes of the gradients? Ecography 22:593–613
Rohde K, Heap M (1998) Latitudinal differences in species and community richness and in community structure of metazoan endo- and ectoparasites of marine teleost fish. Int J Parasitol 28:461–474
Rosenzweig ML (1992) Species diversity gradients: We know more and less than we thought. J Mammal 73:715–730
Rosenzweig ML (1995) Species diversity in space and time. Cambridge Univ Press, Cambridge
Ryckman RE (1971) Plague vector studies. Part I. The rate of transfer of fleas among *Citellus*, *Rattus* and *Sylvilagus* under field conditions in southern California. J Med Entomol 8:535–540
Schmid-Hempel P, Ebert D (2003) On the evolutionary ecology of specific immune defence. Trends Ecol Evol 18:27–32
Schradin C, Pillay N (2005) Demography of the striped mouse (*Rhabdomys pumilio*) in the succulent karoo. Mammal Biol 70:84–92
Sheldon BC, Verhulst S (1996) Ecological immunology: Costly parasite defenses and trade offs in evolutionary ecology. Trends Ecol Evol 11:317–321
Shenbrot GI (1992) Spatial structure and niche patterns of a rodent community in the south Bukhara desert (Middle Asia). Ecography 15:347–357

Shurin JB, Havel JE, Leibold MA, Pinel-Alloul B (2000) Local and regional zooplankton species richness: A scale-independent test for saturation. Ecology 81:3062–3073

Shorrocks B, Rosewell J (1986) Guild size in drosophilids: A simulation model. J Anim Ecol 55:527–541

Simkova A, Desdevises Y, Gelnar M, Morand S (2000) Coexistence of nine gill ectoparasites (*Dactylogyrus*: Monogenea) parasitizing the roach (*Rutilus rutilus* L.): History and present ecology. Int J Parasitol 30:1077–1088

Srivastava D (1999) Using local-regional richness plots to test for species saturation: Pitfalls and potentials. J Anim Ecol 68:1–16

Stanko M, Miklisova D, Goüy de Bellocq J, Morand S (2002) Mammal density and patterns of ectoparasite species richness and abundance. Oecologia 131:289–295

Stone L, Roberts A (1991) Conditions for a species to gain advantage from the presence of competitors. Ecology 72:1964–1972

Sukhdeo MVK, Bansemir AD (1996) Critical resources that influence habitat selection decisions by gastrointestinal helminth parasites. Int J Parasitol 26:483–498

Taylor LH, Mackinnon MJ, Read AF (1998) Virulence of mixed-clone and single-clone infections of the rodent malaria *Plasmodium chabaudi*. Evolution 52:583–591

Tella JL (2002) The evolutionary transition to coloniality promotes higher blood parasitism in birds. J Evol Biol 15:32–41

Terborgh JW, Faaborg J (1980) Saturation of bird communities in the West Indies. Amer Nat 116:178–195

Thrall PH, Antonovics J (1997) Polymorphism in sexual versus non-sexual disease transmission. Proc R Soc Lond B 264:581–87

Valone TJ, Hoffman CD (2002) Effects of regional pool size on local diversity in small scale annual plant communities. Ecol Lett 5:477–480

van der Plank JE (1963) Plant diseases: Epidemics and control. Acad Press, New York.

Vitone ND, Altizer S, Nunn CL (2004) Body size, diet and sociality influence the species richness of parasitic worms in anthropoid primates. Evol Ecol Res 6:183–199

Watters GT (1992) Unionids, fishes, and the species-area curve. J Biogeogr 19:481–49

13 Patterns of host specificity in parasites exploiting small mammals

Robert Poulin, Boris R. Krasnov and Serge Morand

1 Introductory remarks

Host specificity is one of the most fundamental properties of parasitic organisms. In simple terms, host specificity can be defined as the number and identity of host species that are used by a parasite population. Parasites that are highly host-specific will occur in a single host species, whereas generalist parasites will be dispersed unequally among individual hosts from several different species. From an evolutionary perspective, host specificity reflects the parasite's historical associations with its hosts (Brooks and McLennan 1993; Page 2003). The range of host species currently used by a parasite provides strong clues about the identity of the animal that served as host to the ancestral parasite, and their number provides an indication of whether the parasite has the ability to expand its host range by colonizing new species. Host specificity also relates to other evolutionary phenomena, such as the probability of parasite extinction (Koh et al. 2004). From an ecological perspective, host specificity mirrors the diversity of resources used by a parasitic organism, or the breadth of its niche (Futuyma and Moreno 1988). Thus, it allows one to make rough predictions about the likelihood that an introduced parasite species will become established and spread in a new ecosystem.

In this chapter, we will first briefly look at some general ecological features of small mammal hosts, and discuss whether these features should favour low or high host specificity in the parasite species exploiting small mammals. We will then briefly examine patterns of host specificity among species of helminth and arthropod parasites of small mammals, using the limited evidence available at present. Then, we will review the evolutionary forces that can select for either high or low host specificity, and the processes that allow a parasite to add new host species to its range. Finally, we will take advantage of a series of recent comparative studies on the fleas parasitic on small mammals, to address some fundamental questions about host specificity. Our aim in this chapter is not to provide an exhaustive review, but rather to highlight general patterns and the key processes that are likely to underpin those patterns.

2 Key ecological features of micromammals

Although they belong to different orders (i.e., Rodentia, Insectivora, Lagomorpha), small mammals do share several ecological traits. These traits are not possessed by every single micromammal species, but as a general rule, they do characterize the vast majority of micromammals.

First, micromammals, as their name implies, are small-bodied and they have short lifespans. In other words, compared to other mammals, they are smaller (generally less than 5 kg) and live short lives (less than 5 years). Second, again when compared with other mammals, they are characterized by high reproduction rates (i.e. high numbers of offspring per unit time) and high population densities. However, population densities of micromammals can fluctuate widely, both seasonally and from year to year. The combination of small body size, short lifespan and fluctuating population density means that as a resource base for parasites, micromammals may be unstable and unpredictable. Studies on fish parasites suggest that high host specificity is favoured only on stable, predictable resources, such as large-bodied and long-lived host species (Sasal et al. 1999; Desdevises et al. 2002). We might thus expect that, as a rule, parasites of micromammals are less host-specific than those exploiting larger, longer-lived mammalian hosts with more stable population densities.

The third main feature of micromammals concerns their habitat use and social structure. Small mammals are generally territorial, living in burrows or in nests within tree cavities. Whereas these burrows or nests represent ideal foci of parasite transmission among members of the same host species, this sort of habitat use limits opportunities for parasite transmission among different species. Unlike some other mammals, such as ungulates of different species that regularly gather around water holes on the African savannah, or bats of different species that roost together every night, most rodents and other micromammals do not come into contact with other species on a regular basis. This would constrain host-switching by parasites, and may lead to generally high levels of host specificity.

Thus, some features of micromammals seem likely to promote low levels of host specificity, relative to parasites of other mammals, whereas others appear likely to favour stricter host specificity. No comparative study to date has attempted to untangle the potential influences of these host traits on the evolution of host specificity. Clearly, other variables will be involved. For instance, the structural complexity of the habitat can influence the dispersal, and thus the colonising abilities, of parasites. Also, features of the parasites themselves, such as their mode of transmission, determine to a large extent whether the parasite will be highly specialized or not. For instance, among parasites of primates, those transmitted by sexual or other physical contact are highly specific, with two-thirds only known to infect a single host species and none capable of infecting host species belonging to different families (Pedersen et al. 2005). In contrast, parasites using intermediate hosts and transmitted via food are much less specific, with less than half restricted to a single host species, and more than a quarter exploiting hosts belonging to different mammalian orders (Pedersen et al. 2005). Nevertheless, it remains to be seen how much the features of small mammal hosts have contributed to the evolution of host specificity in their parasites.

3 General patterns of host specificity

Assessing levels of host specificity shown by parasites in natural systems requires an account of which host species are used among those that are potentially available to a parasite. Three important issues need to be considered before we provide an illustration of patterns of host specificity among parasites of micromammals.

The first issue concerns the exact operational definition of host specificity, the one that tells us exactly how to measure host specificity. The simplest definition is just the number of host species used by a parasite, from the list of host species available within a given area. Because it is easy to compute, this measure of host specificity is by far the most widely employed in the literature. However, it assumes that all host species used by a parasite are equal, whereas in fact they generally differ on two fundamental levels, and the mere number of host species used fails to capture these differences. First, from an ecological perspective, some host species are used more intensely than others. The prevalence, intensity or abundance of infection by a particular parasite usually varies widely among its host species, even within the same locality. Rohde (1994) proposed an index of specificity, based on the number of parasite individuals found in each host species, that takes these ecological differences into account. Second, from a phylogenetic perspective, some of the host species used by a parasite are likely to be closely related, whereas others are only distantly related. A parasite exploiting congeneric host species can be said to be more host-specific than one exploiting the same number of host species but from different families. Parasites with low host specificity are those capable of broad taxonomic "jumps" during their evolutionary history, regularly switching from one host species to a distantly related one. Poulin and Mouillot (2003) have proposed a useful measure of host specificity that takes host relationships into account, focusing on the average taxonomic distinctness of all host species used by a parasite species. It is even possible to combine both ecological and phylogenetic information into a single index of host specificity (Poulin and Mouillot 2005).

The second issue concerns sampling effort. High host specificity can be an artefact of inadequate sampling (Poulin 1998). Among species of parasites of freshwater fish, sampling effort explains much of the variability in host specificity: the number of known host species is strongly, positively correlated with the number of times a parasite species has been recorded in the literature (Poulin 1992). The same is true among tick species parasitic on mammals, and the distinction between highly specific and non-specific ticks may really be a distinction between rarely and frequently collected species (Klompen et al. 1996). Corrections for sampling effort are therefore necessary in any broad survey of host specificity.

The third issue is the potential impact of incorrect parasite species identification on estimates of host specificity. On the one hand, a species of parasite known to exploit several host species in a given area can in fact prove to be a complex of several species of superficially identical, highly host-specific parasites. With the recent application of molecular techniques to parasite systematics, several groups of cryptic species have been recognized where it was once thought there was a single species exploiting several host species (e.g., Hung et al. 1999; Blouin 2002; Leignel et al. 2002). On the other hand, what appears to be several related species of parasites exploiting several different host species can prove to be a single parasite species with low host specificity and whose morphology is influenced by the identity of the host species, with a resulting confusion in taxonomy. There are probably many instances in which "different" parasite species are in fact one and the same (e.g., Dallas et al. 2001), and these synonymies can also affect estimates of host specificity.

These caveats notwithstanding, a clear pattern emerges from any compilation of host specificity measures across any taxon of parasites infecting small mammals: when measured as the number of host species used, the distribution of host specificity values is typically strongly right-skewed. The majority of parasite species are highly host-specific, and there are only few true generalist species. For instance, among helminths parasitic in rodents and insectivores, between one-third and half of known parasite species in a region are strictly host-specific and found in only one host species (Fig. 1). The majority of other helminth species use 5 or fewer host species, and only very few species use 10 or more host species (Fig. 1).

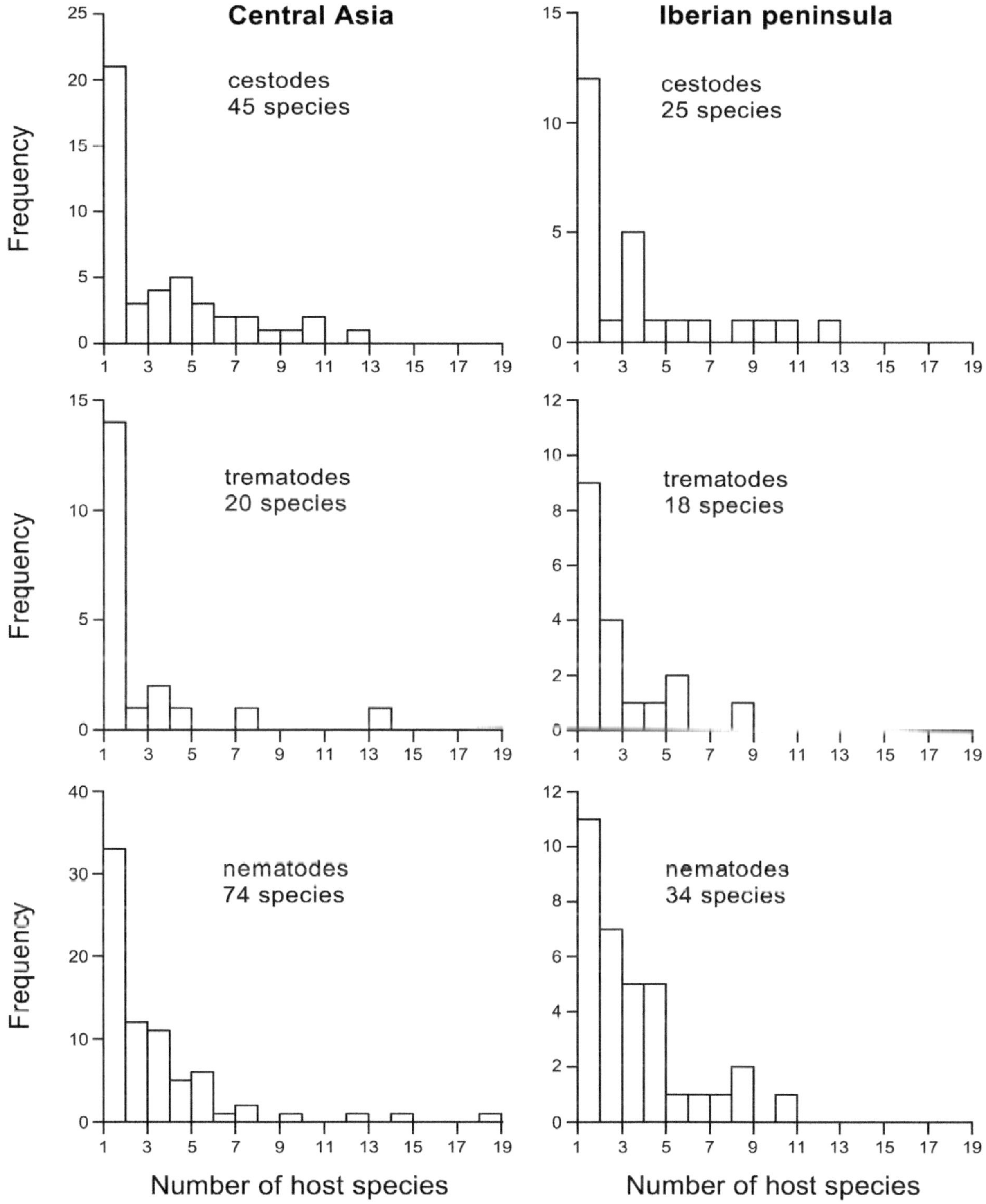

Fig. 1. Frequency distribution of host specificity (number of host species used) among species of cestodes, trematodes and nematodes parasitic in rodents and insectivores from Central Asia, and in rodents from the Iberian peninsula (data from Tokobaev 1976 and Feliu et al. 1997)

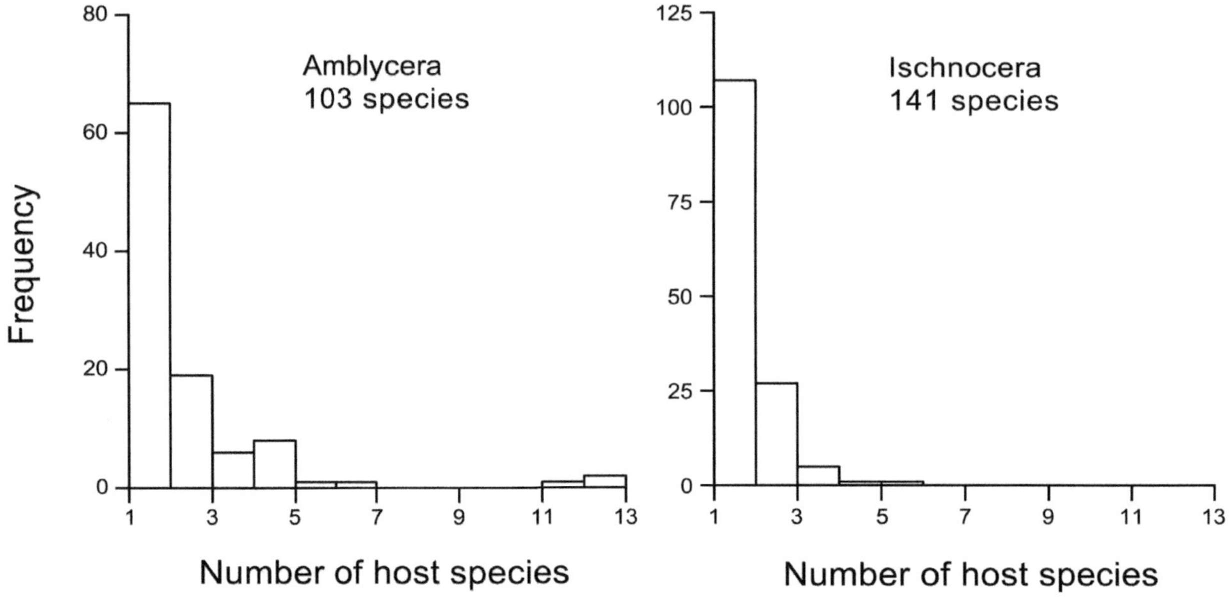

Fig. 2. Frequency distribution of host specificity (number of host species used) among all valid species of chewing lice worldwide known to parasitize rodents (the data, shown separately for the suborders Amblycera and Ischnocera, are from Price et al. 2003)

Among chewing lice ectoparasitic on rodents, the same general pattern is observed (Fig. 2). The data on chewing lice in Figure 2 come from a world checklist of host-parasite associations, as opposed from those on helminths in Figure 1, which originate from regional surveys. By considering the world fauna as the pool of potential hosts, the data on chewing lice should tend to "inflate" the numbers of host species that any given lice population could potentially use. In contrast, the data on chewing lice suggest that they might even be *more* host-specific than helminths. The vast majority of species occur on a single host species, or less frequently on two hosts (Fig. 2).

The apparently greater specificity of lice compared to helminths may be the consequence of their mode of transmission. In general, contact-transmitted parasites such as lice are expected to be more host-specific than parasites acquired via ingestion such as helminths (see Pedersen et al. 2005). This does not appear to apply to fleas, however. Among these ectoparasites of small mammals, the distribution of numbers of host species used is less right-skewed than for other parasite taxa (Fig. 3). Although many flea species are found on only one or two host species, there is a substantial number of flea species that can exploit several host species (Fig. 3). Mode of transmission is thus not necessarily constraining how many micromammal species can be used by a parasite.

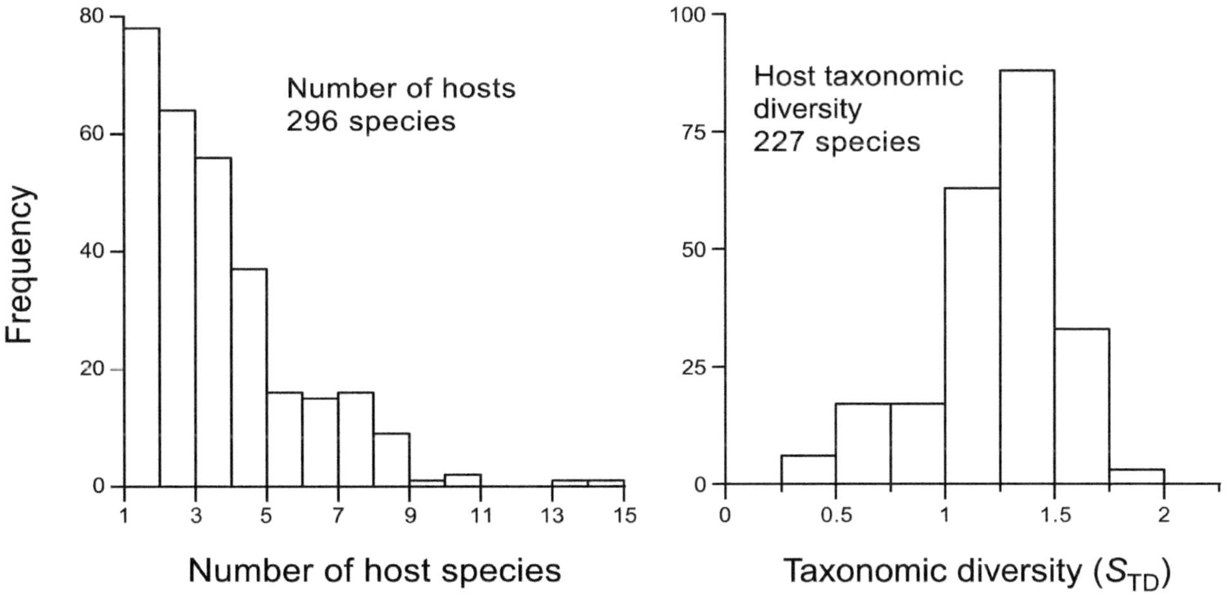

Fig. 3. Frequency distribution of host specificity among flea species parasitic on small mammals (rodents, insectivores and lagomorphs), measured as both the number of host species used and the taxonomic diversity of those host species. The latter measure is expressed as the index S_{TD}, which increases as a function of the average taxonomic distance among host species, and it is only computed for flea species with at least two host species (data from Poulin et al. 2006)

Overall, arthropod and helminth parasites of micromammals show roughly similar patterns of host specificity. Most species are very host-specific, exploiting only one, or maybe two or three, host species; nevertheless, there are also some generalist parasite species capable of exploiting between 4 and 10 host species, sometimes even more (Figs. 1-3). These general patterns are based on host specificity measured as the number of host species used. Other measures of host specificity could produce different patterns. For instance, applying a measure of the average taxonomic distinctness of host species, i.e. the index S_{TD} of Poulin and Mouillot (2003), to the flea data, generates a roughly symmetrical distribution of host specificity values (Fig. 3). This index provides a measure of the average taxonomic distance between host species, computed across all pairs of host species used; in the absence of a complete phylogeny of host species, the index serves as a good surrogate measure of host phylogenetic diversity (Poulin and Mouillot 2003). The most common values, corresponding to the peak of the distribution between values of 1 and 1.5 (Fig. 3), suggest that most flea species capable of exploiting two or more host species occur on hosts belonging either to the same genus, or to different genera within the same subfamily (see Poulin et al. 2006). Estimates of S_{TD} values for other groups of parasites of micromammals are not currently available, but are likely to be of similar magnitude.

How do these patterns of host specificity compare with those displayed by parasites of other groups of mammals? Very little information is available for other taxa of wild mammals, or it simply has not been assembled and compiled in a way that can be used for comparisons. The only group for which there are suitable data are primates. Helminths parasitic in primates show patterns of host specificity that are not too different from those shown by parasites of micromammals. Almost half of the helminth species parasitic in primates are strictly host-specific, i.e. they use a single host species (Pedersen et al. 2005). The data in Pedersen et al. (2005) do not allow the computation of the index S_{TD}, but given that only one helminth species in ten is capable of exploiting host species outside the order Primates, the taxonomic diversity of host species used is probably roughly similar for parasites of primates and parasites of micromammals.

4 Evolutionary processes shaping host specificity

The specificity of a parasite for its host species can be seen as the outcome of both historical events and current ecological conditions. We will examine how host specificity has evolved, first by looking at historical patterns of host-parasite associations, and then at smaller-scale phenomena determining whether new host species can be colonized. We illustrate these processes with examples from parasites of micromammals wherever possible.

4.1 Macroevolutionary processes

Comparisons between the phylogeny of a group of parasites and that of their hosts can shed light on the history of their association (Brooks and McLennan 1993; Page 2003). Mirror-image phylogenies would indicate strict cospeciation between parasites and their hosts. If each time a barrier to gene flow isolates two allopatric subpopulations of hosts, it also prevents gene flow between the two newly-created subpopulations of parasites, then the parasite would be forced to cospeciate with its host. Starting from an ancestral host species with one parasite species, cospeciation will result in n species of hosts and n species of parasites. This simple scenario would produce strictly-host specific parasites. Changes in host specificity occur when there are departures from a strict cospeciation pattern.

Ultimately, there are two ways in which host specificity can decrease over time, i.e. two kinds of evolutionary events through which a parasite can add new host species to its repertoire. First, the original host species can speciate without parallel speciation of the parasite, but with the parasite still capable of exploiting both daughter host species; this would result in the parasite occurring on two related host species instead of only occurring on the single ancestral host species. For example, the parasitic nematode *Longistriata caudabullata* is commonly found in short-tailed shrews of the genus *Blarina* in North America. A mitochondrial DNA phylogeny of nematode populations from different host species shows no subdivision according to host affiliation, suggesting extensive gene flow across host species boundaries (Brant and Orti 2003). In general, however, this is probably a rare situation. Second, and probably much more frequent, the addition of new host species to a parasite's repertoire can also result from host switching, or the colonization of new host species.

Host switching can be detected by comparing host and parasite phylogenies. It causes incongruence between the topologies of the two phylogenetic trees. In the classical example, the evolutionary history of several species of two related genera of chewing lice and their hosts, members of the rodent family Geomyidae (pocket gophers), was shown to be one of rather strict cospeciation with host switching playing a very minor role (Hafner and Nadler 1988, 1990; Hafner and Page 1995). Not only is there a clear congruence between the branching patterns of host and parasite phylogenies, but the timing of speciation events in both host and parasite lineages coincides remarkably well based on evidence from rates of molecular change. Not surprisingly, these lice species display strict host specificity, most being found on a single host species. This specificity is apparent at the morphological level, from the tight coupling between the width of the head groove on the head of lice used to attach to host hair, and the diameter of host hair shafts (Fig. 4). The fit between the groove on a given louse species and the hair of its particular host species resembles that between a lock and key (Morand et al. 2000). This cospeciation pattern may be the outcome of the social structure of pocket gophers and the transmission mode of the lice, both combining to greatly limit opportunities for host switching. In contrast, host switching appears to have been very common and cospeciation almost non-existent in lice parasitic on several species of one genus of rock wallabies in Australia (Barker 1991), and, on a larger scale, across all mammalian taxa (Taylor and Purvis 2003). Thus, the hosts' social structure may be a stronger barrier to host switching than the parasites' mode of transmission in the case of pocket gophers, since lice can switch hosts readily in other mammals.

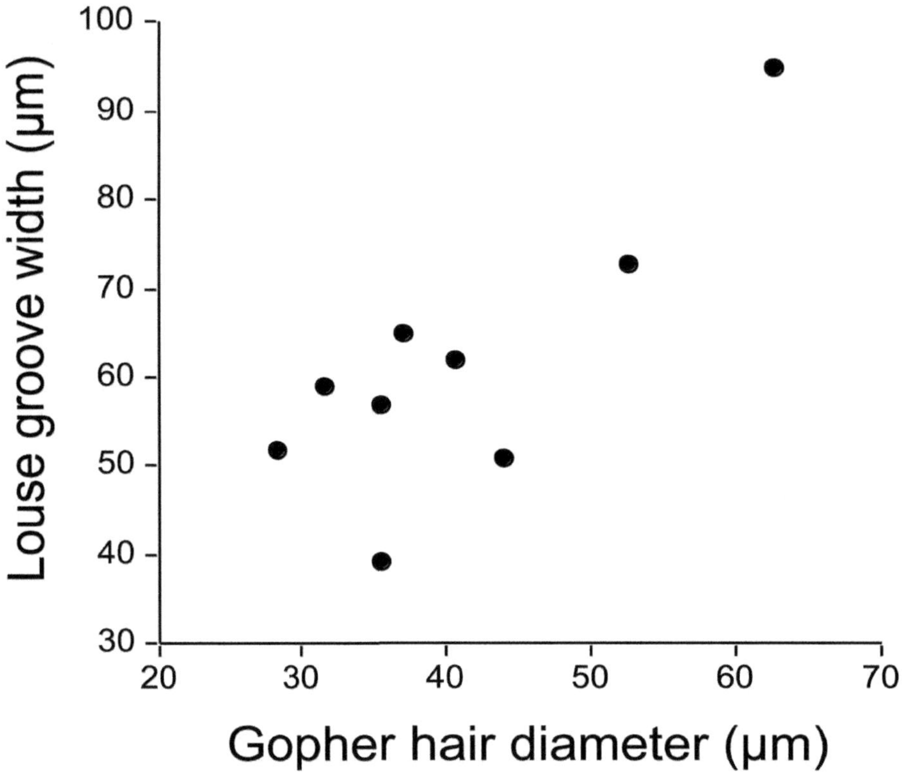

Fig. 4. Relationship between the width of the groove on the head of chewing lice and the average diameter of body hairs from their pocket gopher host species. Each point represents a different louse-gopher species combination (modified from Morand et al. 2000)

There have been few other comparisons of host and parasite phylogenies involving micromammals and their parasites. Krasnov and Shenbrot (2002) tried to reconcile the phylogenies of jerboas and their flea parasites, and concluded that host switching had been common in these host-parasite associations. They proposed that ecological and geographical factors can allow host-switching and override any tendency toward strict cospeciation expected from the transmission mode of these parasites. Brant and Gardner (2000) also concluded that rampant host switching is a better hypothesis to explain the coevolutionary history of filarioid nematodes of the genus *Litomosoides* with their hosts, which include mainly rodents but also bats and marsupials. This is not too surprising, since these nematodes are transmitted by mobile vectors, i.e. blood-sucking dipterans. Phylogeographic studies, although focused on shorter time scales, can also provide information on the evolutionary "fidelity" of parasites to their hosts. For instance, populations of the cestode *Paranoplocephala arctica* form distinct clades across their Holarctic range, showing significant congruence with similar subdivisions existing among the populations and species of their rodent hosts, lemmings of the genus *Dicrostonyx* (Wickström et al. 2003). Similarly, populations of the nematode *Heligmosomoides polygyrus* form three genetic and geographical lineages across their European range, which are congruent with those found among the populations of their rodent host, the field mouse *Apodemus sylvaticus* (Nieberding et al. 2004). In both the nematode and the mouse, postglacial recolonization of northwest Europe came from the Iberian populations, and not from other southern populations (Nieberding et al. 2005). These results suggest that helminth parasites acquired by ingestion, like *P. arctica* and *H. polygyrus*, can evolve with micromammal hosts following a pattern consistent with cospeciation rather than rampant host switching.

To date, the few available studies on the evolutionary history of micromammals and their parasites tend to suggest that strict cospeciation is perhaps a more common pattern than rampant host switching. This may have something to do with the ecological features of small mammals, and could serve to constrain host specificity in these parasites. However, parasites with modes of transmission that can overcome these host features (such as the vector-transmitted nematodes *Litomosoides* spp.; Brant and Gardner 2000) can also evolve following different scenarios. Clearly, we need many more cophylogenetic and phylogeographic studies before any robust conclusion.

4.2 Microevolutionary processes

The above discussion focused on the macroevolutionary history of host-parasite associations and host specificity. On a microevolutionary scale, many phenomena can facilitate host switching and subsequent decreases in host specificity, or, conversely, promote greater specialization on fewer host species. We now discuss the processes by which natural selection may favour changes in host specificity.

The central question concerns the direction of selection: should we expect natural selection to generally favour increases or decreases in host specificity? There are no easy answers. Just as parasites may be selected to increase the range of hosts in which they can successfully develop, they may also sometimes face selection for greater specialization through a narrowing of their range of suitable hosts. The growth and fecundity of any given parasite vary among host species. If selection can fine-tune the mechanisms of host infection to ensure that fewer host species are encountered, then one would predict that host species in which development is suboptimal will eventually be excluded. This would result in a narrow host range comprising only host species on which parasite fitness is high. Although greater specialization on fewer host species can be advantageous, it also links the fate of parasites to that of their hosts and can make highly host-specific parasites more prone to local extinction. There are thus pros and cons associated with both high and low host specificity. We might expect a trade-off between the ability to use many host species and the average fitness achieved in these hosts (Ward 1992). Close adaptation to one host species may only be achieved at the expense of adaptations to other host species. Given that different host species have different defense systems, investing in many counter-adaptations should have a fitness cost for the parasite: a jack-of-all-trades may be a master of none. Different parasite species may achieve greater overall fitness at different points along the continuum of strategies between the high-specificity-low-mean-abundance and low-specificity-high-mean-abundance extremes. This kind of trade-off is often used to explain the host specificity of phytophagous arthropods (Fry 1990). We will further address the trade-off issue using a study on fleas in the next section.

Assuming that lower average fitness is not constraining parasites from expanding to new host species, then what is? On microevolutionary time scales, host specificity is mainly determined by opportunities for colonization and availability of suitable host species. Opportunities can arise in many ways. Hybridisation between host species, for example, can create a genetic and ecological bridge between host species and allow the colonization of one host by parasites from the other (Floate and Whitham 1993). The intermediate ecological and physiological characteristics of hybrids may provide stepping stones facilitating host-switching between two different host species that would otherwise be too distinct to allow parasite colonization. Indeed, two studies have shown that the resistance of rodents to infections by nematodes and cestodes breaks down in hybrid zones (Sage et al. 1986; Moulia et al. 1991). In European areas where the mice *Mus musculus* and *Mus domesticus* hybridise, hybrids acquire higher parasite loads than either parent species (Fig. 5). However, hybridisation between closely related species of micromammals is probably not widespread, and this mechanism may rarely provide opportunities for host switching.

A range of immunological or physiological mechanisms serving to maintain host specificity can be identified by experimental studies. For instance, the nematode *Strongyloides ratti*, a gastrointestinal parasite of rats, has only a limited attachment success and achieves reduced fecundity in mice, even in immunosuppressed mice (Gemmill et al. 2000). Experimental selection, achieved by serial passage in mice for 18 generations, failed to improve the performance of *S. ratti* in this novel host, suggesting that factors stemming from the different physiologies of rats and mice are responsible for maintaining host specificity (Gemmill et al. 2000).

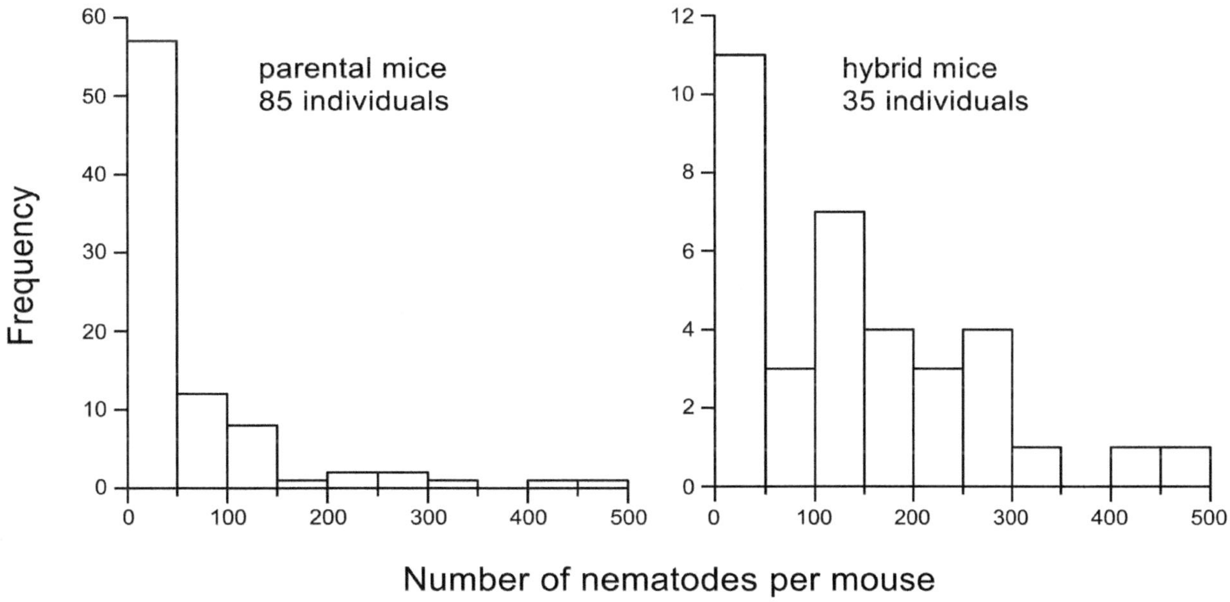

Fig. 5. Frequency distribution of numbers of parasitic nematodes per mouse, among mice belonging to either of two parental species (*Mus musculus* and *M. domesticus*), or among hybrids of these two species. All mice were collected in a Danish hybrid zone. Worms of two nematode species, *Aspiculuris tetraptera* and *Syphacia obvelata*, are combined (data from Moulia et al. 1991)

Strict host specificity is therefore not always easily overcome. Recent models offer reasons for this observation. In these models, adaptation to a particular host species occurs via the fixation of alleles whose beneficial effects are host-specific; this is more rapid and more likely to occur in parasite populations restricted to that host species than in parasite populations spread among several host species (Kawecki 1997, 1998). These models predict that parasite species that begin as generalists gradually lose the ability to exploit seldom-encountered host species and eventually exclude them altogether from their range of suitable alternatives.

Kawecki's (1997,1998) models suggest that local adaptation could maintain host specificity. The selection of greater host specificity in parasites on a local scale would be apparent when comparing the specificity of different populations of the same parasite species exploiting different populations of the same host species. In a review of the literature on local adaptation by parasites, Lajeunesse and Forbes (2002) found that local adaptation is more likely to be observed in parasite species that already show some host specificity, i.e. parasites that only exploit few host species. This makes sense because generalist parasites exploiting many host species would have difficulty simultaneously tracking the changes in genotype frequencies in different local populations of their different host species. In general, though, after generations of isolation from other host genotypes, parasites may lose the ability to infect allopatric hosts in favour of a greater specialization for the local host genotypes. Alternatively, parasites can retain the ability to infect allopatric genotypes but achieve lower fitness when exploiting them.

These ideas have not yet been tested using micromammals and their parasites. The limited evidence available, however, suggests that local adaptation of this nature, serving to promote high levels of host specificity, may not be common in nature. In the trematode *Schistosoma mansoni* parasitic in rats, although some genetic differentiation exists among populations inhabiting fragmented marshy habitats on the island of Guadeloupe, there is also evidence of much gene flow (Sire et al. 2001). This parasite uses two hosts, snail first intermediate hosts and rat definitive hosts. Rats are clearly more vagile than snails, and there is good evidence showing that exchanges of parasites among populations are indeed mediated almost entirely by rat movements (Prugnolle et al. 2005). Whether or not the parasite could infect other hosts, gene flow maintained by one host would prevent local adaptation. Studies of geographical population structure in nematodes parasitic in mammals indicate that, overall, there is only very little genetic structure, a pattern consistent with high levels of gene flow among populations (Anderson et al. 1998). This is true even in situations where genetic structure is expected a priori, such as in the nematode *Strongyloides ratti*, parasitic in wild rats, that reproduces mainly by parthenogenesis (Fisher and Viney 1998). Here again, gene flow prevents local adaptation. Paterson (2005) has tested whether the infectivity of particular genotypes of *S. ratti* depends on the particular host (rat) genotype in which it occurs, and found no evidence of specificity between host and parasite at the genotype level. Therefore, studies to date on micromammals and their parasites do not support the possibility that fine-tuned, local adaptation is promoting tight host specificity.

5 Fleas on micromammals: a case study

There have been rather few studies of host specificity in parasites of micromammals, not nearly enough to allow any general conclusion. Recently, one group of parasites of small mammals has been the focus of several investigations. Fleas (Siphonaptera) are common haematophagous ectoparasites of rodents, insectivores, lagomorphs and other small mammals. They usually alternate between periods when they occur on the body of their hosts and periods when they occur in their hosts' burrows or nests. In most cases, pre-imaginal development is entirely off-host; the larvae are usually not parasitic and feed on organic debris in the burrow or nest of the host. Fleas range from highly host-specific to host opportunistic (Marshall 1981). Here, we use recent comparative studies of fleas parasitic on small mammals to address three fundamental questions about the ecology and evolution of host specificity.

5.1 Is host specificity a species character?

In this chapter, we have treated host specificity as a species character, i.e. a trait that is as characteristic of a species as its morphological features. In fact, host specificity varies among populations of the same parasite species. Whereas the size and shape of a parasite will be more-or-less constant among different populations, host specificity is influenced by the local availability of host species. If variation in host specificity among populations of the same parasite species is less pronounced than variation in host specificity among different parasite species, however, host specificity would still represent a species trait. It would be a variable trait, but one that remains constrained within a range of values.

Krasnov et al. (2004a) investigated geographic variation in host specificity of fleas using data from 21 regional surveys, mainly from the Palearctic. They performed a repeatability analysis using 118 flea species that were recorded in at least two of the regions, to determine whether host specificity showed some constancy across populations of the same flea species. Whether measured as the number of host species used or as the taxonomic distinctness (index S_{TD}) of these hosts, host specificity estimates from the same flea species were more similar to each other than expected by chance, but they varied significantly among flea species (Fig. 6). Although statistically significant, the similarity among host specificity values from different populations of the same flea species is still subject to wide variations (Fig. 6). To some extent, this reflects geographic differences in host availability. Within a given region, the subset of host species used by a flea species tends to be taxonomically constrained, i.e. the host species used by a flea are more closely related to each other than if they were subsets of species drawn at random from the regional pool of available host species (Krasnov et al. 2004a). The absence of one or a few host species from a region can affect the realised host specificity of a flea in that region, and thus contribute to variability in host specificity across regions. In addition, local environment factors, such as mean temperature and precipitation levels, can also affect realised host specificity (Krasnov et al. 2004a). Nevertheless, one can see a certain predictability superimposed over this geographical variation, such that host specificity in fleas, though far from constant, can still be considered as a species character.

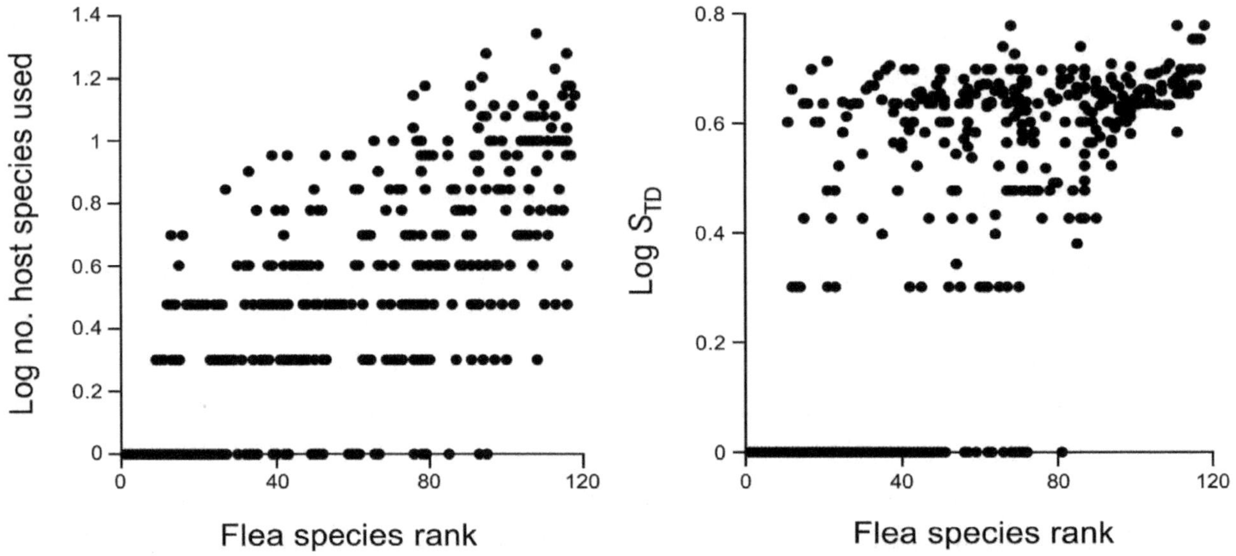

Fig. 6. Rank plots of number of host species used and average taxonomic distinctness, S_{TD}, of these hosts across 118 flea species ranked from lowest to highest mean host specificity. All population estimates are plotted for each species; the number of host species has been corrected for sampling effort, whereas S_{TD} has been corrected for the number of host species used. If geographic variation were small within compared to between flea species, we would expect the points to fall in an area of the plot stretching from the lower left to the upper right corner, with few points in either the upper left or lower right corner (data from Krasnov et al. 2004a)

5.2 Is there a trade-off between number of hosts used and the average fitness achieved in these hosts?

As discussed in the previous section, we might expect a trade-off between the ability to use many host species and the average fitness achieved in these hosts. The rationale behind the trade-off is that close adaptation to one host species may only be achieved at the expense of adaptations to other host species. Given that different host species have different defense systems, investing in many counter-adaptations should have a fitness cost for the parasite: a jack-of-all-trades may be a master of none.

Looking at flea species parasitic on small mammals, it is clear that any given flea species does not do equally well on all its potential host species. Fleas typically achieve much higher abundance (average number of individual parasites per host) on one host species (Krasnov et al. 2004b). If we take this to be the principal host species, then it is also clear that the abundance of a flea on its auxiliary host species decreases with increasing taxonomic distance between an auxiliary host and the principal host species (Krasnov et al. 2004b). The success of a flea following a host switch is thus lower if the newly colonized host is not a close relative of the original host. However, most host species used by a flea tend to fall within the same taxon (e.g., same rodent subfamily), with only rarely one or two host species belonging to other taxa (e.g. another order, like insectivores).

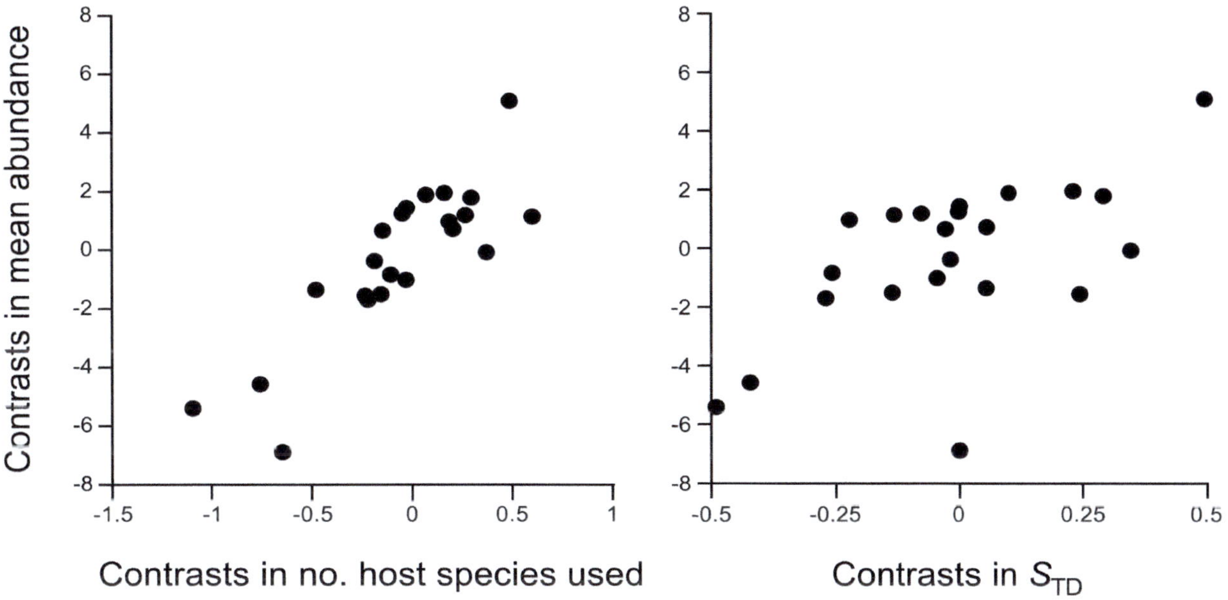

Fig. 7. Relationship between the mean abundance achieved by a flea across all its host species, and either the number of host species used or their taxonomic distinctness, measured by the index S_{TD}. The data are for flea species parasitising rodents in Mongolia; each point represents a phylogenetically independent contrast, with abundance corrected for both sampling effort and host body surface area (data from Krasnov et al. 2004c)

So, is there a trade-off between *average* abundance, i.e. overall fitness, and host specificity among fleas parasitic on small mammals? Using data from 20 regional surveys of fleas on micromammals, Krasnov et al. (2004c) found that there are generally strong *positive* relationships between parasite abundance and either the number of host species used or the index S_{TD}. These relationships were significant in three-quarters of the regions investigated (see Fig. 7 for an example). This finding indicates that there is no general trade-off between how many host species a parasite can use and how well it does on them. In fleas, the opposite happens: whatever features of fleas make them successful on a host also allows them to colonize other host species. In fact, it also appears that fleas using either many host species or taxonomically diverse host species achieve not only greater average abundance, but also a broader geographical range than the more host-specific fleas (Krasnov et al. 2005). Perhaps the epidemiological advantages of having many host species outweigh the physiological costs of adaptations against their immune defences. Mathematical models predict that for parasites with direct, one-host life cycles in which transmission is strongly dependent on host density, such as fleas, the more host species are exploited in a locality, the greater the probability that the parasite population will persist and spread (Dobson 2004). For indirectly transmitted parasites, however, the models predict the exact opposite (Dobson 2004). These different dynamical features of parasite populations with different modes of transmission may explain the complete absence of any trade-off in parasitic fleas. The next step would now be to perform similar analyses for other taxa parasitic on small mammals.

5.3 Is the evolution of host specificity directional and irreversible?

Parasite specialization is generally presumed to be irreversible, leading into evolutionary dead ends that do not give rise to new lineages. On the one hand, specialist taxa, capable of using only a narrow range of host species, should be less likely to colonize new hosts, and therefore the potential of specialists to give rise to new lineages should be limited (Jaenike 1990). If this is so, we might expect that generalists can evolve into specialists, but that the likelihood of specialists evolving into generalists would be much lower. Thus, within a clade, the more specialized species should on average be the more derived, i.e. the more recent ones. On the other hand, specialist taxa should be more prone to extinction than generalists, because of their strict dependence on a narrow range of host species, and thus we might expect generalist taxa to be favoured and to proliferate over evolutionary time. It is therefore not straightforward to predict in which direction host specificity will evolve in a given clade, i.e. whether it will tend to increase or decrease over evolutionary time. Recent studies on other animal groups have challenged the paradigm that specialization is both directional and irreversible. In his review of studies on evolutionary transitions between specialized and generalized host-plant use, Nosil (2002) found that generalist-to-specialist transitions were more frequent overall among phytophagous insects, but that in some groups the opposite was true. Also, Stireman (2005) reported that transitions from specialist to generalist strategies have occurred more frequently than the reverse during the evolutionary history of tachinid flies, a group of endoparasitoids of insect hosts. The result is that generalist tachinid species tend to be the most derived, i.e. they tend to occupy branch tips in the phylogeny of the group (Stireman, 2005).

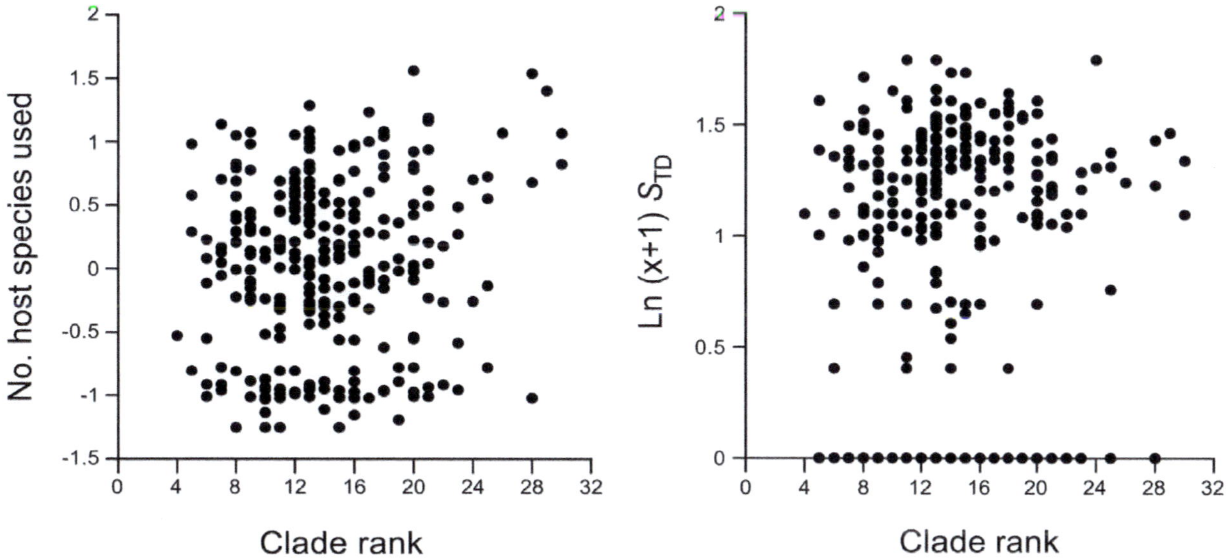

Fig. 8. Relationship between either the number of host species used by a flea species or the taxonomic distinctness of these hosts (measured by the index S_{TD}) and clade rank, among 297 species of fleas parasitic on small mammals. The number of host species used is corrected for sampling effort, i.e. data shown are residuals of the regression of the log-transformed number of host species on which the flea species was found against the log-transformed number of host individuals sampled (data from Poulin et al. 2006)

So what about true parasites? Poulin et al. (2006) tested for directionality in the evolution of host specificity in fleas parasitic on small mammals. They determined whether host specificity, measured both as the number of host species used and their taxonomic diversity, i.e. the index S_{TD}, was related to clade rank of the flea species. Clade rank is the number of branching events between an extant species and the root of the phylogenetic tree; it can be used to distinguish flea species that are basal in the phylogenetic tree from those that are highly derived, i.e. those with low and high clade rank, respectively (Poulin et al. 2006). Both across all flea species in the dataset, and within some families or genera, there were weak positive relationships between clade rank and the number of host species used, but none with the index S_{TD} (Fig. 8). These results suggest a slight evolutionary trend of decreasing host specificity, with many flea lineages increasing over evolutionary time the number of host species they can exploit. However, using a more conservative test, these trends could not be distinguished from a non-directional random walk model, suggesting a lack of directionality in the evolution of host specificity in fleas (Poulin et al.

2006). This can be seen from the scatter of points in Fig. 8. Given the fact that generalist fleas achieve higher abundances on their hosts, as we discussed earlier, it is not surprising that host specificity shows signs, albeit not strong ones, of having loosened over time. Once again, evidence from other parasite taxa would be welcome.

6 Concluding remarks

Host specificity is arguably one of the most important properties of a parasite, because it can determine, among other things, whether a parasite can survive the extinction of a host species, whether a parasite has the potential to invade new habitats such as islands, or whether a parasite can become established and spread following its introduction to a new geographical area. Macroparasites of micromammals have received relatively little attention in this regard. The available evidence suggests that some ecological features of small mammals may interact with parasite transmission mode to determine what levels of host specificity are observed. Still, large-scale patterns of host specificity have only been investigated in fleas, and studies on other parasite taxa are definitely needed. In addition, since many rodent species are now universal laboratory models in many branches of biology, it should prove possible to investigate host specificity in an experimental context. For instance, the mechanisms responsible for the failure or success of a particular parasite species in different host species could be examined using controlled laboratory infections. In addition, selection experiments like that of Gemmill et al. (2000) can be envisaged with host species like mice with short generation times, to track the evolution of host specificity under different selection regimes. The evolution and ecology of host specificity will remain an important research area for years to come. This is particularly true in the light of the global environmental changes occurring at present, and the possibility that, by altering transmission conditions, they will lead to the expansion of the host range of many parasite species.

References

Anderson TJC, Blouin MS, Beech RN (1998) Population biology of parasitic nematodes: Applications of genetic markers. Adv Parasitol 41:219–283

Barker SC (1991) Evolution of host-parasite associations among species of lice and rock-wallabies: Coevolution? Int J Parasitol 21:497–501

Blouin MS (2002) Molecular prospecting for cryptic species of nematodes: Mitochondrial DNA versus internal transcribed spacer. Int J Parasitol 32:527–531

Brant SV, Gardner SL (2000) Phylogeny of species of the genus *Litomosoides* (Nematoda: Onchocercidae): Evidence of rampant host switching. J Parasitol 86:545–554

Brant SV, Orti G (2003) Evidence for gene flow in parasitic nematodes between two host species of shrews. Mol Ecol 12:2853–2859

Brooks DR, McLennan DA (1993) Parascript: Parasites and the language of evolution. Smithsonian Inst Press, Washington

Dallas JF, Irvine RJ, Halvorsen O (2001) DNA evidence that *Marshallagia marshalli* Ransom, 1907 and *M. occidentalis* Ransom, 1907 (Nematoda: Ostertagiinae) from Svalbard reindeer are conspecific. Syst Parasitol 50:101–103

Desdevises Y, Morand S, Legendre P (2002) Evolution and determinants of host specificity in the genus *Lamellodiscus* (Monogenea). Biol J Linn Soc 77:431–443

Dobson AP (2004) Population dynamics of pathogens with multiple host species. Amer Nat 164:S64–S78

Feliu C, Renaud F, Catzeflis F, Hugot J-P, Durand P, Morand S (1997) A comparative analysis of parasite species richness of Iberian rodents. Parasitology 115:453–466

Fisher MC, Viney ME (1998) The population genetic structure of the facultatively sexual parasitic nematode *Strongyloides ratti* in wild rats. Proc R Soc Lond B 265:703–709

Floate KD, Whitham TG (1993) The "Hybrid Bridge" hypothesis: Host shifting via plant hybrid swarms. Amer Nat 141:651–662

Fry JD (1990) Trade-offs in fitness on different hosts: Evidence from a selection experiment with a phytophagous mite. Amer Nat 136:569–580

Futuyma DJ, Moreno G (1988) The evolution of ecological specialization. Ann Rev Ecol Syst 19:207–233

Gemmill AW, Viney ME, Read AF (2000) The evolutionary ecology of host-specificity: Experimental studies with *Strongyloides ratti*. Parasitology 120:429–437

Hafner MS, Nadler SA (1988) Phylogenetic trees support the coevolution of parasites and their hosts. Nature 332:258–259

Hafner MS, Nadler SA (1990) Cospeciation in host parasite assemblages: Comparative analysis of rates of evolution and timing of cospeciation events. Syst Zool 39:192–204

Hafner MS, Page RDM (1995) Molecular phylogenies and host-parasite cospeciation: Gophers and lice as a model system. Phil Trans R Soc Lond B 349:77–83

Hung GC, Chilton NB, Beveridge I, Zhu XQ, Lichtenfels JR, Gasser RB (1999) Molecular evidence for cryptic species within *Cylicostephanus minutus* (Nematoda: Strongylidae). Int J Parasitol 29:285–291

Jaenike J (1990) Host specialization in phytophagous insects. Ann Rev Ecol Syst 21:243–273

Kawecki TJ (1997) Sympatric speciation via habitat specialization driven by deleterious mutations. Evolution 51:1749–1761

Kawecki TJ (1998) Red Queen meets Santa Rosalia: Arms races and the evolution of host specialization in organisms with parasitic lifestyles. Amer Nat 152:635–651

Klompen JSH, Black WC, Keirans JE, Oliver Jr JH (1996) Evolution of ticks. Ann Rev Entomol 41:141–161

Koh LP, Dunn RR, Sodhi NS, Colwell RK, Proctor HC, Smith VS (2004) Species coextinctions and the biodiversity crisis. Science 305:1632–1634

Krasnov BR, Shenbrot GI (2002) Coevolutionary events in the history of association between jerboas (Rodentia: Dipodidae) and their flea parasites. Israel J Zool 48:331–350

Krasnov BR, Mouillot D, Shenbrot GI, Khokhlova IS, Poulin R (2004a) Geographical variation in host specificity of fleas (Siphonaptera) parasitic on small mammals: The influence of phylogeny and local environmental conditions. Ecography 27:787–797

Krasnov BR, Shenbrot GI, Khokhlova IS, Poulin R (2004b) Relationships between parasite abundance and the taxonomic distance among a parasite's host species: An example with fleas parasitic on small mammals. Int J Parasitol 34:1289–1297

Krasnov BR, Poulin R, Shenbrot GI, Mouillot D, Khokhlova IS (2004c) Ectoparasitic "jacks-of-all-trades": Relationship between abundance and host specificity in fleas (Siphonaptera) parasitic on small mammals. Amer Nat 164:506–516

Krasnov BR, Poulin R, Shenbrot GI, Mouillot D, Khokhlova IS (2005) Host specificity and geographic range in haematophagous ectoparasites. Oikos 108:449–456

Lajeunesse MJ, Forbes MR (2002) Host range and local parasite adaptation. Proc R Soc Lond B 269:703–710

Leignel V, Cabaret J, Humbert JF (2002) New molecular evidence that *Teladorsagia circumcincta* (Nematoda: Trichostrongylidea) is a species complex. J Parasitol 88:135–140

Marshall AG (1981) The ecology of ectoparasitic insects. Acad Press, London

Morand S, Hafner MS, Page RDM, Reed DL (2000) Comparative body size relationships in pocket gophers and their chewing lice. Biol J Linn Soc 70:239–249

Moulia C, Aussel JP, Bonhomme F, Boursot P, Nielsen JT, Renaud F (1991) Wormy mice in a hybrid zone: A genetic control of susceptibility to parasite infection. J Evol Biol 4:679–687

Nieberding C, Morand S, Libois R, Michaux JR (2004) A parasite reveals the cryptic phylogeographic history of its host. Proc R Soc Lond B 271:2559–2568

Nieberding C, Libois R, Douady CJ, Morand S, Michaux JR (2005) Phylogeography of a nematode (*Heligmosomoides polygyrus*) in the western Palearctic region: Persistence of northern cryptic populations during ice ages? Mol Ecol 14:765–779

Nosil P (2002) Transition rates between specialization and generalization in phytophagous insects. Evolution 56:1701–1706

Page RDM (ed) (2003) Tangled trees: Phylogeny, cospeciation, and coevolution. Univ Chicago Press, Chicago

Paterson S (2005) No evidence for specificity between host and parasite genotypes in experimental *Strongyloides ratti* (Nematoda) infections. Int J Parasitol 35:1539-1545

Pedersen AB, Altizer S, Poss M, Cunningham AA, Nunn CL (2005) Patterns of host specificity and transmission among parasites of wild primates. Int J Parasitol 35:647–657

Poulin R (1992) Determinants of host-specificity in parasites of freshwater fishes. Int J Parasitol 22:753–758

Poulin R (1998) Evolutionary ecology of parasites: From individuals to communities. Chapman and Hall, London

Poulin R, Mouillot D (2003) Parasite specialization from a phylogenetic perspective: A new index of host specificity. Parasitology 126:473–480

Poulin R, Mouillot D (2005) Combining phylogenetic and ecological information into a new index of host specificity. J Parasitol 91:511–514

Poulin R, Krasnov BR, Shenbrot GI, Mouillot D, Khokhlova IS (2006) Evolution of host specificity in fleas: Is it directional and irreversible? Int J Parasitol 36:185–191

Price RD, Hellenthal RA, Palma RL, Johnson KP, Clayton DH (2003) The chewing lice: World checklist and biological overview. Illinois Natural History Survey Special Publication no. 24, Illinois, USA

Prugnolle F, Théron A, Pointier J-P, Jabbour-Zahab R, Jarne P, Durand P, De Meeûs T (2005) Dispersal in a parasitic worm and its two hosts: Consequence for local adaptation. Evolution 59:296–303

Rohde K (1994) Niche restriction in parasites: Proximate and ultimate causes. Parasitology 109:S69–S84

Sage RD, Heyneman D, Lim K-C, Wilson AC (1986) Wormy mice in a hybrid zone. Nature 324:60–63

Sasal P, Trouvé S, Müller-Graf C, Morand S (1999) Specificity and host predictability: A comparative analysis among monogenean parasites of fish. J Anim Ecol 68:437–444

Sire C, Langand J, Barral V, Théron A (2001) Parasite (*Schistosoma mansoni*) and host (*Biomphalaria glabrata*) genetic diversity: Population structure in a fragmented landscape. Parasitology 122:545–554

Stireman III JO (2005) The evolution of generalization? Parasitoid flies and the perils of inferring host range evolution from phylogenies. J Evol Biol 18:325–336

Taylor J, Purvis A (2003) Have mammals and their chewing lice diversified in parallel? In: Page RDM (ed) Tangled trees: Phylogeny, cospeciation, and coevolution. Univ Chicago Press, Chicago, pp 240–261

Tokobaev MM (1976) Helminths of the wild mammals of Middle Asia (in Russian). Ylim Press, Frunze

Ward SA (1992) Assessing functional explanations of host-specificity. Amer Nat 139:883–891

Wickström LM, Haukisalmi V, Varis S, Hantula J, Fedorov VB, Henttonen H (2003) Phylogeography of the circumpolar *Paranoplocephala arctica* species complex (Cestoda: Anoplocephalidae) parasitizing collared lemmings (*Dicrostonyx* spp.). Mol Ecol 12:3359–3371

14 Coevolution of macroparasites and their small mammalian hosts: Cophylogeny and coadaptation[1]

Jean-Pierre Hugot

1 Introductory remarks

Living organisms permanently interact with each other. Some of these interactions, in particular those that organisms of interest have with organisms belonging to other species, can become necessary. This results in long-standing associations such as commensalism and parasitism. Many authors have suggested that the phylogenetic relationships of highly host specific parasites would provide valuable information about the evolutionary history of their hosts (see review in Brooks and McLennan 1993). This brought to light the general concept of coevolution that has progressively emerged as a particular field of study in evolutionary biology since the beginnings of the 1980s.

Coevolution may be defined as mutual evolutionary influence between two species when each of them exerts selective pressure on the other, so that they evolve together. Although this term is usually attributed to Ehrlich and Raven's (1964) study of butterflies on plants, the idea was already apparent in the "Origin of species" by Darwin. Host-parasite relationships have been particularly often investigated in this regard because of their impact on human health and agricultural production and because they offer numerous different models. However, some particular situations allow more accurate observation of the phenomenon; for example, when coevolution is going on between pairs of species from each group. Patterns of paired coevolution are particularly frequent in the evolution of hosts and parasites.

[1]"What an interesting investigation would be the comparison of the parasites of the closely allied and representative birds of the two countries." (Darwin, 7 November 1844, letter to Henry Denny)

Coevolution between hosts and parasites covers different concepts, which sometimes are confused such as, for instance, cophylogeny (also called cospeciation) and coadaptation. Coadaptation may be defined as the evolution of reciprocal adaptations in hosts and parasites and is probably very frequent. Comparison among species using standard statistical tests is the most common technique for testing hypotheses of how organisms are adapted to each other or to their environments. To discuss the results of such tests we need to be able to control whether related species share traits by common descent rather than through independent adaptation. Thus, the respective phylogenies of the hosts and parasites need to be known and compared.

Life histories of two different lineages are sometimes so intimately linked that speciation in one group induces a parallel speciation event in the other. If cophylogeny were the only process occurring, the host and parasite phylogenies would exactly mirror each other. In such a case, comparison of host and parasite cladograms is crucial. However, cophylogeny does not mean reciprocal phenomena. For example, speciation of the host may induce the speciation of the parasite without parasite-induced speciation of the host. Again, one needs to know the evolutionary history before deciding which type of co-"evolution" is observed.

Finally, whatever coevolutionary processes are considered, phylogenetic analysis is necessary and constitutes the first step. An important component of any study dealing with coevolution and comparing host vs. parasite characters is to distinguish which traits or phylogenies may be considered resulting from transmission by descent and which ones may be attributed to other mechanisms. In this chapter I will present an example of comparison of host and parasite traits, review different concepts used in cophylogenetic studies, highlight the main methods used and briefly examine described cases of cophylogeny. I will use different examples to illustrate the discussion emphasizing the relationships between cophylogeny and coadaptations.

2 Correlated evolution between host immunity and parasite life histories

Adult female body size of nematodes is linked to reproductive output, larger females having higher fecundity. Because growth stops or slows down after maturation, age and size at first reproduction are usually positively correlated and are determinants of reproductive output (Skorping et al. 1991; Morand 1996). On the one hand, delayed maturation might, therefore, be selected for because of the fecundity benefit of larger size at maturity. On the other hand, delayed maturity also entails costs in terms of increased prematurational mortality risks. Therefore, the trade-off between the fecundity benefits of delayed maturity and the mortality costs should ultimately determine the optimal age and size at maturity (Gemmill et al. 1999; Morand and Poulin 2000). Optimality models have explored this trade-off and predicted that age at maturity should be inversely proportional to the prematurational mortality rate (Gemmill et al. 1999; Morand and Poulin 2000). It can thus be expected that nematodes with low prematurational mortality risks should delay maturation, grow longer and have higher fecundity, whereas parasites with high prematurational mortality risks should reproduce earlier at smaller size.

Most macroparasites experience two types of environment, namely the within-host environment and the external environment encountered by infective stages during transmission between hosts. Within the host, a parasite has to establish, grow and reproduce. These fitness components are potentially affected by the control of helminth populations exerted by the host immune response. For example, increases in the numbers of mast cells and eosinophils are commonly observed in response to helminth infections (Maizels et al. 1993). A review of studies on several host-helminth associations suggested a role for eosinophils in killing larval stages rather than adult worms (Meeusen and Balic 2000). In other words, eosinophils might act as an age-dependent mortality factor for parasites and consequently, one might expect that parasites living in a host environment with high larval mortality should reduce age at maturation and have a smaller adult size.

Pinworms are common parasitic nematodes inhabiting the caecum and proximal colon of mammalian hosts. They (a) are highly host-specific (Brooks and Glen 1982; Hugot 1999); (b) generally lack a free-living stage since hosts become infected from ingested eggs (Anderson 2000); (c) infections with larvae in humans have been associated with blood eosinophila and eosinophilic entero and ileocolitis (Liu et al. 1995; Cacopardo et al. 1997); and (d) their reproductive output is tightly correlated with body size (Cho et al. 1985).

Sorci et al. (2003) investigated patterns of correlated evolution between female, male and egg size of 32 pinworm species and basal values of blood circulating lymphocytes, neutrophils and eosinophils of their primate hosts. If the immune system of the host characterizes the environment experienced by the parasite, the expectation is that basal level of circulating eosinophils should induce a correlated evolution towards smaller parasite size, because of the benefits for the parasite of reducing age at maturity. A corollary prediction would also be that small parasites with earlier maturation time might lay smaller eggs to maintain a constant fecundity. This study used the method of phylogenetically independent contrasts (Felsenstein 1985; Harvey and Pagel 1991) implemented with the CAIC statistical package (Purvis and Rambaut 1995). This linear multiple regression program allows controlling for phylogenetic relationships between the hosts or the parasites. This is because within both groups, species share common ancestors and therefore cannot be considered as independent statistical observations. In addition, host body mass is another potential confounding variable because of correlations with cell counts and parasite size (Harvey et al. 1987; Harvey and Keymer 1991), and it was therefore included in the regression model. The stepwise multiple regression model with pinworm female body size as dependent variable and basal values of lymphocytes, neutrophils, eosinophils and host body mass as independent variables revealed that only eosinophil concentration entered the model (Fig. 1).

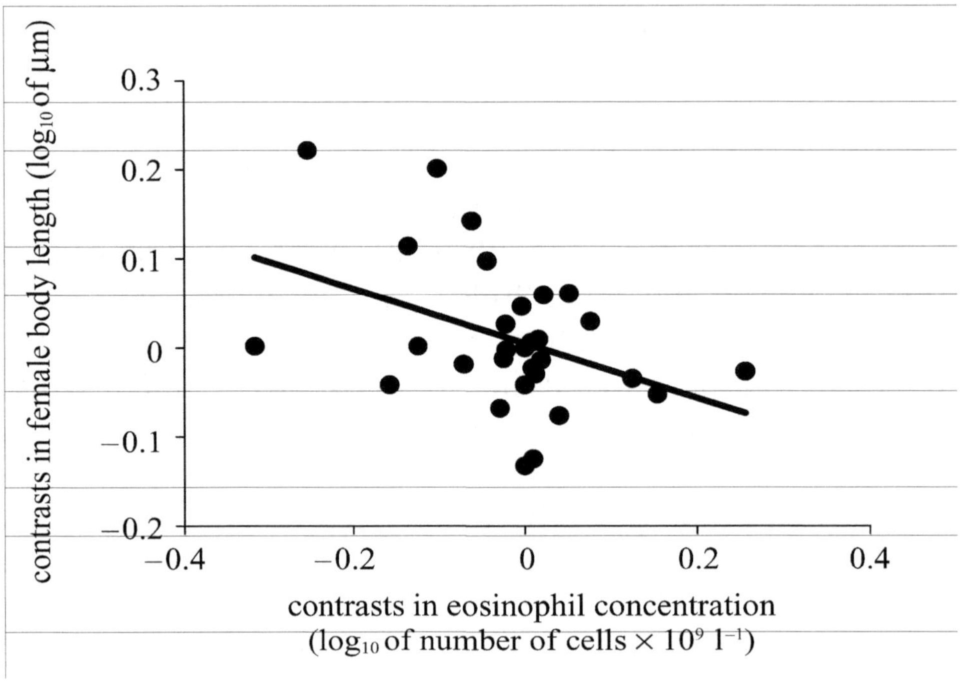

Fig. 1. Negative correlation between female body length of pinworms and eosinophil concentration of primate hosts. Each point represents a phylogenetically independent contrast (after Sorci et al. 2003)

A similar regression model run for male parasite body size revealed that none of the independent variables entered the model. Parasite egg size tended to be positively correlated with female parasite body length, although not significantly. However, when running a stepwise multiple regression model with the immune variables, host body mass and female parasite body size, only eosinophil concentration entered the model.

Again, egg size was negatively correlated with basal eosinophil concentration.

These results show a pattern of correlated evolution between host immunity, assessed as the blood concentration of eosinophils, and life history traits of the pinworms. None of the other white blood cells (neutrophils and lymphocytes) was correlated with the adult or egg size of parasites. Consequently, this study demonstrated the role of eosinophils in the generation of resistance against helminths. The effect of eosinophils on parasite survival (see Meeusen and Balic 2000) and the observed negative correlation between body size of adult female pinworms and eosinophils in the peripheral blood of their hosts are in agreement with the prediction for correlated evolution between host immune defenses and parasite life histories. Reduced age and small size at maturity might be the best options when facing elevated risks of prematurational mortality at the expense of reproductive output.

Another trade-off could, however, intervene to determine the number of eggs laid. Instead of laying fewer eggs, small females might adjust the size of their eggs, maintaining a constant fecundity. As was the case with body size of adult females, eosinophil concentration was the only significant predictor of egg size. In other words, pinworms living in hosts with large numbers of circulating eosinophils lay smaller eggs than pinworms that exploit hosts with a lower eosinophil concentration. These results suggest that small females trade egg size against egg number and that host defenses might affect directly the life-history strategies of their parasites over evolutionary time.

3 Cophylogeny studies: Principles and methods

Closely related organisms show extensive similarities in the parasites associated with them. It is thus possible to base genealogical conclusions on parasite data. This defines the field for the studies of cophylogeny - the comparisons of two groups of organisms for which we can expect a long-standing association over geological times. This means that these groups of associates share (at least partly) a common history during evolution. In practice, this means that we can try using host and parasite trees (cladograms) to reciprocally illuminate each other. During the late 19th century and the first three quarters of the 20th century, host-parasite cophylogeny studies were lacking consistent concepts and methods. As a result, they remained limited to empirical observations, and their interpretations and understanding were mostly intuitive.

Similar ideas were developed in historical biogeography and referred to as component analysis (Nelson and Platnick 1981). In classical component analyses, organisms are used to infer area relationships exactly as parasites may be used to infer host relationships in the parasitological context. The first attempt to accomplish this scored the parasites and their phylogeny as a series of binary characters (Brooks 1985). The parasite data were treated as a character-state tree which was converted into binary characters using additive binary coding. In this new context, this technique was named Brooks Parsimony Analysis (BPA) (Wiley 1988). The similarities between host-parasite studies and cladistic biogeographic analysis led Brooks (1985) to adapt BPA for use in historical biogeography just as component analysis was used to tackle the analysis of coevolving species associations (e.g., Humphries et al. 1986; Humphries and Parenti 1999). However, Brooks' method has been severely criticized (Siddall and Perkins 2003) and accused of circularity. Another flaw of this approach is the difficulty of implementing the method. Recently, a program capable of performing BPA has been announced [Phylogenetic Analysis for Comparing Trees (PACT); Wojcicki and Brooks 2005]. However, to date this software is not yet available.

Ronquist and Nylin (1990) insisted that parsimony analysis in historical biogeography and coevolution should be based on an explicit set of events (transformations). If each of these events is assigned a cost, the minimum-cost (most parsimonious) explanation of the observed data can be sought. A coevolutionary or biogeographic method that satisfies these criteria is an event-based parsimony method (Ronquist 1997). Page (1995) and Ronquist (1997) argued that this type of approach might be applied to different types of associations using the same basic concepts (Table 1). Event-based thinking was introduced under the name "tree reconciliation" by Page (1994). Later, Page (1995) succeeded in introducing switches into the reconciliation framework and suggested that the problem of finding cost-event assignments should be solved by maximizing the number of inferred host-associate codivergences. This approach may be referred to as maximum codivergence (MC).

4 TreeMap: How does that works?

In coevolutionary or biogeographic studies, an event-based parsimony method can be proposed by the comparison of the evolutionary hypotheses concerning the associates summarized in the form of trees. Fig. 2 represents different situations which may be observed using this type of comparison. Page (1995) developed TreeMap: an algorithm to find all reconstructions that maximize the number of codivergences in a particular host-parasite association. TreeMap maximizes the amount of codivergence (shared history), when superimposing the respective trees of two associates on each other in order to reveal the history of a particular association.

Table 1. Equivalent processes in historical associations (according to Page 1995 and Ronquist 1997)

Association type	Codivergence	Duplication	Horizontal transfer	Sorting event
Organism/gene	Interspecific coalescence	Gene duplication	Gene transfer	Gene loss or lineage sorting
Host/parasite	Cospeciation	Within-host speciation	Host switch	Parasite extinction or missing the boat
Area/organism	Vicariance	Sympatry	Dispersal	Extinction

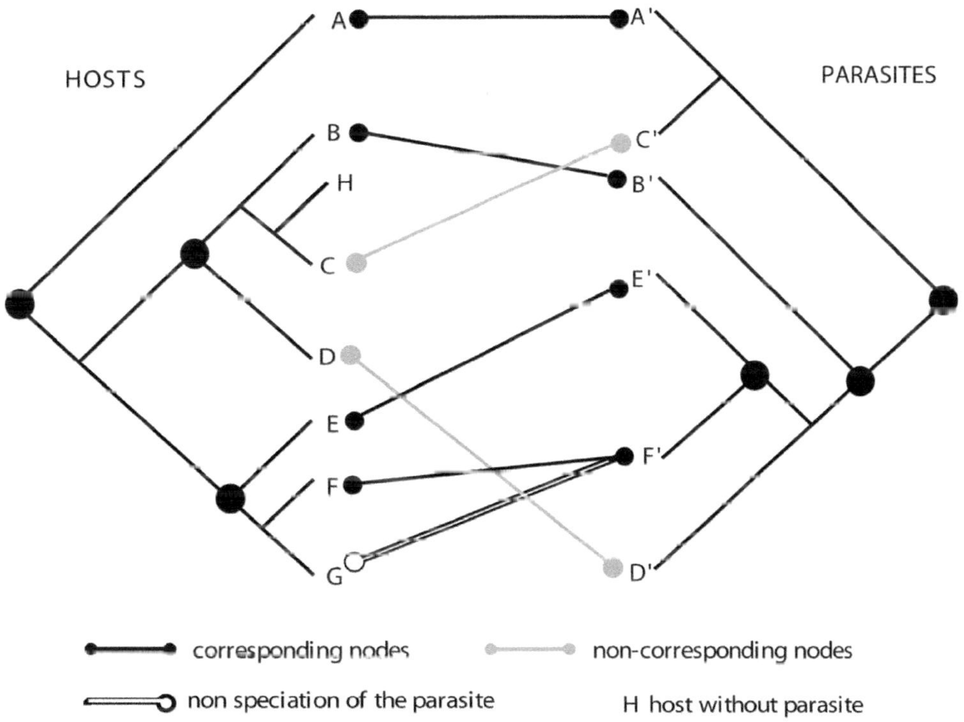

Fig. 2. Comparison of parasite and host trees representing different situations. A' is the specific parasite of host A, B' is the specific parasite of host B, etc. As generally observed, the host and parasite trees are partly similar in their topologies: some nodes in both trees are similar or identical, some others are incompatible. Matching nodes are indicated by black dots. Host H has no parasite. This can be due to different reasons: (a) H lost its parasite; (b) H never had such a parasite; and (c) the parasite of H does exist but is still undiscovered. Finally, the same parasite may be observed on two different hosts (F and G)

The result is a set of diagrams where parasite and host trees are superimposed and nodes of the parasite tree matching with the corresponding node of the host tree are indicated using black dots. The TreeMap exact search algorithm proposes all those scenarios that allow the maximum amount of cospeciations.

TreeMap also allows finding scenario(s) with the smallest total cost. Such scenario(s) is/are most parsimonious in the sense of minimizing the total number of duplications, horizontal transfers, losses and/or sorting events required to explain the evolutionary history of the associates. In the absence of other arguments, which could support or refute a particular scenario, the sole criterion to compare different scenarios is the parsimony of hypotheses. As each scenario exhibits the same number of cospeciations, they have to be compared using the other three kinds of evolutionary events; scenarios with the lowest number of extra hypotheses (evolutionary events other than cospeciations) can be considered the most parsimonious. Fig. 3 shows the design of different types of evolutionary events used to build the diagrams and illustrates the different types of situations which may be encountered. The trees are the simplest possible trees, with only three terminal branches. In Fig. 3.1, the trees are identical, so only one scenario is possible. In Fig. 3.2, parasite B is closest to parasite C, whereas host B is closest to host A. When reconciling the trees, one has to hypothesize a host switch from parasite B (or the common ancestor of parasites A and B) to host B. In Fig. 3.3, the situation is the same, but TreeMap proposes a different solution: a duplication event (=speciation of the parasite without corresponding speciation of the host) occurred in the common ancestor of parasites A, B and C. From these sister-species, two sister parasite lineages might have descended. As a result two different parasite species might be found in every host. However, if we hypothesize two sorting events, then the distribution of the parasites finally fit the observation: one parasite species for one host species. Scenarios 3.2 and 3.3 are two different ways to reconcile the same topologies, but scenario 3.2 needs one hypothesis only (a host switch), whereas scenario 3.3 needs three hypothetical events (a duplication and two sorting events). Consequently, scenario 3.2 is most parsimonious. Fig. 3.4 illustrates yet another situation where the same parasite species is found in two sister host species. This may result from two different explanations. First, because the parasite did not speciate during the host speciation, both host species inherited the parasite from the common ancestor. Second, parasite B was lost (sorting event) and parasite C switched from host C to host B.

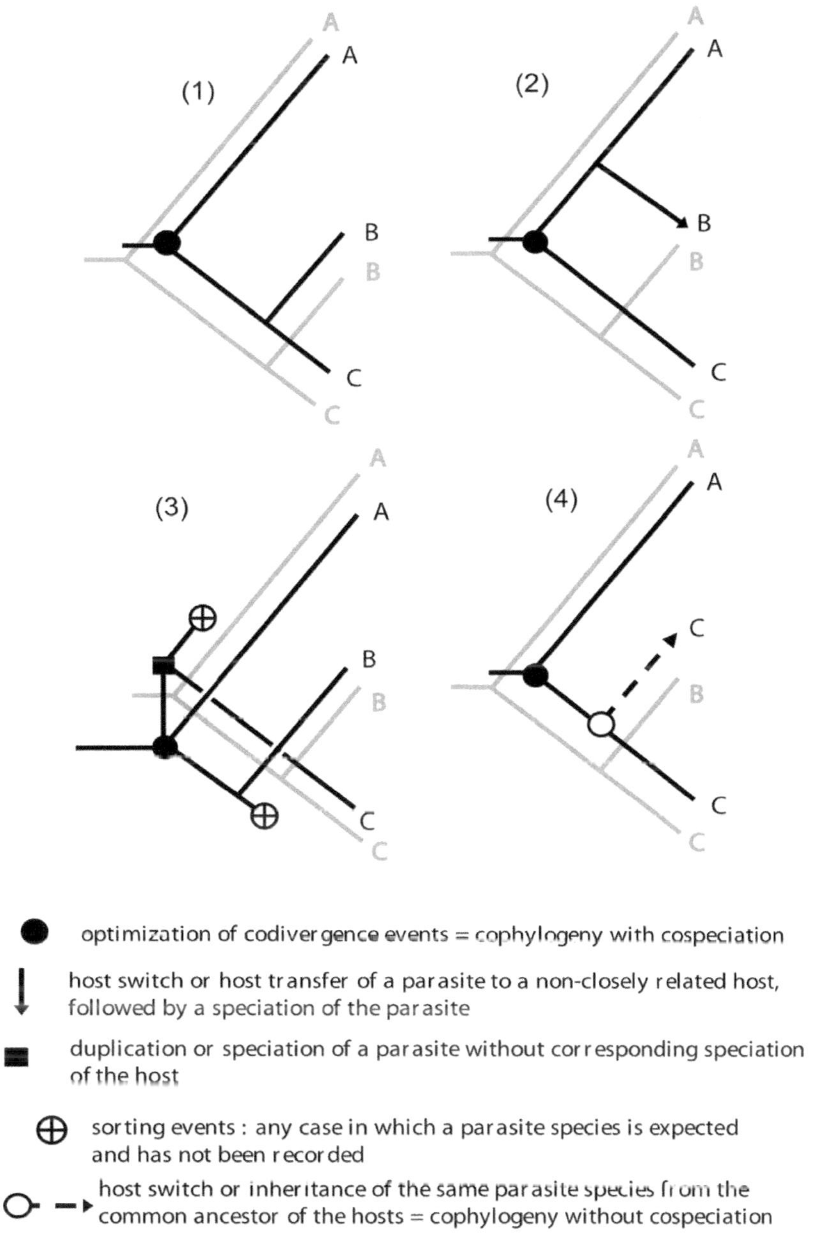

Fig. 3. Host vs. parasite trees superimposition and reconciliation. Following Page (1995), speciation events are divided into three categories: cospeciation, duplication and host switch. To build evolutionary scenarios a fourth category (sorting) is needed. Sorting events cover any case in which a parasite species can be expected on one host species and has not been observed. This can be explained in three different ways: (a) "missing the boat", when the parasites where unable to colonize a particular host, (b) a loss of a parasite during evolution, and (c) the parasite exists but it has not yet been observed. Coevolutionary studies generally have not taken into consideration any additional possibility as follows. When the same parasite species is observed in two closely related hosts, it can be hypothesized that they inherited this parasite from a common ancestor. Such a case of transmission by descent could be called cophylogeny without cospeciation. However, in this situation another, less parsimonious explanation may be suggested as follows. Parasite B was lost before parasite C jumped from C to B (host switching)

Fig. 4 shows the results of the TreeMap analysis using host and parasite trees from Fig. 2. Twelve different scenarios are proposed, from which Fig. 4.1 represents the most parsimonious (five other than cospeciation events) and Fig. 4.2 the least parsimonious (12 other than cospeciation events).

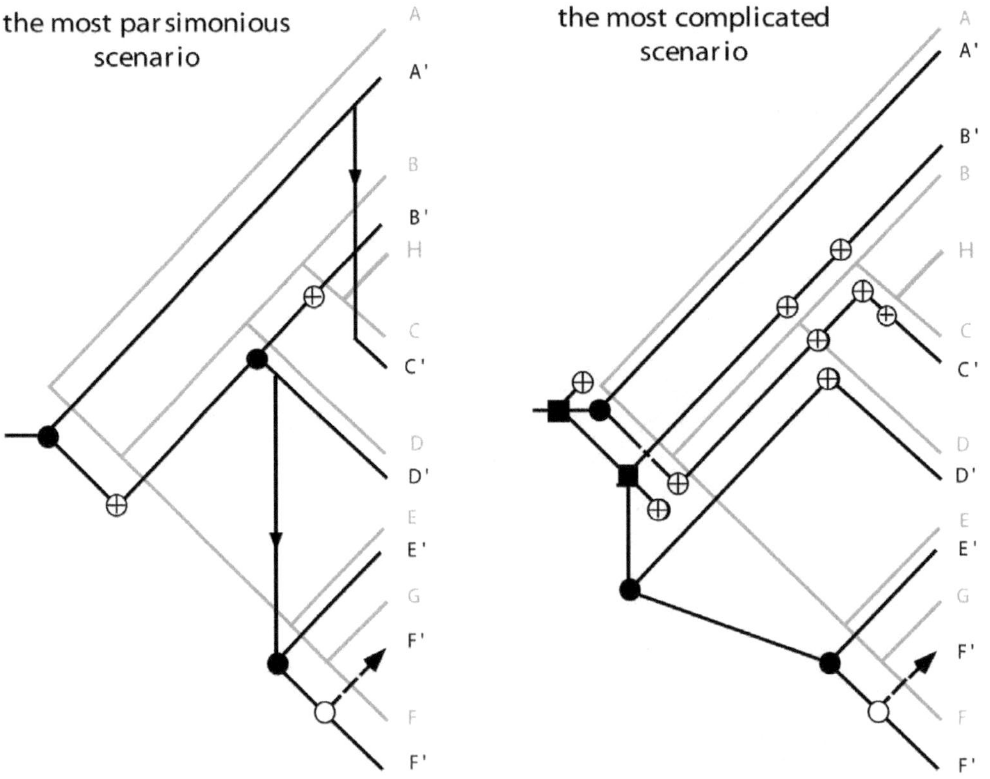

Fig. 4. Reconciliation of the host and parasite trees from Fig. 2. Twelve different scenarios are proposed after TreeMap analysis. All of them have three cospeciation events (black dots).

5 Cophylogeny: Choosing a scenario when different phylogenetic hypotheses may be proposed for the host group

A morphologically based cladistic analysis of Enterobiinae, which includes most of the oxyurid nematodes parasitic in primates, allowed a re-evaluation of the hypothesis of cophylogeny between hosts and parasites (Hugot 1999). Each of three enterobiine genera exploits one of the primate infraorders. Enterobiinae also include two genera (*Xeroxyuris* and *Rodentoxyuris*) parasitic on squirrels (Fig. 5). The cladistic analysis does not support close relationships between the squirrel parasites and suggests an early separation *Xeroxyuris* and a tardy host switch from Platyrrhini to squirrels for *Rodentoxyuris*. Construction of scenarios using TreeMap allows to hypothesize how squirrel parasites originated.

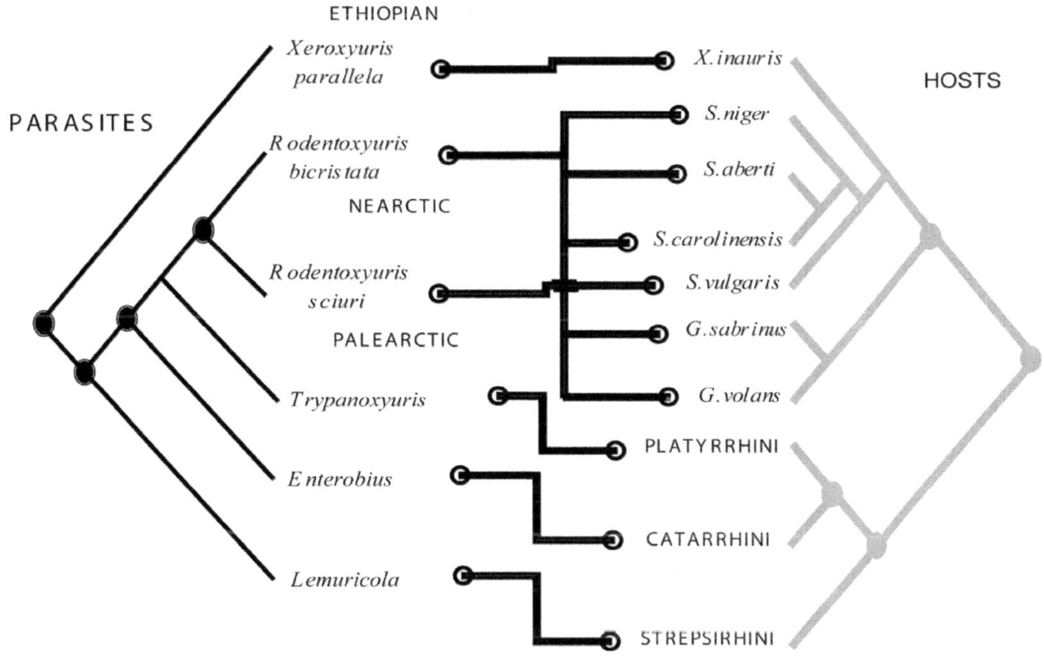

Fig. 5. Reconciliation of Enterobiinae versus their primate and squirrel hosts trees (simplified after Hugot 1999)

5.1 Host and parasite distributions

Squirrel oxyurid parasites belong to two genera. *Xeroxyuris* is a monotypic genus with a single species whose host, *Xerus inauris*, is a ground squirrel living in the dry steppes of southern Africa. The type species of the second genus is *Rodentoxyuris sciuri* parasitic in *Sciurus vulgaris* across most of its geographic range. A second species of this genus, *Rodentoxyuris bicristata*, was recorded from the North American squirrels (*Glaucomys sabrinus*, *Glaucomys volans*, *Sciurus niger*, *Sciurus carolinensis* and *Sciurus aberti*). *S. vulgaris* inhabits the Palearctic, from Spain to Kamchatka. In America, the broadly overlapping geographic ranges of *S. niger, S. carolinensis*, and *G. volans* are from the East Coast to the Rocky Mountains in the west, the Canadian border in the north and the Mexican border in the south. In addition, *G. volans* has mountain populations scattered from Mexico to Honduras. *G. sabrinus* has a very different range extending from Alaska and Canada to the U.S. Northwest. The range of *G. sabrinus* is overlapping with the ranges of other American squirrels around the Great Lakes and in the Appalachian Mountains. *S. aberti* has a very restricted range divided into two unconnected areas (one - parts of Utah,

Colorado, Wyoming, New Mexico, and Arizona, whereas the other - northern Mexico).

5.2 Phylogeny of the hosts

The phylogenetic relationships between *Glaucomys* and other squirrels are still debated. *Glaucomys* has been proposed to belong together with all other flying squirrels in a separate subfamily within Sciuridae (Johnson-Murray 1977; Thorington 1984). In contrast, some authors have suggested that *Glaucomys* could be more closely related to other holarctic squirrels (Sciurini) (Gorgas 1967). These altenative hypotheses can be represented by two different topologies: (a) *Xerus* as sister group of *Sciurus+Glaucomys*; and (2) *Glaucomys* is sister group of *Sciurus+Xerus*.

5.3 Investigating evolutionary scenarios

The position of *Rodentoxyuris* on the cladogram (Fig. 5) implies that the parasites switched from primates to squirrels after Enterobiinae had differentiated into several lineages within Platyrrhini. During most of the Tertiary, Platyrrhini were isolated in South America, and squirrels are presumed not to have been present in the Neotropics during that period. This implies that the contact between the Neotropical monkeys and squirrels occurred after a land connection was reestablished between North and South America. For this reason and because *Xerus* is endemic to Africa, the scenarios that propose host-switching from Platyrrhini to the common ancestor of *Xerus*, *Sciurus* and *Glaucomys* are unlikely and the proposed reconstructions of evolutionary scenarios have to be compatible with host-switching from Platyrrhini to the holarctic squirrels, which chronologically happened in the uppermost Tertiary.

Fig. 6 represents the most parsimonious scenarios obtained using TreeMap with either topology 1 (Fig. 6A), or topology 2 (Fig. 6B). Among all scenarios produced by TreeMap, the reconstructions in Fig. 6 are the most parsimonious and are independent of whatever phyletic relationships between *Sciurus* and *Glaucomys* are hypothesized. To be valid these reconstructions require two assumptions: (1) an initial host switch from Platyrrhini to the ancestor of *Sciurus*, and (2) a later host switch from *Sciurus* to *Glaucomys*. Because *Sciurus* probably migrated into the Neotropics during the late Tertiary, the first assumption seems valid. Considering the present distribution of *Glaucomys* and *Sciurus* in the Nearctic, the second assumption is also acceptable. Consequently, these scenarios look reliable.

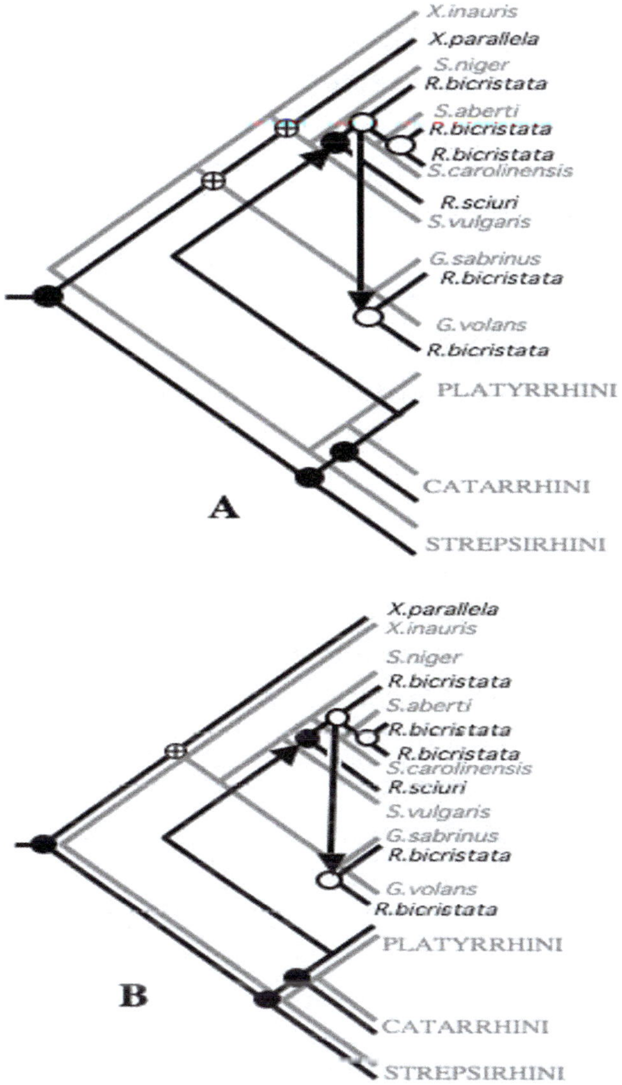

Fig. 6. Reconciliation of trees from Fig. 5 using two alternative host topologies produced numerous different scenarios. Two of them are quite identical and are thus independent of the phylogeny of the hosts. They are also the most parsimonious scenarios for each topology. (A) from topology 1 (*Glaucomys* and the other flying squirrels are monophyletic); (B) from topology 1 (*Glaucomys* and *Sciurus* are closely related) (see text for explanations; modified from Hugot 1999)

6 Cophylogeny: Using TreeMap scenarios to test different phylogenetic hypotheses proposed for the same host group

Rodents are currently divided into two suborders, Sciurognathi and Hystricognathi. Hystricognathi can be further subdivided into the Old World Phiomorpha and the New World Caviomorpha. The evolutionary relationships among hystricognath rodents and the geographical origin of Caviomorpha (which has been endemic to the Neotropics for most of the Tertiary) have generated considerable debate. Some of the arguments used concerned the distribution of the parasite pinworms of these rodents (Quentin 1973a, b). In this example, the cladogram of the pinworms is used as a tool for testing different hypotheses previously proposed for the classification of Hystricognathi. The cladistic analysis includes 19 species of pinworm parasite found in hystricognath rodents. The outgroup includes two species, namely *Hilgertia hilgerti* from a sciurognath rodent belonging to the family Ctenodactylidae and *Acanthoxyurus beecrofti* from a sciurognath rodent belonging to the family Anomaluridae. The cladistic analysis was based on 22 morphological characters from various organ systems. Until now no consensus has been established for the phylogenetic classification of Hystricognathi, and the most recent attempts have given highly divergent results. In this study three different topologies were used (Fig. 7). Two trees were based on molecular data. Host-tree-1 was adapted from Huchon et al.'s (1999; 2000) analysis of von Willebrand factor gene sequences. Host-tree-2 was based on Nedbal et al.'s (1994) analysis of mitochondrial 12S rRNA sequences. Host-tree-3 was constructed using morphological data from Hartenberger (1985), Lavocat and Parent (1985), Bugge (1985), George (1985), Woods and Hermanson (1985), and Bryant and McKenna (1995). In total, this yielded 102 characters which, after exclusion of uninformative characters, resulted in a 55 character matrix.

6.1 Tree comparison

For tree comparison, the parasite tree was considered as an additional phylogenetic hypothesis for the hystricognath rodents. All trees were reduced to these terminal taxa which are shared by the four trees, as represented in Fig. 7. TreeMap was used to determine how many congruent nodes could be found when considering each pair of trees successively. The results are presented in Table 2. The lowest values were found when comparing together Host-tree-1 and Host-tree-2 (only 3 congruent nodes out of six). The best fit was observed between Host-tree-3 and Parasite-tree (5 congruent nodes out of 6), and a very low probability (p =0.001) of getting as much consistency by chance.

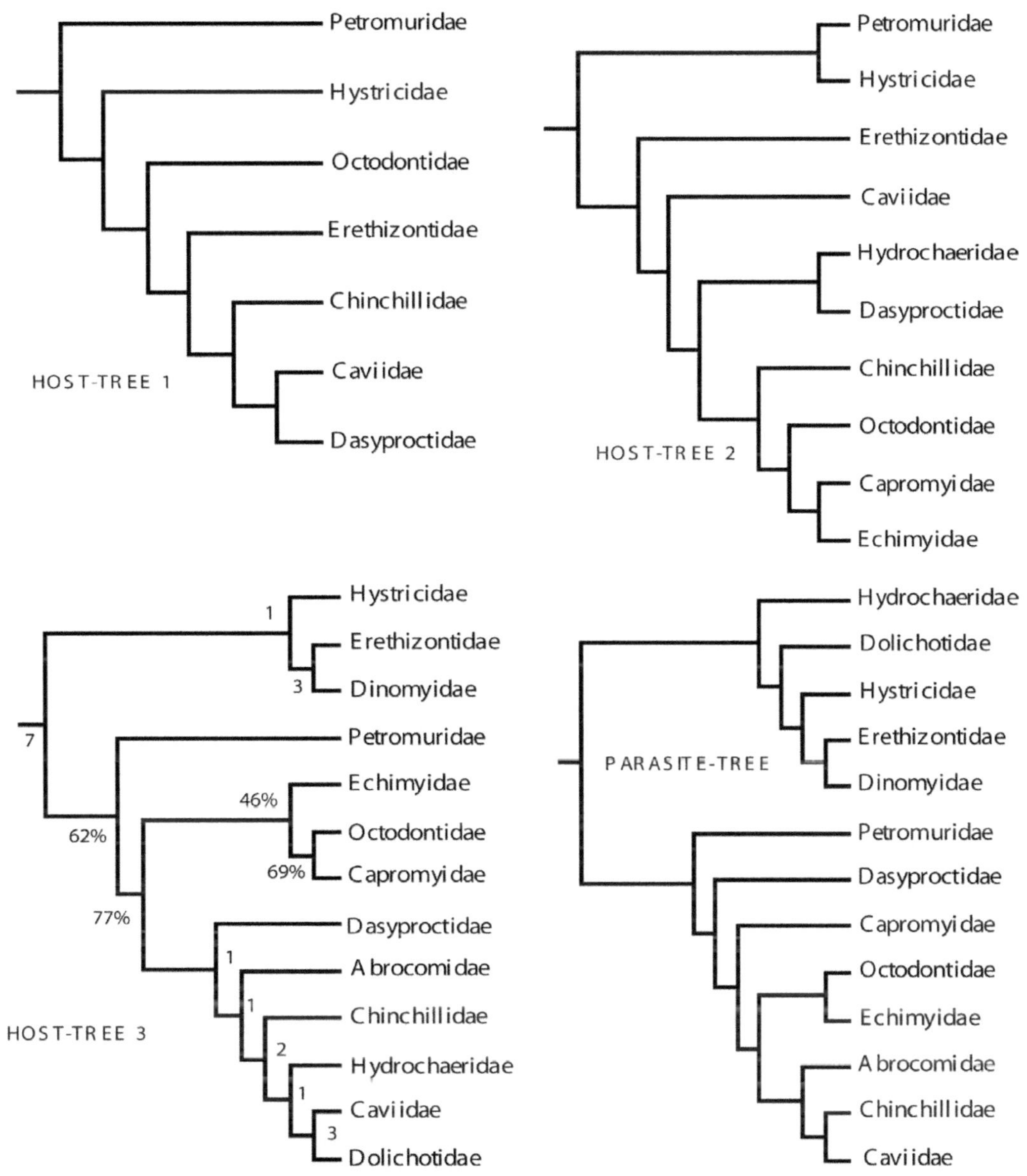

Fig. 7. Parasite and host trees used for tree comparisons (see text for explanations). Numbers on Host-tree-3 are decay indices (when greater than zero); percentages are the percentage occurrence of clades (when different from 100%). In the Parasite-tree the names of the parasites have been substituted with the names of host families (redrawn after Hugot 2003)

Comparison of different cladograms represented in Fig. 7 showed that they generally strongly disagreed with each other (Table 2). Only Host-tree-2 supported the monophyly of the two main subgroups generally distinguished within Hystricognathi (Phiomorpha and Caviomorpha). Host-tree-1 supported the monophyly of Caviomorpha. Host-tree-3 and the Parasite-tree refuted the monophyly of both subgroups. Host-trees-1, -2 and -3 also disagreed on the arrangement of different families, with the exception of the association Octodontidae, Echimyidae, and Capromyidae which is common to Host-trees-2 and -3. This clade corresponds with the superfamily Octodontoidea supported by Woods and Hermanson (1985), Nedbal et al. (1994) and Bryant and McKenna (1995), but recently refuted by Landry (1999). The best fit was observed between the Parasite-tree and Host-tree-3, which both divided Hystricognathi into two subgroups. The first subgroup included Hystricidae as a sister group of the clade Dinomyidae+Erethizontidae. The second group included Petromuridae together with most of the remaining Caviomorpha. Both trees also agreed on the association of Caviidae and Chinchillidae with Abrocomidae. However, the Parasite-tree refuted the existence of Octodontoidea and associated Hydrochaeridae and Dolichotidae with Hystricidae, Erethizontoidae and Dinomyidae.

Table 2. Upper diagonal: number of congruent nodes (maximum number of 6) between different phylogenies of the hystricognath rodents from Fig. 7. Lower diagonal: probability of getting as much congruent nodes between each pair of trees by chance, resulting from a Markovian model generating 1000 random trees

	Host-tree-1	Host-tree-2	Host-tree-3	Parasite-tree
Host-tree-1	6	3	4	4
Host-tree-2	0.424	6	3	4
Host-tree-3	0.056	0.361	6	5
Parasite-tree	0.050	0.052	0.001	6

6.2. Are the Old World and New World porcupines closely related?

The association between pinworms and porcupines provides another example of the usefulness of cophylogenetic studies to answer evolutionary questions. The cladogram of the parasites supported close relationships between the pinworms parasitic on porcupines (Hystricidae and Erethizontidae) and also grouped the parasites of Erethizontidae and Dinomyidae. This posed two different questions: (1) what support can be found for these relationships in the different studies concerning the hosts? and (2) how reliable is the support for the classification of these parasites in the same genus?

The grouping of Dinomyidae and Erethizontidae has been previously proposed (Grant and Eisenberg 1982; Woods and Hermanson 1985). These studies suggested also a close association of these two families with Hystricidae. Recently, Landry (1999) proposed to unite all three families in a superfamily, Erethizontoidea. All these studies were based on morpho-anatomical data. To date, molecular studies have ignored Dinomyidae and consequently have not provided any evidence to support or refute their affinities with porcupines. Similarly, none of the recent studies based on molecular data has given any support to the grouping of Hystricidae and Erethizontidae (Nedbal et al. 1994; Huchon et al. 1999, 2000). However, they generally classified Hystricidae or Erethizontidae as the most divergent branch in their respective groups (Phiomorpha or Caviomorpha). The support for the branching of these two groups on the cladogram of Hystricognathi (decay indices or bootstrap values) repeatedly appeared to be very weak. In addition, molecular studies generally suffer from insufficient taxonomic sampling. Analyzed species are not a comprehensive representation of the different hystrichognath families recognized. Each family included in a particular study was represented by one, rarely two species. Finally, if we also take into account the fact that the topologies proposed by different molecular analyses are highly divergent, the proximity of Hystricidae and Erethizontidae, which has been repeatedly supported by morphological data, cannot be considered seriously challenged by current molecular analyses.

7 Concluding remarks

Parasites are living organisms that comply with taxonomy and ecology as all other living beings in the biosphere do. Their ability to succeed as pathogenic agents or pests seems to a great extent to depend upon their life-history traits, biological traits, ecology, genetic background and ultimately upon their evolution. Research seeking to understand the detailed interaction between parasites and their potential targets has to characterize all the actors and, consequently, deal with the methods and approaches used in the fields of fundamental biological sciences, such as genetics, ecology, phylogenetics and taxonomy.

Although taxonomy is currently considered as a somewhat old-fashioned discipline, it deserves special attention. No advance in related fields may be achieved if we remain or become unable to identify accurately the various biological agents involved in complex ecological systems. Because we cannot define a particular species without studying also most of its relatives, systematics must encompass biodiversity as a whole, because limited studies will not provide robust results. However, modern systematics aims to be phyletic, i.e. a classification aims to reflect relationships between living organisms as a result of evolution. Such a purpose gives more accuracy and strength to the identification of species and the way we define them. In addition, phylogeny is a powerful tool for understanding how life history traits appeared and evolved. Phylogeny also allows the investigation of changes of host and parasite distribution across geological times (biogeography). Moreover, these analyses are useful for construction of hypotheses on the timing of evolutionary events (comparative phylogeny and phylogeography). Altogether these approaches give precious, much-needed information on how complex interactions between different types of organisms emerged and evolved.

References

Brooks DR (1985) Historical ecology: A new approach to studying the evolution of ecological associations. Ann Mo Bot Gard 72:660–680

Brooks DR, Glen DR (1982) Pinworms and primates: A case study in coevolution. Proc Helminthol Soc Washington 49:76–85

Brooks DR, McLennan DA (1993) Parascript. Parasites and the language of evolution. Smithsonian Inst Press, Washington

Bryant JD, McKenna MC (1995) Cranial anatomy and phylogenetic position of *Tsaganomys altaicus* (Mammalia: Rodentia) from the Hsanda Gol formation (Oligocene), Mongolia. Am Mus Nov 3156:1–42

Bugge J (1985) Systematic value of the carotid arterial pattern in rodents. In: Luckett WP, Hartenberger JL (eds) Evolutionary relationships among rodents. A multidisciplinary analysis. Plenum Press, New York, pp 355–380

Cacopardo B, Onorante A, Nigro L, Patamia I, Tosto S, Romano F, Zappala C, Bruno S, Nunnari A (1997) Eosinophilic ileocolitis by *Enterobius vermicularis*: A description of two rare cases. Ital J Gastroenterol Hepatol 29:51–53

Cho SY, Chang JW, Jang HJ (1985) Number of intrauterine eggs in female *Enterobius vermicularis* by body length. Korean J Parasitol 23:253–259

Ehrlich PR, Raven PH (1964) Butterflies and plants: A study of coevolution. Evolution 18:586–608

Felsenstein J (1985) Phylogenies and the comparative method. Amer Nat 125:1–15

Gemmill AW, Skorping A, Read AF (1999) Optimal timing of first reproduction in parasitic nematodes. J Evol Biol 12:1148–1156

George W (1985) Reproductive and chromosomal characters of ctenodactylids as a key to their evolutionary relationships. In: Luckett WP, Hartenberger JL (eds) Evolutionary relationships among rodents. A multidisciplinary analysis. Plenum Press, New York, pp 453–474

Grant TI, Eisenberg JF (1982) On the affinities of the Dinomyidae. Saugetierk Mitt 30:151–157

Gorgas M (1967) Vergleichend-anatomische Untersuchungen am Magen-Darm-Kanal der Sciuromorpha, Hystricomorpha and Caviomorpha (Rodentia). Z Wiss Zool 175:237–404

Johnson-Murray JL (1977) Myology of the gliding membranes of some petauristine rodents (genera: *Glaucomys, Petaurista, Petinomys,* and *Pteromys*). J Mammal 58:374–384

Hartenberger JL (1985) The order Rodentia: Major questions on their evolutionary origin, relationships and suprafamilial systematics. In: Luckett WP, Hartenberger JL (eds) Evolutionary relationships among rodents. A multidisciplinary analysis. Plenum Press, New York, pp 1–34

Harvey PH, Keymer AE (1991) Comparing life histories using phylogenies. Phil Trans R Soc Lond B 332:31–39

Harvey PH, Pagel MD (1991) The comparative method in evolutionary biology. Oxford Univ Press

Harvey PH, Martin RD, Clutton-Brock TH (1987) Life histories in comparative perspective. In: Smuts BR, Cheney DL, Seyfarth RM, Wragham RW, Struhsaker TT (eds) Primate societies. Univ Chicago Press, pp 181–196

Huchon D, Catzeflis FM, Douzery EJP (1999) Molecular evolution of the nuclear von Willebrand factor gene in mammals and the phylogeny of rodents. Mol Biol Evol 16:577–589

Huchon D, Catzeflis FM, Douzery EJP (2000) Variance of molecular datings, evolution of rodents and the phylogenetic affinities between Ctenodactylidae and Hystricognathi. Proc R Soc Lond B 267:393–402

Hugot J-P (1999) Primates and their pinworm parasites: Cameron hypothesis revisited. Syst Biol 48:523–546

Hugot J-P (2003) New evidence of hystricognath rodents monophyly from the phylogeny of their pinworms. In: Page RDM (ed) Tangled trees: Phylogenies, cospeciation and coevolution. Chicago Univ Press, pp 144–173

Humphries CJ, Parenti LR (1999) Cladistic biogeography, 2nd edn. Oxford Univ Press, New York

Humphries CJ, Ladiges PY, Roos MC, Zandee M (1988) Cladistic biogeography. In: Myers AA, Giller PS (eds) Analytical biogeography: An integrated approach to the study of animal and plant distributions. Chapman and Hall. London, pp 371–404

Landry Jr SO (1999) A proposal for a new classification and nomenclature for the Glires (Lagomorpha and Rodentia). Mitt Mus Natkd Berlin Zool 75:283–316

Lavocat R, Parent JP (1985) Phylogenetic analysis of middle ear features in fossil and living rodents. In: Luckett WP, Hartenberger JL (eds) Evolutionary relationships among rodents. A multidisciplinary analysis. Plenum Press, New York, pp 333–354

Liu LX, Chi J, Upton MP, Ash LR (1995) Eosinophilic colitis associated with larvae of the pinworm *Enterobius vermicularis*. Lancet 346:410–412

Maizels RM, Bundy DAP, Selkirk ME, Smith DF, Anderson RM (1993) Immunological modulation and evasion by helminth parasites in human populations. Nature 365:797–805

Meeusen ENT, Balic A (2000) Do eosinophils have a role in the killing of helminth parasites? Parasitol Today 16:95–101

Morand S (1996) Life history traits in parasitic nematodes: a comparative approach for the search of invariants. Funct Ecol 10:210–218

Morand S, Poulin R (2000) Optimal time to patency in parasitic nematodes: Host mortality matters. Ecol Lett 3:186–190

Nedbal MA, Allard MW, Honeycutt RL (1994) Molecular systematics of hystricognath rodents: Evidence from the mitochondrial 12S rRNA gene. Mol Phyl Evol 3:206–220
Nelson G, Platnick NI (1981) Systematics and biogeography: Cladistics and vicariance. Colambia Univ Press, New York
Page RDM (1994) Maps between trees and cladistic analysis of historical associations among genes, organisms, and areas. Syst Biol 43:58–77
Page RDM (1995) Parallel phylogenies: Reconstructing the history of host-parasite assemblages. Cladistics 10:155–173
Purvis A, Rambaut A (1995) Comparative analysis by independent contrasts (CAIC): An Apple Macintosh application for analysing comparative data. Comput Appl Biosci 11:247–251
Quentin JC (1973a) Affinités entre les Oxyures parasites de Rongeurs Hystricidés, Erethizontidés et Dinomyidés. Intérêt paléobiogéographique. Comptes-Rendus de l'Académie des Sciences Paris 276:2015–2017
Quentin JC (1973b) Les Oxyurinae de Rongeurs. Bull Mus Natl Hist Nat Paris 112:1045–1096
Ronquist F (1997) Phylogenetic approaches in coevolution and biogeography. Zool Scripta 26:313–322
Ronquist F, Nylin S (1990) Process and pattern in the evolution of species associations. Syst Zool 39:323–344
Siddall ME, Perkins SL (2003) Brooks parsimony analysis: A valiant failure. Cladistics 19:554–564
Skorping A, Read AF, Keymer AE (1991) Life history covariation in intestinal nematodes of mammals. Oikos 60:365–372
Sorci G, Skarstein F, Morand S, Hugot J-P (2003) Correlated evolution between host immunity and parasite life histories in primates and oxyurid parasites. Proc R Soc London B 270:2481–2484
Thorington Jr RW (1984) Flying squirrels are monophyletic. Science 225:1048–1050
Wiley EO (1988) Parsimony analysis and vicariance biogeography. Syst Zool 37:271–290
Wojcicki M, Brooks DR (2005) PACT: An efficient and powerful algorithm for generating area cladograms. J Biogeogr 32:755–774
Woods CA, Hermanson JW (1985) Myology of hystricognath rodents: An analysis of form, function and phylogeny. In: Luckett WP, Hartenberger JL (eds) Evolutionary relationships among rodents. A multidisciplinary analysis. Plenum Press, New York, pp 515–548

15 Comparative phylogeography: The use of parasites for insights into host history

Caroline M. Nieberding and Serge Morand

1 Introductory remarks

Parasites are useful biological tags of the ecology of their hosts. In this chapter, we will show that parasites can also be used as powerful "evolutionary prints" in order to generate new hypotheses about the history of their hosts. By this we mean that genes of a parasite might actually better reflect host history than genes of the host. This can be useful as incongruence between the genealogies of several genes within a species often limits the resolution of the species history. We will focus on the developing field of comparative phylogeography between hosts and parasites and show that parasites can highlight historical events affecting host lineages that can not be detected by the study of the host itself, such as past host migration or differentiation events. Congruence in host-parasite phylogeographies relies on long-term host specificity, which is favoured by limited dispersal abilities, direct life cycles, high abundance and prevalence of the parasite. Parasites might play the role of an evolutionary print of their host's history if they present reduced ancestral polymorphism, i.e. when parasites have shorter generation times and lower effective sizes than their hosts. Provided that the appropriate parasite species is selected according to these conditions, there appears to be no limitation to the use of parasites as evolutionary prints of the phylogeography of their hosts.

2 Parasites as prints of their hosts in evolutionary biology

The number of extant parasite species is estimated to represent ~30% of the biodiversity of eukaryote species and ~10% of currently known species of Metazoa (DeMeeûs and Renaud 2003; Poulin and Morand 2004). Parasites have been shown to be useful sources of information about various aspects of the ecology of their hosts. For example, parasites can be used as ecological tags to provide insights into different life-history traits of their hosts; parasitic species are used as biological tags to discriminate stocks, migration routes and nursery grounds of commercially exploited marine fish populations (Olson and Pratt 1973; Bouillon and Dempson 1989; Mackenzie 2002). Parasites can also help monitor accurately the health of the host ecosystem and human impacts on that environment, as the tissues of some acanthocephalan and cestode species accumulate up to 2700 times more heavy metal compounds compared with the tissues of their final hosts (Sures 2004).

Whereas most current studies focus on extant host-parasite relationships, parasites can also be used as "evolutionary prints" to better characterize the evolutionary history of their hosts (Thomas et al. 1996; Nieberding et al. 2004; Whiteman and Parker 2005). Indeed, parasites are unique in that their evolutionary history is tightly influenced by their intimate relationships with their hosts (Nadler 1995). As a correlate, it can be assumed that the genetic structure of parasites might be generated by the evolutionary history of their hosts. Provided that hosts and parasites share a common history of speciation, comparing the genetic structure of parasites with that of their hosts could provide information about the evolutionary history of the host that is not detectable by studying the host directly (Thomas et al. 1996; Nieberding et al. 2004). Parasites could therefore be used to reveal cryptic traits of the evolutionary history of their hosts (as a biological "magnifying glass" effect). This approach has been used successfully in several evolutionary biology studies. For example, parasitological data from hybrid or contact zone populations of the hosts have been used to detect host hybrids and to quantify gene flow among host lineages (Cloutman 1988; Moulia et al. 993; Hafner et al. 1998; Derothe et al. 2001). In other studies, comparison of the phylogenies of hosts with those of their specific parasites has allowed researchers to better understand relationships among host species, and to detect the existence of cryptic host species (Thomas et al. 1996; Haukisalmi et al. 2001).

Recently, it was suggested that the parasite magnifying glass effect could prove highly useful in the field of phylogeography: by comparing phylogeographical data from parasites with that of their hosts, one could reveal cryptic phylogeographical traits of the hosts, i.e. cryptic historical gene flow or differentiation events between host lineages. Here we give first a general overview of the increasing number of studies comparing host-parasite phylogeographies, highlighting how the genetic structure of the parasite gave insight into its host phylogeography. Second, we examine the traits of the host and the parasite that can determine whether the phylogeographies of parasites can act as magnifying glasses of the evolutionary history of their hosts. We also provide an overview of the host-parasite characteristics that should be considered to select a parasite useful as a biological magnifying glass of its host.

3 Case studies in host-parasite comparative phylogeography

An increasing number of studies have identified new phylogeographical traits of different host species based on the analysis of the phylogeographical pattern of their parasites. These included the identification of cryptic host lineages, Quaternary refuges, migration routes or migration epochs, and past gene flow between host populations. In contrast, similar other studies have failed to show any magnifying glass effect of the parasite. These different studies are described briefly below.

3.1 Cryptic contacts among host lineages

The phylogeography of the modern human louse *Pediculus humanus* reveals that the species is formed by two ancient lineages that co-diverged 0.7 to 1.2 Myrs ago. As both *P. humanus* lineages have the same ecological niche, they must have diverged in allopatry on ancient *Homo* populations. The presence of both louse lineages on modern humans strongly suggests that a direct physical contact existed between archaic and modern lineages of *Homo* (Reed et al. 2004).

3.2 Cryptic host refuges during the Quaternary cold stages

The phylogeographical patterns of the bast scale *Matsucoccus feytaudi* and of its specific host, the maritime pine *Pinus pinaster*, were compared over the Mediterranean region (Burban et al. 1999; Burban and Petit 2003). Both species display near congruent genetic and geographical groups in Western Europe, along the European Atlantic coast and in North Africa. The phylogeography of *M. feytaudi* revealed the presence of a cryptic evolutionary unit in Punta Cires (Morocco) and multiple cryptic refuges in the Iberian refuge during ice ages from which northward European recolonisation occurred for both host and parasite species.

3.3 Cryptic epochs and routes of host migrations

The phylogeographical patterns of the nematode *Heligmosomoides polygyrus* and its rodent host, *Apodemus sylvaticus*, were compared across Europe. Both species co-differentiated in southwest Europe, North Africa, Italy and on seven Mediterranean islands (Nieberding et al. 2004, 2005, 2006). Moreover, the parasite phylogeography revealed (a) the existence of supplementary distinct allopatric refuges in Iberia and Italy for both species during the Quaternary; (b) that the colonisation of North Africa occurred from southern Spanish populations in both species; and (c) that occasional contacts persisted between the Sicilian and southern Italian populations of both species until ~200 000 years after the differentiation of host populations (Fig. 1).

The umbelliferous *Bowlesia incana* has currently a disjunctive distribution north and south of the tropical regions of Central America. The phylogeography of its herbivore butterfly parasite *Greya powelli* showed that the introduction of *B. incana* in North America from South America was several orders of magnitude older than was previously thought (250 years) (Pellmyr et al. 1998). Similarly, the phylogeographical pattern of *Puumala* hantaviruses (PUUV) across Europe and Russia confirmed the accuracy of one of the two immigration routes proposed for its host, the bank vole *Clethrionomys glareolus* across the Russian plain towards Fennoscandia at the end of the last glacial period (Asikainen et al. 2000; Dekonenko et al. 2003).

3.4 Cryptic host lineages

The phylogeographical structure of the parasitic angiosperm *Arceuthobium americanum* and of its conifer *Pinus* hosts were compared in north-western America (Jerome and Ford 2002a, b). Three genetic races of *A. americanum* each parasitize specifically one allopatric host lineage. This revealed the accuracy of host taxonomic differentiation into *P. contorta* var. *latifolia* and var. *murrayana* and *P. banksania*, in spite of low host genetic and morphological differentiation among them.

3.5 Incongruent phylogeographies

In spite of the many examples of parasite phylogeographies matching closely those of their host, there are also examples where matches do not occur. The phylogeographical patterns of *Greya* butterflies were broadly compared across America with those of their Saxifragacea host plants and their parasitoid wasps *Agathis* (Brown et al. 1997; Althoff and Thompson. 1999). The generalist *Greya politella*, which has achieved numerous host species shifts over its recent evolutionary history, has a phylogeographical pattern showing no co-differentiation with a particular host. Similarly, the phylogeographical structures of the generalist parasitoid wasps *Agathis thompsoni* and *Agathis nsp.* are incongruent with those of their main hosts *G. subalba* and *G. enchrysa*.

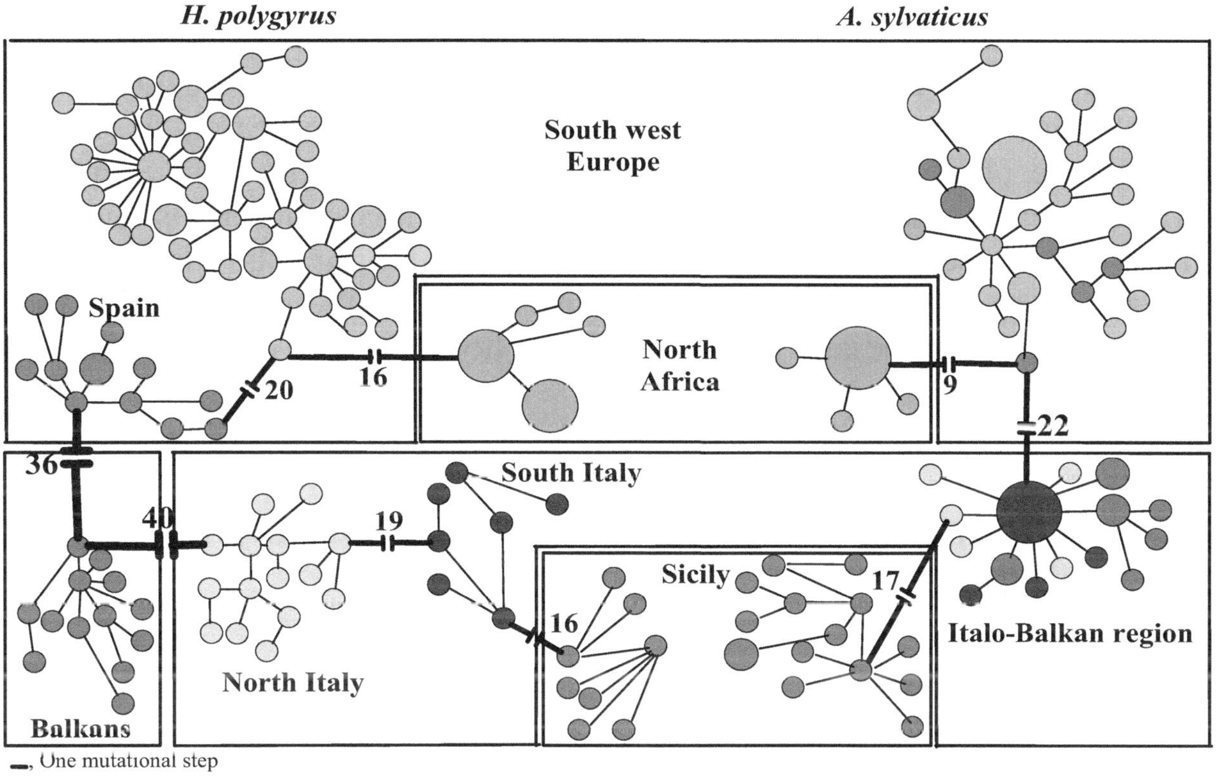

Fig. 1. Minimum spanning networks of the nematode *H. polygyrus* (on the left) and its host, the woodmouse *A. sylvaticus* (on the right), on the basis of cytochrome b gene sequences. Branch lengths correspond to the number of mutational steps (given in bold characters) separating haplotypes, represented by circles. Populations from Spain, southwest Europe, North Africa, the Balkan region, north Italy, south Italy and Sicily, are compared in both species. Partial congruence is observed because both species present a south west European, a north African, an Italian and a Sicilian clade, linked by a high number of mutational steps. However, partial incongruence between the networks of both species is also observed: (a) *H. polygyrus* Italian and Balkan populations form two distinct clades, and (b) *H. polygyrus* Italian and south west European are further divided into several subclades, although the corresponding host populations cannot be distinguished. These particular (sub)clades in *H. polygyrus* highlight the fact that the phylogeography of the parasite is more diversified, differentiated than that of its host (modified from Nieberding et al. 2004)

Similarly, the Australian brood-parasite Horsfield's bronze-cuckoo *Chalcites basalis* lays eggs in the nests of its hosts, *Malurus* fairy-wrens and *Acanthiza* thornbills. The lack of phylogeographical structure of *C. basalis* denies the existence of evolutionarily long-term stable host races in this species, which is likely to have expanded its geographical and host species range within the past few tens of thousands of years, following climatic amelioration during the late Pleistocene (Joseph et al. 2002).

The phylogeographies of the cestode *Paranoplocephala arctica* and of its lemming *Dicrostonyx* hosts were compared over the Holarctic region (Wickstrom et al. 2003). Both taxa present two differentiated clades in the Nearctic and Palearctic regions, separated by the Bering Strait. However, the parasite phylogeography presents additional differentiation events on Wrangle Island around the Bering Strait and on Canadian Artic islands that do not reflect the phylogeography of the host lemming. The discrepancies between *P. arctica* and *Dicrostonyx* could be due to the passage of *P. arctica* through intermediate soil invertebrate hosts, which could reduce dependence on a definitive lemming host.

4 Common traits in hosts and parasites displaying co-differentiating phylogeographies

Here we examine the life-history traits and other features of the host and the parasite that might help determine whether the phylogeographies of parasites can act as magnifying glasses of the evolutionary history of their hosts.

4.1 Congruence versus incongruence

A survey of the existing host-parasite comparative phylogeographies highlights that parasites can indeed reveal cryptic evolutionary events of their host, such as the existence of past cryptic migration or differentiation events, and of past cryptic refuges or lineages across the last few millions years.

Moreover, these studies also provide clues to understand how and why co-differentiation occurs on a phylogeographical evolutionary timescale (i.e. a few million years). Indeed, the congruence in host-parasite phylogeographies discussed above appears to be determined by the duration and degree of intimacy of host and parasite relationships, i.e. long-term host specificity (Johnson et al. 2003). Long-term host specificity should indeed limit *host switching* opportunities (Blouin et al. 1995). As a correlate, the absence of long-term host-parasite specificity explains most incongruent host-parasite phylogeographies: the phylogeographies of generalist parasites are not related to those of their hosts (Althoff and Thompson 1999; Brown et al. 1997; Joseph et al. 2002). Similarly, intermediate hosts in the parasite life cycle reduce the congruence of the parasite with the phylogeographical history of the definitive host (Wickstrom et al. 2003).

Table 1. Survey of the host-parasite congruent comparative phylogeographies according to the following traits of the parasites: dispersal ability, reproduction mode, prevalence and abundance. Importantly, in all the studies cited, the level of host specificity is high and the life cycle is direct (no intermediate host) for the parasite

Host	Parasite	Geographic distribution	Parasite dispersal ability	Parasite reproduction mode	Prevalence, abundance
PLANTS					
Pinus pinaster	*Matsucoccus feytaudi*	Mediterranean region	Higher than its host	Sexual	Variable
P. contorta, *P. banksania*	*Arceuthobium americanum*	North-West America	Lower than its host	Sexual	High
Bowlesia incana	*Greya powelli*	Nearctic region	Lower than its host	Sexual	High
ANIMALS					
Apodemus sylvaticus	*Heligmosomoides polygyrus*	Europe	Lower than its host	Sexual	High
Clethrionomys glareolus	*Hantavirus*	Palearctic region	Lower than its host	Clonal	High
Homo sapiens	*Pediculus humanus*	Worldwide	Lower than its host	Sexual	High

Host-parasite species pairs showing congruent phylogeographies appear to share common life-history, ecological, or demographic traits that favour long-term host specificity (Table 1). A direct life cycle of the parasite (i.e. no intermediate host) is one important trait shared by congruent host-parasite phylogeographies. This trait makes the gene flow of the parasite determined primarily by the contacts and movements of its host, which limits failure of the parasite to speciate in response to host speciation, i.e.

parasite release (Johnson et al. 2003). In addition, high abundance and high prevalence of the parasite on or within its host reduces the risks of *parasite extinction* and of "*missing the boat*" (Clayton et al. 2003). Together, these three factors increase the probability that a parasite matches the differentiation of its host and the probability that congruent host-parasite phylogeographies are observed.

4.1.1 Reproductive mode

Parasite sexual reproduction maintains gene flow between populations because adults have to meet in/on the host in order to produce the next generation. By contrast, parasite parthenogenetic and clonal reproduction induce genetic drift among populations, because adult individuals do not need to meet to produce the next generation (Gow et al. 2004). Therefore, sexually reproducing parasites should better follow the mixing of their host populations and consequently better reflect their history of differentiation or migration events. This will limit the risk that the parasite does not speciate in response to host speciation. An example is given by the nematode *H. polygyrus* parasitizing the wood mouse *A. sylvaticus*: gene flow between *H. polygyrus* populations depends on that of *A. sylvaticus*, which favoured the partial congruence observed between the phylogeographies of both species (Nieberding et al. 2004, 2005, 2006).

4.1.2 Abundance and prevalence

Levels of prevalence and abundance of parasite species also determine the congruence level in host-parasite phylogeographies, because parasite populations whose abundance and prevalence are high in their hosts have a better probability to follow migration and differentiation events between host populations, reducing the probability of parasite extinction.

4.2 Parasites as a "biological magnifying glass"

Although incongruent host-parasite phylogeographies are useful for better understanding the dynamics of host-parasite specificity, these studies can not reveal anything about cryptic historical host events, because the phylogeographical history of the parasite was not generated by the evolutionary history of its current host. Therefore, (partial) congruent host-parasite phylogeographies are a prerequisite for the use of parasites as evolutionary prints. In other words, similar geographical and genetic groups must be present in the corresponding host and parasite populations, and result from the same historical events.

Current studies inferring hypotheses on past historical events of a focal species on the basis of the history of one of its symbiont usually lack an estimation of temporal congruence between the phylogeographical data of the interacting species. However, the presence in host and parasite phylogeographies of similar geographic lineages (spatial congruence) does not necessarily imply real host and parasite historical co-differentiation because similar geographic patterns do not obligatorily result from the same historical events. Therefore, in order to assess co-differentiation, it is necessary to show that similar geographic lineages in the host and its parasite differentiated simultaneously in the past and therefore display temporal congruence of their phylogeographies (Page 2003). This requires the dating of the epoch of differentiation of lineages of both organisms. However, it is generally not possible to directly quantify the absolute rate of molecular evolution of parasite lineages, as parasites generally do not display fossil records. In order to fill this methodological gap, a method was recently proposed to estimate the level temporal congruence between host and parasite lineages (Box 1).

In case of host-parasite (partial) congruence, parasites might display a biological magnifying glass effect, providing information about the evolutionary history of the host that is not detectable at the host level. This is possible if the phylogeography of the parasite is more diversified than that of its host. The genetic diversification of a species is a balance between genetic drift and gene flow: genetic drift counteracts the effects of gene flow by eroding genetic diversity within populations and by increasing differentiation between them (Gow et al. 2004). Consequently, a parasite should present lower gene flow and higher genetic drift between populations than its host in order to act as an evolutionary print of its host. Gene flow and genetic drift appear to be determined primarily by the dispersal ability, the effective population size (N_e), and the DNA mutation rate (μ) of the species (Frankham 2002). From a macroevolutionary perspective, this would promote parasite speciation in the absence of host speciation, i.e. duplication events (Clayton et al. 2003).

4.2.1 Dispersal ability

A lower dispersal ability of the parasite compared with that of its host is likely to be one important factor favouring a higher genetic differentiation between parasite populations, in comparison with its corresponding host populations. Indeed, several studies devoted to the comparison of host-parasite population genetics reveal that high dispersal ability of the parasite homogenizes its population structure (McCoy et al. 2003; Baer et al. 2004), whereas lower dispersal ability enhances local differentiation of the parasite (Blouin et al. 1995; Johnson et al. 2002; Reed and Hafner. 1997; Bohonak. 1999; Bucheli et al. 2001). For example, a lack of congruence is observed between the phylogeographies of the liver fluke *Fascioloides magna* and its specific deer host *Odocoileus virginianus* in northwest America. In this system, the long-distance dispersal of the parasite counteracted local differentiation of its host populations (Mulvey et al. 1991). By contrast, the deer intestinal nematode *Mazamastrongylus odocoilei*, with limited dispersal abilities, presents substantial population differentiation (Blouin et al. 1995). Therefore, the lower dispersal abilities of parasites compared with those of their hosts is a key factor that enables them to represent an evolutionary print of their host.

4.2.2 Effective population size (N_e)

Low parasite N_e increases genetic drift, which in turn increases the probability that the parasite will play the role of a magnifying glass in its host phylogeography. Low prevalence, low abundance and parthenogenetic reproduction are three important factors that reduce N_e in parasites (Anderson et al. 1998; Blouin et al. 1998). For example, the parthenogenetic nematode *Heterorhabditis marelatus* parasitizing soil-dwelling insects in America has a low N_e which has led to high population differentiation (Blouin et al. 1999). However, as high values of prevalence and abundance of parasite species also determine the congruence level in host-parasite phylogeographies by reducing the risks of parasite extinction and of "missing the boat", intermediate levels of prevalence and abundance should be the appropriate compromise to detect a biological magnifying glass effect of the parasite in the phylogeography of its host.

4.2.3 DNA mutation rate

Accelerated substitution rate in the parasite DNA in comparison to homologous genes of their hosts make parasites ideal independent markers of their hosts because they would amplify the evolutionary history of their hosts in their genes (Blouin et al. 1995; Whiteman and Parker 2005). For example, in the *A. sylvaticus - H. polygyrus* host-parasite system, the rate of molecular evolution of the cytochrome b gene is ~1.5 fold higher in the parasite than in its host (Nieberding et al. 2004). This result agrees with those of other host-parasite studies that found a faster rate of molecular evolution in parasitic genes compared with the homologous genes of their specific hosts (Hafner et al. 1994; Page et al. 1998; Paterson et al. 2000).

Box 1. In order to assess whether similar geographic lineages in the host and the parasite differentiated simultaneously in the past (temporal congruence), one method consists of plotting the genetic distances between pairs of host individuals and pairs of parasite individuals, such that each parasite must be related to a host from the same population (see Page 2003 for more details). In case of significant historical codivergence between host and parasite lineages, a linear correlation is observed between the genetic distances of pairs of hosts and pairs of corresponding parasites, and the y-intercept of the resulting correlation line passes through the origin (0,0). If a correlation line is observed but its y-intercept is significantly different from zero, this means that the parasite lineages differentiated before or after the host lineages. This method can be applied to both sequence or multiloci genetic data. In case of sequence data, this methodology was successfully applied to show significant codivergence between *H. polygyrus* and *A. sylvaticus* phylogeographies over south west Europe, using the program TreeMap (Page 1994; Nieberding 2004). Interestingly, if temporal congruence is assessed, the absolute rate of parasite molecular evolution can then be estimated on the basis of the host molecular data, provided that homologous genes were sequenced in both organisms (Avise 2000). In contrast to sequence data, multi loci data (AFLP, RFLP, or microsatellites) can also be used to estimate temporal congruence between host and parasite lineages (Beaumont 2005). Assuming mutation and migration are low, pairwise Fst among populations or individuals can be used as an estimator of the level of genetic differentiation between populations (Rousset 1997; Bohonak 1999). Therefore, the correlation between the genetic structures of two organisms can be estimated using Mantel tests on pairwise genetic distance matrices, using Fst/(1-Fst) values. The program CADM (congruence among distance matrices (Legendre and Makarenkov 2002) allows statistical comparison of full distance matrices. Significance of the Mantel test can then be assessed over large number (e.g., 10000) of permutations. Similarly to sequence data, temporal congruence will be confirmed if a correlation line passing through (0,0) is observed. This methodology has been applied successfully (Jerome 2002b; Anderson et al. 2004).

5 Conclusions and perspectives

Incongruence among gene genealogies within a single species is often observed, and can be due to (a) differences in evolutionary processes, (b) differences in history, or (c) sampling error (reviewed in Nichols 2001). This limits the confidence we can have in the results or the resolution of the organism's history. Recent studies have thus considered the possibility that genes of other, unrelated organisms might actually provide a better picture of the history of the focal organism. These studies showed that indeed gene trees of other, symbiotic, organisms might actually better reflect the true history of a given (focal) organism, compared to gene trees of the focal organism itself (see Whiteman and Parker 2005 for a review in the context of animal parasites and conservation)

This idea is currently emerging in comparative phylogeography. Even if additional studies are needed to better characterize and quantify the relative effects of the different traits discussed here and to identify possible new ones, it appears that numerous studies have already shown the utility and sensitivity of this approach: parasites can successfully be used to generate new hypotheses about the phylogeographic history of their hosts, highlighting possible cryptic historical gene flow, differentiation events, host refuges and lineages, which could not be detected by the phylogeographical study of the host itself.

When focusing on micro-mammal hosts and their macro-parasites, helminths as endoparasites, ticks and lice as ectoparasites have been the most regularly used parasite tags in wild host-parasite comparative phylogeographies (Anderson et al. 1998; Blouin et al. 1995, 1998, 1999; Clayton et al. 2003; Johnson et al. 2003; McCoy et al. 2003; Nieberding et al. 2004; Wickstrom et al. 2003). These parasitic groups have proved to be highly useful as these taxa are harbored by every micro-mammal and present a rapid rate of molecular evolution (Page et al. 1998; Johnson et al. 2003; Nieberding et al. 2004). However, there should be no taxonomic limitation to the use of other macro-parasites as evolutionary prints of the phylogeography of their micro-mammal hosts, provided that the parasite species is selected according to the life history traits and other features mentioned in this review. The appropriate parasite should display current host specificity, a direct life cycle, as well as high abundance and prevalence levels on its host, in order to favour long-term host-parasite specificity. Moreover, the dispersal ability of the selected parasite should be limited by that of its host, while its effective population size and molecular mutation rate should be respectively lower and higher than those of its host. Provided that the appropriate parasite species is chosen according to these conditions, there appears to be no limitation to the use of micro-parasites as evolutionary prints of the phylogeography of their micro-mammal hosts.

References

Agostini HT, Ryschkewitsch CF, Stoner GL (1996) Genotype profile of human polyomavirus JC excreted in urine of immunocompetent individuals. J Clin Microbiol 34:159–164

Althoff DM, Thompson JN (1999) Comparative phylogeographical structure of two parasitoid-host interactions. Evolution 53:818–825

Anderson RM, Jackson HC, May RM, Smith AM (1981) Population dynamics of fox rabies in Europe. Nature 289:765–771

Anderson TJ, Blouin MS, Beech RN (1998) Population biology of parasitic nematodes: Applications of genetic markers. Adv Parasitol 41:219–283

Anderson B, Olivieri I, Lourmas M, Stewart BA (2004) Comparative population genetic structures and local adaptation of two mutualists. Evolution 58:1730–1747

Asikainen K, Hanninen T, Henttonen H, Niemimaa J, Laakkonen J, Andersen HK, Bille N, Leirs H, Vaheri A, Plyusnin A (2000) Molecular evolution of Puumala hantavirus in Fennoscandia: Phylogenetic analysis of strains from two recolonization routes, Karelia and Denmark. J Gen Virol 81:2833–2841

Avise JC (2000) Phylogeography. The history and formation of species. Harvard Univ Press, Cambridge

Baer CF, Tripp DW, Bjorksten TA, Antolin MF (2004) Phylogeography of a parasitoid wasp (*Diaeretiella rapae*): No evidence of host-associated lineages. Mol Ecol 13:1859–1869

Beaumont MA (2005) Adaptation and speciation: What can F_{st} tell us? Trends Ecol Evol 20:435–440

Blouin MS (1998) Mitochondrial DNA diversity in nematodes. J Helminthol 72:285–289

Blouin MS, Yowell CA, Courtney CH, Dame JB (1995) Host movement and the genetic structure of populations of parasitic nematodes. Genetics 141:1007–1014

Blouin MS, Liu J, Berry RE (1999) Life cycle variation and the genetic structure of nematode populations. Heredity 83:253–259

Bohonak AJ (1999) Dispersal, gene flow, and population structure. Q Rev Biol 74:21–45

Bouillon DR, Dempson JB (1989) Metazoan parasite infections in landlocked and anadromous Arctic charr (*Salvelinus alpinus*, Linnaeus), and their use as indicators of movement to sea in young anadromous charr. Can J Zool 67:2478–2485

Bourhy H Kissi B, Audry L, Smreczak M, Sadkowska-Todys M, Kulonen K, Tordo N, Zmudzinski JF, Holmes EC (1999) Ecology and evolution of rabies virus in Europe. J Gen Virol 80:2545–2558

Brown JM, Leebens-Mack JH, Thompson JN, Pellmyr O, Harrison RG (1997) Phylogeography and host association in a pollinating seed parasite *Greya politella* (Lepidoptera: Prodoxidae). Mol Ecol 6:215–224

Bucheli E, Gautschi B, Shykoff JA (2001) Differences in population structure of the anther smut fungus *Microbotryum violaceum* on two closely related host species, *Silene latifolia* and *S. dioica*. Mol Ecol 10:285–294

Burban C, Petit RJ (2003) Phylogeography of maritime pine inferred with organelle markers having contrasted inheritance. Mol Ecol 12:1487–1495

Burban C, Petit RJ, Carcreff E, Jactel H (1999) Range-wide variation of the maritime pine bast scale *Matsucoccus feytaudi* Duc. (Homoptera: Matsucoccidae) in relation to the genetic structure of its host. Mol Ecol 8:1593–1602

Clayton DH, Al-Tamimi S, Johnson KP (2003) The ecological basis of coevolutionary history. In: Page RDM (ed) Tangled trees: Phylogeny, cospeciation, and coevolution. Univ Chicago Press, Chicago, pp 330–350
Clayton DH, Bush SE, Johnson KP (2004) Ecology of congruence: Past meets present. Syst Biol 53:165–173
Cloutman DG (1988) Ancyrocephalids (Monogenea) of redbreast sunfish, bluegill, and their hybrids from Lake Norman, North Carolina: Remarks on monogeneans as indicators of parent species of hybrids. Proc Helminthol Soc Wash 55:108–110
Dekonenko A, Yakimenko V, Ivanov A, Morozov V, Nikitin P, Khasanova S, Dzagurova T, Tkachenko E, Schmaljohn C (2003) Genetic similarity of Puumala viruses found in Finland and western Siberia and of the mitochondrial DNA of their rodent hosts suggests a common evolutionary origin. Infect Genet Evol 3:245–257
De Meeûs T, Renaud F (2003) Parasites within the new phylogeny of eucaryotes. Trends Parasitol 18:247–251
Derothe JM, Le Brun N, Loubes C, Perriat-Sanguinet M, Moulia C (2001) Susceptibility of natural hybrids between house mouse subspecies to *Sarcocystis muris*. Int J Parasitol 31:15–19
Falush D, Wirth T, Linz B, Pritchard JK, Stephens M, Kidd M, Blaser MJ, Graham DH, Vacher S, Perez-Perez GI, Yamaoka Y, Megraud F, Otto K, Reichard U, Katzowitsch E, Wang X, Achtman M, Suerbaum S (2003) Traces of human migrations in *Helicobacter pylori* populations. Science 299:1582–1585
Frankham R, Ballou JD, Briscoe DA (2002) Introduction to conservation genetics. Cambridge Univ Press, Cambridge
Gow JL, Noble LR, Rollinson D, Mimpfoundi R, Jones CS (2004) Breeding system and demography shape population genetic structure across ecological and climatic zones in the African freshwater snail, *Bulinus forskalii* (Gastropoda, Pulmonata), intermediate host for schistosomes. Mol Ecol 13:3561–3573
Hafner MS, Sudman PD, Villablanca FX, Spradling TA, Desmates JW, Nadler SA (1994) Disparate rates of molecular evolution in cospeciating hosts and parasites. Science 265:1087–1090
Hafner MS, Demastes JW, Hafner DJ, Spradling TA, Sudman PD, Nadler SA (1998) Age and movement of a hybrid zone: Implications for dispersal distance in pocket gophers and their chewing lice. Evolution 52:278–282
Haukisalmi V, Wickström LM, Hantula J, Henttonen H (2001) Taxonomy, genetic differentiation and Holarctic biogeography of *Paranoplocephala* spp. (Cestoda: Anoplocephalidae) in collared lemmings (*Dicrostonyx*; Arvicolinae). Biol J Linn Soc 74:171–196
Holmes EC (2004) The phylogeography of human viruses. Mol Ecol 13:745–756
Jerome CA, Ford BA (2002a) Comparative population structure and genetic diversity of *Arceuthobium americanum* (Viscaceae) and its *Pinus* host species: Insight into the host-parasite evolution in parastic angiosperms. Mol Ecol 11:407–420

Jerome CA, Ford BA (2002b) The discovery of three genetic races of dwarf mistletoe *Arceuthobium americanum* (Viscaceae) provides insight into the evolution of parasitic angiosperms. Mol Ecol 11:387–405

Johnson KP, Williams BL, Drown DM, Adams RJ, Clayton DH (2002) The population genetics of host specificity: Genetic differentiation in dove lice (Insecta: Phthiraptera). Mol Ecol 11:25–38

Johnson KP, Adams RJ, Page RD, Clayton DH (2003) When do parasites fail to speciate in response to host speciation? Syst Biol 52:37–47

Joseph L, Wilke T, Alpers D (2002) Reconciling genetic expectations from host specificity with historical population dynamics in an avian brood parasite, Horsfield's Bronze-Cuckoo *Chalcites basalis* of Australia. Mol Ecol 11:829–837

Korber B, Muldoon M, Theiler J, Gao F, Gupta R, Lapedes A, Hahn BH, Wolinsky S, Bhattacharya T (2000) Timing the ancestor of the HIV-1 pandemic strains. Science 288:1789–1796

Legendre P, Makarenkov V (2002) Reconstruction of biogeographic and evolutionary networks using reticulograms. Syst Biol 51:199–216.

Lemey P, Pybus OG, Wang B, Saksena NK, Salemi M, Vandamme AM (2003) Tracing the origin and history of the HIV-2 epidemic. Proc Natl Acad Sci USA 100:6588–6592

MacKenzie K (2002) Parasites as biological tags in population studies of marine organisms: An update. Parasitology 124:S153–S163

McCoy KD, Boulinier T, Tirard C, Michalakis Y (2003) Host-dependent genetic structure of parasite populations: Differential dispersal of seabird tick host races. Evolution 57:288–296

Moulia C, Le Brun N, Dallas J, Orth A, Renaud F (1993) Experimental evidence of genetic determinism in high susceptibility to intestinal pinworm infection in mice: A hybrid zone model. Parasitology 106:387–393

Mulvey M, Aho JM, Lydeard C (1991) Comparative population genetic structure of a parasite (*Fascioloides magna*) and its definitive host. Evolution 45:1628–1640

Nadler SA (1995) Microevolution and the genetic structure of parasite populations. J Parasitol 81:395–403

Nichols R (2001) Gene trees and species trees are not the same. Trends Ecol Evol 16:358–364

Nieberding C, Morand S, Libois R, Michaux J (2004) A parasite reveals cryptic phylogeographical history of its host. Proc R Soc Lond B 271:2559–68

Nieberding C, Morand S, Douady CJ, Libois R, Michaux J (2005) Phylogeography of a nematode (*Heligmosomoides polygyrus*) in the western Palearctic region: Persistence of northern cryptic populations during ice ages? Mol Ecol 14:765–779

Nieberding C, Morand S, Libois R, Michaux J (2006) Parasites and the insular syndrome: The colonisation of the western Mediterranean islands by *Heligmosomoides polygyrus* (Dujardin, 1845). J Biogeogr (in press)

Olson RE, Pratt I (1973) Parasites as indicators of English sole (*Parophrys vetulus*) nursery grounds. Trans Am Fish Soc 2:405–411

Page RDM (1994) Parallel phylogenies: Reconstructing the history of host-parasite assemblages. Cladistics 10:155–173

Page RDM (ed) (2003) Tangled trees: Phylogeny, cospeciation, and coevolution. Univ Chicago Press, Chicago

Page RDM, Lee PL, Becher SA, Griffiths R, Clayton DH (1998) A different tempo of mitochondrial DNA evolution in birds and their parasitic lice. Mol Phyl Evol 9:276–293

Paterson AM, Wallis GP, Wallis LJ, Gray RD (2000) Seabird and louse coevolution: Complex histories revealed by 12S rRNA sequences and reconciliation analyses. Syst Biol 49:383–399

Pellmyr O, Leebens-Mack J, Thompson JN (1998) Herbivores and molecular clocks as tools in plant biogeography. Biol J Linn Soc 63:367–378

Poulin R, Morand S (2004) Parasite biodiversity. Smithsonian Inst Press, Washington

Pybus OG, Holmes EC, Harvey PH (1999) The mid-depth method and HIV-1: A practical approach to testing hypotheses of viral epidemic history. Mol Biol Evol 186:953–959

Reed DL, Hafner MS (1997) Host specificity of chewing lice on pocket gophers: A potential mechanism of cospeciation. J Mammal 78:655–660

Reed DL, Hafner MS (2002) Phylogenetic analysis of bacterial communities associated with ectoparasitic chewing lice of pocket gophers: A culture-independent approach. Microb Ecol 44:78–93

Reed DL, Smith VS, Hammond SL, Rogers AR, Clayton DH (2004) Genetic analysis of lice supports direct contact between modern and archaic Humans. PLOS Biol 2:1972–1983

RicoHesse R, Harrison LM, Salas RA, Tovar D, Nisalak A, Ramos C, Boshell J, deMesa MTR, Nogueira RMR, daRosa AT (1997) Origins of dengue type 2 viruses associated with increased pathogenicity in the Americas. Virology 230:244–251

Rousset F (1997) Genetic differentiation and estimation of gene flow from *F*-statistics under isolation by distance. Genetics 145:1219–1228

Saruwatari L, Sugimoto C, Kitamura T, Ohno N, Sakai E, Shresta P, Hoa BK, Phi PTP, An HPH, Tuyet NTA, Honjo T, Kobayashi N, Zheng HY, Takasaka T, Yogo Y (2002) Asian domains of four major genotypes of JC virus, Af2, B1-b, CY and SC. Arch Virol 147:1–10

Sugimoto C, Hasegawa M, Kato A, Zheng HY, Ebihara H, Taguchi F, Kitamura T, Yogo Y (2002) Evolution of human polyomavirus JC. Implications for the population history of humans. J Mol Evol 54:285–297

Sures B (2004) Environmental parasitology: Relevancy of parasites in monitoring environmental pollution. Trends Parasitol 20:170–177

Thomas F, Verneau O, De Meeûs T, Renaud F (1996) Parasites as host evolutionary prints: Insights into host evolution from parasitological data. Int J Parasitol 26:677–686

Twiddy SS, Farrar JJ, Chau NV, Wills B, Gould EA, Gritsun T, Lloyd G, Holmes EC (2002) Phylogenetic relationships and differential selection pressures among genotypes of denge-2 virus. Virology 298:63–72

Whiteman NK, Parker PG (2005) Using parasites to infer host population history: A new rationale for parasite conservation. Anim Conserv 8:175–181
Wickström LM, Haukisalmi V, Varis S, Hantula J, Fedorov VB, Henttonen H (2003) Phylogeography of the circumpolar *Paranoplocephala arctica* species complex (Cestoda: Anoplocephalidae) parasitizing collared lemmings (*Dicrostonyx* spp.). Mol Ecol 12:3359–3371

16 Insularity and micromammal-macroparasite relationships

Elodie Magnanou and Serge Morand

1 Introductory remarks

1.1 Micromammals and the insular syndrome

Island biogeography has been extensively studied since the pioneering studies of MacArthur and Wilson (1963, 1967), followed by numerous studies showing the magnitude and direction of ecological and evolutionary processes that operate within and among insular populations, species and communities. Insular trends have been extensively studied in terrestrial vertebrates, especially in lizards, birds, and mammals (e.g., Stamps and Buechner 1985). The cascade of changes that affect the life-history traits of insular populations is sometimes called "insular syndrome". Despite the fact that a great number of environmental factors may vary among different islands, the main components of the insular syndrome can be summarized as morphological, behavioural, demographic, ecological, physiological and genetic shifts exhibited by organisms living in isolation (Blondel 1995). Indeed, all levels of biological organization, from individuals to interactive communities, are under the effects of the insular environment.

Insular faunas are characterized by an impoverishment in species number. Species with good dispersal ability, being generalist or abundant on the mainland are favoured on islands (Sarà and Morand 2002). As a consequence, there is a decrease in interspecific competition and in predation pressure often associated with some demographic and ecological changes: density increase, ecological niche widening, resistance to invasions, and greater vulnerability to perturbations (Alder and Levins 1994; Blondel 1995). Insular small mammals show also a decrease in litter size (Fons et al. 1997), an increase in adult survival and often also a decrease in aggressiveness (Alder and Levins 1994).

At the same time, morphological changes divide insular mammals into two main evolutionary groups so that small species (<100g) tend to become larger (gigantism), whereas large species (>100g) tend to become smaller (dwarfism) (Van Valen's rule; Van Valen 1973; Lomolino 1985). Patterns of variation in body size of insular mammals seem to support the hypothesis that, based on physiological arguments, the optimal body mass for a terrestrial mammal is approximately 100 g (Brown 1995; Meiri et al. 2004). Finally, changes in energy expenditure in insular mammals have also been observed (Arends and McNab 2001; Magnanou et al. 2005).

1.2 Macroparasites on islands

Studies analysing selective forces that may drive evolutionary changes on islands generally consider resource abundance or competition and predation pressure release but too often ignore parasitism (Michaux et al. 2002). However, the "insular syndrome" should also affect parasites and host-parasite interactions.

First, new environmental conditions linked to an impoverishment in free-living species richness could obviously have consequences on the structure of parasite assemblages. Several studies showed that the macroparasite fauna of a given host species can strongly differ between mainland and island populations (Thomas 1953; Mas-Coma and Feliu 1984; Mas-Coma et al. 1988; Goüy de Bellocq et al. 2002). This difference involves the richness in parasite species, their identity (depending on mainland abundance or life cycle) but also their specificity (see review in Combes 1995).

Second, parasites may be affected by the life-trait shifts of their mammal hosts. Changes in mammal density, mammal body size and/or behaviour should have direct consequences on the epidemiological parameters (prevalence and intensity) (Mas-Coma et al. 2000). The effects of the decrease in mammal genetic diversity (Frankham 1997) on their parasites are even less known (Table 1).

As a consequence, changes in selective forces may affect the evolution of insular host-parasite interactions as theoretically predicted by Hochberg and Møller (2001). An increasing number of studies have demonstrated that parasites play an important role in host ecology, immune investment, population dynamics, behaviour, energy allocation, etc (Gregory 1991; Hudson et al. 1992; Newborn and Hudson 1992; Lindstrom et al. 2004), so it seems relevant to identify and quantify the effect of modified parasite assemblages on the ecology and evolution of insular micromammals.

Table 1. Interactions between hosts and parasites in relation to the insular syndrome. (M: micromammals, P: parasites)

Micromammals	Parasites	References
Decrease in species number	Decrease in species number	M: MacArthur and Wilson 1963, 1967 P: Thomas 1953; Mas-Coma & Feliu 1984; Goüy de Bellocq et al. 2002, 2003
Nested patterns	Nested patterns	M: Sarà and Morand 2002 P: Goüy de Bellocq et al. 2003
Generalist species	Direct life cycle species	M: Blondel 1995 P: Gouy de Bellocq et al. 2002
Density increase	Abundance increase	M: Alder and Levins 1994 P: Mas-Coma et al. 2000
Decrease in intraspecific competition	Decrease in intraspecific competition	M: Alder and Levins 1994 P: Mas-Coma et al. 2000
Ecological niche widening	Host capture	M: Cheylan 1988 P: Théron and Pointier 1995
Dwarfism and gigantism (100g threshold)	Shifts in parasite body size	M: Van Valen 1973 P: Valéro et al. 1996
Decrease in fecundity and increase in longevity	Parasite demography?	M: Fons et al. 1997; Alder and Levins 1994
Decrease in genetic diversity	Decrease in genetic diversity	M: Frankham 1997 P: Nieberding et al. 2006
Decrease in immune defence investment	Decrease in parasite diversity	Lindstrom et al. 2004
Impact on host fitness	Decrease in virulence	Hochberg and Møller 2001

1.3 Macroparasite contributions

This chapter illustrates to what extent parasites have to be considered in island biogeography, as parasites may have an influence on host evolution in addition to other selective pressures. We present an overview of the knowledge concerning micromammal parasite faunas on islands, focusing on two rodent species, namely the woodmouse *Apodemus sylvaticus* and the black rat *Rattus rattus*. Specifically, we first discuss the issues related to species richness and parameters that may influence it. The second part of the chapter deals with parasite host specificity and focuses on two well-known examples of parasite transfer occurring in an invasive host species on an island (*Fasciola hepatica* and *Schistosoma mansoni* in *R. rattus*). These examples may be relevant for (1) understanding host and parasite invasion processes; (2) management as some parasites considered here have veterinary and public health importance; and (3) conservation as endemic free-living species are especially vulnerable to perturbations caused by invaders.

2 Patterns of macroparasite species richness and abundance on islands

2.1 Nestedness

A nested pattern in the assemblages of both free-living organisms and parasites is often the rule. Species in depauperate assemblages generally constitute subset samples of richer assemblages (Patterson and Atmar 1986; Poulin 1996; Poulin and Guégan 2000; Krasnov et al. 2005). Nested patterns, mainly investigated in species assemblages living in fragmented or insular habitats, are considered to be a result of colonisation and extinction processes. Goüy de Belloq et al. (2003) showed that nestedness does occur in the parasite community of *A. sylvaticus*, and that this pattern can be related to the parasite life cycles and to the host geographic distribution (mainland versus islands, large versus small islands) (Fig. 1). Goüy de Bellocq et al. (2003) suggested that parasite species at the top of the presence/absence matrix are those with high probabilities of colonization and low probabilities of extinction, whereas parasite species at the bottom of the matrix are those with properties resulting in low probabilities of colonization and high probabilities of extinction. These properties reflect the parasite life-style: direct versus indirect life cycle, high versus low host specificity.

Parasite species found on small islands are subsets of larger assemblages found on larger islands or on the mainland. Hosts that have succeeded in establishing themselves on an island do not harbour the whole parasite community that is observed in the mainland host populations, but a specific subset of parasite species. Indeed, even if hosts succeed on an island, their parasites have to meet the conditions allowing them to survive. These include, for example, presence of suitable intermediate or definitive hosts and/or environmental conditions that permit the survival of free-living stages. Thomas (1953), Combes (1995) and Goüy de Bellocq et al. (2003)

showed that indirectly-transmitted parasites have a lesser chance of establishing themselves on islands than directly-transmitted parasites because of the poverty of free-living species encountered there. Parasites that need three hosts with an aquatic phase for their transmission are generally considered to have the lowest chance to invade an island. For example, Goüy de Bellocq et al. (2003) observed a lack of trematode and cestode species with complex life cycles on both continental and oceanic islands. Trematodes and acanthocephalans are usually found on large but not on small islands (Mas-Coma et al. 1984, 2000; Jiménez Piqueras 1992; Goüy de Bellocq et al. 2002, 2003).

Host populations live on islands that differ in biodiversity and ecological stability. Hence, according to the matrix model, islands at the left of the matrix may be characterized by their low extinction rates, high biodiversity and/or large area size and islands at the right may be characterized by their high extinction rates, low biodiversity and/or small area size (Fig. 1).

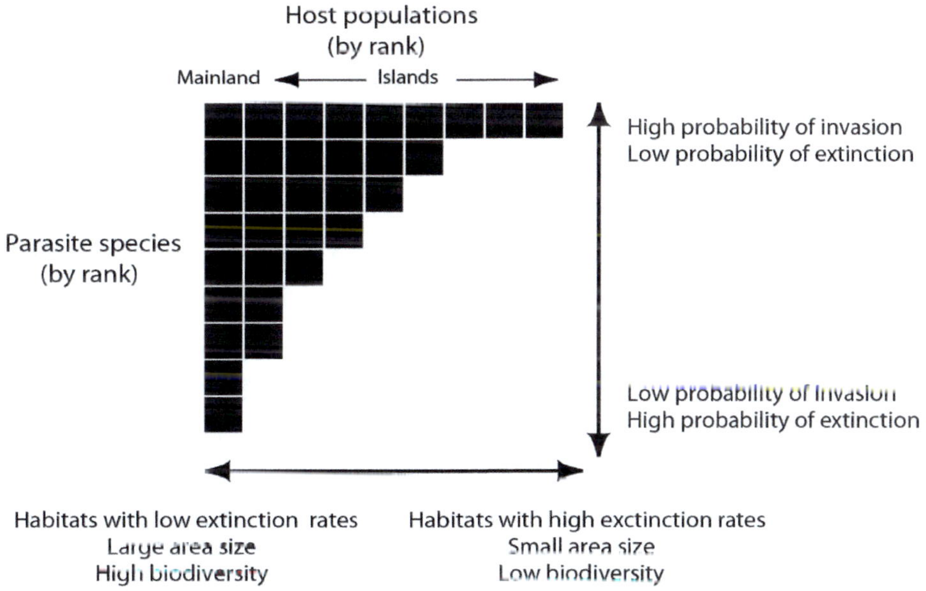

Fig. 1. Nested patterns of parasites in relation to insularity (adapted after Goüy de Bellocq et al. 2003)

2.2 The effect of island size

As mentioned above, islands are characterized by a dramatic reduction in species richness compared to mainland areas of similar size (MacArthur and Wilson 1967). Likewise, parasite faunas exhibit a marked decrease in species number on islands (Thomas 1953; Mas-Coma and Feliu 1984; Goüy de Bellocq et al. 2002). However, the relationship between parasite species richness and island size is far from being universal. The non human-associated *A. sylvaticus* follows this rule: the smaller the island, the lower the parasite species richness. However, this relationship is not observed with the human-associated and invasive *R. rattus* (Fig. 2).

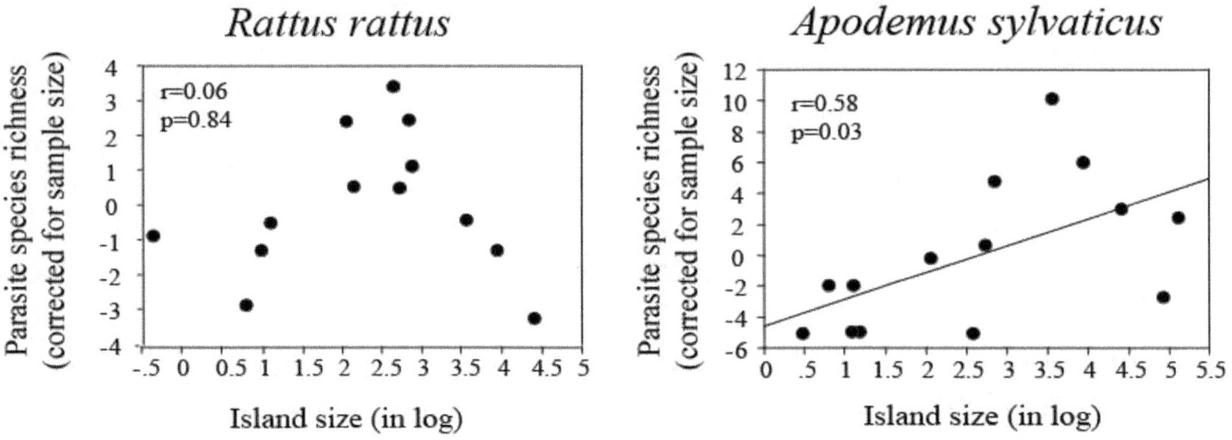

Fig. 2. Relationship between the size of the island and parasite species richness in two rodent species, *Rattus rattus* and *Apodemus sylvaticus* (modified after Goüy de Bellocq et al. 2002, 2003 and Goüy de Bellocq and Morand, unpubl. data)

2.3 The effect of insular host diversity

Insular biotope diversity, biological diversity and ecosystem stability should play important roles in the richness, structure and composition of parasite assemblages. Esteban Sanchis (1983) recorded the parasite fauna of the lesser white-toothed shrew *Crocidura suaveolens* on the mainland and Mediterranean islands. The impoverishment in larval cestodes was lower on Corsica than on Minorca or Porquerolles, reflecting a lower abundance of predators of the shrew (which are the definitive hosts of these cestodes) on Minorca than on Corsica (Esteban Sanchis 1983).

Nieberding et al. (2005b) observed a positive relationship between the mammal species richness and the parasite species richness of *A. sylvaticus* on Mediterranean islands. However, when mammal species richness was controlled for island size (i.e. for the area/diversity allometry), this relationship disappeared (Fig. 3). The parasite species richness of *A. sylvaticus* is independent of the residual variation in mammal diversity, which may indicate a saturation of the parasite community.

In contrast to *A. sylvaticus*, a negative relationship is observed between the residuals of mammal species richness and the parasite species richness of *R. rattus*. This may indicate that the parasite communities of insular *R. rattus* were far from being saturated, and that niches are open for new parasites. Empty niches may explain why parasite capture and lateral transfer are often recorded from *R. rattus* when colonizing new islands (see below).

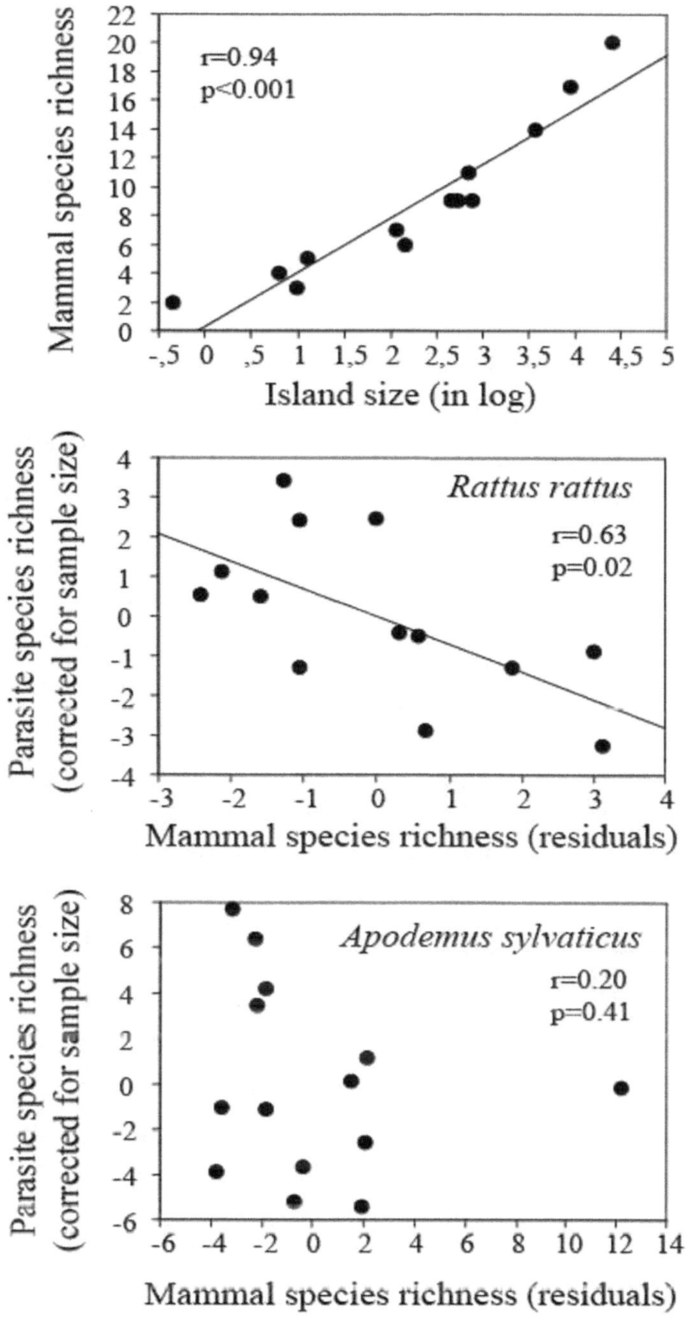

Fig. 3. Mammal diversity in relation to island size, and relationship between parasite species richness and mammal species richness (controlled for island size) in two rodents, *Rattus rattus* and *Apodemus sylvaticus* (data from Goüy de Bellocq et al. 2002, 2003 and Goüy de Bellocq and Morand, unpubl.).

2.4. Insularity and host specificity

Many parasites cannot become established on an island because of the complexity of their life cycles, and/or the environmental conditions and the hosts needed for their survival. It has been repeatedly shown that host specificity, i.e. the number of host species per parasite species, significantly decreases on islands (see Combes 1995 for review).

Goüy de Bellocq et al. (2002) investigated mean host specificity of parasites exploiting four species of hosts from the insular versus mainland populations. They found that parasites of three host species demonstrated a significant decrease in their host specificity on islands. This was explained by the decrease in the richness of mammal communities that may prevent the installation of host specific parasites. This, however, was not the case for parasites of *R. rattus* that demonstrated no difference in their host specificity on islands compared to the mainland, suggesting that *R. rattus* is able to exchange many parasites with other insular mammals as well as to acquire new parasites from them. The reason for this could be the broadening of the rat's ecological niche on islands (Cheylan 1988).

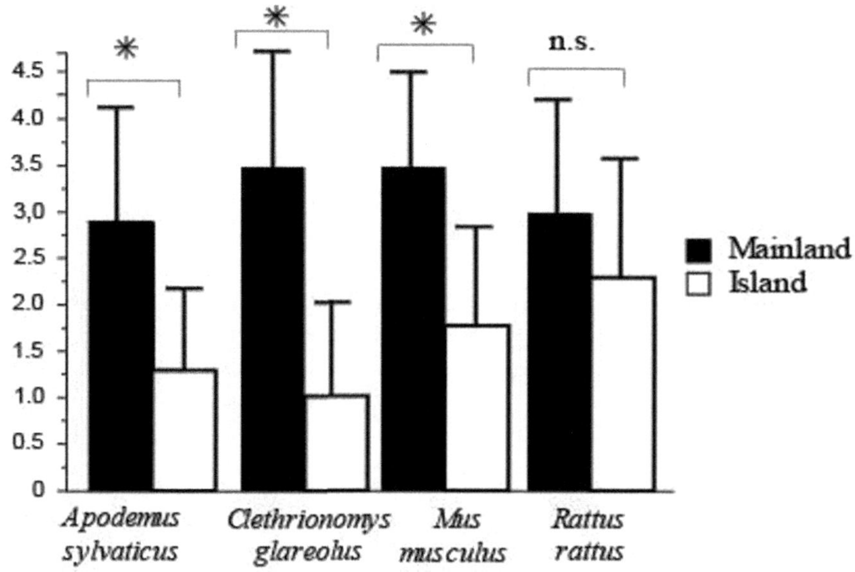

Fig. 4. Mean number of host species exploited by a parasite species harboured by individuals of four rodent species from mainland and island populations (data from Goüy de Bellocq et al. 2002)

2.5 Anthropic factor

The above examples show that the pattern of parasite species richness does not always fit with the predictions of the island biogeography theory related to species-area relationship, isolation, or life cycle. On the one hand, patterns demonstrated by parasites of the non-human associated *A. sylvaticus* could be well explained by the insular paradigm. On the other hand, this appeared not to be the case for the parasites of synanthropous rodents (*Mus musculus* and *R. rattus*) (Goüy de Bellocq et al. 2002).

Mediterranean islands were and are still at the crossroad of human migrations and commercial routes. All civilizations that succeeded on islands have introduced many plants and animals, including micromammals (Dobson 1998). Several studies have shown that insular species may come from farther regions as a consequence of human activities (Lo Brutto et al. 2004; Cosson et al. 2005). As humans have facilitated the introduction of mammals, the effect of the distance to the closest mainland should be drastically diminished. In other words, geographic distance is unlikely a potential factor that may influence the parasites associated with synanthropous species. Moreover, because of these human introductions, mammal species richness could be higher on some Mediterranean islands than predicted by the species-area equilibrium (Vigne 1998) due to a bias towards invasion, even if the overall diversity has decreased with several extinctions of insular endemic species due to these introductions. This implies that many more parasites, especially generalist parasites, can find the ecological conditions (i.e. hosts) necessary to their establishment and spread. Moreover, inferring modifications of the insular parasite faunas, based on a comparison with the closest mainland, may introduce bias as the nearest mainland is not necessarily the source of the invaders to a given island.

2.6 Macroparasite richness and composition illustrate ecological and historical phenomena

Parasites help understand host ecology, including trophic, competitive, phylogenetic or phylogeographic relationships between species (Mas-Coma and Feliu 1984; Nieberding et al. 2005a, 2006a; this volume).

For example, the parasite fauna of the greater white-toothed shrew *Crocidura russula* is markedly impoverished on the Mediterranean island of Ibiza compared to the closest mainland. Some parasites abundant in the European mainland, such as *Parastrongyloides winchesi*, are absent on Ibiza or restricted to few localities (Mas-Coma and Feliu 1984). On the contrary, *Brachylaima simoni*, *Gongylonema* spp., which are supposed to be of North African origins are common and abundant on Ibiza. Based on the parasitological data, Mas-Coma and Feliu (1984) concluded the insular population of this shrew is of African rather than European origin. More recently, molecular studies confirmed this hypothesis (Lo Brutto et al. 2004; Cosson et al. 2005).

3 Parasite invasion and lateral transfers on islands

A lateral transfer implies a host change, or switch to a new host that was never before infected by a given parasite, and occurs usually after an introduction into a new environment (Combes 1995). Several cases of lateral transfers have been observed on islands. For example, *Gongylonema brevispiculum* and *Streptopharagus kutassi* are found respectively in *A. sylvaticus* and *R. rattus*, on the Mediterranean island of Ibiza. The geographical origin of these two parasites is North Africa, where they parasitize Gerbillidae absent in Ibiza (Mas-Coma and Feliu, 1984).

The best-documented examples of lateral transfer on islands concern two trematode species: the liver fluke *Fasciola hepatica* on Corsica and S*chistosoma mansoni* in the Caribbean archipelago. The adult stages of these parasites were introduced by their definitive hosts, Bovidae and humans, respectively.

3.1 *Fasciola hepatica* and *Rattus rattus* in Corsica

The liver fluke *F. hepatica* has a very broad spectrum of definitive hosts including livestock, wild animals (Lagomorpha, Marsupialia and Cervidae) and humans (Spratt and Presidente 1981; Menard et al. 2000; Shimalov and Shimalov 2000; Rondelaud et al. 2001). This parasite has a complex life cycle and uses freshwater molluscs as intermediate hosts. Host specificity of the fluke varies geographically (Hurtrez-Bousses et al. 2001). Although rats are only occasionally infested by the liver fluke all over the world (Li 1952; Molan and Hussein 1988), high prevalence of this parasite in *R. rattus* on the Mediterranean island of Corsica has been reported (Mas-Coma et al. 1988).

While typically commensal and omnivorous on the mainland, the black rat becomes more definitely wild (less human-associated) and markedly herbivorous on islands (Cheylan 1988). On islands, this species is a habitat opportunist. Furthermore, *R. rattus* is the most abundant small mammal on Corsica. Thus, the black rat represents one of the extreme examples of insular enlarging of an ecological niche so far reported for a mammal. The infestation by *F. hepatica* appears to be related to these changes in ecological characteristics and feeding behaviour of the black rat.

3.1.1 Epidemiological context

On Corsica, all localities with *F. hepatica* present in black rats are humid natural habitats that are used as pastures for cattle and sheep. These pastures also constitute favourable habitats for both *R. rattus* and the intermediate snail host (Fons and Magnanou 2004). In these localities, the prevalence of the liver fluke in the black rat is high (on average, 58% of rats were found to be infested by the parasite) without seasonal variation

(Magnanou 2005). Prevalences of various trematodes in small mammals in other regions are usually much lower (Ribas et al. 2005).

3.1.2 The effect of lateral transfer on host physiology

Helminth parasites sometimes have a sharp negative effect on their vertebrate hosts (Connors and Nickol 1991; Kristan and Hammond 2000, 2001). This is also the case for *F. hepatica* (Dan et al. 1981; Smithers 1982; Mas-Coma and Bargues 1997). In particular, parasitism by the liver fluke increases the energy expenditure of the black rats (Magnanou et al. 2006). Resting metabolic rate of infected rats was always higher than that of unparasitized rats independently of the ambient temperature. Differences in mass specific energy expenditure between infected and uninfected rats were maximal at the lowest ambient temperature (when thermoregulation constraints were also at their maximum). There can be various mechanisms explaining the increase in the energy requirements of parasitized hosts (see Degen, this volume). Kristan and Hammond (2000, 2001) demonstrated that the presence of an intestinal helminth decreases the intestinal glucose transport capacity. Likewise, bile duct inflammations and necroses of liver tissues due to *Fasciola* infection have been reported. The trematode alters the function of the gall bladder and ducts and reduces digestion capacity (Chen and Mott 1990; Mas-Coma and Bargues 1997).

The most important finding of the study on the metabolic effect of liver fluke parasitism is that the increase in energy requirements caused by the fluke infection in black rats appeared to be extremely high (56%; Magnanou et al. 2006) compared with that of, for example, mice infected with *Heligmosomoides polygyrus* (9%;Kristan and Hammond 2000, 2001). This unexpectedly high increase in energy requirements may reflect the unusual situation experienced by both host and parasite.

3.1.3 Adaptations of the parasite

Although the liver fluke can develop in *R. rattus* (Mas-Coma et al. 1987, 1988; Valero et al. 1998, 2002), murids are the smallest natural definitive hosts known for this parasite. Their small size may account for the numerous constraints experienced by the fluke. Indeed, *F. hepatica* eggs shed by murids are smaller than those shed by infected cattle (Valero et al. 2002). Adult body size at sexual maturity is smaller in flukes harboured by Corsican black rat than that of flukes harboured by cattle (Valero et al. 1996). Finally, liver flukes coil up in the bile duct of murids, whereas they do not coil in other, larger definitive hosts (Valero et al. 1998, 2002).

3.1.4 Delay in the co-adaptation

Cattle and sheep were introduced to Corsica by humans about 7 000 years ago. *R. rattus* was introduced more recently (Vigne 1992). If this scenario is correct, the interaction *R. rattus–F. hepatica* from an evolutionary point of view is relatively young. This interaction seems to be established on Corsica due to precluded compatibility and shifts in host behaviour that, in turn, allowed the transfer of the parasite from cattle to the black rat. At present, an adaptation of the parasite to its new hosts seems to be already established, whereas adaptation of the host to its new parasite seems to be delayed. Indeed, the resistance of black rat against the liver fluke seems to be relatively low as suggested by high prevalences and high energetic costs imposed by the parasite (Magnanou et al. 2006).

3.2 *Schistosoma mansoni* and *Rattus rattus* in the Caribbean islands

S. mansoni is a causative agent of human intestinal schistosomiasis. It is widely distributed in the tropical zone, and affects several hundred million people. *S. mansoni* was introduced to America during the slave trade. The parasite found an appropriate local intermediate mollusc, *Biomphalaria glabrata*. This allowed *S. mansoni* to become established and to spread in various parts of Central America, South America and in the Caribbean archipelago.

3.2.1 Epidemiological processes in a heterogeneous environment

The epidemiology of *S. mansoni* on the island of Guadeloupe (West Indies) has been studied in great details by Théron and Pointier (1995). The transmission dynamics differed among three main eco-epidemiological systems, namely the urbanized, the marshy forest and the sylvatic foci (Théron and Pointier 1995). In the urbanized foci, humans were principal definitive hosts and the black rats were exploited by the parasite only occasionally, in localities where prevalence in humans was high. In the localities where both rats and snails were present, but humans absent, the black rats were never infected. In the marshy forest focus, characterized by scattered human settlements, prevalences of infection in the black rat were high (50-87%). The same was true for the intensity of infestation attaining 500 worms per rat. In other words, both humans and rats were equally involved in the transmission dynamics and the black rat was able to maintain the infection. In the sylvatic focus, where the rat population density was especially high, black rats were heavily infected with prevalences higher than 60%. Thus, the rodent was the only definitive host able to maintain the infection in the focus where humans were absent (Morand et al. 1999).

3.2.2 Origin of lateral transfer

S. mansoni has increased its host range (from 1 to 2 hosts) on the island of Guadeloupe following the introduction of the black rat. However, the Norwegian rat *Rattus norvegicus* was also introduced to the island without any implication of this rodent in the parasite transmission. Indeed, both rat species can be infected by *S. mansoni* in the laboratory. Whereas *R. rattus* permits the complete development of the parasite with egg release in the urine, *R. norvegicus* was found unable to expel the parasite eggs (Combes et al. 1975). A precluded compatibility of the black rat was then necessary for the transfer to occur.

3.2.3 Parasite adaptation

Théron and Pointier (1995) suggested that epidemiological processes affected the genetic structure of the parasite populations. Their study highlighted the fact that two host-adapted populations of schistosomes co-exist, one adapted to the human host, the other to the murine host. Each can be distinguished by several characters as follows.

1. Three egg morphs of schistosomes have been distinguished, based on the shape of the egg and the lateral spine. Egg polymorphism was correlated with the level of participation of the murine host in schistosome circulation.
2. The frequency variation of the ndh-1a allele has allowed identification of three different groups of schistosomes, which also indicated a strong correlation between genetic differentiation and the implication of the black rat in the epidemiology of the disease.
3. The chronobiological polymorphism of the cercariae shed by snails also helped to define three schistosome groups, with cercarial shedding correlating with rodent activity in the sylvatic foci.

3.3 Relevance of these two models

These two examples of lateral transfer have occurred as a consequence of human activities, in particular, species introduction on islands. The invasion success is supposed to have some links with parasites. Torchin et al. (2004) showed that invasive species are usually less parasitized in their new areas than in their native areas but that they are also less parasitized than other species in the new areas. Indeed, rats are less parasitized on islands than on the mainland. Moreover, invasive hosts may also successfully establish themselves and spread in new habitats because of high immuno-competence (Møller and Cassey 2004), which may explain why they are more tolerant of local parasites (Lee and Klasing 2004). The immunocompetence hypothesis remains to be tested in the case of the black rat.

The recent confrontation of parasites and rodents offers interesting models for co-evolutionary studies, i.e. the evolution of parasite virulence and host resistance. Both parasite species respond to their new host (the black rat) and their new insular environment by changes in their life-history traits (egg size, size at maturity) and life cycle (shedding behaviour). However, adaptation and, in particular, resistance of the black rat to their new parasites appear to be limited. One provisional conclusion is that the co-evolutionary processes in the recently established host-parasite associations seem to be asymmetric and biased in favour of the parasites. These two examples highlight the importance of evaluating the role of parasites when dealing with invasion and biological conservation (Prenter et al 2004; Christe et al., this volume).

4 Conclusion: the vulnerability of insular communities

Human alteration of the global environment has triggered the sixth major extinction event in the history of life and caused widespread changes in the global distribution of organisms. These changes in biodiversity alter ecosystem processes and resilience of ecosystems to environmental perturbation (Chapin et al. 2000).

Invasive species are the second cause, after habitat fragmentation, of species extinction. The number of species introductions is expected to increase dramatically in the coming years as a result of accelerating international trade, making the control of biological invasions a priority (Lee and Klasing 2004).

Island populations have a much higher risk of extinction than mainland populations. Recorded extinctions since 1600 show that a majority of extinctions concerns insular animals and plants, although island species represent only a fraction of total species richness. For example, only 20% of all bird species are endemic on islands, but 90% of bird species driven to extinction in historic times were island dwellers (Myers 1979; Frankham 1997). Human activities (over-exploitation, habitat loss, and introduced species) have been the major cause of species extinction on islands in the past 50 000 years. The relative importance of each factor varies according to the taxonomic group, over-exploitation and introduced species being the most important causes for vertebrates (Olson 1989). There is also a growing suspicion that new and modified diseases represent a significant factor (Frankham 1997).

The reasons why insular species are particularly prone to extinction or endangerment are still controversial:

1. Frankham (1997) has suggested that most island populations have low genetic variation because of founder effects and low population size. Genetic variation is the raw material for evolutionary change and allows populations to evolve in response to environmental changes such as human perturbations. Thus genetic impoverishment could favour greater sensitivity to perturbations.
2. Endemic insular species are unable to cope with stochastic effects (human perturbations, climatic events) because of their small population size, especially on small islands.
3. Many endemic island species are facing an increased risk of encountering new pathogens that are introduced through human activities. Insular species are especially vulnerable to these non-native parasites.

The decrease in parasite species richness could be the first explanation for this higher vulnerability of insular populations to invasive pathogens. Hochberg and Møller (2001) argued that the insular epidemiological context should reduce host resistance to parasites and parasite virulence. Ecological immunology can provide crucial clues concerning this question (Lee and Klasing 2004; Lindström et al 2004). Specifically, a major challenge for future researchers will be to understand how immunology and physiology make an animal population more susceptible to pathogens (see Degen, Barnard and Benhke, Weil et al. in this volume).

Some recent studies assessed genetic diversity for a locus of the major histocompatibility complex (MHC) in insular small mammals (mainly murid) populations compared to mainland populations (Seddon and Baverstock 1999; Goüy de Bellocq et al. 2005; see Charbonnel et al. in this volume), suggesting that MHC diversity is not directly linked to insularity but rather to parasite diversity (Goüy de Bellocq et al. unpubl.).

Lindström et al. (2004) tested the hypothesis that investment in immune defence is influenced by parasite-mediated selection. They analyzed immune response of birds in relation to island size and parasite load and found that parasite prevalences and infection intensities increase with island size. Moreover, birds on large islands have increased concentrations of natural antibodies and raised a strong specific antibody response faster than birds on smaller islands. In contrast, the magnitude of cell-mediated immune responses decreases with increasing parasite pressure, i.e. on larger islands. Lindström et al. (2004) report results that are consistent with the hypothesis that different immunological defence strategies are optimal in parasite-rich and parasite-poor environments.

Lindström et al. (2004) illustrated how various immunological strategies depend on the parasite context, but this does not allow us to determine if parasite-poor environments or parasite-rich ones are more vulnerable to a non-native parasite introduction. Changes in parasite assemblages, species richness, population densities and genetic diversity, all have numerous consequences for the evolution (or extinction) of insular species. Some of these changes have been quantified but much of them must be evaluated in further investigations.

The increasing problem of host invasion and parasite invasion is not restricted to island situations. Island populations constitute the most extreme case of species vulnerability, and the processes that operate on islands are of great significance to managing the risks due to human activities. Invasion and extinctions on islands are the first steps of what may occur everywhere else.

References

Alder GH, Levins R (1994) The island syndrome in rodent populations. Q Rev Biol 69:473–490

Arends A, McNab BK (2001) The comparative energetics of "caviomorph" rodents. Comp Biochem Physiol A 130:105–122

Blondel J (1995) Biogéographie: Approche écologique et évolutive. Masson, Paris

Brown JH (1995) Macroecology. Univ Chicago Press, Chicago

Chapin FS, Zavaleta ES, Naylor RL, Vitousek PM, Reynolds HL, Hooper DU, Lavorel S, Sala OE, Hobbie SE, Mack MC, Diaz S (2000) Consequences of changing biodiversity. Nature 405: 242–264

Chen MG, Mott KE (1990) Progress in assessment of morbidity due to *Fasciola hepatica* infection: A review of recent literature. Trop Dis Bull 87:1–38

Cheylan G (1988) Les adaptations écologiques à la survie de *Rattus rattus* dans les îlots méditerranéens (Provence et Corse). Bull d'Ecol 19:417–426

Combes C (1995) Intéractions durables – écologie et évolution du parasitisme. Masson, Paris

Combes C, Leger N, Golvan Y (1975) Le rôle du rat dans la dynamique de l'endémie shistosomienne en Guadeloupe. CR Acad Sci 281:1059–1061

Connors VA, Nickol BB (1991) Effects of *Plagiorhynchus cylindraceus* (Acanthocephala) on the energy metabolism of adult starlings, *Sturnus vulgaris*. Parasitology 13:395–402

Cosson JF, Hutterer R, Libois R, Sarà M, Taberlet P, Vogel P (2005) Phylogeographical footprints of the Strait of Gibraltar and Quaternary climatic fluctuations in the western Mediterranean: A case study with the greater white-toothed shrew, *Crocidura russula* (Mammalia: Soricidae). Mol Ecol 14:1151–1162

Dan M, Lichteinstein D, Lavochkin J, Stavorowsky M, Jedwab M, Shibolet S. (1981) Human fasciolosis in Israel, an imported case. Israel J Med Sci 17:430–432

Dobson M (1998) Mammal distributions in the western Mediterranean: The role of human intervention. Mammal Rev 28:77–88

Esteban Sanchis JG (1983) Contribucion al conocimiento de la helmintofauna de micromamiferos (Insectivoros y Roedores) de las islas Gimnesicas y Pitiusas (Archipielago Balear, Espana). Ph. D. Thesis, University of Valencia

Fons R, Magnanou E (2004) Processus évolutifs lies à l'insularité: Le cas des micromammifères et de leurs helminthes parasites en Corse. In: Anonymous (ed) Encyclopaedia Corsicae, vol 7. Dumane Sabiani, Corte, pp 418–455

Fons R, Croset H, Poitevin F, Catalan J (1997) Decrease in litter size in the shrew, *Crocidura suaveolens* (Mammalia, Insectivora) from Corsica: Evolutionary response to insularity? Can J Zool 75:954–958

Frankham R (1997) Do island populations have less genetic variation than mainland populations? Heredity 78:311–327

Goüy de Bellocq J, Morand S, Feliu C (2002) Patterns of parasites species richness of western Palaearctic region micro-mammals: Island effects. Ecography 25:173–183

Goüy de Bellocq J, Sarà M, Casanova J-C, Feliu C, Morand S (2003) A comparison of the structure of helminth communities in the woodmouse, *Apodemus sylvaticus*, on islands of the western Mediterranean and continental Europe. Parasitol Res 90:64–70

Goüy de Bellocq J, Delarbre C, Gachelin G, Morand S (2005) Allelic diversity at the Mhc-DQA locus of woodmouse populations (*Apodemus sylvaticus*) present in the islands and mainland of the north Mediterranean. Global Ecol Biogeogr 14:115–122

Gregory RD (1991) Parasite epidemiology and host population growth: *Heligmosomoides polygyrus* (Nematoda) in enclosed wood mouse populations. J Anim Ecol 60:805–821

Hochberg ME, Møllller AP (2001) Insularity and adaptation in coupled victim-enemy associations. J Evol Biol 14:539–551

Hudson PJ, Newborn D, Dobson AP (1992) Regulation and stability of a free-living host-parasite system *Trichostrongylus tenuis* in red grouse. 1. Monitoring and parasite reduction experiments. J Anim Ecol 61:477–486

Hurtrez-Bousses S, Meunier C, Durand P, Renaud F (2001) Dynamics of host-parasite interactions: The example of population biology of the liver fluke (*Fasciola hepatica*). Microb Infect 3:841–849

Jimenez Piqueras AM (1992) Contribucion al conicimiento de la parasitofauna de micromamiferos de la isla de Corcega (Francia). Ph. D. Thesis, University of Valencia

Krasnov BR, Shenbrot GI, Khokhlova IS, Poulin R (2005) Nested pattern in flea assemblages across the host's geographic range. Ecography 28:475–484

Kristan DM, Hammond KA (2000) Combined effects of cold exposure and sub–lethal intestinal parasites on host morphology and physiology. J Exp Biol 203:3495–504

Kristan DM, Hammond KA (2001) Parasite infection and caloric restriction induce physiological and morphological plasticity. Am J Physiol Reg Int Comp Physiol 281:R502–R5110

Lee KA, Klasing KC (2004) A role for immunology in invasion biology. Trends Ecol Evol 19:523–529

Li SY (1952) On the new host, black rat, of *Fasciola hepatica*. J Parasitol 28:88

Lindstrom KM, Foufopoulos J, Parn H, Wikelski M (2004) Immunological investments reflect parasite abundance in island populations of Darwin's finches. Proc R Soc Lond B 271:1512–1519

Lo Brutto S, Arculeo M, Sarà M (2004) Mitochondrial simple sequence repeats and 12S-rRNA gene reveal two distinct lineages of *Crocidura russula* (Mammalia, Soricidae). Heredity 92:527–533

Lomolino MV (1985) Body size of mammals on islands: The island rule reexamined. Amer Nat 125:310–316

MacArthur RH, Wilson EO (1963) An equilibrium theory of insular zoogeography. Evolution 17:373–387

MacArthur RH, Wilson EO (1967) The theory of island biogeography. Princeton Univ Press, Princeton

Magnanou E (2005) Micromammifères, helminthes et insularité: Evolution des traits d'histoire de vie du rat noir (Muridae) et de deux musaraignes (Crocidurinae) sur les îles ouest-méditerranéennes. Ph. D thesis, University of Montpellier

Magnanou E, Fons R, Blondel J, Morand S (2005) Energy expenditure in Crocidurinae shrews (Insectivora): Is metabolism a key component of the insular syndrome? Comp Biochem Physiol A 142:276–285

Magnanou E, Fons R, Feliu C, Morand S (2006) Physiological responses of insular wild black rat (*Rattus rattus*) to natural infection by the digenean trematode *Fasciola hepatica*. Parasitol Res 99:97–101

Mas-Coma S, Bargues MD (1997) Human liver flukes: A review. Res Rev Parasitol 57:145–218

Mas-Coma S, Feliu C (1984) Helminthfauna of small mammals (Insectivores and Rodents) on the Pityusic islands. In: Kubbier H (ed) The Pityusic islands. Monographie Biologicae, Junck Publishers, Hague, pp 469–525

Mas-Coma S, Feliu C, Beaucournu JC (1984) Parasitofauna de micromamifers de les illes Medes. In: Ros J, Olivella I, Gili JM (eds) Els sistems naturals de les illes Medes. Institut d'estudis Catalans, Barcelona, pp 259-272

Mas-Coma S, Fons R, Feliu C, Bargues MDV, Valero MA, Galan-Puchades MT (1987) Conséquences des phénomènes liés à l'insularité dans les maladies parasitaires. La grande douve du foie (*Fasciola hepatica*) et les Muridés en Corse). Bull Soc Neuchat Sci Nat 110:57–62

Mas-Coma S, Fons R, Feliu C, Bargues MD, Valero MA, Galan-Puchades T (1988) Small mammals as natural definitive hosts of the liver fluke *Fasciola hepatica* (Linnaeus, 1758) (Trematoda: Fasciolidae); a review and two new records of epidemiologic interest on the island of Corsica. Riv Parasitol 5:73–78

Mas-Coma S, Esteban JG, Fuentes MV, Bargues MDV, Galan-Puchades MT (2000) Helminth parasites of small mammals (Insectivores and Rodents) on the Pityusic island of Eivissa (Balearic Archipelago). Res Rev Parasitol 60:41–49

Meiri S, Dayan T, Simberloff D (2004) Body size of insular carnivores: Little support for the island rule. Amer Nat 163:469–479

Menard A, L'Hostis M, Leray G, Marchandeau S, Pascal M, Roudot N, Michel V, Chauvin A (2000) Inventory of wild rodents and lagomorphs as natural hosts of *Fasciola hepatica* on a farm located in a humid area in Loire Atlantique (France). Parasite 7:77–82

Michaux J, Goüy de Bellocq J, Sarà M, Morand S (2002) Body size increase in insular rodent populations: A role for predators? Global Ecol Biogeogr 11:427–436

Molan AL, Hussein MM (1988) A general survey of blood and tissue parasites of some rodents in Arbil province, Iraq. Acta Pathol Microbiol Scan 3:47–49

Møller AP, Cassey P (2004) On the relationship between T-cell mediated immunity in bird species and the establishment success of introduced populations. J Anim Ecol 73:1035–1042

Morand S, Pointier J-R, Theron A (1999) Long term persistence of a host-parasite system: Population biology of *Schistosoma mansoni* in the black rat. Int J Parasitol 29:673–684

Myers N (1979) The thinking ark. A new look at the problem of disappearing species. Pergamon Press, New York

Newborn D, Hudson PJ (1992) Regulation and stability of a free-living host–parasite system *Trichostrongylus tenuis* in red grouse. 2. Population models. J Anim Ecol 61:487–498

Nieberding C, Libois R, Douady CJ, Morand S, Michaux JR (2005a) Phylogeography of a nematode (*Heligmosomoides polygyrus*) in the western Palearctic region:persistence of northern cryptic populations during ice ages? Mol Ecol 14:765–779

Nieberding C, Goüy de Bellocq J, Morand S (2005b) The biogeography of host–parasite interactions: From nested assemblages to comparative phylogeography. Mammal Study 30:S87–S93

Nieberding C, Libois R, Morand S, Michaux JR (2006) Parasites and the insular syndrome: The colonisation of the western Mediterranean islands by *Heligmosomoides polygyrus* (Dujardin, 1845). J Biogeogr 33:1212–1222

Olson SL (1989) Extinction on islands: Man as a catastrophe. In: Western D, Pearl MC (eds) Conservation for the twenty-first century. Oxford Univ Press, New York, pp 50–53

Poulin R (1996) Richness, nestedness, and randomness in parasite infracommunity structure. Oecologia 105:545–551

Poulin R, Guégan J-F (2000) Nestedness, antinestedness, and relationship between prevalence and intensity in ectoparasite assemblages of marine fish: A spatial model of species co-existence. Int J Parasit 30:1147–1152

Prenter J, Mac Neil C, Dick JTA, Dunn AM (2004) Roles of parasites in animal invasions. Trends Ecol Evol 19:385–390

Ribas A, Casanova J-C, Miquel J, Fons R, Guisset C, Feliu C (2005) On the fauna of digeneric trematodes parasites of small mammals in the natural reserves of Py and Mantet (Eastern Pyrenees – France). Helminthologia 42:71–75

Rondelaud D, Vignoles P, Abrous M, Dreyfuss G (2001) The definitive and intermediate hosts of *Fasciola hepatica* in the natural watercress beds in central France. Parasitol Res 87:475–478

Sarà M, Morand S (2002) Island incidence and mainland population density: Mammals from Mediterranean islands. Divers Dist 8:1–9

Seddon JM, Baverstock PR (1999) Variation on islands: Major histocompatibility complex (Mhc) polymorphism in population of the bush rat. Mol Ecol 8:2071–2079

Shimalov VV, Shimalov VT (2000) Findings of *Fasciola hepatica* Linnaeus, 1758 in wild animals in Belorussian Polesie. Parasitol Res 86:527

Smithers SR (1982) Fascioliasis and other trematode infections. In: Cohen S, Warren KS (eds) Immunology of parasitic diseases. Blackwell Sci, Oxford, pp 608–621

Spratt DM, Presidente PJ (1981) Prevalence of *Fasciola hepatica* infection in native mammals in southeastern Australia. Austr J Exp Biol Med Sci 59:713–721

Stamps JA, Buechner M (1985) The territorial defense hypothesis and the ecology of insular vertebrates. Q Rev Biol 60:155–181

Théron A, Pointier JP (1995) Ecology, dynamics, genetics and divergence of trematode populations in heterogeneous environments: The model of *Schistosoma mansoni* in the insular focus of Guadeloupe. Res Rev Parasitol 55:49–64

Thomas RJ (1953) On the nematode and trematode parasites of some small mammals from the Inner Hebrides. J Helminthol 27:143–168

Torchin ME, Lafferty K, Dobson AP, McKenzie VJ, Kuris AM (2003) Introduced species and their missing parasites. Nature 421:628–630

Valero MA, Marcos MD, Fons R, Mas-Coma S (1996) *Rattus rattus* (Rongeur, Muridae) réservoir de la fasciolose en Corse: Étude comparative de populations naturelles de *Fasciola hepatica* (Linnaeus, 1758) (Trematoda:Fasciolidae) adultes, chez le rat noir et les bovins. Vie & Milieu 46:379–380

Valero MA, Marcos MD, Fons R, Mas-Coma S (1998) *Fasciola hepatica* development in the experimentally infected black rat *Rattus rattus*. Parasitol Res 84:188–194

Valero MA, Panova M, Comes AM, Fons R, Mas-Coma S (2002) Patterns in size and shedding of *Fasciola hepatica* eggs by naturally and experimentally infected murid rodents. J Parasitol 88:308–313

Van Valen L (1973) A new evolutionary law. Evol Theory 1:1–30

Vigne JD (1992) Zooarcheology and the biogeographical history of the mammals of Corsica and Sardinia since the last ice age. Mammal Rev 22:87–96

Vigne JD (1998) The large "true" Mediterranean islands as a model for the Holocene human impact on the European vertebrate fauna? Recent data and new reflections. In: Benecke N (ed) The Holocene history of the European vertebrate fauna. Modern aspects of research. Verlag Marie Leidorf GmbH, Berlin, pp 295–321

Part IV. Processes

17 Models for host-macroparasite interactions in micromammals[1]

Roberto Rosà, Annapaola Rizzoli, Nicola Ferrari and Andrea Pugliese

1 Introductory remarks

Modelling the spread of infectious diseases has proved to be useful for elucidating the conditions necessary for parasite persistence, the role of disease in population regulation and the expected patterns of host-pathogen coevolution.

Initially, models were applied to human infections (see, for instance, Bailey 1975). Anderson and May (1978) were among the first to use this type of model for natural animal populations, introducing also a different formulation for micro- and macro-parasites, to accommodate their diverse biological and epidemiological features. Microparasites characteristically increase rapidly in number when introduced into a susceptible host, and the precise count of infective agents is not only difficult to estimate, but irrelevant. In this case compartmental models are traditionally used that classify individuals in the population as either susceptible or infected or immune.

Reproduction in macroparasites usually includes the production of free living stages that pass from one host to the next after being exposed to a series of environmental constraints. Direct reproduction rarely occurs within the definitive host, although asexual reproduction can occur in the intermediate hosts of the digenean trematodes and some cestodes. Compared to microparasites, they are relatively large, have long generation times and are immunologically characterised by a diversity of antigens. Infections tend to be chronic, leading to morbidity rather than mortality, the

[1] RR was supported by the Autonomous Province of Trento under Grant no. 3479 (30th December 2003) "MOSTWIN - Modelling the spatio-temporal dynamics of zoonotic wildlife infections"

severity of which tends to increase with the number of parasites harboured. Therefore, in order to understand epidemiological patterns in macroparasites, it is important to measure and consider not only the prevalence of infection (i.e. the proportion of infected hosts), but also the mean parasite burden as well as the entire distribution of parasite among host, since fertility, mortality and behaviour of the host population will depend on how parasites are distributed within it (Hudson and Dobson 1995).

An important feature of macroparasite infections is their aggregation within host populations. In human communities, for example, less than 20% of individuals generally harbour 80% of the helminth parasites present in that population. Thus, a relatively small number of individuals in the "tail" of the parasite distribution are responsible for the majority of parasite transmission and play an essential role in the persistence of the parasite (Anderson and May 1985; Woolhouse et al. 1997). This pattern has also been recently observed for the tick *Ixodes ricinus* and one of its rodent hosts, the yellow-necked mouse, *Apodemus flavicollis* (Perkins et al. 2003). Heterogeneities such as these are generated by variation between individual hosts in exposure to parasite infective stages, and by differences in susceptibility once an infectious agent has been encountered, which in turn depend on a series of intrinsic and extrinsic factors. A detailed review is presented by Wilson et al. (2001).

Host age is among the most important factors affecting parasite burden. In many host-parasite systems, especially those including rodent hosts, an increase in parasite abundance with age is reported (Abu-Madi et al. 1998; Behnke et al. 1999), although other patterns are possible. Hudson and Dobson (1995) classify the shape of the age-intensity curves as Type I (if parasite burden increases monotonically with age), Type II (if parasite burden reaches an asymptote), and Type III (if the parasite burden peaks at an intermediate age). Patterns of Types II and III have also been observed in populations of yellow-necked mice (*A. flavicollis*) and woodmice (*Apodemus sylvaticus*) (Quinnell 1992; Gregory et al. 1992).

There are a number of mechanisms that may affect age-intensity curves. These include parasite-induced host mortality, acquired immunity, age-related changes in predisposition to infection, age-dependent changes in exposure, and age-related probabilities of accurately determining parasite loads (Hudson and Dobson 1995; Wilson et al. 2001). It is believed that acquired immunity develops in response to accumulated experience of infection and acts to decrease parasite establishment, survival, reproduction and/or maturation. Thus, in populations where transmission rates are high, the level of parasitic infection will rise rapidly, followed by a rapid increase in the level of acquired immunity causing a subsequent decline in parasite loads. In contrast, in populations where parasite transmission rates are low, parasite loads (and acquired immunity) will increase at a slower rate, and the age at peak infection will be greater. This will result in a negative correlation between peak levels of infection and the age at which the peak occurs, a phenomenon known as the "peak shift" (Anderson and May 1985). Peak shift has now been demonstrated in a number of human helminth infections, in several experimental infections of laboratory mice (Wilson et al. 2001), and in wild rabbit populations (Cattadori et al. 2005).

Host gender also has significant effects on parasite infection patterns. Epidemiologists have long recognized that males of vertebrate species, including humans, tend to exhibit higher rates of parasitism and disease than females (Wilson et al. 2001; Skorping and Jensen 2004). Sex biases in parasitism rates may be caused by physiological differences between males and females, such as in the levels of a number of steroid hormones, including testosterone, progesterone and estrogens. All of these hormones are known to have direct or indirect effects on components of the immune system and/or on parasite growth and development. Other mechanisms generating sex bias include differences in behaviour, diet composition and body size. In mammals, males are generally larger than females and there is good evidence that parasite load correlates with host size in a number of systems (Arneberg et al. 1998). Each sex may play a different role in the dynamics of parasitism, even if sex bias in parasitism rates does not occur; the example of *A. flavicollis* is discussed at end of sub-chapter 2.

In sub-chapter 2, we review the main results emerging from models of infections with helminths having direct life-cycles. Reviews focusing on nematode parasites can be found in Roberts et al. (1995), Hudson et al. (2001) and Cornell (2005). Host-macroparasite interaction models were first applied to rodent hosts by Scott (1990), who combined theoretical and experimental studies of the dynamics of a laboratory mouse population infected with nematodes (*Heligmosomoides polygyrus*): at the end of sub-chapter 2, we discuss the application of this type of model to a wild population of *A. flavicollis*.

Models of ectoparasite population dynamics seem to be less widespread, and are often motivated by an interest in the diseases that these parasites can transmit. For example, several models have recently been developed for tick-borne diseases, such as Lyme disease and tick-borne encephalitis (TBE), since these have become problematic in human populations of more temperate regions. Tick-borne disease systems are highly complex due to the presence of a number of heterogeneities coupled with non-linear phenomena operating in the transmission processes between tick, host and pathogen (Randolph et al. 2001). In sub-chapter 3, first we discuss models of ectoparasite population dynamics, in particular, tick dynamics; then, we briefly consider the tick-borne infection models that include rodent species as the principal pathogen reservoirs.

2 Models for rodent-nematode interactions

2.1 Models of nematode infection

2.1.1 Aggregation in parasite distribution

In order to model macroparasitic infections it is crucial to quantify the parasite burden in the host population, since host mortality and morbidity depend on the number of parasites harboured. As well as prevalence (the proportion of hosts carrying at least one parasite), and mean parasite burden, it is also important to quantify the overall parasite distribution among hosts.

A key feature of macroparasitic infections is the aggregated distribution of parasites among hosts, meaning that most parasites are concentrated in a small number of hosts. Parasite distributions have traditionally been fitted with a negative binomial (Crofton 1971), for which the parameter k is considered an index of aggregation. k is relatively easy to estimate empirically: the closer k is to 0, the more aggregated the distribution; in contrast, when k goes to infinity, the distribution approaches a Poisson. More recently, the Weibull distribution has been used as an alternative to the negative binomial and in some cases seems to be a more appropriate fit to macroparasite distributions (Gaba et al. 2005).

The causes of aggregation and its consequences on the dynamics of host-parasite interactions are still hotly debated. Several factors are thought to contribute to aggregation, including heterogeneities in hosts, predisposition to infection (Anderson and May 1985; Wilson et al. 2001), multiple infection (Isham 1995; Quinnel et al. 1995; Rosà and Pugliese 2002) and immunoepidemiological interactions (Grenfell et al. 1995). Several of these will be discussed briefly in the models below. The consequences of aggregation on the dynamics of host-parasite interactions are one of the main focus of the present chapter.

2.1.2 The Anderson and May model

In a seminal series of papers, Anderson and May (1978) modelled the effect of aggregation by assuming that, regardless of the causes, parasite distribution can be represented by a negative binomial, with a fixed aggregation parameter, k, and by studying the effect of k on parasite dynamics. In their model, the distribution of adult macroparasites is described entirely by the variable $x(t)$, representing the mean parasite burden, and the constant k (estimated in the field). All free-living stages are usually grouped into a single stage, $L(t)$, and any time delay between egg release and development of infectious stages is neglected. Hence the variables of the model are $x(t)$, the mean parasite burden, $L(t)$, the density of free-living larvae, and $N(t)$, the density of hosts. The resulting system is:

$$\frac{dN}{dt} = N\left((b-d)\left(1-\frac{N}{K}\right) - b\left(1-\left(\frac{1}{1+\frac{\xi}{k}x}\right)^{k}\right) - \alpha x \right) \tag{1}$$

$$\frac{dx}{dt} = \beta\psi L - x\left(b\left(\frac{1}{1+\frac{\xi}{k}x}\right)^{k} + \sigma + \alpha + \frac{\alpha}{k}x \right)$$

$$\frac{dL}{dt} = hNx - \delta L - \beta LN.$$

The equations in system (1) are based on very simple assumptions: hosts are born and die according to a logistic demographic model (b and d are birth and death rates; K is the carrying capacity). Moreover, if a host carries i parasites, its death rate increases linearly with i (α is the proportionality constant) and its birth rate decreases multiplicatively (ξ measures the intensity of decrease). Adult parasites die at rate σ and give birth to larvae at rate h. Finally, larvae die at rate δ and randomly encounter hosts and are ingested at rate β; ψ is the probability that they then develop into adult parasites. However, the translation of these assumptions about individual hosts and parasites into equations for aggregated variables, such as $x(t)$, is not obvious. In fact, system (1) can be considered a type of moment closure of an infinite system, whose variables are $p_i(t)$, the number of hosts carrying i adult parasites at time t, first proposed by Kostizin (1934, cited via Scudo and Ziegler 1978) and rather difficult to study (see Pugliese and Tonetto 2004 for a recent rigorous treatment). Note that the term in brackets describing parasite-induced reduction in fertility is rather complex, and is different from the simple linear term present in Anderson and May (1978). Indeed, their linearity assumption is untenable, and the term considered here is due to work by Diekmann and Kretzschmar (1991); however, in the linear approximation it is the same, and indeed it is simply equal to 0 when ξ=0 (the assumption used in the case study found in subchapter 2).

2.1.3 Qualitative predictions of the model

One of the most important contributions of modelling host-pathogen interactions has been introduction of the concept of the basic reproduction number (R_0). R_0 is a quantity that describes the average number of infective agents produced by one infective agent in the initial phase; if R_0>1 a parasite species can invade a host population, and persist in it. This concept can also be used for macroparasites, and R_0 can be computed for model (1) as

$$R_0 = \frac{\psi\beta K}{\delta + \beta K}\frac{h}{b + \sigma + \alpha} \tag{2}$$

The first factor in (2) expresses the probability that a larva finds a host and develops into an adult; the second factor, the average number of larvae produced by an adult parasite during its lifetime. From the expression (2) one can see that successful parasites will have a high fecundity (h) and a small effect on host mortality (α), which is the basis of theories on parasite evolution (Anderson and May 1982; Pugliese 2002). On the other hand, it can be seen that, for model (1), there is no effect on R_0 of the aggregation k, and of the parasite-induced fertility decrease ξ.

When R_0>1, system (1) reaches an equilibrium where parasites are at a positive level, and the host population is below carrying capacity, and thus, being regulated by parasites. An analysis of the model clearly shows that a host population can be regulated by a parasite to a density significantly lower than the carrying capacity, even though parasite density is low and parasite-related deaths are rare.

It is also possible to analyse the stability of the equilibrium where hosts and parasites coexist. In an ideal case, with Malthusian host demography, no effect of parasites on host mortality or fertility, infinitely quick larval dynamics, and no parasite aggregation, there would be a continuum of neutrally stable equilibria. Against this neutral baseline, all factors can be seen as stabilizing or destabilizing: host logistic demography, parasite-induced host mortality, and parasite aggregation are stabilizing factors, whereas parasite-reduced host fertility and long-lived larvae are destabilizing. If only stabilizing factors are present in the model, host-parasite dynamics will tend towards a unique equilibrium; if only destabilizing factors are included, the equilibrium is unstable and solutions will tend to periodic cycles; if both stabilizing and destabilizing factors are present, the result will depend on the relative strength of the contrasting factors.

A particularly good example of this approach can be found in Dobson and Hudson (1992) on the population cycles of red grouse (*Lagopus lagopus scoticus*). They showed that many patterns of population cycles were in accordance with model predictions, and were also able to confirm experimentally the role of parasites in maintaining the cycles (Hudson et al. 1998).

2.1.4 Extensions of the model

The Anderson and May model is based on an empirical moment closure; however, it is obviously possible to consider other approximations. Such a model was proposed by Adler and Kretschmar (1992) to enable the degree of aggregation to evolve dynamically. Other moment closure methods have also been developed, such as the normal approximations used in a stochastic setting by Herbert and Isham (2000).

More interestingly, it is possible to extend model (1) by considering other phenomena such as multiple infections and host heterogeneity, age structure and the dynamics of the immune response. For example, Rosà and Pugliese (2002) analysed models where aggregation in parasite distribution was generated by including multiple infections or host heterogeneity (due, for instance, to sex differences in susceptibility). They found that, with either mechanism, parasite aggregation always has a stabilizing effect; however, the quantitative strength of the effect is rather different depending on the biological mechanism that produces them.

It is well recognized that individual immune response has a strong influence on the dynamics of parasitism, and cannot be ignored, especially when analysing parasite aggregation. Most models describe immunological status as a single variable, m, varying according to the equation

$$\frac{dm}{dt} = X - \mu_I m \tag{3}$$

where X represents parasite abundance (larval challenge, adult burden, or another factor, according to the case study), and μI allows for finite immunological memory (Woolhouse 1992). It is extremely difficult to integrate such an equation for individual immunity at the population level (Grenfell et al., 1995). However, Roberts and Heesterbeek (1995) used (3) for measuring average immunity in a model of domestic animals, restocked every year.

2.1.5 Age-related parasite intensity

Individual immune response has been considered most often in models for explaining the age patterns of parasitism. In this case, individual parasite load is assumed to change with age according to some stochastic process. The simplest case is to assume constant rates of parasite acquisition and death, so that parasite load increases monotonically with age, and parasite distribution at any fixed age should be Poisson. Since these results are generally in contrast with observed patterns, more complex models have been developed that yield predictions in better agreement with data. In particular, Type III age intensity curves (see Introduction) seem to indicate a role for immune response in decreasing susceptibility to infection and increasing parasite death rates, although it is not easy to detect the role of the immune response in available data (Chan et al. 2000). However, clumped infections, and intrinsic differences in host susceptibility seem to be necessary to explain the relatively high aggregation found in parasite distributions, even when age and sex are discounted (Duerr et al. 2003).

As discussed above, it is very difficult to integrate these models at the level of an individual host with the models at the population level considered above. A simpler approach is to use a phenomenological model that assumes age difference in susceptibility (without attempting to derive them from a model of immunity development). Such an approach has been used by Rosà et al. (2000) to model the interaction of a chamo*is (Rupicapra rupicapra*) population and Trichostrongildae parasites. The results of this study clearly demonstrate a difference between models that consider age differences in susceptibility and those that do not. All other factors being equal, the model including age differences predicts a higher density for the host population, a lower parasite load, and a more aggregated distribution; in addition, parasites are concentrated in fewer hosts, and have a lower effect on the host population.

2.1.6 The stochastic approach

Models based on differential equations are justified when populations are large, and can capture a relevant part of the phenomenology of population interactions. However, small populations (and related phenomena, such as local extinctions) can only be described by stochastic models; moreover, it is conceivable that a stochastic model in which interactions are modelled at the individual level yields different results than a deterministic model that is based on some sort of averaging. The recent, marked increase in computing power now makes it possible to run simulations of complex individual-based models with hundred of thousands of individuals, so that the predictions of deterministic models can be compared to outcomes of simulations.

For example, Rosà et al. (2003a) have analysed simulations of a stochastic model, based on the same assumptions as the deterministic models, and focusing on two aspects: population cycles and extinctions. Whereas deterministic models predict damped population cycles, stochastic simulations often exhibit reasonably regular and sustained cycles for the same parameter values, in agreement with the general idea that stochastic perturbations can sustain otherwise damped cycles (Kaitala et al. 1996). Population extinctions have been examined for parameter values suitable for the *Ascaridia compar* infection in rock partridge (*Alectoris graeca saxatilis*), a species disappearing from the Alps. It was found that, since partridge population were small, parasite populations could not persist on only a single host, but had to rely on additional hosts, such as the black grouse (*Tetrao tetrix*), that has a much higher population density. That a generalist parasite may drive to extinction a small host population is widely accepted (McCallum and Dobson 1995); therefore, an intriguing result of the simulation analysis of these host-parasite interactions is that, for intermediate population sizes, the extinction probability is more or less constant or even decreases with increasing external infection (that is, with an increase in the degree to which parasites are shared). In simulation models, many other factors may easily be integrated, from the dynamics of individual immune responses, to the genetic structure of populations, to behavioural or spatial differences among individuals (Hess 1996; Keeling 1999).

In conclusion, there is great potential for understanding the role of various factors in host-parasite dynamics, and consequently, for designing appropriate control measures. We firmly believe that simple deterministic models will provide useful qualitative predictions, while complex simulation models are best used for exploring the qualitative effect of more realistic assumptions, rather than as predictive tools.

2.2 Case study: *Apodemus flavicollis – Heligmosomoides polygyrus* interactions

Among the numerous macroparasites infecting micromammals, *Heligmosomoides polygyrus* is often investigated because of its high prevalence and abundance in wild populations. Furthermore, its biological and immunological features make it a good laboratory model for studying gastrointestinal parasite infections. *H. polygyrus* (Family Trichostrongylidae) has a direct life cycle which is completed in 13-15 days with some variation among subspecies (Keymer, 1985). Infection occurs by ingestion of the third stage of free-living larvae (L_3) (Slater and Keymer 1988; Scott 1990). Once ingested, larvae colonize the small intestine and develop into adult parasites that live up to 3 months depending on the *H. polygyrus* strain and immunological status of the host (Gregory et al. 1990). Eggs produced by adult female worms are deposited in the host's faeces where hatch and develop to infective L_3 (Anderson 2000). *H. polygyrus* has a broad infective spectrum, with possible hosts including *Apodemus* spp., *Mus musculus musculus*, *Mus musculus domesticus*, and *Peromyscus maniculatus,* while its occurrence in two species of *Clethrionomys* spp. is debated (Lewis 1987; Behnke et al. 1991).

Studies of the effects of *H. polygyrus* on host population dynamics led to the first empirical evidence of the regulatory role of parasites on host populations confirming the theoretical predictions formulated by Anderson and May (1978). The discovery that inbred mice reared on damp peat maintained a self-sustained infection cycle of *H. polygyrus* without any artificial intervention provided the basis of a free-running experimental system for investigations on this host-parasite interaction. Scott (1987) placed three yellow-necked mouse populations on this type of arena, and by supplying a fixed amount of food, and created density dependence in host population growth (Scott 1987). After an exponential increase, the mouse populations reached an equilibrium density of 320 mice/m^2 (Scott 1987). Afterward *H. polygyrus* was introduced, the mice 10% of their density with respect to control arenas. Following pharmacological removal of the parasite from the populations, host density was restored to its original level. Scott (1990) then carried out a series of detailed analyses with the help of mathematical models, to identify the consequences of parasite infection on host mortality. The fact that the output of the model matched observed data confirmed the theoretical predictions that the main effect of *H. polygyrus* was reduction on mice survival while the effect on host fecundity seems to be negligible (Scott 1990). Subsequent research in semi-natural populations in outdoor enclosures confirmed that the negative effect of *H. polygyrus* on mouse population growth was mainly a consequence of parasite induced mortality (Gregory 1991; Quinnell 1992).

Other studies on *H. polygyrus* focused on the mechanisms promoting the maintenance of infection. Sex bias in parasite load has been documented in most host-parasite systems and although factor generating this bias have been analysed, the consequences for parasite persistence have been poorly investigated (Poulin 1996; Schalk and Forbes 1997; McCurdy et al. 1998; Wilson et al. 2001; Moore and Wilson 2002). The *Apodemus* spp. – *H. polygyrus* systems are an exception to this rule, suggesting that both sexes contribute equally to parasite maintenance (Gregory 1992; Gregory et al. 1992; Abu-Madi et al. 1998; Behnke et al. 1999; Skorping and Jensen 2004). Experimental field manipulation of parasite load in male or female mice indicated that males play a major role in driving the infection in the whole host population: while removal of *H. polygyrus* from male mice resulted in a reduction in parasite load even in untreated hosts, removal of parasites from females had no effect on the parasite load of untreated hosts (Ferrari et al. 2004). Results of the field study did not allow the identification of the underlying parasitological mechanism involved, but mathematical modelling suggests the pattern was caused by sexual differences in immunological response or behaviour (according to spatial arrangement of free infective larvae). The following simple model considers a separate dynamics for male and female hosts; the equations are the following:

$$\frac{dM}{dt} = bF + M[-d_M - (b - d_M)(M + F)/K - \alpha_M x_M] \tag{4}$$

$$\frac{dF}{dt} = F[b - d_F - (b - d_F)(M + F)/K - \alpha_F x_F]$$

$$\frac{dx_M}{dt} = x_M[-\sigma_M - b - \alpha_M(x_M/k_M + 1)] + \beta_M \psi_M L$$

$$\frac{dx_F}{dt} = x_F[-\sigma_F - b - \alpha_F(x_F/k_F + 1)] + \beta_F \psi_F L$$

$$\frac{dL}{dt} = h_M M x_M + h_F F x_F - \delta L - \beta_M ML - \beta_M FL,$$

where M and F are densities of male and female hosts, respectively, x_M and x_F are the average load of parasites harboured by males and females, respectively and L represents the common free-living infective pool.

Fig. 1 shows a simulation obtained with model (4) under the assumption that parasite fertility (h) is different for parasites harboured by hosts of a different sex. Specifically, eggs expelled by parasites in males have a higher hatching rate and larval survival than eggs expelled by parasites in females (i.e. $h_M > h_F$, Table 1). All other parameter values are the same for both sexes and are those reported in Table 1. *A. flavicollis – H. polygyrus* system reaches a stable equilibrium through damped oscillations, with no sex-bias in parasite burden and host density between the two host sexes (Fig. 1).

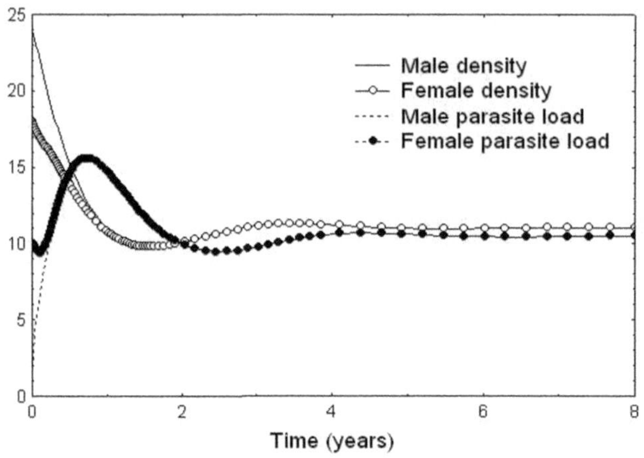

Fig. 1. Temporal dynamics of male and female *A. flavicollis* and their average load of *H. polygyrus*, for the case where fertility of parasites found on male hosts is greater than those found on female hosts ($h_M > h_F$). Parameters values are those in Table 1

Table 1. Numerical values of population parameters for *A. flavicollis* and *H. polygyrus*

Symbol	Parameter	Value	References
d	Host death rate	$3.7 \cdot 10^{-3}$	Flowerdew, 1984
b	Host birth rate	$8.21 \cdot 10^{-3}$	Flowerdew, 1984
K	Host population carrying capacity	50	Unpublished data on *A. flavicollis* in Trentino
σ	Mortality rate of adult parasite	$1.3 \cdot 10^{-2}$	Gregory et al. 1990
ψ	Proportion of ingested infective larvae that develop to the adult stage	$8 \cdot 10^{-2}$	Slater and Keymer 1988; Enriquez et al. 1988; Gregory et al. 1990
α	Mortality rate of host due to the parasite	$2.4 \cdot 10^{-4}$	Keymer and Hiorns 1986
K	Aggregation parameter of the negative binomial distribution	0.36	Ferrari et al. 2004
δ	Mortality rate of free-living infective stages	$1.6 \cdot 10^{-2}$	Dobson and Hudson 1992; Fernàndez et al. 2001
h_M	Rate of production of infective larvae per adult parasite in male mice	1.67	Based on observed data on *A. flavicollis* in Trentino
h_F	Rate of production of infective larvae per adult parasite in female mice	0.33	Based on observed data on *A. flavicollis* in Trentino
β	Rate of ingestion of free-living larvae	$4 \cdot 10^{-4}$	Based on observed data on *A. flavicollis* in Trentino

3 Models for host-tick interactions

Tick-borne infections transmitted by ixodid ticks are increasing in many parts of Eurasia and North America. Several investigations of the epidemiology of tick-borne diseases have been performed in recent years, including that of Lyme disease, tick-borne encephalitis (TBE) and human granulocytic anaplasmosis (HGA). The increase in prevalence of these diseases is associated with different ecological and sociological changes such as the abandonment of fields and pastures coupled with the expansion of woodland, which favour an increase in the range and densities of suitable hosts for ticks; and hence, the potential for disease transmission by ticks.

The complexity of tick-borne infection dynamics has required the development of specific mathematical models. Clearly, these models require the description of tick population dynamics, reviewed by Kitron and Mannelli (1994). Models of tick population dynamics, as well as for tick-borne infections, can be classified as computer-based models, which often include many details of the interactions, and simple models, mainly suited for qualitative results. The latter generally assume continuous time, despite the fundamental importance of seasonality on tick dynamics in a temperate area. Hudson et al. (1995), Norman et al. (1999) and Rosà et al. (2003b) have developed different models with the specific aim of exploring the effect of host abundance and host community composition on persistence of ticks and pathogens. For example, detailed field investigations of louping-ill disease coupled with large scale experimental studies showed that reducing the population size of some species can have a profound effect on the dynamics and persistence of infection in other species (Hudson et al. 1995; Gilbert et al. 2001). Other studies have focused instead on the deer tick-*Borrelia* system in northeastern America examining how the species richness of the host community influences the persistence of Lyme disease (Ostfeld and Keesing 2000; LoGiudice et al. 2003). These authors found that the percentage of nymphal ticks infected (and hence risk to humans) was dependent on the abundance of the white-footed mouse *Peromyscus leucopus*, relative to other non-rodent hosts. This suggests that the preservation of vertebrate biodiversity and community composition can reduce the incidence of Lyme disease (Lo Giudice et al. 2003).

Norman et al. (1999) computed R_0 for tick-borne infections and introduced the so-called dilution effect: i.e. when two alternative hosts exist for ticks, only one of which is competent for transmission, an increase in the density of the incompetent host may shift R_0 from above to below 1, and thus cause pathogen extinction. Qualitatively similar results have been obtained in computer-based models (Van Buskirk and Ostfeld 1998). Rosà et al. (2003b) and Rosà and Pugliese (unpublished) have extended the model by computing R_0 in several cases and exploring the dilution effect in greater detail. Randolph et al. (1996, 1999) have studied the effect of seasonality, and have obtained a rough estimate of R_0, for TBE. They found $R_0>1$ only where nymphs and larvae follow a similar pattern of emergence. Moreover, they related this pattern to climatic factors, and, using satellite data, showed that certain climatic patterns could explain the occurrence of TBE in many part of Europe (Randolph et al. 2000). In fact, the TBE virus only persists in those geographic areas where a combination of biotic and abiotic factors favours co-feeding of larvae and nymphs on rodent hosts, predominantly the yellow-necked mouse, permitting the non-systemic transmission of the virus (Labuda et al. 1993, 1997; Jones et al. 1997; Labuda and Randolph 1999; Randolph et al. 1999, 2000; Randolph 2001). This transmission mechanism affords a greater degree of virus amplification than the conventional viraemic route of transmission (Randolph et al. 1996, 1999, 2001). It appears that the co-feeding activity of larvae and nymphs occurs in areas where a rapid decrease in daily autumnal temperature (rapid autumnal cooling) delays the host-seeking activity of summer-born larvae until the following spring (Randolph 2001), when they then feed together with nymphs on hosts (Fig. 2).

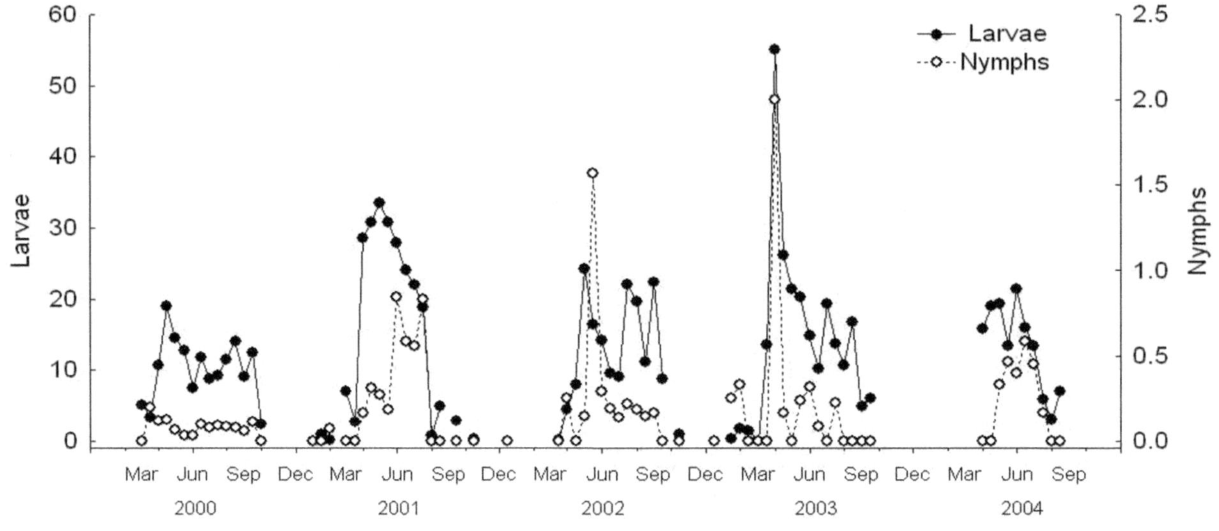

Fig. 2. Observed values of the average load of feeding larvae and nymphs on *A. flavicollis* during the period 2000-04 in a TBE endemic area in Trentino (northern Italy)

Given this dependence on climate, the current distribution of TBEV foci is expected to shift towards higher latitudes and altitudes over the next 50 years as the climate becomes warmer (Randolph and Rogers 2000; Randolph 2001; Zeman and Benes 2004).

A sophisticated tick population model that incorporates the effect of climate variables such as temperature in the demography of tick population was developed by Randolph and Rogers (1997) for the African tick *Rhipicephalus appendiculatus*. The model includes temperature- and density-dependent rates together with climate-driven density-independent laws for different tick stages, and describes both the seasonality and the annual range of variation in numbers of each stage in different sites in Africa. The model is potentially applicable to other tick species for predicting tick abundance and seasonality as risk factor for tick-borne diseases.

Finally, more complex models use multi-dimensional state variables and ticks are usually classified not only on the basis of stages but also on age. The complexity of these models requires the application of appropriate mathematical techniques such as Leslie matrix models and dynamic life tables. For example, Mount and Haile (1987, 1989) developed computer simulation models for different American tick species. They built age-specific life-table tick population models, where each tick stage is divided into discrete age classes. Density-dependent and temperature-dependent constraints were included in these models and the growth and the generation time for tick populations were simulated for several input levels of day length, weather, habitat and host density.

In this sub-chapter, we discuss a simple model for tick population dynamics recently introduced by Rosà and Pugliese (unpublished), then add the dynamics of the pathogen. Throughout, we focus on the effect of the relative abundance of the main host species on the persistence and the dynamics of tick populations and the pathogen

3.1 A simple tick population model

The life cycle of ixodid ticks includes three post embryonic developmental stages: larva L, nymph N and adult A. Each stage can be subdivided in turn according to the phases of activity: "questing", in which the unfed tick seeks a host and 'feeding' in which the attached tick feeds, becomes engorged and drops off. After dropping off their hosts, ticks complete a period of development, after which they emerge as questing ticks at the next stage (or lay eggs, in the case of adult females). Ticks are found on many vertebrate hosts; usually adults have a more restricted host range than larvae and nymphs (Eisen and Lane 2002). Nevertheless, in many natural systems, the dynamics of ticks and tick-borne diseases (e.g. Lyme disease and TBE) depends largely on two classes of hosts: small rodents such as mice and voles and larger mammals, especially ungulates. Rodents, which will be indicated in the following model as H_1, are the most common host species for immature stages of ticks (larvae and nymphs) while adults are generally found on medium-sized and large mammals (H_2), especially deer.

A simple model for the dynamics of tick populations has been developed assuming that host populations are fixed at given densities H_1 and H_2. This model, as for most models of tick-borne infections, is continuous, i.e. it disregards seasonality. A different approach has been used by Ghosh and Pugliese (2004) which introduces seasonality through a semi-discrete model. The parameters included in the model are summarized in Table 2 and the resulting equations that describe the tick population dynamics are the following (Rosà and Pugliese, unpublished):

$$\begin{aligned}
\frac{dL_Q}{dt} &= \sigma^A a_T(T) A_F - d^L L_Q - (\beta_1^L H_1 + \beta_2^L H_2) L_Q \\
\frac{dL_F}{dt} &= (\beta_1^L H_1 + \beta_2^L H_2) L_Q - \sigma^L L_F \\
\frac{dN_Q}{dt} &= m^L \sigma^L L_F - d^N N_Q - (\beta_1^N H_1 + \beta_2^N H_2) N_Q \\
\frac{dL_F}{dt} &= (\beta_1^N H_1 + \beta_2^N H_2) N_Q - \sigma^N N_F \\
\frac{dA_Q}{dt} &= m^N \sigma^N N_F - d^A A_Q - (\beta_1^A H_1 + \beta_2^A H_2) A_Q \\
\frac{dA_F}{dt} &= (\beta_1^A H_1 + \beta_2^A H_2) A_Q - \sigma^A A_F .
\end{aligned} \tag{5}$$

Encounters between questing ticks and hosts of either class are governed by mass-action; for example the corresponding encounter rate with questing nymphs is given by the product $(\beta_1^N H_1 + \beta_2^N H_2) N_Q$. A tick-host encounter results in the transition of the tick to the feeding stage. Questing larvae, nymphs and adults die at rate d^L, d^N and d^A, respectively (Table 2). Mortality in the feeding period, which lasts on average $1/\sigma$ days, is neglected. The parameters m^L and m^N represent the probability of moulting success for larvae and nymphs after feeding, respectively. Finally, we assume that the production of larvae per feeding adult tick is density-dependent, and is represented by a decreasing function a_T(T); below, we consider different choices for this function.

Through the study of the local stability of the tick-free equilibrium (Rosà et al. 2003b) the following basic reproduction number for the tick population can be derived:

$$R_{0,ticks} = a_T(0)\frac{m^L(\beta_1^L H_1 + \beta_2^L H_2)}{d^L + \beta_1^L H_1 + \beta_2^L H_2}\frac{m^N(\beta_1^N H_1 + \beta_2^N H_2)}{d^N + \beta_1^N H_1 + \beta_2^N H_2}\frac{\beta_1^A H_1 + \beta_2^A H_2}{d^A + \beta_1^A H_1 + \beta_2^A H_2}. \quad (6)$$

The quantity in (6) represents the threshold condition for the persistence of ticks in the system. When $R_{0,ticks}>1$ ticks will persist and tick and host populations will settle to a positive coexistence equilibrium. The quantity $R_{0,ticks}$ has a rather simple biological interpretation in that, if the product of the losses from each tick stage is greater than the product of the gains to each stage, then the ticks will die out, otherwise they will persist.

Table 2. Notation and values for variables and parameters included in the models

Symbol	Description
Ticks and host densities	
L_Q	Density of questing larvae
L_F	Density of feeding larvae
N_Q	Density of questing nymphs
N_F	Density of feeding nymphs
A_Q	Density of questing adults
A_F	Density of feeding adults
T	Density of total tick population
H_1	Density of hosts 1 (rodents)
H_2	Density of hosts 2 (roe deer)
Demography and encounter parameters for ticks and hosts	
d^i	Natural death rate of hosts H_i (i=1,2)
d_z	Natural death rate of questing ticks in stage z (z=L,N,A)
a_T	Average number of larvae produced per fed adult tick
σ^z	Detachment rate of feeding ticks in stage z (z=L,N,A)
M^z	Moulting success probability for ticks of stage z (z=L,N)
β_i^z	Encounter rates between questing ticks of stage z (z=L,N,A) and hosts H_i (i=1,2)
Infection parameters	
p_1^L	Probability of becoming infected for a larva feeding on an infected host 1
q_1^N	Probability of becoming infected for a host 1 bitten by a nymph
λ_{LN}	Co-feeding probability between larvae and infected nymphs
λ_1	Recovery rate for hosts 1
α_1	Disease related death rate of host 1

3.2 Effect of host densities on tick persistence and dynamics

The relative densities of hosts that allow ticks to persist can be shown by a persistence-extinction boundary in the plane H_1 (rodents) - H_2 (deer); the curve R_0=1 that divides the region of host densities where tick population persists from the region in which ticks go extinct (Fig. 3).

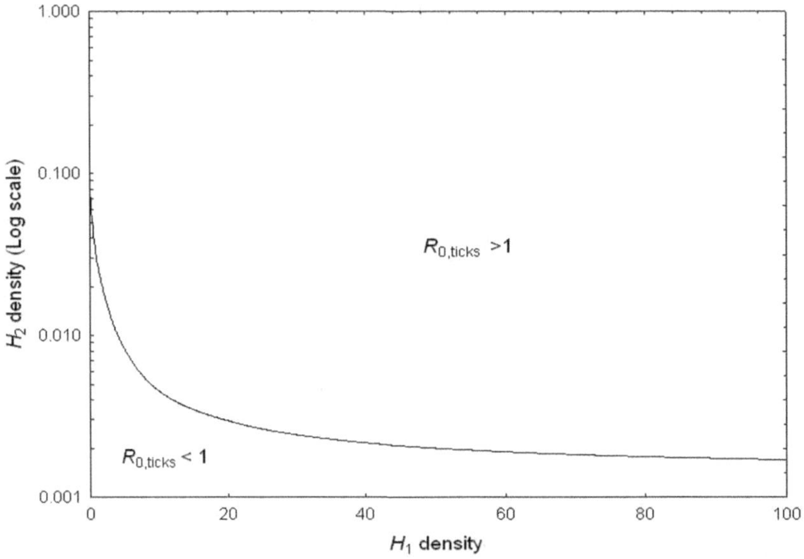

Fig. 3. The effect of host densities on the persistence of tick populations. The parameter values are: $\beta_1^L = 0.03$, $\beta_1^N = 0.001$, $\beta_1^A = 0$, $\beta_2^L = 0.05$, $\beta_2^N = 0.05$, $\beta_2^A = 0.28$, d^L=d^N=d^A=0.02, r_T=1300.

Tick populations persist as long as there are a few roe deer for the adult ticks to complete their life cycle. The ticks can persist with very low densities of rodents as long as this is compensated for by some increase in deer density. Note that the curve R_0=1 in Fig. 3 does not cross the rodent axis because we assume no adult ticks feed on rodents ($\beta_1^A = 0$).

The basic reproductive number is usually defined assuming there are no density-dependent constraints acting anywhere in the tick life cycle (Hudson et al. 2001). However, for model simulations it is important to introduce some density-dependent factor into tick life cycle to avoid unrealistic exponential tick population growth (or decrease). Norman et al. (1999) and Rosà et al. (2003b) assumed that density-dependence occurs in the production of larvae per feeding adult tick, indicated with the decreasing function $a_T(T)$. More precisely, they assumed that the production of larvae per feeding adult tick is a linear function of the total number of ticks present in the system:

$$a_T(T) = r_T - s_T T. \tag{7}$$

It may be more reasonable to assume that egg production depends on the average tick load of hosts, rather than on the absolute density of ticks. Since individual tick fecundity varies directly with meal size, a possible biological explanation for density dependence in tick fecundity is that hosts develop an immune response after being exposed to several tick bites (Wikel 1982; Randolph 1994; Hughes and Randolph 2001) which may decrease the average blood meal taken by adult female ticks (Hudson and Dobson 1995). Without considering individual tick loads and immune histories, this mechanism can be modelled as adult fecundity depending on average tick load; using again a linear model the following form for the production of larvae per feeding adult is:

$$a_T(T) = r_T - s_T\, T/(uH_1 + vH_2). \quad (8)$$

The two parameters u and v are weights associated with the two host species that take into account their different contributions to tick population dynamics.

Fig. 4 shows the effect of host 1 density on tick equilibria using different choices for density-dependence in tick fecundity, $a_T(T)$. The same effect (not shown here) occurs for increasing density of host 2. When hosts are abundant, ticks are more likely to attach to a host, hence there is an increase in the rates at which ticks progress from stage to stage and reproduce. However, as a result of negative density-dependence in fecundity, the reproductive success of ticks decreases with tick density, so that equilibrium density will saturate with increasing host density. Panel *A* of Fig. 4 illustrates the case using the density-dependence function in (7). Under this assumption, it generally follows that the equilibrium density of larvae will depend in a non-monotonous way on the density of host 1 (the type on which larvae mainly feed). In fact, at high host densities, the total number of ticks at equilibrium is almost independent of host density; hence, the rate at which new larvae are recruited is practically constant. On the other hand, the rate at which larvae feed (and thus leave the stage) is still a strongly increasing function. An almost constant recruitment of questing larval ticks combined with a rate of removal that increases with host density must result in an equilibrium density of questing larvae that decreases with increasing density of host 1.

Instead, if tick fecundity depends on average tick load [$a_T(T)$ in (8)], the effect of host density on tick equilibrium density changes substantially. The resulting effect of host 1 density on equilibrium tick densities is shown in panel *B* of Fig 4. In this case, tick density increases almost linearly with host density without reaching a plateau as in the previous case, because ticks fecundity is regulated by tick:host ratio, so that ticks' carrying capacity increases almost linearly with host densities.

Density-dependence in ticks may also occur in different periods of their life cycle; for example, in Randolph and Rogers (1997), density dependence was detected in moulting probabilities. The same qualitative effect of host densities on tick equilibria was observed with model (5), considering density dependence in moulting success (Rosà and Pugliese, unpublished).

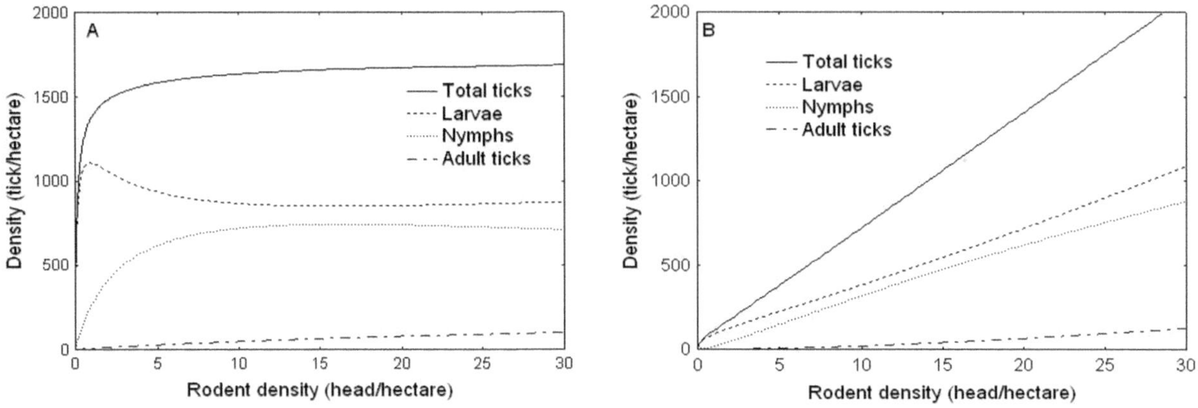

Fig. 4. Effect of rodent density on tick equilibrium densities. In panel *A* a_T (*T*) in (7) is used while in panel *B* we used a_T (*T*) in (8). Parameter values are as in Fig. 3; the others are m_L=m_N=0.2,σ=0.25, r_T=1300, s_T=0.73, *u*=0.04, *v*=0.4

3.3 An example of a model for tick-borne infection

Various tick-borne infections have different competent hosts, and diverse infection pathways. For instance, the dynamics of Lyme disease involves the spirochaete *Borrelia burgdorferi* s.l., the ticks that carry the bacteria and different hosts (rodents and deer). Some hosts, such as rodents, act as reservoirs of the infection, meaning that they can acquire the pathogen from infected ticks and transmit it to other ticks. Other hosts, like deer, are classified as tick maintenance hosts and they simply amplify the tick population without amplifying the pathogen. For Lyme disease, the main route of transmission is from an infected tick to a susceptible host and vice versa: this type of transmission is usually called systemic transmission. Recently it has been discovered (Gern and Rais 1996) that pathogens can be also transmitted from an infected tick to a non-infected tick while they *co*-feed on the same host: this process is known as non-systemic transmission. This co-feeding transmission seems to be very efficient for tick-borne encephalitis (TBE). Many workers have demonstrated that certain tick hosts, rodents in particular, which do not produce a viraemic response to TBE virus will permit non-viraemic transmission between co-feeding ticks (Jones et al. 1987; Labuda et al. 1993). Randolph et al. (1996, 1999) have shown the importance of co-feeding and temporal coincidence of different tick stages in the maintenance of TBE.

The classical method of deriving a model for tick-borne infection is to modify a tick population model by introducing the infection status of hosts and the various tick stages following the classical approach of SIR (susceptible, infected, recovered) models. Rosà and Pugliese (unpublished), modifying model (5), have developed several tick-borne infection models that differ in their assumptions of tick dynamics, the competence of the various host species and different infection pathways occurring between hosts and ticks. Here we report some results obtained with one of those models that includes both systemic and non-systemic transmission, and considers two classes of hosts: hosts of type 1 acting both as reservoirs for the pathogens and as hosts for ticks, and hosts of type 2 which act only as feeding hosts for ticks. Questing and feeding tick stages are explicitly modelled and all stages are divided between susceptible and infected. Furthermore, hosts of class 1, are divided into susceptible, infective and immunes and their densities will change following infections and recoveries. No trans-ovarial transmission is assumed, hence questing larvae can only be susceptible, on the other hand, the pathogen is transmitted interstadially, so once an immature stage is infected, the subsequent stages can transmit the pathogen to a susceptible host (for the details of all processes included in the model see Rosà and Pugliese, unpublished). This case appears adequate to describe the transmission of borreliosis or TBE, assuming host 1 to be small rodents, and host 2 to be mainly deer (Randolph *et al.* 2001).

The threshold condition for pathogen persistence found with the model is the following (Rosà and Pugliese, unpublished):

$$R_{0,pathogen} = \frac{m^L p_1^L \beta_1^L L_Q}{d_1 + \gamma_1 + \alpha_1} \frac{q_1^N \beta_1^N H_1}{d^N + \beta_1^N H_1 + \beta_2^N H_2} + \frac{m^L \beta_1^L (\beta_1^N / \sigma^N) \lambda_{LN} L_Q H_1}{d^N + \beta_1^N H_1 + \beta_2^N H_2} > 1. \quad (9)$$

The expression for R_0 in (9) can be read as the expected number of infected larvae produced (in a wholly susceptible population) by a newly infected larva over its infectious period: the first term computes those infected through the systemic route (i.e. the probability of surviving, moulting, finding a host 1 and infecting it multiplied by the average number of larvae infected by that host); the second term computes those infected through the non-systemic route (i.e. the probability of surviving, moulting, finding a host 1 and infecting multiplied by times the average number of co-feeding larvae infected over that host). As for most macroparasite species (see sub-chapter 2) ticks are aggregated on hosts. In addition, the aggregated distributions of different tick stages are coincident rather than independent: those hosts feeding large number of larvae were simultaneously feeding the greatest number of nymphs (Craine et al. 1995; Randolph et al. 1996). It has been surmised that this pattern of tick infestation facilitates transmission via co-feeding and thus significantly increases the basic reproductive number R_0 of the pathogen (Randolph et al. 1999). To explore this, Rosà et al. (2003b) incorporated the effects of tick aggregation and correlation of different tick stages in the non-systemic route of transmission. They assumed that both immature tick stages distributions follow a negative binomial distribution and that larvae and nymphs are positively correlated on hosts. Specifically, they used the following expression for the co-feeding probability between larvae and infected nymphs:

$$\lambda_{LN} = \theta_{LN}\left(1 + \rho_{LN} / \sqrt{k^L k^N}\right), \qquad (10)$$

where k^L and k^N are the negative binomial aggregation parameters for larvae and nymphs, respectively, while ρ_{LN} represents the correlation coefficient between larvae and nymphs (see Rosà et al. 2003b for details).

As for ticks, we focus on the effect of host densities on the persistence of the pathogen. In Fig. 5 the curves R_0=1 divides the region of host densities where pathogen persists from the region in which the pathogen goes extinct. The curve with a solid line represents one possible output of the model with systemic transmission only, while the dashed line shows a simulation when non-systemic transmission is also included in the model (Fig. 5). Clearly, the insertion of an additional route of transmission causes an increase of R_0, and the region where pathogen persists becomes larger (Fig. 5). A higher level of aggregation in larvae and nymphs distributions (lower values of k^L and k^N) and a stronger correlation between larvae and nymphs (higher values of ρ_{LN}) increase the co-feeding transmission between larvae and infected nymphs [see the expression of λ_{LN} in (10)] with a consequent rise of the basic reproductive number of the pathogen (Fig. 5).

The pathogen persists when H_2 density is in a range above a minimum density, needed for tick persistence, and below a maximum. Above the latter density, the incompetent hosts (roe deer) prevent the transmission of the disease, by acting as a pathogen sink that loses more pathogens from the system than the competent rodent hosts may produce. In other words, an increase in roe deer density has a positive effect on tick abundance but causes a dilution effect through "wasted bites"; in this way, at high deer density, R_0 falls below unity (Fig. 5).

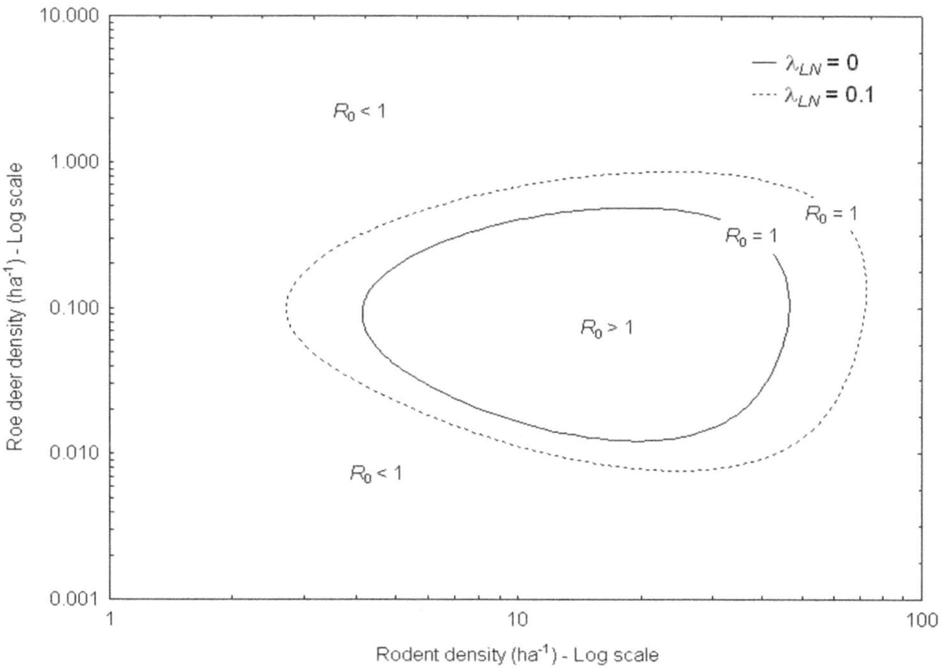

Fig. 5. Effect of host densities density on the $R_{0,pathogen}$ in (8) when only systemic infection is considered (solid line) an when non systemic transmission is added (dashed line). Parameters values are as in Fig. 4 with $a_T(T)$ in (6); the others are d_1=0.003, p_1^L=1, q_1^N=0.025, γ_1=0.01, α_1=0.005

As for rodent density, both curves in Fig. 5 show that a minimum rodent density is needed for the pathogen to persist (R_0>1). This is obvious, since infection is assumed to be transmitted only through rodents. The shape of the persistence curve also shows a dilution effect due to rodents: when rodent density is too high, the infection cannot persist. This is because for this simulation, the function $a_T(T)$ in (6) is used. In this case, ticks density (and especially larvae) does not increase much with increasing H_1 (Fig. 4 panel *A*), so that tick:host ratio will strongly decrease with increasing H_1. Thus, when hosts are abundant, each host will have the opportunity to infect only a few larvae; although each infected larva will then have a high probability of finding a host as a nymph, the overall effect is to decrease the reproduction ratio below 1.

However, Rosà and Pugliese (unpublished) showed that the negative effect of high rodent density strongly depends on the structure of density-dependence in ticks, and more generally, that tick population dynamics and its interaction with hosts plays a crucial role in the transmission of tick-borne pathogens. Hence, understanding how tick population dynamics depends on host densities and the biodiversity of the environment is a necessary step for drawing conclusions about the dynamics and persistence of tick-borne infections.

4 Concluding remarks

Recent years have seen a dramatic increase in the mathematical modelling of epidemics and an increasing recognition of the need to view such problems in their proper ecological context as host-parasite interactions. Experiments with infectious diseases in natural populations are often unethical, very expensive or impractical, and modelling provides the means of making explorative predictions.

One of the advantages of using epidemiological models is to develop explicit formulae for determining thresholds, equilibria, and periodic solutions and to provide a clear understanding of disease dynamics. Thus, models are also essential tools for identifying possible disease control strategies.

Throughout this chapter we focus on the threshold condition for parasite persistence, examining the effect of a specific aspect or variable, such as host densities and/or aggregation in parasite distribution, on parasite persistence. One important result is that the level of aggregation in tick distributions may influence the basic reproduction number of tick-borne infections, and consequently, the persistence of the disease. Not including tick aggregation among hosts on which non-systemic transmission takes place might cause an underestimate of R_0 of the infection, possibly leading to inappropriate conclusions.

This result emphasizes the importance of considering heterogeneities in the modelling of host-parasite interactions. Even more important is the modelling of biological mechanisms that produce aggregation in parasite distributions, rather than describing aggregation with particular population parameters, such as k of the negative binomial distribution, which not correspond to any biological process, but are simply population statistics.

Some factors which may have a profound effect on infection transmission are missing in the models reviewed in this chapter, both for nematode and tick-borne infections, such as seasonality (White et al. 1996), multispecies and/or trophic levels (Grenfell 1992; Begon and Bowers 1995) and immunity (Woolhouse 1992; Grenfell et al. 1995). The insertion of these factors would certainly improve the models, although explicit expressions for thresholds and equilibria will probably no longer be computable.

On the whole, wildlife disease modelling remains an essential tool for understanding the increasing flood of data on hosts and pathogens. However, before using models as management tools for planning control and prevention programs, detailed empirical studies are needed to assess model results.

References

Abu-Madi MA, Behnke JM, Lewis JW, Gilbert FS (1998) Descriptive epidemiology of *Heligmosomoides polygyrus* in *Apodemus sylvaticus* from three contrasting habitats in south-east England. J Helmintol 72:93–100

Adler FR, Kretzschmar M (1992) Aggregation and stability in parasite-host models. Parasitology 104:199–205

Anderson RC (2000) Nematode parasites of vertebrates: Their development and transmission, 2nd edn. CABI, Wallingford

Anderson RM, May RM (1978) Regulation and stability of host-parasite population interactions. I. Regulatory processes. J Anim Ecol 47:219–247

Anderson RM, May RM (1982) Coevolution of hosts and parasites. Parasitology 85:411–426

Anderson RM, May RM (1985) Age-related changes in the rate of transmission: Implications for the design of vaccinations programmes. J Hyg 94:365–436

Arneberg P, Skorping A, Read AF (1998) Parasite abundance, body size, life histories and the energetic equivalent rule. Amer Nat 151:497–513

Bailey NTJ (1975). The mathematical theory of infectious diseases and its applications. Griffin, London

Begon M, Bowers RG (1995) Beyond host-pathogen dynamics. In: Grenfell BT, Dobson AP (eds) Ecology of infectious diseases in natural populations. Cambridge Univ Press, Cambridge, pp 478–509

Behnke JM, Keymer AE, Lewis JW (1991) *Heligmosomoides polygyrus* or *Nematospiroides dubius*? Parasitol Today 7:177–179

Behnke JM, Lewis JW, Mohd Zain SN, Gilbert FS (1999) Helminth infections in *Apodemus sylvaticus* in southern England: Interactive effects of host age, sex and year on the prevalence and abundance of infections. J Helmintol 73:31–44

Cattadori IM, Boag B, Björnstadt ON, Cornell SJ, Hudson PJ (2005) Peak shift and epidemiology in a seasonal host-nematode system. Proc R Soc Lond B 272:1163–1169

Chan MS, Mutapi F, Woolhouse MEJ, Isham VS (2000) Stochastic simulation and the detection of immunity to schistosome infections. Parasitology 120:161–169

Cornell S (2005) Modelling nematodes populations: 20 years of progress. Trends Parasitol 11:483–546

Craine NG, Randolph SE, Nuttall PA (1995) Seasonal variation in the role of grey squirrels as hosts of *Ixodes ricinus*, the tick vector of the Lyme disease spirochaete, in a British woodland. Folia Parasitol 42:73–80

Crofton HD (1971) A quantitative approach to parasitism. Parasitology 63:179–193

Diekmann O, Kretzschmar M (1991) Patterns in the effects of infectious diseases on population growth. J Math Biol 29:539–570

Dobson AP, Hudson PJ (1992) Regulation and stability of a free-living host-parasite system: *Trichostrongylus tenuis* in red grouse. II. Population models. J Anim Ecol 61:487–498

Duerr HP, Dietz K, Eichner M (2003) On the interpretation of age-intensity profiles and dispersion patterns in parasitological surveys. Parasitology 126:87–101

Eisen L, Lane RS (2002) Vectors of *Borrelia burgdorferi* sensu lato. In: Gray O, Kahl RS, Lane RS, Stanek G (eds) Lyme borreliosis: Biology, epidemiology and control, CABI, New York, pp 91–115

Enriquez FJ, Scarpino V, Cypress RH, Wassom DL (1988) In vivo and in vitro egg production by *Nematospiroides dubius* during primary and challenge infections in resistant and susceptible strains of mice. J Parasitol 74:262–266

Fernàndez S, Šarkunas M, Roepstorff A (2001) Survival of infective *Ostertagia ostertagi* larvae on pasture plots under different simulated grazing conditions. Vet Parasitol 96:291–299

Ferrari N, Cattadori IM, Nespereira J, Rizzoli A., Hudson PJ (2004) The role of host sex in parasite dynamics: Field experiments on the yellow-necked mouse *Apodemus flavicollis*. Ecol Lett 7:88–94

Flowerdew J (1984) Wood mice. Anthony Nelson, Oswestry, Shropshire

Gaba S, Ginot V, Cabaret J (2005) Modelling macroparasite aggregation using a nematode-sheep system: The Weibull distribution as an alternative to the negative binomial distribution? Parasitology 131:393–401

Gern L, Rais O (1996) Efficient transmission of *Borrelia burgdorferi* between cofeeding *Ixodes ricinus* ticks (Acari: Ixodidae). J Med Entomol 33:189–192

Ghosh, M, Pugliese A (2004) Seasonal population dynamics of ticks, and its influence on infection transmission: A semi-discrete approach. Bull Math Biol 66:1659–168

Gilbert L, Norman R, Laurenson KM, Reid HW, Hudson PJ (2001) Disease persistence and apparent competition in a three-host community: An empirical and analytical study of large-scale, wild populations. J Anim Ecol 70:1053–1061

Gregory RD (1991) Parasite epidemiology and host population growth: *Heligmosomoides polygyrus* (Nematoda) in enclosed wood mouse populations. J Anim Ecol 60:805–821

Gregory, RD (1992) On the interpretation of host-parasite ecology: *Heligmosomoides polygyrus* (Nematoda) in wild wood mouse (*Apodemus sylvaticus*) populations. J Zool Lond 226:109–121

Gregory RD, Keymer AE, Clarke JR (1990) Genetics, sex, and exposure: The ecology of *Heligmosomoides polygyrus* (Nematoda) in the wood mouse. J Anim Ecol 59:363–378

Gregory RD, Montgomery SSJ, Montgomery WI (1992) Population biology of *Heligmosomoides polygyrus* in the wood mouse. J Anim Ecol 61:749–757

Grenfell BT (1992) Parasitism and the dynamics of ungulate grazing systems. Amer Nat 139:907–929

Grenfell BT, Dietz K, Roberts MG (1995) Modelling the immuno-epidemiology of macroparasites in wildlife host populations In: Grenfell BT, Dobson AP (eds) Ecology of infectious diseases in natural populations. Cambridge Univ Press, Cambridge, pp 362–383

Herbert J, Isham V (2000) Stochastic host-parasite interaction models. J Math Biol 40:343–371
Hess G (1996) Disease in metapopulation models: Implications for conservation. Ecology 77:1617–1632
Hudson PJ, Dobson AP (1995) Macroparasites: Observed patterns. In: Grenfell BT, Dobson AP (eds) Ecology of infectious diseases in natural populations. Cambridge Univ Press, Cambridge, pp 144–176
Hudson PJ, Norman R, Laurenson MK, Newborn D, Gaunt M, Jones L, Reid H, Gould E, Bowers R, Dobson AP (1995) Persistence and transmission of tick-borne viruses: *Ixodes ricinus* and louping-ill virus in red grouse populations. Parasitology 111:S49–S58
Hudson PJ, Dobson AP, Newborn D (1998) Prevention of population cycles by parasite removal. Science 282:2256–2258
Hudson PJ, Rizzoli A, Grenfell BT, Heesterbeek H, Dobson AP (eds) (2001) The ecology of wildlife diseases. Oxford Univ Press, Oxford
Hughes VL, Randolph SE (2001) Testosterone depresses innate and acquired resistance to ticks in natural rodent hosts: A force for aggregated distributions of parasites. J Parasitol 87:49–54
Isham V (1995) Stochastic models of host-macroparasite interaction. Ann Appl Prob 5:720–740
Jones LD, Davies CR, Steele GM, Nuttall PA (1987) A novel mode of arbovirus transmission involving a nonviraemic host. Science 237:775–777
Jones LD, Gaunt M, Hails RS, Laurenson K, Hudson PJ, Reid H, Henbest P, Gould EA (1997) Transmission of louping-ill virus between infected and uninfected ticks co-feeding on mountain hares (*Lepus timidus*). Med Vet Entomol 11:172–176
Kaitala V, Ranta E Lindstroem J (1996) Cyclic population dynamics and random perturbations. J Anim Ecol 65:249–251
Keeling MJ (1999) The effects of local spatial structure on epidemiological invasions. Proc R Soc Lond B 266:859–867
Keymer AE (1985) Experimental epidemiology: *Nematospiroides dubius* and laboratory mouse. In: Rollison D, Anderson RM (eds) Ecology and genetics of host-parasite interactions. Acad Press, London, pp 55–75
Keymer AE, Hiorns RW (1986) *Heligmosomoides polygyrus* (Nematoda): The dynamics of primary and repeated infection in outbred mice. Proc R Soc Lon B 229:47–67
Kitron U, Mannelli A (1994) Modeling the ecological dynamics of tick-borne zoonoses. In: Mather TN, Sonenshine DE (eds) Ecological dynamics of tick-borne zoonoses. Oxford Univ Press, Oxford, pp 198–239
Kostizin VA (1934) Symbiose, parasitisme et èvolution (ètude mathèmatique). Hermann, Paris. Translated in: Scudo F, Ziegler J (eds) (1978) The golden age of theoretical ecology. Lecture notes in biomathematics, vol 52. SpringerVerlag, Berlin, pp 369–408
Labuda M, Randolph SE (1999) Survival strategy of tick-borne encephalitis virus: Cellular basis and environmental determinants. Zentralbl Bakteriol 289:513–524

Labuda M, Jones LD, Williams T, Nuttal P (1993) Enhancement of tick borne encephalitis virus transmission by tick salivary gland extracts. Med Vet Entomol 7:193–196

Labuda M, Kozuch O, Zuffova E, Eleckova E, Hails RS, Nuttal PA (1997) Tick-borne encephalitis virus transmission though ticks co-feeding on specific immune natural rodent hosts. Virology 235:138–143

Lewis JW (1987) Helminth parasites of British rodents and insectivores. Mammal Rev 17:81–93

LoGiudice K, Ostfeld RS, Schmidt KA, Keesing F (2003) The ecology of infectious disease: Effects of host diversity and community composition on Lyme disease risk. Proc Natl Acad Sci USA 100:567–571

McCallum H, Dobson AP (1995) Detecting disease and parasite threats to endangered species and ecosystems. Trends Ecol Evol 10:190–194

McCurdy DG, Shutler D, Mullie A, Forbes MR (1998) Sex-biased parasitism of avian host: Relations to blood parasite taxon and mating system. Oikos 82:303–312

Moore SL, Wilson K (2002) Parasites as a viability cost of sexual selection in natural populations of mammals. Science 297:2015–2018

Mount GA, Haile DG (1987) Computer management of area-wide management strategies for the lone star tick, *Ambllyomma americanum* (Acari: Ixodidae). J Med Entomol 24:523–531

Mount GA, Haile DG (1989) Computer simulation of population dynamics of the America dog tick (Acari: Ixodidae). J Med Entomol 26:60–76

Norman R, Bowers RG, Begon M, Hudson PJ (1999) Persistence of tick-borne virus in the presence of multiple host species: Tick reservoirs and parasite mediated competition. J Theor Biol 200:111–118

Ostfeld RS, Keesing F (2000) Biodiversity and disease risk: The case of Lyme disease. Cons Biol 14:722–728

Perkins SE, Cattadori IM, Tagliapietra V, Rizzoli A, Hudson PJ (2003) Empirical evidence for key hosts in persistence of a tick-borne disease. Int J Parasitol 33:909–917

Poulin R (1996) Sexual inequalities in helminth infections: A cost of being male? Amer Nat 147:287–295

Pugliese A (2002) Virulence evolution in macro-parasites. In: Castillo-Chavez C, Blower S, Kirschner D, Van Den Driessche P, Yakubu A (eds) Mathematical approaches for emerging and reemerging infectious diseases: Models, methods and theory, part 2. IMA Series vol. 126, SpringerVerlag, New York, pp 193–213

Pugliese A, Tonetto L (2004) Thresholds for macroparasite infections. J Math Biol 49:83–110

Quinnell RJ (1992) The population dynamics of *Heligmosomoides polygyrus* in an enclosure population of wood mice. J Anim Ecol 61:669–679

Quinnell RJ, Grafen A, Woolhouse MEJ (1995) Changes in parasite aggregation with age: A discrete infection model. Parasitology 111:635–644

Randolph SE (1994) Density-dependent acquired resistance in natural and unnatural hosts. Parasitology 79:141–156

Randolph SE (2001) The shifting landscape of tick-borne zoonoses: Tick-borne encephalitis and Lyme borreliosis in Europe. Phil Trans R Soc Lond B 356:1045–1056

Randolph SE, Rogers DJ (1997) A generic population model for the African tick *Rhipicephalus appendiculatus*. Parasitology 115:265–279

Randolph SE, Rogers DJ (2000) Fragile transmission cycles of tick-borne encephalitis virus may be disrupted by predicted climate change. Proc R Soc Lond B 267:1741–1744

Randolph SE, Gern L, Nuttal PA (1996) Co-feeding ticks: Epidemiological significance for tick-borne pathogens transmission. Parasitol Today 12:472–479

Randolph SE, Miklisova D, Lysy J, Rogers DJ, Labuda M (1999) Incidence from coincidence: Patterns of tick infestations in rodents facilitate transmission of tick-borne encephalitis virus. Parasitology 118:177–186

Randolph SE, Green RM, Peacey MF, Rogers DJ (2000) Seasonal synchrony: The key to the tick-borne pathogen transmission. Parasitology 121:15–23

Randolph SE, Chemini C, Furlanello C, Genchi C, Hails RA, Hudson PJ, Jones LD, Medley G, Norman R, Rizzoli AP, Smith G, Woolhouse MEJ (2001) The ecology of tick-borne infections in wildlife reservoirs. In: Hudson PJ, Rizzoli A, Grenfell BT, Hesterbeek H, Dobson AP (eds) The ecology of wildlife diseases. Oxford Univ Press, Oxford, pp 119–138

Roberts MG, Heesterbeek JAP (1995) The dynamics of nematode infections of farmed ruminants. Parasitology 110:493–502

Roberts MG, Smith G, Grenfell BT (1995) Mathematical models for macroparasites of wildlife. In: Grenfell BT, Dobson AP (eds) Ecology of infectious diseases in natural populations. Cambridge Univ Press, Cambridge, pp 177–208

Rosà R, Pugliese A (2002) Aggregation, stability and oscillations in different models for host-macroparasite interactions. Theor Pop Biol 61:319–334

Rosà R, Rizzoli A, Pugliese A, Genchi C (2000) Managing chamois (*Rupicapra rupicapra*) populations: A model with macroparasites infection and host dynamics. Hystrix 11:103–114

Rosà R, Pugliese A, Villani A, Rizzoli A (2003a) Individual-based vs. deterministic models for macroparasites: Host cycles and extinction. Theor Pop Biol 63:295–307

Rosà R, Pugliese A, Norman R, Hudson PJ (2003b) Thresholds for disease persistence in models for tick-borne infections including non-viraemic transmission, extended feeding and tick aggregation. J Theor Biol 224:359–376

Schalk G, Forbes MR (1997) Male biases in parasitism of mammals: Effects of study type, host age and parasite taxon. Oikos 78:67–74

Scott ME (1987) Regulation of mouse colony abundance by *Heligmosomoides polygyrus*. Parasitology 95:111–124

Scott ME (1990) An experimental and theoretical study of the dynamics of a mouse-nematode (*Heligmosomoides polygyrus*) interactions. Parasitology 101:75–92

Skorping A, Jensen KH (2004) Disease dynamics: All caused by males? Trends Ecol Evol 19:219–220

Slater AF, Keymer AE (1988) Epidemiology of *Heligmosomoides polygyrus* in mice: experiments on natural transmission. Parasitology 93:177–187
Van Buskirk J, Ostfeld RS (1995) Controlling Lyme disease by modifying the density and species composition of tick hosts. Ecol Appl 5:1133–1140
White KAJ, Grenfell BT, Hendry RJ, Lejeune O, Murray JD (1996) Effect of seasonal host reproduction on host-macroparasite dynamics. Math Biosci 137:79–99
Wikel SK (1982) Immune response to arthropods and their hosts. Ann Rev Entomol 27:21–48
Wilson K, Bjørnstad ON, Dobson AP, Merler S, Poglayen G, Randolph SE, Read AF, Skorping A (2001) Heterogeneities in macroparasite infections: Patterns and processes. In: Hudson PJ, Rizzoli A, Grenfell BT, Heesterbeek H, Dobson AP (eds) The ecology of wildlife diseases. Oxford Univ Press, Oxford, pp 6–44
Woolhouse MEJ (1992) A theoretical framework for the immunoepidemiology of helminth infection. Parasite Immunol 14:563–578
Woolhouse MEJ, Dye C, Etard JF, Smith T, Charlwood JD, Garnett GP, Hagan P, Hii JLK, Ndhlovu PD, Quinnel RJ, Watts CH, Chandiwana SK, Anderson RM (1997) Heterogeneities in the transmission of infectious agents; implications for the design of control programs. Proc Natl Acad Sci USA 94:338–342
Zeman P, Benes C (2004) A tick-borne encephalitis ceiling in Central Europe has moved upwards during the last 30 years: Possible impact of global warming? Int J Med Microbiol 37:48–54

18 Transmission ecology and the structure of parasite communities in small mammals

Isabella M. Cattadori, Voitto Haukisalmi, Heikki Henttonen and Peter J. Hudson

1 Introductory remarks

The majority of individual hosts are inhabited by more than one parasite species during their life span. These communities of parasites are dynamically structured and the patterns observed are known to be shaped by both intrinsic and extrinsic factors. Intrinsic factors involve host susceptibility, for instance host immunity, hormones or physiological conditions, while extrinsic factors are related to host exposure, such as seasonality, habitat characteristics, host behaviour or the interaction with other sympatric host species. In addition, parasite communities can be shaped by inter-specific interactions among parasites.

Studies on the parasite community dynamics of rodents based on a comparative approach have shown that parasite communities are more or less random assemblages (Haukisalmi and Henttonen 1993a, b; Montgomery and Montgomery 1990; Behnke et al. 2001a, b; 2005). Once the effects of host sex, age and reproductive status as well as seasonality, habitat characteristics and other confounding factors, have been taken into account inter-specific interactions among parasites are weak or irrelevant in affecting parasite community dynamics. In contrast, laboratory experiments that have tested different combinations of concomitant infections in mice have identified the importance of parasite interactions (Phillips and Wakelin 1976; Jenkins 1977; Behnke et al. 1978, 1984; Christiansen et al. 1987; Adams et al. 1989; Norozianamiri and Behnke 1993; Rose et al. 1994). These studies have reported both antagonistic and synergistic interactions and suggested that host immunity often plays a crucial role in modulating such interactions either through cross-immunity or immuno-suppression (Patrick 1991). The majority of work on community ecology of parasites has been based on macroparasites in small mammals and yet, it is apparent that there is no general consensus about the role of inter-specific interactions between parasites and how they shape the dynamics of other parasite species and their host. To explain these contradictions we need to understand the within-host mechanisms affecting parasite community structure and host-mediated parasites interactions. In a broader context, an understanding of these processes is important if we wish to understand the dynamics of a single parasite species and explain the epidemiological consequences for host dynamics. In this respect we feel the central question that needs addressing is: Do we need to incorporate the interactions between parasites to explain the dynamics of any single component parasite species?

In this chapter, we shall initially describe the structure of parasite communities within individual hosts. We shall then explore the relative importance of two fundamental components affecting the dynamics of host-parasite interaction, host susceptibility and host exposure. From here, we shall move to describe the pattern of common and rare parasites and finally, we shall examine the parasite community dynamics and how susceptibility and exposure modulate inter-specific parasite interactions. We shall use some of our work on shrews, mice, voles and rabbits to describe these patterns, concentrating on macroparasites (helminths and arthropods) but also considering microparasites when interactions between micro- and macroparasites have been detected.

2 Host characteristics and parasite community composition

The assemblage of all parasite species within an individual host is commonly referred to as the infra-community of parasites of that host (Poulin 2001). The composition and intensity of an infra-community of parasites in a natural animal population is not constant but varies markedly over time and space (Boag et al. 2001; Wilson et al. 2001). A comparison of different hosts and their parasite species showed that the number and abundance of parasite species increase with host density and host body mass (Haukisalmi 1989; Haukisalmi and Henttonen 1994; Arneberg et al. 1998; Arneberg 2002). Field studies on the infra-community of parasites in mice have suggested that hosts with widespread geographical distribution tend to harbour more parasite species than hosts with restricted geographical ranges (Gregory 1990). However, a look at the parasite communities in populations of wood mice (*Apodemus* spp.) and bank vole (*Clethrionomys glareolus*) suggests that this is not necessarily the case. Parasite species richness was highly variable across the distribution range of these small mammals, such that the number of species and their intensity differed between Northern, Central and Southern Europe (Montgomery and Montgomery 1989, 1990; Behnke et al. 1999; Abu-Madi et al. 2000; Haukisalmi and Henttonen 2000; Goüy de Bellocq et al. 2003; Ferrari et al. 2006). It has also been suggested that host species with large body size (indicative of efficient uptake of parasite transmission stages and voluminous alimentary tract) and high population density (compared to other host species sharing the same parasite assemblage) are exposed to abundant and diverse macroparasite infections (Haukisalmi 1989). More generally, a strong positive relationship between helminth species richness and host body size has been detected in vertebrate hosts (Gregory et al. 1996).

A sex biased parasitism where males have higher nematode prevalence and intensity than females has been observed in many mammal species (Zuk and McKean 1996; Poulin 1996; Wilson et al. 2001). For example, Perkins et al. (2003) found that sexually mature and large-bodied males of yellow-necked mice, *Apodemus flavicollis*, were responsible for the majority of tick larval infestation and tick co-feeding groups. This is extremely important since it suggests that sexually mature yellow-necked mouse males are also the critical hosts in the maintenance of Tick Borne Encephalitis (TBE). Sex-bias has also been observed in macroparasite transmission, as shown in a field manipulation of a yellow-neck mouse population in Italy. Ferrari et al. (2004) experimentally reduced the helminth community of either males or females in *A. flavicollis* population using an anthelmintic and found that reducing parasites in males caused a consistent reduction of parasitic intensity in females, estimated through faecal egg counts, but the removal of parasites in females had no significant influence on the parasites in males. This finding suggests that males are responsible for driving the parasite infection in the host population and females may play a relatively trivial role. Interestingly a female biased parasitism is observed in some "rare" helminths of voles, which occur predominantly in old, over-wintered female hosts (Haukisalmi and Henttonen 1993a; see also section 5), although this may be because the infected females survive better than the males. Parasite community structure can also be affected by heterogeneities in host population age-structure, such that species richness and intensity are low in young age classes but increase with age and host exposure (more details in section 4). Finally, since hosts are often inhabited by more than one parasite, inter-specific parasite interactions (direct or mediated by host immunity) can also cause changes in parasite community structure (Boag et al. 2001; more details in section 6). An examination of the community pattern of large-sized cestodes can be very different from the patterns observed in the small nematodes (Haukisalmi 1989; Haukisalmi and Henttonen 1994). In Northern Europe five sympatric shrew species (*Sorex* spp.) share a common pool of some twenty helminth species, but shrew species with larger body and intestine size harbour more parasite species and have higher intensities than the smaller shrew species. Moreover, when comparing voles with shrews, the relative average body size of helminths (worm/host length) is dramatically higher in voles than shrews. The cestode *Paranoplocephala batzli* in the vole *Microtus miurus* can be 25 cm long and 0.65 cm wide, and one individual can be so large as to fill the host's small intestine. In contrast, most of the shrew helminths have dimensions of the order of millimetres and this allows the establishment of more than one parasite, often tens or hundreds per host (Haukisalmi 1989; Haukisalmi and Henttonen 1994). In some circumstances, the difference in parasite size between voles and shrews can reach two orders of magnitude.

Since there are variations between hosts in their exposure and susceptibility, the parasite infra-communities are not random assemblages of species but complex entities that change not only through time and space but also with host age, sex and status. While it would be straightforward to apply the traditional theory from community ecology, this approach does not work when investigating parasite community dynamics. In fact, many inter-specific interactions can be immuno-mediated and so we need to take into account that parasite intensity is not just the result of host exposure to the current infection but is the sum of both present and previous infections the individual has been exposed to (Woolhouse 1992, 1998).

We shall now examine in more detail the relative roles of host exposure and susceptibility on the dynamics of host-parasite interaction and how they affect parasite community dynamics.

3 Host exposure

In general, the relative abundance of parasite species in any individual host is a consequence of two dominant factors: the exposure of the host to the infective stages and the susceptibility of the host once it has been infected. Host exposure is influenced by the survival, development and transmission of infective stages, which in turn are affected by environmental conditions such as seasonality and habitat characteristics (Soulsby 1982; Marquardt et al. 1999; Anderson 2000). This is a consistent pattern both for parasites with free-living infective stages and parasites that utilise intermediate hosts or vectors (Young 1980; Epstein 2000). For instance, nematodes need a warm and moist environment to survive, eggs commence development when temperatures exceed 10°C, and migration of infective larvae onto vegetation ceases below 2.7°C (Anderson 2000). If we examine the ectoparasites, researchers have proposed that the rapid cooling of air temperature during the late summer-autumn period results in larvae delaying development until the following spring. This effect causes the coincidental emergence of both nymphs and larvae and consequently provides an opportunity for co-feeding to occur on small mammals, mostly mice and voles (Randolph et al. 2000). These co-feeding tick groups are important since they permit transmission of the TBE virus.

Seasonal climatic fluctuations are important in driving parasites dynamics and community composition. The availability of parasites (either as free-living infective stages or through vectors or intermediate hosts) is often highly seasonal, and most of the transmission occurs during the warmest months of the year that also coincide with the seasonal host reproduction. However, there is also evidence of distinct, predictable winter peaks both in helminths and ectoparasites (e.g., Haukisalmi et al. 1988). As such, we may expect seasonality to have an important role in synchronizing the development and transmission of infective stages and therefore infection outbreaks. The pattern of infection of two macroparasite species in a wild population of rabbits (*Oryctolagus cuniculus*) in Scotland, sampled monthly from 1977 to 2002, illustrates well this seasonal pattern (Fig. 1 and Cattadori et al. 2005). The relationship between mean parasite intensity and host age was examined in cohorts of rabbits born in spring, summer, autumn and winter, and two clear patterns emerged. First, parasite transmission was high in spring and summer and low in the autumn and winter periods, suggesting that transmission rate increased as temperature rose (due to the grouping by seasons this pattern is not clear for *Trichostrongylus retortaeformis* in Fig 1). Second, the age-intensity profiles were consistent for each cohort of rabbits born in different seasons, supporting the hypothesis that for each parasite species the mechanism regulating the host-parasite interaction must be the same throughout the year. In another way, seasonality and host exposure are important in affecting the availability of parasites, while changes in host susceptibility determine the shape of these age-intensity profiles. On the other hand, in Finnish Lapland where snow cover prevails for 8 months per year, the pattern of parasite infection in *C. glareolus* is still very seasonal, but the prevalence and intensity of common helminths, both nematodes and cestodes, increases from late summer throughout the winter (Haukisalmi et al. 1988).

Generally speaking, there is a minimum temperature below which free-living stages of parasites do not develop, and as temperature increases so parasite development rate increases linearly. This is a well documented pattern in a wide range of nematode species where workers have incubated free living stages at set temperatures and recorded development rates

(Anderson 2000). In natural conditions development rate can then be modelled using a day degree approach, based on the assumption that a fixed amount of thermal energy is required by a parasite to develop, and this accumulates daily according to the local temperature (e.g. Grenfell and Smith 1983). Implicit in this assumption is that variation in temperature is not important but development is achieved after the thermal energy is accumulated. However, some recent work on *Heterakis gallinarum*,

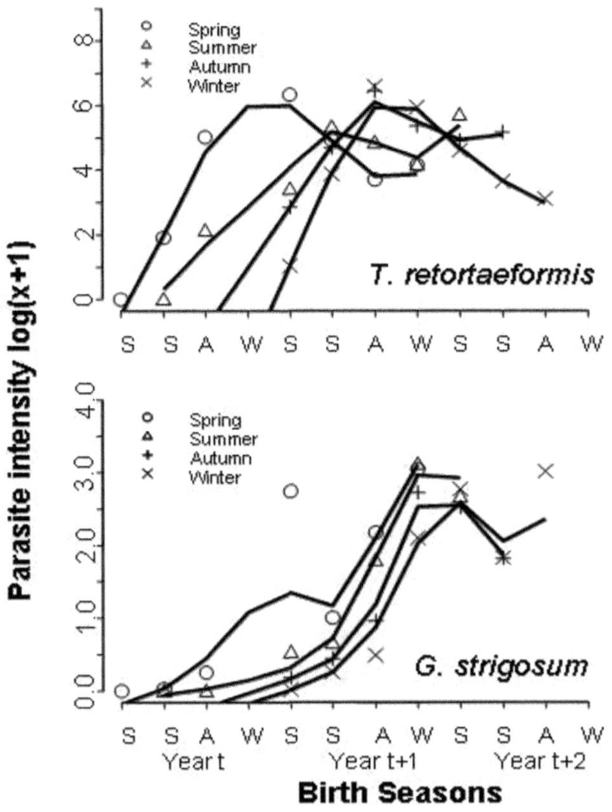

Fig. 1. Host age-parasite intensity profiles for each cohort of rabbits born in the four main seasons (Spring, Summer, Autumn and Winter). The smooth profiles represent a cubic spline curve fitted to the relationship between geometric mean of parasite intensity and host cohorts, averaged over 26 years of data (symbols)

a nematode that infects galliform birds, has shown that such models do not apply to the free living egg stages of the parasite when the rate of development increases with variability in climatic conditions (Saunders 2000). Therefore, we need to take into account that stochastic variations in climate, or unexpected extreme weather conditions, can produce non-linear effect on parasite transmission and modify the way changes in climatic conditions will alter the availability and transmission of parasites.

While we can expect climatic conditions to have a profound and direct influence on parasite abundance and distribution, habitat changes, such as changes in composition and structure of vegetation, can also influence the spatial distribution of free living stages, the intermediate hosts or the vectors, and may well cause additional non-linearities in parasite transmission. For example, changes in the matt layer, the microclimate in which many free-living helminth stages develop, can influence parasite survival, development and vertical migration of infective stages, and so affect larval availability (Crofton 1948a, b). Also, the predicted climate change can widely affect the development rates of parasite intermediate forms, especially in the high latitudes (Kutz et al. 2005).

4 Host susceptibility

Insight into the dynamics of parasite communities requires not only the understanding of the mechanism of parasite transmission but also how susceptibility varies between individual hosts and how individuals vary in their responses to infection. Since the majority of small mammal host species exhibit highly seasonal birth production, the naïve hosts are only available to the parasites during the breeding season, and for much of the remaining year the host population consists of individuals with experience of prior infection that usually exhibit signs of acquired immunity or age-related changes in parasite intensity. In the circumstances of a host that acquires immunity to a parasite infection, the interaction between a parasite and the host is not simply dependent on the current parasite abundance but an integral of past exposures and the competence/maturity of the host's immune response to secondary infections. Theoretical modelling of the host age-parasite intensity relationship suggests that acquired immunity can be modelled as a function of the cumulative exposure to parasites (Anderson and May 1985; Woolhouse 1992; 1998). When parasite transmission rates are high, intensity will rise rapidly reaching a peak in relatively young hosts. This peak will be followed by a reduction in intensity as acquired immunity increases. In contrast, when the rate of transmission is low, intensity rises more slowly and peaks in older age individuals; in this case the peak intensity is lower than the one observed when the rate of transmission is higher. The cumulative exposure at the point of turnover will be roughly the same but the age-at-peak depends on the force of infection. This pattern, known as the "peak shift" (Woolhouse 1998), predicts a negative relationship between the peak of infection and the age at which the peak occurs. Evidence of peak shift in small mammals has been supported by laboratory studies on mice (inbreed CBA/CA mice) infected with macroparasites (Crombie and Anderson 1985; Berding et al. 1986) and the analysis of seasonal changes in parasite intensity in a free-living natural population of rabbits (Cattadori et al. 2005). In the latter study the mean intensity of the nematode *T. retortaeformis* was examined in relation to host age in cohorts of rabbits born each month from February to August. The results revealed that peak intensity was higher and occurred in younger rabbits in the summer months, when the force of infection was high, and was lower and occurred in the older rabbits in the late winter months, when worm availability was low (see also Fig. 1). This seasonal nematode-rabbit system supports the assumption that acquired immunity can be modelled using the cumulative exposure to infection and also suggests that the pattern observed is the combination of three seasonal components, seasonality in host immune response, seasonality in host condition, and seasonality in cohort-dependent quality (Cornell et al., unpublished data). Acquired immunity has also been suggested to be the principal cause of host-parasite interaction in *Heligmosomoides polygyrus* infection in wood mice (Gregory et al. 1992).

While acquired immunity can be expected to have a relevant role in host-parasite interactions, other mechanisms can also be involved. For example, *Graphidium strigosum*, a nematode with direct life cycle – and a free living infective stage – inhabiting the rabbits' stomach, does not cause an apparent acquired immune response in rabbits but intensity exponentially increases with host age, and this pattern is consistent throughout the year (Fig. 1). Similarly, *Heligmosomum mixtum* and *Catenotaenia henttoneni*, the common parasites of *C. glareolus* in Finnish Lapland, do not seem to cause an acquired host immune response (Haukisalmi et al. 1988). One possible explanation is that ingestion and survival of infective stages exceed parasite mortality rate and the high parasite burdens in adults do not significantly harm host condition or host survival. Other mechanisms may also play a role in influencing the host-parasite interaction. Physiological conditions may overwhelm the immunological status of the host and increase host susceptibility to parasite infection. The detrimental role of malnutrition on host susceptibility was clearly demonstrated in a laboratory experiment where *H. mixtum* was allowed to circulate in cages with bank voles that were placed on diets that varied in their protein levels. At low protein levels (3%), high numbers of the third stage larvae could penetrate the intestinal wall of the host causing massive intestinal infections (Haukisalmi and Henttonen 2000). If the nutrition-dependent components become relevant, and allow for a quick and massive invasion of parasites, then host death can be fast (Haukisalmi and Henttonen 2000).

One of the inferences from these studies is that only a few infected individuals within the host population can be responsible for parasite persistence and transmission. Heterogeneities between individual hosts can have a significant effect on the dynamics of parasite infection and the persistence of infections at the host population level. One typical consequence of host heterogeneities is a highly aggregated distribution of parasites within the host population, such that the majority of individuals harbour very few or no parasites while a minority of hosts carry large parasite loads. There is an increasing number of empirical studies that have identified these groups of highly infected hosts and evidence suggests that they do not necessarily belong to this category all year round but move in and out according to the change in their level of infection across the year (Perkins, unpublished data; Cattadori et al. 2005).

A clear example of seasonal change in host infection rate is observed in adult rabbit females that exhibit a strong periparturient rise in *T. retortaeformis* infection (Cattadori et al. 2005). This phenomenon occurs as a consequence of relaxation of immunity in females during late pregnancy and lactation, which results in an increase in the intensity of parasite infection, worm egg production, and is expected to lead to a pulse of infective stages on the pasture coinciding with the birth of new naïve hosts available for infection. High prevalence of "rare" nematodes in old, post-reproductive female *C. glareolus* could be due to a similar mechanism (Haukisalmi et al. 1988). Indeed, if the hosts that become more susceptible are also producing more infective stages, then these two features co-vary and introduce a non-linearity that can potentially have large effects on the parasite infection rate. One of the consequences of this is that the heavily parasitized hosts are also the individuals that contribute to the majority of the transmission. Empirical evidence suggests that this can well be the case. As previously reported, sexually mature *A. flavicollis* males are responsible for the majority of larval and co-feeding tick infestation (Perkins et al. 2003) and again, adult males are responsible for the majority of *H. polygyrus* transmission while females have a marginal role (Ferrari et al. 2004). Host heterogeneities may also play a fundamental role in shaping parasite community structure and parasite species distribution. An investigation of wild populations of *C. glareolus* found that the aggregated parasite distribution observed was more imputable to individual difference in host exposure/susceptibility rather than large scale spatial variability (Haukisalmi and Henttonen 1999).

We shall now investigate the role of host heterogeneities on the pattern of common and rare species in parasite communities.

5 Host population structure and helminth commonness/rarity

When sampling parasites in small mammals it is most essential to understand the heterogeneity of small mammal populations, and particularly, how the heterogeneity changes continuously from one season to the next in terms of changes in the population structure of the hosts. Various population categories, often called functional groups (Haukisalmi et al. 1988; see also Prévot-Julliard et al. 1999), behave differently, have different hormonal and immune states and consequently often show very different parasite infection patterns. In rodents, these functional groups include old overwintered breeding or post-breeding 1) males and 2) females, breeding 3) males and 4) females, born in the current year, 5) sub-adults (animals which have delayed their breeding to the following summer even if they could breed in their first summer on the basis of their age, typically late summer and overwintering animals) and 6) juveniles. Depending on the population density, subsequent years can differ a lot in respect of the seasonal dynamics of various functional groups. In fact, when a small mammal population is screened for macroparasites, a detailed analysis of each functional group should be carried out, and their seasonal occurrence identified. For example, Haukisalmi et al. (1995) described pronounced differences in cestode infection patterns with respect to sex and breeding status in *Microtus* voles. Interestingly, the observed patterns differed significantly between the two common cestode species examined. In contrast to breeding voles, infection parameters did not significantly differ between non-breeding sub-adult voles. Similarly, both sexes of non-breeding shrews of the year showed similar levels of helminth infection (Haukisalmi et al. 1994), while the old over-wintered breeding male shrews were much more heavily infected than females.

In a longitudinal monitoring of an over-wintering *C. glareolus* cohort (born in late summer, and breeding the next summer after the overwintering), Haukisalmi et al. (1988) showed the contrasting dynamic patterns of so-called common and rare helminths. Common helminths occurred already in young voles in late summer and their prevalence and intensity increased with host age until late winter, and then decreased in early spring when hosts matured and their hormonal status changed; the drop being steeper in maturing females, but increasing again in the aging over-wintered voles. In contrast, the rare helminths were almost only observed in old breeding and post-breeding voles in late summer, when this senescing cohort was about to disappear from the population. Rare helminth species were common at a specific time of the year, the late summer, in this specific host functional group, which usually makes up only a small proportion (5-10%) of the whole population. Thus, rare helminths were rare at the host population level, but they concentrated into a specific sub-group of individuals, with a prevalence that could reach more than 50% of the individuals. Because of the sampling schedules, it is possible that rare helminths are missed due to their restricted occurrence. Haukisalmi and Henttonen (1990, 1999) subsequently characterized general features of common and rare helminths in *C. glareolus*. Common helminths occur year round in all host subgroups, have high prevalences, their long-term dynamics are predictable, and their sub-habitat distribution in forests is wide. In contrast, rare species tend to occur only in a restricted subgroup of the host population for a limited time of the year, their long-term dynamics are unpredictable, and despite an even distribution of the hosts among habitat types, rare helminths occur in patches. These patches, however, seem to be rather permanent, at least in the 20 year follow-up by Haukisalmi and Henttonen (1999). This persistence could be due to some microhabitat and/or microclimatic conditions that affect the occurrence of possible intermediate hosts of the rare helminths. Rare helminths also appeared to be competitively superior to common helminths. An analogy to rare and common species could be derived from the core and satellite species hypothesis (Hanski 1982). This hypothesis suggests that core species have high prevalence and intensity, while satellite species have low prevalence and intensity. In a detailed analysis of helminth communities of two *Sorex* shrews, *Sorex araneus* and *Sorex caecutiens*, Haukisalmi and Henttonen (1994) found a dichotomy fitting this hypothesis to the common rare hypothesis. However, this pattern was detected only in the large species, *S. araneus*. The absence of a significant prevalence-intensity relationship in small-sized *S. caecutiens* may be due to the fact that such patterns are hard to observe when the assemblage consists of a small number of species. Moreover, most of the shrew helminth transmission occurs through the most abundant, large-sized host species, *S. araneus*. This suggests that community patterns in general may be more discernible in the common and possibly larger-sized host species of a guild.

Recently, it has become increasingly clear, with the development of modern molecular technologies, that many of the "classic" tapeworm species in small mammals represent complexes of cryptic species with poorly-defined morphological characters (Haukisalmi et al. 2004; 2006; Cook et al. 2005). As such, understanding the true nature of helminth species diversity is fundamental for understanding their community patterns. In this respect, it could be supposed that this problem is greater for rare species, emphasizing the importance of a proper sampling approach.

6 Parasite coexistence and interspecific interactions

Within the infra-community of parasites, species may show synergistic, antagonistic, or independent associations with other parasite species. When interactions take place, they do occur either through a direct relationship between parasite species for resources, space or other niche constraints, or are mediated by the host's characteristics, like immunity, physiology, sex, age or reproductive status. Holmes and Price (1986) suggested that in a host population the strongest inter-specific interactions should be expected between the most common parasite species (core species), while the more rare species should show weak or no interactions (satellite species). Researchers who have examined parasite community dynamics have done so either by analysing field surveys or undertaking mixed infections in the laboratory. Workers that have applied the comparative approach using small mammals have suggested that parasite communities are little more than random assemblages, implying that variation in exposure drives the pattern observed (Montgomery and Montgomery 1990; Haukisalmi and Henttonen 1993a, b; Ellis et al. 1999; Behnke et al. 2001, 2005). In contrast, laboratory studies on concomitant infections have identified the importance of species interactions, and indicated that host immuno-mediated responses are important (Phillips and Wakelin 1976; Jenkins 1977; Behnke et al. 1978, 1984; Christiansen et al. 1987; Adams et al. 1989; Norozianamiri and Behnke 1993; Rose et al. 1994). Clearly this field of epidemiology is still undeveloped and these conflicting results need to be carefully explained. If inter-specific interactions among parasites are common and contribute to host susceptibility to infection by another species, then the classic studies of "single-species" systems can potentially be misleading.

There are several alternative statistical ways to detect random and non-random patterns of species associations (Gotelli and Graves 1996). Haukisalmi and Henttonen (1998) used real and simulated data for helminth species associations in *Sorex* shrews and, in general, the analyses indicated a low probability of Type I statistical error. However, all the methods used were more likely to detect positive non-randomness than negative non-randomness of comparable strength and this may explain the predominance of positive overall associations in the empirical data sets. The models indicated slight differences between the methods in their ability to detect non-randomness of known strength (Type II error). Some methods failed to detect strong overall association when the simulated assemblages consisted of roughly equal numbers of positive and negative pair-wise interactions. Importantly, the structure of simulated data sets always disappeared when the expected distribution was constrained to account for sampling heterogeneity, i.e. varying prevalence of species in subsamples. In all methods used, detecting positive non-randomness was "inherently" much more probable than detecting negative non-randomness. Ignoring various sampling heterogeneities could partly explain the commonness of overall positive associations in field studies. For example, ignoring the age, or functional groups of hosts will bias the results towards positive association, because most helminths are more prevalent in adult hosts than in young ones. Host immuno-mediated interactions could also bias the results towards an overall positive association. The helminth immune reaction can be both homologous and heterologous (Christensen et al. 1987). In the latter case, one key species can be sufficient to create the overall positive structure if other species are affected by its immunosuppressive capacity.

In a long term study on the community structure of helminths in natural *C. glareolus* populations in Finland, Haukisalmi and Henttonen (1993a) found that inter-specific parasite interactions, as determined by changes in intestinal distribution, were relatively scarce. Interactions were more likely to occur between parasite species that showed large intestine overlap, and coexistence was apparently favoured by niche segregation and the trade off between transmission efficiency and competitive ability (Haukisalmi and Henttonen 1993b). A series of possible mechanisms were suggested to explain the positive co-occurrence patterns observed, such as mutualistic effects, similarities in life-cycles and habitat requirements, and decreased host resistance. As an example of obvious direct interference, consider the competitively inferior *H. mixtum*, which inhabits the first third of the small intestine in *C. glareolus*, but clearly shifts to the second third of the intestine when the competitively dominant *Ancotheca sp.* (= "*Capillaria*" *sp.*) occupies the very beginning of the small intestine (Haukisalmi and Henttonen 1993b).

The negative correlation between helminth body size and intensity found in many small mammal species (e.g. Haukisalmi and Henttonen 1994), suggests that intra-specific competition for resources and space seems to play a relevant role in affecting the average density of the infracommunity of helminths. Crowding effects are obviously more drastic for cestodes that are on average larger than nematodes, and often more sensitive to direct intra-specific competition, than nematodes. A lack of statistical significance, after correcting for habitat and host characteristic constraints, was reported in a similar study on helminth community structure in wild *Apodemus sylvaticus* populations in the UK (Behnke et al. 2005). In contrast, the study of the gastrointestinal parasite community in a natural population of rabbits in Scotland (Lello et al. 2004) found a number of interspecific interactions. The statistical analysis of more than 20 years of individual based data on adult rabbits suggested a series of parsimonious positive and negative pair-wise interactions between the gastrointestinal helminths. Parasite community structure was not necessarily the same between hosts and the intensity and direction of the interactions among parasites differed between male and female rabbits. There was reasonable evidence to suppose that host-mediated competition between parasites was the dominant feature in shaping the structure of this community. The putative explanation suggested that this acts through the immune system, either through cross-immunity or immuno-suppression (Lello et al. 2004).

Detailed laboratory work on mice and some of their most common helminths (for instance *H. polygyrus*, *Trichinella spiralis* and *Nippostrongylus brasiliensis*) have suggested that host immuno-mediated regulation plays a key role in affecting parasites interactions (Behnke et al. 2001). We now have increasing evidence of co-infections and immunological cross-reactions between different infectious agents in human, i.e. worms and malaria (Nacher 2002) and helminths and HIV (Wolday et al. 2002), but this type of interaction is one that has not been investigated in detail in the ecological context using wild animal species. As an example of a micro-macroparasite interaction consider the effects of the myxoma virus on the community of macroparasites of wild rabbits. Myxoma is known to be immuno-suppressive within the host (Gerin et al. 2002). Since myxoma is a strong immuno-compromising agent, when a rabbit infected by a macroparasite is also infected with myxoma we may expect a strong reduction of the host immune response to the macroparasite but not the virus. In this respect, Boag et al. (2001) found that rabbits co-infected with myxoma virus had greater mean macroparasite burdens and reduced aggregation and an overall increase in macroparasite species richness and intensity (Fig. 2). Moreover, immunity is influenced by predisposing factors and it is important to take into account age, status and host sex when analysing such interactions.

So a question that arises is: why do we not find any strong and consistent inter-specific parasite interactions from field studies as we do in laboratory experiments? One possible explanation is that laboratory experiments are often based on host and parasite species that have gone through a large number of laboratory passages and have developed host-parasite qualities that differ from the corresponding wild species. For example, the parasite *H. polygyrus bakeri*, which has been used in many of the concomitant infection experiments, has been passed through mouse strains for half a century and can well be considered a laboratory strain (Behnke et al. 1991). Similarly, strains of the same species of mice have been generated in different laboratories through inbreeding or low re-assortment of laboratory populations and this could cause further differences between laboratory and field results.

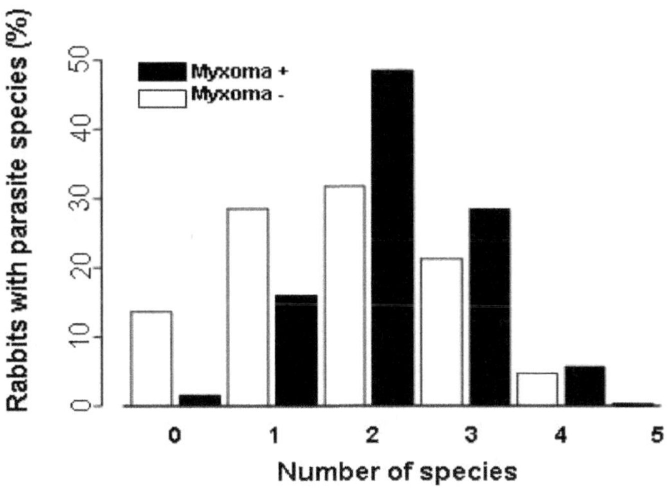

Fig. 2. Differences in the species richness of gastrointestinal macroparasite species in rabbits infected and uninfected with the myxoma virus. Rabbits carry a richer macroparasite community when infected by the immuno-suppressive myxoma virus

Another possibility is that laboratory experiments often use high parasite doses that rarely mimic the environmental rate of infection and the rate of parasite intake observed in the field. Host characteristics, age and breeding status, can profoundly affect the rate of infection and can generate non-linearities in the pattern of infections, but this aspect is never considered when performing single or even trickle dose infections and can indirectly affect the final conclusions. We should also be aware that field studies are based on data of parasite presence/absence or intensity that may not necessarily be the correct variables to use for detecting parasites interactions. In fact, if inter-specific interactions occur between the larval stages at the beginning of an infection, and we concentrate our analysis on adult stage intensities then we may fail to detect certain types of interaction. Similarly, if inter-specific interactions occur at a particular phase of the infection process and we miss to record parasite abundance during this frame, or we average out our data over an expanded time frame, again, we shall not be able to detect interactions between parasite species.

7 Concluding remarks

One of the major challenges in epidemiology is to understand the processes that regulate parasite transmission ecology and community structure and identify whether it is important to consider the whole community of parasites when studying the interaction of the host with individual parasite species. Small mammals have been largely used as study systems in the field and laboratory, and different parasite pair-wise combinations have been examined including both macro- and micro-parasites. Nevertheless, studies on parasite communities have not agreed about the role of parasite interactions at the infra-community level, and how they are modulated by host immunity or other host characteristics. To clarify this conflict and reveal the fundamental mechanisms of regulation, we need to integrate field studies and laboratory experiments and examine not just the transmission ecology but also the immunological mechanisms involved in the host-parasite, and consequently, in parasite-parasite interactions. More broadly, identifying the mechanisms of parasite interactions is an important issue in disease ecology, and in fact, if such interactions are important, we could use one species of parasite to help to control another. We may thus start to understand why diseases and parasites suddenly emerge or why some parasites are more common than others and start to consider ways of predicting where the infection will occur. These considerations are also pertinent with issues related to conservation of animal species and the more recent interest on the effect of environmental changes on parasite community and host-parasite interactions. Changes in environmental features (land use practices, global climatic changes, etc.) have altered the community of wildlife and its parasites and may play an important role in the emergence and persistence of pathogens and parasites that may infect humans or their livestock, thus reducing the quality of life for people.

References

Abu-Madi MA, Behnke JM, Lewis JW, Gilbert FS (2000) Seasonal and site specific variation in the component community structure of intestinal helminths in *Apodemus sylvaticus* from three contrasting habitats in south-east England. J. Helminthol 74:7–15

Anderson RC (2000) Nematode parasites of vertebrates. Their development and transmission, 2nd edn. CABI, Wallingford

Anderson RM, May RM (1985) Herd immunity to helminth infection and implications for parasite control. Nature 315:493–496

Arneberg P (2002) Host population density and body mass as determinants of species richness in parasite communities: Comparative analyses of directly transmitted nematodes of mammals. Ecography 25:88–94

Arneberg P, Skorping A, Grenfell B, Read A (1998) Host densities as determinants of abundance in parasite communities. Proc R Soc Lond B 265:1283-1289

Adams DB, Anderson BH, Windon RG (1989) Cross-immunity between *Haemonchus contortus* and *Trichostrongylus colubriformis* in sheep. Int J Parasitol 19:717–722

Behnke JM, Wakelin D, Wilson MM (1978) *Trichinella spiralis*: Delayed rejection in mice concurrently infected with *Nematospiroides dubius*. Exp Parasitol 46:121–130

Behnke JM, Ali NMH, Jenkins SN (1984) Survival to patency of low-level infections with *Trichuris muris* in mice concurrently infected with *Nematospiroides dubius*. Ann Trop Med Parasitol 78:509–517

Behnke JM, Keymer AE, Lewis JW (1991) *Heligmosomoides polygyrus* or *Nematospiroides dubius*. Parasitol Today 7:177–179

Behnke JM, Lewis JW, Mohd Zain SN, Gilbert FS (1999) Helminth infections in *Apodemus sylvaticus* in southern England: Interactive effects of host age, sex and year on the prevalence and abundance of infections. J Helminthol 73:31–44

Benke JM, Bajer A, Sinski E, Wakelin D (2001a) Interactions involving intestinal nematodes of rodents: Experimental and field studies. Parasitology 122:S39–S49

Behnke JM, Barnard CJ, Bajer A, Bray D, Dinmore J, Frake K, Osmond J, Race T, Sinski E (2001b) Variation in the helminth community structure in bank voles (*Clethrionomys glareolus*) from three comparable localities in the Mazury Lake District region of Poland. Parasitology 123:401–414

Behnke JM, Gilbert FS, Abu-Madi MA, Lewis JW (2005) Do the helminth parasites of wood mice interact? J Anim Ecol 74:982–993

Berding C, Keymer AE, Murray JD, Slater AFG (1986) The population dynamics of acquired immunity to helminth infection. J Theor Biol 122:459–471

Boag B, Lello J, Fenton A, Tompkins DM, Hudson PJ (2001) Patterns of parasite aggregation in the wild European rabbit *Oryctolagus cuniculus*. Int J Parasitol 31:1421–1428

Cattadori IM, Boag B, Bjørnstad ON, Cornell S, Hudson PJ (2005) Immuno-epidemiology and peak shift in a seasonal host-nematode system. Proc R Soc Lond B 272:1163–1169

Christensen NO, Nansen, P, Fagbemi BO, Monrad J (1987) Heterologous antagonistic and synergistic interactions between helminths and protozoans in concurrent experimental infection of mammalian hosts. Parasitol Res 73:387–410

Cook JA, Hoberg EP, Koehler A, Henttonen H, Wickström L, Haukisalmi V, Galbreath K, Chernyavski F, Dokuchaev N, Lahzuhtkin A, MacDonald SO, Hope A, Waltari E, Runck A, Veitch A, Popko R, Jenkins E, Kutz S, Eckerlin R (2005) Beringia: Intercontinental exchange and diversification of high latitude mammals and their parasites during the Pliocene and Quaternary. Mammal Study 30:S33–S44

Cox FEG (2001) Concomitant infections, parasites and immune responses. Parasitology 122:S23–S38

Crofton HD (1948a) The ecology of immature phases of trichostrongyle nematodes: I. The vertical distribution of infective larvae of *Trichostrongylus retortaeformis* in relation to their habitat. Parasitology 39:17–25

Crofton HD (1948b) The ecology of immature phases of trichostrongyle nematodes: II. The effect of climatic factors on the availability of the infective larvae of *Trichostrongylus retortaeformis* to the host. Parasitology 39:26–38

Crombie JA, Anderson RM (1985) Population dynamics of *Schistosoma mansoni* in mice repeatedly exposed to infection. Nature 315:491–493

Denby J, McKillop G, Fox S,Western G, Langton S (2003) The development of a model to assess the effects of rabbit grazing on grass. Ann Appl Biol 142:317–322

Ellis RD, Pung OJ Richardson DJ (1999) Site selection by intestinal helminths of the Virginia opossum (*Didelphis virginiana*). J Parasitol 85:1–5

Epstein PR (2000) Is global warming harmful to health? Sci Amer 283:50–57

Ferrari N (2006) Macroparasite transmission and dynamics in *Apodemus flavicollis*. Ph. D. Thesis, University of Stirling

Ferrari N, Cattadori IM, Hudson PJ (2004) The role of host sex in parasite dynamics of the yellow necked mice (*Apodemus flavicollis*). Ecol Lett 7:88–94

Gotelli NJ, Graves GR (1996) Null models in ecology. Smithsonian Inst Press, Washington

Goüy de Bellocq J, Sara M, Casanova J-C, Feliu C, Morand S (2003) A comparison of the structure of helminth communities in the wood mouse, *Apodemus sylvaticus*, on islands of the western Mediterranean and continental Europe. Parasitol Res 90:64–70

Gregory RD (1990) Genetics, sex and exposure - the ecology of *Heligmosomoides polygyrus* (Nematoda) in the wood mouse. J Anim Ecol 59:363–378

Gregory RD, Montgomery SSJ, Montgomery WI (1992) Population biology of *Heligmosomoides polygyrus* (Nematoda) in the wood mouse. J Anim Ecol 61:749–757

Grenfell B, Smith G (1983) Population biology and control of ostertagiasis in first year grazing calves. Proc Soc Vet Epid Prevent Med 1983:54–61

Guerin JL, Gelfi J, Boullier S, Delverdioer M, Bellanger FA, Bertagnoli S, Drexler I, Sutter G, Messud-Petit F (2002) Myxoma virus leukemia-associated protein is responsible for MHC complex Class 1 and Fas-CD95 down regulation and defines scrapins, a new group of surface cellular receptor abductor proteins. J Virol 76:2912–2923

Hanski I (1982) Dynamics of regional distribution: The core and satellite species hypothesis. Oikos 38:210–221

Haukisalmi V (1989) Intestinal helminth communities of *Sorex* shrews in Finland. Ann Zool Fenn 23:141–150

Haukisalmi V, Henttonen H (1990) The impact of climatic factors and host density on the long-term population dynamics of vole helminths. Oecologia 83:309–315

Haukisalmi V, Henttonen H (1993a) Coexistence in helminths of the bank vole - *Clethrionomys glareolus*. 1. Patterns of co-occurrence. J Anim Ecol 62:221–229

Haukisalmi V, Henttonen H (1993b) Coexistence in helminths of the bank vole - *Clethrionomys glareolus*. 2. Intestinal distribution and interspecific interactions. J Anim Ecol 62:230–238

Haukisalmi V, Henttonen H (1994) Distribution patterns and microhabitat segregation in gastrointestinal helminths in *Sorex* shrews. Oecologia 97:236–242
Haukisalmi V, Henttonen H (1998) Analysing interspecific associations in parasites: Alternative methods and effects of sampling heterogeneity. Oecologia 116:565–574
Haukisalmi V, Henttonen H (1999) Determinants of helminth aggregation in host populations: Individual differences or spatial heterogeneity? Ecography 22:629–636
Haukisalmi V, Henttonen H (2000) Helminth dynamics and community structure in the bank vole *Clethrionomys glareolus*. Polish J Ecol 48:S219–S230
Haukisalmi V, Henttonen H, Tenora F (1988) Population dynamics of common and rare helminths in cyclic vole populations. J Anim Ecol 57:807–826
Haukisalmi V, Henttonen H, Mikkonen T (1994) Parasitism by gastrointestinal helminths in the shrews *Sorex araneus* and *S. caecutiens*. Carnegie Mus Nat Hist Spec Publ 18:97–102
Haukisalmi V, Henttonen H, Batzli GO (1995) Helminth parasitism in the voles *Microtus oeconomus* and *M. miurus* on the North Slope of Alaska: Host specificity and the effects of host sex, age and breeding status. Ann Zool Fenn 32:193–201
Haukisalmi V, Wickström LM, Henttonen H, Hantula J, Gubányi A (2004) Molecular and morphological evidence for multiple species within *Paranoplocephala omphalodes* (Hermann, 1783) (Cestoda: Anoplocephalidae) in *Microtus*-voles (Arvicolinae). Zool Scripta 33:277–290
Haukisalmi V, Henttonen H, Hardman LM (2006) Taxonomy, diversity and zoogeography of *Paranoplocephala* spp. (Cestoda: Anoplocephalidae) in voles and lemmings of Beringia, with a description of three new species. Biol J Linn Soc (in press)
Holmes JC, Price PW (1986) Communities of parasites. In: Kikkawa J, Anderson DJ (eds) Community ecology: Patterns and processes. Blackwell Sci, London, pp 187–213
Jenkins SN (1977) Impairment of primary expulsion of *Trichuris muris* in mice concurrently infected with *Nematospiroides dubius*. Parasitology 75:71–78
Kutz S, Hoberg EP, Polley L, Jenkins E (2005) Global warming is changing the dynamics of Arctic host-parasite systems. Proc R Soc Lond B 272:2571–2576
Lello J, Boag B, Fenton A, Stevenson I, Hudson PJ (2004) War of the worms. Competition and mutualism amongst the gut helminths of a mammalian host. Nature 428:840–844
Marquardt W, Demaree R, Grieve R (1999) Parasitology and vector biology, 2nd edn. Harcourt/Acad Press, San Diego London Boston New York Sydney Tokyo Toronto
Montgomery SSJ, Montgomery WI (1989) Spatial and temporal variation in the intracomunitary structure of helminths of *A. sylvaticus*. Parasitology 98:145–150
Montgomery SSJ, Montgomery WI (1990) Structure, stability and species interactions in helminth communities of wood mice *Apodemus sylvaticus*. Int J Parasitol 20:225–242

Nacher M (2002) Worms and malaria: Noisy nuisances and silent benefits. Parasite Immunol 24:391–393

Norozianamiri SMB, Behnke JM (1993) Density-dependent effects on establishment of *Necator americanus* and *Ancylostoma ceylanicum*. J Helminthol 67:151–157

Patrick MJ (1991) Distribution of enteric helminths in *Glaucomys volans* L. (Sciuridae): A test for competition. Ecology 72:755–758

Perkins SE, Cattadori IM, Tagliapietra V, Rizzoli AP, Hudson PJ (2003) Disease transmission through functional groups, non linear transmission of tick bone encephalitis through yellow necked mice. Int J Parasitol 33:909–917

Phillips RS, Wakelin D (1976) *Trichuris muris* - effect of concurrent infections with rodent piroplasms on immune expulsion from mice. Exp Parasitol 39:95–100

Poulin R (1996) Sexual inequalities in helminth infections: A cost of being a male? Amer Nat 147:287–295

Poulin R (2001) Interactions between species and the structure of helminth communities. Parasitology 122:S3–S11

Prévot-Julliard A-C, Henttonen H, Yoccoz NG, Stenseth NC (1999) Delayed maturation in female bank voles: Optimal decision or social constraint? J Anim Ecol 68:684–697

Randolph SE, Green RM, Peacey MF, Rogers DJ (2000) Seasonal synchrony: The key to tick-borne encephalitis foci identified by satellite data. Parasitology 121:15–23

Rose ME (1994) Interactions between infections with *Eimeria* spp. and *Trichinella spiralis* in inbred mice. Parasitology 108:69–75

Saunders LM (2000) Infection strategies of gamebird macroparasites. Ph. D. thesis, University of Stirling

Silver BB, Dick TA, Welch HE (1980) Concurrent infections of *Hymenolepis diminuta* and *Trichinella spiralis* in the rat intestine. J Parasitol 66:786–791

Soulsby EJL (1982) Helminths, arthropods and protozoa of domesticated animals. Lea and Febiger, Philadelphia

Wilson K, Bjørnstad ON, Dobson AP, Merler S, Poglayen G, Randolph SE, Read AF, Skorping A (2001) Heterogeneities in macroparasite infections: Patterns and processes. In: Hudson PJ, Rizzoli A, Grenfell BT, Heesterbeek H, Dobson AP (eds) The ecology of wildlife diseases. Oxford Univ Press, Oxford, pp 6–44

Wolday D, Mayaan S, Maraiamm ZG, Berhe N, Seboxa T, Britton S, Galai N, Landay A, Bentwich Z (2002) Treatment of intestinal worms is associated with decreased HIV plasma viral load. J AIDS 31:56–62

Woolhouse MEJ (1992) A theoretical framework for the immuno-epidemiology of helminth infection. Parasite Immunol 14:563–578

Woolhouse MEJ (1998) Patterns in parasite epidemiology: The peak shift. Parasitol Today 14:428–434.

Young RR, Anderson N, Overend D, Tweedie RT, Malafant KWJ, Preston GAN (1980) The effect of temperature on times to hatching of eggs of the nematode *Ostertagia circumcita*. Parasitology 81:477–491

Zuk M, McKean KA (1996) Sex differences in parasite infections: Patterns and processes. Int J Parasitol 26:1009–1024

19 Effect of macroparasites on the energy budget of small mammals

A. Allan Degen

1 Introductory remarks

Parasites are organisms that live in or on one or a few hosts from which they derive their nutrients and other biological needs (Kim 1985). Their biological impact is substantial as more than 50% of the species on earth and more than 50% of all individual organisms are parasites or pathogens. They can play a major role in the life history of their hosts in that they can affect such variables as fecundity, survival rate, mate selection, population dynamics, and growth of the host (Tompkins and Begon 1999; Forchhammer and Asferg 2000; Lochmiller and Deerenberg 2000; Zuk and Stoehr 2002).

Parasites normally cause harm to their hosts, but not immediate death. This definition of parasitism implies a reduction of host fitness while parasitized (Arnold and Lichtenstein 1993; Clayton and Moore 1997), including an increase in host mortality and/or morbidity as well as a decrease in host fecundity (Tompkins and Begon 1999; Newey and Thirgood 2004). The effects of parasites can be direct, such as using energy and nutrients of hosts (Khokhlova et al. 2002) and indirect, such as increasing activity of the immune system (Wedekind 1992; Lochmiller and Deerenberg 2000), modifying behaviour (Barnard et al. 1998; Kavaliers et al. 1998; Poulin 2000; ter Hofstede and Fenton 2005) and decreasing food intake of hosts (Tripet and Richner 1997; Kyriazakis et al. 1998; Simon et al. 2004).

However, evidence of the negative effects of parasites is equivocal. Although many studies have demonstrated detrimental effects of parasites (Munger and Karasov 1989; Alves 1997), others have failed to do so (Munger and Slichter 1995; Pacejka et al. 1998; Bouslama et al. 2001; Kristan 2004). These contradictory findings have lead to a number of studies on the physiological and ecological impacts of parasites on hosts in different systems. Most of these studies examined endoparasites, especially helminths (Munger and Karasov 1991; Meagher 1998; Behnke et al. 2001; Kristan 2002a, b; Meagher and Dudek 2002; Kristan and Hammond 2004). Ectoparasites have been examined mainly on avian hosts (Arendt 1985; Brown and Brown 1986; Fauth et al. 1991; Møller 1990; Hudson et al. 1992; De Lope et al. 1993; Richner et al. 1993; Richner and Tripet 1999), with very few studies on mammals (Dzieciolowski and Clarke 1990; Butler and Roper 1996; Chekchak et al. 2000; Khokhlova et al. 2002, 2004a, b; Krasnov et al. 2004, 2005). In fact, as late as in 2003, in a study of the effects of ectoparasites on Columbian ground squirrels, Neuhaus (2003) stated that “To my knowledge, this is the first experimental attempt to look at the impact of ectoparasites on reproductive success and body condition in a wild mammal”.

In this chapter, I will discuss the effects of macroparasites on the energy budget of small mammals and start by brief definition.

1.1 Energy budget of the host

Energy is provided to the animal from its food intake, which is digested in the gastrointestinal tract and absorbed via the blood system. The fraction of dry matter that the animal extracts from its diet is known as the apparent dry matter digestibility (ADMD, fraction) of the diet and is determined from the dry matter intake (DMI, g) and dry matter faecal output (DMFO, g) as

$$ADMD = \frac{DMI - DMFO}{DMI}. \qquad (1)$$

The amount of apparent digestible energy the animal obtains from its diet is calculated from the gross energy of the diet minus the energy lost in faeces, and the apparent metabolizable energy of a diet is the digestible energy minus the energy lost in urine plus energy lost in combustible gases.

The metabolizable energy intake of an animal is the energy available for maintenance and production (growth, milk production, etc). If the metabolizable energy intake of an animal equals its maintenance energy requirements, then there is no change in the energy content of the animal and the entire chemical energy intake is lost as heat. If the metabolizable energy intake is below maintenance requirements, then the animal is forced to catabolize tissue from its body energy to compensate for this lack of energy. If metabolizable energy intake is above maintenance requirements, however, the animal can then add to its body reserves (Degen 1997). These relationships may be somewhat more complicated when, for example, the intake of a parasitized animal is below energy requirements for both maintenance and immunity costs, but nonetheless, the immunity system is activated by the mobilization of body energy. This, of course, can only continue for a limited period of time. The relationship between metabolizable energy intake (MEI, kJ/d), heat production (HP, kJ/d) and energy retention (ER, kJ/d) can be presented as:

$$MEI = HP + ER, \tag{2}$$

where ER can be either positive or negative.

Basal metabolic rate represents a baseline of minimal energy expenditure for the body functions of an animal while awake, and when almost zero energy is expended in movement, thermoregulation, combating diseases and food absorption. It is a very widespread measure of energy expenditure in animals, and is usually determined by their rate of oxygen (O_2) uptake. In fact, the use of oxygen uptake to estimate basal metabolic rate is so common that it is often expressed in terms of oxygen uptake. Resting metabolic rate, fasting metabolic rate and standard metabolic rate are all similar to basal metabolic rate and these terms are often used interchangeably. Average daily metabolic rate is the metabolizable energy intake required by a caged, laboratory animal to maintain constant body energy content. It includes basal metabolic rate, heat increment of feeding for maintenance, some minimal locomotory costs and possibly some thermoregulatory costs (Degen et al. 1998). The field metabolic rate of an animal is its energy expenditure under free-living conditions, which includes average daily metabolic rate, locomotory costs and thermoregulatory costs (Degen 1997).

2 Trade-offs between immune costs and other energetic costs

Parasitized hosts can be confronted with trade-off decisions between energy costs of the immune defence system and other energy demanding processes such as maintenance, reproduction, growth and thermoregulation. Consequently, parasites can affect their hosts through the diversion of resources (Candolin and Voigt 2001; Zuk and Stoehr 2002).

Immunity can be innate, that is, present in a host irrespective of diseases or can be acquired, that is, activated in response to a challenge. Both require energy from the host during parasite infection. In general, metabolic costs of: 1) mounting an immune response; and 2) maintaining a competent immune system have been assessed by observing physiological changes in the host, as these costs are extremely difficult to measure. As expressed by Lochmiller and Deernberg (2000), "These two physiological traits are not easily addressed quantitatively, particularly with respect to the former, because of the integrated and organizational characteristics of the immune system with other physiological systems". In addition, "severity, type, and duration of infection, ambient temperature, and gender, age, and nutritional status of the host all influence the cost of mounting an immune response".

Nonetheless, it is generally accepted that activation of an immune response and even maintenance of a competent immune system are very energetically demanding processes (Sheldon and Verhulst 1996; Lochmiller and Deernberg 2000; Zuk and Stoehr 2002). This is further supported by studies showing that energy restriction of a host can lead to a suppression of the immune system and an increase in the risk of infection, in particular in opportunistic pathogens (Lochmiller and Dabbert 1993). However, counter-intuitively, immune challenges often suppress food intake of a host (sepsis-induced anorexia) at a time when additional energy is vitally needed. Immune cells require high levels of glucose and glutamine, in particular. The body mobilizes protein and energy reserves to support the initial acute-phase immune response with an accelerated lipolysis, proteolysis and glycolysis and, to do so, increases metabolic rate. This can often lead to a negative nitrogen balance and loss in body mass in the host (Hasselgren and Fischer 1998).

The costs of mounting and maintaining immune responses were examined in wild white-footed mice (*Peromyscus leucopus*) (Derting and Compton 2003). To assess the energetic costs of mounting an immune response, mice were injected with sheep red blood cells and phytohaemagglutinin (in foot pads), so that their humoral and cell-mediated immune responses, respectively, would be stimulated. There was no significant difference in dry matter intake, apparent dry matter digestibility, resting metabolic rate and average daily metabolic rate between immunochallenged and control mice. Also, white blood cells concentration did not differ between immunochallenged and control mice after injection of either sheep red blood cells or phytohaemagglutinin but, feet of mice injected with phytohaemagglutinin were 57% heavier than in controls, showing a significant cell-mediated immune response. The wet and dry masses of the small intestine (by 33% and 22%, respectively) and testes (by 82% and 74%, respectively) and wet mass of lungs (by 19%) were significantly greater in control than immunochallenged mice. Consequently, it was concluded that "mounting an immune response to a mild immunochallenge was associated with change in patterns of energy allocation ... specifically, energy allocation to the small intestine and testes were reduced. White-footed mice appeared to accommodate the cost of mounting an immune response through trade-offs in energy allocation to other physiological systems rather than through increased ingestion of food energy". To assess the energetic costs of maintaining an immune response, immune responses were suppressed by injection of cyclophosphamide, a drug that suppresses humoral immunocompetence but not macrophages or cell-mediated immunity (Allison 2000), and then the same measurements as above were made on the mice. There was an immunosuppressive effect as white blood cells in control mice were 225% higher than in injected mice. However, there was no difference between immunosuppressed and control mice in any of the other variables measured. It was concluded that "the cost of maintaining a normally functioning immune system was minimal in wild adult-male white-footed mice. Significant suppression of the immune system was not associated with a significant change in RMR (resting metabolic rate) or in DMR (daily metabolic rate) or any measurable change in energy allocation to the vital, intestinal, or reproductive organs". In summary, from this study, it was concluded that the immune response could be energetically costly, but that the cost of its maintenance was negligible.

3 Basal metabolic rate, average daily metabolic rate and field metabolic rate of parasitized hosts

Basal metabolic rate, cold-stress maximum oxygen consumption (MR_{peak}), metabolic scope (MR_{peak}/BMR), and thermogenic endurance were determined in two populations of deer mice (*Peromyscus maniculatus gracilis*) when infected by the nematode *Capillaria hepatica* (Meagher and O'Conner 2001). *C. hepatica* inhabits the liver in a large number of mammals, but most commonly in rodents. Adult female worms lay eggs in the host's liver which remain there till the host dies and, therefore, this nematode requires the death of its host for transmission. The two mice populations in the study originated from different geographic areas in Michigan in which the nematode was present in one area (Upper Peninsula – UP) but not in the other (Lower Peninsula – LP). In laboratory studies, mice collected from each area were infected by oral gavage with approximately 750 nematode eggs and compared with non-infected mice. Basal metabolic rate was not affected by nematode infection. However, LP infected mice had a reduced maximum oxygen consumption, metabolic scope, and thermogenic endurance (Fig. 1) compared with the other groups. The authors concluded that the "decreased metabolic performance caused by *C. hepatica* infection, as displayed by LP mice" could "diminish their overwinter survival". In addition, while infected LP mice had lower maximum oxygen consumption than uninfected LP mice, infected UP mice, in contrast, had higher maximum oxygen consumption than uninfected UP mice (Fig. 1).

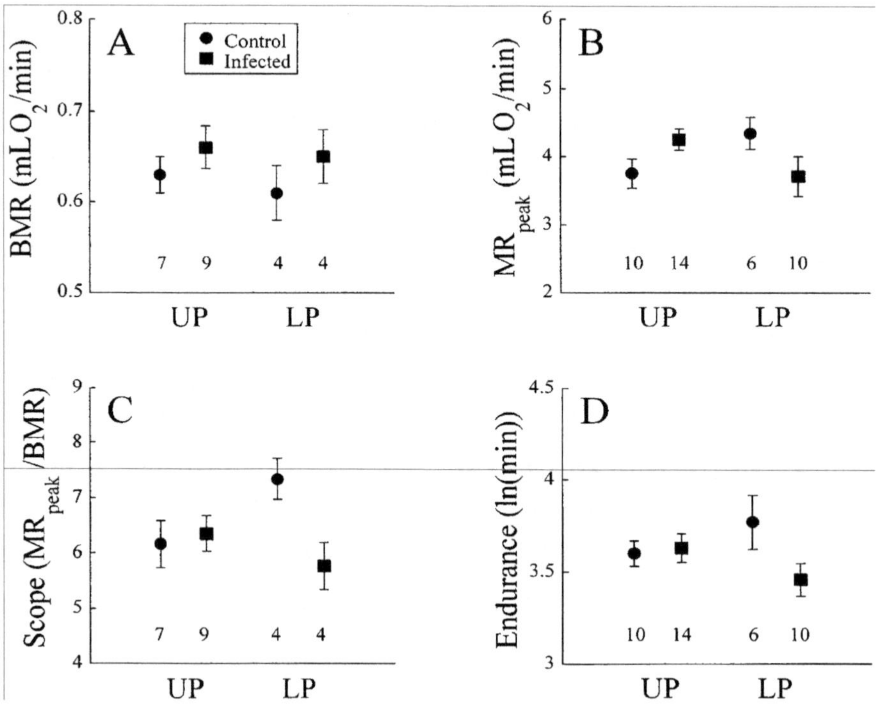

Fig. 1. Effect of host locality and *C. hepatica* infection on mass-adjusted metabolic measurements (mean and standard error) in *P. m. gracilis* in the Upper Peninsula (UP) and Lower Peninsula (LP) of Michigan. (A) Basal metabolic rate (BMR); (B) Cold-induced peak metabolic rate (MR_{peak}); (C) Metabolic scope (MR_{peak} /BMR); (D) Thermogenic (cold-stress) endurance time (ln-transformed). The measurement for each mouse was adjusted to that predicted for an animal with a body mass of 23.25 g. Numbers below the symbols are sample sizes (after Meagher and O'Conner 2001, reprinted with permission from the Canada Institute for Scientific and Technical Information)

These results suggest that there are "genetically determined differences in host response to this parasite" and that the UP "hosts have evolved to cope with the detrimental consequences of *C. hepatica* infection".

Resting metabolic rate and maximal exercise-induced oxygen consumption were also determined in wild-derived house mice (*Mus musculus*) (Kristan and Hammond 2004) and in laboratory mice (Kristan and Hammond 2000) when infected by the intestinal nematode *Heligmosomoides polygyrus* and maintained at room temperature (23°C) or exposed to cold (5°C). These mice were infected with the infective stage larvae (L3) of *H. polygyrus* and, after 14 days, the mature *H. polygyrus* occupied the small intestine. In the wild-derived house mice, resting metabolic rate increased on average by 12% in parasitized animals and by 9% when exposed to cold, but these differences were not significantly different from controls. These measurements were also made on lactating mice and here too, infection and cold exposure had no effect. The authors concluded that "wild-deprived mice are unaffected by exogenous (temperature) and endogenous (*H. polygyrus*) demands, and therefore, wild-deprived mice respond to these demands without incurring potential costs associated with changes in aerobic performance". But, parasitism and cold exposure did affect these measurements in laboratory mice. Parasitized mice had a significantly greater average resting metabolic rate than unparasitized mice, by 4% at 5°C and by 9% at 23°C, and cold-exposed mice had a greater resting metabolic rate than room temperature mice, by 19% and 14% for unparasitized and parasitized mice, respectively. However, there was no difference in metabolizable energy intake between parasitized and unparasitized mice, in spite of the increased resting metabolic rate, but cold exposed mice consumed more energy than room temperature mice. Apparent dry matter digestibility was similar for all groups. Furthermore, body mass did not differ among treatment groups, but parasitized mice had less fat than unparasitized mice. It is possible then that the main source of energy for the increased resting metabolic rate for the parasitized mice was their body fat. If that is the case, the question that arises then is why these infected mice did not increase their metabolizable energy intake and conserve their body energy? If it is not the case, then what was the source of the energy for the increased resting metabolic rate for the parasitized mice?

Resting metabolic rate was also measured in another study of laboratory mice when infected by *H. polygyrus* and when energy intake was restricted (Kristan and Hammond 2001). Resting metabolic rate, when corrected for whole body mass was significantly greater in parasitized than unparasitized mice, by 9%, but there was no significant difference between groups when corrected for lean body mass. Energy restricted mice had a lower resting metabolic rate than *ad libitum* fed mice when corrected for either of the covariates. Of interest, "caloric restriction and parasitism were independent for physiological responses", that is, "changes in RMR (resting metabolic rate) associated with parasite infection and caloric restriction remained independent when both demands (were presented) simultaneously". However, in yet another study on laboratory mice parasitized by *H. polygyrus*, resting metabolic rate did not differ between infected and noninfected rodents (Kristan 2002a).

The energetic costs of larval bot fly (*Cuterebra fontinella*) infection were measured in white-footed mice (*P. leucopus*) (Munger and Karasov 1994). Infected hosts generally harbour one or two larvae. Adult bot flies lay eggs near rodent burrow entrances and eggs hatch in response to body heat of a host. The first-instar larvae adhere to a passing host, penetrate the skin in the oral or nasal region and migrate subdermally where the larva grows within a subdermal capsule for 21 days. Then, at about 1 g mass, it emerges through its respiratory pore, burrows into the ground, pupates and undergoes diapause until emergence the next summer. Mice naturally infected with one or two larvae were used in laboratory studies, while deinfected mice had larvae removed and acted as controls. Metabolic rate (measured as oxygen uptake) of the host increased by 2.9% for each larva harboured. This energetic cost corresponded well with the estimated energy consumed by each larva, which was 3-4% of the energy intake of the host (6-8% for two larvae). However, there was no difference in either food consumption or apparent dry matter digestibility due to the larval bot fly. In a field study of this host – parasite relationship using doubly labelled water, there was no significant difference in field metabolic rate between infected and deinfected mice, but infected mice lost significantly more body mass than deinfected mice during the 2 days of measurement (-1.1 g versus -0.1 g, respectively). This would suggest that the infected mice consumed less metabolizable energy than the deinfected mice and, consequently, catabolized more body tissue than or that, although the parasitized mice lost more body mass, their loss in body energy was similar to that of deinfected mice.

Ectoparasites that do not enter the host, such as fleas, do not compete with the host for nutrients as do intestinal or burrowing endoparasites, however they obtain energy from their hosts and can affect the energy budget of the host by diverting energy to fight the infection. Fleas are blood-sucking, usually spending time either on the body of their host or in its burrow or nest. The feeding rate of fleas varies among species as well as within-species and is dependent on such factors as flea age, ambient temperature, relative humidity and host behaviour (Marshall 1981; Krasnov et al. 2001). Potential negative effects on the host due to flea parasitism, however, is not limited only to blood sucking as fleas also damage the skin of hosts, give irritating bites, inject salivary toxins into the wound and inoculate pathogens (Nelson et al. 1977; Marshall 1981).

The effect of the flea *Xenopsylla ramesis* while parasitizing the desert gerbil *Gerbillus dasyurus* was studied by determining the average daily metabolic rate of parasitized and unparasitized rodents (Khokhlova et al. 2002). Metabolizable energy consumption and body mass change were determined for each individual within each group of rodents and average daily metabolic rate was estimated for each group from the linear regression of body mass change of the gerbil on its metabolizable energy intake, with average daily metabolic rate taken at the point where there was no change in body mass (Fig. 2). Average daily metabolic rate of the parasitized gerbils (7.75 kJ $g^{-0.54}d^{-1}$) was 16% higher than that of unparasitized gerbils (6.69 kJ g $g^{-0.54}d^{-1}$). In addition, at zero metabolizable energy intake, the parasitized gerbils lost body mass at a faster rate than the unparasitized gerbils, 4.34 and 3.95% body mass d^{-1}, respectively. Therefore, the increased average daily metabolic rate of parasitized gerbils was compensated by catabolism of body tissue.

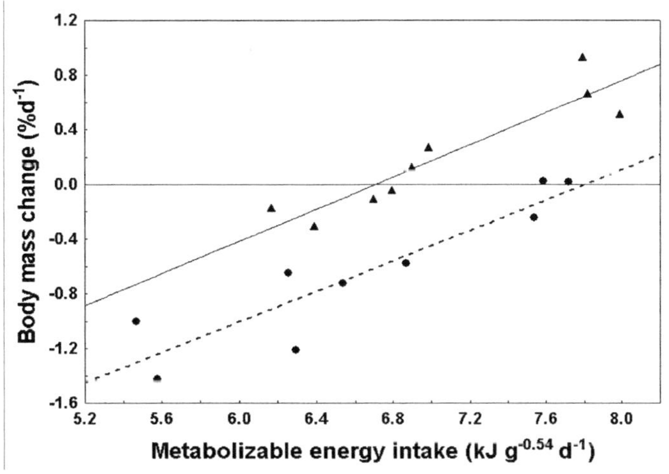

Fig. 2. The effect of metabolizable energy intake on body mass change in parasitized (circles, dashed line) and unparasitized (triangles, solid line) *G. dasyurus* consuming different levels of energy. Body mass to the exponent 0.54 was used to compensate for different body masses (redrawn from Khokhlova et al. 2002, reprinted with permission from Cambridge University Press)

4 Dry matter and energy intakes, digestibility and body mass change of parasitized hosts

Macroendoparasites inhabiting the gastrointestinal tract of their hosts: 1) compete for nutrients and thus reduce the amount available to their hosts: and 2) can damage the absorptive surfaces of the tract and thus reduce the amount of nutrients absorbed by the gut. Both these actions could potentially reduce the apparent dry matter digestibility of the diet, that is the amount of dry matter absorbed from the gut. A reduction in apparent dry matter digestibility results in a reduction in the amount of energy the hosts obtain from their diet. To compensate for such energy losses, hosts can increase their metabolizable energy intake and/or decrease their heat production and/or reduce their body temperature. Otherwise, the lost energy has to be obtained from the mobilization of body energy reserves, which can usually be indicated by losses in body mass of the hosts. This assumes that the proportionate body composition and the body temperature (heat storage) of the animal remain constant (Degen 1997).

Munger and Karasov (1989) examined the effects of *Hymenolepis citelli*, a tapeworm that inhabits the intestine of its host, on the apparent dry matter digestibility of the diet and energy budget of the white-footed mouse (*P. leucopus*). Treated animals received tapeworm cysts and one of two powdered diets: a high-protein ration, similar in nutrient composition and digestibility to an insect-rich diet (57% arthropods, 31% seeds and 12% vegetation - composed of 45:20:23:7:5; protein: a carbohydrate: fiber: vitamins + minerals) or high-carbohydrate ration, similar in composition to a seed-rich diet (56% seeds, 23% vegetation and 22% arthropods – 20:45:23:7:5). Infected mice had a significant 2% decrease in apparent dry matter digestibility when compared to non-infected mice. Furthermore, protein digestibility in the infected group was decreased by 3.3%, a significant reduction, and fiber digestibility by 4.3%, a non-significant reduction. Two possible explanations were offered for the reduced apparent dry matter digestibility by the infected mice. Firstly, "digestive efficiency might be decreased simply by loss of tapeworm eggs and gravid proglottids. Nutrients normally available for assimilation would be tied up in the tapeworms and thus appear in the faeces". Secondly, "tapeworms might damage the intestinal wall, thereby decreasing rates of digestion and absorption". However, there was no significant effect on either dry matter intake or on body mass change due to tapeworm infection, although mice on the high-protein diet tended to gain body mass but mice on the high carbohydrate diet tended to lose body. Heat production (metabolic rate) and body temperature were not measured in these laboratory kept mice. However, in a field study in which doubly labelled water was used to measure field metabolic rates and water turnover fluxes and temperature-sensitive transmitters were used to measure body temperatures in these mice, there was no effect of the tapeworms on any of these variables. Thus, there was no difference in heat production nor in body temperature of the infected mice and, consequently, no compensation was detected for the reduced apparent dry matter digestibility. How can this occur? Possible explanations that were offered by the authors included :1) an error in apparent dry matter digestibility measurements and, actually, there was no difference between groups; and 2) there was compensation by the parasitized group but this was not detected by the measurements employed.

Food intake and body mass change were examined in mice infected with metacestodes of the tapeworm *Taenia crassiceps* (Crompton et al. 1985) over a period of 140 days. A significantly higher proportion of uninfected mice survived the study than infected mice but there was no effect on food intake due to parasitism. Although body mass of infected mice was higher than uninfected mice, mass of dry carcasses of uninfected mice was higher than that of infected mice. This suggests that body energy was being mobilized at a faster rate in infected than in uninfected mice and that *T. crassiceps* caused a shift in body composition with parasitized mice retaining more body water volume than unparasitized mice.

The effect of the whipworm, *Trichuris dipodomys*, a nematode that inhabits the caecum of a number of rodents, was studied in two species of kangaroo rats (Munger and Slichter 1995). *Dipodomys microps* (72–91 g) and *Dipodomys ordii* (42–72 g) both harbour *T. dipodomys* but its prevalence is much greater in the former species (53% versus 14%). Three groups of both rodent species were examined: 1) naturally uninfected (control); 2) naturally infected; and 3) naturally infected but then deinfected with an anthelminthic drug. The animals were offered whole millet seeds, which had an apparent dry matter digestibility above 95% in both *Dipodomys* spp. The deinfected *D. microps* had a significantly higher apparent dry matter digestibility than the other two *D. microps* groups (by 1.9%), but there was no effect in *D. ordii*. The authors concluded that the difference in response between the two rodent hosts could be due to the natural higher intensity of parasite infection in *D. microps* and "if fewer worms were present in *D. ordii*, the effect of eradicating those worms would have been less apparent." They also concluded that "although a statistically significant effect was shown, its small magnitude indicates that whipworm infection is unlikely to have a biologically significant impact on the energy budgets of host kangaroo rats".

Wild-derived *M. musculus* that were infected with *H. polygyrus* also showed a reduced apparent dry matter digestibility (Kristan and Hammond 2003). Mice were offered a high carbohydrate diet and were subjected to either cold (5°C) or room (23°C) air temperatures. Apparent dry matter digestibility of the parasitized mice was 2% lower than that of the unparasitized mice; there was no difference due to air temperature. Parasitized mice consumed 7% less food than unparasitized mice, but this difference was not significant (P=0.072); cold exposed mice increased intake significantly by 51% compared to room temperature mice. Whole body mass, lean body mass, and total body fat of the mice were similar in all treatments, but parasitized mice had a 15% longer small intestine than unparasitized mice and the dry mass of the small intestine in parasitized mice was 46% greater than in unparasitized mice. Of importance, glucose uptake rate by the parasitized mice was 41% lower than for the unparasitized mice over the entire small intestine (calculated per unit tissue); however, since the parasitized mice had a greater amount of intestinal tissue, there was no difference between groups in total glucose uptake capacity. Here again, there was no evident compensation for the reduced apparent dry matter digestibility due to the parasites. The authors concluded that "the diminished digestive efficiency when parasitized was probably not directly biologically relevant to the mouse (at least under the laboratory conditions of our experiment) because of the phenotypic augmentation of small intestinal mass which resulted in similar nutrient uptake capacity of parasitized and unparasitized mice". They also concluded that the "effects of cold exposure and parasite infection were largely independent of each other for the morphological and physiological parameters we measured".

However, in contrast to the findings in wild-derived house mice, laboratory mice infected by *H. polygyrus* showed no difference in apparent dry matter digestibility when compared with uninfected mice (Kristan and Hammond 2000; 2001). But, in common with the wild-derived house mice, there was no difference in body mass and energy intake between infected and uninfected laboratory mice. However, there was a difference in body composition in that infected mice had less body fat than uninfected mice. The decrease in glucose uptake and increase of the mass of the small intestine, as reported in wild-derived house mice, was also found in these laboratory mice.

Body mass was affected when laboratory oldfield mice (*Peromyscus polionotus*) were infected with *Trichinella spiralis* (Meagher and Dudek 2002), a widespread, gastrointestinal nematode that can be lethal. *T. spiralis* infects a wide variety of mammalian hosts, including *P. leucopus*, but *P. polionotus* has not been shown to be a compatible host for this parasitic nematode. Unlike most macroparasites, *T. spiralis* is intracellular and the same animal serves as both the definitive and intermediate host. In the study, mice received doses of juvenile worms ranging in number from 70 to 960. All mice that received up to 210 juvenile worms survived the 120 days of the study; however, 29% (2/7) of mice given 600 juvenile worms and 78% (5/7) of mice given 960 juvenile worms died. Control mice gained body mass but parasitized mice lost body mass, the change in body mass being negatively correlated with the infection intensity (Fig. 3). In addition, final body mass of the mice was not affected by initial mass, but was negatively affected by nematode intensity and testes mass and seminal vesicles were associated positively with body mass, but negatively with nematode intensity (Fig. 4). Three possibilities were proposed for the reduced gonad size in infected mice: 1) reallocation of energy from reproduction to immunity; 2) high cortisone levels due to stress response because of the infection; and 3) gonadal damage due to inflammatory responses. These authors concluded that "*T. spiralis* infection causes significant negative physiological changes in *P. polionotus*. These include death, mass loss, and decreased gonad size, and these nonlethal effects could lead to reduced reproductive success by the loss of social dominance".

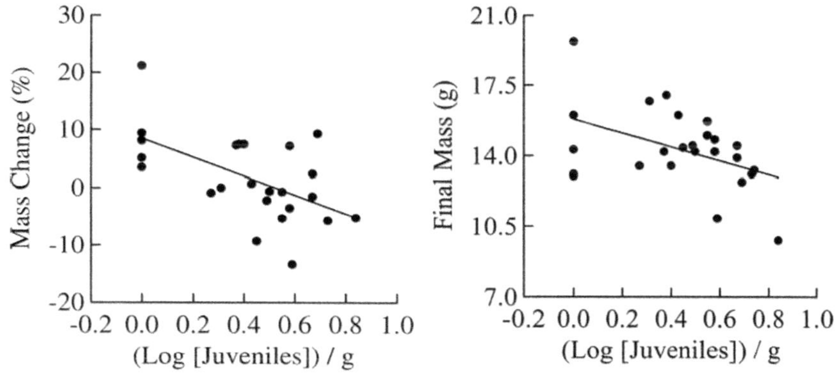

Fig. 3. Effect of *T. spiralis* infection intensity ($\log_{10}$[juveniles]/g) on percentage of mass change at 30 days and on final mass of *P. polionotus* (after Meagher and Dudek 2002, reprinted with permission from the American Society of Parasitologists)

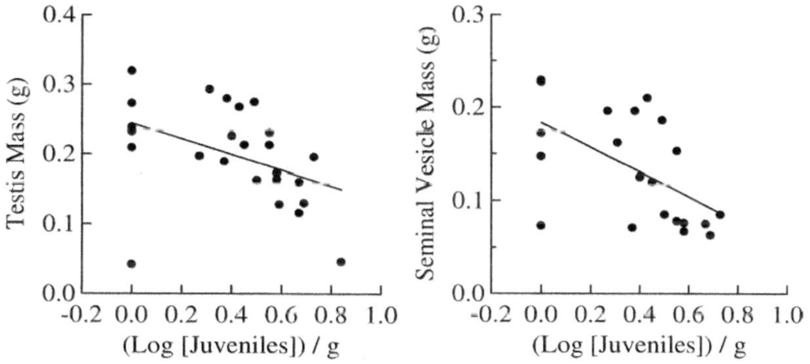

Fig. 4. Effect of *T. spiralis* infection intensity ($\log_{10}$[juveniles]/g) on paired testes and on paired seminal vesicles masses of *P. polionotus* (after Meagher and Dudek 2002, reprinted with permission from the American Society of Parasitologists)

In contrast to these results, male bank voles (*Clethrionomys glareolus*) showed significantly increased adrenal glands (Fig. 5), testes and seminal vesicles (Fig. 5) with greater endoparasite intensities, including helminths (in particular the nematodes, *Heligmosomum mixtum*, *Heligmosomoides glareoli* and *Aspiculuris tetraptera*) (Barnard et al. 2002). As a result of these larger glands, there was an increase in glucocorticoid hormones and androgens. Concomitantly, there was an increased worm and protozoan intensity in these voles, indicating a reduced resistance to helminths and protozoan infection, which is "consistent with the hypothesis that steroid activity will generally correlate with reduced resistance and thus greater parasite intensities". Furthermore it was postulated that "steroid hormones, particularly glucocorticoids and sex steroids, provide a plausible mechanism for mediating trade-offs ... between the immune system and other components of life history".

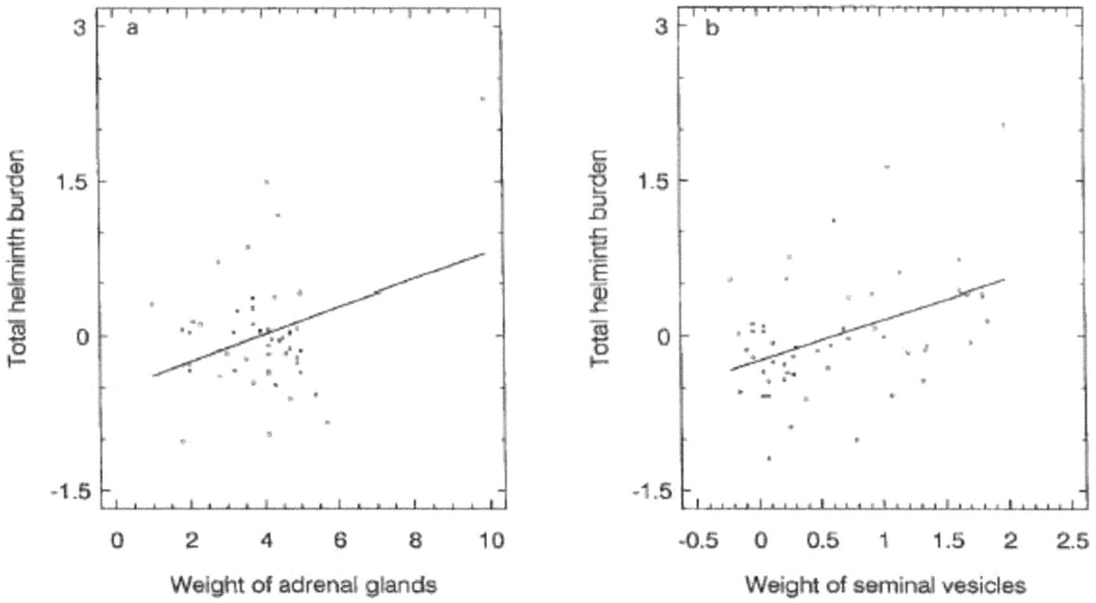

Fig. 5. Component effect (departure from mean value of dependent variable) from multiple regression analysis for the relationship between abundance ($\log_{10}$number of individuals + 1) of helminth infection and mass (mg) of (a) adrenal glands and (b) ($\log_{10}$) seminal vesicles in male *C. glareolus* (after Barnard et al. 2002, reprinted with permission from CAB International)

Body mass change, metabolizable energy intake and apparent dry matter digestibility were determined in *G. dasyurus* when parasitized by the flea *X. ramesis* (Khokhlova et al. 2002) and offered millet seeds that satisfied 85%, 100% and 115% of maintenance energy requirements. Fifty fleas consumed, in total, 3.68 mg (ranging from 3.05 to 4.54 mg) of blood per 3 h meal. The 0.074 mg of blood consumed per flea amounted to 34.3% of its unfed body mass, but the total blood lost by the host was only 0.17% of its blood volume. Apparent dry matter digestibility and metabolizable energy intake (as a percentage of gross energy intake) did not differ between parasitized and unparasitized gerbils and averaged 92.7% and 93.0%, respectively. But, body mass change in *G. dasyurus* was strongly affected by parasitism. Unparasitized gerbils maintained steady state body mass when offered food at maintenance (100% of energy requirements) or above maintenance (115% of energy requirements) energy levels and lost body mass when offered food at submaintenance (85% of energy requirements) energy levels. In contrast, parasitized gerbils maintained steady state body mass only when offered food at the highest level (115% of energy requirements) and lost body mass when offered food at 100% and 85% levels of energy requirements. However, the efficiency of utilization of energy in the parasitized and unparasitized gerbils was similar, which suggests that the utilization of energy to combat parasitism and for maintenance is similar (Kam and Degen 1997).

5 Reproduction and energy allocation

Many studies have shown that parasitism increases during reproduction and also that an experimentally increased effort during reproduction increases parasitism (Norris et al. 1994; Ots and Horak 1996; Allander 1998; Nordling et al. 1998; Wiehn et al. 1999). According to life-history theory, individuals are expected to maximize their fitness by adjusting their investment in current reproduction to a level where the sum of the fitness contributions from future and current reproduction is maximized (Stearns 1976).

Reproduction, in particular lactation, is the most costly activity in terms of energy expenditure in female mammals. Females attempt to maximize reproductive success by producing a large number of offspring and often use their own energy reserves to achieve this (Degen et al. 2002). However, energy demands of the reproducing female may not be met if other energy demanding activities are occurring concomitantly (Kam and Degen 1993).

In reproducing mammals, the cost of parasitism can be paid by the offspring (reduced litter mass and/or growth rate) (Arendt 1985; Richner et al. 1993; Hollmen et al. 1999) and/or the parents (reduced mass and/or survival and/or future reproductive performance) (Brown et al. 1995; Richner and Tripet 1999; Fitze et al. 2004), and/or by increased levels of resource acquisition by parents (Tripet and Richner 1997; Thomas and Shutler 2001; Tripet et al. 2002). While many studies on parasitism and reproductive success have been done on birds and fishes (Møller 1990; Tompkins et al. 1996; Allander 1998; Kopachena et al. 2000), little is known about the interactions between parasites and reproductive success in mammals (Lehmann 1993; Murray et al. 1998; Neuhaus 2003).

A model describing energy allocation of the total energy expenditure (TEE) of the reproducing female during reproduction could be presented as:

$$\text{TEE} = \text{MEI} - (\text{ER} + \text{energy production}), \quad (3)$$

where energy production could be related to pregnancy (foetal energy) or lactation (milk energy). Parasitism, which can cause an increase in energy expenditure and/or a decrease in metabolizable energy intake (MEI) can lead to a decrease in energy production (foetus or milk). Alternatively, the female could increase mobilization of energy reserves (ER) to compensate for the increased energy demands.

If it is assumed that parasitism can have either a negative or negligible effect on the reproductive success of a host, than nine possible scenarios (Table 1) are envisioned for the parasitized reproducing females (P) compared to the unparasitized female (N).

In these scenarios, three dependent variables are examined: metabolizable energy intake (MEI); body mass changes of the dam, (ΔM_b); and growth rate of the litter (Δm_b). To compensate for energy used in combating parasitism, infested females can increase energy consumption and/or increase the mobilization of body energy reserves (body mass loss) to the extent that they can either fully or partially support growth of their litter as compared with controls. Differences in growth rate (Δm_b) of the litter between treatment and control are indicative of differences in milk energy production. The effect of parasitism on the reproducing female can, therefore, result in one of four responses: 1) no effect of parasitism; 2) no compensation by the reproducing female; 3) partial compensation; 4) and full compensation. I did not consider a situation of improved reproductive success due to parasitism.

Table 1. Nine possible scenarios on the effect of parasitism on the reproducing female

Scenario #:	Dependent variables			Effect of parasites
	MEI	ΔM_b	Δm_b of litter	
1	P=N	P=N	P=N	No effect
2	P=N	P=N	P<N	No compensation
3	P=N	P<N	P=N	Full compensation
4	P=N	P<N	P<N	Partial compensation
5	P<N	P=N	P<N	No compensation
6	P<N	P<N	P<N	Partial compensation
7	P<N	P<N	P=N	Full compensation
8	P>N	P=N or P<N	P<N	Partial compensation
9	P>N	P=N or P<N	P=N	Full compensation

In scenarios one to four (Table 1), there is no difference in energy intake between parasitized and unparasitized reproducing females. In the first, there is no effect on either the mass of the female or the litter and, therefore, no effect due to parasitism. In the second, there is no effect on the female, but the litter mass is reduced and, therefore, there is no compensation from the female for the added energy costs of parasitism. In the third, there is an effect on the female but not on the litter and, therefore, the female is compensating fully for the costs of parasitism by catabolizing body tissue. In the fourth, there is an effect on both the female and the litter and, therefore, the female is compensating partially for the added energy costs of parasitism.

In scenarios five to seven (Table 1), energy intake of parasitized reproducing females is less than unparasitized females. In the fifth, there is no effect on the dam but an effect on the offspring and, therefore, there is no compensation. In the sixth, there is an effect on both the dam and the offspring and, therefore, there is partial compensation, whereas in the seventh, there is an effect on the dam but not on the offspring and, therefore, there is full compensation.

In scenarios eight and nine (Table 1), the parasitized females consume more energy than the unparasitized females. In the eighth, there is (or is not) an effect on the dam and an effect on the offspring and, therefore, there is partial compensation, whereas in ninth, there is (or is not) an effect on the dam but no effect on the offspring and, therefore, there is full compensation.

5.1 Effects of macroparasites on reproductive success of free-living mammals

The effect of ectoparasites, mainly fleas, on reproductive success was studied in free-living Columbian ground squirrels (*Spermophilus columbianus*), a social rodent living in colonies, in Alberta, Canada (Neuhaus 2003). These squirrels emerge from hibernation in late April when females breed. Gestation is approximately 24 days and young remain in the burrow for about 27 days. By mid to late August, all animals hibernate. All individuals in the study area carried some fleas and prevalence was high. Animals were trapped before mating and ectoparasites were removed from half the females using a commercially available flea and tick powder. Treatment was then applied weekly until the end of lactation. All females mated with between three to five males, and there was no difference due to treatment. Body mass did not differ significantly between treated and untreated females when emerging from spring hibernation (458 g versus 474 g, respectively) or just after parturition (541 g versus 557 g, respectively), but was higher in treated than untreated females at weaning (560 g versus 518 g, respectively). During lactation, that is, from parturition to juvenile emergence, treated females gained 19 g while untreated females lost 40 g. Furthermore, treated females gained more mass than untreated females from emergence in spring to the emergence of young, both when litter mass was and was not included (Fig. 6).

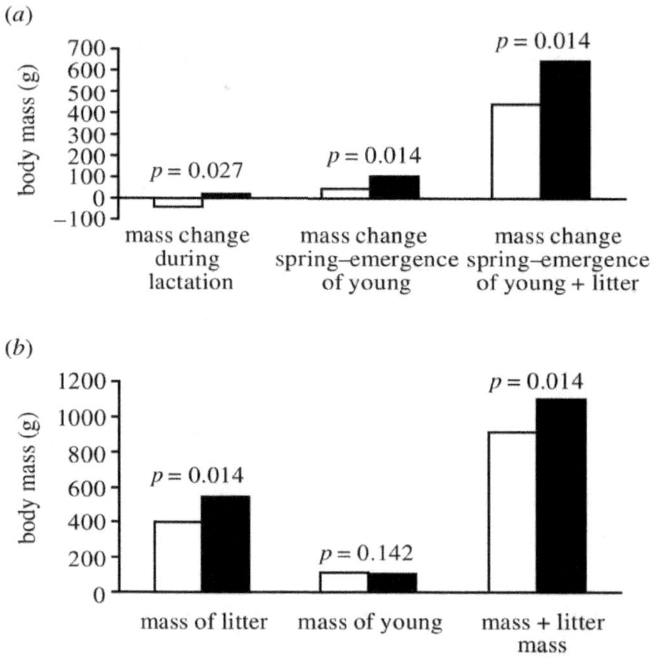

Fig. 6. Comparisons of mass gain or mass loss between Columbian ground squirrels (*S. columbianus*) treated against ectoparasites (filled bars) and untreated (open bars). (a) Average female mass change between birth and emergence of young, mass change of females from emergence in spring to the emergence of their young and total mass produced from adult emergence in spring to the emergence of the young, including litter mass. (b) Comparison of average litter masses, average mass per young and of the female mass and litter mass combined, from left to right (after Neuhaus 2003; reprinted with permission from the Royal Society of London)

Treated females weaned 5.25 young whereas untreated females weaned 3.6 young per litter and the number of young surviving to yearling age was higher for treated than untreated females (3.5 versus 2.0 young per litter, respectively). There was no difference between treated and untreated groups in body mass per young at birth, but because of the larger litters in treated than untreated females, mass of litter was higher in treated females. Neuhaus (2003) concluded that “it seems obvious that ectoparasites in my study area have a profound negative effect on individual reproductive success and on body mass”. In addition, he concluded that “the constant presence of ectoparasites in these ground squirrels leads to a constantly lower reproductive success and survival of females and their offspring and probably affects the whole population, since all animals are infected to a certain degree”. Untreated females lost body mass whereas treated females gained body mass and total litter mass was greater in the treated than untreated females. Consequently, scenarios 4, 6 and 8 could possibly describe these results. Energy intake of the females would be required to refine the prediction of the right scenario.

The effect of endoparasites (nematodes, mainly *Trichostrongylus retortaeformis* and *Graphicum strigosum*) on the fecundity of free-living mountain hares (*Lepus timidus*; adult body mass = 2-4 kg) in the Highlands of Scotland (Newey and Thirgood 2004) was examined. Three groups of female mountain hares were studied: 1) treated with an anti-helminthic drug (subcutaneous injection of Ivermectin); 2) handled but untreated and 3) not handled and untreated. Following treatment, the treated group had less *Trichostrongylus retortaeformis* nematodes than the other two groups but, there was no difference among groups in survival. In addition, there was no difference in body condition, measured as kidney fat index, among the three groups.

However, fecundity, measured by the number of corpora albicantia in both ovaries, was significantly higher in the treated group than in the other two groups, showing an improved fecundity in the treated group (Fig. 7). Scenarios 2, 5, and 8 could possibly describe these results as there is no effect on the reproducing female but there are reduced offspring. Again, energy intake of the females is not available to choose the exact scenario.

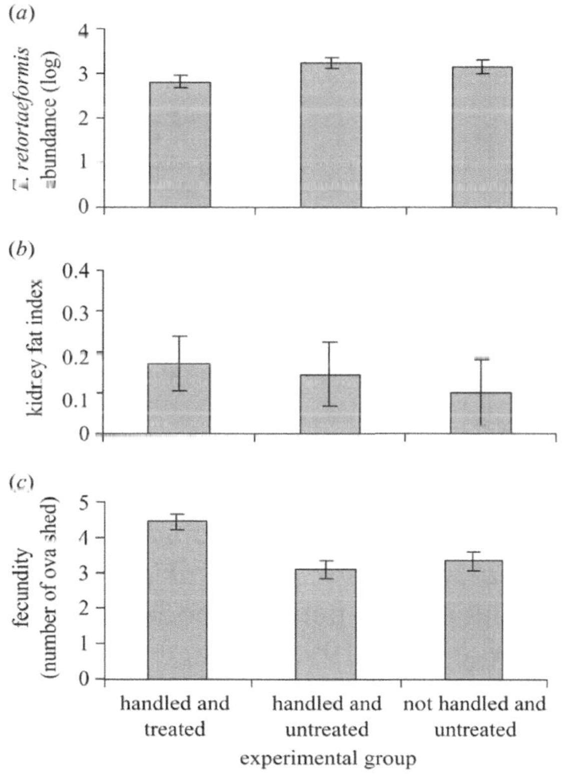

Fig. 7. The effect of experimental treatment of female mountain hares (*L. timidus*) with Ivermectin on (a) *T. retortaeformis* abundance (logged), (b) kidney fat index and (c) fecundity. Whiskers are standard errors (after Newey and Thirgood 2004; reprinted with permission from the Royal Society of London)

The effect of nematodes and nutrition on seasonal body condition and reproduction was studied in snowshoe hare (*Lepus americanus*; adult body mass = 1200 g) in south-central Manitoba (Murray, Keith and Cary 1998). These hares harbour three nematodes (*Obeliscoides cuniculi*, *Nematodirus triangularis* and *Trichuris leporis*) that infect the digestive tract and have direct life cycles and two nematodes (*Dirofilaria scapiceps* and *Protostrongylus boughtoni*) that infect the ankles and lungs, respectively, and have indirect life cycles. Nematodes in both males and females were reduced by a subcutaneous injection of an anti-helminthic drug (Ivermectin) and nutrition was manipulated by supplementation of commercial rabbit pellets. Also, the hares were divided into low, medium and high density groups. The treated hares had significantly less nematodes for four of the five nematode species and the fifth nematode species (*T. leporis*) was 35% lower in the treated group, but this difference was not significant. Body mass of non-treated male and female hares was significantly negatively correlated to nematode abundance in May to June only (Fig. 8), indicating that the nematodes had a negative effect on the hares at that time.

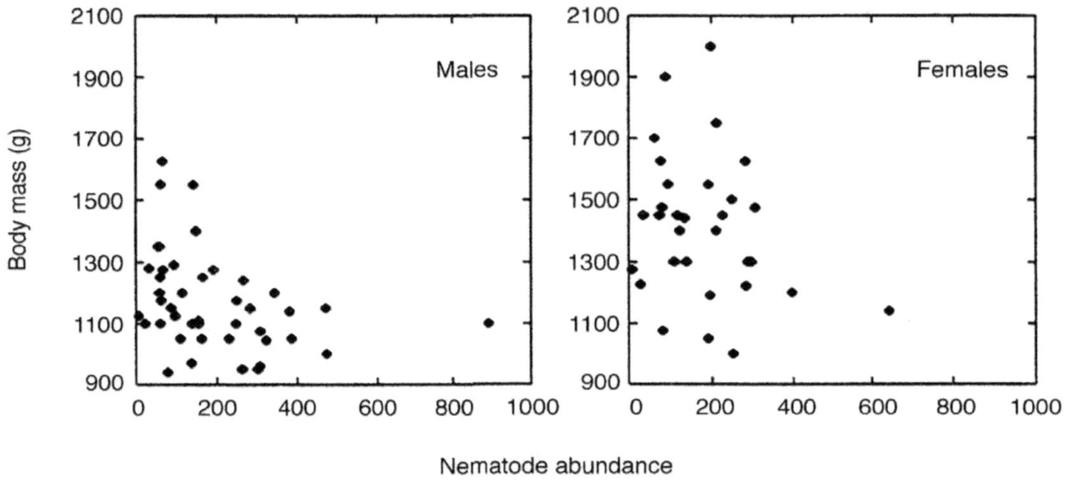

Fig. 8. Relationship between nematode abundance and body mass of 75 adult snowshoe hares (parasite-normal and food-normal controls) collected during May-June (1992-1993) in Manitoba. Nematodes include *Obeliscoides cuniculi*, *Nematodirus triangularis*, *Trichuris leporis*, *Dirofilaria scapiceps*, and *Prostrongylus boughtoni*. The negative relationship between body mass and nematode abundance was significant (after Murray et al 1998, printed with permission from the Ecological Society of America)

However, overall, neither parasite reduction nor food supplementation affected body mass of the hares although food x density interaction was significant from November to April. Food supplementation increased marrow fat level of males by 16%; however parasite reduction and parasite reduction x food supplementation interaction had no effect. In females, neither food, parasitism nor their interaction had an effect on marrow fat level.

Reproductive status of males was determined by testes palpation mainly during May-August (95 of 142 examinations). The proportion of males with descended testes was similar among food treatments (73% and 63% for supplemented and non-supplemented, respectively) and parasite treatments and the interaction between them was not significant. Females had three to four litters during the breeding season that lasted between April and August and were palpated for embryos from March to August. Food supplementation did not increase consistently the proportion of pregnant females but food supplementation x hare density interaction was significant, being higher in the moderate and high density groups. In addition, food supplementation resulted in a more rapid onset of pregnancy: in March – April, 47% of supplemented females had palpable embryos while in the non-supplemented group, it was only 13%. Neither parasite treatment nor parasite x food interaction had an effect on pregnancy rates. Furthermore, examination of 50 euthanized females in May – June showed that food supplementation increased the number of embryos, but the number of corpora lutea and conception dates were similar between treatment groups. Neither parasite reduction nor its interaction with food had an effect on either number of embryos, number of corpora lutea or conception dates. The authors concluded that "the nonsignificant effect of nematodes on hare production may indicate the small overall cost of such parasitism to hares". They also concluded that the effects of parasitism might be evident with severe food restriction, which did not occur during the study. It seems as if there was no effect on the reproducing female due to parasitism in this study and if I assume that there was no effect on the litter mass, then either scenario 1 or 9 could possibly describe the results.

5.2 Effects of macroparasites on reproductive success of laboratory mammals

Kristan (2004) examined the effects of *H. polygyrus* on reproduction and on offspring growth in wild-derived *M. musculus* in which young were weaned at 20 days. Half the pups of the parasitized and unparasitized mothers were infected with nematodes and growth rate of the pups were followed to 60 days. Parasitized females had 45% larger litter sizes at birth (8.4 versus 5.8 pups) and 51% larger litters at weaning than unparasitized mothers (8.3 versus 5.5 pups) (Fig. 9). Pup loss between birth and weaning at 20 days did not differ between groups and there was no effect due to parasites on average time to first litter, on inter-litter intervals and on the success of weaning at least one pup per litter (93% and 89% for parasitized and unparasitized females, respectively).

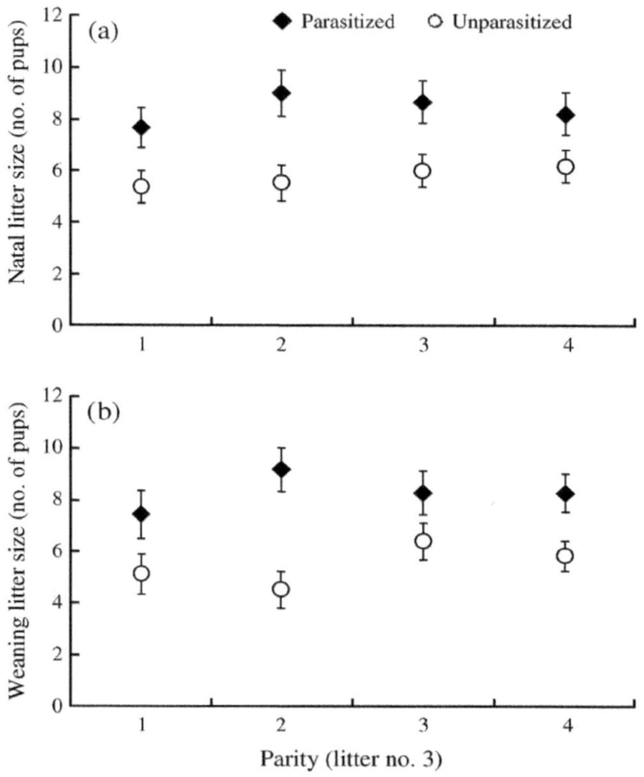

Fig. 9. Natal litter size (a) and weaning size (b) of parasitized and unparasitized wild-derived house mice for parity ranging from 1 to 4. Whiskers are standard errors. Sample size for natal litter size is parasitized =9, unparasitized =14 and for weaning litter size is parasitized =7, unparasitized =10 (after Kristan 2004, reprinted with permission from Blackwell Publishing)

Pups at birth from parasitized mothers were larger than those from unparasitized mothers (by 1.2% after accounting for effects of litter size; 9.26 versus 9.05 g; $P = 0.03$); however, pup growth to weaning was not affected by mother parasite infection. There was no difference in body mass or lean mass at 60 days between parasitized and unparasitized pups, but parasitized pups had 4% more fat than unparasitized pups (1.7 vs 1.6 g; P=0.04). On a dry matter basis, parasitized pups had 5% greater liver mass, 40% greater small intestine mass but 8% smaller stomach mass and similar spleen mass than unparasitized pups. Interestingly, parasitized females produced more and larger offspring than unparasitized females in four consecutive litters; that is, improved reproductive success. In the scenarios that were presented, it was assumed that there was no effect or a negative effect on reproductive success and, therefore, results from this study do not fit any of the possible scenarios envisioned.

A similar experiment by Kristan (2002) was done on laboratory mice in which the pups were also weaned at 20 days. Pups from parasitized and unparasitized mothers received either low intensity nematode infection, high intensity infection or no infection (as controls). Body mass growth, maximum rate of body mass gain, tail length and foot growth of the pups were not affected by maternal parasite condition; however, pups from parasitized mothers were 4% heavier than from unparasitized mothers (11.10 versus 10.93 g, respectively) when pups were growing at their fastest. Parasitized pups grew 5% faster (1.041 versus 0.988 g/day, respectively) and reached their maximum growth rate (20.2 versus 20.6 days, respectively) 0.5 days earlier than unparasitized pups. Nonetheless, at 60 days of age, there was no effect on body mass or on body energy of the pups due to maternal or pup parasite treatment. However, in general, high intensity parasitized pups had larger livers, spleens and small intestines, but smaller kidneys than low intensity parasitized pups and unparasitized pups. This may suggest a "change in energy allocation to organs during growth or may simply reflect systematic morphological changes owing to parasite pathology". Here, again there appears to be a slight positive effect on reproductive success due to parasitism in that the offspring from parasitized dams tended to be a larger than offspring from unparasitized dams and there was no effect on the parasitized female.

No effect on current reproduction was found in the laboratory rat (*Rattus rattus*) parasitized by the intestinal tapeworm, *Hymenolepis diminuta* (Willis and Poulin 1999). Rats were infected with eight cysticercoids (larval stage, which grew into adult worms in 3-5 weeks) and were then mated. Food pellets and water were available *ad libitum*, but were not measured. Litter size and mass did not differ between parasitized and unparisitized mothers either at birth or at weaning (21 days). In addition, body mass change of parasitized and unparisitized mothers were similar during lactation, both increasing in body mass. In this study, either scenario 1 or 9 could satisfy the findings.

Willis and Poulin (1999) concluded that "It would appear that parasite increases the relative value of the current litter and current levels of maternal investment, possibly because it reduces future reproductive success" and it is possible that this holds true for the studies mentioned above. It would be interesting, then to determine the source of the added energy required for the current litter? However, this theory should be taken with caution as there was an increased litter production in parasitized wild-deprived house mice in four consecutive litters when compared to unparasitized females, and not only in the first litter (Kristan 2004).

6 Concluding remarks

Seeing the importance of parasites on the life history of their hosts, it is surprising to find the paucity of studies on this subject. In this chapter I attempted to divide the studies into different subjects, however, results among and within host species were so variable, that it was difficult to generalize responses to parasitism. For example, responses ranged from improvement in reproduction to relatively high mortality of the host, from a decrease in apparent dry matter digestibility to no effect on apparent dry matter digestibility, from an increase in metabolic rate to no effect on metabolic rate and from an increase in size in some body organs to a decrease in size in the same body organs. Furthermore, reproductive success in parasitized free-living and parasitized laboratory maintained small mammals appear to be quite different. It is evident that the impact of parasites on life-history traits in mammals is still poorly understood.

References

Allander K (1998) The effects of an ectoparasite on reproductive success in the great tit: A 3-year experimental study. Can J Zool 76:19–25

Allison AC (2000) Immunosuppressive drugs: The first 50 years and a glance forward. Immunopharmacol 47:63–83

Alves MAS (1997) Effects of ectoparasites on the sand martin *Riparia riparia* nestlings. Ibis 139:494–496

Arendt WJ (1985) *Philornis* ectoparasitism of pearly-eyed thrashers. 1. Impact on growth and development of nestlings. Auk 102:270–280

Arnold W, Lichtenstein AV (1993) Ectoparasite loads decrease the fitness of alpine marmots (*Marmota marmota*) but are not a cost of sociality. Behav Ecol 4:36–39

Barnard CJ, Behnke JM, Gage AR, Brown H, Smithurst PR (1998) The role of parasite-induced immunodepression, rank and social environment in the modulation of behaviour and hormone concentration in male laboratory mice (*Mus musculus*). Proc R Soc Lond B 265:693–701

Barnard CJ, Behnke JM, Bajer A, Bray D, Race T, Frake K, Osmond J, Dinmore J, Sinski E (2002) Local variation in endoparasite intensities of bank voles (*Clethrionomys glaroelus*) from ecologically similar sites: Morphometric and endocrine correlates. J Helminthol 76:103–112

Behnke JM, Bajer A, Sinski E, Wakelin D (2001) Interactions involving intestinal nematodes of rodents: Experimental and field studies. Parasitology 122:S39–S49

Bouslama Z, Chabi Y, Lambrechts MM (2001) Chicks resist high parasite intensities in an Algerian population of blue tits. Ecoscience 8:320–324

Brown CR, Brown MB (1986) Ectoparasitism as a cost of coloniality in cliff swallows (*Hirundo pyrrhonota*). Ecology 67:1206–1218

Brown CR, Brown MB, Rannala B (1995) Ectoparasites reduce long-term survival of their avian host. Proc R Soc Lond B 262:313–319

Butler JM, Roper TJ (1996) Ectoparasites and set use in European badgers. Anim Behav 52:621–629

Candolin U, Voigt HR (2001) No effect of a parasite on reproduction in stickleback males: A laboratory artefact? Parasitology 122:457–464

Chekchak T, Chapuis JL, Pisanu B, Bousses P (2000) Introduction of the rabbit flea, *Spilopsyllus cuniculi* (Dale), to a subantarctic island (Kerguelen Archipelago) and its assessment as a vector of myxomatosis. Wildl Res 27:91–101

Clayton DH, Moore J (1997) Introduction. In: Clayton DH, Moore J (eds) Host-parasite evolution: General principles and avian models. Oxford Univ Press, Oxford, pp 1–6

Crompton DW, Arnold SE, Walters DE, Whitfield PJ (1985) Food intake and body weight changes in mice infected with metacestodes of *Taenia crassiceps*. Parasitology 90:449–456

Degen AA (1997) Ecophysiology of small desert mammals. SpringerVerlag, Berlin Heidelberg New York

Degen AA, Kam M, Khokhlova IS, Krasnov BR, Barraclough T (1998) Average daily metabolic rate of rodents: habitat and dietary comparisons. Funct Ecol 12:63–73

Degen AA, Khokhlova IS, Kam M, Snider I (2002) Energy requirements during reproduction in female common spiny mice (*Acomys cahirinus*). J Mammal 83:645–651

De Lope, F, Gonzalez G, Perez JJ, Moller AP (1993) Increased detrimental effects of ectoparasites on their bird hosts during adverse environmental conditions. Oecologia 95:234–240

Derting TL, Compton S (2003) Immune response, not immune maintenance, is energetically costly in wild white-footed mice (*Peromyscus leucopus*). Physiol Biochem Zool 76:744–752

Dzieciolowski RM, Clarke CMH (1990) Physical condition and ectoparasite infestation of feral pigs in New Zealand. Acta Theriol 35:299–309

Fauth PT, Krementz DG, Hines JE (1991) Ectoparasitism and the role of green nesting material in the European starling. Oecologia 88:22–29

Fitze PS, Tschirren B, Richner H (2004) Life history and fitness consequences of ectoparasites. J Anim Ecol 73:216–226

Forchhammer MC, Asferg T (2000) Invading parasites cause a structural shift in red fox dynamics. Proc R Soc Lond B 267:779–786

Hasselgren PO, Fischer JE (1998) Sepsis: Stimulation of energy-dependent protein breakdown resulting in protein loss in skeletal muscle. World J Surg 22:203–208

Hollmen T, Lehtonen JT, Sankari S, Sovery T, Hario M (1999) An experimental study on the effects of polymorphiasis in common eider ducklings. J Wildl Dis 35:466–473

Kam M, Degen AA (1993) Energetics of lactation and growth in the fat sand rat, *Psammomys obesus* – new perspectives of resource partitioning and the effect of litter size. J Theor Biol 162:353–369

Kam M, Degen AA (1997) Energy requirements and the efficiency of utilization of metabolizable energy in free-living animals: Evaluation of existing theories and generation of a new model. J Theor Biol 184:101–104

Kavaliers M, Colwell DD, Choleris E (1998) Parasitized female mice display reduced aversive responses to the odours of infected males. Proc R Soc Lond B 265:1111–1118

Khokhlova IS, Krasnov BR, Kam M, Burdelova NI, Degen AA (2002) Energy cost of ectoparasitism: The flea *Xenopsylla ramesis* on the desert gerbil *Gerbillus dasyurus*. J Zool Lond 258:349–354

Khokhlova IS, Spinu M, Krasnov BR, Degen AA (2004a) Immune responses to fleas in two rodent species differing in natural prevalence of infestation and diversity of flea assemblages. Parasitol Res 94:304–311

Khokhlova IS, Spinu M, Krasnov BR, Degen AA (2004b) Immune response to fleas in a wild desert rodent: Effect of parasite species, parasite burden, sex of host and host parasitological experience. J Exp Biol 207:2725–2733

Kim KC (1985) Evolutionary relationships of parasitic arthropods and mammals. In: Kim KC (ed) Coevolution of parasitic arthropods and mammals. John Wiley, New York, pp 3–82

Krasnov BR, Khokhlova IS, Fielden LJ, Burdelova NV (2001) The effect of temperature and humidity on the survival of pre-imaginal stages of two flea species (Siphonaptera: Pulicidae). J Med Entomol 38:629–637

Krasnov BR, Khokhlova IS, Burdelova NV, Mirzoyan NS, Degen AA (2004) Fitness consequences of host selection in ectoparasites: Testing reproductive patterns predicted by isodar theory in fleas parasitizing rodents. J Anim Ecol 73:815–820

Krasnov BR, Khokhlova IS, Arakelyan MS, Degen AA (2005) Is a starving host tastier? Reproduction in fleas parasitizing food-limited rodents. Funct Ecol 19:625–631

Kristan DM (2002a) Effects of intestinal nematodes during lactation: Consequences for host morphology, physiology and offspring mass. J Exp Biol 205:3955–3965

Kristan DM (2002b) Maternal and direct effects of intestinal nematode *Heligmosomoides polygyrus* on offspring growth and susceptibility to infection. J Exp Biol 205:3967–3977

Kristan DM (2003) Physiological and morphological responses to simultaneous cold exposure and parasite infection by wild-derived house mice. Funct Ecol 17:464–471

Kristan DM (2004) Intestinal nematode infection affects host life history and offspring susceptibility to parasitism. J Anim Ecol 73:227–238

Kristan DM, Hammond KA (2000) Combined effects of cold exposure and sublethal intestinal parasites on host morphology and physiology. J Exp Biol 203:3495–3504

Kristan DM, Hammond KA (2001) Parasite infection and caloric restriction induce physiological and morphological plasticity. Am J Physiol 281:R502–R510

Kristan DM, Hammond KA (2004) Aerobic performance of wild-derived house mice does not change with cold exposure or intestinal parasite infection. Physiol Biochem Zool 77:440–449

Lehmann T (1993) Ectoparasites: Direct impact on host fitness. Parasitol Today 9:8–13

Lochmiller RL, Dabbert CB (1993) Immunocompetence, environmental stress, and the regulation of animal populations. Trends Comp Biochem Physiol 1:823–855

Lochmiller RL, Deerenberg C (2000) Trade-offs in evolutionary immunology: Just what is the cost of immunity? Oikos 88:87–98

Marshall AG (1981) The ecology of ectoparasite insects. Acad Press, London

Meagher S (1998) Physiological responses of deer mice (*Peromyscus maniculatus*) to infection with *Capillaria hepatica* (Nematoda). J Parasitol 84:1112–1118

Meagher S, Dudek SN (2002) Effects of *Trichinella spiralis* on survival, total mass, and organ mass of oldfield mice (*Peromyscus polionotus*). Parasitology 88:833–838

Meagher S, O'Connor T (2001) Population variation in the metabolic response of deer mice to infection with *Capillaria hepatica* (Nematoda). Can J Zool 79:554–561

Møller AP (1990) Effects of parasitism by a haematophagous mite on reproduction in the barn swallow. Ecology 71:2345–2357

Munger JC, Karasov WH (1989) Sublethal parasites and host energy budgets: Tapeworm infection in white-footed mice. Ecology 70:904–921

Munger JC, Karasov WH (1991) Sublethal parasites in white-footed mice – impact on survival and reproduction. Can J Zool 69:398–404

Munger JC, Karasov WH (1994) Costs of bot fly infection in white-footed mice – energy and mass-flow. Can J Zool 72:166–173

Munger JC, Slichter TA (1995) Whipworm (*Trichuris dipodomys*) infection in kangaroo rats (*Dipodomys* spp.) – effects on digestive efficiency. Great Basin Nat 55:74–77

Murray DL, Keith LB, Cary JR (1998) Do parasitism and nutritional status interact to affect production in snowshoe hares? Ecology 79:1209–1222

Nelson WA, Bell JF, Clifford CM, Keirans JE (1977) Interactions of ectoparasites and their hosts. J Med Entomol 13:389–428

Neuhaus P (2003) Parasite removal and its impact on litter size and body condition in Columbian ground squirrels (*Spermophilus columbianus*). Proc R Soc Lond B 270:S213–S215

Newey S, Thirgood S (2004) Parasite-mediated reduction in fecundity of mountain hares. Proc R Soc Lond B 271:S413–S415

Newey S, Thirgood SJ, Hudson, PJ (2004) Do parasite burdens in spring influence condition and fecundity of female mountain hares *Lepus timidus*? Wildl Biol 10:171–176

Nordling D, Andersson M, Zohari S, Gustafsson L (1998) Reproductive effort reduces specific immune response and parasite resistance. Proc R Soc Lond B 265:1291–1298

Norris K, Anwar M, Read A (1994) Reproductive effort influences the prevalence of haematozoan parasites in great tits. J Anim Ecol 63:601–610

Ots I, Horak P (1996) Great tits *Parus major* trade health for reproduction. Proc R Soc Lond B 263:1443–1447

Pacejka AJ, Gratton CM, Thompson CF (1998) Do potentially virulent mites affect house wren (*Troglodytes aedon*) reproductive success? Ecology 79:1797–1806

Poulin R (2000) Manipulation of host behaviour by parasites: A weakening paradigm? Proc R Soc Lond B 267:787–792

Richner H, Tripet F (1999) Ectoparasitism and the trade-off between current and future reproduction. Oikos 86:535–538

Richner H, Oppliger A, Christe P (1993) Effect of an ectoparasite on reproduction in great tits. J Anim Ecol 62:703–710

Sheldon BC, Verhulst S (1996) Ecological immunology: Costly parasite defences and trade offs in evolutionary ecology. Trends Ecol Evol 11:317–321

Simon A, Thomas D, Blondel J, Perret P, Lambrechts MM (2004) Physiological ecology of Mediterranean blue tits (*Parus caeruleus* L.): Effects of ectoparasites (*Protocalliphora* spp.) and food abundance on metabolic capacity of nestlings. Physiol Biochem Zool 77:492–501

Stearns SC (1976) Life history tactics: A review of the ideas. Q Rev Biol 51:3–47

ter Hofstede HM, Fenton MB (2005) Relationships between roost preferences, ectoparasite density and grooming behaviour of neotropical bats. J Zool Lond 266:333–340

Thomas K, Shutler D (2001) Ectoparasites, nestling growth, parental feeding rates, and begging intensity of tree swallows. Can J Zool 79:346–353

Tompkins DM, Jones T, Clayton DH (1996) Effect of vertically transmitted ectoparasites on the reproductive success of swifts (*Apus apus*). Funct Ecol 10:733–740

Tompkins DM, Begon M (1999) Parasites can regulate wildlife populations. Parasitol Today 15:311–313

Tripet F, Richner H (1997) Host responses to ectoparasites: Food compensation by parent blue tits. Oikos 78:557–561

Tripet F, Glaser M, Richner H (2002) Behavioural responses to ectoparasites: time-budget adjustments and what matters to blue tits *Parus caeruleus* infested by fleas. Ibis 144:461–469

Wedekind C (1992) Detailed information about parasites revealed by sexual ornamentation. Proc R Soc Lond B 247:169–174

Wiehn J, Korpimäki E, Pen I (1999) Haematozoan infections in the Eurasian kestrel: Effects of fluctuating food supply and experimental manipulation of parental effort. Oikos 84:87–98

Willis C, Poulin R (1999) Effects of the tapeworm *Hymenolepis diminuta* on maternal investments in rats. Can J Zool 77:1001–1005

Zuk M, Stoehr AM (2002) Immune defense and host life history. Amer Nat 160:S9–S22

20 Immunogenetics of micromammal-macroparasite interactions

Nathalie Charbonnel, Joelle Goüy de Bellocq and Serge Morand

1 Introductory remarks

During the last 20 years, the concepts of population biology and evolutionary ecology have been applied to the studies of immune defences, in an attempt to explain the variation of immune responses observed across individuals and species. Among these approaches, immunogenetics focuses on immunity genes and on the variability of outcomes generated by genotype-genotype interactions between and within host and parasite species.

The recent explosion of genetic data emerging from sequencing and genomics has highlighted the importance of immunity because a considerable fraction of the genome appears to be dedicated to defence against parasites. In most mammals, 5% of genes are involved in this role (Trowsdale and Parham 2004). Such advances allow investigation of the mechanisms associated with immune-related coding and non-coding mRNA transcripts, and definition of the genetic basis of susceptibility to complex diseases. Identifying the genes underlying adaptation, how these genes change in response to natural selection, and what selective pressures drive their evolution are now at the core of evolutionary biology and immunogenetics, and will be discussed here.

2 Relevance of immunogenetics for small mammals and macroparasites interactions

The profound influence of the host genetic background on resistance against infections has been established in numerous animal studies, which mainly focused on human infections such as malaria, HIV and hepatitis (see review in Cooke and Hill 2001). The emphasis of these studies is understandable given the potential implication arising from the importance of the evolutionary mechanisms of pathogen resistance for medicine (Sorci et al. 1997). Because of their medical or veterinary importance, the genetic bases of resistance to helminths and ectoparasites have also been investigated, although to a lesser extent.

2.1 Evidence for the genetic variability of susceptibility and infectivity

2.1.1 Helminth infections

It is now well recognized that resistance to helminth infections in mammals is under genetic control, and that helminths themselves show genetic variation in their infectivity characteristics, i.e. intensity of infection or level of susceptibility to host responses (see review in Wakelin et al. 2002). Epidemiological and genetic studies in human and cattle populations have demonstrated that susceptibility to helminths frequently limited to some families, and that it differs between ethnic groups or breeds, suggesting the possible involvement of genetic factors (Quinnell 2003). On the other hand, epidemiological studies of a number of parasite species have shown that the intensity of infection (worm burden) is a host heritable character. This was first found by Wakelin (1975) in mice parasitized by *Trichuris muris*, where selection experiments enhanced host resistance after six generations from 70 % to 100 % (reviewed in Behnke et al. 2003). Heritability of worm burden varying from 0.21 to 0.44 has now been reported for many gastrointestinal helminths (Behnke et al. 2003; Quinnell 2003). The values of the heritabilities of helminth burden can provide a measure for the cumulative effect of all genes involved in helminth resistance, but cannot provide any information about the specific genes involved. Most of our current understanding of the regulation of the protective mechanisms mediating resistance to intestinal nematodes has come from laboratory studies that used rat and mouse models. These studies used well-defined inbred strains and genetic manipulation (transgenes, gene targeting; see review in Abel and Desein 1997) revealing the effects of host genes on anti-parasitic resistance. They concentrated on a limited number of intestinal nematode species such as *Ascaris* spp., *Nippostrongylus brasiliensis*, *Trichinella spiralis*, *Heligmosomoides polygyrus* and *Trichuris muris* (Grencis 1997), tissue-dwelling helminths *(Brugia malayi*) and various *Schistosoma*. Although genetic control of the worm burden is likely to be polygenic, recent studies have focussed on genes of major effect, and whether the same genes of major effect are involved in controlling different helminth species.

2.1.2 Ectoparasite infections

Although ectoparasites are abundant in natural communities, are vectors of many important diseases and negatively affect host condition and fitness (e.g., Lehmann 1992), very few studies have estimated the magnitude of heritable genetic variation underlying resistance against them. Most of what we know about genetic variation and the proximate basis of resistance against ectoparasites comes from studies on domesticated vertebrates. Natural resistance exists in ruminants and varies among breeds (Wakelin 1991). Resistance to ticks, estimated from the mortality of female ticks, has been found to be highly heritable and responsive to selection (Utech and Wharton 1982; Mattioli et al. 1993). Resistance against ticks in the Illawara Shorthorn herd increased from 89 to 99% after just three generations of selection. The genetic basis of tick resistance has further been supported in natural bird populations by cross-fostering experiments (Sorci et al. 1997) and parent-offspring regression of ectoparasite load (Boulinier et al. 1997). A quantitative genetic study (Kerr et al. 1994) reported the presence of a major gene for average tick number in a line of British cattle. The complete genome sequencing of the Southern cattle tick, *Boophilus microplus* and that of the deer tick, *Ixodes scapularis*, will offer new opportunities for immunogenetics in the future.

2.2 Importance of immunogenetics in natural populations

Many recent advances in molecular genetic techniques have led to an explosion of information on gene and genome structures and functions, which facilitate the understanding of the molecular basis of genetic resistance to infections of veterinary or medical importance. Consequently, the candidate species used to assess the role of and evolutionary processes acting on immunogenetics in natural populations are likely to be related to laboratory or domestic animals, such as wild rodents. Small mammals and macroparasites, including helminths and ectoparasites, are thus promising biological models to investigate immune defence genes from ecological and evolutionary perspectives.

2.2.1 Immunogenetics and the outcome of the interactions

There is increasing evidence that genetically determined variation in host capacity to express resistance to a given parasite plays a major role in determining the intensity of infection (Wakelin et al. 2002). Analysing the fitness consequences of these polymorphisms and the patterns of evolutionary change over time and space are the second step towards an understanding of host-parasite interactions. Thus, without a detailed knowledge of the genetics of both host and parasite, the outcome of the host-parasite interaction in natural populations, either in terms of population dynamics or evolutionary interactions, may be difficult to predict (Sorci et al. 1997). In particular, immunogenetics might provide essential information to disentangle the effects of genetic variation and environmental factors on the differences observed in the impact of macroparasites on individual hosts or populations.

2.2.2 Immunogenetics and epidemiology

One of the main goals of immunogenetics is to understand the link between genetics and immune related diseases (Geraghty 2002). There is no doubt that this has major implications for the design of new vaccines and immunotherapeutics in humans (Hill 2001), and that it provides a highly sensitive means of detecting anti-helminthic or anti-ectoparasitic resistance in livestock (Frisch et al. 2000). Another, but less developed, potential application of immunogenetics concerns the assessment of emergent or re-emergent disease risks in natural populations. Immunogenetics may provide key insights into epidemiology and transmission ecology, and may contribute to the identification of zoonotic potential in previously unidentified agents.

2.2.3 Immunogenetics and conservation

Immunogenetics may also be of major relevance in conservation biology. Infectious diseases have become increasingly important factors in wildlife conservation. Introduced diseases have caused the decline and, in some cases, extinction of several species (Altizer et al. 2003). Knowledge of the genetic variability of immune defence genes in natural populations may help to predict the evolutionary potential of wild host populations in response to native or novel parasites (see Christe et al. in this volume). Preserving networks of coevolving populations could maintain host-parasite interactions as an evolutionary process important to both biodiversity and conservation (e.g., Seddon and Baverstock 1999). Moreover, some families of immune defence genes are among the most polymorphic genes. Such highly variable loci can provide insights into the recognition of species, evolutionarily significant units (ESUs) and management units (MUs). Immune defence genes may also present the advantage of being reflective of adaptive differences between units (Hedrick 2001).

3 Immune defence genes and their patterns of variability in small mammals

3.1 How to find immune defence genes against macroparasites?

The results of studies of infectious diseases have indicated that much of the genetic component of resistance is specified by many minor susceptibility genes (sometimes termed polygenes) rather than by a few major loci (Hill 2001). The characterization of the immune response to macroparasites is thus a prerequisite to determine the genetic factors of the host involved in its resistance.

3.1.1 Gene identification

Due to the size of the host genome and the complexity of the immune responses developed against helminth and ectoparasite infections, identifying resistance genes is a difficult job and has rarely been attempted for diseases in populations of wild animals. There are two general approaches that may be adopted to identify genes whose polymorphism might affect immune response: genome scanning and candidate gene studies. In genome scanning, markers spaced across the genome, usually "neutral" microsatellites, are typed in the crossbred populations derived from two strains that exhibit different phenotypes regarding the experimental infection, and evidence for the correlation between each marker and the trait of interest is examined. In rodents, mapping of the regions of the genome controlling genetic variation in resistance and identification of genes underlying these loci is facilitated by the availability of a high resolution murine linkage map (the mouse database genome – http://www.genome.wi.mit.edu). Genome scans have thus identified quantitative trait loci responsible for the control of infections involving *Angiostrongylus costaricensis* (Ohno et al. 2002) and *H. polygyrus* (Iraqi et al. 2003; Menge et al. 2003) in laboratory mice. Alternatively, candidate genes may be identified through previous genetic studies or knowledge of the biological processes involved. These methods are usually powerful although the causative relationship between the gene/allele tested and the phenotype remains to be demonstrated (e.g., Quinnell 2003). Indeed, an association can result from linkage disequilibrium with a causative gene/allele at another locus.

3.1.2 Different consequences of gene polymorphism

The phenotypic differences related to the host immune response may vary genetically in several ways (Frank 2002). Polymorphism may affect specific recognition (variation allows the recognition of different spectra of parasite epitopes) and regulation of gene expression (variation leads to differences in the intensity of particular immune effectors deployed against parasite attack).

Nonsynonymous coding substitutions change the protein structure. In the case of a cytokine molecule or its receptor, this may induce drastic change or loss of function, such as reduction in high affinity binding. On the other hand, polymorphism in promoters, introns and surrounding regions will affect mRNA stability and gene expression, thus influencing the level of the produced proteins. Some substitutions in intronic regions may also affect mRNA splicing and thus lead to the production of splice variants (Geraghty 2002). Single nucleotide polymorphisms (SNP) and microsatellite repeat numbers are current measures used to analyse to analyse gene variation in immune defence (Ollier 2004).

The last feature of immune defence gene polymorphism concerns their polygeny. Some families of immune defence genes comprise multiple copies that appeared via duplication (Trowsdale and Parham 2004). The number of duplications, which may vary among individuals, is likely to provide polymorphism in the proteins coded and/or in their degree of expression. We now review different immune defence genes involved in small mammal resistance to helminths and ectoparasites, by considering their role and their importance in immunogenetics and the patterns of genetic diversity observed between species or populations in the field.

3.2 Polymorphism in specificity

3.2.1 MHC

The major histocompatibility complex (MHC) is a multigene family controlling immunological self/nonself recognition in vertebrates. Some of these genes encode cell surface glycoproteins that present peptides of foreign and self proteins to T-cells, thereby controlling all specific immune responses, both humoral and cell-mediated.

The genetic structure of the MHC has been characterized to various degrees of details in all vertebrate classes. In mammals, the MHC is generally divided into regions including class I, class II, class III, extended class I, and extended class II (review in Kelley et al. 2005). The number of genes and the presence of each region vary among species. The class I region encodes classical I and non-classical I molecules. The classical I molecules present peptide antigens to CD8 T-lymphocytes, through T-cell receptors, while the functions of non-classical class I genes are diverse. The class II region encodes molecules that present antigens to CD4 T-lymphocytes but also the TAP 1 and TAP 2 genes (see below). The class III region harbours a diverse array of structurally unrelated genes; among them several genes involved in innate immunity (e.g., complement components and cytokines).

Two micro-mammal MHC genetic structures are available: the H-2 system of *Mus musculus* and the TR1 complex of *Rattus norvegicus* (Günther and Walter 2001; Hurt et al. 2004). The H-2 system is located on chromosome 17 and only class II, III and incomplete class I regions have been determined on this chromosome (Hurt et al. 2004). In contrast to the H-2 system, the RT1 complex has been completely sequenced. It is a 3.8 Mb sequence located on chromosome 20 (Hurt et al. 2004).

Class I molecules are monomers with four extracellular domains (α 1–3 and β 2m). They present antigenic peptides (usually 9-mers) derived from the intracellular proteins and are expressed on the surface of all nucleated somatic cells (Wegner et al. 2004). In contrast, class II molecules have much more restricted expression pattern as they are primarily expressed on antigen-presenting cells such as dendritic cells, B-cells and macrophages. Class II molecules are heterodimers of two peptide chains, α and β , encoded by A and B genes that often occur in tandem arrangement. The diversity of potential class II molecules is a quadratic function of the number of class IIA and B tandems, because all combinations of α and β chains to form a functional class II molecule are possible (Wegner et al. 2004). Class II molecules present antigenic peptides (usually 9–12 mers) derived from exogenous and membrane proteins. The ability of both class I and II genes to face various pathogens is believed to be mainly related to sequence variation among the MHC alleles in the antigen binding site or ABS (also named peptide binding region or PBR). Both classes of MHC genes have only a limited number of anchor residues for peptide binding. ABS is coded by exons 2 and 3 of the class I loci and exon 2 of the class II loci. Thus, most research in micromammals focuses on the second exon of the class II DQA, DQB and DRB loci. Genes of the MHC are generally in strong linkage disequilibrium, thus the pattern observed for a locus should be a good indicator of the genetic variation in other MHC genes (Sommer 2005).

Associations between pathogens and certain MHC haplotypes or even single alleles, cover a wide range of host as well as pathogen taxa (see review in Apanius et al. 1997). However, taxonomic bias is shifted towards mammals, especially humans where a lot of studies have established correlative evidence for interactions between the MHC genotype and the course of in-fections with various pathogens. The role of MHC genes in the control of resistance to infection has been extensively studied in several experi-mental mouse-helminth systems. Associations between levels of resistance to helminths such as *H. polygyrus*, *T. muris*, *T. spiralis*, *Schistosoma mansoni* and H-2 genotypes have been observed using inbred and congenic strains of mice (Sher et al. 1984; Else et al. 1990; Keymer et al. 1990; Behnke and Wahid 1991; Wahid and Behnke 1993). Studies in wild animal populations are still very limited. However, recent studies on rodents have shown association between specific alleles of the class II DRB locus and susceptibility to helminths (Froeschke and Sommer 2005; Harf and Sommer 2005; Meyer-Lucht and Sommer 2005). In *Apodemus flavicollis*, specific DRB alleles showed significant associations to both susceptibility to and resistance against nematodes (Meyer-Lucht and Sommer 2005). Associations between susceptibility to ectoparasites and the MHC are rare. However, a study of the mole rat *Spalax ehrenbergi* demonstrated positive and significant correlation between average MHC class II heterozygosity and species richness of gamasid mites (Nevo and Beiles 1992).

Inter-species polymorphism

The evolution of the MHC alleles in murids has been studied mostly in regard to DQA and DRB loci. Allelic diversity at the MHC genes is thought to be maintained by balancing selection over long periods of time, even across multiple speciation events (Klein 1987). Thus, alleles of a given species are often more closely related to alleles of distantly related species than they are to other alleles from the same species ("trans-species" evolution). Trans-species sharing of the MHC alleles among genera has previously been supported by many studies of mammal and fish species (see review in Hedrick 2001). However, evidence of trans-species persistence of allelic lineages in murids is still ambiguous. Although the ancestral polymorphism of MHC class II genes has been supported by some authors (Figueroa et al. 1988; Edwards et al. 1997; Seddon and Baverstock 2000; Musolf et al. 2004; Bryja et al. 2006), others did not find support for trans-species persistence of the MHC alleles among different murid taxa (Pfau et al. 1999). Ambiguity in trans-species polymorphism might depend on the investigated locus with varying selection pressures acting on different loci (even if they are in strong linkage disequilibrium, as mentioned above).

Intra-species polymorphism

Genes of the MHC are the most polymorphic loci known for vertebrates (Geraghty et al. 2002; Trowsdale and Parham 2004) but it is still unclear how natural selection maintains this extreme polymorphism. The leading hypotheses to explain this polymorphism are (a) selection from parasites, either through overdominant selection or frequency-dependent selection, and (b) disassortative mating preferences (Apanius et al. 1997; Penn and Potts 1999).

During the last few years, many studies on MHC polymorphism in micro-mammal populations were carried out. The goals of these studies were for the most part to test the type of selection at the MHC loci (e.g., Harf and Sommer 2005; Meyer-Lucht and Sommer 2005), to study the allelic diversity within the framework of conservation biology (Smulders et al. 2003) and to understand phylogeographic patterns (Berggren et al. 2005). Among these studies, there is a strong bias in favour of rodents, whereas there are no studies of MHC polymorphism in lagomorphs and only one study on insectivores (Berggren et al. 2005). The three main loci chosen for these studies were DQA, DQB and DRB (Table 1). In general, the DQA locus appears less variable than the DQB or DRB loci.

Studies of wild populations have documented different aspects of MHC variability that were not apparent from previous studies on domesti-cated or laboratory animals. Almost all species exhibit extremely high le-vels of MHC polymorphism (Table 1). The action of balancing selection can be noted by examining the ratio of non-synonymous (dn) to synonymous (ds) substitutions in the ABS and non-ABS regions. In general, comparisons of dn and ds in ABS reveal a significantly higher rate of dn/ds (e.g., Musolf et al. 2004; Babik et al. 2005; Meyer-Lucht and Sommer 2005).

Table 1. Studies showing the number of MHC class II alleles in wild populations of micromammals. *locus duplicated.

Species	Individuals (populations)	Locus	Alleles (total)	References
Apodemus flavicollis	146 (7)	DRB exon 2	27	Meyer-Lucht and Sommer 2005
Apodemus sylvaticus	119 (9)	DRB exon 2	38	Musolf et al. 2004
	164 (8)	DQA exon 2	13	Goüy de Bellocq et al. 2005
Arvicola terrestris	96 (5)	DQA exon 2*	7	Bryja et al. 2006
Castor fiber	76 (7)	DRB exon 2	10	Babik et al. 2006
Clethrionomys glareolus	16 (2)	DQA exon 2*	7	Bryja et al. 2006
Cricetus cricetus	51 (4)	DQA exons 2,3	1	Smulders et al. 2003
		DRB exon2	14	
Cryptomys damarensis	13 (6)	DQα1exons 2,3	13, 5	Kundu and Faulkes 2004

Cryptomys hottentotus hottentotus	10 (3)	DQα1exons 2,3	8, 5	Kundu and Faulkes 2004
Ctenomys haigi	34	DQB exon 2	5	Hambuch and Lacey 2002
Ctenomys sociabilis	33 (14)	DQB exon 2	5	Hambuch and Lacey 2002
Eliurus myoximus	44	DQA exons 2,3 DRB exon 2	11 9	Sommer et al. 2002
Erinaceus concolor	22	DQA exon2 DQB exon2	2 10	Berggren et al. 2005
Erinaceus europaeus	62	DQA exon 2 DQB exon 2	2 6	Berggren et al. 2005
Gerbillurus paeba	40	DRB exon 2*	34	Harf and Sommer 2005
Heliophobius argenteocinereus	10 (5)	DQα1exons 2,3	10, 6	Kundu and Faulkes 2004
Heterocephalus glaber	10 (5)	DQα1exons 2,3	9,4	Kundu and Faulkes 2004
Hypogeomys antimena	191	DQA exon 2 intron 2 exon 3	2	Sommer and Tichy 1999
H. antimena	44	DQA exons 2,3 DRB exon 2	2 5	Sommer et al. 2002
Macrotarsomys bastardi	22	DQA exons 2,3 DRB exon 2	3 6	Sommer et al. 2002
Microtus arvalis	16 (2)	DQA exon 2*	8	Bryja et al. 2005
Peromyscus californicus		DRB	5	Richman et al. 2003
Peromyscus eremicus	20	DRB	14	Richman et al. 2003
Peromyscus leucopus		DRB	6	Richman et al. 2003
Peromyscus maniculatus	25 (1)	DRB	16	Richman et al. 2003
P. maniculatus	22	DRB	16	Richman et al. 2003
Rattus fuscipes greyii	390 (16)	DQA (RT1.Ba) exon 2	36	Seddon and Baverstock 1999
Rhabdomys pumilio	58	DRB exon 2	20	Froeschke and Sommer 2005
Sigmodon hispidus	180 246 (13)	DQA exon 2	11 25	Pfau et al. 1999 Pfau et al. 2001
Spalax ehrenbergi	121 (13)	Palpha1(=DPA) Q beta (=DQB)	11 14	Ben-Shlomo et al. 1988
Thomomys bottae	44 (3)	DQA	5	Sanjayan et al. 1996

However this pattern differs depending on the loci and species investigated (Pfau et al. 1999), indicating that balancing selection has different effects depending on loci and species.

Although most natural populations of micromammals have high levels of diversity at the MHC loci, there are some populations or species that exhibit little variation such as the beaver *Castor fiber* (Ellegren et al. 1993) or the Malagasy giant jumping rat *Hypogeomys antimena* (Sommer et al. 2002). This indicates that balancing selection may not always be the predominant force shaping MHC variation and that behavioural and demographic factors may also play an important role. The effects of balancing selection and genetic drift on the genetic diversity of the MHC class II DQA locus were recently investigated in 14 island and two mainland populations of *Rattus fuscipes*. Although balancing selection may slow the loss of variation in small populations, the effect of genetic drift has been shown to play a dominant role in the majority of island populations leading to a decrease in the number of the MHC alleles (Seddon and Baverstock 1999). However, although MHC allelic diversity is low in some species, a high degree of divergence is often observed between alleles (Sommer 2005).

Transporters associated with antigen processing (TAP)

The transporters associated with antigen processing (TAP) genes code for proteins which form a heterodimer that transports short peptides from the cytosol into the endoplasmic reticulum lumen where they are associated with the MHC class I molecules (McCluskey et al. 2004). Binding of the peptide stabilizes the complex and induces the export to the cell surface for presentation to T-cell receptors. A recent study indicates that variability at the coding region of the TAP2 gene could influence susceptibility to the cestode *Echinococcus multilocularis* infection in humans, independently of HLA status (Zhang et al. 2003).

TAP genes are polymorphic in rats, chickens, hamsters and *Xenopus* but conserved in most other species including humans. In rats, one of the subunits for the TAP molecule exhibits allelic variability at up to 30 different amino acid sites. Among these rat alleles, two have been identified which have distinct substrate preferences in transport assays (Powis et al. 1992). Such influence of the TAP polymorphism on the substrate selectivity of the peptide transporter has also been observed in the Syrian hamster *Mesocricetus auratus* (Lobigs et al. 1999). Analysis of TAP polymorphism between mouse strains revealed a structurally unique TAP transporter per strain, suggesting the possibility of functional polymorphism. Allelic variations were predominantly located in or adjacent to membrane-spanning domains, but no significant bias in the ratio of nonsynonymous to synonymous substitutions was observed (Marusina et al. 1997).

3.2.2 B-cell receptor germline

Immunoglobulins are proteins expressed on the surface of B-cells. They may circulate as antibodies or form the B-cell receptors (BCRs). The BCRs have specificity for particular epitopes. When a BCR binds an antigen, it may pull the antigen into the cell. If the antigen is a protein, the B-cell processes the antigen into smaller peptides, binds some of those peptides to the MHC class II molecules, and presents the peptide class II complexes on the cell surface (Frank 2002).

Each molecule is composed of two identical heavy chains and two identical light (κ or λ) chains. Heavy chains have three domains that affect recognition, namely the variable (V), diversity (D) and joining (J) domains. Light chains have only the V and J domains. The V domains, which interact with foreign antigens, can be divided into the complementarity-determining regions (CDRs) and the framework regions (FRs). CDRs are antigen-binding sites and are known to be highly variable. The constant region is sometimes referred to as the Fc fragment.

In mice, there are 10 groups of immunoglobulin genes (IG): (1) Genes belonging to the immunoglobulin heavy locus (IGH) on chromosome 12 include IGHV, IGHD, IGHJ and IGHC. The IGHV genes encode the antigen-binding regions of antibodies (Su and Nei 1999). These genes play an important role in the diversification of the primary antibody repertoire. (2) The IGKV, IGKJ and IGKC groups comprise genes belonging to the immunoglobulin κ locus (IGK) on chromosome 6. (3) The genes belonging to the immunoglobulin λ locus (IGL) on chromosome 16 include IGLV, IGLJ and IGLC groups.

Although the immunoglobulin heavy chain region (IGH) is a well-established candidate region for disease susceptibility, there are still considerable doubts about its role in protective immunity against helminths (see reviews in Else and Finkelman 1998; Lawrence 2003). Furthermore, high antibody production after infection is not antigen-specific (Grencis 1997). A recent genome-wide search for QTLs influencing mouse resistance to *H. polygyrus* identified a region on chromosome 12 close to the one where the genes for several immunoglobulin heavy chains are encoded (Menge et al. 2003).

On the other hand, antibodies secreted at mucosal surfaces may yet prove to be important in protective immunity to intestinal helminths. Evidence supporting the genetic contributions of immunoglobulins in resistance against macroparasites comes from human lymphatic filariasis studies (Choi et al. 2003). A small case-control study in Ecuador revealed an association between the immunoglobulin *κ* chain allotypes and suscepti-bil-ity to onchocerciasis in Afro-Ecuadorians, but not in Amerindians

(Quinnell 2003). Moreover, Rajan et al. (2005) confirmed a critical role for IgM in mice protection against filarial infection.

During ectoparasite infections, acquired resistance may require humoral immunity although failure of effective humoral protection has been reported in several host-ectoparasite systems (Wikel and Bergman 1997; Mattioli et al. 2000). For example, infestation induces synthesis of tick-reactive homocytotropic antibodies that bind Fc receptors on basophils and mast cells in rabbits and guinea pigs (Wikel and Bergman 1997).

Inter-species polymorphism

Phylogenetic analyses of the immunoglobulin heavy and light chain variable region gene families have been conducted in vertebrates as sequences for many taxa have become available. FRs sequences are preferentially used rather than CDRs, these latter presenting high rates of somatic mutations. Mice and rabbits are often used in these studies, whereas rats were used more rarely.

The IGHV genes form three clusters that are not necessarily represented in the genome of every species. Mice possess IGHV genes from all three groups but rabbits possess a more restricted IGHV repertoire, with only one IGHV cluster (Sitnikova and Su 1998). It has been shown that at least ten mouse IGHV subgroups are homologues in the rat genome (Dammers and Kroese 2001). The phylogenetic tree for IGLV sequences forms six clusters. In mice, IGLV genes belong to two out of these six clusters, and in rabbits, all sequences are restricted to one cluster including mouse and human sequences (Sitnikova and Su 1998). Finally, the IGKV gene sequences show the same clustering pattern as IGHV and IGLV: the human, mouse and rat sequences are intermingled with each other and belong to many clusters, whereas the rabbit IGVK genes form one cluster (Sitnikova and Nei 1998; Sitnikova and Su 1998). Furthermore, the rabbit sequences are quite divergent from other mammalian species (Sitnikova and Nei 1998).

Mice and rats (at least for the immunoglobulin κ and heavy chain variable region genes) contain diverse repertoires of IGHV, IGKV and IGLV genes. In contrast, rabbits have a restricted diversity. These taxa have thus different strategies to generate diverse antibodies: (a) combinatorial rearrangement of various IGHV, IGHD and IGHJ genes for heavy chains and IGLV and IGLJ genes for light chains in mice and rats, and (b) extensive somatic hypermutation or gene conversion in rabbits (Sitnikova and Su 1998).

These analyses reveal that the long-term evolution of the immunoglobulin multigene families occurred through a birth-death process with the birth of genes occurring by gene duplication with subsequent divergent evolution and the death of genes being caused by either gene deletion or loss of functionality (Nei et al. 1997). As closely related genes in the phylogenetic trees are not located closely in the physical maps of IGHV, IGKV or IGLV genes on the chromosome, translocation events are also an important factor involved in IG gene evolution (Nei et al. 1997). Recently, some evidence of trans-species polymorphism has been found at the IGHV genes. Some alleles are found in species belonging to different genera, for example, in an Alaskan population of the snowshoe hare *Lepus americanus* and in domestic and wild rabbits *Oryctolagus cuniculus*. This polymorphism has thus persisted through speciation events at least from the time of divergence of rabbits and hares (Su and Nei 1999). The mechanism of maintenance of this polymorphism remains to be investigated, although it may be due to the specific adaptive significance of these alleles, which could have adapted to cope with different sets of pathogens (Su and Nei 1999).

Intra-species polymorphism
Tanaka and Nei (1989) found that the rate of nonsynonymous nucleotide substitution in mouse sequences is significantly higher than that of synonymous substitution at the CDR of IGHV and IGLV genes. The antigen-binding regions of mice are thus likely to be subject to positive Darwinian selection. When considering the FRs, the rate of nonsynonymous nucleotide substitution is lower than that of synonymous substitution in mice and rats (Dammers and Kroese 2001). Furthermore, the extent of genetic polymorphism is much lower at the IGHV genes than that of other mammalian MHC loci, suggesting that directional selection rather than overdominance or balancing selection is likely to cause this observed pattern of polymorphism in mice (Nei et al. 1997; Su and Nei 1999). Contrasting results have been observed in rabbits. Su and Nei (1999) showed no difference between the rate of nonsynonymous and synonymous nucleotide substitution at the CDRs of the IGHV rabbit genes, and higher rates of nonsynonymous compared to synonymous nucleotide substitution at the FRs, suggesting purifying selection (i.e. for the maintenance of canonical structure of the antibody-binding site). Population genetics studies conducted on the European rabbit with IGKC genes provide evidence of overdominance-type selection as the patterns of allelic diversity were very similar to those reported for MHC loci (see review in Van der Loo et al. 1996). Indeed, by increasing the allelic persistence through time, overdominance may have contributed to the exceptional divergence observed between alleles and can account for the high heterozygosity levels found in the original range of the species.

Population genetic studies provide information on the influence of the number of IG genes on individual fitness, and also on the possible association between diseases such as filariasis and human IG haplotypes. Immunoglobulin markers were also used to study the genetics of wild populations of the European rabbit. These studies have been conducted by Van der Loo since the 80s and on the IGKC and IGHV genes. They focused on between-population genetic differentiation and within-population diversity, and compared the aboriginal range of the species (Iberian peninsula) with the more "recent" distribution range (Western Europe, Great Britain, Australia) (Van der Loo 1993; Van der Loo et al. 1996), or subspecies (*O. cuniculus cuniculus* and *O. c. algirus*) in the Iberian peninsula. They provided evidence for balancing selection enhancing gene diversity at some of the heavy and light κ chain loci among populations and individuals (Van der Loo 1993; Van der Loo et al. 1996; Esteves et al. 2004), but revealed that this genetic variability at IG heavy and light κ chain constant regions was sustained by compensatory diversity enhancing selection (Van der Loo et al. 1996).

3.2.3 T-cell receptor germline

The T-cell receptors (TCR) are cell surface molecules which co-recognize antigen and the MHC during antigen presentation. They have many similarities with immunoglobulins. Successful cross-linking of the TCRs leads to T-cell activation and cytokine production, which is an essential step in the generation of a specific immune response.

The TCR is a heterodimeric receptor composed of either αβ chains or γδ chains. αβ T-cells recognize antigenic peptides linked to MHC molecules, whereas γδ T-cells recognize native peptide or non-peptide antigens independently of the MHC. The peripheral T-cell population in mice and humans is dominated by αβ T-cells, whereas only 5 % is composed of γδ T-cells. The γδ T cell repertoire in mice is partly characterized by the predominant expression of different genes in different tissues (Cho et al. 2005). In other species including rabbits, γδ T-cells compose 20 to 50 % of the peripheral T-cell pool (Cho et al. 2005).

The α and δ chains are encoded by a variable (V) domain, constant (C) domain and joining (J) domain, whereas the V, C, J domains together with the diversity (D) domains are required in the case of β and γ chains (Mackelprang et al. 2002). Each V gene segment in the TCR loci is composed of three complementary determining regions (CDR) which are responsible for antigen binding. CDR1 and CDR2 regions interact primarily with the more conserved elements of the MHC molecules, while the CDR3 regions interact more with the central region of the bound peptide. Conserved framework regions (FR) flank CDR regions in the V gene segments and are important for the structure of the TCR.

While there are only a few copies of the D and J genes in the genome of higher vertebrates, the number of V genes is large, and the sequence diversity among these gene copies is largely responsible for the generation of diversity of TCRs. These V genes are usually clustered in a certain region of a chromosome and form a multigene family (Su and Nei 2001).

The TCR loci have long been considered important candidates for common disease susceptibility. Although specific recognition of antigens depends primarily on somatic mechanisms to create variability, the germline alleles are the basis on which somatic processes are built. It is thus likely that germline polymorphisms influence individual tendencies to react to particular antigens (Frank 2002). In support of this hypothesis, there is evidence that the TCRV polymorphism affects the expressed TCR repertoire in mice (e.g. Jaeger et al. 1998), and that a functional role for the human TCRV gene polymorphism in defence against pathogens exists (Jaeger et al. 1998).

Going back to the early 1980s, many studies have attempted to find associations between markers in the TCR regions and common diseases. Although these studies have mainly focused on autoimmune and allergic diseases, some of them explore the resistance or susceptibility to infection with the cestode *Taenia crassiceps* (Lopez-Briones et al. 2001; Lopez-Briones et al. 2003). It was found that the absence of TCR αβ T-cells greatly increases mice susceptibility in terms of parasite load (Lopez-Briones et al. 2001). Moreover, Lopez-Briones et al. (2003) established a relationship between T-cell repertoire and resistance or susceptibility to *T. crassiceps* infection in mice. In addition, a recent genome-wide search for quantitative trait loci (QTLs) influencing immunological mouse response to infection with the gastrointestinal nematode *H. polygyrus* have identified the TCR γ chain on chromosome 13 as a potential candidate gene located within the QTL mapped region (Behnke et al. 2003; Menge et al. 2003).

Finally, the presence/absence of specific V segments could have consequences in terms of disease susceptibility. Changes in the V gene segment number can have a great impact on the overall T-cell repertoire. For example, one additional BV gene segment can result in an increase of more than 10^{14} TCR variants (Funkhouser et al. 1997).

Inter-species polymorphism

Initially, variations in the TCR loci were discovered as restriction en-zyme length polymorphisms (RFLPs). Because variation in TCR loci is generated through multiple copies of gene segments, variation discovery has also included the characterization of the position and copy number of the V gene segments (see review in Mackelprang et al. 2002). Now, most approaches search for SNPs at the TCR loci (Geraghty 2002). The database IMGT (http://imgt.cines.fr:8104) provides a useful resource for structure, nomenclature, and allelic variation in this multigenic family.

A striking level of similarity has been observed between the TCRV genes of humans, mice and rabbits (Livak 2003). In the TCRB locus, every mouse and human BV genes has at least one recognizable orthologous pair, with the exception of the mouse BV2 locus. Similarly, the nucleotide sequences of the rabbit TCRBV segments form nine families, which are homologous to nine human families (Isono et al. 1994). In addition, the majority of mouse species contain a similar content of TCRBV genes although some gene families are missing from *Rattus* and *Mus shortridgei* (Huppi et al. 1988). The importance and function of the conservation of this primary TCR combinatorial repertoire may reflect the generation of an optimally shaped antigen-receptor repertoire which can support the most efficient adaptive host defence against diverse pathogens (Livak 2003).

Phylogenetic trees of humans and mice based on the entire TCRBV gene segments do not form two separate clusters. Rather, they intermingle extensively (Su and Nei 2001). This suggests that many gene duplications occurred both before and after primates and rodents diverged, and that duplicate TCRBV genes have not been subject to any significant interlocus homogenization of sequences within either of the two species. TCRBV genes can be classified into six groups. This had previously been proposed by Su et al. (1999), where the framework regions rather than the entire V domain were analysed in humans, mice, rabbits, sheep, cattle and poultry. Human and mouse genes are found in all six subgroups, whereas rabbit TCRBV genes belong to only five of the subgroups. The phylogenetic tree based on the TCRAV genes of six species well characterized for their TCRV repertoire (including human, mouse and rabbit) is composed of four clusters. Human and mouse TCRAV genes are found in four subgroups, whereas rabbit TCRAV genes are found in only one subgroup. This suggests that the common ancestor of amniotes must have possessed the TCRAV genes from all four subgroups and that three different subgroups have been lost in the rabbit lineage (Su and Nei 1999). The evolutionary analysis of the TCRGV subgroups in rabbits indicates duplication and gene conversion events (Cho et al. 2005).

Intra-specific polymorphism

The study of interpopulation differences in TCR genes may identify genes subjected to selection (Ibberson et al. 1998). However, there is still little information on the contribution of allelic variation to molecular variability of the TCRs, and on the degree of TCRV allelic variation between populations (Ibberson et al. 1998). The few reported studies concern humans and mice.

In mammals, the TCR constant domains are relatively invariant, though a few amino acid differences are known to exist between products of different TCRBC or TCRGC genes (Criscitiello et al. 2004). In mice, the relatively stable TCRAC locus from inbred strains has fostered the idea that the TCR constant regions are relatively invariant. However, RFLP studies of the TCR among wild mice indicate that significant polymorphisms are detectable, particularly in the TCRBC where three alleles exist (Nobuhara et al. 1989). It is likely that sequencing of wild *Mus* TCR constant genes would reveal polymorphisms beyond those detected with RFLP (Criscitiello et al. 2004). The significance of these population-specific variations is currently unknown.

Extensive polymorphism has been observed for the TCRAV gene segments for both inbred strains and wild mice species using RFLP (Jouvin-Marche et al. 1989). According to the IMGT database, among the 100 gene segments involved, 62 are polymorphic and present between two and nine allele variants. To our knowledge, there is no population genetic study of wild micromammals based on such loci. The genomic diversity observed at the TCRBV gene segments in colony-bred rat and mouse species revealed much more limited variation than that seen for the TCRAV loci (Huppi et al. 1988). Most of the loci are monomorphic or bi-allelic, and the maximum number of allelic variants recorded is five. In natural populations, evidence of positive selection was found in two TCRBV gene segments. A population genetic study of the Indian house mouse conducted on the TCRBV17 gene revealed differences in the allelic frequencies observed among localities (Awasi et al. 1999). Amino acid substitutions in this region are known to alter T-cell receptor specificity, and could potentially increase the range of peptide/MHC complexes that can be recognized by the TCRs. Genetic differentiation among wild mouse populations could thus reflect differences in selective pressure. Another source of TCRBV variability has been reported in humans and comes from the repertoire of TCRBV segments which greatly varies between individuals and populations (Mackelprang et al. 2002).

Finally, although less studied, the TCRGV gene polymorphism has revealed interpopulation specificities in mouse populations. Roger et al. (1993) described seven TCRGV haplotypes among 23 wild mice corresponding to four *Mus musculus* subspecies (*M. m. domesticus*, *M. m. castaneus*, *M. m. musculus*, and *M. m. molossinus*). These haplotypes were unequally distributed among the four subspecies.

3.2.4 CD1

The CD1 family consists of the antigen presenting molecules mainly located on the surface of the immune system's antigen-presenting cells, such as macrophages and dendritic cells. CD1 molecules bind and present lipid- and glycolipid-based antigens to the specific types of T-cells, which results in the activation of cell-mediated responses. Sequence comparison has led to the classification of CD1 proteins into: CD1-a, CD1-b, and CD1-c (group 1), and significantly divergent CD1-d and CD1-e (group 2) (see review in Porcelli et al. 1998).

Lipid-based antigens presented by the group 2 CD1 molecules in humans and mice stimulate an important subset of T-cells called natural killers (NKs) (see review in Dutronc and Porcelli 2002). Upon primary stimulation through their T-cell receptors, these lymphocytes rapidly release considerable amounts of immunoregulatory cytokines such as interleukin 4 (IL-4), which is critical in the control of intestinal helminth infections (Bendelac et al. 1995). Moreover, human and murine group 2 CD1 molecules are expressed at high levels on the intestinal epithelium (Dutronc and Porcelli 2002). CD1 molecules are thus good candidates for presenting antigen from intestinal dwelling parasites (Bendelac et al. 1995; Else and Finkelman 1998; Trottein et al. 2006). To date, association studies involving the CD1 family have mainly focused on protozoan and mycobacterial infections (i.e. tuberculosis, leprosy). A role for CD1-d during nematode infections is still controversial although it has been established in murine schistosomiasis (Faveeuw et al. 2002; Trottein et al. 2006).

Polymorphism

The CD1 proteins are encoded by genes that lack the extensive polymorphism of the classical MHC encoded antigen presenting molecules (Dutronc and Porcelli 2002). Allelic variants found among mouse and rat strains are rare and of a conserved nature, and therefore unlikely to af-fect binding properties substantially (Katabami et al. 1998). Among 11 strains of rat analysed, only two allelic variants were detected in the exons encoding the extracellular domains. CD1 could have evolved for the presentation of a very restricted set of structures (Calabi and Milstein 2000). These results have yet to be confirmed for wild populations. On the other hand, the numbers of CD1 loci are extremely variable among different species of mammals. CD1 genes have been cloned from seven different species: human, mouse, rat, rabbit, sheep, pig and guinea pig. The size of the CD1 family varies from mice which harbour 1-2 closely related genes (CD1d homologues) to rabbits, sheep, guinea pigs, dogs and cattle which have more than seven genes (Calabi and Milstein 2000). Rabbits possess both group 1 and group 2 CD1 genes (Hayes and Knight 2001), and guinea pigs possess clear homologues of the group 1 CD1 genes (Dascher et al. 2002; Hiromatsu et al. 2002). This raises questions concerning the evolution of the gene family. First, it is unlikely that group 1 CD1 genes arose independently in guinea pigs and rabbits. The most parsimonious explanation is that CD1 genes have undergone duplication events that generated a series of linked genes (Dascher et al. 1999; Han et al. 1999; Hayes and Knight 2001). The common ancestor of rodents and lagomorphs must have had both CD1 groups and the group 1 CD1 genes have been lost from mouse and rats during evolution (Dascher and Brenner 2003). Second, it would be interesting to analyse whether other muroid rodents lack group 1 CD1 genes as mice do, and to estimate how prevalent the absence of these CD1 genes is among different rodent species. One hypothesis for the presence of multiple isoforms of CD1 in a given species is to facilitate sampling of antigens from different compartments within the antigen-presenting cells. This has gained support from studies carried out on humans and guinea pigs (see review in Hiromatsu et al. 2002). Parasites may have exerted selective pressure to alter the number or expression of different CD1 genes in different species (Dascher et al. 1999). This hypothesis suggests that associations between the number of isomorphs and susceptibility/resistance could be explored.

3.2.5 Toll like receptors (TLR)

Toll-like receptors (TLRs) are pathogen recognition molecules that are present on many antigen-presenting cells such as macrophages and dendritic cells (Akira et al. 2001). So far, six major families have been described in vertebrates (Roach et al. 2005). Each TLR family recognizes distinct pathogen-associated molecular patterns (PAMPS). The TLR-delivered signals ultimately culminate in the production of pro-inflammatory cytokines, which, on the one hand, mediate direct defence responses and, on the other hand, alert adaptive immune cells to the presence of a pathogen (see review in Kawai and Akira 2005). Some TLRs, such as TLR4, constitute a direct interface with the microbial world. The primary structure of these molecules thus determines ligand specificity (Poltorak et al. 2000).

Recent studies have demonstrated that the TLRs play a key pathogenic role in chronic gastrointestinal nematode infections. Helmby and Grencis (2003) found that mice with a natural defect in TLR4, as well as TLR4 KO mice, are resistant to chronic *T. muris* infection. In regard to the origin of the antigens involved in activation through TLR4, it is possible that *T. muris* secretes products that can bind to TLR4 and activate the innate response. This hypothesis has also been suggested for *T. crassiceps* (Dissanayake et al. 2004), and for *S. mansoni*, which secretes a molecule capable of activating and altering dendritic cell function through TLR2 (Van der Kleij et al. 2002).

Although associations between TLR gene polymorphism and infectious diseases have been broadly investigated (see review in Schroder and Schumann 2005), there are yet no case-control studies revealing associations between TLR gene polymorphism and macroparasites.

Inter-species polymorphism
The TLR genes have been recognized in a number of vertebrate genomes and many sequences are now available. A recent phylogeny proposed by Roach et al. (2005) revealed that vertebrate TLR genes are not fast evolving genes, and are highly conserved among species (Smirnova et al. 2000). The observed "star-like" pattern indicated that the TLR genes are evolving at about the same rate. The relative absence of discrepancies in molecular distance observed between species with different generation times implied that selection overwhelmed mutation in governing the evolution of the TLR genes. These results provided evidence of major selective pressure acting for the maintenance of the specific PAMP recognition of the TLR gene families. Evaluation of synonymous versus non-synonymous substitution ratios yielded no support for positive selection in the vertebrate TLR phylogeny (Roach et al. 2005). On the contrary, purifying selection was detected in the TLR1 and TLR4 gene families (Smirnova et al. 2001; Roach et al. 2005).

Intra-species polymorphism
Most of the studies of within population polymorphism of the TLR genes concern humans. The only study conducted on small mammals was performed by Smirnova et al. (2000), who investigated the intra-species polymorphism of the TLR4 gene in 35 strains of *M. musculus*. They identified 10 alleles on the basis of mutations observed at 22 sites. Most of these mutations occupy portions of the gene corresponding to the extracellular domain. This high diversity could be a result of selective evolutionary pressure driven by parasites.

3.3 Polymorphism in regulation

3.3.1 MHC promoters

Polymorphism is conspicuous in the promoters of the class II MHC genes in both human and mouse. It has been suggested that this may result in differential expression of these molecules in diverse antigen-presenting cells, and may thus influence the balance of the Th1 and Th2 cytokines (Daser et al. 1996). A tight and precise control of MHC expression is required for a functional immune system. Deviations from normal expression patterns have been associated with severe immunodeficiency and autoimmune diseases (Berggren and Seddon 2005). However, in mice, MHC class II gene promoter polymorphism has frequently been observed (Mitchison and Roes 2002). Studies of regulation of MHC class II expression in species other than humans and mice are rare (but see Berggren and Seddon 2005 for canids).

3.3.2 Cytokine gene polymorphism

With regard to helminth infections, the balances of the Th1- and Th2-associated cytokines are the major agents of the final nature of the host response. Individual cytokines may produce opposing effects depending upon dose and timing of their participation in the immune response (Allen and Maizels 1997). Genetic predispositions to express cytokines that influence this dichotomy between the Th1 and Th2 subsets are likely, therefore, to be a major factor in determining the susceptible or resistant phenotype (Abbas et al. 1996). The more frequent SNP and microsatellite polymorphism observed in the regulatory regions of cytokine genes compared to the exonic regions of these genes is consistent with this hypothesis (Daser et al. 1996; Mitchison 1997).

Many studies have identified cytokine candidate gene polymorphisms for helminth resistance/susceptibility using either genetic linkage or case-control association studies for the most part based on human pedigrees, sheep flocks and mouse strains. Experiments conducted with knock-out and transgenic mouse strains have revealed that the cytokines encoded by the identified genes are required for helminth expulsion (Else and Finkelman 1998; Grencis 2001; Lawrence 2003; Hayes et al. 2004). However, pronounced difficulties in interpretation of associations between the cytokine gene expression and resistance/susceptibility arise from the confounding influences of gene-gene, gene-environment and cytokine-cytokine interactions (Ollier 2004).

Information acquired from genome scanning and from case/control studies hints that the main form of variation outside the MHC is in the cytokine genes, and especially in those of the pro-inflammatory cytokines (Mitchison 1997). Studies aiming at describing cytokine gene polymorphism in human populations, or phylogeny in vertebrates or mammals, are now too numerous to be described in detail. We only focus on a few genes encoding the main cytokines involved in the immune response against helminth infection.

Inter-species polymorphism

Cytokine genes have been used in phylogenetic studies to provide a better understanding of the evolutionary relationships within families such as Muridae (Herbst et al. 2002). In these studies, a high rate of sequence conservation has been observed for the TNF-α and IFN-γ loci among different orders, probably due to the highly selective pressure linked with the fundamental role of these cytokines in the immunological function. Sequence comparisons of mammalian species have also been carried out to analyse the evolutionary pressures acting on the cytokine genes. Interspecies polymorphism was first investigated in the interleukin genes (IL-6 and IL-4) in murine rodents extending phylogenetically from *M. m. domesticus* to rat using random fragment length polymorphism. The ratio between synonymous and non-synonymous mutations in the IL-4 gene suggested a complex pattern of selective forces involved, which included both purifying and positive selection (Richter et al. 1990). Such a complex evolutionary scenario has been invoked to explain the polymorphism patterns observed at the IFN α, β and ω genes (Hughes 1995). More recent studies have focused on the IL-2 gene, which exhibits a high rate of evolution (Zelus et al. 2000; Zhang and Nei 2000).

An interesting issue concerning these phylogenetic analyses is the detection of differences in the substitution rates between lineages. Positive selection, detected as significantly higher rates of non-synonymous compared to synonymous substitution rates, was observed in artiodactyls, but not in primates, carnivores or rodents (Zelus et al. 2000; Zhang and Nei 2000). The IL-2 gene evolved more rapidly in the ruminants than in any other mammalian taxon. In rodents, nucleotide substitution rates indicated purifying selection (Zelus et al. 2000).

Intra-species polymorphism

TNF-α

Since the tumor necrosis factor α (TNF-α) gene was found to be located in the central MHC, it has been the epicentre of intense attention concerning the relevance of its promoter diversity to susceptibility to various infectious diseases. Although this work remains controversial in large part (Hauptmann and Bahram 2004), it seems that this pro-inflammatory cytokine is involved in the intestinal pathology of *T. spiralis* infection of mice (Lawrence et al. 1998), and is responsible for the expulsion of *T. muris* (Grencis 2001). On the basis of this rationale, a number of the SNPs in the promoter region of the TNF-α gene have been investigated for their effects on gene transcription as well as for their possible correlation with particular diseases. Pernthaner et al. (2005) showed that the TNF-α mRNA was more expressed in resistant than in susceptible sheep lines, which might account for the more acute inflammatory response to natural nematode challenge in the resistant lines. To our knowledge no TNF-α population genetic study has been conducted in wild micromammal populations.

IFN-γ
IFN-γ is associated with chronicity in *T. muris* and *N. brasiliensis* infections (Hayes et al. 2004). To our knowledge, the only study conducted in a natural population was that of Soay sheep on St Kilda (Coltman et al. 2001). Reduced faecal egg counts of the nematode *Teladorsagia circumcincta* were associated with an allele at a microsatellite locus located in the first intron of the INF-γ gene in lambs and yearlings.

Interleukins
Genetic variation patterns at some human interleukin loci reveal substantial locus-specific population differentiation (e.g., IL-4; Rockman et al. 2003). These loci are known to produce critical cytokines for protection against helminth infections (Lawrence et al. 1998; Artis et al. 1999; Finkelman et al. 1999). Selection on interleukin promoters could act to alter the balance of interactions within the immune system, which offers the advantage of being difficult to counteract by parasites (Rockman et al. 2003).

3.3.3 iNOs promoter polymorphism

Nitric oxide (NO) was initially described as an effector generated by endothelial cells. It exerts multiple modulating effects on inflammation and plays a key role in the regulation of the immune responses. Its biological roles have expanded to include activity against viruses, bacteria, fungi and helminths (see review in Bogdan 2001). The isoform of the enzyme relevant to immune function is iNOS (inducible nitric oxide synthetase), which is expressed in a number of cell types, including macrophages. The expression of this isoform is tightly regulated by cytokines (IFN- γ , IL-1h, IL-6and TNF- α in murine and rat cells). Transcriptional activation of the gene results in rapid accumulation of the mRNA, and induced macrophages are capable of producing large amounts of NO over a prolonged time period (Rajan et al. 1996).

Rajan et al. (1996) found a consistent role for NO in host defence against filarial infection. The parasite *B. malayi* lives in the lymphatic system of the mammalian host, in close proximity to endothelial cells, which are known to express iNOS and to generate prodigious quantities of NO in response to cytokines. Taylor et al. (1996) have shown that NO is toxic to *Onchocerca linealis* microfilariae. Oswald et al. (1994) suggested similar effects of NO on *Schistosoma* and Kanazawa et al. (1993) against *Echino-*

coccus multilocularis. However, other studies on extra-cellular blood dwelling parasite species did not find antiparasitic effects of NO *in vivo* despite well-documented susceptibility *in vitro* (Kanazawa et al. 1993).

Polymorphism
A large number of SNPs are described in the database PubMed for humans (334 SNPs), mice (139 SNPs), rats (22 SNPs) and dogs (58 SNPs). Recently, Johannesen et al. (2003) analyzed iNOS gene promoter polymorphism in two strains of rats and tested the functionality of the observed nucleotide differences. They identified three polymorphisms in two separate areas and estimated that the homology between all published rat iNOS sequences was 96%. Phylogenies based on these polymorphisms would give a better understanding of the evolutionary pressure acting on these genes.

3.3.4 Other potential candidate genes

A few other potential candidate genes have been identified using case-control or genome linkage studies, and require further attention. Among these, we note the Ly49 receptor (Korten et al. 2002), the mannose-binding lectin 2 (Choi et al. 2001; Hise et al. 2003), the CCR3 chemokine (He et al. 2004, and the mucin protein (Lawrence 2003) genes.

4 Virulence genes in parasites

4.1 Immune evasion strategies in parasites

Parasites require time in their host in order to complete development, to reproduce and to ensure their transmission. Consequently, chronic infections of macroparasites, ranging from a few months to many years are the rule (Morand 1996; Trouvé et al. 1998). Parasites have to avoid immune elimination by hosts with good immune defences and thus have evolved immune evasion strategies, which may be passive, or may involve active intervention with the host's immune regulation. Some strategies are quite simple, for example migration of hookworms within the host' s gut to avoid local inflammatory reactions. However, most strategies are more sophisticated, and can be categorized as immune evasion, immune exploitation and molecular piracy.

Evasion and auto-immunity are potentially connected as the immune system has to maintain the delicate balance between effective defence and auto-immunity. The strategy of immunomodulation displayed by many macroparasites may have some beneficial effects on their hosts by regulating both mucosal inflammation in the case of gut parasites and Th1/Th2 responses (Weinstock et al. 2004).

4.2 Evasion in helminths

Most parasitic helminths are extracellular and are too large to be removed by phagocytosis. Hosts have developed inflammation and hypersensitivity responses for the larger worms such as gastrointestinal nematodes. Acute response after previous exposure can involve an IgE and eosinophil mediated systemic inflammation which results in the expulsion of the worms from the intestine. These reactions are similar to allergic reactions. Chronic exposure to worm antigens can cause chronic inflammation with delayed type hypersensitivity. The Th2/B-cell responses increase IgE levels, whereas mast cells and eosinophils activate inflammation. Helminths commonly induce Th2 responses characterised by specific cytokines, eosinophils and antibody responses, in particular IgE. Characteristic antibody-dependent cell-mediated cytotoxicity (ADCC) reactions, i.e. killer cells (e.g. macrophages, neutrophils, eosinophils) are directed against target parasites by specific antibodies (e.g., eosinophils killing parasite larvae by IgE, or by some IgG subclasses; Gause et al. 2003).

Several helminth immune evasion mechanisms in the vertebrate host are known (Maizels et al. 2004):

1. Molecular mimicry, when a parasite is able to mimic a host structure or function. For example, schistosomes have E-selectin that may help in adhesion or invasion. A parasite may also coat itself with host proteins. The tegument of cestodes and trematodes is able to adsorb host components, thus giving the worm the immunological appearance of host's tissue. Molecular mimicry has been updated in the light of recent discoveries concerning degeneracy and plasticity of the TCR/MHC-peptide recognition. For example, schistosomes take up host blood proteins, e.g. blood group antigens and the MHC class I and II molecules. As a result, the worms are seen as "self".

2. Anatomical seclusion in order to escape the host's immune response, i.e. larvae of *Trichinella* live inside mammalian muscle cells for many years.

3. Shedding or replacement of surface. This is a characteristic of trematodes and hookworms.

4. Immunosuppression or manipulation of the host's immune response. High burdens of nematode infection are often carried with no outward sign of infection. This may happen if a parasite secretes products that include

anti-inflammatory agents which act to suppress the recruitment and activation of the effector leukocytes. For example, a hookworm protein which binds the β integrin CR3 inhibits neutrophil activities. There is other evidence of secreted products which block the chemokine-receptor interactions. For example, an acetylhydrolase from *N. brasiliensis* inactivates the pro-inflammatory molecule Platelet-activating Factor. Liver fluke larvae secrete an enzyme that cleaves antibodies. Filarial parasites secrete a number of anti-oxidant enzymes such as glutathione peroxidase and superoxide dismutase which most likely contribute to their observed resistance to antibody-dependent cellular cytotoxicity and oxidative stress. Many nematodes which colonise the alimentary tract of the host secrete acetylcholinesterases (AChEs), enzymes generally associated with the termination of neuronal impulses via hydrolysis of acetylcholine at synapses and neuromuscular junctions. Host cytokines may serve as parasite growth factors such as helminth pseudocytokines. Finally, another strategy used by a number of parasites to regulate the immune response is parasite-induced T-cells apoptosis (Jenkis et al. 2005).

4.3 Helminth genomes and genes

In comparison to viruses which devote a large portion of their genomes to immune evasion (Maizels et al. 2001b), helminth genomes (1–3 108 bp, approximately 20 000 protein-coding genes) are repositories of novel mediators of potential importance in the regulation of immune responses (Maizels et al. 2001a). These helminth immunomodulators are generally analogous of the mammalian immune system genes, and are unlikely to be genes transferred from mammals. Most of the gene products are members of ancient gene families that have evolved in parallel in the vertebrate and invertebrate lineages. The genome projects offer a great opportunity to discover genes of interest. For example, *B. malayi* expresses sequence tags (ESTs) on 6822 partial gene sequences, 11,6% were homologues to the nematode *Caenorhabditis elegans* and 62.7% had no database homolog, i.e. they are novel genes. For *N. brasiliensis*, 36.6% of ESTs are novel genes and 27.4% were similar only to other nematode sequences (Maizels et al. 2004). The higher proportion of *N. brasiliensis* genes with homologues in *C. elegans* reflects a close phylogenetic relationship between these two species.

In both *B. malayi* and *N. brasiliensis*, a substantial number of new genes have been discovered with no similarity to existing database sequences. A considerable proportion of these new genes are directly involved in host-parasite interactions. However, the diversity and function of these genes still remain to be analysed.

4.4 Ectoparasitic arthropods and ectoparasite genes

The ability of ectoparasitic arthropods to suppress innate and acquired host immunity is also well documented (Wikel and Bergman 1997; Singh and Girschick 2003). For example, host immune-effector elements can be suppressed by ticks. It has been observed that salivary gland extracts of several tick species (including, for example, *Ixodes scapularis*) profoundly inhibit the proliferative response of the mitogen stimulated T-lymphocytes. Tick saliva contains a protein named soluble IL-2-binding, which suppresses T-cell proliferation and the activity of immune effector cells that are responsive to IL-2 stimulation.

Ticks have evolved mechanisms to suppress the production of T-lymphocyte cytokines. Salivary gland extracts of *Dermacentor andersonii* suppressed the production of TH1 lymphocyte cytokines IL-1b and IFN-g by lymphocytes from uninfested mice (Wikel and Bergman 1997).

5 Evolutionary factors maintaining variability of parasite resistance and epidemiological consequences

5.1 Mechanisms maintaining immunogenetic variability

The immune system genes are amongst the most variable in the genome. The loci that exhibit extreme levels of polymorphism are those whose products interact directly with parasites. Evolutionary biologists have tried to disentangle the potential mechanisms maintaining this genetic diversity, and a number of possible explanations have been advanced.

5.1.1 Birth/death processes

Diversification of the immune genes often occurs through gene duplication, and underpins many of their innovative properties. Duplication provides a way of retaining, through conservation of one duplicate, the currently useful function of the encoded protein, whilst its twin is liberated to mutate and possibly acquire a novel function. The immune genes tend to duplicate *in cis*, so that they lie next to each other in the genome, forming clusters in which a coordinated evolution of molecules is possible. Gene conversion, i.e. genetic exchange of short sequences, is frequent in recently duplicated genes, and also plays a major role in the diversification of some immune gene families (e.g., the TCR genes; Funkhouser et al. 1997). Examples are the MHC class I and II, immunoglobulin superfamily, cytokines, Toll-like receptors and mouse Ly49 genes (Trowsdale and Parham 2004). On the other hand, the duplicated immune genes may also become non-functional pseudogenes. They may remain intact but expressed at negligible levels. Although clear roles of these pseudogenes remain problematic, they remain useful for tracing the source of the donor sequences gene conversion (Trowsdale and Parham 2004).

5.1.2 Positive balancing selection

Evidence indicating that some immune genes such as MHC (Hughes and Yeager 1998) or IG genes (Tanaka and Nei 1989) evolve through balancing selection comes from their uniform allelic frequency distribution, their pattern of nucleotide substitution in coding exons, differences in evolutionary pattern between exons and introns, and the patterns of trans-species polymorphism observed in phylogenetic analyses of sequences (see Hughes 2002). It is also a hallmark of their counterpart genes in pathogens. Two non-exclusive mechanisms have been proposed to explain the evolution of MHC polymorphism in natural populations (which has been the most extensively studied immune gene family) and concern sexual selection and parasite-driven selection pressures (see Hughes and Yeager 1998). This latter mechanism invokes the strong selective pressures created by the need to respond to rapidly evolving pathogens with short generation times. Parasite-driven selection is expected to operate when specific alleles are favoured because of their ability to provide protection from different parasite species or strains.

5.1.3 Co-evolutionary arms races

The biochemical specificity of host susceptibility and parasite infectivity underlies the potential for genetic diversity (Frank 2002). High levels of polymorphism might be maintained through negative frequency-dependent selection, in which parasites and hosts are continually responding and counter-responding to selection pressures imposed by the other, or overdominance (Takahata and Nei 1990; Slade and McCallum 1992). Parasites evolve so that their constituent peptides cannot be bound by the most prevalent immune receptors. Common host alleles are the primary targets of parasite evolution, which gives a disadvantage to these alleles and an advantage to rare host alleles. This selective force can lead to rapid coevolutionary dynamics over timescales of a few generations (Peters and Lively 1999). These ideas are formalized in the Red Queen metaphor, but to our knowledge, they have not been tested ex-perimentally in the small mammal-macroparasite systems.

5.1.4 Trade-offs

The evolution of the genetic diversity of immune genes is constrained by physiological and evolutionary trade-offs between the fitness costs associated with resistance and those associated with parasitism. First, regulatory polymorphism may be limited by trade-offs, as a more intense immune response clears parasite more effectively but also causes more collateral tissue damage to the host (Frank 2002). Second, several trade-offs may involve immune receptor genes: The trade-offs between resistance to disease and susceptibility to autoimmunity may be important (Trowsdale and Parham 2004). However, owing to a limited number of genes coding for the recognition alleles, there are constraints on the particular recognition alleles an individual can have. Although variation in recognition alleles is likely to be cost-free, there could be exceptions if the recognition allele is negatively pleiotropic or if the allele is accompanied by increased expression of the receptor. Third, the evolution of genetic resistance to parasite infection is often assumed to be traded-off against other fitness components such as fecundity and growth, although a positive correlation has also been found between nematode resistance and body size in the feral Soay sheep (Coltman et al. 2001).

From an evolutionary point of view, the lack of theory including both concepts of coevolution between host and parasite immunogenetics, and trade-offs has been repeatedly pointed out (e.g., Schmid-Hempel and Ebert 2003). Theoretical models and empirical analyses are still sorely lacking.

5.2 Epidemiological consequences of immunogenetic variability

5.2.1 Evolution of specificity

Immunogenetics can be used to determine the evolution of host and parasite specificity. The reasons why hosts differ in their susceptibility to different parasite types may be based on the genetic diversity or the degree of matching between immune genes and parasite antigens.

5.2.2 Local adaptation

Knowing the potential amount of immune gene diversity in small mammals and their macroparasites, a future central goal should be concern the understanding of the factors determining the appearance, spread and distribution of resistance/immuno-modulating alleles within populations and across geographical landscapes. Spatial variation in the interactions between parasites and their hosts is thought to be a major force in the coevolutionary process and in generating biological diversity. Under particular conditions of gene flow and genetic drift, it can result in local adaptation of parasites (see a recent review in Kawecki and Ebert 2004); that is, greater infectivity of local parasites than foreign parasites on local hosts. Although this concept has motivated a large numbers of theoretical models in the last decade, the ubiquity of this phenomenon remains to be determined.

Given the importance of gene flow and demographic patterns to local adaptation, it has been suggested that studying the genetic structure of populations in both hosts and parasites is a prerequisite for understanding the evolution of pathosystems. Such studies have mainly focused on plant-parasite or invertebrate models. Relatively little is known about the structure of genetically determined susceptibility-infectivity in natural micromammal-helminth or micromammal-ectoparasite associations, especially with regard to potential interactions between host and parasite genotypes. Recently, Prugnolle et al. (2005) studied the population genetics of *S. mansoni* and its two hosts (*R. rattus* and *Biomphalaria glabrata*). Their study showed genetic differentiation in both intermediate and definitive hosts and parasite subpopulations at a regional scale, which is an essential feature for the evolution of local adaptation. Variable selection pressures between sites could then lead to the adaptation of hosts and/or parasites to local environmental conditions. Similar surveys have been conducted by McCoy et al. (2003) on ticks and seabirds.

6 Conclusions and the future

Immunogenetics, the analysis of genetic polymorphisms in specific recognition and immune regulation, is at the core of the study of host-parasite coevolution. Given the well studied laboratory micromammal systems, broad research into micro-mammal immunogenetics now has a firm basis. Unfortunately, most studies continue to focus on MHC genes in lab mouse strains. We crucially lack further investigations considering genetic variation in natural populations, across biogeographical areas or at local scales, and exploring a wider diversity of immune genes. In addition, macroparasite immunogenetics remains scarcely explored, even though it is a topic of major importance to evaluate the effects of host immune gene diversity, from both specificity and regulation aspects, on antigenic variation. Such information will provide significant insights into the processes of adaptation at the molecular levels in these host-parasite systems. It will also provide the basis for future promising research areas including experimental evolution, which still mainly concerns viruses, and is essential to determine the nature of selection acting on the host or parasite immune genes or to reconcile the hypotheses of coevolution and genetic trade-offs.

References

Abbas AK, Murphy KM, Sher A (1996) Functional diversity of helper T lymphocytes. Nature 383:787–793

Abel L, Desein AJ (1997) The impact of host genetics on susceptibility to human infectious diseases. Curr Opin Immunol 9:509–516

Akira S, Takeda K, Kaisho T (2001) Toll-like receptors: Critical proteins linking innate and acquired immunity. Nat Immunol 2:675–80

Allen JE, Maizels RM (1997) ThI-Th2: Reliable paradigm or dangerous dogma? Immunol Today 18:387–392

Altizer S, Harvell D, Friedle E (2003) Rapid evolutionary dynamics and disease threats to biodiversity. Trends Ecol Evol 18:589–596

Apanius V, Penn D, Slev PR, Ruff LR, Potts WK (1997) The nature of selection on the major histocompatibility complex. Crit Rev Immunol 17:179–224

Artis D, Humphreys NE, Bancroft AJ, Rothwell NJ, Potten CS, Grencis RK (1999) Tumor necrosis factor is a critical component of interleukin 13-mediated protective T helper cell type 2 responses during helminth infection. J Exp Med 190:953–962

Awasthi M, Cazenave PA, Rath S, Anand RK (1999) TCR V beta 17 gene polymorphism in wild populations of Indian *Mus musculus*. J Biosci 24:163–169

Babik W, Durka W, Radwan J (2005) Sequence diversity of the MHC DRB gene in the Eurasian beaver (*Castor fiber*). Mol Ecol 14:4249–4257

Behnke JM, Wahid FN (1991) Immunological relationships during primary infection with *Heligmosomoides polygyrus*: H-2 linked genes determine worm survival. Parasitology 103:157–64

Behnke JM, Iraqi F, Menge BM, Baker LR, Gibson J, Wakelin D (2003) Chasing the genes that control resistance to gastrointestinal nematodes. J Helminthol 77:99–109

Bendelac A, Lantz O, Quimby ME, Yewdell JW, Bennink JR, Brutkiewicz RR (1995) CD1 recognition by mouse NK1+ T lymphocytes. Science 268:863–865

Ben-Shlomo R, Figueroa F, Klein J, Nevo E (1988) MHC class II DNA polymorphisms within and between chromosomal species of the *Spalax ehrenbergi* superspecies in Israel. Genetics 119:141-149

Berggren KT, Seddon JM (2005) MHC promoter polymorphism in grey wolves and domestic dogs. Immunogenetics 57:267–272

Berggren KT, Ellegren H, Hewitt GM, Seddon JM (2005) Understanding the phylogeographic patterns of European hedgehogs, *Erinaceus concolor* and *E. europaeus* using the MHC. Heredity 95:84–90

Bogdan C (2001) Nitric oxide and the immune response. Nature Immunol 2:907–916

Boulinier T, Sorci G, Monnat JY, Danchin E (1997) Parent-offspring regression suggests heritable susceptibility to ectoparasites in a natural population of kittiwake *Rissa tridactyla*. J Evol Biol 10:77–85

Bryja J, Galan M, Charbonnel N, Cosson J-F (2006) Duplication, balancing selection and trans-species evolution explain the high levels of polymorphism of the DQA MHC class II gene in vole (Arvicolinae). Immunogenetics 58:191–202

Calabi F, Milstein C (2000) The molecular biology of CD1. Seminar Immunol 12:503–509

Cho KS, Zhai S-K, Esteves PJ, Knight KL (2005) Characterization of the T-cell receptor gamma locus and analysis of the variable gene segment expression in rabbit. Immunogenetics 57:352–363

Choi EH, Zimmerman PA, Foster CB, Zhu S, Kamaraswami W, Nutman TB, Chanok SJ (2001) Genetic polymorphisms in molecules of innate immunity and susceptibility to infection with *Wuchereria bancrofti* in South India. Genes Immun 2:248–253

Choi EH, Nutman TB, Chanock SJ (2003) Genetic variation in immune function and susceptibility to human filariasis. Expert Rev Mol Diagn 3:367–74

Coltman DW, Wilson K, Pilkington JG, Stear MJ, Pemberton JM (2001) A microsatellite polymorphism in the gamma interferon gene is associated with resistance to gastrointestinal nematodes in a naturally-parasitized population of Soay sheep. Parasitology 122:571–582

Cooke GS, Hill AVS (2001) Genetics of susceptibility to human infectious disease. Nature Rev Genet 2:967–977

Criscitiello MF, Wermenstam NE, Pilstrom L, McKinney EC (2004) Allelic polymorphism of T-cell receptor constant domains is widespread in fishes. Immunogenetics 55:818–824

Dammers PM, Kroese FGM (2001) Evolutionary relationship between rat and mouse immunoglobulin IGHV5subgroup genes (PC7183) and human IGHV3subgroup genes. Immunogenetics 53:511–517

Dascher CC, Brenner MB (2003) Evolutionary constraints on CD1 structure: Insights from comparative genomic analysis. Trends Immunol 24:412–419

Dascher CC, Hiromatsu K, Naylor JW, Brauer PP, Brown KA, Storey JR, Behar SM, Kawasaki ES, Porcelli SA, Brenner MB, LeClair KP (1999) Conservation of a CD1 multigene family in the guinea pig. J Immunol 163:5478–5488

Dascher CC, Hiromatsu K, Xiong X, Sugita M, Buhlmann JE, Dodge IL, Lee SY, Roura-Mir C, Watts GF, Roy CJ, Behar SM, Clemens DL, Porcelli SA, Brenner MB (2002) Conservation of CD1 intracellular trafficking patterns between Mammalian species. J Immunol 169:6951–6958

Daser A, Mitchison H, Mitchison A, Müller B (1996) Non-classical MHC genetics of immunological disease in man and mouse. Cytokine 8:S593–597

Dissanayake S, Amith RS, Shahin A (2004) *Taenia crassiceps* carbohydrates stimulate IL-6 expression in naïve murine macrophages via Toll-like receptors (TLRs). Mol Immunol 41:391–398

Dutronc Y, Porcelli SA (2002) The CD1 family and T cell recognition of lipid antigens. Tissue Antigens 60:337–53

Edwards SV, Chesnut K, Satta Y, Wakeland EK (1997) Ancestral polymorphism of MHC class II genes in mice: Implications for balancing selection and the mammalian molecular clock. Genetics 146:655–68

Ellegren H, Hartman G, Johansson M, Andersson L (1993) Major histocompatibility complex monomorphism and low levels of DNA fingerprinting variability in a reintroduced and rapidly expanding population of beavers. Proc Natl Acad Sci USA 90:8150–8153

Else KJ, Finkelman FD (1998) Intestinal nematode parasites, cytokines and effector mechanisms. Int J Parasitol 28:1145–1158

Else KJ, Wakelin D, Wassom DL, Hauda K (1990) The influence of gene mapping within the major histocompatibility complex on resistance to *Trichuris muris* infections in mice. Parasitology 101:61–67

Esteves PJ, Lanning D, Ferrand N, Knight KL, Zhai SK, Van der Loo W (2004) Allelic variation at the VHa locus in natural populations of rabbit (*Oryctolagus cuniculus*, L.). J Immunol 172:1044–1053

Faveeuw C, Angeli V, Fontaine J, Maliszewski C, Capron A, Van Kaer L, Moser M, Capron M, Trottein F (2002) Antigen presentation by CD1d contributes to the amplification of Th2 responses to *Schistosoma mansoni* glycoconjugates in mice. J Immunol 169:906–12

Figueroa F, Gunther E, Klein J (1988) MHC polymorphism pre-dating speciation. Nature 335:265–267

Finkelman FD, Wynn TA, Donaldson DD, Urban JF (1999) The role of IL-13 in helminth-induced inflammation and protective immunity against nematode infections. Curr Opin Immunol 11:420–426

Frank SA (2002) Immunology and evolution of infectious disease. Princeton Univ Press

Frisch JE, O' Neill CJ, Kelly MJ (2000) Using genetics to control cattle parasites: The Rockhampton experience. Int J Parasitol 30:253–264

Froeschke G, Sommer S (2005) MHC class II DRB variability and parasite load in the striped mouse (*Rhabdomys pumilio*) in the southern Kalahari. Mol Biol Evol 22:1254–1259

Funkhouser W, Koop BF, Charmley P, Martindale D, Slightom J, Hood L (1997) Evolution and selection of primate T cell antigen receptor BV8 gene subfamily. Mol Phyl Evol 8:51–64

Gause WC, Urban JF, Stadecker MJ (2003) The immune response to parasitic helminths: Insights from murine models. Trends Immunol 24:269–277

Geraghty DE (2002) Genetic diversity and genomics of the immune response. Immunol Rev 190:5–8

Geraghty DE, Daza R, Williams LM, Vu Q, Ishitani A (2002) Genetics of the immune response: Identifying immune variation within the MHC and throughout the genome. Immunol Rev 190:69–85

Goüy de Bellocq J, Delarbre C, Gachelin G, Morand S (2005) Allelic diversity at the MHC-DQA locus of woodmouse populations (*Apodemus sylvaticus*) present in island and mainland northern Mediterranean area. Global Ecol Biogeogr 14:115–122

Grencis RK (1997) Th2-mediated host protective immunity to intestinal nematode infections. Phil Trans R Soc Lond 352:1377–1384

Grencis RK (2001) Cytokine regulation of resistance and susceptibility to intestinal nematode infection, from host to parasite. Vet Parasitol 100:45–50

Günther E, Walter L (2001) The major histocompatibility complex of the rat (*Rattus norvegicus*). Immunogenetics 53:520–542

Hambuch TM, Lacey EA (2002) Enhanced selection for MHC diversity in social tuco-tucos. Evolution 56:841–845

Han M, Hannick LI, DiBrino M, Robinson MA (1999) Polymorphism of human CD1 genes. Tissue Antigens 54:122–127

Harf R, Sommer S (2005) Association between major histocompatibility complex class II DRB alleles and parasite load in the hairy-footed gerbil, *Gerbillurus paeba*, in the southern Kalahari. Mol Ecol 14:85–91

Hauptmann G, Bahram S (2004) Genetics of the central MHC. Curr Opin Immunol 16:668–672

Hayes KS, Knight KL (2001) Group 1 CD1 genes in rabbit. J Immunol 166:403–410

Hayes KS, Bancroft AJ, Grencis RK (2004) Immune-mediated regulation of chronic intestinal nematode infection. Immunol Rev 201:75–88

He L, Hu CS, Cai GB, Zhang QP, Li Q, Zhang XL, Huang BJ, Zhang LJ, Liu JY, Jiang MS, Tan JQ (2004) Highly up-regulated CXCR3 expression on eosinophils in mice infected with *Schistosoma japonicum*. Immunology 111:107–117

Hedrick PW (2001) Conservation genetics: Where are we now? Trends Ecol Evol 16:629–636

Helmby H, Grencis RK (2003) Essential role for TLR4 and MyD88 in the development of chronic intestinal nematode infection. Eur J Immunol 33:2974–2979

Herbst MM, Prescott J, Palmer AD, Schountz T (2002) Sequence and expression analysis of deer mouse interferon-gamma, interleukin-10, tumor necrosis factor, and lymphotoxin-alpha. Cytokine 17:203–213

Hill AVS (2001) Immunogenetics and genomics. Lancet 357:2037–2041

Hiromatsu K, Dascher C, Sugita M, Gingrich-Baker C, Behar SM, LeClair KP, Brenner MB, Porcelli SA (2002) Characterization of guinea-pig group 1 CD1 proteins. Immunology 106:159–172

Hise AG, Hazlett FE, Bockarie MJ, Zimmerman PA, Tisch DJ, Kazura JW (2003) Polymorphisms of innate immunity genes and susceptibility to lymphatic filariasis. Genes Immun 4:524–527

Hughes AL (1995) The evolution of the type I interferon gene family in mammals. J Mol Evol 41:539–548

Hughes AL (2002) Natural selection and the diversification of vertebrate immune effectors. Immunol Rev 190:161–168

Hughes AL, Yeager M (1998) Natural selection at major histocompatibility complex loci of vertebrates. Ann Rev Genet 32:415–435

Huppi KE, D' Hoostelaere LA, Mock BA, Jouvin-Marchel E, Behlke MA, Chou HS, Berry RJ, Loh DY (1988) T-cell receptor $V_T\beta$ genes in natural populations of mice. Immunogenetics 27:51–56

Hurt P, Walter L, Sudbrak R, Klages S, Muller I, Shiina T, Inoko H, Lehrach H, Gunther E, Reinhardt R, Himmelbauer H (2004) The genomic sequence and comparative analysis of the rat major histocompatibility complex. Genom Res 14:631–639

Ibberson MR, Copier JP, Llop E, Navarrete C, Hill AVS, Cruickshank JK, So AKL (1998) T-cell receptor variable alpha (TCRAV) polymorphisms in European, Chinese, South American, AfroCaribbean, and Gambian populations. Immunogenetics 47:124–130

Iraqi FA, Behnke JM, Menge DM, Lowe AM, Teale AJ, Gibson JP, Baker LR, Wakelin D (2003) Chromosomal regions controlling resistance to gastrointestinal nematode infections in mice. Mammal Genom 14:184–191

Jaeger EEM, Bontrop RE, Parham P, Wickings EJ Kadwell M, Lanchbury JS (1998) Characterization of chimpanzee TCRV gene polymorphism: How old are human TCRV alleles? Immunogenetics 47:115–123

Jenkins SJ, JP Hewitson, GR Jenkins, Mountford AP (2005) Modulation of the host's immune response by schistosome larvae. Parasite Immunol 27:385–393

Johannesen J, Karlsen AE, Pociot F, Roenn SG, Nerup J (2003) Strain dependent rat iNOS promoter activity-correlation to identified WT1 transcription factor binding site. Autoimmunity 36:167–175

Jouvin-Marche E, Morgado MG, Trede N, Marche PN Couez D, Hue I, Gris C, Malissen M, Cazenave PA (1989) Complexity, polymorphism, and recombination of mouse T-cell receptor α gene families. Immunogenetics 30:99–104

Kanazawa T, Asahi H, Hata H, Mochida K, Kagei N, Stadecker MJ (1993) Arginine-dependent generation of reactive nitrogen intermediates is instrumental in the in vitro killing of protoscoleces of *Echinococcus multilocularis* by activated macrophages. Parasite Immunol 15:619–623

Katabami S, Matsuura A, Chen H, Imai K, Kikuchi K (1998) Structural organization of rat CD1 typifies evolutionarily conserved CD1D class genes. Immunogenetics 48:22–31

Kawai T, Akira S (2005) Pathogen recognition with Toll-like receptors. Curr Opin Immunol 17:338–344

Kawecki T, Ebert D (2004) Conceptual issues in local adaptation. Trends Ecol Evol 7:1225–1241

Kelley J, Walter L, Trowsdale J (2005) Comparative genomics of major histocompatibility complexes. Immunogenetics 56:683–695

Kerr RJ, Frisch JE, Kinghorn BP (1994) Evidence for a major gene for tick resistance in cattle. In: Anonymous (ed) Proceedings of the 5th World congress on genetics applied to livestock production, vol 20. Guelph, Canada, pp 265–268

Keymer AE, Tarlton AB, Hiorns RW, Lawrence CE, Pritchard DI (1990) Immunogenetic correlates of susceptibility to infection with *Heligmosomoides polygyrus* in outbred mice. Parasitology 101:69–73

Klein J (1987) Origin of major histocompatibility complex polymorphism: The trans-species hypothesis. Human Immunol 19:155–162

Korten S, Volkmann L, Saeftel M, Fischer K, Taniguchi M, Fleischer B, Hoerauf A (2002) Expansion of NK cells with reduction of their inhibitory Ly-49A, Ly-49C, and Ly-49G2 receptor-expressing subsets in a murine helminth infection: Contribution to parasite control. J Immunol 168:5199–5206

Kundu S, Faulkes CG (2004) Patterns of MHC selection in African mole-rats: The effects of sociality and habitat. Proc Biol Sci 271:273–278

Lawrence CE (2003) Is there a common mechanism of gastrointestinal nematode expulsion? Parasite Immunol 25:271–281

Lawrence CE, Paterson JCM, Higgins LM, MacDonald TT, Kennedy MW, Garside P (1998) IL-4-regulated enteropathy in an intestinal nematode infection. Eur J Immunol 28:2672–2684

Lehmann T (1992) Ectoparasite impacts on *Gerbillus andersoni allenbyi* under natural conditions. Parasitology 104:479–488

Livak N (2003) Evolutionarily conserved pattern of gene segment usage within the mammalian TCRb locus. Immunogenetics 55:307–314

Lobigs M, Mullbacher A, Blanden RV, Hammerling GJ, Momburg F (1999) Antigen presentation in syrian hamster cells: Substrate selectivity of TAP controlled by polymorphic residues in TAP1 and differential requirements for loading of H2 class I molecules. Immunogenetics 49:931–41

Lopez-Briones S, Soloski MJ, Bojalil R, Fragoso G, Sciutto E (2001) CD4+ TCRαβ T cells are critically involved in the control of experimental murine cysticercosis in C57BL/6J mice. Parasitol Res 87:826–832

Lopez-Briones S, Lamoyi E, Fragoso G, Soloski MJ, Sciutto E (2003) *Taenia crassiceps* cysticercosis: Immune response in susceptible and resistant BALB/c mouse substrains. Parasitol Res 90:236-242

Mackelprang R, Carlson CS, Subrahmanyan L, Livingston RJ Eberle DA, Nickerson MA (2002) Sequence variation in the human T-cell receptor loci. Immunol Rev 190:26–39

Maizels RM, Blaxter ML, Scott AL (2001a) Immunological genomics of *Brugia malayi*: Filarial genes implicated in immune evasion and protective immunity. Parasite Immunol 23:327–344

Maizels RM, Gomez-Escobar N, Gregory WF, Murray J, Zang X (2001b) Immune evasion genes from filarial nematodes. Int J Parasitol 31:889–898

Maizels RM, Balic A, Gomez-Escobar N, Nair M, Taylor MD, Allen JE (2004) Helminth parasites – masters of regulation. Immunol Rev 201:89–116

Marusina K, Iyer M, Monaco JJ (1997) Allelic variation in the mouse Tap-1 and Tap-2 transporter genes. J Immunol 158:5251–5256

Mattioli RC, Bah M, Faye J, Kora S, Cassama M (1993) A comparison of field tick infestation on N'Dama, Gobra zebu and N'Dama x Gobra crossbred cattle. Vet Parasitol 47:139–148

Mattioli RC, Pandey VS, Murray M, Fitzpatrick JL (2000) Immunogenetic influences on tick resistance in African cattle with particular reference to trypanotolerant N'Dama (*Bos taurus*) and trypanosusceptible Gobra zebu (*Bos indicus*) cattle. Acta Trop 75:263–277

McCluskey J, Rossjohn J, Purcell AW (2004) TAP genes and immunity. Curr Opin Immunol 16:651–659

McCoy KD, Boulinier T, Tirard C, Michalakis Y (2003) Host-dependent genetic structure of parasite populations: Differential dispersal of seabird tick host races. Evolution 57:288–96

Menge BM, Behnke JM, Lowe A, Gibson JP, Iraqi FA, Baker RL, Wakelin D (2003) Mapping of chromosomal regions influencing immunological responses to gastrointestinal nematode infections in mice. Parasite Immunol 25:341–349

Meyer-Lucht Y, Sommer S (2005) MHC diversity and the association to nematode parasitism in the yellow-necked mouse (*Apodemus flavicollis*). Mol Ecol 14:2233–2243

Mitchison A (1997) Partitioning of genetic variation between regulatory and coding gene segments: The predominance of software variation in genes encoding introvert proteins. Mol Biol Evol 46:46–51

Mitchison NA, Roes J (2002) Patterned variation in murine MHC promoters. Proc Natl Acad Sci USA 99:10561–10566

Morand S (1996) Life-history traits in parasitic nematodes: A comparative approach for the search of invariants. Funct Ecol 10:210–218

Musolf K, Meyer-Lucht Y, Sommer S (2004) Evolution of MHC-DRB class II polymorphism in the genus *Apodemus* and a comparison of DRB sequences within the family Muridae (Mammalia: Rodentia). Immunogenetics 56:420–426

Nei M, Gu X, Sitnikova T (1997) Evolution by the birth-and-death process in multigene families of the vertebrate immune system. Proc Natl Acad Sci USA 94:7799–7806

Nevo E, Beiles A (1992) Selection for class II MHC heterozygosity by parasites in subterranean mole rats. Experientia 48:512–515

Nobuhara H, Kuida K, Furutani M, Shiroishi T, Moriwaki K, Yanagi Y, Tada T (1989) Polymorphism of T-cell receptor genes among laboratory and wild mice: Diverse origins of laboratory mice. Immunogenetics 30:405–413

Ohno T, Ishih A, Anaka T, Nishimura M, Terada M (2002) Chromosomal mapping of host susceptibility loci to *Angiostrongylus costaricensis* nematode infection in mice. Immunogenetics 53:925–929

Ollier WER (2004) Cytokine genes and disease susceptibility. Cytokine 28:174–178

Oswald IP, Eltoum I, Wynn TA, Schwartz B, Caspar P, Paulin D, She, A, James S L (1994) Endothelial cells are activated by cytokine treatment to kill an intravascular parasite, *Schistosoma mansoni*, through the production of nitric oxide. Proc Natl Acad Sci USA 91:999–1003

Penn DJ, Potts WK (1999) The evolution of mating preferences and major histocompatibility complex genes. Amer Nat 153:145–164

Pernthaner A, Cole SA, Morrison L, Hein WR (2005) Increased expression of interleukin-5 (IL-5), IL-13, and tumor necrosis factor alpha genes in intestinal lymph cells of sheep selected for enhanced resistance to nematodes during infection with *Trichostrongylus colubriformis*. Inf Immunol 73:2175–2183

Peters AD, Lively CM (1999) The Red Queen and fluctuating epistasis: A population genetic analysis of antagonistic coevolution. Amer Nat 154:393–405

Pfaff AW, Schulz-Key H, Soboslay PT, Geiger SM, Hoffman WH (2000) The role of nitric oxide in the innate resistance to microfilariae of *Litomosoides sigmodontis* in mice. Parasite Immunol 22:397–405

Pfau RS, Van Den Bussche RA, McBee K, Lochmiller RL (1999) Allelic diversity at the Mhc-DQA locus in cotton rats (*Sigmodon hispidus*) and a comparison of DQA sequences within the family Muridae (Mammalia: Rodentia). Immunogenetics 49:886–93

Pfau RS, van den Bussche RA, McBee K (2001) Population genetics of the hispid cotton rat (*Sigmodon hispidus*): Patterns of genetic diversity at the major histocompatibility complex. Mol Ecol 10:1939–1945

Poltorak A, Ricciardi-Castagnoli P, Citterio S, Beutler B (2000) Physical contact between lipopolysaccharide and Toll-like receptor 4 revealed by genetic complementation. Proc Natl Acad Sci USA 97:2163–2167

Porcelli SA, Segelke BW, Sugita M, Wilson IA, Brenner MB (1998) The CD1 family of lipid antigen presenting molecules. Immunol Today 19:289–294

Prugnolle F, Manica A, Charpentier M, Guégan J-F, Guernier V, Balloux F (2005) Pathogen-driven selection and worldwide HLA class I diversity. Curr Biol 15:1022–1027

Quinnell RJ (2003) Genetics of susceptibility to human helminth infection. Int J Parasitol 33:1219–1231

Rajan TV, Porte P, Yates JA, Keefer L, Schultz LD (1996) Role of nitric oxide in host defense against an extracellular, metazoan parasite, *Brugia malayi*. Inf Immunol 64:3351–3353

Rajan B, Ramalingam T, Rajan TV (2005) Critical role for IgM in host protection in experimental filarial infection. J Immunol 175:1827–1833

Richman AD, Herrera LG, Nash D (2001) MHC class II beta sequence diversity in the deer mouse (*Peromyscus maniculatus*): Implications for models of balancing selection. Mol Ecol 10:2765–2773

Richman AD, Herrera LG, Nash D (2003) Evolution of MHC class II beta diversity within the genus *Peromyscus*. Genetics 164:289–297

Richter G, Blankenstein T, Diamantstein T (1990) Evolutionary aspects, structure, and expression of the rat interleukin 4 gene. Cytokine 2:221–228

Roach JC, Glusman G, Rowen L, Kaur A, Purcell MK, Smith KD, Hood LE, Aderem A (2005) The evolution of vertebrate Toll-like receptors. Proc Natl Acad Sci USA 102:9577–9582

Rockman MV, Hahn MW, Soranzo N, Goldstein DB, Wray GA (2003) Positive selection on a human-specific transcription factor binding site regulating IL4 expression. Curr Biol 13:2118–2123

Roger T, Pepin LF, Jouvin-Marche E, Cazenave PA, Seman M (1993) New T-cell receptor gamma haplotypes in wild mice and evidence for limited Tcrg-V gene polymorphism. Immunogenetics 37:161–169

Sanjayan MA, Crooks K, Zegers G, Foran D (1996) Genetic variation and the immune response in natural populations of pocket gophers. Cons Biol 10:1519–1527

Schmid-Hempel P, Ebert D (2003) On the evolutionary ecology of specific immune defence. Trends Ecol Evol 18:27–32

Schroder NW, Schumann RR (2005) Single nucleotide polymorphisms of Toll-like receptors and susceptibility to infectious disease. Lancet Inf Dis 5:156–164

Seddon JM, Baverstock PR (1999) Variation on islands: Major histocompatibility complex (MHC) polymorphism in populations of the Australian bush rat. Mol Ecol 8:2071–2079

Seddon JM, Baverstock PR (2000) Evolutionary lineages of RT1.Ba in the Australian *Rattus*. Mol Biol Evol 17:768–772

Sher A, Hieny S, James S (1984) Mechanisms of protective immunity against *S. mansoni* infection in mice vaccinated with irradiated cercariae. VI. Influence of the major histocompatibility complex. Parasite Immunol 6:319–328

Singh SK, Girschick HJ (2003) Tick-host interactions and their immunological implications in tick-borne diseases. Curr Sci 85:1284–1298

Sitnikova T, Nei M (1998) Evolution of immunoglobulin kappa chain variable-region genes in vertebrates. Mol Biol Evol 15:50–60

Sitnikova T, Su C (1998) Coevolution of immunoglobulin heavy and light chain variable region gene families. Mol Biol Evol 15:617–625

Smirnova I, Poltorak A, Chan EKL, McBride C, Beutler B (2000) Phylogenetic variation and polymorphism at the Toll-like receptor 4 locus (TLR4). Genome Biol 1:1–10

Smirnova I, Hamblin MT, McBride C, Beutler B, Di Rienzo A (2001) Excess of rare amino acid polymorphisms in the Toll-like receptor 4 in humans. Genetics 158:1657–1664

Smulders MJM, Snoek LB, Booy G, Vosman B (2003) Complete loss of MHC genetic diversity in the common hamster (*Cricetus cricetus*) population in the Netherlands. Cons Genetics 4:441–451

Sommer S (2005) The importance of immune gene variability (MHC) in evolutionary ecology and conservation. Front Zool 2:16

Sommer S, Tichy H (1999) Major histocompatibility complex (MHC) class II polymorphism and paternity in the monogamous *Hypogeomys antimena*, the endangered, largest endemic Malagasy rodent. Mol Ecol 8:1259–1272

Sommer S, Schwab D, Ganzhorn JU (2002) MHC diversity of endemic Malagasy rodents in relation to geographic range and social system. Behav Ecol Sociobiol 51:214–221

Sorci G, Møller AP, Boulinier T (1997) Genetics of host-parasite interactions. Trends Ecol Evol 12:196–200

Su C, Nei M (1999) Fifty-million-year-old polymorphism at an immunoglobulin variable region gene locus in the rabbit evolutionary lineage. Proc Natl Acad Sci USA 96:9710–9715

Su C, Nei M (2001) Evolutionary dynamics of the T-cell receptor VB gene family as inferred from the human and mouse genomic sequences. Mol Biol Evol 18:503–513

Su C, Jakobsen I, Gu X, Nei M (1999) Diversity and evolution of T-cell receptor variable region genes in mammals and birds. Immunogenetics 50:301–308

Tanaka T, Nei M (1989) Positive darwinian selection observed at the variable-region genes of immunoglobulins. Mol Biol Evol 6:447–459

Taylor MJ, Cross HF, Mohammed AA, Trees AJ, Bianco AE (1996) Susceptibility of *Brugia malayi* and *Onchocerca lienalis* microfilariae to nitric oxide and hydrogen peroxide in cell-free culture and from IFN gamma-activated macrophages. Parasitology 112:315–322

Trottein F, Mallevaey T, Faveeuw C, Capron M, Leite-de-Moraes M (2006) Role of the natural killer T lymphocytes in Th2 responses during allergic asthma and helminth parasitic diseases. Parasite Allerg 90:113–127

Trouvé S, Sasal P, Jourdane J, Renaud F, Morand S (1998) Revisited views on the evolution of life-history traits in parasitic and free-living platyhelminthes. Oecologia 115:370–378

Trowsdale J, Parham P (2004) Defense strategies and immunity-related genes. Eur J Immunol 34:7–17

Utech KBW, Wharton RH (1982) Breeding for resistance to *Boophilus microplus* in Australian Illawarra Shorthorn and Brahman x Australian Illawarra Shorthorn cattle. Austral Vet J 58:41–46

Van der Kleij D, van Remoortere A, Schuitemaker JH, Kapsenberg ML, Deelder AM,Tielens AG, Hokke CH, Yazdanbakhsh M (2002) Triggering of innate immune responses by schistosome egg glycolipids and their carbohydrate epitope GalNAc beta 1-4GlcNAc. J Infect Dis 185:531–539

Van der Loo W (1993) Variance analysis of immunoglobulin alleles in natural populations of rabbit (*Oryctolagus cuniculus*): The extensive interallelic divergence at the b locus could be the outcome of overdominance-type selection. Genetics 134:171–187

Van der Loo W, Boussès P, Arthur CP, Chapuis JL (1996) Compensatory aspects of allele diversity at Ig loci: Gene correlations in rabbit populations devoid of light chain diversity. Genetics 144:1181–1194

Wahid F, Behnke JM (1993) Immunological relationships during primary infection with *Heligmosomoides polygyrus*. Regulation of response phenotype by H2 and non-H2 genes. Parasitology 107:343–50

Wakelin D (1975) Genetic control of immune responses to parasites: Immunity to *Trichuris muris* in inbred and random-bred strains of mice. Parasitology 70:393–405

Wakelin D (1991) Model systems on the genetic basis of disease resistance. In: Owen JB, Axford RFE (eds) Breeding for disease resistance in farm animals. CABI, Wallingford, pp 54–70

Wakelin D, Farias SE, Bradley JE (2002) Variation and immunity to intestinal worms. Parasitology 125:S39–S50

Wegner KM, Kalbe M, Schaschl H, Reusch TBH (2004) Parasites and individual major histocompatibility complex diversity - an optimal choice? Microb Inf 6:1110–1116

Weinstock JV, Summers R, Elliott D (2004) Helminths and harmony. Gut 53:7–9

Wikel SK, Bergman D (1997) Tick-host immunology: Significant advances and challenging opportunities. Parasitol Today 13:383–389

Zelus D, Robinson-Rechavi M, Delacre M, Auriault C, Laudet V (2000) Fast evolution of interleukin-2 in mammals and positive selection in ruminants. J Mol Evol 51:234–244

Zhang J, Nei M (2000) Positive selection in the evolution of mammalian interleukin-2 genes. Mol Biol Evol 17:1413–1416

Zhang S, Penfornis A, Harraga S, Chabod J Beurton I, Bresson-Hadni S, Tiberghien P, Kern P, Vuitton DA (2003) Polymorphisms of the TAP genes in human alveolar echinococcosis. Eur J Immunogenet 30:133–139

21 Interactions among immune, endocrine, and behavioural response to infection

Zachary M. Weil, Lynn B. Martin II and Randy J. Nelson

1 Introductory remarks

Macroparasite infection is common among small mammals. As such, extensive coevolution between parasite and host species has occurred. In this chapter, we consider several aspects of the physiology of host-parasite relationships. In the first section, we discuss the generalities of host immune defences against macroparasites and the myriad tactics that parasites employ to avoid or minimize these responses. In the second section, we describe known effects of macroparasites on host behaviour and how hosts behave to eliminate or control macroparasite infections. In the final section, we consider the role of hormones in modulating behavioural and immune responses to parasites, highlighting the seasonal interplay between changes in host immunophysiology and macroparasite abundance and diversity that has been described in some species. Throughout the chapter, we discuss phenomena from both ultimate and proximate perspectives. However, we focus primarily on the regulatory processes that determine how micromammals avoid or resist macroparasite infections in natural contexts.

2 The vertebrate immune system

The vertebrate immune system is a complex network of cells, tissues, and soluble substances that either prevents macroparasite infections or eliminates or controls parasitic agents once infection occurs. Although much has been learned about host-parasite immune interactions in model species of domestic rodents (e.g., rats, mice, and guinea pigs), it is not apparent how relevant these studies are for understanding how free-living small mammals defend themselves against macroparasite attacks. The scant evidence available indicates that wild animals rarely control and/or eliminate infection as readily as domesticated species. Further, wild animals rarely exhibit sterilizing immunity; i.e., they generally attempt to control but not eliminate most macroparasite infections (Wikel 2002). Below, we discuss how domestic and wild small mammals defend themselves immunologically against the various macroparasites discussed elsewhere in this volume. Instead of discussing the minutiae of each macroparasite-host interaction however, we highlight generalities among these interactions, as the immunological activities involved in each host-parasite relationship vary slightly.

Macroparasites preceded the phylogenetic development of the vertebrate immune system by millions of years, so it is to be expected that the immune defences of small mammals have been shaped over time to control these infections (Barriga 1999). The vertebrate immune system can be coarsely divided into two components: innate and adaptive (Janeway et al. 1999). The major distinction between these two lines of defence is that the innate system non-specifically destroys non-self substances whereas the adaptive system generates immunological memory of encounters with non-self substances and effects targeted attacks against these substances upon secondary interactions. The innate immune system includes cells such as macrophages, basophils, granulocytes, and other soluble compounds such as complement, acute phase proteins, and a host of enzymes and lytic molecules, all of which are often managed by the adaptive immune system. The adaptive immune system is comprised of two groups of lymphocytes, one of which is derived predominantly from bone marrow (B-cells). These cells either produce antibodies (immunoglobulins: IgG, IgM, IgE, IgA etc.) that control infection directly or more often target parasites or parasite-derived molecules for destruction (through a process called opsonization). The other group of lymphocytes (T-cells) originates in the thymus. These cells can also be divided into classes. T helper cells are responsible for managing immune responses, particularly the activity and movement of innate system effector cells. T killer cells, or cytotoxic T lymphocytes, attack and destroy self cells infected by intracellular pathogens. A third group of T cells, T regulatory cells, has been proposed, but their function and identity remains in debate. The actions of all of these cells are regulated by cytokines and chemokines. Although cytokines and chemokines are diverse in structure and function, they predominantly orchestrate one of two immune defence strategies. These strategies are referred to as Th1 versus Th2 responses, with Th representing the predominant T helper cell type involved in the immune response. Generally, Th1 responses induce inflammatory and/or cytotoxic cell activity and are best at controlling intracellular infections, such as those caused by viruses. Th2 responses on the other hand drive humoral, or B-cell mediated immune responses, and hence are better at controlling extra-cellular parasites. Specific cytokine mediators of each response are variable and depend on the parasite and location within the body where infection occurs (Mahida 2003). However, some general patterns occur during responses to particular parasite types.

3 Immune responses to macroparasites

Macroparasites differ from bacteria, viruses, and protists in terms of how the vertebrate immune system must deal with them to prevent or control infection. First, macroparasites are generally larger and hence contain many more antigens to which immune responses can be generated. Intestinal helminthes in particular have been estimated to possess 7 to 20 thousand protein encoding genes, giving their hosts ample targets for immune attack (Pearce and Tarleton 2002). From the host's perspective, this characteristic can make macroparasite control difficult because parasites change antigenicity over the period of host colonization (i.e. – across developmental stages). In other words, hosts must constantly generate and mount immune responses against "moving targets" (Grzych et al. 1991). Second, parasites in most cases have an interest in promoting the survival of their host. Although macroparasites must subvert some immune defences in order to reproduce successfully, their tendency to replicate outside hosts generally precludes high virulence. The optimal strategy for a chronically infective macroparasite therefore is to suppress components of the immune system that prevent parasite persistence and reproduction but do not compromise the host to the point that it is killed by another infectious agent (Pearce and Tarelton 2002).

Several traits of macroparasites allow them to be successful in these efforts. First, they can often migrate to other areas of the body and hence avoid some immune responses generated against them. Second, they typically progress through multiple developmental stages within hosts, so immune defences generated against them may lose effectiveness over time. Finally, some parasites possess traits that allow them to avoid immune detection altogether. Cysts of *Echinococcus granulosus*, for instance, incorporate host complement regulatory factors into their outer membranes, which allow them to avoid innate-mediated immune attacks (Finkelson 1995).

From the perspective of hosts, these traits make macroparasites better candidates for control rather than eradication. Indeed, in the wild, it is rare to find animals that engage sterilizing levels of immune activity (i.e., complete clearance) of intestinal parasites (Viney 2002). The more common pattern is maintenance of some low level of infection. By maintaining such a modest infection, hosts keep intact resistance they generate against primary infections, which some have suggested may provide them with increased defence against secondary infections (Viney 2002). This concomitant immunity, as it is commonly referred, effectively represents the hosts use of the parasite itself as a first line of defence against further infection.

A second and less obvious benefit of parasitism is the avoidance of autoimmune diseases that animals may experience late in life if they are infected (with low levels of parasites) early in life. Non-obese diabetic (NOD) mice, for example, show signs of diabetes soon after they reach adulthood. If they are experimentally infected early in life, onset of diabetes in adulthood never occurs. These early-life infections, which are often associated with elevated proinflammatory cytokine levels, are believed to oppose autoimmune cytokine production and induce greater T-regulatory cell activity in adulthood (Thomas et al. 2004).

Finally, individual hosts vary in their ability to control parasite infections depending on the time of year or current life stage through which they are progressing. For example, after experimental infection with *Nippostrongylus brasiliensis* pregnant female rats harboured more adult nematodes in their intestines and more eggs in their colons than age-matched virgin rats (Houdijk et al. 2003). Although it is not apparent what specific immune mediators produced this outcome, other work suggests that multiple aspects of immune activity vary with physiological state and time of year in small mammals (Nelson 2004).

3.1 Endoparasites

Helminths represent the most prevalent macroparasite group. The most common infectious genera for small mammals include *Ascaris*, *Trichuris*, *Strongyloides*, and *Trichinella*. Endoparasitic macroparasites can enter the host three ways: directly through the skin, via food or water consumption, or through the bite of blood-feeding insects. Unlike bacterial and viral infections, the number of macroparasites that hosts harbour reflects the number of times they came into contact with parasites; rarely do endoparasites replicate within hosts (Scott and Grencis 2002). Once parasites enter the host's body, the host's basic strategy is to make the local environment intolerable. Depending on the parasite, this strategy results either in expulsion (*Nippostrongylus*, *Trichinella*, and *Strongyloides*) or chronic mild infection (*Trichuris* and *Heligosomoides*). Regardless of the outcome, similar strategies of defence are enacted. Indeed, although hosts are generally able to recognize intestinal parasites, they show little ability to distinguish among different types (at least upon primary infection) to orchestrate more coordinated immunological attacks (Finkelman et al. 1997).

Four generalizations can be made about the immune processes engaged to combat intestinal parasites: (1) CD4+ T-cells are critical for protection, (2) IL-12 and IFNγ can counteract protective immunity generated by CD4+ cells and the effector mechanisms they induce, (3) IL-4 is either imperative for protection, important for limiting severity of infection, or pivotal in inducing redundant forms of protection, and (4) other cytokines increase in circulation during infection, but do not directly affect parasites (Finkelman et al. 1997). In terms of the specifics of immune control, Th1 mediated immune responses predominantly control initial infection whereas Th2 responses provide defence against chronic infection (Tarleton et al. 2000). Th1 responses are characterized by abundant IL-2 and IFNγ production, which leads to high cytolytic activity and increased complement activity. Such activities are effective at eliminating infiltrating larval stages of parasites. However, if a Th1 bias persists for long within a host, that host can become more susceptible to a persistent infection. High IL-12 production, which occurs late during Th1 responses, can increase helminth susceptibility (Else and Finkelman 1998).

Th1 biased responses activate many innate effector cells. Mast cells in particular are important for nematode resistance, and their activity is T-cell-dependent (Else and Finkelman 1998). Eosinophils are also important for resolving helminth infections (Butterworth 1984), but their efficacy depends on the developmental stage and parasite identity. Generally, eosinophils prevent larval establishment (Meeusen and Balic 2000). However, no studies have demonstrated that these cells affect adult parasite stages *in vivo*. Further, eosiniphilia is only effective against some endoparasites; *Taenia taeniformis, Trichuris muris*, and *N. brasiliensis* resistance is positively correlated with capacity to generate eosinophilia in mice, but other parasites show no such relationships. Also, elimination of eosinophils during *T. spiralis* infections delays expulsion of parasites, but the same treatment has no effect for *N. brasiliensis* infections (Meeusen and Balic 2000). Eosinophilia is mediated by IL-5 secretion by T-cells or NK-cells, but B-cells serve important roles in these responses as antibodies must label parasites for eosinophils to have an effect (Hunter and Sher 2002).

Although larval stages of endoparasites are susceptible to innate effector cells, adult stages are better controlled by antibody-mediated defences, especially once they have established (Hunter and Sher 2002). Th2 cytokines, such as IL-4, IL-5, IL-10, and IL-13 are produced in abundance in response to adult parasite infections, and they are effective at mast cell activation, inducing eosinophilia, and suppression of Th1 responses (Mahida

2003). In some cases, Th2 cytokines can induce other parasite defences that are not immunological in nature. For instance, some Th2 cytokines alter gut mucosa structure and composition, including the intestinal cell phenotype composition and smooth muscle structure (Mahida 2003). Although Th1 versus Th2 biases affects host success against controlling infections, the effector mechanisms that are induced by these cytokine cascades are not well understood. For instance, production of IL-4 and IL-13 is related to control of *Heligosomoides* infection in mice, but not *Trichuris* or *Nippostrongylus* (Scott and Grencis 2002).

Until recently, these mechanisms were thought to be the primary pathways through which endoparasitic infections were fought. Recently, two other cell types, CD5+ B1-cells and γδ T cells, were discovered to have anti-helminth properties. CD5+ B1-cells combat helminths by producing low affinity poly-reactive antibodies that recognize a variety of polysaccharides. These B cells respond to IL-10, a Th2 cytokine, and are present in large numbers in the body cavity. CD5+ B1-cells proliferate and produce IL-10 in response to *Schistosoma mansoni* eggs in mice (Vellupillai and Harn 1994). γδ T cells express a diverse array of receptors, and are abundant in epithelial and mucosal tissue. In rats, *N. brasiliensis* infection induced proliferation of these cells and production of IL-4 (Rosat et al. 1995) initiating further Th2 mediated immune activity.

In common with these cell types, other soluble compounds have roles in parasite defence that were only recently recognized. One particularly promising line of research involves mannan-binding protein, which is produced by hosts and is capable of detecting mannose residue on parasite surfaces (Hunter and Sher 2002). Upon contacting mannose, these proteins initiate lectin-mediated complement activity, which allows hosts to eliminate parasites before they become established in tissue. Additional comparable mechanisms of detection may also occur in mammals. Some have even predicted that immune cells may possess receptors for particular components of parasite tissue much like Toll-like receptors are receptive to components of bacteria (Medzhitov et al. 1997).

Although not all of the mechanisms that small mammals use to prevent or control intestinal parasite infection have been identified, their immune defences are effective against the parasites that infect them. In rats, *Strongyloides ratti* become shorter and less fecund over the course of an infection. If the same individual parasites are transplanted into a naïve host rat, then parasites re-grow to their original size and produce eggs at higher rates, although these changes last only until their new hosts generate immune responses once again (Schad et al. 1997). In other words, changes in parasite fitness are directly related to the activity of the hosts' immune system, even in captive-housed animals. Indeed, shrinkage and decreased fecundity of parasites in the case above was caused by a plug being generated over the oral cavity of worms; this plug, which is partly composed of host immunoglobulins, effectively starves the worm. If hosts are treated with immunosuppressive glucocorticoids, these effects are eliminated (Wilkes et al. 2004).

3.2 Ectoparasites

Ectoparasitic macroparasites take one of two general forms: (1) individuals attach and feed on blood or other bodily fluids of the host over an extended period (e.g. – ticks and lice), or (2) they attack and feed rapidly (e.g. – biting flies and most fleas). Because these two groups use different strategies to obtain resources from hosts, the tactics they use to subvert host immune defence and the immune defences that hosts engage to combat them vary. Much like endoparasitic helminths, ectoparasites that use an attachment strategy tend to down-regulate host immune defences to a level that promotes feeding but not host mortality. Rapid feeding ectoparasites also depress some aspects of the hosts' immune system to obtain a blood meal. Even though they encounter components of the host immune system for brief periods of time, they expose themselves to pre-programmed immune cells and more importantly cells primed by exposure to arthropod-specific antigens from previous ectoparasitic attacks.

Hosts use a variety of immune defences to combat ectoparasites. These mechanisms include up regulation of antigen-presenting cell and T and B cell activity, antibody mediated complement activity, and mast cell and granulocyte activity. Ectoparasites do not passively avoid these barriers however. Almost all ectoparasites possess anti-hemostatic compounds in their saliva. In some cases, such as the ixodid ticks, parasites maintain substances in their saliva (e.g. kininase), which decreases host grooming activity by destroying bradykinin, a mediator of skin irritation (Wikel 2002).

3.2.1 Ticks

Ticks are one of the best studied groups of ectoparasites. As in other ectoparasites, salivary products produced by these arthropods change over the course of engorgement, making immune defence difficult for the host because of changes in the antigen environment. Indeed, ticks may use this salivary antigen variability as a defence mechanism itself. High production of several different proteins forces the host to make antibodies that are often ineffective for parasite control. In other words, by forcing the host to generate abundant ineffective antibodies, generation of sufficient protective antibodies is prevented (Barriga 1999). Besides antibody responses, hosts can engage many other immune defences against macroparasites, which can lead to impaired feeding, reduced engorgement weight, a diminished number and decreased viability of ova, and even parasite death. Further, once hosts have developed immunity to ticks, this immunity can be transferred to naïve animals either by infusing immune serum or lymphocytes (Wikel and Allen 1976). Avoidance or resistance of tick infestation is important to hosts, as tick parasitization can have strong negative consequences. Guinea pigs experimentally infected with *Dermacentor andersoni,* for instance, were less able to produce antibodies to another novel antigenic substance compared to uninfected control animals (Wikel 1982). Although this effect disappeared once ticks stopped feeding, these data indicate that tick infections may suppress overall immune activity in hosts, which could lead to infections by more virulent disease agents.

To successfully feed on hosts, ticks must overcome the host's immune responses and other defences such as blood coagulation agents, platelet aggregation, and pain and/or irritation responses. Ticks are not passive in their efforts to avoid these defences. Their saliva can affect the host immune system in many ways including inhibition of complement activity, depression of pro-inflammatory cytokine production, T lymphocyte proliferation, Th1 cytokine secretion, antibody production, and natural killer (NK) cell function (Wikel 1999; Trinchieri 1995). One of the main targets of tick-induced immunosuppression is T lymphocyte activity, as salivary gland extracts reduce T- but not B-cell lymphocyte proliferation *in vitro* (Wikel 1999). Interestingly, tick saliva often contains prostaglandins (particularly PGE2). Currently, no data exist to indicate that this substance has immunomodulatory effects in these contexts, but prostaglandins are generally well-known modulators of vertebrate immune activity (Wikel 1999). Surprisingly, little is known about the immune activity that takes place at the site of tick bites. Some evidence indicates that Langerhans cells, which serve as the major antigen presenting cells in skin, are less able to recognize parasite antigen post-bite (Brossard and Wikel 2004).

3.2.2 Fleas

Fleas are another common parasite of small mammals. Most fleas are solenophagous, meaning that they feed on blood from small blood vessels of hosts, but there is extensive variability among species in terms of how long they remain on hosts and how they obtain blood meals (Jones 1996). Only one genus (*Tunga* sp.) remains attached to its host for an extended period, although many remain on the skin for weeks or longer. Flea bites incite multiple immune activities in hosts including mast cell and basophil infiltration and IgE production. Generally, in the skin, a stereotyped progression of immune activity occurs. Upon the first infection, inflammatory immune activity is induced at the site of the bite, but outward signs are not obvious. After a second bite, a delayed hypersensitivity response is induced (T cell-mediated inflammation) characterized by infiltration of mononuclear leukocytes within 24 hours. With a third bite, both immediate and delayed hypersensitivity reactions are induced, and eosinophils infiltrate as rapidly as 20 minutes after the bite. A fourth bite generates only immediate hypersensitivity, but a fifth bite induces no skin reactivity (Larrivee et al. 1964). As in ticks, substances in flea saliva change over the course of the feeding process, with different substances being released during blood vessel probing versus feeding.

3.2.3 Anoplura lice

Lice (Anoplura) resemble ticks in terms of the strategies they use to exploit and evade their hosts. They evade grooming activity by hiding among hairs, and their elongate, flattened body form allows them to avoid easy removal. Overall, they feed rapidly and rarely remain attached to hosts for long (Jones 1996). In mice, resistance to Anoplura is correlated with increased numbers of multiple immune cell types at site of bite (Nelson et al. 1972). Sometimes, hosts control lice infection by modifying the structure and immunological access to their skin. In mice, lice infection increased epidermal thickness over the four weeks of infestation. During this period, neutrophils, eosinophils, and lymphocytes increased in number in tissue, followed by degranulation and subsequent tissue destruction at the site of infection over this period (Nelson et al. 1972).

3.2.4 Mosquitoes and flies

Although more often recognized as vectors for viral and bacterial parasites, biting flies themselves induce changes in the immune systems of their hosts. For instance, bites of the sand fly, *Simulium vittatum*, induces hosts to make IgM, IgE, and IgG reactive saliva antigens (Cross et al. 1993). Such immune responses serve a protective purpose; female sand flies (*Phlebotomus argentipes*) were less able to obtain blood meal from repeatedly bitten versus unbitten hamster hosts (Ghosh and Mukhopadyay 1998). In common with other ectoparasites, fly attacks are chemically aggressive. Extracts from salivary glands of multiple species can affect the immune system by changing antigen presentation capacity and decreasing T and B cell proliferation (Titus 1998). Some of these processes are beneficial to the infectious agents that flies carry in saliva, as evidenced by increased transmission of disease causing agents when saliva components are inoculated into animals in addition to infectious agents (Titus and Ribeiro 1990). However, the main purpose of these salivary compounds seems to be host immunosuppression. In rats, salivary gland extract of female but not male *Aedes* mosquitos limit TNFα release from mast cells (Bissonette et al. 1993). Because males do not feed on blood in this mosquito species, they probably do not need, and thus do not produce, immunosuppressive salivary substances.

4 Behavioural responses to infection

Animal models of infection and parasitism have been studied for years. However, until recently, the role of behaviour in host defence and pathogen transmission was largely ignored. In this section we consider host behaviour from two broad perspectives: (1) behaviours mediated by the parasite and (2) behaviours mediated by the host. First we consider the phenomenon of parasite modulation of host behaviour because of mounting evidence that parasites can (and do) alter host behaviour to facilitate their own fitness (Klein 2003). In addition, we consider some of the underlying neuroendocrinological and immunological mechanisms by which this occurs. Second, some host-mediated behaviours that have evolved in order to prevent or control parasitic infections will be described. Behavioural responses can be complementary to immune function in the avoidance and regulation of macroparasites (Hart 1997). Hart (1990) described two conditions for which a behavioural pattern has a parasite defence function: (1) the parasite in question must have a detrimental effect on host fitness and (2) the behaviour must have the effect of removing, avoiding, or other wise controlling the parasite. Specifically, behavioural mechanisms that are complementary to immune function in preventing or managing parasites, such as grooming behaviour and sickness responses, are presented. In addition, the social recognition and avoidance of parasitized conspecifics are considered. Finally, we discuss the role of major histocompatibility complex in resistance to parasites and the phenomenon of disassortive mating that occurs in some rodent populations.

4.1 Parasite modulation of host behaviour

The study of the modulation of host behaviour by parasites has received enormous interest in recent years (Moore 2002). Parasitic modulation of behaviour can occur through several different pathways including: (1) directly cellular infection, (2) immunologically mediated changes in the nervous system, or (3) alteration of the chemical messengers modulating behaviour (Klein 2003). The type of behavioural alterations that parasites induce is often dependent on the life cycle of the parasite in question. Although the specific responses vary among parasite-host responses, some generalities have emerged. For instance, parasites with direct life cycles are more likely to alter host behaviour to increase contact between the infected organism and vulnerable conspecifics. On the other hand, parasites with intermediate hosts often act to increase the probability of predation in order to facilitate transmission into the definitive host. Most of the work on parasitic alterations in behaviour as focused on microparasites (Klein 2003). However, some evidence exists for macroparasitic infections altering behaviour.

Parasites alter host neurochemistry as a proximate mechanism to alter behaviour. Parasitization and parasite-associated cues are able to induce analgesia in some rodent species. Analgesia, a reduction in pain thresholds, is part of a larger suite of defensive responses associated with real or potential danger. For instance, parasitization with *S. mansoni* and *N. brasiliensis,* but not *H. polygyrus,* induces a state of analgesia. Although some of these responses may be mediated by the host brain, *S. mansoni* produces and releases proopiomelanocortin (POMC), a peptide precursor to opiate molecules, as well as other opiate-like peptides (Duvaux-Miret et al. 1992). For parasites with intermediate hosts, natural selection would presumably favour organisms that could facilitate the intermediate host's depredation. The best known example of this phenomenon is that ants infected with the trematode "brain worm" (*Dicrocoelium dendriticum)* are more likely to ascend blades of grass, which increases the possibility that they will be consumed by sheep. Several lines of evidence suggest that this sort of manipulation may be occurring in mammals. As some other stressors, exposure to predator odours induces analgesia as well as number of other physiological alterations. Remarkably, animals infected with *H. polygyrus* and *Taenia crassiceps* fail to exhibit the normal analgesic responses to predatory stress (Kavaliers and Colwell 1995b; Gourbal et al. 2001). Future research should focus on identifying parasite induced susceptibility to predation and the proximate mechanisms underlying them.

Other behavioural responses are mediated via alterations in peripheral tissues. For example, a tape worm (*T. crassiceps)* induces a behavioural and physiological feminization in its male hosts. The disruption in normal neuroendocrine signaling interferes with the expression of reproductive behaviour. However, tapeworms (e.g., *T. crassiceps* and *T. taeniaformis)* inhibit mating behaviour in parasitized animals by adjusting testosterone signaling rather than acting on brain circuits that mediate mating. Reduced mating behaviour in these animals can be restored with exogenous testosterone (Morales et al. 1996). Other androgen dependent behaviours including aggression are also reduced (Gourbal et al. 2002).

4.2 Behavioural avoidance of parasitization

Simple motor behaviours can be powerful defences against parasites. Lab rats may spend up to 1/3 of their waking time grooming (Bolles 1960). Also, biting and blood-sucking flies can be repelled with twitches, tail and ear flipping. Key largo wood rats (*Neotoma floridana*) exposed to mosquitoes demonstrate ~3-fold increase in the number of fly-repelling behaviours per hour (Edman and Kale II 1971). Mice infected with malaria fail to exhibit normal anti-mosquito behaviours resulting in increased feeding success for the mosquito vector (Day and Edman 1983) and potentially greater transmissibility for the malarial parasite.

4.2.1 Recognition of parasitized individuals

Physical contact represents a major mechanism by which parasites can be transmitted from infected to uninfected individuals (Kavaliers et al. 2005a). Thus, hosts have evolved a number of mechanisms to minimize their exposure to parasitized individuals. In rodents, the olfactory system is paramount among sensory systems. Chemical signals can provide key information about a conspecific (e.g., sex, reproductive condition, social status, etc.). As such, the accessory olfactory or vomeronasal systems are critical in detecting and avoiding parasitized conspecifics (Kavaliers et al. 2005b). Avoidance of parasitized conspecifics confers a fitness advantage to individuals in two distinct ways: (1) avoidance of close social contact with parasitized conspecifics can prevent transmission of the parasite, and (2) because resistance to many types of parasites is genetic, females select for males that are apparently resistant to parasites (for discussion see next section).

Rodents can often discriminate between parasitized and unparasitized conspecifics (Kavaliers and Colwell 1995a). The nematode, *H. polygyrus*, has been used extensively to study the effects of parasitic infection on social behaviours (Ehman and Scott 2002). This gastrointestinal nematode is shed in the feces and then after a short period is infective of other mice. Importantly, infected animals do not display classical sickness behaviours (see next section), so the number of interactions in which they engage with uninfected conspecifics does not change post-infection. Unparasitized individuals can discriminate between parasitized and unparasitized conspecifics however, and not surprisingly, they prefer to mate with unparasitized animals (Ehman and Scott 2002). In addition, mice also prefer the urine of uninfected animals relative to parasitized ones despite the lack of parasite components contained in the urine (Kavaliers et al. 2004). The phenomenon of female avoidance of parasitized males also occurs in rats infected with the nematode *Hymenolepsis diminuta* (Willis and Poulin 2000) and meadow voles (*Microtus pennsylvanicus)* harboring *T. spiralis* (Klein et al. 1999). The presence of ectoparasites also evokes avoidance responses. Female mice could discriminate between uninfected males and males infected with the louse, *P. serrata* (Kavaliers et al. 2003).

Although much of the interest in detection of parasitized conspecifics has focused on female choice, there is evidence that males prefer uninfected conspecifics and benefit from this choice. Female house mice harbouring *T. crassiceps*, *Echinostoma revolutum,* and *Echinostoma caproni* had smaller litter sizes than uninfected mice (Moore 2002), so males may do well to choose uninfected females, particularly if they exhibit a tendency towards monogamy and/or extended parental care. Indeed, male mice avoid females infected with *T. spiralis*, *T. crassiceps,* and *H. polygyrus* (Edwards and Barnard 1987; Kavaliers et al. 1998; Gourbal and Gabrion 2004). Male mice also demonstrate aversive responses to the odours of other infected males (Kavaliers et al. 2004), indicating that these behavioural tendencies may be driven by factors other than mate choice.

4.2.2 MHC diversity

Another area wherein olfactory cues may be important in regulating social behaviour is in relationship to the major histocompatibility complex (MHC). The MHC genes are among the most variable loci in the vertebrate genome (Penn and Potts 1998). MHC genes encode two types of large glycoproteins (class I and II molecules) involved in presenting peptide antigens to T cells and hence initiating some immune responses (Falk et al. 1991). MHC class I molecules present antigens from infected or cancerous cells. MHC II molecules present antigens from extracellular pathogens and parasites and are expressed on phagocytes and antigen presenting cells (Janeway et al. 1999).

The relationship between macroparasites and MHC is an area that has received much interest for two reasons: (1) MHC diversity is associated with greater resistance to parasitism (possibly due to greater T cell diversity (Dyall et al. 2000). For instance, specific MHC alleles were negatively associated with nematode burdens in wild yellow-necked mice (*Apodemus flavicollis;* (Meyer-Lucht and Sommer 2005); (2) Parasites appear at least in part to drive MHC variation to greater diversity. For instance, MHC diversity was positively correlated with prevalence of macroparasites in blind mole rats (*Spalax ehrenbergi*, Nevo and Beiles 1992).

Laboratory mice prefer to mate with individuals with dissimilar MHC alleles in the laboratory (Yamazaki et al. 1976; Egid and Brown 1989) and in semi-natural enclosures (Potts et al. 1991). The major hypotheses underlying the preference for this disassortive mating are increased pathogen/parasite resistance for offspring and avoidance of inbreeding (Brown and Eklund 1994). Importantly, mice can distinguish between genetically identical conspecifics that differ only in MHC haplotype (Yamazaki et al. 1979). The exact mechanism by which MHC proteins alter olfactory cues is not known, although this appears to be volatile components in the urine that may be components of MHC glycoprotein itself, or specific peptides that bind MHC molecules (Singer et al. 1997).

4.2.3 Sickness behaviour

Many behavioural responses to parasites are evoked by the parasite in an effort to enhance its own fitness. However, sickness responses are primarily mediated by the host. Sick or infected animals display a coordinated suite of physiological and behavioural responses (Exton 1997) collectively termed the acute phase response (Baumann and Gauldie 1994). The behavioural sequelae of bacterial infections are particularly salient and include lethargy, anorexia, adipsia, anhedonia, and reduced social interactions (Hart 1988). Additionally, the reproductive neuroendocrine axis is inhibited at multiple physiological levels (Rivier and Vale 1990; Avitsur and Yirmiya 1999). These responses collectively termed "sickness behaviour", along with the induction of fever, are thought to be part of a coordinated, adaptive effort to aid in recovery from infection (Hart 1988; Kent et al. 1992). Indeed, interference with components of the sickness response can negatively impact recovery from infection. Force-feeding of mice infected with *Listeria monocytogenes* (e.g. – preventing anorexia) resulted in a nearly 100% increase in mortality over mice that were allowed to eat *ad libitum* (Murray and Murray 1979). The primary mediators of the sickness response are the proinflammatory cytokines, IL-1β, IL-6, and TNFα (Kent et al. 1992; Dantzer 2001). Glucocorticoid secretion is potently activated by cytokines that feed back to inhibit cytokine gene expression (Turnbull and Rivier 1995; Goujon et al. 1997).

So far, sickness behaviours have not been shown to occur in response to macroparasitic infection. One component of sickness behaviour, anorexia, in response to parasitic infection is fairly common. The best studied example involves rats infected with the nematode, *N. brasiliensis*. These animals display a biphasic pattern of anorexia followed by a hyperphagic period that occurs once the parasite has been cleared. The anorectic period is associated with the increased expression of mRNA for neuropeptide Y, which is a potent stimulator of food intake (Horbury et al. 1995). Additionally, the proinflammatory cytokine IL-6 is released and corticosterone concentrations are elevated during the early period of infection with *T. spiralis* (Roberts et al. 1999). Suppression of cytokine signaling with the synthetic glucocorticoid betamethasone reduced the anorectic affects of *T. spiralis,* although the authors attributed this effect to alterations in intestinal rather than neural inflammation (Faro et al. 2000).

Although specific examples of host-mediated sickness behaviours can be found in the macroparasite literature, they are not as common as in bacterial infections. Classical sickness behaviours may not be adaptive responses to chronic parasitic infections. Many animals face some parasite burden and as such the generalized behavioural depression associated with sickness responses would compete with other critical behaviours (Kavaliers et al. 2000). In addition, there is mounting evidence that sickness responses are plastic and can be modulated by the proximate social or physical environment (Aubert 1999; Bilbo et al. 2002a; Weil et al. 2006). Therefore, some chronic parasitization might not induce the sickness responses indicative of other types of infections. It would be interesting to determine if some parasites are interfering with the signal transduction pathways associated with sickness behaviours in order to enhance their own fitness.

5 Neuroendocrine regulation of immune function

Most individuals in a population are infested with small numbers of macroparasites whereas a minority of individuals within the same population harbours large numbers of macroparasites (Wilson et al. 2001). What accounts for this variation within populations? Variation in macroparasite infection may be partly due to the demographic composition of a population. For instance, prepubertal and aged individuals are generally more susceptible to macroparasites than adults (Klein 2004; Gryseels 1994). Thus the distribution of age classes within a population may affect the diversity and intensity of infection within a population. Part of this variation in infection status of the population may be reflective of differences in endocrine regulation among age classes. In a two-year study, male field voles (*Microtus agrestis*) displayed higher number of fleas than female cohorts (Smith et al. 2005). Flea numbers were maximal in autumn, likely reflecting the increased social densities (Smith et al. 2005). Importantly, this population of voles displayed a convex age-prevalence pattern implicating the development of immunity for flea-borne protozoan (*Trypanosoma microti*) (Smith et al, 2005).

Host genetic differences in macroparasite susceptibility may also exist (Wilson et al. 2001). In part, such differences may reflect differences in hormone-modulated immune function. Individuals that perceive stressors tend to compromise immune function as compared to individuals that do not perceive stressors in their environment (Ader et al. 1995; Padgett and Glaser 2003). Glucocorticoids are released during stress responses and chronic exposure to these adrenal steroid hormones suppresses immune function (Ader et al. 1995; Padgett and Glaser 2003). Other possible proximate sources of variation in macroparasite prevalence include genetic differences among parasites, variation in host behaviour, and the time of year (Wilson et al. 2001).

5.1 Sex steroid hormones

One proximate mechanism underlying variation in macroparasite infection is differential endocrine modulation of behaviour and physiology, especially immune function. For example, male mammals and birds tend to be more infected than female conspecifics, which may be due to steroid-mediated differences in immune function (e.g., Grossman 1985; Klein 2000a, b). Generally, androgens compromise immune function whereas estrogens promote immune function (Klein 2000a, b). Other proximate explanations include the observation that the sex difference in body size favours male infection by macroparasites simply because they are easier to locate than the smaller females (Klein 2004).

The evolutionary pressures and trade-offs underlying sex differences in parasite load and immune activity have been previously described (e.g., Zuk 1990; Folstad and Karter 1992; Norris and Evans 2000); briefly, the original Hamilton-Zuk hypotheses stated that elaborate secondary sex characteristics were preferred by females because they provided evidence for the presence of genes that conferred resistance to parasites (Hamilton and Zuk 1982). Later Folstad and Karter (1992) expanded on this concept by proposing that the high androgen concentrations necessary for the expression of many sexually selected traits could suppress immune function. Thus, parasitized males would have to suppress their androgen secretions in order to enhance their defences and inversely males that can survive potential infection despite high testosterone concentration would be selected for.

Males are more susceptible to most parasitic infections than females. In a recent review of the sex differences in parasite literature of the 58 parasite species listed, 49 had male-biased infection patterns including the macroparasite genera nematodes, trematodes, and arthropods (Klein 2004). In already cited study, males of *M. agrestis* were heavier parasitized by fleas than females (Smith et al. 2005).The study of the proximate actions of androgens on the immune system is an active area of research. Androgen receptors are present on many immune cells (Abraham and Buga 1976). Much of the sexually dimorphic immune responses may be accounted for by a Th2 dominated response in males that is mediated in part by androgens. Testosterone also slows cutaneous wound healing, and inhibits delayed-type hypersensitivity responses (Mendenhall et al. 1990; Ashcroft and Mills 2002)

In contrast to testosterone, estradiol, the primary estrogen secreted by females, generally enhances immune function (Grossman 1985; Klein 2000a, b). Consistent with the finding that estrogen receptors are on lymphoid tissue, estrogen may have a direct action on lymphocytes. T-cell function can be modified *in vitro* by estrogen treatment (Paavonen et al. 1981). For example, *in vitro* B-cell activity is enhanced after the addition of estrogen, presumably due to estrogen inhibition of T-cell-mediated suppression (Paavonen et al. 1981). Similarly, macrophages increase lysosomal enzyme activity and phagocytosis in response to the addition of estrogen in tissue culture (Stimson 1983).

5.1.1 Glucocorticoids

The hypothalamic-pituitary adrenal (HPA) axis regulates the production of glucocorticoids from the adrenal glands. Glucocorticoids are steroid hormones that are important regulators of several physiological systems including carbohydrate metabolism, neuronal function, and immune function. Further, glucocorticoids have been conceptualized as "stress" hormones because they are elevated in responses to stressors.

Glucocorticoid hormones (corticosterone in most rodents and cortisol in primates) are intimately involved in regulating the immune system. The relationship between glucocorticoids and immune activity has been studied in two parallel but separate contexts. In general, immunological activation is associated with potent activation of the HPA axis. Proinflammatory cytokines are capable of activating the HPA axis at all three anatomical levels (Turnbull and Rivier 1995). Inversely, glucocorticoids feed back to inhibit the expression of proinflammatory cytokines.

Because of their effects on cytokines, as well as other cellular signaling pathways, glucocorticoids tend to suppress inflammation and because of their induction by inflammatory stimuli they have been conceptualized as "brakes" on the immune system having evolved to prevent runaway inflammation. Indeed, clinically, synthetic corticosteroids are used to treat inflammatory conditions.

The other major area in which glucocorticoids and the immune system have been studied is in response to stressors. Both psychological and physiological stressors are potent activators of the HPA axis. A relatively new field (psychoneuroimmunology) has developed to study the complex interactions between psychological variables and the immune system (for a review see McEwen et al. 1997; Kiecolt-Glaser et al. 2002). Generally, acute stressors and the associated rise in glucocorticoids enhance several aspects of immune function via alterations in immune cell distribution. In contrast, chronic stressors (or chronically high glucocorticoid exposure) tend to suppress immune function (Dhabhar and McEwen 1997). Glucocorticoid suppression of inflammation can also be characterized as a biasing towards a Th1 mediated immune response (Padgett et al. 1995) and thus shifting away from the Th2 response involved defences against many parasites.

The relationship between glucocorticoids and parasites is likely to be complex. For instance, elevated glucocorticoids may be associated with reduced resistance to infection. However, parasitization and the associated immune responses are likely to induce glucocorticoid production. In addition, the effects of glucocorticoids on the host-parasite system can be mediated in at least two ways: (1) increased glucocorticoid concentrations can be associated with immunosuppression and thus enhanced parasite survival, reproduction, or both, or (2) glucocorticoids may be important for limiting parasite-induced inflammation and tissue damage.

The direct evidence for glucocorticoid interactions with macroparasites is somewhat limited. In general, synthetic glucocorticoids, which often have high affinity for glucocorticoid receptors, tend to increase the susceptibility to parasites. For instance, rats infected with *S. ratti* and treated with betamethasone had more surviving worms and increased parasite fecundity as compared to uninfected rats (Wilkes et al. 2004). Two populations of bank voles *Clethrionomys glareolus* that differed in helminth burdens showed a positive correlation between circulating corticosterone concentrations and parasite numbers (Barnard et al. 2002; Barnard et al. 2003).

5.2 Seasonal influences

The most salient (and best studied) seasonal fluctuation observed among small mammals is seasonal breeding. Small mammals tend to breed in the spring and summer when conditions are most conducive to offspring survival (Prendergast et al. 2002). Although several proximate factors, such as inanition and extreme ambient temperatures, can inhibit breeding, the most reliable cue used by animals to phase breeding to the appropriate season is photoperiod (day length). Photoperiodic information is transduced from the eyes to an endocrine signal in the pineal gland. Pineal melatonin is secreted only at night; as night lengths increase during short days, the duration of melatonin secretion increases. Prolonged exposure to short day lengths (<12.5 h/light/day), or extended duration of nightly melatonin secretion, induces gonadal regression in many species of small mammals. In addition to extended duration of melatonin secretion, other hormones are influences by short days; short days reduced circulating concentrations of prolactin, gonadotropin-releasing hormone (GnRH), gonadotrophins, and sex steroid hormones (Prendergast et al. 2002). Gonadal regression in response to short days leads to diminished gonadal steroidogenesis and spermatogenesis, which results in infertility and cessation of androgen-dependent behaviours.

In response to reduced testosterone, individual males suppress territorial and reproductive behaviours during winter and tend to form group aggregations to improve retention of heat and humidity in communal burrows comprised of mixed species of rodents (Madison et al. 1984). Lab studies have indicated a strong relationship among androgens, aggressive behaviours, and macroparasite infection. For example, dominant male mice display elevated testosterone concentrations and are both more aggressive and more likely to be infected with macroparasites *Babesia microti* and *H. polygyrus* than low ranking individuals (Barnard et al. 1994; Barnard et al. 1998). Also, higher helminth loads were observed in bank voles (*C. glareolus*) in northeast Poland that had heavier adrenal glands, testes, and seminal vesicles (Barnard et al. 2002; Barnard et al. 2003). Thus, the seasonal pattern of macroparasite infection may represent a complicated pattern associated with reduced testosterone concentrations during winter, which would seem to reduce macroparasite infections via improved immune function, as well as a seasonal change in social structure, from largely solitary to group-huddling, which improves the opportunities for macroparasite infection.

Several studies have indicated seasonal changes in macroparasites or in the prevalence and severity of macroparasitic infections (reviewed in Nelson et al. 2002; Klein 2004). As noted, the causes underlying seasonality in macroparasites can range from seasonal changes in climate to seasonal changes in the physiology or behaviour of intermediate or host species. Small mammals, especially rodents and bats, are often the intermediate hosts for many macroparasites that have medical implications for humans. Because many of these mammalian species display robust seasonal fluctuations in breeding, territorial, and other social grouping behaviours, the prevalence of macroparasites varies across the year (Read 1990)

For example, maras (*Dolichotis patagonum*), a hystricomorph rodent species from Argentina, display a unique social organization comprising either monogamous pairs or communal nests (Porteous and Pankhurst 1998). In these animals, intensity and prevalence of Strongyloidae egg counts was highest among the communal family groups as compared to adult pairs of maras living in a zoological park in the UK. These results support the idea that seasonal changes in social group size can contribute variation in macroparasite infection.

In addition to seasonal changes in social organization, the energetic bottleneck during winter results from increased thermoregulatory demands when food availability is scarce; this makes winter a particularly difficult time to breed and survive. Immune function often varies on a seasonal basis; it is generally decreased during the winter in the wild, but is enhanced in the laboratory during short-day conditions when all other factors are held constant (Nelson and Demas 1996; Nelson 2004). Because (1) immune function is compromised by the chronic stressors of winter and (2) winter stressors are seasonally predictable, we have previously proposed that individuals use photoperiodic information to anticipate winter and accordingly redistribute energy among competing reproductive and survival functions (Nelson et al. 2002; Nelson 2004). Obviously, mounting an immune response requires resources that could otherwise be allocated to other biological functions (Sheldon and Verhulst 1996). Thus, it is reasonable to consider immune function in terms of energetic trade-offs. Individuals may partition resources among the immune system and other biological processes, such as reproduction, growth, or thermogenesis. Consequently, animals may maintain the highest level of immune function that is energetically possible given the constraints of processes essential for survival, growth, reproduction, thermogenesis, foraging, and other activities (Festa-Bianchet 1989; Deerenberg et al. 1997; Nelson et al. 2002). The observations that immune function fluctuates seasonally and is compromised during stressful times are consistent with this idea (Zuk 1990; John 1994).

"Stress" is a notoriously ethereal concept that has been used to describe any factor that increases glucocorticoid secretion including injury, pain, infection, overcrowding, harsh ambient temperature, food deprivation, noise, restraint, and aversive social interactions (Nelson et al. 2002). The ecological literature illustrates that environmental factors perceived as stressors, such as low food availability, low ambient temperatures, overcrowding, lack of shelter or increased predator pressure can be seasonal. Oftentimes, seasonal changes in environment correlate with seasonal fluctuations in immune function among individuals and seasonal changes in population-wide disease and death rates (Lochmiller et al. 1994). Thus, winter survival, at least among non-human animals, is hypothesized to require a positive balance between short-day enhanced immune function and glucocorticoid-induced immunosuppression (Nelson and Demas 1996). The balance between short-day enhanced immune function (i.e., to the point where autoimmune disease becomes a danger) and stressor-induced immunosuppression (i.e., to the point where opportunistic macroparasites overwhelm the host) must be met for animals to survive and become reproductively successful (Raberg et al. 1998). Indeed, the stressor of reduced food availability during winter may contribute to macroparasite infections. In some cases, starved hosts are preferred to well-nourished hosts (Krasnov et al. 2005). For example, egg production of fleas (*Xenopsylla ramesis*) was significantly increased when parasitizing underfed as compared to control gerbils (*Meriones crassus*). Although inanition of hosts affected survival of flea eggs and larvae on these rodents, survival of pupae was unaffected. These results suggest nutritional state in combination with the energetic costs of host resistance can affect parasites (Krasnov et al. 2005).

Photoperiod affects the immune system of many rodent species. Short days increase the number of circulating blood leukocytes, lymphocytes, T cells and NK cells, as well as spontaneous blastogenesis in whole blood and isolated lymphocytes and the cytolytic capacity of natural killer cells (Yellon et al. 1999; Bilbo et al. 2002b). Moreover, short days suppress phagocytosis and oxidative burst activities of granulocytes and monocytes (Yellon et al. 1999)). Short days also enhance lymphocyte proliferation in species ranging from mice to primates (Mann et al. 2000; Nelson et al. 2002; Nelson 2004). There are species differences in photoperiodic influences on immune function. In addition, certain specific components of immune function might be more costly to maintain, although methods for precise measurements are generally not available. Finally, the types of immune responses, such as enhanced primary defences in the skin, lymph nodes and gastrointestinal tract, could vary because the types of infectious risks vary seasonally. However, the general pattern is that short day lengths are usually associated with enhanced immune function.

Melatonin transduces photoperiodic information and also influences immune function both directly and indirectly (reviewed in Nelson et al.

2002). For example, melatonin receptors have been localized on lymphocytes, and in vitro melatonin treatment enhances splenocyte proliferation (Pozo et al. 1997) Enhancement of several components of immune function in mice is mediated directly through type 2 melatonin receptors on lymphocytes (Drazen and Nelson 2001). Melatonin also stimulates production of endogenous opioids directly from T cells, which might mediate the immunoenhancing effects of melatonin; melatonin also modulates the effects of stressors on immune function during the winter (Moore and Siopes 2003; Nelson 2004). Finally, melatonin increases survival of mice infected with *Schistosoma mansoni* (El-Sokkary et al. 2002).

Short days inhibit prolactin in all small mammals examined (Goldman and Nelson 1993), and prolactin influences immune function. For example, hypophysectomy suppresses hematopoiesis and immune cell proliferation; these effects are reversed after administration of prolactin (Berczi et al. 1991). Bromocriptine, a drug that inhibits prolactin release, suppresses antibody formation and cell-mediated immune activity; this immunosuppression is reversed with prolactin (Nagy et al. 1983). Finally, prolactin stimulates several immune parameters in untreated animals (Nagy et al. 1983). There is also evidence that prolactin may be produced locally in immune cells to regulate immune parameters. For example, antibodies directed against prolactin inhibit lymphocyte proliferation *in vitro* (Hartmann et al. 1989) suggesting that immune-derived prolactin may enhance lymphocyte proliferation.

Estrogens tend to enhance immune function while androgens tend to suppress it. Therefore, if photoperiodic changes in immune function are due to fluctuations in sex steroid hormones, then female rodents housed in short days, that have low estrogen concentrations, should reduce immune function compared to long-day animals. Alternatively, if the effects of short days or melatonin on immune function are independent of changes in gonadal steroid hormones, then immune enhancement should be observed among both males and females housed in short days regardless of circulating concentrations of gonadal steroids. Most studies support the latter hypothesis. Specifically, both male and female deer mice (*Peromyscus maniculatus*) housed in short days display enhanced lymphocyte proliferation compared to long-day housed animals, regardless of gonadal status (Demas and Nelson, 1998). Gonadectomized male and female animals that lack circulating testosterone and estradiol, respectively, display similar enhancement of immune function compared to intact animals; exogenous hormone replacement does not change these results (Demas and Nelson 1998). Furthermore, the photoperiodic effects on immune function in this species do not appear to be due to changes in glucocorticoids because corticosterone concentrations do not differ between short- and long-day deer mice (Demas and Nelson 1998). Thus, it appears that some of the immunoenhancing effects of short days are probably caused by direct effects of melatonin.

6 Future directions

A few host-macroparasite systems have been extensively investigated, but to date the general principles governing interactions between macroparasites and micromammals have not been identified. In particular, most of the work on host immune responses to parasites has been conducted using *in vitro* models in the laboratory. Studies on immune responses to macroparasites in wild mammals would greatly extend our understanding of these interactions. Such field experiments, although methodologically difficult at present, would provide the ecological validity that exclusively lab-based studies do not.

One particularly promising line of research involves the interplay between macroparasites and host behaviour. Indeed, it remains unclear whether and in what contexts parasites induce sickness behaviour. More specifically, we know little about how animals parse the competing processes of growth, feeding, reproduction and social behaviours when sickness behaviours makes these activities partially incompatible. Have some hosts simply evolved to forgo systemic inflammation typically associated with sickness behaviour? Perhaps some parasites actively suppress the chemical signals that orchestrate these responses. If so, do hosts become more susceptible to chronic infection? From a human health perspective, neuroendocrine-immune interactions have become an important research topic. However, this work can only provide us with direction through which we can begin to understand how wild animals use their immune defences to combat macroparasites. Overall, the physiological interactivity between micromammals and their macroparasites is fascinating, but there is still much to learn.

References

Abraham AD, Buga G (1976) 3H-testosterone distribution and binding in rat thymus cells in vivo. Mol Cell Biochem 13:157–163

Ader R, Cohen N, Felten D (1995) Psychoneuroimmunology: Interactions between the nervous system and the immune system. Lancet 345:99–103

Ashcroft GS, Mills SJ (2002) Androgen receptor-mediated inhibition of cutaneous wound healing. J Clin Invest 110:615–624

Aubert A (1999) Sickness and behaviour in animals: A motivational perspective. Neurosci Biobehav Rev 23:1029–1036

Avitsur R, Yirmiya R (1999) The immunobiology of sexual behavior: Gender differences in the suppression of sexual activity during illness. Pharmacol Biochem Behav 64:787–796

Barnard CJ, Behnke JM, Sewell J (1994) Social behaviour and susceptibility to infection in house mice (*Mus musculus*): Effects of group size, aggressive behaviour and status-related hormonal responses prior to infection on resistance to *Babesia microti*. Parasitology 108:487–496

Barnard CJ, Behnke JM, Gage AR, Brown H, Smithurst PR (1998) The role of parasite-induced immunodepression, rank and social environment in the modulation of behaviour and hormone concentration in male laboratory mice (*Mus musculus*). Proc Biol Sci 265:693–701

Barnard CJ, Behnke JM, Bajer A, Bray D, Race T, Frake K, Osmond J, Dinmore J, Sinski E (2002) Local variation in endoparasite intensities of bank voles (*Clethrionomys glareolus*) from ecologically similar sites: Morphometric and endocrine correlates. J Helminthol 76:103–113

Barnard CJ, Kulis K, Behnke JM, Bajer A, Gromadzka-Ostrowska J, Stachon M, Sinski E (2003) Local variation in helminth burdens of bank voles (*Clethrionomys glareolus*) from ecologically similar sites: Temporal stability and relationships with hormone concentrations and social behaviour. J Helminthol 77:185–195

Barriga OO (1999) Evidence and mechanisms of immunosuppression in tick infestations. Genetic Anal Biomol Eng 15:139–142

Baumann H, Gauldie J (1994) The acute phase response. Immunol Today 15:74–80

Berczi I, Nagy E, de Toledo SM, Matusik RJ, Friesen HG (1991) Pituitary hormones regulate c-myc and DNA synthesis in lymphoid tissue. J Immunol 146:2201–2206

Bilbo S, Drazen D, Quan N, He L, Nelson R (2002a) Short day lengths attenuate the symptoms of infection in Siberian hamsters. Proc R Soc Lond B 269:447–454

Bilbo SD, Dhabhar FS, Viswanathan K, Saul A, Yellon SM, Nelson RJ (2002b) Short day lengths augment stress-induced leukocyte trafficking and stress–induced enhancement of skin immune function. Proc Natl Acad Sci USA 99:4067–4072

Bissonette EY, Rossignol PA, Befus AD (1993) Extracts of mosuqito salivary gland inhibit tumor necrosis factor alpha. J Med Entomol 35:505–509

Bolles RC (1960) Grooming behavior in the rat. J Comp Physiol Psychol 53:306–310

Brossard M, Wikel SK (2004) Tick immunobiology. Parasitology 129:S161–S176

Brown JL, Eklund A (1994) Kin recognition and the major histocompatibility complex – an integrative review. Amer Nat 143:435–461

Cross ML, Cupp MS, Cupp EW, Galloway AL, Enriquez EL (1993) Modulation of murine immunological responses by salivary gland extract of *Simulium vittatum* (Diptera: Simuliidae). J Med Entomol 30:928–935

Dantzer R (2001) Cytokine-induced sickness behavior: Where do we stand? Brain Behav Immun 15:7-24

Day JF, Edman JD (1983) Malaria renders mice susceptible to mosquito feeding when gametocytes are most infective. J Parasitol 69:163–170

Deerenberg C, Arpanius V, Daan S, Bos N (1997) Reproductive effort decreases antibody responsiveness. Proc R Soc Lond B 264:1021–1029

Demas GE, Nelson RJ (1998) Short-day enhancement of immune function is independent of steroid hormones in deer mice (*Peromyscus maniculatus*). J Comp Physiol B 168:419–426

Dhabhar FS, McEwen BS (1997) Acute stress enhances while chronic stress suppresses cell-mediated immunity in vivo: A potential role for leukocyte trafficking. Brain Behav Immun 11:286–306

Drazen DL, Nelson RJ (2001) Melatonin receptor subtype MT2 (Mel 1b) and not mt1 (Mel 1a) is associated with melatonin-induced enhancement of cell-mediated and humoral immunity. Neuroendocrinology 74:178–184

Duvaux-Miret O, Stefano GB, Smith EM, Dissous C, Capron A (1992) Immunosuppression in the definitive and intermediate hosts of the human parasite *Schistosoma mansoni* by release of immunoactive neuropeptides. Proc Natl Acad Sci USA 89:778–781

Dyall R, Messaoudi I, Janetzki S, Nikolic-Zugic J (2000) MHC polymorphism can enrich the T cell repertoire of the species by shifts in intrathymic selection. J Immunol 164:1695–1698

Edman JD, Kale II HW (1971) Host behavior: Its influence on the feeding success of mosquitoes. Ann Entomol Soc Am 64:513–516

Edwards JC, Barnard CJ (1987) The effects of *Trichinella* infection on intersexual interactions between mice. Anim Behav 35:533–540

Egid K, Brown JL (1989) The major histocompatibility complex and female mating preferences in mice. Anim Behav 38:548–550

Ehman KD, Scott ME (2002) Female mice mate preferentially with non–parasitized males. Parasitology 125:461–466

El-Sokkary GH, Omar HM, Hassanein AF, Cuzzocrea S, Reiter RJ (2002) Melatonin reduces oxidative damage and increases survival of mice infected with *Schistosoma mansoni*. Free Radic Biol Med 32:319–332

Exton M (1997) Infection-induced anorexia: Active host defence strategy. Appetite 29:369–383

Falk K, Rotzschke O, Stevanovic S, Jung G, Rammensee HG (1991) Allele-specific motifs revealed by sequencing of self-peptides eluted from MHC molecules. Nature 351:290–296

Faro CJ, Reidelberger RD, Palmer JM (2000) Suppression of food intake is linked to enteric inflammation in nematode-infected rats. Am J Physiol Regul Integr Comp Physiol 278:R118–R124

Festa-Bianchet M (1989) Individual differences, parasites, and the cost of reproduction for bighorn ewes (*Ovis canadensis*). J Anim Ecol 58:785–795

Folstad I, Karter AJ (1992) Parasites, bright males and the immunocompetence handicap hypothesis. Amer Nat 139:603–622

Ghosh KN, Mukhopadhyay J (1998) The effect of anti-sandlfy salive antibodies on *Phlebotmous argentipes* and *Leishmania donovani*. Int J Parasitol 28:275–281

Goldman BD, Nelson RJ (1993) Melatonin and seasonality in mammals. In: Yu HS, Reiter RJ (eds) Melatonin: Biosynthesis, physiological effects, and clinical applications. CRC Press, Boca Raton, pp 225–252

Goujon E, Laye S, Parnet P, Dantzer R (1997) Regulation of cytokine gene expression in the central nervous system by glucocorticoids: Mechanisms and functional consequences. Psychoneuroendocrinology 22:S75–S80

Gourbal BEF, Gabrion C (2004) A study of mate choice in mice with experimental *Taenia crassiceps* cysticercosis: Can males choose? Can J Zool 82:635–643

Gourbal BEF, Righi M, Petit G, Gabrion C (2001) Parasite-altered host behavior in the face of a predator: Manipulation or not? Parasitol Res 87:186–192

Gourbal BEF, Lacroix A, Gabrion C (2002) Behavioural dominance and *Taenia crassiceps* parasitism in BALB/c male mice. Parasitol Res 88:912–917

Grossman CJ (1985) Interactions between the gonadal steroids and the immune system. Science 227:257–261

Hamilton W, Zuk M (1982) Heritable true fitness and bright birds: A role for parasites? Science 218:384–387

Hart BL (1988) Biological basis of the behavior of sick animals. Neurosci Biobehav Rev 12:123–137

Hart BL (1990) Behavioral adaptations to pathogens and parasites: Five strategies. Neurosci Biobehav Rev 14:273–294

Hart BL (1997) Effects of hormones on behavioral defenses against parasites. In: Beckage NE (ed) Parasites and pathogens: Effects on host hormones and behavior. Chapman and Hall, New York, pp 210–230

Hartmann DP, Holaday JW, Bernton EW (1989) Inhibition of lymphocyte proliferation by antibodies to prolactin. FASEB J 3:2194–2202

Horbury SR, Mercer JG, Chappell LH (1995) Anorexia induced by the parasitic nematode, *Nippostrongylus brasiliensis*: Effects on NPY and CRF gene expression in the rat hypothalamus. J Neuroendocrinol 7:867–873

Hunter CA, Sher A (2002) Innate immunity to parasitic infections. In: Kaufmann SHE, Sher A, Ahmed R (eds) Immunology of infectious diseases. ASM Press, Washington, pp 111–126

Janeway C, Travers P, Walport M, Capra J (1999) Immunobiology: The immune system in health and disease. Elsevier, Garland

John JL (1994) The avian spleen: A neglected organ. Q Rev Biol 69:327–351

Jones CJ (1996) Immune responses to fleas, bugs, and sucking lice. In: Wikel SK (ed) The immunology of host-ectoparasite arthropod relationships. CABI, Wallingford, pp 150-174

Kavaliers M, Colwell DD (1995a) Discrimination by female mice between the odours of parasitized and non-parasitized males. Proc Biol Sci 261:31–35

Kavaliers M, Colwell DD (1995b) Reduced spatial learning in mice infected with the nematode, *Heligmosomoides polygyrus*. Parasitology 110:591–597

Kavaliers M, Colwell DD, Choleris E (1998) Analgesic responses of male mice exposed to the odors of parasitized females: Effects of male sexual experience and infection status. Behav Neurosci 112:1001–1011

Kavaliers M, Colwell DD, Choleris E (2000) Parasites and behaviour: An ethopharmacological perspective. Parasitol Today 16:464–468

Kavaliers M, Fudge MA, Colwell DD, Choleris E (2003) Aversive and avoidance responses of female mice to the odors of males infected with an ectoparasite and the effects of prior familiarity. Behav Ecol Sociobiol 54:423–430

Kavaliers M, Choleris E, Agmo A, Pfaff DW (2004) Olfactory-mediated parasite recognition and avoidance: linking genes to behavior. Horm Behav 46:272–283

Kavaliers M, Choleris E, Pfaff DW (2005a) Genes, odours and the recognition of parasitized individuals by rodents. Trends Parasitol 21:423–429

Kavaliers M, Choleris E, Pfaff DW (2005b) Recognition and avoidance of the odors of parasitized conspecifics and predators: Differential genomic correlates. Neurosci Biobehav Rev 29:1347–1359

Kent S, Bluthe R, Kelley K, Dantzer R (1992) Sickness behavior as a new target for drug development. Trends Pharmacol Sci 13:24–28

Kiecolt-Glaser JK, McGuire L, Robles TF, Glaser R (2002) Psychoneuroimmunology: Psychological influences on immune function and health. J Consult Clin Psychol 70:537–547

Klein SL (2000a) The effects of hormones on sex differences in infection: From genes to behavior. Neurosci Biobehav Rev 24:627–638

Klein SL (2000b) Hormones and mating system affect sex and species differences in immune function among vertebrates. Behav Process 51:149–166

Klein SL (2003) Parasite manipulation of the proximate mechanisms that mediate social behavior in vertebrates. Physiol Behav 79:441–449

Klein SL (2004) Hormonal and immunological mechanisms mediating sex differences in parasite infection. Parasite Immunol 26:247–264

Klein SL, Gamble HR, Nelson RJ (1999) *Trichinella spiralis* infection in voles alters female odor preference but not partner preference. Behav Ecol Sociobiol 45:323–329

Krasnov BR, Khokhlova IS, Arakelyan MS, Degen AA (2005) Is a starving host tastier? Reproduction in fleas parasitizing food-limited rodents. Funct Ecol 19:625–631

Larivee DH, Benjamini E, Feingold BF, Shimizu BF (1964) Histological studies of guinea pig skin: Different stages of allergic reaction to flea bites. Exp Parasitol 15:491–502

Lochmiller R, Vesty M, Boren M (1994) Temporal variation in humoral and cell-mediated immune response in a *Sigmodon hispidus* population. Ecology 75:236–245

Madison DM, FitzGerald RW, McShea WJ (1984) Dynamics of social nesting in overwintering meadow voles (*Microtus pennsylvanicus*): Possible consequences for population cycling. Behav Ecol Sociobiology 15:9–17

Mahida YR (2003) Host-parasite interactions in rodent nematode infections. J Helminthol 77:125–131

Mann DR, Akinbami MA, Gould KG, Ansari AA (2000) Seasonal variations in cytokine expression and cell-mediated immunity in male rhesus monkeys. Cell Immunol 200:105–115

McEwen BS, Biron CA, Brunson KW, Bulloch K, Chambers WH, Dhabhar FS, Goldfarb RH, Kitson RP, Miller AH, Spencer RL, Weiss JM (1997) The role of adrenocorticoids as modulators of immune function in health and disease: neural, endocrine and immune interactions. Brain Res Brain Res Rev 23:79–133

Medzhitov R, Preston-Hurlbert P, Janeway CA (1997) A human homologue of the *Drosophila* Toll protein signals activation of adaptive immunity. Nature 388:394–397

Mendenhall CL, Grossman CJ, Roselle GA, Hertelendy Z, Ghosn SJ, Lamping K, Martin K (1990) Anabolic steroid effects on immune function: Differences between analogues. J Steroid Biochem Mol Biol 37:71–76

Meyer-Lucht Y, Sommer S (2005) MHC diversity and the association to nematode parasitism in the yellow-necked mouse (*Apodemus flavicollis*). Mol Ecol 14:2233–2243

Moore CB, Siopes TD (2003) Melatonin enhances cellular and humoral immune responses in the Japanese quail (*Coturnix coturnix japonica*) via an opiatergic mechanism. Gen Comp Endocrinol 131:258–263

Moore J (2002) Parasites and the behavior of animals. Oxford Univ Press, New York

Morales J, Larralde C, Arteaga M, Govezensky T, Romano MC, Morali G (1996) Inhibition of sexual behavior in male mice infected with *Taenia crassiceps* cysticerci. J Parasitol 82:689–693

Murray MJ, Murray AB (1979) Anorexia of infection as a mechanism of host defense. Am J Clin Nutr 32:593–596

Nagy E, Berczi I, Wren GE, Asa SL, Kovacs K (1983) Immunomodulation by bromocriptine. Immunopharmacol 6:231–243

Nelson RJ (2004) Seasonal immune function and sickness responses. Trends Immunol 25:187–192

Nelson RJ, Demas GE (1996) Seasonal changes in immune function. Q Rev Biol 71:511–548

Nelson RJ, Demas GE, Klein SL, Kriegsfeld L (2002) Seasonal patterns of stress, immune function and disease. Cambridge Univ Press, New York

Nelson WA, Clifford CM, Bell JF, Hestekin B (1972) *Polyplax serrata*: Histopathology of the skin of louse-infected mice. Exp Parasitol 31:194–202

Nevo E, Beiles A (1992) Selection for class II Mhc heterozygosity by parasites in subterranean mole rats. Experientia 48:512–515

Norris K, Evans MR (2000) Ecological immunology: Life history trade-offs and immune defense in birds. Behav Ecol 11:19–26

Paavonen T, Andersson LC, Adlercreutz H (1981) Sex hormone regulation of in vitro immune response. Estradiol enhances human B cell maturation via inhibition of suppressor T-cells in pokeweed mitogen-stimulated cultures. J Exp Med 154:1935–1945

Rosat JP, Conceicao-Silva F, Waanders GA, Beerman F, Wilson A, Owen MJ, Hayday AC, Huang S, Aguet M, MacDonald HR, Louis JA (1995) Expansion of δγ+ T-cells in BALB/c mice infected with *Leishmania major* is dependent on Th2-type CD4+ T cells. Infect Immunol 63:3000–3004

Padgett DA, Glaser R (2003) How stress influences the immune response. Trends Immunol 24:444–448

Padgett DA, Sheridan JF, Loria RM (1995) Steroid hormone regulation of a polyclonal TH2 immune response. Ann N Y Acad Sci 774:323–325

Pearce, EJ, Tarelton RL. 2002. Overview of the parasitic pathogens. In: Kaufmann SHE, Sher A, Ahmed R (eds) Immunology of infectious diseases. ASM Press, Washington, pp 39–52

Penn D, Potts WK (1998) Untrained mice discriminate MHC-determined odors. Physiol Behav 64:235–243

Porteous IS, Pankhurst SJ (1998) Social structure of the mara (*Dolichotis patagonum*) as a determinant of gastrointestinal parasitism. Parasitology 116:269–275

Potts WK, Manning CJ, Wakeland EK (1991) Mating patterns in seminatural populations of mice influenced by MHC genotype. Nature 352:619–621

Pozo D, Delgado M, Fernandez-Santos JM, Calvo JR, Gomariz RP, Martin–Lacave I, Ortiz GG, Guerrero JM (1997) Expression of the Mel1a-melatonin receptor mRNA in T and B subsets of lymphocytes from rat thymus and spleen. FASEB J 11:466–473

Prendergast BJ, Nelson RJ, Zucker I (2002) Mammalian seasonal rhythms: Behavior and neuroendrocrine substrates. In: Pfaff DW (ed) Hormones, brain, and behavior. Acad Press, San Diego, pp 93–156

Read A (1990) Parasites and the evolution of host sexual behaviour. In: Barnard C, Behnke J (eds) Parasitism and host behaviour. Taylor and Francis, London, pp 117–157

Rivier C, Vale W (1990) Cytokines act within the brain to inhibit luteinizing hormone secretion and ovulation in the rat. Endocrinology 127:849–856

Roberts HC, Hardie LJ, Chappell LH, Mercer JG (1999) Parasite-induced anorexia: Leptin, insulin and corticosterone responses to infection with the nematode, *Nippostrongylus brasiliensis*. Parasitology 118:117–123

Schad GA, Thompson F, Talham G, Holt D, Nolan TJ, Ashton FT, Lange AM, Bhopale VM (1997) Barren female *Strongyloides stercoralis* from occult chronic infections are rejuvenated by transfer to parasite-naïve recipient hosts and give rise to an autoinfective burst. J Parasitol 83:785–791

Scott P, Grencis RK (2002) Adaptive immune effector mechanisms against intracellular protozoa and gut-dwelling nematodes. In: Kaufmann SHE, Sher A, Ahmed R (eds) Immunology of infectious diseases. ASM Press, Washington, pp 235–246

Sheldon B, Verhulst S (1996) Ecological immunology: Parasite defences and trade-offs in evolutionary ecology. Trends Ecology Evol 11:317–321

Singer AG, Beauchamp GK, Yamazaki K (1997) Volatile signals of the major histocompatibility complex in male mouse urine. Proc Natl Acad Sci USA 94:2210–2214

Smith A, Telfer S, Burthe S, Bennett M, Begon M (2005) Trypanosomes, fleas and field voles: Ecological dynamics of a host-vector-parasite interaction. Parasitology 131:355–365

Titus RG (1998) Salivary gland lysate from the sandfly *Lutzomyia longipalpis* has an inhibitory effect on macrophage function *in vitro*. Exp Parasitol 89:133–136

Titus RG, Ribiero JMG (1990) The role of vector saliva in transmission of arthropod borne disease. Parasitol Today 6:157–160

Trinchieri G (1989) Interleukin 12: A proinflammatory cytokine with immunoregulatory functions that bridge innate resistance and antigen-specific adaptive immunity. Ann Rev Immunol 13:251–276

Turnbull AV, Rivier C (1995) Regulation of the HPA axis by cytokines. Brain Behav Immun 9:253–275

Velupillai P, Harn DA (1994) Oligosaccharide-specific induction of interleukin-10 production by B220+ cells from schistosome-infected mice – a mechanism for regulation of CD4+ T-cell subsets. Proc Natl Acad Sci USA 91:18–22

Weil ZM, Bowers SL, Pyter LM, Nelson RJ (2006) Social interactions alter proinflammatory cytokine gene expression and behavior following endotoxin administration. Brain Behav Immun 20:72–79

Wikel SK (1982) Influence of *Dermacentor andersoni* infestation on lymphocyte responsiveness to mitogens. Ann Trop Med Parasitol 76:627–632

Wikel SK (1996) Immunology of the tick-host interface. In: Wikel SK (ed) The immunology of host-ectoparasite arthropod relationships. CABI, Wallingford, pp 204-231

Wikel SK (1999) Modulation of the host immune system by ectoparasitic arthropods. Bioscience 49:311–320

Wikel SK, Allen JR (1976) Acquired resistance to ticks: I. Passive transfer of resistance. Immunology 30:311–316

Wilkes CP, Thompson FJ, Gardner MP, Paterson S, Viney ME (2004) The effect of the host immune response on the parasitic nematode *Strongyloides ratti*. Parasitology 128:661–669

Willis C, Poulin R (2000) Preference of female rats for the odours of non–parasitised males: The smell of good genes? Folia Parasitol 47:6–10

Wilson K, Bjørnstad ON, Dobson AP, Merler S, Poglayen G, Randolph SE, Read AF, Skorping A (2001) Heterogeneities in macroparasite infections: Patterns and processes. In: Hudson PJ, Rizzoli A, Grenfell BT, Heesterbeek H, Dobson AP (eds) The ecology of wildlife diseases. Oxford Univ Press, New York, pp 6–44

Yamazaki K, Yamaguchi M, Baranoski L, Bard J, Boyse EA, Thomas L (1979) Recognition among mice. Evidence from the use of a Y-maze differentially scented by congenic mice of different major histocompatibility types. J Exp Med 150:755–760

Yamazaki K, Boyse EA, Mike V, Thaler HT, Mathieson BJ, Abbott J, Boyse J, Zayas ZA, Thomas L (1976) Control of mating preferences in mice by genes in the major histocompatibility complex. J Exp Med 144:1324–1335

Yellon SM, Fagoaga OR, Nehlsen-Cannarella SL (1999) Influence of photoperiod on immune cell functions in the male Siberian hamster. Am J Physiol 276:R97–R102

Zuk M (1990) Reproductive strategies and disease susceptibility: An evolutionary viewpoint. Parasitol Today 6:231–233

22 Behaviour, life history strategies and parasite infection in rodents

Christopher J. Barnard and Jerzy M. Behnke

1 Introductory remarks

The idea that susceptibility to infection, and thus the prevalence and intensity of discase in host populations, reflect adaptive variation in investment in the immune system (so-called "ecological immunology", Sheldon and Verhulst 1996) has gained considerable support in recent years (Behnke et al. 1992; Folstad and Karter 1992; Sheldon and Verhulst 1996; Barnard and Behnke 2001; Barnard ct al. 2005). Evidence from both field and laboratory studies suggests that immunity may be traded off against other components of life history, or, conversely, constrain investment in other components, and that, in birds and mammals at least, steroid hormones play an important mediating role (Grossman, 1985; Alexander and Stimson, 1988; Folstad and Karter, 1992; Wedekind and Folstad, 1994; Poiani et al 1997). Attention in recent times has focused in particular on the role of immunity trade offs in the evolution of secondary sexual characters and their suggested function as a diagnostic of health in potential mates (the so-called immunocompetence handicap hypothesis; Hamilton and Zuk 1982; Folstad and Karter 1992). Such a view has profound implications for the evolution of host-parasite relationships and the epidemiology of parasite-related disease since it suggests that resistance may vary strategically and not simply through chance (genetic or environmental) variation in immunocompetence or exposure to infection.

As Maier et al (1994) have pointed out, the clearest basis for arguing that the immune system competes for metabolic resources with other physiological and behavioural systems derives from the stress response-like consequences of immune activation. Studies have long indicated increased autonomic nervous system activity and serum concentrations of pituitary-adrenal "stress" hormones during immune responses to antigen

(Besedovsky et al. 1975), implying a peripheral physiological equivalent to a classical stress response. Indeed, the pituitary-adrenal response here is activated by the same mechanisms that trigger responses to classical environmental stressors, a chain of events mediated by cytokines released by cells of the immune system during immune response (Berkenbosch et al. 1987).

Of particular interest in the present context, however, is the fact that immune responses have been shown to be associated with behavioural changes on the part of the challenged individual, especially increased somnolence and a reduction in activity, social and sexual interaction, exploration of novel objects and food and water intake, that imply a redirection of energy away from muscular activity and mimic changes that often follow a period of "fight-flight"-inducing stress (Maier et al. 1994). All these stress-like changes can be induced by the administration of interleukin (IL)-1 or substances that stimulate immune cells to secrete IL-1 and other cytokines (Dantzer et al. 1993; Cirulli et al. 1998) which may be instrumental in coordinating behavioural and other (e.g. slowed digestion, hyperalgesia) changes via cells of the immune system and CNS (Maier et al. 1994). The obvious functional interpretation of the changes is that redirected energy can be made available for the acute phase (involving dramatic cellular proliferation), inflammation, fever and other energy-demanding components of the immune response, and attention is focused (via hyperalgesia) on sites of injury, infection or other distress (Maier et al. 1994). Conversely, it makes functional sense at times of acute danger or conflict to direct resources into muscular activity and analgesia to facilitate aggressive or escape responses and reduce distraction by injuries (Kelly 1986; Maier et al. 1994). Thus changing priorities can, in principle, dictate shifts in the pattern of energy investment between behaviour and other responses and the immune system. Glucocorticoid hormones, which modulate immunity in several ways (Wilckens and de Rijk 1997), provide a convincing potential mechanism for mediating such shifts in resource allocation between immune function and behaviour (Maier et al. 1994). The characteristic elevation in glucocorticoid levels during stress responses may thus (at least partly) reflect such immunity trade-offs. A similar argument can be made for the immunomodulatory effects of other steroid hormones, such as testosterone (Alexander and Stimson 1988; Grossman 1985; Grossman and Roselle 1986; Roberts et al. 1996), in which downregulation of some aspects of immunity may reflect adaptive shifts of resources away from the immune system into, in the case of testosterone, elaborate sexual signals and competitive behaviour (Mooradian et al. 1987; Wedekind and Folstad 1994, Penn and Potts 1998; though see Braude et al. 1999 for an alternative view of the origin of the immunodepressive effects of testosterone).

1.1 Local variation in the risk of infection

One prediction of the adaptive immunity trade-off hypothesis, as it relates to the maintenance of steady ("resting") state immune responsiveness (see below), is that trade-offs should reflect the risk and hazard of infection. That is, individuals in populations subject to high risk and/or hazard should be expected to conserve responsiveness and either downregulate physiological and behavioural processes that impact negatively on immunocompetence, or modulate their levels in relation to current immune status. In contrast, those in low risk/hazard populations might be expected to relax investment in immunity, and be more likely to risk long term survival for current reproduction. While any such differences may reflect selection for different trade-offs or reaction norms (Stearns and Koella 1986), or different states within the same reaction norm, depending on the magnitude and temporal stability of local differences in parasite pressure, the spatial and temporal unpredictability of many parasite infections (Kennedy et al. 1991; Hartvigsen and Kennedy 1993) mean that local risk and hazard are likely to vary, even within populations where infection levels are generally high or low. We might expect potential hosts to be sensitive to this and to adjust steady state immune responsiveness accordingly. One way they could do this is by monitoring social information about infection and/or immune system activity.

Studies of mice and other rodent species have shown that both parasites and the immune system influence social odours and behaviour of subjects and that this can lead to physiological and behavioural changes in social companions or other individuals in the vicinity (e.g. Edwards and Barnard 1987; Edwards 1988; Kavaliers and Colwell 1995; Kavaliers et al. 1998; Penn and Potts 1998; Fernandes 2000). For example, Kavaliers and coworkers have shown that both male and female mice respond to the odours of conspecifics infected with protozoan or helminth parasites with changes in social discrimination and opioid-mediated analgesia (Kavaliers and Colwell 1995; Kavaliers et al. 1998). These changes are also influenced by the infection status of the subject (Kavaliers et al. 1998b) and manipulation of circulating levels of steroid hormones (Kavaliers and Ossenkopp 2001). Infection and immune system activity are known to influence the olfactory qualities of urine in mice, for instance via increased expression of the MHC (Penn and Potts 1998; Ehman and Scott 2001; but see Hurst et al. 2001), and thus the complex system of urine-based chemical communication that underpins sociosexual relationships in mouse populations (Hurst 1990; 1993; Hurst et al. 2001). In keeping with this, recent studies of rats (*Rattus norvegicus*) have shown that immune responses to benign antigen in one animal can produce parallel changes in body weight and steroid hormone activity in unchallenged conspecifics sharing its environment (Fernandes 2000). This suggests that unchallenged animals respond to the increased immune activity in their companions in order to prepare physiologically for imminent stress (Fernandes 2000). We argue that this may reflect a more general capacity for cueing immune responsiveness, and its hormonal and behavioural modulators, to ambient risk. Information about risk may come from social companions, including littermates, other group members or transient encounters, or from in utero or early postnatal experience of the mother, perhaps leading to foetal programming of immune responsiveness (Barker 1995; Hales 1997). In the first part of this chapter, we shall look at associations between helminth infections, behaviour and underlying physiology in naturally fragmented populations of three species of rodent to see whether there is any evidence for local population differences in the modulation of immune responsiveness.

1.2 Life history variation and immunity modulation

A second prediction of the adaptive immunity trade-off hypothesis, however, is that immunity modulation should differ between individuals within populations according to the relative priority they attribute to the different principal components of life history (growth, survival and reproduction).

Immune function is a vital component of survivorship for most animal species. In general, therefore, we should expect animals to attach a high priority to maintaining immunocompetence. However, if the immune system competes for resources with other activities important for survival and reproduction, investment in immunity must be prioritised. Our expectation would be that activities with a depressing effect on immune function would be downregulated, and those with an enhancing effect (e.g. through releasing resources) would be upregulated, in situations where antigen challenge increased the demand for immune response, or when the animal was immunodepressed (Maier et al. 1994; Barnard and Behnke 2001). Since the risk of infection is likely to be an ever-present one in the real world, however, we should expect animals to maintain a minimum level of immune responsiveness even in the absence of a challenge (we generally refer to this as steady-state responsiveness; Barnard and Behnke 2001).

The importance attached to future survival, and thus investment in the behavioural and physiological mechanisms contributing to it, are likely to depend on the extent to which survival translates into reproductive success. Thus sex, and factors such as age and competitive ability that affect reproductive value, might be expected to play an important role in modulating immune responsiveness by affecting the lost opportunity costs of curtailed survival. In the second part of the chapter, we shall focus on, among other things, social status in male laboratory mice (*Mus musculus*) as a likely correlate of reproductive opportunity (e.g. Hurst 1987, 1993; Barnard et al. 1991) and discuss evidence for adaptive immunity modulation in relation to life history strategy.

2 Local population differences in helminth infection, and relationships with steroid hormone secretion and behaviour

We predicted above that the extent and nature of immunity modulation should vary with the selection pressure imposed by parasites during the host's evolutionary history (Behnke et al. 1992; Sheldon and Verhulst 1996; Barnard and Behnke 2001). However, the outcomes that might be expected under this prediction are varied. One plausible outcome is that individuals in populations with a high risk of infection respond to this directly by maintaining a high steady state level of immune responsiveness at the expense of some lower priority function. Alternatively, individuals may respond indirectly via compensatory trade-offs, such as generally investing less in those aspects of physiology or behaviour that might impinge on any immune response should it be called upon to respond to challenge. Candidates here might include investment in reproductive development and behaviour, insofar as these rely on the secretion of steroid hormones that impact on immune responsiveness (Folstad and Karter 1992; Wedekind and Folstad 1994; Barnard and Behnke 2001). The result in either case may be that both the mean and variance in infection intensity become damped relative to populations at lower risk, so leading to a paradoxical inverse relationship between risk and level of infection, but the two will differ markedly in their consequences for overall host life history strategy.

However, we can also imagine trade-offs involving reproductive effort occurring in the opposite direction. In relatively short-lived species, for example, or where future survival is uncertain for other environmental reasons, a high perceived risk of infection may indicate a severe constraint on future reproductive potential, thus generating an incentive for investment in reproduction in the immediate to short term. In this case, reproductive effort may increase susceptibility and enhance, rather than dampen, associations between risk and intensity of infection. In either case, any associations between reproductive investment and infection levels might be expected to occur at the level of the population rather than the individual, since we are suggesting adaptive responses to ambient social information rather than to individual challenge.

2.1 Local variation in infection in bank voles (*Clethrionomys glareolus*) and Egyptian spiny mice (*Acomys dimidiatus*)

While we have argued that the risk and reproductive costs of infection, and the value of current versus future reproductive opportunity are likely to vary within and between populations, little direct information bearing on this has come from studies of populations in the field. In a series of recent studies, however, we have begun to look at associations between relevant components of life history in a number of rodent species inhabiting fragmented landscapes, and thus occurring as apparently discrete local populations (Behnke et al. 2001; Barnard et al. 2002; Hassaneen 2005; Bajer et al. 2006). Those studied most extensively are the European bank vole (*C. glareolus*) from the fragmented forest region of Mazury in northeast Poland, and the Egyptian spiny mouse (*A. dimidiatus*) from the network of wadis (dry valleys) in the St Katherine region of the Sinai, Egypt. The following discussion comes from those of Barnard et al. (2002, 2003a, b).

2.1.1 Bank voles

In the case of *C. glareolus*, we found that animals inhabiting ecologically similar (in terms of vegetation structure), but mutually isolated, mixed woodland habitats differed in the component community structure and intensity of helminth parasite infections (Behnke et al. 2001), and that these differences were associated with anatomical (organ weight) and morphometric measures relating to endocrine function and apparent developmental constraint (overall body size and fluctuating asymmetry; e.g. Palmer and Strobeck 1986; Møller and Swaddle 1997; Mallard and Barnard 2003, 2004) (Barnard et al 2002). In general, sites with greater parasite burdens were those in which male and female *C. glareolus* had significantly larger adrenal glands, and males had larger testes and seminal vesicles for their age and body size. Voles from higher intensity sites also had smaller thymus glands, implying some degree of reduced immune capacity, and showed greater fluctuating asymmetry (FA) in hind foot length, suggestive of underlying developmental instability (Palmer and Strobeck 1986; Markow 1994; Barnard et al. 2002). These associations are consistent with the kind of site-specific relationships between infection risk, immunocompetence, steroid hormone activity and life history variation predicted above, suggesting, in this case, that animals in populations under a degree of stress (high glucocorticoid activity, developmental instability, and pressure from parasites) put more into current reproductive opportunity (as evidenced by greater testis and seminal vesicle size). However, the picture is not quite as clear cut as this, as we shall now see.

A follow-up study two years later by Barnard et al. (2003a), at two of the sites used by Barnard et al. (2002), confirmed the previous differences in infection intensity and host morphometrics and organ weights (Fig. 1), suggesting the site differences were temporally stable. In addition, measures of circulating hormone concentrations confirmed that voles at the site with the higher helminth burdens and adrenal weights also had higher levels of corticosterone. As this second study was carried out slightly later in the season (October), however, males were non-scrotal, with low circulating levels of testosterone that, along with testis and seminal vesicle weights, showed no difference between sites.

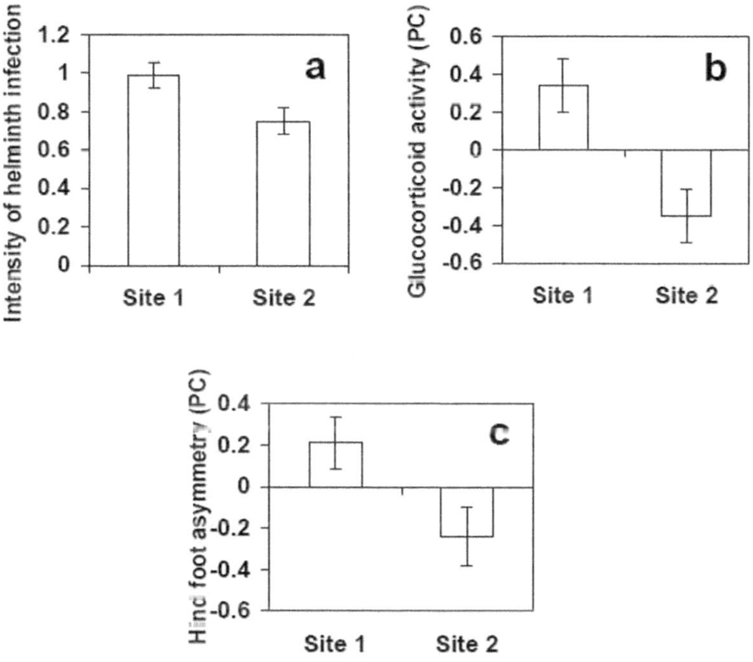

Fig. 1. (a) The mean intensity of helminth infections in bank voles (*C. glareolus*) at two forest sites in northeast Poland; (b) mean glucocorticoid activity (composite measure derived from principal components analysis [PCA] of adrenal gland weights and corticosterone concentrations) in voles from the two sites in (a); (c) mean hind foot asymmetry at the two sites (principal component contrasting foot asymmetry [positive loading] and body size [negative loading]). Bars are least squares deviations (based on data from Barnard et al. 2003a)

As well as confirming the relative stability of site differences in associations between parasite burdens, morphometrics and steroid hormone activity, the second study looked at behavioural differences between animals from the high and low burden sites, focusing on aggression among males, as aggressive behaviour has been shown in studies of other rodents to be associated with depressed immune responsiveness and increased susceptibility to infection (Barnard et al. 1996a), and, in turn, to be down-regulated under conditions of immunodepression (Barnard et al. 1997a, b).

Male *C. glareolus* in the field appear to establish aggressive dominance relationships on the basis of their sequence of recruitment to the breeding population (Gliwicz and Rajska-Jurgiel 1983) and form stable hierarchies when maintained in groups in the laboratory (Gustafsson et al. 1980; Hoffmeyer 1982). Aggressive relationships are underpinned by a chemical communication system involving urinary and faecal odour cues denoting, among other things, prior residence and social status (e.g. Rozenfeld and Rasmont 1991). The test procedure was developed from the method of Hurst et al. (1996) for resident-intruder dyads of aboriginal house mice (*Mus spretus*). Males were allocated as "residents" for the purposes of testing and paired with one "intruder" male from each of three subsites within the main sampling sites, so that each resident experienced an intruder from its own subsite, another subsite at its own location and one of the subsites at the other location. Pairings were arranged so that males within dyads encountered each other only once. Males in the present experiment showed little escalated aggression (chasing, biting). Instead, disputes were generally settled by a combination of vocalization and stylized "boxing" in which winners and losers were identified by which retreated in response to the other (see e.g. Rutovskaya 1998). In some cases, aggressive interactions simply involved one animal pushing the other back. We therefore adapted Hurst et al.'s procedure of testing dyads in a clear Perspex tube in which advances and retreats could readily be measured (Barnard et al. 2003a).

As aggression appears to respond negatively to immunodepression (see above), our expectation was that it would be reduced in males from sites with high parasite burdens and high glucocorticoid activity [since the latter is also potentially immunodepressive (Folstad and Karter 1992; Wilckens and de Rijk 1997)] and evidence of developmental stress (FA). This expectation was borne out in three ways: first, "resident" males from the high burden site showed less aggression overall (Fig. 2a); second, site differences in aggression were most pronounced when "residents" were confronting "intruders" from the low burden site, and, third, the amount of aggression shown in tests was negatively associated with foot asymmetry, individual parasite burden and terminal (end of experiment) corticosterone concentration (Fig. 2b). Interestingly, the relationship with infection was apparent only at the low burden site (Site 2 in Figs. 1 and 2), which at first sight seems paradoxical. While this may simply reflect a floor effect imposed by the chronically higher infection and apparent stress levels at the high burden site (i.e. sick animals don't fight), it is also consistent with animals at this site downregulating aggression independently of their present individual infection status. This is as might be expected if potentially immunodepressive behaviours were modulated in relation to perceived infection risk in the current environment rather than in response to infection itself, as we predicted above. However, there is arguably a degree of inconsistency between the trends in gonadal measures (larger organs at higher burden sites) and aggression (less at higher burden sites), if both are taken as reflecting investment in current reproductive effort. On the other hand, the negative relationship between aggression and corticosterone concentration suggests that animals showing aggression tended to be those that were less stressed, a trend in keeping with the suggestion above that stress underlay the broad associations between life history characters across local populations.

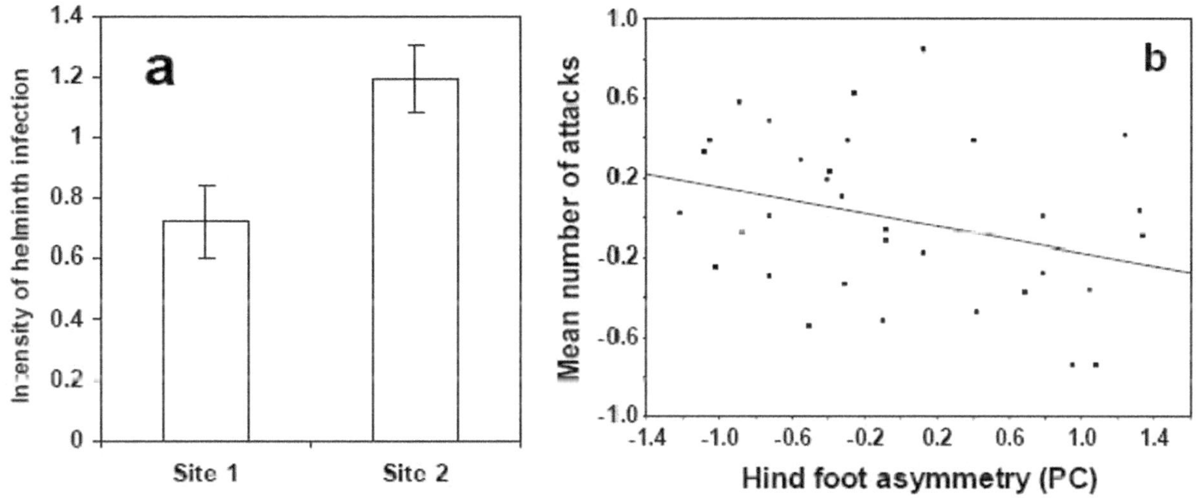

Fig. 2. (a) The mean number of attacks initiated by male voles (*C. glareolus*) from the two sites in Figure 1; (b) component effects plot from stepwise partial regression showing a significant relationship between hind foot asymmetry and the number of attacks initiated (based on data from Barnard et al. 2003a)

2.1.1 Spiny mice

Studies of local populations of *A. dimidiatus* in different wadis in the St Katherine region of the Sinai, showed a number of similarities with those of voles in Poland, but also some crucial differences (Barnard et al. 2003b). Some of the latter are likely to reflect differences in basic life history: for instance, differences in longevity and length of breeding season, and producing few, precocial young (*A. dimidiatus*) versus many, altricial young (*C. glareolus*) during the season (e.g. Osborn and Helmy 1980; Brunjes 1983; Mappes et al. 1995; Hassaneen 2005).

Population-level comparisons between wadis showed that local differences in the intensity of helminth infection (here dominated by spirurid and oxyuroid nematodes) were associated positively with a composite index of androgen activity (Fig. 3a, b). This is consistent with the voles above, and other studies of rodents demonstrating positive relationships with androgen activity and infection (e.g. Klein 2000; Barnard and Behnke 2001; Barnard et al. 2002, 2003a). The negative relationship between measures of androgen activity (organ weights and testosterone concentrations) and thymus weight in the *A. dimidiatus* study is also consistent with gonadal depression of immune responsiveness (Barnard et al 2003b). However, while measures of androgen and glucocorticoid activity were positively correlated, again as found in many other studies (e.g. Folstad and Karter 1992; Poiani et al. 2000), there was no evidence of any association between local differences in helminth intensity and glucocorticoid activity [gauged as adrenal weight and, in this species, cortisol, rather than corticosterone, concentration (Lamers et al. 1986)]. Neither was there any association between hind foot asymmetry and helminth infection. The results for the spiny mice thus suggest that variation in infection does not reflect differences in stress between local populations, an outcome that contrasts with our findings in bank voles, where both putative stress indicators (glucocorticoid activity and hind foot asymmetry) covaried with helminth intensity (Barnard et al. 2002, 2003a). One reason for the difference between the two host species may lie in the temporal stability of selection pressures acting on local populations. In the voles, local differences in helminth intensity, androgen and glucocorticoid activity and morphological asymmetry were, as we saw, consistent in samples separated by two years, suggesting chronic differences in selection pressure. While relative infection levels showed temporal consistency across some wadis in spiny mice, it is clear that other wadis differ between years (Behnke et al. 2000; Barnard et al. 2003a; Hassaneen 2005). Thus local conditions facing spiny mice in the St Katherine's region may be temporally more variable than those facing voles in different forest fragments in Poland. Of course, the cause and effect reasons for the association between helminth infection and androgen activity in any of these studies cannot be gleaned directly from the results, though one possibility in the case of the spiny mice is that the timing or magnitude of investment in reproductive activity differed between local populations because of differences in resource distribution or availability, so leading to differences in exposure or susceptibility to infection at the time of sampling.

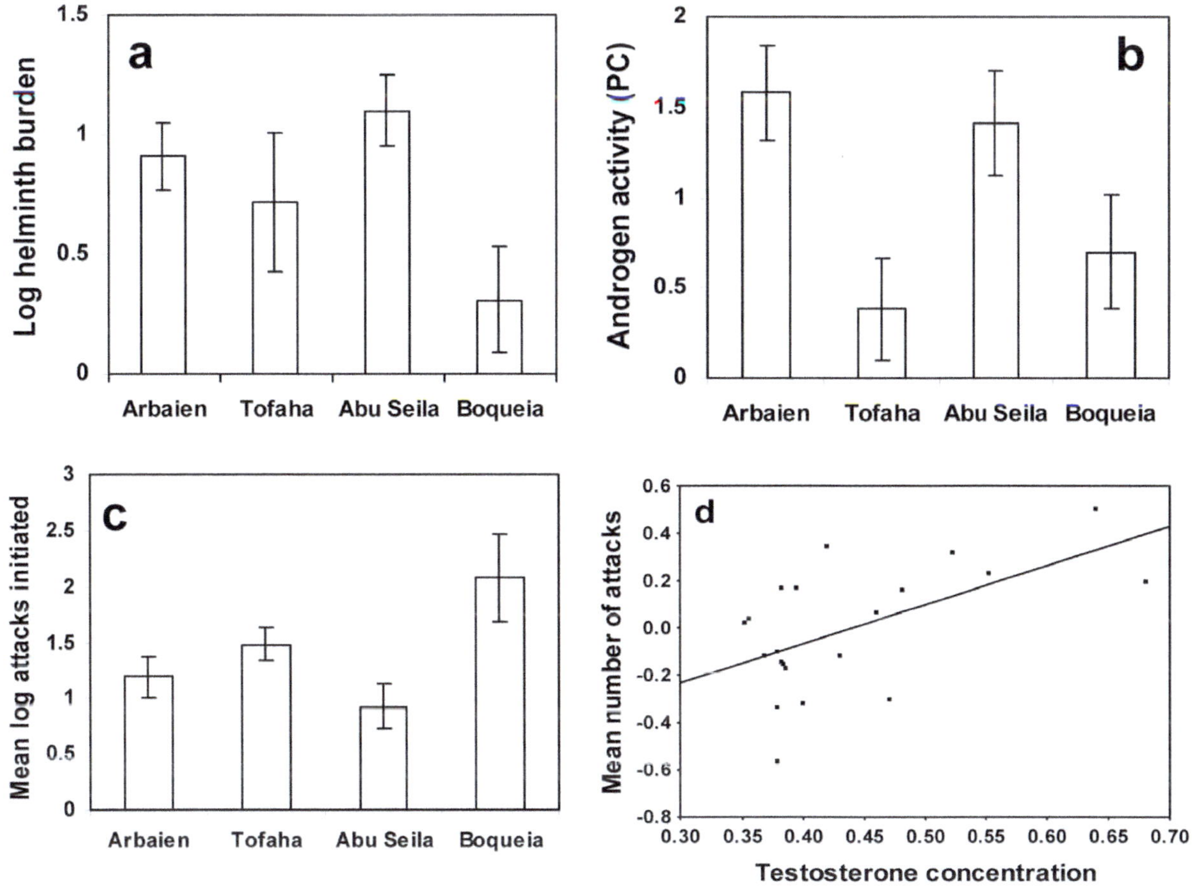

Fig. 3. (a) The mean intensity of helminth infection in Egyptian spiny mice (*A. dimidiatus*) from four different wadis in the St Katherine region of the Sinai; (b) mean androgen activity (from PCA of testis and seminal vesicle weights and testosterone concentrations) in males from the four wadis; (c) mean number of attacks initiated by males from each of the wadis; (d) component effects plot from stepwise partial regression showing a significant relationship between testosterone concentration and the number of attacks initiated (redrawn after Barnard et al 2003b; reprinted with permission from CAB International)

As in the study of voles, complementary laboratory experiments, looking at social interactions between males from selected wadis, focused on aggression, but also responses to social odours (soiled sawdust), between individuals from the same, neighbouring or distant sites, this time using an open field arena test environment. Unlike the voles, but in keeping with our general prediction earlier, aggression in experimental males was not associated with the helminth burdens of the individuals themselves, but with the average intensity of infection in their source population (Fig. 3c). This is consistent with potentially immunodepressive behaviours being modulated in relation to perceived infection risk in the environment rather than in response to individual challenge. Like the voles, however, aggression was reduced in animals from the higher burden sites despite a significant positive association between aggression and testosterone concentration in experimental males (Fig. 3d). Interestingly, there was also a positive association between aggression and cortisol concentration in male *A. dimidiatus*, the opposite of the relationship with corticosterone in voles, and one that again appears to support the idea that stress is not a significant driver of the relationships between helminth burdens, hormone activity and behaviour in spiny mice from these populations.

As well as showing less aggression, males from high burden wadis were significantly less interested in the odours of "intruder" males than those from low burden wadis. "Residents" in general showed more interest towards the odours of "intruders" from distant wadis compared with those of individuals from home or neighbouring wadis. This outcome was also mirrored in laboratory-bred stock, where males were established as "home" or 'distant' on the basis of whether or not they shared a cage (Hassaneen 2005). However, when "residents" later encountered the odour donors, they showed least aggression towards distant "intruders", thus appearing to set a higher priority on gleaning low risk social information about those individuals they deemed more serious competitors. As with aggression, the tendency to investigate odours showed no association with the individual helminth burden of either "resident" or "intruder".

3 Life history priorities and immunity trade-offs in laboratory mice

While evidence from populations of rodents in the field is strongly suggestive of adaptive covariation between parasite infection and physiological and behavioural attributes, it is correlational and so indicative rather than conclusive of cause and effect. Laboratory mice, however, have provided a productive model system for exploring some these relationships experimentally (e.g. Barnard and Behnke 2001). We shall focus on two aspects of these studies: first, relationships between immunity trade-offs and social rank, and, second, relationships between immune function and learning. The discussion of relationships with social rank is taken directly, with minor modification, from the summary review in Barnard and Behnke (2001).

3.1 Social rank and immunity trade-offs

While trade-offs may vary more or less continuously within a population, they can frequently be characterised in terms of adaptive suites and thus more or less discrete categories of life history strategy (Rohwer and Ewald 1981; Arak 1984; Hutchings and Myers 1994; but see Gosling et al. 1996). Social rank classifications are a good example. Rank classifications are usually based on measures of competitive ability, but are often associated with differences in several other important life history traits, such as growth and body size, disease resistance and reproductive status (Freeland 1981; Schur 1987; Meikle et al. 1996; Komers et al. 1997). In an extensive series of studies, we have shown that male mice (*M. musculus*) of the outbred CFLP/BKW strain can be classified into two discrete rank categories ("high" and "low" rankers) based on the relative amount of aggression initiated and received and that ranks are associated with different strategies of immunity modulation and susceptibilty to experimental infections (Barnard et al. 1994, 1996a, b; 1997a, b; 1998a; Smith et al. 1996). Moreover, the fact that these differences precede the emergence of clear rank categories in random groups and the development of rank is influenced by maternal condition (Barnard et al. 1998a), suggests deeper differences in life history strategies between individuals than responses to current aggression (see below). A range of physiological and behavioural evidence from other studies also suggests that rank categories in mice are associated with different strategies of immune response (e.g. Bigi et al. 1994; Aloe et al. 1995).

From the arguments above, we might expect individuals of high social rank, and thus high competitive ability, to invest in energy-demanding social and sexual activities at the expense of immune responsiveness as they are able to take advantage of short-term mating opportunities and command the resources that are likely to lead to them. Some evidence for this, and for a mediating role of glucocortocoid hormones (see above), comes from our studies of mice infected with the trichostongyloid nematode *Heligmosomoides polygyrus*.

Barnard et al. (1998b) infected male CFLP mice with *H. polygyrus* and maintained them in single-sex groups where their social interactions were recorded. Overall, high rankers (males with a disproportionately high attack rate and high attack initiation:receipt ratios within their group) showed reduced immune responsiveness compared with low rankers, the difference between ranks being accountable in terms the interrelationship between aggression, plasma corticosterone concentration, peripheral immune responsiveness (measured as haemagglutination titre to sheep red blood cells [SRBCs]) and resistance to *H. polygyrus* among high rankers.

Among high ranking males, post-infection aggressive behaviour correlated positively with corticosterone concentration and negatively with both haemagglutination titre to SRBCs and resistance to *H. polygyrus*. These post-infection phase relationships compounded a tendency for high rankers to maintain corticosterone levels during the preceding period of infection itself, while corticosterone concentrations dropped among low rankers and uninfected controls. Partial regression analyses revealed that the change in corticosterone levels during the period of infection (rather than the post infection period) was the best hormone-measure predictor of eventual worm burden, a relationship in keeping with the impact of glucocorticoids on the secretion of Th2 cytokines (Padgett et al. 1995; Rook and Zumia 1997) and depression of the Th2 arm of the immune response important in resistance to helminth infections (Wakelin and Selby 1974; Behnke and Parish 1979; Quinnell et al. 1991). The results are thus consistent with our expectation that individuals of greater competitive ability will be more likely to trade off immune responsiveness for reproductive benefit under conditions of challenge.

The important point to stress about Barnard et al.'s (1998b) results is that it was the *change* in corticosterone concentration over the period of infection, rather than differences in absolute concentration, that accounted for the difference in resistance to *H. polygyrus* between high and low ranking males. The conclusion that this reflected a tendency for high competitive ability males to trade off worm burden against short term investment in aggressive competition is strengthened by the fact that the negative relationship between corticosterone concentration and resistance among high rankers emerged as a specific component effect within a multivariate analysis that controlled for differences in absolute corticosterone concentration and the concentration of potentially immunodepressive testosterone that have been associated with rank differences in immune responsiveness and infection in other studies (e.g. Vessey 1964; Chapman et al. 1969; Brain and Nowell 1970; Barnard et al. 1994, 1996a, b; Poiani et al. 2000). The rank differences here contrast with the generally held view that associations between corticosterone and immunodepression in circumstances of social stress usually arise among low rankers (e.g. Vessey 1964; Chapman et al. 1969; Leshner and Politch 1979; Beden and Brain 1985; Maestripieri et al. 1990), where it might arguably reflect a trade-off between the present priority for escaping attack and immune responsiveness to a potential future challenge (Barnard and Behnke 2001). The question, of course, remains as to whether the apparent tendency for high rankers to trade off resistance to helminth infection in Barnard et al.'s (1998b) study reflects a sacrifice of future survivorship in favour of short term competitiveness or an ability of higher quality individuals to withstand greater parasite burdens at little cost to survivorship (see later). However, this distinction does not alter the putative role of corticosterone in shifting resources between inflammatory immune responses and other systems among high rankers (Barnard and Behnke 2001).

3.2 Social rank and steady state responsiveness

The males infected with *H. polygyrus* above appeared to trade off resistance to a current infection according to their competitive ability and the likelihood that the hormonal mechanisms driving the trade off would lead to short-term reproductive success. But even in the absence of a current challenge, we might expect males of different competitive ability to modulate the trade-off between immunity and current reproductive effort with different priority. An obvious focus here might be on the modulation of sex steroids, especially testosterone, which, like glucocorticoids, serve a number of important metabolic functions, but also have direct and indirect influences on the immune system (see Folstad and Karter 1992). The evolution of elaborate secondary sexual characters as honest signals of mate quality under the immuncompetence handicap principle may be one manifestation of immunity trade-offs involving androgens (but see Owens and Short 1996), but other recent work suggests such trade-offs may be a more fundamental feature of physiological and behavioural decision-making. Our work with laboratory mice again provides a good example. In a series of studies, Barnard et al. (1993, 1994, 1996a, b) exposed male CFLP mice to varying conditions of social stress in randomly constituted single sex groups. Plasma concentrations of testosterone, corticosterone and total IgG were measured before and after periods of social grouping and mice were finally subjected to an experimental infection of *Babesia microti*, a piroplasmid protozoan infecting the erythrocytes of its vertebrate host (Cox and Young 1969; Clark and Howell 1978). The time course of *B. microti* infection allows a number of measures of the host's ability to respond immunologically to the parasite, the main ones being the magnitude of the peak of infection (% cells infected, usually peaking around 8-10 days post infection) and the time to clear the infection (usually around 17-22 days post infection). The social rank of males during the random grouping phase was calculated on the basis of attacks initiated and received, as in Barnard et al.'s (1998b) study above, and allowed comparisons of hormone levels, immunocompetence, responses to infection and interrelationships between the three in males of different competitive ability.

As in the studies of voles and spiny mice above, interest focused in particular on relationships between aggressive behaviour, changing testosterone, corticosterone and IgG levels during the period of grouping, and their impact on later resistance to *B. microti*. Several lines of evidence point to an immunodepressive effect of androgens and glucocorticoids in the context of intra-erythrocytic protozoan parasites, including *B. microti* (e.g. Benten et al. 1991; Wunderlich et al. 1992; Davies 1998). Many more indicate that immunodepressive effects of both groups of hormones are associated with aggression and social stress (Broom and Johnson 1993; Poiani et al. 1997). One expectation, therefore, was that testosterone would be modulated in relation to current immunocompetence, of which total IgG was a peripheral bystander measure [Davies (1998) has since demonstrated that total IgG correlates well with the production of *B. microti*-specific antibody during infection], and the level of corticosterone, a stress-associated immunodepressant. However, on the basis of differences in competitive life history strategy, we might expect such modulation to be more apparent among high ranking, rather than low ranking, males. In addition, if modulating testosterone influences resistance, we should expect an effect of testosterone on resistance to be more apparent in individuals which fail to modulate. Both expectations have been borne out in a series of studies by Barnard et al. (1994, 1996a, b; 1998a; Smith et al. 1996).

3.3 Social rank and modulation of testosterone

Barnard et al. (1994, 1996a) found that an independent effect of change in testosterone concentration over the period of random grouping on subsequent resistance to *B. microti* was confined to high ranking males. When relationships between changing hormone and total IgG concentrations were partialled out, it emerged that high rankers tended to decouple changes in testosterone from those in total IgG whereas low rankers showed a significant positive correlation between the two (Barnard et al. 1996a; Fig. 4a, b). High rankers subsequently showed a significant negative effect of testosterone concentration on resistance to *B. microti*, an effect absent among low rankers (Barnard et al. 1994, 1996a; Fig. 4c, d). However, while high rankers appeared to decouple changes in testosterone and IgG, both rank categories showed some evidence of downregulating testosterone as corticosterone concentration increased (Barnard et al. 1994, 1996a). It is important to emphasize here that absolute testosterone concentration did not differ between rank categories, the effect of testosterone on resistance appeared to be due entirely to the tendency or otherwise to modulate levels relative to current immunocompetence. While aggression resulted in increased levels of corticosterone and reduced IgG concentration and resistance to *B. microti*, these relationships were restricted to low ranking males when ranks were analyzed separately. However, in both Barnard et al.'s (1994) and (1996a) studies, it was high rankers that experienced the greatest reduction in immunocompetence over the period of grouping. These results therefore suggest that high ranking males, by decoupling testosterone secretion from current immunocompetence at a time when they appear to be immunocompromised, actively trade off future survival against the presumed competitive and reproductive benefits of maintaining testosterone levels.

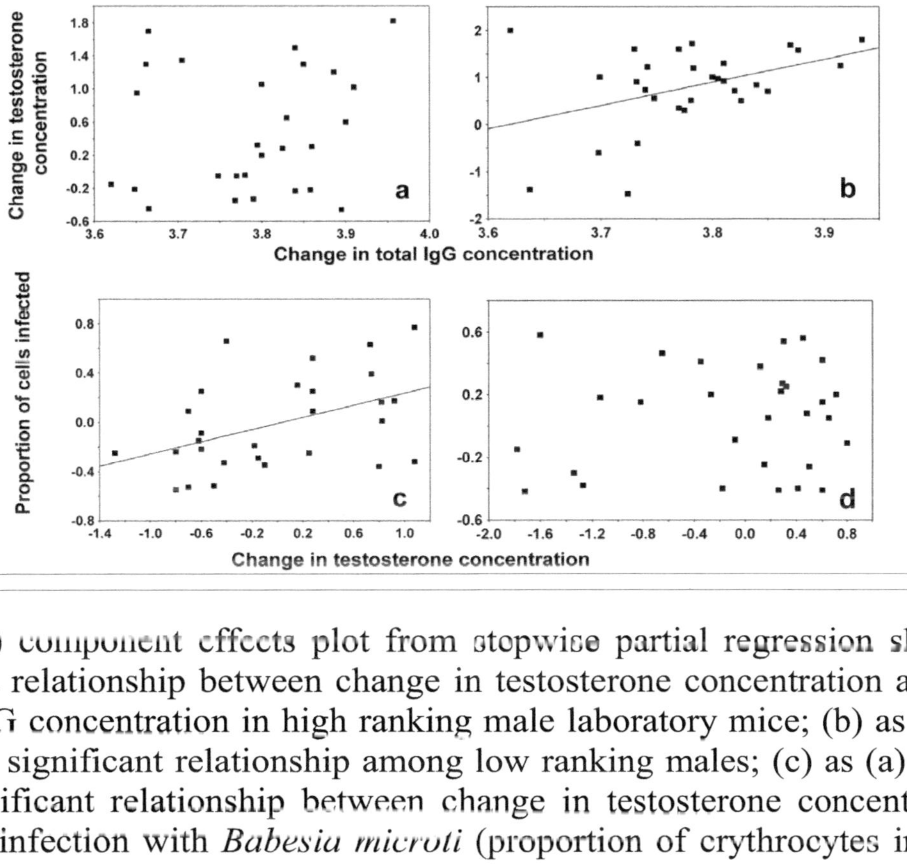

Fig. 4. (a) component effects plot from stepwise partial regression showing no significant relationship between change in testosterone concentration and change in total IgG concentration in high ranking male laboratory mice; (b) as (a) expect showing a significant relationship among low ranking males; (c) as (a) but showing a significant relationship between change in testosterone concentration and degree of infection with *Babesia microti* (proportion of erythrocytes infected) in high ranking males; (d) as (c) but showing a nonsignificant relationship among low ranking males (data from Barnard et al. 1996a)

The role of modulation in these rank differences in resistance has received further support from experiments in which hormone levels were manipulated experimentally (Davies 1998). Administration of testosterone and glucocorticoids accelerated the time course and increased the peak of *B. microti* infection, as expected from previous work with *Babesia* and other blood-borne protozoan infections (e.g. Hussein 1984; Wunderlich et al. 1992). If modulation of testosterone, rather than absolute concentration, is what determines the hormone-related resistance to *B. microti*, however, our prediction would be that increasing testosterone levels, and thus overriding any tendency to modulate, would reduce resistance among low rankers, but not among high rankers (because high rankers do not generally modulate). The prediction was borne out by Davies's results: time to reach peak infection was reduced, and the magnitude of the peak increased in testosterone-treated low rankers compared with vehicle controls, while there was no significant difference in the time course of infection among high rankers.

Interestingly, mice that turned out to be high rankers in Barnard et al.'s randomly constituted groups showed a tendency to enter their groups with higher testosterone and corticosterone, but lower IgG, levels than eventual low rankers (Barnard et al. 1993, 1994, 1996a), implying pre-existing differences in their natal litters. Barnard et al. (1998a) showed that the rank difference in testosterone modulation referred to above is present in natal litters, with eventual high rankers lacking the correlation between testosterone levels and IgG shown by eventual low rankers, but in the absence of any polarised aggressive social relationships within litters. These early differences may have their origin in the effects of maternal condition on suckling and weight gain, or in its effects on the prenatal environment (Barker 1995). Extensive work in humans and rodents (Barker 1995; Phillips 1996; Rao 1996) has identified a crucial role of nutritional constraints and other mother/foetus conflicts (Haig 1993) in utero in determining a suite of life history attributes in resultant offspring, including patterns of growth and organ development, immune function, menopause and longevity (Barker 1995, 1996; Cresswell et al. 1997; Hales 1997; Langley-Evans 1997), processes that are underpinned by various endocrine changes involving many different hormones, but particularly glucocorticoids, insulin and growth hormone (Barker 1995; Clark et al. 1996; Phillips 1996; Rohner-Jeanrenaud and Jeanrenaud 1997). Such fundamental shifts in development and metabolism, mediated by maternal condition, could account for the early differences in Barnard et al.'s (1998a) rank categories prior to the aggressive social environment in which rank is normally expressed.

3.4 Constraints on immunity trade-offs

While these results support the idea of life history-dependent immunity trade-offs, there is likely to be a limit to the benefits of discounting future survivorship. If immunocompetence becomes severely depressed it may pay all individuals to downregulate potentially immunodepressive activities and physiological responses, regardless of short term reproductive incentives. Both correlational evidence and evidence from experimental manipulation of immunocompetence among laboratory mice lend support to this expectation.

Correlational evidence emerges from hormonal responses of animals at times of increased social stress, such as on first introduction into an unfamiliar group or maintenance in environments that encourage aggressive interaction. Comparisons of pre-grouping and immediately post-grouping testosterone levels among male CFLP mice consistently showed a sharp drop in hormone concentration when individuals are introduced into new groups (Barnard et al. 1996a, b; 1997a, b) and the frequency of aggressive encounters is high (Barnard et al. 1993, 1996a; see also Poole and Morgan 1973). In some cases the marked decline removed pre-existing differences in testosterone concentration between high and low ranking males (Barnard et al. 1996a), which then decayed back towards pre-grouping levels in relation to apparent peripheral immune responsiveness (low rankers) or independently of immune responsiveness (high rankers) with the consequences for resistance discussed earlier. However, whether or not a rank difference in testosterone modulation emerges depends on the degree of social stress in the group environment. One potential, if somewhat counter-intuitive, source of increased social stress is so-called environmental enrichment.

Environmental enrichment has been an important element in the drive to improve the welfare of captive animals, including standard laboratory species (Markowitz 1982; Chamove 1989; Wurbel 2001). In caged environments, this often means the provision of objects that create spatial heterogeneity or opportunities for different activities, shelter and so on (Chamove 1989). Such initiatives can have unexpected results, however. In laboratory mice, the addition of various objects, such as bricks, flower-pots and labyrinths, to cages can result in increased aggression and plasma corticosterone concentrations (Brain 1988; McGregor and Ayling 1990; Haemisch and Gartner 1994; Barnard et al. 1996b), probably because they provide defendable resources for dominants (Hurst 1987, 1990).

Barnard et al. (1996b) equipped the unfurnished cages used in some of their earlier studies (Barnard et al. 1993, 1994) with shelving and nest-boxes to act as refuges where mice could escape or avoid aggressive encounters. The rest of the experimental design repeated that of Barnard et al. (1994). In comparison with the earlier experiment, mice in the furnished cages showed significantly elevated aggression and reduced subsequent resistance to *B. microti*. Moreover, both total IgG concentration and resistance to *B. microti* decreased as the number of attacks received by mice increased, implying that it was the change in the aggressiveness of the social environment that was instrumental in the reduction in immunocompetence and resistance. From a welfare perspective, it is interesting to note that, when time spent on the shelves or in the nestboxes was partialled out of the analyses, it showed a significant enhancing effect on both IgG levels and resistance to *B. microti*. The availability of the refuges thus appeared to offset to some extent their negative effects on overall social stress within the furnished cage environment. The crucial point in terms of the adaptive modulation hypothesis, however, was the lack of any hormone-related reduction in resistance. In part this appeared to be due to a general downregulation of testosterone and corticosterone levels in comparison with mice in the unfurnished cages, but it also appeared to be due to the fact that *both* rank categories now showed significant positive correlations between change in testosterone concentration and levels of total IgG. Among low rankers there was a similar correlation between corticosterone and IgG concentrations, the only study of ours to date in which there has been evidence for the modulation of corticosterone levels in relation to immunocompetence (see Smith et al. 1996 for some discussion of this). The conclusion from this seems to be that, when environmental stressors depress immunity, other potential immunodepressants become more tightly regulated, despite their important role in various functional systems, so as to reduce further impact on the animal's immune capability.

This conclusion is reinforced by a very different study in which singly-housed male mice were exposed to the odours of unfamiliar males and/or females (Smith et al. 1996). In Smith et al.'s (1996) study, male mice, previously identified as high or low rankers within small, randomly constituted groups were presented with the substrate odours of unfamiliar individuals in their home cage, assayed for changing hormone and IgG concentrations during the period of exposure, then infected with *B. microti*. Exposure to odours resulted in a marked decrease in IgG levels relative to clean sawdust controls, with levels being reduced most when male and female odours were presented together. As in Barnard et al.'s (1998) experiment with *H. polygyrus*, previously high ranking males showed a severer infection than previous low rankers, and infections among high rankers were severest after they had been presented with male and female odours combined. However, when relationships between resistance and hormone concentrations were analysed, resistance related to corticosterone concentration but not testosterone. Reduced resistance across ranks and odour treatments was associated with elevated corticosterone relative to pre-treatment concentrations and post-treatment corticosterone concentrations were higher among previous high rankers and among high rankers when female odours (especially when combined with those of a male) had been presented. Importantly, as in all our other experiments, corticosterone showed no evidence of being modulated with respect to IgG concentration. Testosterone concentration, on the other hand, showed a significant positive correlation with IgG concentration across all males. Once again, therefore, the absence of an independent effect of testosterone, on resistance to

B. microti was associated with a tendency to modulate levels of the hormone in relation to current immunocompetence.

The correlational evidence is thus in keeping with the idea of constraints on adaptive immunity trade-offs. If this conclusion is robust, however, we should be able to manipulate the tendency to modulate potential immunodepressants by reducing immunocompetence experimentally. Barnard et al. (1997a, b) tested this by temporarily depressing immunocompetence with anti-thymocyte serum (ATS). ATS treatment was selected because it acts primarily on T-lymphocytes that are essential for efficient antibody responses and cell-mediated immunity and because it is relatively innocuous in other respects, without the general side-effects of other forms of immunodepressive therapy (e.g. cytotoxic drugs, whole-body irradiation, ablation of lymphoid tissue) (see references in Barnard et al. 1997a). In addition, there are known feedback mechanisms mediating the secretion of sex steroids in relation to immunocompetence via the thymus (Grossman 1985; Alexander and Stimson 1988). Our expectation from the adaptive immunity trade-off hypothesis was that ATS-treated mice would show a downregulation of testosterone and aggressive and other energy-consuming behaviours commensurate with the degree of immunodepression experienced. From our previous results, however, we did not expect to see a reduction in corticosterone concentration.

Male CFLP mice were observed in small groups prior to separation and treatment with ATS or naïve rabbit serum (NRS) control, then reallocated to their pre-treatment groups for further observation. Blood samples for hormone and IgG concentrations were taken at the beginning and end of the two periods of grouping, and, terminal blood samples were assayed for SRBC haemagglutination titres. Following treatment, ATS-treated animals showed an absence of detectable antibody response to SRBCs and a pronounced reduction in IgG concentration compared with NRS controls. The change in IgG was reflected in a simultaneous decline in testosterone. However, when relationships were partialled out, the decline in testosterone among ATS males correlated with a reduction in thymus weight rather than IgG concentration, supporting the idea of thymus-mediated regulation of testosterone secretion. As expected, there was no associated change in corticosterone concentration.

The downregulation of testosterone among ATS animals was accompanied by a reduction in the amount of aggression and general activity relative to both pre-treatment levels and NRS controls, the reduction being particularly striking because overall activity, and especially aggression, are usually greatly increased when groups are first introduced or reconstituted after a period of separation (Poole and Morgan 1973; Barnard et al. 1993). Time spent sleeping, however, was maintained among ATS mice but fell sharply among controls. The fact that the reduction in aggression correlated with that in IgG concentration across both ATS mice and controls strengthened the conclusion that the treatment effect was the result of increased immunodepression among ATS mice. Relationships between immunocompetence measures and sleep were confined to ATS animals and involved thymus weight rather than IgG. Interestingly, neither the changes in activity and aggression, nor those in sleep (or other behaviours), correlated with testosterone concentration. The recurrent finding in our experiments that aggressive (and other) behaviour show little correlation with testosterone suggests that hormonal and behavioural changes are independent responses to immunodepression.

Of course, if the modulation of hormones and behaviour by immunodepressed ATS males reflected an adaptive trade-off between survival and reproduction, an increased reproductive incentive might be expected to reduce their tendency to modulate. Barnard et al. (1997b) repeated the ATS experiment, but this time introduced the substrate odours of females into the males' cages during the post-treatment phase. Under these conditions, ATS-treated males failed to show any of the behavioural changes observed in Barnard et al.'s (1997a) experiment. Indeed, there was a significant increase in aggression and general locomotory activity, and a reduction in sleep, during the post-treatment phase relative to pre-treatment levels with no difference between ATS-treated and control mice. The presence of female cues therefore appeared to override the downregulation of active behaviours in immunodepressed males, as we should predict if future survival was discounted against the prospect of immediate reproductive opportunity.

4 Immunity trade-offs and learning

While many studies looking at relationships between behaviour, immunity and parasite infection, and their mediation by steroid hormones, have focused on trade-offs in the context of sexual selection and the immunocompetence handicap hypothesis (Folstad and Karter 1992; Sheldon and Verhulst 1996; Zuk and Stoehr 2002), we have seen above that other aspects of behaviour can have bidirectional interactions with the immune system that may be mediated by steroid hormones. So far, however, we have looked only at social behaviour and how this relates to differences in competitive ability and thus life history priorities, but the kind of trade-offs apparent there can, in principle, apply to any aspect of behaviour where underlying mechanisms impinge on immunity and susceptibility to infection.

One possible example is learning, and we shall finish by discussing briefly some recent evidence for this.

There is good evidence that learning ability is influenced by immunity and/or infection status. Studies of several species, using various learning paradigms, have shown reduced learning ability when subjects are immunodepressed or suffering from infection (Dolinsky et al. 1981; Nokes et al. 1992; Kavaliers et al. 1995; but see Braithwaite et al. 1998). Learning can also be improved experimentally by boosting components of the immune system, such as certain cytokines (e.g. Gibertini 1998; Brennan et al. 2003; but see e.g. Fiore et al. 2000; Song 2002). A number of mechanisms could account for these relationships, including stress, metabolic burden and parasitic manipulation (Bloom 1979; Ali and Behnke 1984; Nokes et al. 1992; Bodnoff et al. 1995; Maier and Watkins 1998). Some or all of these may operate, at least partly, through opposing effects of glucocorticoids and androgens on immunocompetence (Grossman 1985; Alexander and Stimson 1988; Barnard and Behnke 2001) and attention and memory (Andrew et al. 1981; Rousse et al. 1997; Kritzer et al. 2001). Thus, if hormone activity is boosted to underpin learning effort, there may be a concomitant impact on immune responsiveness, the corollary being reduced learning effort (and thus immunodepressive hormone secretion) under conditions of immune depression or challenge (when the premium on maintaining immune responsiveness is high). There are good reasons for supposing that both learning and maintaining immune responsiveness are metabolically expensive (e.g. Dukas 1999; Lochmiller and Deerenberg 2000; Martin et al. 2003) and thus likely to be subject to constraints from competing demands. Just as with secondary sexual behaviours and associated characters, therefore, we might expect a shifting investment trade-off between learning and immune function, mirrored by underlying changes in steroid hormone secretion, that reflects the relative reproductive consequences of the learning outcome and its associated immunity costs. Barnard et al. (2005, 2006), from which the following discussion comes, have tested this using two different learning paradigms in BKW mice: radial maze performance and odour learning. These approaches were chosen because of their salience in terms of the behavioural ecology of house mice which depend heavily on spatial learning and olfactory social communication (e.g. Olton et al. 1981; Hurst 1993; Nevison et al. 2003; Latham and Mason 2004).

4.1 Radial maze performance and immunity

Barnard et al. (2005) exposed single male mice to a radial maze task in which they were required to learn the location of a food reward in a 7-arm maze. The degree of variation in the correct location, and thus the difficulty of the learning task, differed between treatments. In one treatment, food was always available in the same arm across trials (constant location), in the others it changed randomly after every six 10-min trials (low change location), or after every three trials (high change location). A significant decline in learning performance (percent correct entries) across the three treatments confirmed the increase in task difficulty. As in the study of social rank effects above, circulating testosterone, corticosterone and total IgG were assayed intercurrently before and after experience in the maze, and mice were subsequently infected with *B. microti* and their response to the infection monitored.

The results provided some support for a relationship between learning effort and immune function in mice in that an experimental blood protozoan infection was severest in animals that had been exposed to the most difficult spatial learning task, and in those that performed better under more difficult learning conditions. Total locomotory activity, which increased with task difficulty, as might be expected, was included as a covariate in the analysis to control for any confounding effect, but showed no independent association with infection. Inclusion of learning performance (percentage correct responses), however, did show a significant independent relationship with infection (Fig. 5), and a significant interaction with task difficulty.

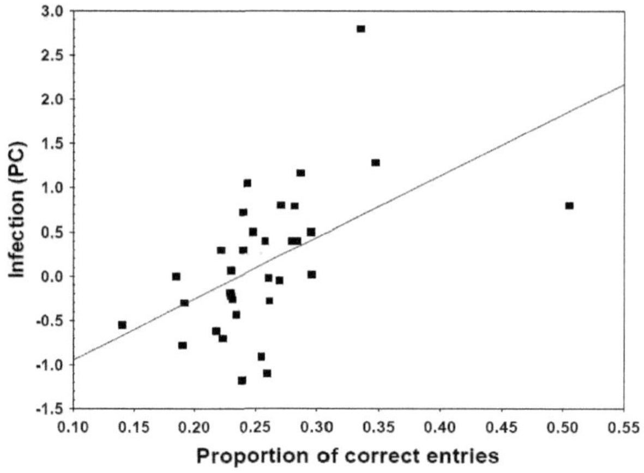

Fig. 5. Component effects plot from stepwise partial regression showing a significant relationship between learning performance in a radial maze and degree of infection with *B. microti* (composite measure from PCA of time to reach peak infection, intensity at peak infection and relative time to clear infection) in male laboratory mice (based on data from Barnard et al. 2005)

Post hoc stepwise regression analysis within each treatment showed that the correct entries x treatment interaction was due to a significant positive relationship between infection and correct entries in the low and high change treatments but not when food location was kept constant (no variables entered the equation). Thus, mice that appeared to put more effort into learning when the task was difficult, and overall performance poor, suffered greater infection afterwards.

While infection was not a simple function of task difficulty, as mice experiencing the low change task showed the weakest infection, the relative levels of infection across treatments was mirrored by the change in total IgG over the period of testing in the maze. Total IgG generally increased over the period of the experiment, but the increase was greatest in the treatment showing the weakest infection (low change) and least in the treatment where mice were affected most severely (high change). Post-experimental infection therefore reflected changes in circulating antibody titre during the learning task.

The association between learning performance and infection was consistent with a specific effect of the cost of learning between treatments, rather than more general differences such as stress (de Kloet et al. 2002) and increased overall activity. Neither increased number of visits nor weight loss had any independent effect on response to infection or decline in IgG. There was also no increase in corticosterone or evidence of adrenal hypertrophy associated with treatments or learning performance. Stress thus seems unlikely as a cause of reduced resistance in this case.

Given the lack of change in corticosterone concentration, it is not surprising that no relationship emerged between change in corticosterone and either learning performance or infection. While there was no association with intercurrent measures of corticosterone, however, there was a positive relationship between pre-experimental corticosterone concentration and infection. The fact that pre-experimental corticosterone levels showed only a weak positive relationship with learning performance, and learning performance maintained a significant positive relationship with infection independently of pre-experimental corticosterone suggests that *both* starting hormonal predisposition and learning effort contributed to the response to infection but were not causally linked in their effects. This somewhat ambiguous outcome is consistent with other studies that have shown that relationships between glucocorticoids and learning depend heavily on experimental context (Kloet et al. 2002). However, a weak positive association between corticosterone and learning performance in Barnard et al.'s (2005) experiment is arguably in keeping with other studies showing that non stress-associated elevations in corticosterone facilitate the retention of spatial learning tasks (e.g. Grootendorst et al. 2002; Kent et al. 2002; Beylin and Shors 2003), but not consistent with stress-induced elevation in corticosterone levels which have been found to depress spatial learning performance (e.g. Bodnoff et al. 1995; McClay et al. 1995).

Overall, therefore, Barnard et al.'s (2005) results suggest some predisposition towards spatial learning and infection among mice with higher starting concentrations of corticosterone, but no modulation of corticosterone in relation to the learning experience itself. That mice with high pre-experimental concentrations of corticosterone were more susceptible to infection is at least consistent with the idea that elevated corticosterone levels reflect underlying individual differences in metabolic investment and thus life history strategy. In contrast to the studies of social rank-related immunity trade-offs in BKW mice summarized above, however, there was no evidence for effects of pre-experimental or intercurrent concentrations of testosterone on learning performance or infection.

4.2 Odour learning and immunity

In a second study, Barnard et al. (2006) used a standard habituation procedure to test the ability of mice to learn different numbers of social odours (presented as soiled sawdust) and relate this to changes in circulating hormone and antibody titres and subsequent response to infection with *B. microti*. Mice were exposed to odours from two, four or eight unfamiliar males and then tested with paired samples of one of the odours previously encountered and a novel odour, the degree of learning being assessed as the extent to which mice attended to the novel as opposed to familiar sample. Subjects showed clear habituation to odour samples over successive periods of exposure, suggesting they had been learned, and significantly greater interest in the novel odour sample in subsequent tests, suggesting recall of familiar odours.

Interestingly, the number of odours had no significant effect in itself on attention paid to either novel or familiar odours. This could imply one of two things: either the difference in difficulty between the learning tasks was trivial, so there was no effect on learning, or mice put more effort into learning when larger numbers of odours were presented. Relationships between learning and the outcome of experimental infection suggested the learning tasks were not trivial. To begin with, *B. microti* infection increased with odour number, with infection being lowest when mice were required to learn two odours and highest when required to learn eight. This implies a greater cost to the learning effort in multiple odour treatments. Second, within-treatment relationships between infection and attention to novel versus familiar odours in tests suggested that, while overall learning was maintained across treatments, those subjects that did less well (made more "attend to familiar" errors) in the most demanding treatment (eight odours) suffered a greater cost. In other words, it was the relative failure to learn, rather than maintaining learning performance, in the most difficult treatment that led to greater infection – a contrast with results from the maze-learning experiment above.

This conclusion is supported further by the within-treatment relationships between learning performance, infection and change in corticosterone concentration. While there was an overall decline in corticosterone across the period of the experiment, infection and "attend to familiar" errors during tests were greater in those mice showing least reduction in corticosterone. This is consistent with other studies of rodents showing that corticosterone levels reflect an association between learning performance and stress (e.g. Bodnoff et al. 1995; McClay et al. 1995) and that prolonged social memory is associated with reduced basal concentration of corticosterone (Tang et al. 2003), a trend reinforced in Barnard et al. (2006) by an upward drift of testosterone and total IgG concentrations, which tend to be depressed under stress in BKW mice (Barnard et al. 1996a, b), during the course of the experiment. Just as importantly, no difference emerged between odour treatments in either absolute concentrations, or changes in concentration, of hormones or IgG, again supporting the conclusion that there was nothing intrinsically stressful about treatments with more odours that might confound inferences about the effect of odour number on learning effort. The results certainly did not support a role of corticosterone in enhancing learning (e.g. Grootendorst et al. 2002; Kent et al. 2002; Beylin and Shors 2003; Barnard et al. 2005; but see Sandi 2003; Brush 2003).

The different apparent roles of corticosterone in two learning paradigms (maze versus odour) using the same strain of mouse are in keeping with evidence across studies that relationships between corticosterone and learning depend heavily on the learning task and experimental context (e.g. Grootendoorst et al. 2002; Kloet et al. 2002). While corticosterone has widely been found to enhance performance in maze-learning tasks (e.g. Rousse et al. 1997; Beylin and Shors 2003; Barnard et al. 2005), there is evidence that corticosterone levels tend to be inversely related to performance and retention in olfactory tasks (e.g. Kavaliers and Ossenkopp 2001; Tang et al. 2003; Moriceau and Sullivan 2004).

5 Concluding remarks

Like the field studies in the first part of this chapter, the results from experiments with social rank and learning suggest that interactions between behavioural and physiological systems, and the way animals prioritize them, are likely to vary considerably from context to context, and that sweeping generalizations about the nature of any trade-offs will usually be misleading. Indeed, analyses at the level of experimental treatments or convenient social categories may obscure crucial information about individual differences (e.g. Gosling et al. 1996) that may be best approached as a phenotypic continuum. Motivational changes within individuals are a further factor to consider. Barnard and Luo (2002), for example, looked to the effects of acquiring a particular social status on learning performance in mice. Animals that had previously shown no difference in maze performance when housed singly, performed very differently when again tested singly after having been housed in pairs for a few days and allowed to develop polarized rank relationships with their cagemate; mice that became dominant performed significantly better than those that became subordinate. Importantly, the effects on learning of adopting a given status can persist for a long period of subsequent solitary housing, suggesting that certain kinds of experience can modify investment in different behaviours well into the future (Fitchett et al. 2005).

The crucial element in any approach to the problem of competing priorities is a clear understanding of the selection pressures acting on different kinds of individual, something that, in most cases, requires comprehensive analysis of ecological relationships in the field coupled with manipulative experimentation based on clear predictions arising from it. In very few cases, are we anywhere close to such a position. Much exciting work therefore lies in store.

References

Alexander J, Stimson WH (1988) Sex hormones and the course of parasitic infection. Parasitol Today 4:189–193

Ali NMH, Behnke JM (1984) Non-specific immunodepression by *Nematospiroides dubius*: Of concurrent responses to oxazolone and lipopolysaccharide. J Helminthol 58:301–311

Aloe L, Musi B, Micera A, Santucci D, Tirassa P, Alleva E (1995) NGFantibody production as a result of repeated psychosocial stress in adult mice. Neurosci Res Comm 16:19–28

Andrew RJ, Clifton PG, Gibbs ME (1981) Enhancement of effectiveness of learning by testosterone in domestic chicks. J Comp Physiol Psychol 95:406–417

Arak A (1984) Sneaky breeders. In: Barnard CJ (ed) Producers and scroungers: Strategies of exploitation and parasitism. Chapman and Hall, New York, pp 154–194

Bajer A, Harris PD, Behnke JM, Bednarska M, Barnard CJ, Sherif N, Clifford S, Gilbert FS, Sinski E, Zalat S (2006) Marked local variation in haemoparasites and arthropod vectors, and intestinal Protozoa in spiny mice (*Acomys dimidiatus*) from four montane wadis in the St. Katherine Protectorate, Sinai, Egypt. J Zool Lond (in press)

Barker DJP (1995) The foetal and infant origins of disease. Eur J Clin Inv 25:457–463

Barker DJP (1996) The foetal origins of hypertension. J Hypertension 14:S117–S120

Barnard CJ, Behnke JM (2001) From psychoneuroimmunology to ecological immunology: Life history strategies and immunity trade-offs. In: Ader RL, Felten DL, Cohen N (eds) Psychoneuroimmunology, vol 2, 3rd edn. Acad Press, New York, pp 35–47

Barnard CJ, Luo N (2002) Acquisition of dominance status affects maze learning in mice. Behav Proc 60:53–59

Barnard CJ, Hurst JL, Aldhous P (1991) Of mice and kin: The functional significance of kin bias in social behaviour. Biol Rev 66:379–430

Barnard CJ, Behnke JM, Sewell J (1993) Social behaviour, stress and susceptibility to infection in house mice (*Mus musculus*): Effects of duration of grouping and aggressive behaviour prior to infection on susceptibility to *Babesia microti*. Parasitology 107:183–192

Barnard CJ, Behnke JM, Sewell J (1994) Social behaviour and susceptibility to infection in house mice (*Mus musculus*): Effects of group size, aggressive behaviour and status-related hormonal responses prior to infection on resistance to *Babesia microti*. Parasitology 108:487–496

Barnard CJ, Behnke JM, Sewell J (1996a) Social status and resistance to disease in house mice (*Mus musculus*): Status-related modulation of hormonal responses in relation to immunity costs in different social and physical environments. Ethology 102:63–84

Barnard CJ, Behnke JM, Sewell J (1996b) Environmental enrichment, immunocompetence and resistance to *Babesia microti* in male laboratory mice. Physiol Behav 60:1223–1231

Barnard CJ, Behnke JM, Gage AR, Brown H, Smithurst PR (1997a) Modulation of behaviour and testosterone concentration in immunodepressed male laboratory mice (*Mus musculus*). Physiol Behav 61:907–917

Barnard CJ, Behnke JM, Gage AR, Brown H, Smithurst PR (1997b) Immunity costs and behavioural modulation in male laboratory mice (*Mus musculus*) exposed to the odour of females. Physiol Behav 62:857–866

Barnard CJ, Behnke JM, Gage AR, Brown H, Smithurst PR (1998a) Maternal effects on the development of social rank and immunity trade-offs in male laboratory mice (*Mus musculus*). Proc R Soc Lond B 265:2087–2093

Barnard CJ, Behnke JM, Gage AR, Brown H, Smithurst PR (1998b) The role of parasite-induced immunodepression, rank and social environment in the modulation of behaviour and hormone concentration in male laboratory mice (*Mus musculus*). Proc R Soc Lond B 265:693–701

Barnard CJ, Behnke JM, Bajer A, Bray D, Race T, Frake K, Osmond J, Dinmore J, Sinski E (2002) Local variation in endoparasite burdens of bank voles

(*Clethrionomys glareolus*) from ecologically similar sites: Morphometric and endocrine correlates. J Helminthol 76:103–112

Barnard CJ, Kulis K, Behnke JM, Bajer A, Gromadzka-Ostrakowska J, Stachon E, Sinski E (2003a) Local variation in helminth burdens of bank voles (*Clethrionomys glareolus*) from ecologically similar sites: Temporal stability and relationships with hormone concentrations and social behaviour. J Helminthol 77:185–195

Barnard CJ, Sayed E, Barnard LE, Behnke JM, Abdel Nabi I, Sherif N, Shutt A, Zalat S (2003b) Local variation in helminth burdens of Egyptian spiny mice (*Acomys cahirinus dimidiatus*) from ecologically similar sites: Relationships with hormone concentrations and social behaviour. J Helminthol 77:197–207

Barnard CJ, Collins SA, Daisley JN, Behnke JM (2005) Maze performance and immunity costs in mice. Behaviour 142:241–263

Barnard CJ, Collins SA, Daisley JN, Behnke JM (2006) Odour learning and immunity costs in mice. Behav Proc 72:74–83

Beden SN, Brain PF (1985) Effects of attack-related stress on the primary immune response to sheep red blood cells in castrated mice. IRCS Med Sci 12:675

Behnke JM, Parish HA (1979) *Nematospiroides dubius*: Arrested development of larvae in immune mice. Exp Parasitol 47:116–127

Behnke JM, Barnard CJ, Wakelin D (1992) Understanding chronic nematode infections: Evolutionary considerations, current hypotheses and the way forward. Int J Parasitol 22:861–907

Behnke JM, Barnard CJ, Mason N, Harris PD, Sherif NE, Zalat S, Gilbert FS (2000) Intestinal helminths of spiny mice (*Acomys cahirinus dimidiatus*) from St Katherine's Protectorate in the Sinai, Egypt. J Helminthol 74:31–43

Behnke JM, Barnard CJ, Bajer A, Bray D, Dinmore J, Frake K, Osmond J, Race T, Sinski E (2001) Variation in the helminth community structure in bank voles (*Clethrionomys glareolus*) from three comparable localities in the Mazury Lake District region of Poland. Parasitology 123:401–414

Benten WPM, Bettenhauser U, Wunderlich F, van Vliet F, Mossman H (1991) Testosterone-induced abrogation of self-healing of *Plasmodium chabaudi* malaria in B10 mice: Mediation by spleen cells. Infect Immun 59:4486–4490

Berkenbosch F, van Oers J, del Rey A, Tilders F, Besedovsky H (1987) Corticotropin-releasing factor-producing neurons in the rat activated by interleukin-1. Science 238:524–526

Besedovsky HO, Sorkin E, Keller M, Muller J (1975) Changes in blood hormone levels during immune response. Proc Soc Exp Biol Med 150:466–470

Beylin AV, Shors TJ (2003) Glucocorticoids are necessary for enhancing the acquisition of associative memories after acute stressful events. Horm Behav 43:124–131

Bigi S, Huber C, DeAcetis L, Alleva E, Dixon AK (1994) Removal of the submaxillary salivary glands first increases then abolishes the agonistic response of male mice in social encounters. Physiol Behav 55:13–19

Bloom BR (1979) Games parasites play: How parasites evade immune surveillance. Nature 279:21–26

Bodnoff SR, Humphreys AG, Lehman JC, Diamond DM, Roe GM, Meaney MJ (1995) Enduring effects of chronic corticosterone treatment on spatial learning, synaptic plasticity and hippocampal neuropathology in young and mid-aged rats. J Neurosci 15:61–69

Brain PF (1988) Social stress in laboratory mouse colonies. In: Laboratory animal welfare research: Rodents. Universities' Federation for Animal Welfare, Potters Bar, pp 49–61

Brain PF, Nowell NW (1970) The effects of differential grouping on endocrine function of mature albino mice. Physiol Behav 5:907–910

Braithwaite VA, Salkeld DJ, McAdam HM, Hockings CG, Ludlow AM, Read AF (1998) Spatial and discrimination learning in rodents infected with the nematode *Strongyloides ratti*. Parasitology 117:145–154

Braude S, Tang-Martinez Z, Taylor GT (1999) Stress, testosterone and the immunredistribution hypothesis. Behav Ecol 10:345–350

Brennan FX, Beck KD, Servtius RJ (2003) Low doses of interleukin-1 improve the leverpress avoidance performance of Sprague-Dawley rats. Neurobiol Learn Mem 80:168–171

Broom DM, Johnson K (1993) Stress and animal welfare. Chapman and Hall, London

Brunjes P (1983) Olfactory bulb maturation in *Acomys cahirinus*: Is neural growth similar in precocial and altricial murids? Brain Res 284:335–341

Brush FR (2003) The Syracuse strains, selectively bred for differences in active avoidance learning, may be models of genetic differences in trait and state anxiety. Stress 6:77–85

Chamove AS (1989) Environmental enrichment: A review. Anim Technol 40:155–178

Chapman VM, Desjardins C, Bronson FH (1969) Social rank in male mice and adrenocortical response to open field exposure. Proc Soc Exp Biol Med 130:624–627

Clark PM, Hindmarsh PC, Shiell AW, Law CM, Honour JW, Barker DJP (1996) Size at birth and adrenocortical function in childhood. Clin Endocrinol 45:721–726

Cirulli F, DeAcetis L, Alleva E (1998) Behavioural effects of peripheral interleukin-1 administration in adult CD-1 mice: Specific inhibition of the offensive components of intermale agonistic behavior. Brain Res 791:308–312

Cox FEG, Young AS (1969) Acquired immunity to *Babesia microti* and *Babesia rodhaini* in mice. Parasitology 75:189–196

Cresswell JL, Egger P, Fall CHD, Osmond C, Fraser RB, Barker DJP (1997) Is the age of menopause determined in utero? Early Human Dev 49:143–148

Dantzer R, Bluthe RM, Kent S, Goodall G (1993) Behavioral effects of cytokines: An insight into mechanisms of sickness behavior. In: De Souza DB (ed) Neurobiology of cytokines. Acad Press, San Diego, pp 130–151

Davies IB (1998) The effects of hormone treatment on social status-related differences in infection with *Babesia microti* in adult male CFLP mice. Ph. D. thesis, University of Nottingham

de Kloet ER, Grootendorst J, Karssen AM, Oitzl ME (2002) Gene x environment interaction and cognitive performance: Animal studies on the role of corticosterone. Neurobiol Learn Mem 78:570–577
Dolinsky ZS, Mitchell DF, Grove DI (1981) Behavioral effects of lead and *Toxocara canis* infection in mice. Science 231:1142–1144
Dukas R (1999) Costs of memory: Ideas and predictions. J Theor Biol 197:41–50
Edwards JC (1988) The effects of *Trichinella spiralis* infection on social interactions in mixed groups of infected and uninfected male mice. Anim Behav 36:529–540
Edwards JC, Barnard CJ (1987) The effects of *Trichinella* infection on intersexual interactions between mice. Anim Behav 35:533–540
Ehman KD, Scott ME (2001) Urinary odour preferences of MHC congenic female mice, *Mus domesticus*: Implications for kin recognition and detection of parasitized males. Anim Behav 62:781–789
Fernandes GA (2000) Immunological stress in rats induces bodily alterations in saline-treated conspecifics. Physiol Behav 69:221–230
Fitchett AE, Collins SA, Barnard CJ, Cassaday HJ (2005) Subordinate mice show long-lasting differences in spatial learning that persist when housed alone. Neurobiol Learn Mem 84:247–251
Folstad I, Karter AJ (1992) Parasites, bright males and the immunocompetence handicap. Amer Nat 139:603–622
Freeland WJ (1981) Parasitism and behavioural dominance among male mice. Science 213:461–462
Gibertini M (1998) Cytokines and cognitive behavior. Neuroimmunomod 5:160–165
Gliwicz J, Rajska-Jurgiel E (1983) Social organization. Acta Theriol 28:S134–S140
Gosling LM, Atkinson NW, Dunn S, Collins SA (1996) The response of subordinate male mice varies in relation to their own competitive ability. Anim Behav 52:1185–1191
Grootendorst J, Kempes MM, Lucassen PJ, Dalm S, de Kloet ER, Oitzl MS (2002) Differential effect of corticosterone on spatial learning in apolipoprotein E knockout and C57BL/6J mice. Brain Res 953:281–285
Grossman CJ (1985) Interactions between the gonadal steroids and the immune system. Science 227:257–261
Grossman CJ, Roselle GA (1986) The control of immune response by endocrine factors and the clinical significance of such regulation. Progr Clin Biochem Med 4:9–56
Gustafsson T, Andersson B, Meurling P (1980) Effect of rank on the growth of the preputial glands in male bank voles, *Clethrionomys glareolus*. Physiol Behav 24:689–692
Haemisch A, Gartner K (1994) The cage design affects intermale aggression in small groups of male laboratory mice: Strain specific consequences of social organization, and endocrine activations, in two inbred strains (DBA/2J and CBA/J). J Exp Anim Sci 36:101–116
Haig D (1993) Genetic conflicts in human pregnancy. Q Rev Biol 68:495–532

Hales CN (1997) Metabolic consequences of intrauterine growth retardation. Acta Paediatr 86:184–187

Hamilton WD, Zuk M (1982) Heritable true fitness and bright birds: A role for parasites? Science 218:384–387

Hartvigsen R, Kennedy CR (1993) Patterns in the composition and richness of helminth communities in brown trout, *Salmo trutta*, in a group of reservoirs. J Fish Biol 43:603–615

Hassaneen ESMA (2005) Investigating population dynamics, social behaviour and immunophysiology of *Acomys cahirinus dimidiatus* populations in the naturally fragmented habitat of the St. Katherine Protectorate, South Sinai, Egypt. M. Phil. Thesis, University of Nottingham

Hoffmeyer I (1982) Responses of female bank voles (*Clethrionomys glareolus*) to dominant vs subordinate conspecific males and to urine odors from dominant vs subordinate males. Behav Neur Biol 36:178–188

Hurst JL (1987) Behavioural variation in wild house mice (*Mus domesticus* Rutty): A quantitative assessment of female social organization. Anim Behav 35:1846–1857

Hurst JL (1990) Urine marking in population of wild house mice (*Mus domesticus* Rutty). I. Communication between males. Anim Behav 40:209–222

Hurst JL (1993) The priming effects of urine substrate marks on interactions between male house mice, *Mus musculus domesticus* Schwarz and Schwarz. Anim Behav 45:55–81

Hurst JL, Hall S, Roberts R, Christian C (1996) Social organization in the aboriginal house mouse, *Mus spretus* Lataste: Behavioural mechanisms underlying the spatial dispersion of competitors. Anim Behav 51:327–344

Hurst JL, Payne CE, Nevison CM, Marie AD, Humphries RE, Robertson DHL, Cavaggioni A, Beynon RJ (2001) Individual recognition in mice mediated by major urinary proteins. Nature 414:631–634

Hussein HS (1984) *Babesia hylomysci* and *Babesia microti*: Dexamethasone treatment of infected mice. Exp Parasitol 57:165–171

Hutchings JA, Myers RA (1994) The evolution of alternative mating strategies in variable environments. Evol Ecol 8:256–268

Kavaliers M, Colwell DD (1995) Odours of parasitized males induce aversive responses in female mice. Anim Behav 50:1161–1169

Kavaliers M, Ossenkopp KP (2001) Corticosterone rapidly reduces male odor preferences in female mice. Neuroreport 12:2999–3002

Kavaliers M, Colwell DD, Galea LAM (1995) Parasitic infection impairs spatial learning in mice. Anim Behav 50:223–229

Kavaliers M, Colwell DD, Choleris E (1998) Parasitized female mice display reduced aversive responses to the odours of infected males. Proc R Soc Lond B 265:1111–1118

Kelly DD (1986) Stress-induced analgesia. Ann NY Acad Sci 467:426–445

Kennedy CR, Hartvigsen R, Halvorsen O (1991) The importance of fish stocking in the dissemination of parasites throughout a group of reservoirs. J Fish Biol 38:541–552

Kent WDT, Cross-Mellor SK, Kavaliers M, Ossenkopp KP (2002) Acute effects of corticosterone on LiCl-induced rapid gustatory conditioning in rats: A microstructural analysis of licking patterns. Behav Brain Res 136:143–150

Klein SL (2000) Hormones and mating system affect sex and species differences in immune function among vertebrates. Behav Proc 51:149–166

Komers PE, Pelabon C, Stenstrom D (1997) Age at first reproduction in male fallow deer: Age-specific versus dominance-specific behaviours. Behav Ecol 8:456–462

Kritzer MF, McLaughlin PJ, Smirlis T, Robinson JK (2001) Gonadectomy impairs T-maze acquisition in adult male rats. Horm Behav 39:167–174

Lamers WH, Mooren PG, Griep H, Endert E, Degenhart HJ, Charles R (1986) Hormones in perinatal rat and spiny mouse: Relation to altricial and precocial timing of birth. Am J Physiol 14:E78–E85

Langley-Evans S (1997) Fetal programming of immune function and respiratory disease. Clin Exp Allergy 27:1377–1379

Latham N, Mason G (2004) From house mouse to mouse house: The behavioural biology of free-living *Mus musculus* and its implications in the laboratory. Appl Anim Behav Sci 86:261–289

Leshner AI, Politch JA (1979) Hormonal control of submissiveness in mice: Irrelevance of the androgens and relevance of the pituitary-adrenal hormones. Physiol Behav 22:531–534

Lochmiller RL, Deerenberg C (2000) Trade-offs in evolutionary immunology: Just what is the cost of immunity? Oikos 88:87–98

Maier SF, Watkins LR (1998) Cytokines for psychologists: Implications of bidirectional immune-to-brain communication for understanding behavior, mood, and cognition. Psych Rev 105:83–107

Maier SF, Watkins LR, Fleshner M (1994) Psychoneuroimmunology: The interface between behavior, brain and immunology. Am Psychol 49:1004–1017

Maestripieri D, Desimone R, Aloe L, Alleva E (1990) Social status and nerve growth factor serum level affect agonistic encounters in mice. Physiol Behav 47:161–164

Mallard ST, Barnard CJ (2003) Competition, fluctuating asymmetry and sperm transfer in male gryllid crickets (*Gryllus bimaculatus* and *Gryllodes sigillatus*). Behav Ecol Sociobiol 53:190–197

Mallard ST, Barnard CJ (2004) Food stress, fluctuating asymmetry and reproductive performance in the gryllid crickets *Gryllus bimaculatus* and *Gryllodes sigillatus*. Behaviour 141:219–232

Mappes T, Koskela E, Ylonen H (1995) Reproductive costs and litter size in the bank vole. Proc R Soc Lond B 261:19–24

Markow TA (ed) (1994) Developmental instability: Its origins and evolutionary implications. Kluwer, Amsterdam

Markowitz H (1982) Behavioral enrichment in the Zoo. van Nostrand, New York.

Martin II LB, Scheuerlein A, Wikelski M (2003) Immune activity elevates energy expenditure of house sparrows: A link between direct and indirect costs? Proc R Soc Lond B 270:153–158

McClay RN, Freeman SM, Zadina JE (1995) Chronic corticosterone impairs memory performance in the Barnes maze. Physiol Behav 63:933–937
McGregor PK, Ayling SJ (1990) Varied cages result in more aggression in male CFLP mice. Appl Anim Behav Sci 26:277–281
Meikle DB, Drickamer LC, Vessey SH, Arthur RD, Rosenthal TL (1996) Dominance rank and parental investment in swine (*Sus scrofa domesticus*). Ethology 102:969–978
Møller AP, Swaddle J (1997) Asymmetry, developmental stability and evolution. Oxford Univ Press, Oxford
Mooradian AD, Morley JE, Korenman SG (1987) Biological action of androgens. Endocrine Rev 8:1–28
Moriceau S, Sullivan RA (2004) Corticosterone influences mammalian neonatal sensitive-period learning. Behav Neurosci 118:274–281
Nevison CM, Barnard CJ, Hurst JL (2003) The consequences of inbreeding for modulating social relationships between competitors. Appl Anim Behav Sci 81:387–398
Nokes C, Gratham-McGregor SM, Sawyer AW, Cooper ES, Bundy DAP (1992) Moderate to heavy infections of *Trichuris trichiura* affect cognitive function in Jamaican school children. Parasitology 104:539–547
Olton DS, Handelmann GE, Walker JA (1981) Spatial memory and food searching strategies. In: Kamil AC, Sargent TD (eds) Foraging behavior: Ecological, ethological and psychological approaches. Garland, New York, pp 333–354
Osborn DJ, Helmy I (1980) The contemporary land mammals of Egypt (including Sinai). Fieldiana (Zool) 5:1–579
Owens IPF, Short RV (1995) Hormonal basis of sexual dimorphism in birds: Implications for new theories about sexual selection. Trends Ecol Evol 10:44–47
Padgett DA, Sheridan JF, Loria R (1995) Steroid hormone regulation of a polyclonal Th2 immune response. Ann NY Acad Sci 774:323–325
Palmer AR, Strobeck C (1986) Fluctuating asymmetry: Measurement, analysis, pattern. Ann Rev Ecol Syst 17:391–421
Penn D, Potts WK (1998) Chemical signals and parasite-mediated sexual signals. Trends Ecol Evol 13:391–396
Phillips DIW (1996) Insulin-resistance as a programmed response to fetal undernutrition. Diabetologia 39:1119–1122
Poiani A, Goldsmith AR, Evans MR (2000) Ectoparasites of house sparrows (*Passer domesticus*): An experimental test of the immunocompetence handicap hypothesis and a new model. Behav Ecol Sociobiol 47:230–240
Poole TR, Morgan HDR (1973) Differences in aggressive behaviour between male mice (*Mus musculus* L.) in colonies of different sizes. Anim Behav 21:788–795
Quinnell RJ, Behnke JM, Keymer AE (1991) Host specificity of and cross-immunity between two strains of *Heligmosomoides polygyrus*. Parasitology 102:419–427
Rao RH (1996) Experimental evidence for the thrifty phenotype hypothesis in rats. Diabetes 45:911

Roberts CW, Satoskar A, Alexander J (1996) Sex steroids, pregnancy-associated hormones and immunity to parasitic infections. Parasitol Today 12:382–388
Rohner-Jeanrenaud F, Jeanrenaud B (1997) Central nervous system and body weight regulation. Ann D Endocrinol 58:137–142
Rohwer S, Ewald P (1981) The cost of dominance and advantage of subordination in a badge signalling system. Evolution 35:441–454
Rook GAW, Zumia A (1997) Gulf War syndrome: Is it due to a systemic shift in cytokine balance towards a Th2 profile? Lancet 349:1831–1833
Rousse I, Beaulieu S, Rowe W, Meaney M, Barden N, Rochford J (1997) Spatial memory in transgenic mice with impaired glucocorticoid receptor function. Neuroreport 8:841–845
Rozenfeld FM, Rasmont R (1991) Odour cue recognition by dominant male bank voles, *Clethrionomys glareolus*. Anim Behav 41:839–850
Rutovskaya V (1998) Acoustic activity and social behaviour of the bank vole (in Russian). In: Sokolov VE, Rozhnov VV, Serbenyuk MA (eds) Behaviour, communication and ecology of mammals. Inst Ecol Evol RAS, Moscow, pp 177–188
Sandi C (2003) Glucocorticoid involvement in memory consolidation. Rev Neurol 37:843–848
Schur B (1987) Social structure and plasma corticosterone level in female albino mice. Physiol Behav 40:698–693
Sheldon BC, Verhulst S (1996) Ecological immunology: Costly parasite defences and trade-offs in evolutionary ecology. Trends Ecol Evol 11:317–321
Smith FV, Barnard CJ, Behnke JM (1996) Social odours, hormone modulation and resistance to disease in male laboratory mice (*Mus musculus*). Anim Behav 52:141–153
Song C (2002) The effect of thymectomy and IL-1 on memory: Implications for the relationship between immunity and depression. Brain Behav Immun 16:557–568
Stearns SC, Koella J (1986) The evolution of phenotypic plasticity in life-history traits: Predictions for norms of reaction for age- and size-at-maturity. Evolution 40:893–913
Tang AC, Reeb BC, Romeo RD, McEwen BS (2003) Modification of social memory, hypothalamic-pituitary-adrenal axis, and brain asymmetry by neonatal novelty exposure. J Neurosci 23:8254–8260
Vessey SH (1964) Effects of grouping on levels of circulating antibodies in mice. Proc Soc Exp Biol Med 115:252–255
Wakelin D, Selby GR (1974) The induction of immunological tolerance to the parasitic nematode *Trichuris muris* is cortisone-treated mice. Immunology 26:1–10
Wedekind C, Folstad I (1994) Adaptive or nonadaptive immunosuppression by sex hormones? Amer Nat 143:936–938
Wilckens T, de Rijk R (1997) Glucocorticoids and immune function: Unknown dimensions and new frontiers. Immunol Today 18:418–424

Wunderlich F, Benten WPM, Bettenhauser U, Schmitte-Wrede H, Mossmann H (1992) Testosterone unresponsiveness of existing immunity against *Plasmodium chabaudi*. Parasite Immunol 14:307–320

Würbel H (2001) Ideal homes? Housing effects on rodent brain and behaviour. Trends Neurosci 24:207–211

Zuk M, Stoehr AM (2002) Immune defense and host life history. Amer Nat 160:S9–S22

Part V. Management and Case Studies

23 Nematode zoonoses

Juan Carlos Casanova, Alexis Ribas and Juan Matías Segovia

1 Introductory remarks

Small mammals are hosts of numerous nematode species, some of which, if transmitted to humans and domestic animals, may cause diseases often associated with high morbidity. This is the case with three important zoonotic nematodes in humans: *Trichinella* spp, *Toxocara canis* and *Angiostrongylus* spp. Trichinellosis, toxocarosis and angiostrongylosis are geographically widespread diseases. The roles of small mammals, and particularly rodents, in the life cycles of *T. canis*, *Trichinella* spp. and *Angiostrongylus* spp. are undoubtedly crucial.

These nematode species are remarkable models to explore the relationships between humans and small mammals in wild, rural and urban areas. For instance, in the natural environment, trichinellosis and toxocarosis are maintained by wild carnivores and their small mammalian prey, whereas in urban environments, the main roles are played by humans, rats, dogs and pigs.

2 *Toxocara canis*

2.1 *T. canis* and rodents

Toxocara canis is a widely distributed and directly transmitted ascarid. The direct life cycle of this nematode was interpreted as an evolutionary simplification, with an indirect life cycle being an ancestral attribute of Ascarididae (Anderson 1988). Nevertheless, the biological cycle of *T. canis* involves internal migration and a great variety of hosts.

The adult stage usually parasitizes wild and domestic canids, and less commonly other families of carnivores, such as Felidae, Mustelidae and Viverridae (Torres et al. 1996; Casanova et al. 2000; Scholtz et al. 2003;

Segovia et al. 2004; Borecka 2005). Eggs are released with the faeces of the definitive host into the environment, and become embryonated in 9-15 days under optimal humidity and temperature. Lower temperatures can delay the development for months to years (Anderson 2000).

Eggs are uptaken by a host with contaminated food and hatch in its digestive tract. Larvae dwell in tissues and undergo complex migrations within the host, with trans-placental and trans-mammary transmission in female hosts. The third stage larvae can also be found in tissues of a wide range of animals other than definitive hosts. Some species of vertebrates (including humans) and invertebrates may act as paratenic hosts. Indeed, this nematode species has been extensively studied as it is implicated in the human syndrome of visceral larva migrans, i.e. human toxocarosis (Overgaauw et al. 2000). Rodents are common paratenic hosts with third-stage larvae found in their liver, kidneys, heart and brain (Brunaska et al. 1995; Anderson 2000).

Dubinsky et al. (1995) studied the epidemiological role of small mammals in the Slovak Republic, where human toxocarosis is highly prevalent compared to the epidemiological situations in other European countries. These authors demonstrated that several small mammals in urban and rural areas are suitable hosts for *T. canis* larvae. Among them, they reported *Mus musculus*, *Apodemus agrarius* and *Micromys minutus* with *T. canis* prevalence of 32, 30.4 and 25%, respectively. Later, Lescano et al. (2000) experimentally infected *Rattus norvegicus* with the nematode embryonated and demonstrated that the rat can be a susceptible paratenic host for *T. canis*. The larval migration of the nematode was found to be similar to what was previously observed in mice. After the oral ingestion of eggs, the larvae hatch in the mouse stomach, then they migrate and penetrate the posterior half of the small intestine, which is considered as their preferred site (Abo-Shehada et al. 1984). During the first week post infection, the internal migration begins by a hepato-pulmonary phase, when the larvae reach the liver and lungs, followed by a myotrophic-neurotrophic phase, when the larvae migrate throughout the body including the brain (Abo-Shehada and Herbert 1984; Skerrett and Holland 1997).

2.2 Rodents as experimental models for *T. canis*

The use of rodents as experimental models for studying the nematode-induced diseases in humans has a long history. Rodent models were routinely used to study pathological, epidemiological, immunological, biochemical and physiological aspects of diseases. In addition, rodents experimentally infected with *T. canis* have been used for evaluating the efficiency of anti-helminthic drugs (Boes and Helwigh 2000).

Small mammals suffer from various pathological alterations due to the presence and migration of *T. canis* larvae. Clinical signs are variable, depending on several factors such as the number of embryonated eggs ingested, the number of migrating larvae, the organ affected, the frequency of re-infections and the immune response of the host. In general, pathogenic mechanisms observed in rodents are similar to those developed in humans (Guardis et al. 2002), justifying, thus, the use of rodent experimental models.

Experimental studies carried out on wild canids (silver and artic foxes) have demonstrated that the number of *T. canis* larvae migrating to pulmonary and renal tissues depends on the infective dose (Saeed et al. 2005), similarly to what was observed in rodents (Cox and Holland 2001).

The brain seems to be one of the main organs of a mouse host where *T. canis* larvae accumulate (Guardis et al. 2002). The number of larvae localized in the brain may influence the behaviour of mice. In general, mice infected with *T. canis* larvae showed less aggressiveness (Anderson, 2000). The effects of *T. canis* larvae on some behavioural activities such as spatial exploration, anxiety, learning and memory were studied by Cox and Holland (2001). In particular, they described a reduction in response to novelty and an increase in exploratory behaviour in heavily infected mice, which could be interpreted as enhancing predation and facilitating transmission of the nematode towards the definitive carnivore hosts. On the other hand, marked effects on learning, memory and anxiety were also observed in moderately and heavily infected mice (Cox and Holland 2001).

Trans-mammary and trans-placental transmission of larvae constitutes a peculiar feature of the biology of *T. canis* that causes early infections in young definitive hosts. This mode of transmission is of great veterinary importance because puppies experience the disease with higher severity than adult individuals (Overgaauw and Knapen 2000). The phenomenon of larval migration was also intensively studied in mice. Larvae enhance expression of secretory granules in the Paneth cells and crypts of Leiberkuhn, i.e. the main larval penetration site (Chiakwung et al. 2004). Recent research has been aimed at identifying the molecules associated with the inflammatory reaction caused by *T. canis* larvae, with the metalloproteinase-9 being a useful marker of *T. canis* larvae migration (Lai et al. 2005).

Some studies have focused on the cellular immune response. The interaction of mouse granulocytes with the nematode, with particular emphasis on the role of the eosinophils, has been studied by Sobota et al. (1988). Eosinophilia is normally associated with all clinical signs of the disease, with the exception of ocular infection. This was confirmed by Guardis et al. (2002) who found no association between eosinophilia and the number of *T. canis* larvae lodged in eyes or in the encephalon of mice. The role of T lymphocytes (CD4+ and CD8+) was also evaluated in post-parturition changes induced by *T. canis* in mice (Reiterova et al. 2004). The role of mast cells in the phenomenon of hyper-reactivity during *T. canis* infection was recently revealed in mice (Sa-Nunes et al. 2005).

Cellular immune response does not provide a clear diagnosis for dog and human toxocarosis, and hence the identification and quantification of humoral response is now preferred by clinicians. The humoral immune response in mice was studied by Havaiova-Reiterova et al. (1995), who found that the occurrence of antibodies against infection of *T. canis* was more expressed in mice infected with high doses of eggs. Several techniques of serological diagnoses have been developed (Dubinky et al. 1995).

However, techniques of serological diagnosis may have several limitations due to the existence of cross-reactions. Perteguer et al. (2003) identified such a cross-reaction between the immune response against secretory-excretory antigens of *T. canis* and the immune response against the third-stage larvae of *Anisakis simplex* in C57BL/10 mice.

3 *Trichinella* in small mammals

3.1 How many species?

Nematodes of the genus *Trichinella* (Nematoda, Trichinellidae) are probably the most intensively studied of all helminths (Anderson 2000). The reason is that this parasite is the causative agent of a human disease known as trichinellosis. The re-emergence of *Trichinella* as an important risk for human health has lead to the creation of the International Trichinellosis Commission with the main objective being to exchange information on the biology, physiopathology, epidemiology, immunology and clinical aspects of trichinellosis in humans and animals. Trichinellosis in humans is an old story as testified by the discovery of the parasite larvae in an Egyptian mummy. The encysted larvae were found in the 1820's, followed by experimental infestations in animals (dogs, pigs, rabbits, mice) in the 1850's.

The first difficulty was to identify which species are involved. At present, the genus *Trichinella* is thought to comprise a total of 8 species and 3 "genotypic species": *T. spiralis*, *T. nativa*, *T. britovi*, *T. murreli*, *T. nelsoni*, *T. pseudospirallis*, *T. papuae* and *T. zimbawensis* and the genotypes *Tri-*

chinella T6, *Trichinella* T8 and *Trichinella* T9. The use of molecular techniques greatly facilitates the identification.

3.2 Life cycle

The life cycle of *T. spiralis* was the first nematode life cycle ever described. Adults of *T. spiralis* are located in the mucosal epithelium of the intestine at the basis of the villi in the glandular crypts and, less commonly, on the tips of villi. The entire body of a nematode is embedded in the epithelium. Females lay active first-stage larvae that invade the veins or the lymphatics of the intestine. It is estimated that each female produces 500 larvae. Larvae invade the cells of the striated muscles, and may survive for prolonged periods as intracellular parasites. The infected cell is modified into a nurse cell, which nourishes and protects the parasite from the host immune response. The nurse cell, which develops a double-membrane around the larvae, and the larvae stay together for the entire life of the host. *T. spiralis* is transmitted to a new host during predation or scavenging. Larvae liberated in the stomach and intestine of a new host invade the region between the lamina propia and the columnar epithelium of the small intestine, giving rise to males and females after a series of moults.

Trichinellosis is mainly a disease of carnivores. The infection of small mammals suggests that they may feed upon infected carcasses. The effects of *Trichinella* infection on the behavioural activity of a host depends on both the *Trichinella* species involved and the host species (Pozio 2005). For example, infected *Peromyscus maniculatus* showed a decrease in behavioural activity proportional to the number of *T. spiralis* larvae recovered from their tissues.

3.3 The parasite in wild vertebrates

The majority of studies on *Trichinella* deal with the genetics and molecular aspects of the host-parasite interactions. For example, more than 200 papers on *Trichinella* were published in 2005. However, only a small portion of these studies dealt with wild mammals, reporting *Trichinella* spp. incidences in several species such as polar bears (Moller et al. 2005; Rah et al. 2005), wild boars (Gamble et al. 2005; Heper et al. 2005), brown bears, wolves and wolverines (Mörner et al. 2005), viverrids (Pozio et al. 2005) and red foxes (Rafter et al. 2005). Only one paper focused on small mammals, namely rodents from Mexico (Pulido et al. 2005), in which the authors reported *Trichinella* infection in *Mus musculus*, *Peromyscus maniculatus* and *Rattus rattus*.

More than 150 mammalian species belonging to many orders have been reported as harbouring *Trichinella* spp. (Pozio 2005). No small marsupials have been found infected. Ten species of insectivores have been reported as hosts of *Trichinella*, but no nematode isolates have been identified at the species level. In Lagomorpha, larvae of *Trichinella* have been detected in hares and rabbits in North America and Europe (Beack 1970; Rausch 1970; Zimmermann 1971). Many species of rodents have been found to be naturally infected with *Trichinella* larvae (Pozio 2005). However, several authors argued that most reports were based on incorrect identifications of other encysted nematodes. Biochemical and molecular methods allow now to distinguish *Trichinella* species, and isolates have been characterized (Pozio 2005). Another problem with these studies is related to the lack of information on rodent trapping locations, whether these were close to human settlements or in the wild. This lack of information reduces their relevance for understanding the role of small mammals in the sylvatic and domestic foci of trichinellosis (Pozio 2005).

In the Iberian peninsula, *Trichinella* spp. have been detected in *R. norvegicus* and *M. musculus* (Rodentia), and *Sorex araneus* (Insectivora) (Cordero del Campillo et al. 1994). This number of known hosts appears to be extremely low when the high effort in the investigation of the helminth fauna of Iberian small mammals is taken into account (Cordero del Campillo et al. 1994; Feliu et al. 1997). However, it is possible that one of the reasons for the low number of known host species for *Trichinella* is related to the necessity to use specific and appropriate technique (tissue compression techniques, digestion). For example, the adequate screening of wild mammals in Lithuania by Senutaité and Grikieniené (2001) showed that six of 10 examined host species were infected by *Trichinella* spp.

A real effort should be done for a better survey of wild animals and, especially, small mammals in order to better understand the sylvatic cycle of *Trichinella* spp. and better contribute to the control of the disease.

4 Human angiostrongyloidiosis

Nematodes of the subgenus *Parastrongylus* of the genus *Angiostrongylus* belong to the family Metastrongyloidae that use rodents as definitive hosts and gastropods as intermediate hosts all around the world (Anderson 2000). Bhaibulaya (1979) and Ohbayashi et al. (1979) considered that at least 20 species of *Angiostrongylus* were described from rodents, carnivores and insectivores, but only two of them have been confirmed in humans, namely *Angiostrongylus cantonensis* and *Angiostrongylus costaricensis*.

Angiostrongylus cantonensis is the causative agent of a human neurological disease. *Angiostrongylus costaricensis* inhabits the mesenteric arteries and causes abdominal angiostrongyliasis in tropical America (Beaver et al. 1984; Morera and Cespedes 1971), but this species can also invade the central nervous system and the lungs. It was experimentally proved that *Angiostrongylus malaysiensis* can cause neurological disease in monkeys, whereas *Angiostrongylus mackerrasae* also has zoonotic potential (Cross 1979).

4.1 Life cycle

Angiostrongylus parasites are not highly specific for either definitive or intermediate hosts. Adults occur in the pulmonary or mesenteric arteries of naturally infected rodents. Eggs hatch in the arteries and arterioles and are expelled via the faeces. A number of molluscs may serve as intermediate hosts, including slugs, aquatic and terrestrial snails. Moreover, several aquatic arthropod species may serve as paratenic hosts. Humans and other mammals become infected when they accidentally ingest larvae of the third stage.

4.2 Host parasite interaction

Chabaud (1972) proposed a hypothesis on the origin of the Angiostrongylinae, with an ancestral group parasitizing Insectivora and Lemuroidea, an intermediate group exploiting Carnivora and the most derived group parasitizing Rodentia.

The geographical distribution of the subgenus *Parastrongylus* matches the distribution and migration of rodents of the genus *Rattus*, with the dispersal of some of them (*R. norvegicus* and *R. rattus*) being facilitated by humans. Drozdz (1970) argued that the original definitive hosts of *A. cantonensis* should be members of the genus *Rattus* from south and south-east Asia. The parasite originated and migrated from Asia to Europe and America with its rodent hosts along the human commercial routes. The restricted presence of *A. mackerrasae* in Australia and Tasmania, and its specificity for *Rattus fuscipes* and *Rattus lutreolus* in Australia and other native rat species in Tasmania, seem to indicate an older historical interaction (Prociv et al. 2000).

4.3 Emerging zoonoses

Angiostrongylus cantonensis has been added to the list of food-borne parasites (Anonymous 2004). Since 1961, it is known that human infections are usually acquired by accidental ingestion of infective larvae in raw or poorly cooked gastropod intermediate hosts. Other sources of infection are the ingestion of paratenic hosts or contaminated food (Cross 1998). The parasite has a worldwide distribution and has been found in small mammals other than *Rattus*, such as different species of rodents in Madagascar (Kliks and Palumbo 1992) and opossums in Lousiana (Kim et al. 2002).

Angiostrongylus cantonensis is the most common cause of human eosinophilia meningitis, primarily in Southeast Asia and the Pacific Basin (Polley 2005). According to Alicata (1966, 1969), *A. cantonensis* originated from Madagascar and spread to the Indian Ocean and South Pacific with the rapid spread of the invasive giant African snail *Achatina fulica*. Although *A. fulica* could play an important role in the dispersion of *A. cantonensis* in Asia and America, it seems likely that the parasite was introduced to new territories more efficiently by rats (Prociv et al. 2000).

5 Concluding remarks

Small mammals play significant roles in the life-cycles of *T. canis*, *Trichinella* spp. and *Angiostrongylus* spp, three important agents of zoonotic human diseases. Although the number of human cases is still low compared to the major infectious diseases, incidence of these helminthoses is expected to increase in the face of the changes that affect the environment and the interactions between wild animals, domestic animals and humans. Hence, a great effort will be essential to improve our knowledge of the relationships between these nematodes and small mammals.

References

Abo-Shehada MN, Herbert IV (1984) The migration of larval *Toxocara canis* in mice. II. Post-intestinal migration in primary infections. Vet Parasitol 17:75–83

Abo-Shehada MN, Al-Zubaidy BA, Herbert IV (1984) The migration of larval *Toxocara canis* in mice. I. Migration through the intestine in primary infections. Vet Parasitol 17:65–73

Alicata J (1966) The presence of *Angiostrongylus cantonensis* in Islands of the Indian Ocean and probable role of the giant African snail, *Achatina fulica*, in dispersal of the parasite to the Pacific Islands. Can J Zool 44:1041–1049

Alicata J (1969) Present status of *Angiostrongylus cantonensis* in man and animals in the tropics. J Trop Med Hyg 72:53–63

Anderson RC (1988) Nematode transmission patterns. J Parasitol 74:30–45

Anderson RC (2000) Nematode parasites of vertebrates. Their development and transmission, 2nd edn. CABI, Wallingford, UK

Anonymous (2004) Centers for Disease Control and Prevention. Diagnosis and management of foodborne illnesses: A primer for physicians and other health care professionals. MMWR, Recommendations and Reports, 53 (RR04)

Beack JW (1970) Trichinosis in domesticated and experimental animals. In: Gould SE (ed) Trichinosis in man and animals. C.C. Thomas, Springfield, pp 61–80

Beaver PC, Jung RC, Cupp EW (1984) The Strongylida: Hookworms and other bursate nematodes. In: Beaver PC (ed) Clinical parasitology, 9th edn. Lea and Febiger, Philadelphia, pp 269–301

Bhaibulaya M (1979) Morphology and taxonomy of major *Angiostrongylus* species of eastern Asia and Australia. In: Cross JH (ed) Studies on angiostrongyliasis in eastern Asia and Australia, Special Publication No. 2 (NAMRU-2-SP-44). US Naval Medical Research Unit, Taipei, pp.4–13

Boes J, Helwigh AB (2000) Animal models of intestinal nematode infections of humans. Parasitology 121:S97–S111

Borecka A (2005) Prevalence of intestinal nematodes of dogs in the Warsaw area, Poland. Helminthologia 42:35–39

Brunaska M, Dubinsky P, Reiterova K (1995) *Toxocara canis*: Ultrastructural aspects of larval moulting in the maturing eggs. Int J Parasitol 25:683–690

Casanova JC, Feliu C, Miquel J, Torres J, Spakulova M (2000) Faunistic and ecological trends on the helminthic community of *Genetta genetta* Linnaeus, 1758 (Carnivora: Viverridae) in the Iberian Peninsula. Helminthologia 37:223–228

Chabaud AG (1972) Description de *Stefanskostrongylus dubosti* n. sp. parasite du *Potomogale* et essai de classification des nématodes Angiostrongylinae. Ann Parasitol Hum Comp 47:735–744

Chiakwung F, Chienching H, Yunho L, Minghsien L, Kuacyre S (2004) Enhanced expression of transforming growth factor-beta 1 in inflammatory cells and secretory granules in Paneth cells in the small intestine of mice infected with *Toxocara canis*. Parasitol Res 94:397–404

Cordero del Campillo M, Castañón Ordóñez L, Reguera Feo A (1994) Indice-Catálogo de zooparásitos Ibéricos. Universidad de León, Secretariado de Publicaciones, León

Cox DM, Holland CV (2001) Relationship between three intensity levels of *Toxocara canis* larvae in the brain and effects on exploration, anxiety, learning and memory in the murine host. J Helminthol 75:33–41

Cross JH (1979) Experimental studies on *Angiostrongylus* species and strains in monkeys and laboratory animals. In: Cross JH (ed) Studies on angiostrongyliasis in eastern Asia and Australia, Special Publication No.2 (NAMRU-2-SP-44). US Naval Medical Research Unit, Taipei, pp118–137

Cross JH (1998) Angiostrongylosis. In: Palmer SR, Soulsby EJL, Simpson DIH (eds) Zoonoses: Biology, clinical practice and public health control. Oxford Univ Press, Oxford, pp 773–782

Dick TA, Pozio E (2001) *Trichinella* spp. and trichinellosis. In: Samuel WM, Pybus MJ, Kocan AA (eds) Parasitic dseases of wild mammals, 2nd edn. Iowa State Univ Press, Ames, pp 380–396

Drozdz J (1970) Révision de la systématique du genre *Angiostrongylus* Kamensky, 1905 (Nematoda: Metastrongyloidea). Ann Parasitol Hum Comp 45:597–603

Dubinsky P, Havasiora-Reiterova K, Petko B, Hovorka I, Tomasovicova O (1995) Role of small mammals in the epidemiology of toxocariasis. Parasitology 110:187–193

Feliu C, Renaud F, Catzeflis F, Hugot J-P, Durand P, Morand S (1997) A compara-tive analysis of parasite species richness of Iberian rodents. Parasitology. 115:453–66

Gamble HR, Pozio E, Lichtenfels JR, Zarlenga DS, Hill DE (2005) *Trichinella pseudospiralis* from a wild pig in Texas, USA. Vet Parasitol 132:147–150

Guardis MV, Radman NE, Burgos L, Fonrouge RD, Archelli SM (2002) *Toxocara canis*: Larval migration and eosinophilia in the paratenic host. Parasitol Latinoam 57:46–49

Havasiova-Reiterova K, Tomasovicova O, Dubisnky P (1995) Effect of various doses of infective *Toxocara canis* and *Toxocara cati* eggs on the humoral response and distribution of larvae in mice. Parasitol Res 81:13–17

Heper Y, Yilmaztepe F, Komitova R, Akalin H, Vutova K, Helvaci S (2005) A trichinosis outbreak caused by wild boar meat in Turkey. Parasite 12:191–192

Kliks MM, Palumbo NE (1992) Eosinophilic meningitis beyond the Pacific Basin. The global dispersal of a peridomestic zoonosis caused by *Angiostrongylus cantonensis*, the nematode lungworm of rats. Social Sci Med 34:199–212

Lai SC, Chen KM, Chen HC, Lee HH (2005) Induction of matrix metalloproteinase-9 in mice during *Toxocara canis* larvae migration. Parasitol Res 95:193–200

Lescano SZ, Queiroz ML, Chieffi PP (2004) Larval recovery of *Toxocara canis* in organs and tissues of experimentally infected *Rattus norvegicus*. Mem Inst Oswaldo Cruz 99:627–628

Møller LN, Peterson E, Kapel CMO, Melbye M, Kock A (2005) Outbreak of trichinellosis associated with consumption of game meat in West Greenland. Vet Parasitol 132:131–136

Morera P, Céspedes R (1971) *Angiostrongylus costaricencis* n. sp. (Nematoda: Metastrongyloidea), a new lungworm occurring in man in Costa Rica. Rev Biol Trop 18:173–185

Mörner T, Eriksson H, Bröjer C, Nilsson K, Uhlhorn H, Ågren E, Hård af Segerstad C, Jansson DS, Gavier-Widén D (2005) Diseases and mortality in free-ranging brown bear (*Ursus arctos*), grey wolf (*Canis lupus*), and wolverine (*Gulo gulo*) in Sweden. J Wildl Dis 41:298–303

Ohbayashi M, Kamiya M, Bhaibulaya M (1979) Studies on the parasite fauna of Thailand. I. Two new metastrongylid nematodes, *Angiostrongylus siamensis* sp. n. and *Thaistrongylus harinasutai* gen. et sp. n. (Metastrongyloidea: Angiostrongylidae) from wild rats. Japan J Vet Res 27:5–10

Overgaauw PAM, Knapen F (2000) Dogs and nematode zoonoses. In: Macpherson CNL, Meslin FX, Wandeler A (eds) Dogs, zoonoses and public health. CABI, Wallingford, pp 213–256

Perteguer MJ, Aguila C, Fenoy S, Guillen JL, Cuellar C (2003) Cross-reactivity induced by *Anisakis simplex* and *Toxocara canis* in mice. J Helminthol 77:331–334

Polley L (2005) Navigating parasite webs and parasite flow: Emerging and re-emerging parasitic zoonoses of wildlife origin. Int J Parasitol 35:1279–1294

Pozio E (2001) Taxonomy of *Trichinella* and the epidemiology of infection in the Southeast Asia and Australian regions. Southeast Asian J Trop Med Public Health 32:129–132

Pozio E, Foggin CM, Marucci G, La Rosa G, Sacchi L, Corona S, Rossi P, Mukaratirwa S. (2002) *Trichinella zimbabwensis* n. sp. (Nematoda), a new non-encapsulated species from crocodiles (*Crocodylus niloticus*) in Zimbabwe also infecting mammals. Int J Parasitol 32:1787 1799

Pozio E, Pagani P, Marucci G, Zarlenga DS, Hoberg EP, De Meneghi D, La Rosa G,Rossi L (2005) *Trichinella britovi* etiological agent of sylvatic trichinellosis in the Republic of Guinea (West Africa) and re-evaluation of geographical distribution for encapsulated species in Africa. Int J Parasitol 35:955–960

Prociv P, Spratt DM, Carlisle MS (2000) Neuro-angiostrongyliasis: Unresolved issues. Int J Parasitol 30:1295–1303

Pulido-Flores G, Moreno-Flores S, Monks S (2005) Helminths of rodents (Rodentia: Muridae) from Metztitlan, San Cristobal, and Rancho Santa Elena, Hidalgo, Mexico. Comp Parasitol 72:186–192

Rafter P, Marucci G, Brangan P, Pozio E (2005) Rediscovery of *Trichinella spiralis* in red foxes (*Vulpes vulpes*) in Ireland after 30 years of oblivion. J Infect 50:61–65

Rah H, Chomel BB, Follmann EH, Kasten RW, Hew CH, Farver TB, Garner GW, Amstrup SC (2005) Serosurvey of selected zoonotic agents in polar bears (*Ursus maritimus*). Vet Rec 156:7–13

Rausch RL (1970) Trichinosis in the Arctic. In: Gould SE (ed) Trichinosis in man and animals. C.C. Thomas, Springfield, pp 348–373

Reiterova K, Tomasovicova O, Dubinsky P (2004) Post-parturitional changes in the proportion of CD4+ and CD8+ T lymphocytes in *Toxocara canis*-infected mice and their offspring. Vet Med 49:103–108

Saeed I, Taira K, Kapel CMO (2005) *Toxocara canis* in experimentally infected silver and arctic foxes. Parasitol Res 97:160–166

Sa-Nunes A, Bolzoni RMF, Medeiros AI, Jamur MC, Oliver C, Baruffi MD, Faccioli LH (2005) Ultrasonically nebulized distilled water prevents exogenous histamine hyperreactivity in *Toxocara canis* infected mice. Inflamm Res 54:243–248

Scholz T, Uhlirova M, Ditrich O (2003) Helminth parasites of cats from the Vientiane province, Laos, as indicators of the occurrence of causative agents of human parasitoses. Parasite 10:343–350

Segovia JM, Torres J, Miquel J (2004) Helminth parasites of the red fox (*Vulpes vulpes* L., 1758) in the Iberian Peninsula: An ecological study. Acta Parasitol 49:67–79

Senutaité J, Grikienienė J (2001) Prevalence of *Trichinella* in muscles of some domestic and wild mammals in Lithuania and their impact in the organism. Acta Zool Lituanica 11:395–404

Skerrett H, Holland CV (1997) Variation in the larval recovery of *Toxocara canis* from the murine brain: Implication for behavioural studies. J Helminthol 71:253–255

Sobota K, Kotuliakova M, Sobotova O, Krcmery V (1988) Our experiences in the clinic and treatment of larval toxocarosis. Helminthologia 25:61–67

Torres J, Feliu C, Miquel J, Casanova J-C, García-Perea R, Gisbert J (1996) Helminth fauna of *Mustela putorius* Linnaeus, 1758 (Carnivora: Mustelidae) in the Iberian peninsula. Boll Soc Hist Nat Bal 39:155–165

Zimmermann, WJ, Hubbard ED (1969) Trichiniasis in wildlife of Iowa. Am J Epidemiol 90:84–92

24 Rodents as definitive hosts of *Schistosoma*, with special reference to *S. mansoni* transmission

Jean-Marc Duplantier and Mariama Sene

1 Introductory remarks

Schistosomes are digenetic trematodes belonging to the superfamily Schistosomatoidea. They are blood parasites responsible for schistosomiasis in man and domestic and wild animals. After malaria, human schistosomiasis is the most important parasitic disease in the world. It is widespread in the tropics and subtropics and has a major socio-economic impact in terms of public health. In the rural areas of developing countries, Doumenge et al. (1987) ranked schistosomiasis as the first professional risk in terms of prevalence of the water-borne diseases. About 200 million people living in 74 countries are infected with schistosomes and some 500 to 600 million are exposed to infection (Doumenge et al. 1987; Chitsulo et al. 2000). Approximately 80% of people inhabiting sub-Saharan Africa are infected with schistosomiasis, and the annual mortality rate is high (Southgate et al. 2005).

About twenty different species of schistosomes are recognised today. They occur mainly in the Old World, while only one species (*Schistosoma mansoni*) is present in the Neotropics. These species have been classified into five major groups according to their geographical distribution, egg morphology, intermediate snail hosts and, more recently, DNA analyses (Rollinson et al. 1997; Morgan et al. 2003a).

2 Occurrence of different *Schistosoma* species in rodents

2.1 *S. japonicum* group

This group is composed of 5 species, with *S. japonicum* being the most widespread and most important agent of the human disease. This species has the largest known spectrum of definitive hosts (more than 40 mammal species, including cattle, pigs, dogs, cats, rodents and humans). At present, *S. japonicum* is distributed in China, Thailand, Philippines and Indonesia, and has recently been eradicated from Japan (last human case in 1977; Tanaka and Tsuji 1997). Although the list of publications on rodent hosts of *S. japonicum* is very long (see review in Imbert-Establet 1986), most of these hosts belong to the genus *Rattus*.

S. mekongi is also an agent of human disease and parasitizes pigs and dogs, but has never been found in rodents (Kitikoon et al. 1975). Its distribution is restricted to a small area in Lao and Cambodia (Urbani et al. 2002). *S. malayensis*, described in Malaysia by Greer et al. (1988), is primarily a parasite of rats, but has also been found in humans. Finally, two closely related species, *S. sinensium* (China) and *S. ovuncatum* (Thailand; Attwood et al. 2002), are specific parasites of rats.

2.2 *S. indicum* group

The geographic distribution of this group is restricted to South-East Asia and the Indian sub-continent. It includes at least three species, namely *S. indicum, S. spindale* and *S. nasale*. The taxonomic position of *S. incognitum*, which was previously included in this group, has not been confirmed by DNA analysis (Morgan et al. 2003a). All these species parasitize mainly cattle, and have been found sometimes in rodents. For example, in Malaysia, *S. spindale* has been found in *Bandicota indica, Rattus argentiventer*, *Rattus diardii* and *Rattus tiomanicus* (Singh et al. 1997).

2.3 *S. mansoni* group

Two species (*S. mansoni* and *S. rodhaini*) belong to this group. Both have eggs with lateral spines and use *Biomphalaria* snails as intermediate hosts. These species are so closely related that they can hybridize in the laboratory (Taylor 1970; Brémond et al. 1989) and in the field (Morgan et al. 2003b). Close relationships between *S. rodhaini* and *S. mansoni* led Combes (1990) to suggest a lateral transfer and to advocate that humans acquired *S. mansoni* from a lineage of parasites that had evolved in rodents.

S. rodhaini parasitizes rodents in Central Africa, while *S. mansoni* infects rodents and humans in Africa, Arabian Peninsula, Caribbean Islands and South America (Brazil and Venezuela). Special attention is given below to *S. mansoni* because (a) it is an agent of the most important human schistosomiasis and (b) the importance of rodents in its transmission is well-documented.

Several rodent species were found to be infected by *S. rodhaini* in Central Africa. These included *Lophuromys flavopunctatus, Praomys jacksoni, Mastomys natalensis, Pelomys fallax, Malacomys longipes, Dasymys incomtus* and *Oenomys hypoxanthus* in the Democratic Republic of Congo; *M. natalensis, P. fallax* and *O. hypoxanthus* in Rwanda-Burundi; *Praomys morio* and *M. longipes* in Uganda; and *L. flavopunctatus* and *Thallomys paedulcus* in Kenya (Pitchford 1977). Some of these species (*M. longipes, D. incomtus* and *P. fallax*) live in close contact with bodies of water, but, surprisingly, there are also arboreal (*P. jacksoni* and *T. paedulcus*) and commensal (*M. natalensis*) species. The spectrum of rodent hosts of *S. mansoni* will be described later in this chapter.

In addition, Schwetz (1953) isolated another *Schistosoma* from *Dasymys bentleyae*, *Mastomys coucha* and *L. flavopunctatus*. This species also had eggs with lateral spines, but was morphologically distinct from *S. mansoni*. He named the species *S. mansoni var. rodentorum*. It is still unknown whether these were *S. rodhaini* x *S. mansoni* hybrids or simply rodent-adapted *S. mansoni*.

2.4 *S. hippopotami* group

This group includes 3 species specific to hippopotami [*S. hippopotami*, *S. edwardiensis* and still an unnamed species described by Morgan et al. (2003a)]. None of these has been found in other mammals.

2.5 *S. haematobium* group

Seven species of this African group are characterised by eggs with terminal spines and having snails of the genus *Bulinus* as intermediate hosts. Two species (*S. haematobium* and *S. intercalatum*) are agents of important human diseases (urinary and rectal schistosomiases, respectively). Three species (*S. bovis*, *S. curassoni* and *S. mattheei*) cause important cattle diseases, whereas the two remaining species (*S. leiperi* and *S. margrebowiei*) are specific to antelopes in southern Africa and have not been reported in small mammals. The transmission of schistosomes of this group by rodents seems very doubtful as the only reports of rodents infected by *S. intercalatum* (Schwetz 1956) and by *S. haematobium* (Pitchford 1959) were later challenged (Pitchford 1961). The double infestation of a Nile rat by *S. mansoni* and *S. haematobium* from Egypt (Mansour 1973) is also doubtful (Pitchford 1977; Imbert-Establet 1986). Imbert-Establet et al. (1997) demonstrated the susceptibility of *Mastomys huberti* and *Arvicanthis niloticus* to *S. intercalatum* in the laboratory, but the geographic distributions of these rodents and *S. intercalatum* do not overlap. The only observation of *S. bovis* in rodents was reported in Kenya (in *M. natalensis* and *L. flavopunctatus*; Nelson et al. 1962).

3 The role of rodents in *S. mansoni* transmission

Human intestinal schistosomiasis caused by *S. mansoni* occurs in the Arabian Peninsula, Africa, the West Indies and South America. DNA analyses suggested a recent colonisation of the New World by *S. mansoni* (Morgan et al. 2005). The definitive hosts of this species are usually humans, but many other mammal species, especially rodents, have been found to be infected with this parasite. Data on the presence of *S. mansoni* in non-human mammals have been collected for years, yet it remains unclear whether the occurrence of *S. mansoni* in animals is related to a transmission cycle or whether the parasites in these hosts reach a dead end.

3.1 Field studies

The infestation of rodents with S. *mansoni* was first reported in the early 1950s by Kuntz (1952) and Schwetz (1953) in Africa, and then by Amorin (1953) and Barbosa et al. (1953) in South America. The role of two rodent species (*Rattus rattus* and *Rattus norvegicus*) in the transmission of *S. mansoni* on Guadeloupe Island has been thoroughly studied both in the field and in the laboratory by Combes and co-workers (see reviews in Combes et al. 1975; Combes and Delattre 1981; Imbert-Establet 1986). In South America, especially in Brazil, several foci involving rodents have been studied, but this work was more descriptive than experimental. Data from Africa are still inadequate. Apart from the surveys by Pitchford (1959) and Pitchford and Visser (1962) in South Africa and by Duplantier and Sene (2000) in a recently emerged focus in Senegal, there are only a few sporadic reports of the presence of *S. mansoni* in various rodent species. Three very comprehensive reviews of the definitive hosts of S. *mansoni* were published by Pitchford (1977) for the Middle East and Africa, by Imbert-Establet (1986) for the entire geographic range of S. *mansoni* and by Rey (1993) for Brazil. Very often, though, the rodent taxonomy used in the original publications was not up-to-date. Here, we have used the names of rodent species according with Wilson and Reeder (1993).

3.1.1 African foci

The first reference to an African rodent infected with *S. mansoni* is anecdotal. In the foci with very high human prevalences (40 to 80%), *S. mansoni* was reported in a gerbil, *Gerbillus pyramidum*, that inhabits arid environments, while no infection was found in approximately 100 examined *A. niloticus* that occur in wet habitats (Kuntz 1952). *S. mansoni* has not been reported from gerbillines since. Other rodents infected with *S. mansoni* in Africa belong mainly to Murinae.

Rats of the genus *Rattus* infected by *S. mansoni* were found in the Democratic Republic (DR) of Congo (Schwetz 1955) and Egypt (Arafa and Massoud 1990). However, prevalence of infection in *R. rattus* was relatively low and attained only 1.5% in the DR Congo and 3.7-5.4% in Egypt. The prevalence of infection in *R. norvegicus* in Egypt was substantially higher (18.4%; Gunther 1979 cited by Arafa and Massoud 1990). Non-commensal rodents infected with *S. mansoni* included *D. incomtus* (Schwetz 1956; Nelson et al. 1962), *P. fallax* (Kawashima et al. 1978), *L. flavopunctatus* (Schwetz 1956), *Lemniscomys griselda* (Pitchford and Visser 1962), *Otomys angoniensis* (Nelson et al. 1962; Pitchford and Visser 1962), *Mastomys sp.* (Nelson et al. 1962; Pitchford and Visser 1962), *Mastomys huberti* (Duplantier and Sene 2000), and *A. niloticus* (Mansour 1973; Arafa and Massoud 1990; Mbieleu Nkouedeu 1990; Duplantier and Sene 2000). Prevalence of infection in these rodents ranged from 1.5 to 41.3%, whereas individual rodents harboured from 1 to 91 parasites. Some animals had *S. mansoni* eggs in their faeces. In some cases, miracidia were obtained from viable eggs collected from the tissues of hosts. These miracidia appeared to be infective for the snails *Biomphalaria alexandrina* and *Biomphalaria sudanica*. In the Richard-Toll schistosomiasis focus in Senegal, two rodent species, *A. niloticus* and *M. Huberti*, had similar high prevalences of *S. mansoni* infection (Talla *et al.* 1991). This focus is very recent as the first human cases were reported as late as in 1988 (Talla et al. 1990).

3.1.2 American and West Indies foci

Schistosoma mansoni is the only *Schistosoma* species of the New World. It occurs from northern Venezuela to Brazil and in the Caribbean, being introduced in these areas only recently.

Brazilian foci

All non-human hosts of *S. mansoni* in Brazil were reviewed by Rey (1993). The majority of infected rodent species belong to Sigmodontinae. In different states of Brazil, different rodents are infected with *S. mansoni*. Prevalence and intensity of infection vary greatly among host species as well as among geographic locations. In particular, *S. mansoni* was reported in *Oryzomys subflavus*, *Oligoryzomys nigripes*, *Oligoryzomys microtis* (=*Oryzomys mattogrossae*), *Bolomys lasiurus* (=*Akodon arviculoides*, *Zygodontomys lasiurus*, *Z. pixuna* and *Z. brachyurus*), *Oxymecturus angularis*, *Oxymecturus hispidus*, *Calomys tener* (=*Calomys expulsus*), *R. norvegicus*, *R. rattus*, *Nectomys squamipes*, *Holochilus brasiliensis*, *Holochilus sciureus*, and *Cavia aperea* (Caviidae). The highest prevalence of *S. mansoni* was found in hosts with the most pronounced aquatic habits. For example, prevalences in *N. squamipes*, *H. brasiliensis* and *H. sciureus* ranged from 16 to 75%. The highest intensity of infection was reported in *C. aperea* (up to 782 worms per rodent; Baretto et al. 1964), although viable *S. mansoni* eggs were not found in the faeces of this rodent.

Venezuela

In Venezuela, only two *R. rattus* individuals were found to be infected with *S. mansoni* (Gonzales et al. 1976), although human cases in the country are rather frequent (Alarcon de Noya et al. 1999).

Guadeloupe Island

R. rattus and *R. norvegicus* had very high prevalences (approaching 100% in some areas) of *S. mansoni*, with loads up to 600 worms. Nevertheless, prevalence of infection in *R. rattus* was consistently higher than that in *R. norvegicus* [e.g., mean prevalences were 52% and 37%, respectively (Combes and Delattre 1981) and 72% and 28%, respectively (Imbert-Establet 1986)]. Furthermore, local *R. norvegicus* did not shed *S. mansoni* eggs but, in contrast, many viable eggs of *S. mansoni* were excreted by *R. rattus* (e.g., about 2000 eggs were observed in 1g of faeces of a single rat). The viable miracidia were infective for the intermediate host *Biomphalaria glabrata*. This difference between *R. rattus* and *R. norvegicus* is surprising, since the latter is more aquatic in its habits and would, therefore, be expected to become infected more easily. It seems that some other factors (e.g., immunological) make *R. norvegicus* a less suitable host for *S. mansoni* than *R. rattus*.

Investigations in Guadeloupe suggest the existence of three types of schistosomiasis foci (Théron et al. 1980) as follows:

- Urban foci that are characterized by running water in the irrigation canals. Here, prevalence of infection in humans is very high (50-75%) and in rodents is low. In this case, human-to-human transmission route is the most important, whereas the role of rodents in *S. mansoni* transmission appears to be limited.

- Foci on the fringes of mangrove swamps that are characterized by stagnant bodies of water. Here, prevalence of infection in humans is low and, at least seasonally, in rodents is high (up to 80% in the rainy season). Consequently, here rodents play an important role in *S. mansoni* transmission.
- Sylvatic foci that are characterized by stagnant bodies of water not frequented by humans. Here, prevalence of infection in rodents is very high and they play an important role in the transmission of *S. mansoni* due to the absence of alternative (=human) routes.

3.2 Experimental transmission

The fact that a particular rodent is infected naturally with *S. mansoni* is not sufficient to consider this rodent as a reservoir host. A reservoir host is not only infected, but must also pass viable and infective eggs and maintain the life cycle of a parasite. Many experimental infection studies have been made with *S. mansoni*, using both laboratory and wild rodents. However, these experiments often aimed to find a suitable laboratory model rather than understand the role of a rodent species in the dynamics of *S. mansoni* transmission. For example, some rodents such as *Gerbillus* sp. (Kuntz and Malakatis 1955) and *Mus musculus* (Leigh and Alarcon de Noya 1978; Sène *et al.* 1996) have proven to be good laboratory hosts although they do not inhabit wet areas and, thus, cannot play any substantial role in the natural transmission of schistosomes.

3.2.1 African rodents

In the laboratory, Kuntz and Malakatis (1955) tested the susceptibility of different rodent species to *S. mansoni*, including rodents that occupy arid habitats only. They recorded the presence of adult worms in *M. musculus, R. rattus, A. niloticus, Acomys cahirinus, G. pyramidum, Jaculus jaculus, Meriones shawi, Psammomys obesus* and *Nesokia indica*. The excretion of viable eggs in faeces was observed only in *M. musculus, R. rattus, A, niloticus, A. cahirinus, Meriones shawi* and *N. indica*. These results were later confirmed by Kuntz (1961). However, most of these species (except *R. rattus* and *A. niloticus*) cannot be infected naturally because they occupy habitats where schistosomes are not present.

In South Africa, experimental transmissions of *S. mansoni* to multimammate rats were carried out repeatedly (Lurie and De Meillon 1956; Pitchord and Visser 1962; Cheever 1965). However, results of many of these studies are equivocal because of problems related to identification of the experimental hosts. Thus, it is unclear whether *Mastomys natalensis* or *Mastomys coucha* was used (Dettman et al. 1987; Higgins-Opitz et al. 1987). In some experiments, however, the specific affinity of a host such as *M. coucha* was determined (Higgins-Opitz et al. 1987; Higgins-Opitz and Dettman 1991). This species proved to be susceptible to *S. mansoni*, although the rate of egg excretion in faeces appeared to be very low.

In Senegal, Sène et al. (1996) demonstrated that naturally infected rodents (*A. niloticus*, *M. huberti* and *Mastomys erythroleucus*) could be infected and pass viable and infective eggs. The susceptibility of *A. niloticus* to infection with *S. mansoni* was also shown by Stirewald et al. (1951), Kuntz and Malakatis (1955), Dumon and Quilici (1956), Karoum and Amin (1985) and Mbieuleu-Nkouedeu (1990).

3.2.2 Brazilian rodents

Experimental infections of Brazilian rodents by *S. mansoni* were determined by the presence of adult worms in the mesenteric venous system and viable eggs in faeces and were successful for *N. squamipes*, *B. lasiurus*, *H. sciureus*, *H. brasiliensis*, and *R. rattus* (Rey 1993). However, the ability to maintain the life cycle of *S. mansoni* (infection of snails by exposure to the miracidia and then infection of rodents by the cercariae) was found only in *N. squamipes* (Silva 1988; Antunes et al. 1971; both cited by Rey 1993). In addition, Ribeiro et al. (1998) found that two sibling species, *N. squamipes* and *Nectomys rattus* showed high similarity in their abilities to be infected by and to maintain the life cycle of *S. mansoni*.

3.2.3 Guadeloupe Island

In the Guadeloupean foci, rats were infected experimentally with *S. mansoni* to examine their role in the transmission of the parasite. It was demonstrated that *R. rattus* and *R. norvegicus*, although both being naturally infected with *S. mansoni* in Guadeloupe, have different relationships with this parasite (see above). *R. rattus* shed fertile eggs containing infectious miracidia, while *S. mansoni* exploiting *R. norvegicus*, produced non-fertile eggs that were not shed by the host (Imbert-Establet 1982a, b).

3.3 Comparison between strains of *S. mansoni* of human and murine origin

3.3.1 Rhythms of cercarial emission and activity rhythms of rodents

S. mansoni normally reach a peak of cercarial emission near midday, whereas the rodent parasite *S. rodhaini* shows a crepuscular peak of emission (Théron 1985). This suggests that the cercarial emission depends on the activity patterns of the definitive host. This was further supported by comparison between the emission rhythms of cercariae of human and murine origins on Guadeloupe Island (Théron 1984). Here, two chronobiological types (early and late) of cercarial emission that correlated with the activity of the definitive hosts were found. In urban foci, where transmission is almost exclusively human-to-human, the peak emission time was near 11:00 (= early type). In the mangrove fringe foci, where both humans and rats are involved in transmission, an intermediate peak time was observed. In the sylvatic foci, where transmission is supported by rodents only, peak emission time was near 16:00 (= late type).

3.3.2 Comparison of eggs and cercariae

Comparisons of egg morphology and patterns of cercarial emission have shown a certain polymorphism in *S. mansoni* from Guadeloupe Island. Théron (1986) found three different types of eggs and demonstrated that the relative proportion of these three types is different among the three types of foci that differ in the degree of participation of *R. rattus* in the transmission of *S. mansoni*. In populations of schistosomes transmitted mainly by rats, eggs were very similar morphologically to those of experimental hybrids between *S. mansoni* and *S. rodhaini*. Théron (1986) suggested a double origin of the Guadeloupean population. However, comparison of enzymes between isolates of *S. mansoni* from Guadeloupe and from the Burundi isolate of *S. rodhaini* demonstrated that this is unlikely (Rollinson et al. 1986).

Combes and Imbert-Establet (1980) found that there was no difference in the probability of maturation in lab mice and wild black rats between the cercariae of human and murine origins. This led them to suggest that the cercariae of human and murine origin have the same rate of success at infecting humans. The parasite thus circulates freely between humans and rodents in Guadeloupe.

In Brazil, Neves et al. (1998) observed significant size differences of eggs and adult worms between *S. mansoni* isolated from *N. squamipes* and from humans. However, this observation was not supported by Freire et al. (2002) who did not find any other morphological differences between the cercariae from this rodent and from humans and concluded that they belong to the same population of *S. mansoni*.

3.3.3 Biochemical comparisons

Rollinson et al. (1986) compared *S. mansoni* isolated from rats and humans from Guadeloupe Island using isoelectric focusing and came to the conclusion that "there is no suggestion from the present enzyme data of genetic divergence or separate gene pools; the isolates from rats proved indistinguishable from those of man for the seven enzymes studied. "

Théron and Combes (1988) carried out experimental cross-breeding between schistosomes with early and late cercarial shedding to investigate the genetic determinants of the cercarial emergence rhythms. They found that individuals of the first generation had an emission rhythm of cercariae characterised by a single peak, intermediate between those of the parents. In the second generation, they observed a variety of chronobiological phenotypes (early, intermediate and late patterns) with emergence peaks between 10:00 and 16:00. These results suggest that the cercarial emergence rhythms of schistosomes are genetically determined and that the intermediate pattern described by Théron (1985) in the swamp mangrove foci could result from a natural crossing between schistosomes with early chronobiology, adapted to a human type of transmission, and schistosomes with late chronobiology, adapted to a rat type of transmission. They interpreted this genetic variability as a consequence of the selective pressure exerted by the two different hosts involved in the life cycle of *S. mansoni* from the Guadeloupean schistosomiasis focus. The selective pressure exerted by murine hosts after successive passages were also reported by Fletcher et al. (1981), LoVerde et al. (1985) and Brémond et al. (1993).

In Senegal, human and murine *S. mansoni* isolates were compared using isoelectric focusing (Sene et al. 1996). Results showed that these two isolates were genetically similar. In addition, each isolate showed a low genetic variation, suggesting that little intraspecific variation occurred in the isolate either from humans or rodents. However, very rare phenotypes identified at several loci were found in the isolates of *S. mansoni* recovered from *A. niloticus*. Nevertheless, it is still difficult to consider the appearance of these phenotypes as an indicator of independent evolution of *S. mansoni* in *A. niloticus*.

Among the analysed enzymatic systems, the malate dehydrogenase (MDH) patterns observed in Senegal (Sène et al. 1997) were similar to those described for Guadeloupe (Rollinson et al. 1986; Théron and Combes 1988; Théron et al. 1989; Brémond et al. 1993). Théron et al. (1989) found that "MDH-1a frequency, always increases when one passes: (1) from *S. mansoni* from urbanised foci to *S. mansoni* from mangrove foci and thence to *S. mansoni* from sylvatic foci; (2) from *S. mansoni* strains with early cercarial chronobiology to strains with intermediate and then with late chronobiology." An integrated analysis of these genetic results with epidemiological and ecological considerations led Théron et al. (1989) to conclude that the greater the role played by the rat in maintaining the parasite cycle in a transmission site, the higher the occurrence of the MDH-1a frequency among *S. mansoni* populations of this focus.

In Guadeloupe, Sire et al. (2001) showed a highly significant degree of geographical genetic variation among *S. mansoni* populations from naturally infected *R. rattus* at a regional and at a micro-spatial scale. They suggested that a combination of different factors could promote this genetic differentiation. At a regional scale, these factors are represented by differences in landscape ecology between two areas (e.g., river systems with running water versus marshy forest with standing water) and/or in epidemiology (e.g., highly mobile human hosts versus sedentary murine hosts). At a micro-spatial scale, the factors are restricted to movements of rats, patchy spatial aggregation of infected snails and limited cercarial dispersion in stagnant water.

Finally, the recent DNA analyses of parasites throughout the geographic range of *S. mansoni* and originating from different hosts (snails, humans and rodents) showed that rodent isolates do not constitute a monophyletic assemblage (Morgan et al. 2005).

4 Concluding remarks

Considering that rodent species, naturally infected with schistosomes, have not always been correctly identified, it is quite likely that different species were reported as the same species and vice versa. In Guadeloupe, sharp differences between two congeneric rodent hosts in their relationships with *S. mansoni* were found. It is, therefore, very important to identify correctly a rodent host.

Prevalence of *S. mansoni* seems independent of sex of the rodent host (Combes and Delattre 1981, Duplantier and Sene 2000). However, prevalence increases significantly with the age of rodents, as was observed in *Mastomys sp.* in South Africa (Pitchford and Visser 1962), *R. rattus* and *R. norvegicus* on Guadeloupe Island (Combes and Delattre 1981) and in *M. huberti* and *A. niloticus* in Senegal (Duplantier and Sene 2000). In spite of this, the age of a rodent host has not been taken into account in the majority of studies of schistosome infections.

Epidemiological and experimental data reveal entirely different situations concerning schistosomes between African and American endemic areas. In general, African foci are characterized by low prevalences of infections in rodent hosts (usually, less than 10%,), while in the New World, prevalences in rodents are much higher. Perhaps, South American sigmodontine are more compatible physiologically with *S. mansoni* than African murine hosts. However, De Jong et al. (2001) pointed out the importance of the intermediate host *Biomphalaria glabrata* in the successful introduction of *S. mansoni* in the New World. This snail is more closely related to the African than to the neotropical species of the genus and probably is more compatibile with *S. mansoni*. A substantial difference in the prevalences of the same host, *R. rattus*, in Africa and in America should be noted. Imbert-Establet and Combes (1986) provided genetic evidence of the profound differences in the compatibility of Caribbean and African strains of *S. mansoni* with *R. rattus*.

Viable eggs in the faeces of rodents is the first indication of a possible role of the host species in schistosome transmission. This has been supported for South American rodents, but only a few African studies examined this parameter. This (together with the low levels of prevalence observed) can be the reason why many authors concluded that the role of rodents in schistosome transmission is less important in Africa than in South America. Two interesting cases were reported for *Rattus norvegicus* in Guadeloupe and *Cavia aperea* in Brazil. Although the prevalence and the intensity of *S. mansoni* infection in these hosts are high, the emission of viable eggs from them has not been reported. Although naturally infected due to their aquatic habits, these rodents do possess sufficient physiological or immunological compatibility with *S. mansoni* to be its reservoir hosts.

Finally, current understanding of the role of a rodent in the transmission of schistosomes is limited. For example, in Africa, representatives of only three genera (*Mastomys, Arvicanthis* and *Pelomys*) seem to participate in the transmission of *S. mansoni*. However, it is still difficult to assess the role of multimammate rats in this transmission. *M. huberti* and *M. erythroleucus* are able to maintain the life cycle of the parasite in the laboratory, but only the former was naturally infected in Senegal. *A. niloticus* appeared to be a good experimental host that maintained the life cycle of *S. mansoni*, but the prevalence and intensity of infection of this host in the field was low. In contrast, *P. fallax* demonstrated the highest prevalence of infection and, thus, is expected to participate actively in the transmission of the parasite. However, no experimental study has been undertaken to assess the ability of this rodent to maintain the life cycle of *S. mansoni*.

In the New World, surveys in the field and experimental studies showed that at least five rodent species were strongly involved in the transmission of *S. mansoni*: *Holochilus brasiliensis* and *H. sciureus*, *Nectomys squamipes* and *N. rattus* in Brazil, *R. rattus* on Guadeloupe Island. However, it should be pointed out that there is no known race of *S. man-*

soni that is exclusively zoonotic, as is the case with *S. japonicum* in Formosa.

More information concerning population dynamics (lifespan, turnover and reproductive periods, fecundity) and ecology (habitat and habits, diet, spatial organisation, density, diversity) is needed to determine the importance of other rodents in *S. mansoni* transmission.

References

Alarcon de Noya B, Balzan C, Arteaga C, Cesari I, Noya O (1999) The last fifteen years of schistosomiasis in Venezuela: Features and evolution. Mem Inst Oswaldo Cruz 94:139–146

Amorin JP (1953) Infestaçao experimental e natural de murideos pelo *Schistosoma mansoni*. Revista Brasileira de Malariologia e Doenças Tropicais 5:219–222

Arafa MAS, Massoud MM (1990) Natural *Schistosoma mansoni* infection in *Arvicanthis niloticus* in Ismailia, Egypt. J Egypt Soc Parasit 20:775–778

Attwood SW, Panasoponkul C, Upatham ES, Meng XH, Southgate VR (2002) *Schistosoma ovuncatum* n. sp. (Digenea: Schistosomatidae) from northwest Thailand and the historical biogeography of Southeast Asian *Schistosoma* Weinland, 1858. Syst Parasitol 51:1–19

Barbosa FS (1972) Natural infection with *Schistosoma mansoni* in small mammals trapped in the course of a schistosomiasis control project in Brazil. J Parasitol 58:405–407

Barbosa FS, Dobbin JE, Coelho MV (1953) Infestaçao naturral de *Rattus rattus frugivorus* por *Schsitosoma mansoni* em Pernambuco. Publ Av Inst Aggeu Magalhaes 2:42–43

Barretto AC, Santos I, Oliveira VS (1964) Epidemiologia da esquistossomose masonica e infectaçao natural de *Cavia aperea aperea*. Rev Inst Med Trop Sao Paulo 6:233

Bastos OC, Sadigursky M, Nascimento MDSB, Brazil RP, Holanda JC (1984) *Holochilus brasiliensis nanus* Thomas 1897; Sugestao de modelo experimental para filariose, leishmaniose e esquistossomose. Rev Inst Med Trop Sao Paulo 26:307–315

Brémond P, Pasteur N, Combes C, Renaud F, Théron A (1993) Experimental host-induced selection in *Schistososma mansoni* strains from Guadeloupe and comparisons with natural observations. Heredity 70:33–37

Brumpt E (1931) Description de deux bilharzies de mammifères africains, *Schistosoma mansoni* sp. inqu. et *Schistosoma rodhaini* n. sp. Ann Parasitol 9:325–338

Chitsulo L, Engels D, Montresor A, Savioli L (2000) The global status of schistosomiasis and its control. Acta Tropica 77:41–51

Combes C (1990) Where do human schistosomes comes from? An evolutionary approach. Trends Ecol Evol 5:334–337

Combes C, Delattre P (1981) Principaux paramètres de l'infestation des rats (*Rattus rattus* et *Rattus norvegicus*) par *Schistosoma mansoni* dans un foyer de Schistosomose intestinale de la région Caraïbe. Acta Oecol 2:63–79

Combes C, Imbert-Establet D (1980) Infectivity in rodents of *Schistosoma mansoni* cercariae of human and murine origin. J Helminthol 54:167–171

Combes C, Léger N, Golvan H (1975) Le rôle du rat dans la dynamique de l'endémie schistosomienne en Guadeloupe. Comptes Rendus de l'Académie des Sciences de Paris série D 281:1059–1061

D'Andrea PS, Fernandes FA, Cerqueira R, Rey L (2002) Experimental evidence and ecological perspectives for the adaptation of *Schistosoma mansoni* Sambon 1907 (Digenea: Schistosomatidae) to a wild host, the water-rat, *Nectomys squamipes* Brants, 1827 (Rodentia, Sigmodontinae). Mem Inst Oswaldo Cruz 97:11–14

Dettman CD, Higgins-Opitz SB, Bronner GN (1987) A comment on the taxonomic status and correct nomenclature of laboratory colonies of *Mastomys*. S Afr J Sci 83:395

Doumenge J, Mott KE, Cheung C, Villenave D, Chapuis O, Perrin MF, Reaud-Thomas G (1987) Atlas de la répartition mondiale des Schistosomiases. CEGET/OMS, Presses Universitaires de Bordeaux, Bordeaux

Dumon H, Quilici M (1976) Réceptivité d'*Arvicanthis niloticus* a l'infestation par *Schistosoma mansoni.* CR Société de biologie 170:642–645

Duplantier J-M, Sene M (2000) Rodents as reservoir hosts in the transmission of *Schistosoma mansoni* in Richard-Toll, Senegal, West Africa. J Helminthol 74:129–135

Fletcher M, Lo Verde PT, Woodruff DS (1981) Genetic variation in *Schistosoma mansoni*: Enzyme polymorphisms in populations from Africa, South West Asia, South America and the West Indies. Am J Trop Med Hyg 30:406–421

Freire N, Machado-Silva JR, Rodrigues-Silva R, Rey L (2002) Cercarial chaetotaxy and sex differentiation of *Schistosoma mansoni* deriving from humans and *Nectomys squamipes* (Muridae: Sigmondontinae) in Brazil. Mem Inst Oswaldo Cruz 97:127–132

Freire N, Rodrigues-Silva R, Machado-Silva JR, Rey L (2003) A comparative parasitoloc study on *Biomphalaria glabrata* snail and C3H/He mice infected with human and murine isolates of *Schistosoma mansoni* from Sumidouro, Rio de Janeiro, Brazil. Mem Inst Oswaldo Cruz 98:783–787

Gonzalez NR, Torrealba JW, Comach GA, Afaro O (1976) *Rattus rattus*, hospedador natural de *Schistosoma mansoni* en Venezuela. Boletin de la Dirreccion de Malariologia y Saneamiento Ambiental 16:126–128

Greer GJ, Ow-Yang CK, Yong HS (1988) *Schistosoma malayensis* n. sp.: A *Schistosoma japonicum*-complex schistosome from Peninsular Malaysia. J Parasitol 74:471–480

Higgins-Opitz SB, Dettman CD (1991) The infection characteristics of a South African isolate of *Schistosoma mansoni*: A comparison with a Puerto Rican isolate in BALB/c mice and Mastomys coucha. Parasitol Res 77:142–151

Higgins-Opitz SB, Bhoola KD, Dettman CD (1987) *Mastomys coucha* as a host for experimental schistosomiasis. I. Some pathological responses to *Schistosoma mansoni.* S Afr J Sci 83:696–699

Imbert-Establet D (1982a) Infestation naturelle des rats sauvages par *S. mansoni* en Guadeloupe. Ann Parasitol 57:573–585

Imbert-Establet D (1982b) Approche expérimentale du rôle de *Rattus rattus* et *Rattus norvegicus* dans le foyer de *Schistosoma mansoni* de Guadeloupe. Ann Parasitol Human Comp 57:271–284

Imbert-Establet D (1986) Approche synthétique du rôle des rongeurs sauvages dans la transmission de *Schistosoma mansoni*: Résultats démographiques, écologiques, immunologiques et génétiques. Ph. D. Thesis, University of Perpignan

Imbert-Establet D, Moné H, Tchuem-Tchuenté LA, Jourdane J (1997) Permissiveness of two African wild rodents *Mastomys huberti* and *Arvicanthis niloticus*, to *Schistosoma intercalatum*: Epidemiological consequences. Parasitol Res 83:569–573

Karoum KO, Amin MA (1985) Domestic and wild animals naturally infected with *Schistosoma mansoni* in the Gezira Irrigated Scheme, Sudan. J Trop Med Hyg 88:83–89

Kawashima K, Katamine D, Sakamoto M, Shimada M, Nojima H, Miyahara M (1978) Investigations on the role of wild rodents as reservoirs of human schistosomiasis in the Taveta area of Kenya, East Africa. Japancsc J Trop Med Hyg 3:195–203

Kitikoon V, Schneider CR, Bhaibulaya M, Sittilerd S, Thirachantra S (1975) Mekongi schistosomiasis. IV. A parasitological survey of wild rodents, domestic pigs and cattle on Khong island, Laos. Southeast Asian J Trop Med Pub Health 6:223–229

Kuntz RE (1952) Natural infection of an Egyptian gerbil with *Schistosoma mansoni*. Proc Helminth Soc Wash 19:123–124

Kuntz RE (1961) Passage of eggs by hosts infected with *Schistosoma mansoni*, with emphasis on rodents. J Parasitol 47:905–909

Kuntz RE, Malakatis GM (1955) Susceptibility studies in schistosomiasis. II Susceptibility of wild mammals to infection by *Schistosoma mansoni* in Egypt with emphasis on rodents. Am J Trop Med Hyg 4:75–89

Lo Verde PT, De Wald J, Minchella DJ, Bosshart SC, Damian RT (1985) Evidence for host-induced selection in *Schistosoma mansoni*. J Parasitol 71:297–301

Mansour NS (1973) *Schistosoma mansoni* and *Schistosoma haematobium* found as a natural double infection in the Nile rat, *Arvicanthis niloticus*, from a human endemic area in Egypt. J Parasitol 59:424

Mbieleu-Nkouedeu B (1990) Rôle potentiel de deux familles de rongeurs Nigériens dans la transmission naturelle des Schistosomoses humaines et animale. Ph. D. Thesis, University of Namey

Morgan JAT, De Jong RJ, Kazibwe F, Mjoki GM, Loker ES (2003a) A newly-identified lineage of *Schistosoma*. Int J Parasitol 33:977–985

Morgan JAT, De Jong RJ, Lwambo NJS, Mungai BN, Mjoki GM, Loker ES (2003b) First report of a natural hybrid between *Schistosoma mansoni* and *S. rodhaini*. J Parasitol 89:416–418

Morgan JAT, Dejong RJ, Adeoye GO, Ansa EDO, Barbosa CS, Brémond P, Cesari IM, Charbonnel N, Corrêa LR, Coulibaly G, D'andrea PS, De Souza CP, Doenhoff MJ, File S, Idris MA, Nino Incani R, Jarne P, Karanja DMS,

Kazibwe F, Kpikpi J, Lwambo NJS, Mabaye A, Magalhães LA, Makundi A, Moné H, Mouahid G, Muchemi GM, Mungai BN, Séne M, Southgate V, Tchuenté LAT (2005) Origin and diversification of the human parasite *Schistosoma mansoni*. Mol Ecol 14:3889–3902

Nelson GS, Teesdale C, Highton RB (1962) The role of animals as reservoirs of bilharziasis in Africa. In: Wolstenholme GEW, O'Connor M (eds) Chiba foundation symposium: Bilharziasis. London, pp 127–149

Neves RH, Pereira MJS, Oliveira RMF, Gomes DC, Machado-Silva JR (1998) *Schistosoma mansoni* Sambon, 1907: Morphometric differences between adult worms from sympatric rodent and human isolates. Mem Inst Oswaldo Cruz 93:309–312

Pitchford RJ (1959) Natural schistosome infections in South African rodents. Trans R Soc Trop Med Hyg 53:2123

Pitchford RJ (1961) Observations on a possible hybrid between the two schistosomes, *S. haematobium* and *S. mattheei*. Trans R Soc Trop Med Hyg 55:44–51

Pitchford RJ (1977) A check list of definitive hosts exhibiting evidence of the genus *Schistosoma* Weinland, 1858 acquired naturally in Africa and the Middle-East. J Helminthol 51:229–252

Pitchford RJ, Visser PS (1962) The role of naturally infected wild rodents in the epidemiology of schistosomiasis in the Eastern Transvaal. Trans R Soc Trop Med Hyg 56:126–135

Rey L (1993) Non-human vertebrate hosts of *Schistosoma mansoni* and schistosomiasis transmission in Brazil. Res Rev Parasitol 53:13–25

Ribeiro AC, Maldonado AJ, D'Andrea PS, Vieira GO, Rey L (1998) Susceptibility of *Nectomys rattus* (Pelzen, 1883) to experimental infection with *Schistosoma mansoni* (Sambon, 1907): A potential reservoir in Brazil. Mem Inst Oswaldo Cruz 93:295–299

Rollinson D, Imbert-Establet D, Ross GC (1986) *Schistosoma mansoni* from naturally infected *Rattus rattus* in Guadeloupe: Identification, prevalence and enzyme polymorphism. Parasitology 93:39–53

Rollinson D, Kaukas A, Johnston DA, Simpson AJG, Tanaka M (1997) Some molecular insights into schistosome evolution. Int J Parasitol 27:11–28

Schwetz J (1953) On a new schistosome of wild rodents found in the Belgian Congo, *Schistosoma mansoni* var *rodentorum* var nov. Ann Trop Med Parasit, 47:183–186

Schwetz J (1954) On two schistosomes of wild rodents of the Belgian Congo: *Schistosoma rodhaini* Brumpt 1931; and *Schistosoma mansoni* var *rodentorum* Schwetz 1953; and their relationship to *S. mansoni* of man. Trans R Soc Trop Med Hyg 48:89–100

Schwetz J (1956) Role of wild rats and domestic rats (*Rattus rattus*) in schistosomiasis of man. Trans R Soc Trop Med Hyg 50:275–282

Sène M, Duplantier J-M, Marchand B, Hervé JP (1996) Susceptibility of rodents to infection with *Schistosoma mansoni* in Richard-Toll (Senegal). Parasite 4:321–326

Sène M, Brémond P, Hervé JP, Southgate VR, Sellin B, Marchand B, Duplantier J-M (1997) Comparison of human and murine isolates of *Schistosoma man-*

soni from Richard-Toll, Senegal, by isoelectricfocusing. J Helminthol 71:175–181

Silva TM, Andrade ZA (1989) Infecçao natural de roedores silvestres pelo *Schistosoma mansoni*. Mem Inst Oswaldo Cruz 84:227–235

Singh I, Krishnasamy M, Ambu S, Rasul R, Chong NL (1997) Studies on animal schistosomes in Peninsular Malaysia: record of naturally infected animals and additional hosts of *Schistosoma spindale*. Southeast Asian J Trop Med Public Health 28:303–307

Sire C, Durand P, Pointier JP, Théron A (2001) Genetic diversity of *Schistosoma mansoni* within and among individual hots (*Rattus rattus*): Infrapopulation differentiation at microspatial scale. Int J Parasitol 31:1609–1616

Southgate VR, Rollinson D, Tchuem-Tchuenté LA, Hagan P (2005) Towards control of schistosomiasis in sub-Saharan Africa. J Helminthol 79:181–185

Talla I, Kongs A, Verlé P, Belot J, Sarr S, Coll AM (1990) Outbreak of intestinal schistosomiasis in the Senegal river basin. Ann Soc Belg Med Trop 70:173–180

Talla I, Kongs A, Verlé P (1991) Preliminary study of the prevalence of human schistosomiasis in Richard-Toll (the Senegal river basin). Trans R Soc Trop Med Hyg 86:182

Tanaka H, Tsuji M (1997) From discovery to eradication of schistosomiasis in Japan: 1847-1996. Int J Parasitol 27:1465–1480

Taylor MG (1970) Hybridisation experiments on five species of African schistosomes. J Helminth 44:253–314

Théron A (1984) Early and late shedding patterns of *Schistosoma mansoni* cercariae: Ecological significance in transmission to human and murine hosts. J Parasitol 70:652–655

Théron A (1985) Polymorphisme du rythme d'émission des cercaires de *S. mansoni* et de ses relations avec l'écologie de la transmission du parasite. Vie & Milieu 35:23–31

Théron A (1986) Polymorphisme des oeufs de *Schistosoma mansoni* dans le foyer Guadeloupéen (Antilles françaises): Présence de *S. rodhaini*-like parasites? Acta Tropica 43:335–342

Théron A, Combes C (1988) Genetic analysis of cercarial emergence rhythms of *Schistosoma mansoni*. Behav Genetics 18:201–209

Théron A, Pointier JP, Combes C (1978) Approche écologique de la responsabilité de l'homme et du rat dans le fonctionnement d'un site de transmission de *Schistosoma mansoni* en Guadeloupe. Ann Parasitol Human Comp 53:223–234

Théron A, Brémond P, Imbert-Establet D (1989) Allelic frequency variations at the MDH 1 locus within *Schistosoma mansoni* strains from Guadeloupe (French West Indies): Ecological interpretation. Comp Biochem Physiol B 93:33–37

Urbani C, Sinoun M, Socheat D, Pholsena K, Strandgaard H, Odermatt P, Hatz C (2002) Epidemiology and control of mekongi schistosomiasis. Acta Tropica 82:157–158

Wilson DE, Reeder DM (1993) Mammal species of the world: A taxonomic and geographic reference. Smithsonian Inst Press, Washington

25 Towards understanding the impacts of environmental variation on *Echinococcus multilocularis* transmission[1]

David R. J. Pleydell, Francis Raoul, Amélie Vaniscotte, Philip S. Craig and Patrick Giraudoux

1 Introductory remarks

The cestode *Echinococcus multilocularis* exploits predator-prey relationships between canid definitive hosts and small mammal intermediate hosts (Thompson and McManus 2001). Human infection arises from accidental ingestion of *E. multilocularis* eggs which leads to the zoonotic liver disease alveolar echinococcosis (AE) in one in ten cases (Vuitton et al. 2003). This disease can be fatal. No parasiticide is currently available and only radical surgery of the liver lesion, when possible, offers a definitive cure. During a long and asymptomatic incubation period in humans, metastasis of the larvae produces a complex multilocular network of larval material causing necrosis and malfunction of infected tissue. This disperse multilocular nature of infection makes treatment extremely difficult once symptomatic (Pawlowski et al. 2001).

The red fox (*Vulpes vulpes*) is regarded as the principal definitive host responsible for sustaining transmission in a wildlife cycle throughout much of Paleartic Eurasia (Eckert et al. 2001). A southern limit of this endemic area is imposed by the susceptibility of *E. multilocularis* eggs to desiccation in hot or dry conditions. Within Europe, rising fox densities have coincided with observations of both increases in prevalence rates and range extensions (Romig 2002) including into cities (Deplazes et al. 2004). These current trends are naturally of public health concern although it is

[1] Supported by the US National Institutes of Health and National Science Foundation "Ecology of infectious diseases" (TWO1565-02 and 2 R01 TW001565-05)

currently unclear how emergence in wildlife and human populations might be linked.

Here the variables which give rise to variation in the life cycle of *E. multilocularis* are reviewed with particular attention focused on those sources of variation of an ecological or environmental origin. Early models on *Echinococcus* transmission dynamics mainly concerned *E. granulosus*. They will be discussed with other models in the perspective of explorating environmental influences on transmission. Finally, the chapter draws contrast between transmission in various biogeographical areas and draws attention to the diversity of ecological systems which may sustain transmission. The chapter highlights commonalities of those various biogeographical areas for the purpose of identifying potential global parameters and then addresses local idiosyncrasies which force us to address geographical heterogeneity in the strength of environmental forcings. By this contrasting of various study sites in Europe and China this chapter attempts to provide a framework within which the influences of biogeographical diversity on parasite transmission might be studied in a global context.

2 Sources of variation in transmission intensity

The life cycle of *E. multilocularis* consists of three stages: egg, larvae and adult. Growth or decline in the parasites population depends upon the rates at which the worm passes successfully from one stage to the next. These transitions and the variables which influence them are outlined in Fig. 1.

Echinococcus multilocularis starts its life cycle as an egg produced by adult worms in the intestine of a definitive host. The time course of adult development in the intestine is affected by immunological interactions between the host and parasite. Consequently, host immunology affects the time course of egg production as well as egg viability. The species of host is therefore of primary importance: foxes are natural hosts and appear to have evolved a limited degree of natural immunity. By comparison, immunity in dogs appears to be weaker and egg production can be greater than in foxes. Cats, on the other hand, are poor hosts and infections result in relatively few infective eggs (Kapel et al. 2006). The eggs are passed with faeces into the external environment such that the pattern of environmental contamination is principally governed by the defecation behaviour of the host. Foxes use faeces as territorial markers and preferentially defecate along linear landscape features or near favoured hunting grounds (Giraudoux et al. 2002). Foxes are generalist predators and hunting patterns adapt to changing abundance of available food resources. Defecation behaviour will therefore also respond to food availability. Once in the environment, degradation in egg viability becomes a function of micro-climatic conditions and egg longevity is seriously reduced in hot or dry conditions (Veit et al. 1995). The probability that a small mammal ingests a viable egg will depend upon the densities of and juxtaposition between eggs and small mammal populations. This relationship will change dynamically in space and time as a result of both seasonal and multi-annual population density variations in small mammal populations.

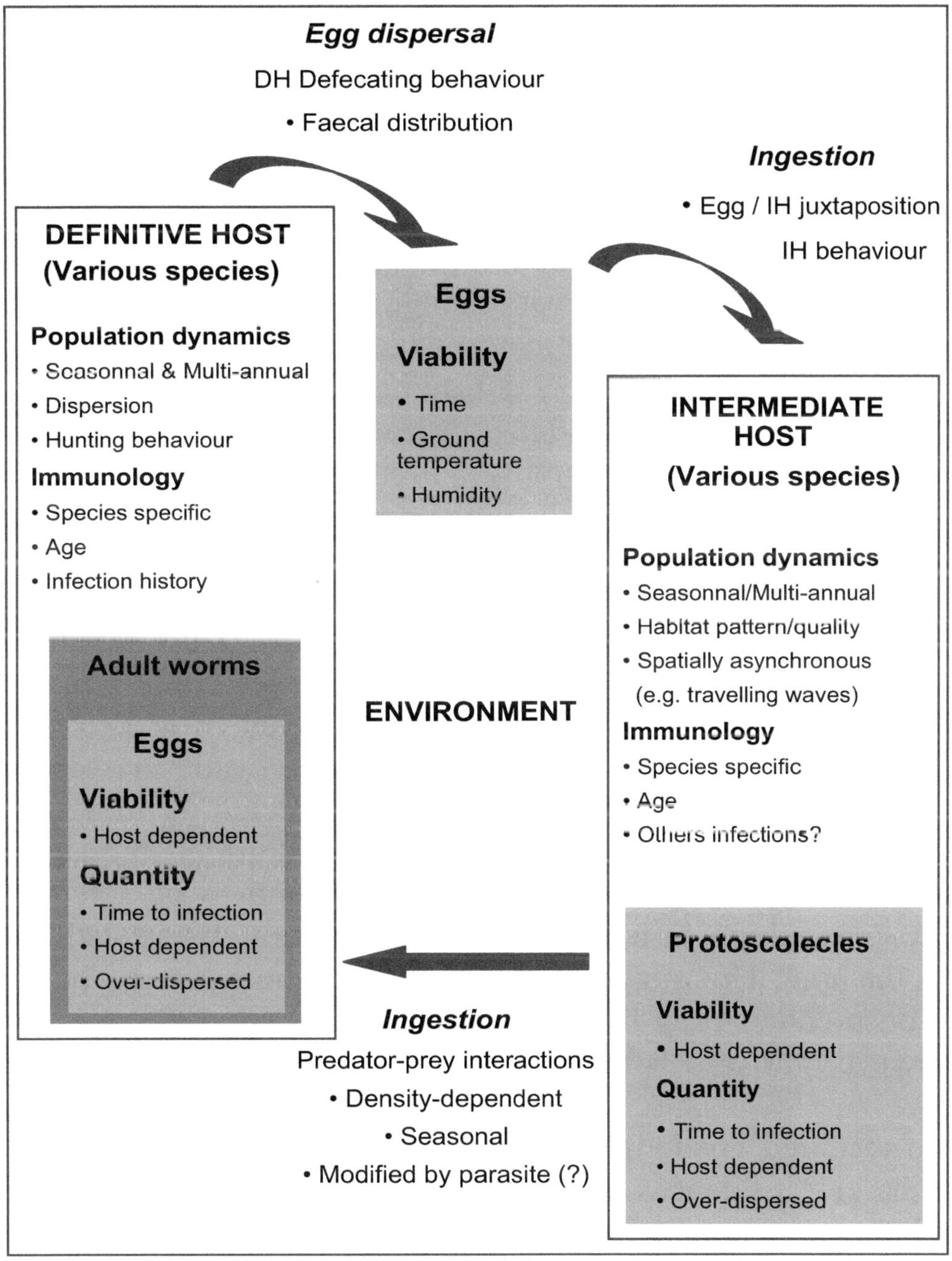

Fig. 1. The transmission cycle of *E. multilocularis* and sources of variation

If an egg is ingested by a small mammal, the parasite can start the second stage of its life cycle. This stage is composed of two sub-stages: development of larval cysts and protoscolex production via asexual reproduction within cysts. Whether or not the parasite becomes established in the host depends upon the viability of the ingested egg(s) and the physiological response of the host. Asexual reproduction in the intermediate host implies that both time and host immunology influence the infectivity of a small mammal at the time of predation. Immunological factors vary greatly between small mammal species and even between individuals (Rausch 1995; Vuitton et al. 2002). It is not always apparent which species within a given rodent assemblage provide key reservoirs for the parasite.

The final transition in the life cycle is from protoscolex to adult and this requires that a carnivore ingests an infective rodent. The infection rate of foxes is dependent upon the number of infected intermediate hosts ingested which is a function of the prevalence in intermediate hosts and the number of animals eaten. The number of animals eaten is the result of predator-prey interaction which can change as a function of prey density. *Vulpes vulpes* is a generalist predator which has been observed to prey preferentially on microtine rodents during high density periods (Weber and Aubry 1993; Giraudoux et al. 2002). Predator-prey interaction at a given time and place can therefore naturally inhibit or promote the probability of moving from the protoscolex to the adult stage. Evidence of the role of rodent density on infection rate in foxes has already been demonstrated in Japan (Saitoh and Takahashi 1998), although the actual predator-prey relationship was not fully documented. The probability that a given larvae is eaten by a fox will depend upon the number of foxes hunting in the vicinity of the infected host. It appears likely that this event is also influenced by the parasite via at least two mechanisms. Large cystic masses impose a physical reduction in mobility which can be suspected to influence predator-prey interactions in cases of heavy infection. Further, the adrenal metabolism of *Meriones unguiculatus* has been found to be affected by *E. multilocularis* resulting in a deprivation of desperately needed source of energy for the host. This suggests that the parasite may exert a biochemical influence on predator-prey interactions (Novak et al. 1993; Kepron et al. 1999). However, the effects of these interactions in the field have yet to be quantified.

It is the received wisdom that the daily predation patterns of resident foxes are the key influences behind transmission. However, the parasite is known to be over-dispersed in foxes (Hofer et al. 2000) and the importance of extreme events is currently understudied. Over-dispersion in small mammals driven by asexual reproduction, the possibility of age-acquired immunity in foxes and the ability of juvenile foxes to migrate hundreds of kilometers (Ables 1965; Artois 1989; Allen and Sargeant 1993; Rosatte 2002) points towards an epidemiological system influenced by rare but extreme events interacting with definitive host dispersal. The relative roles of average and extreme events on *E. multilocularis* transmission dynamics has yet to be quantified in the field.

3 Mechanistic models of transmission dynamics

Several kinds of models have been developed about *Echinococcus* transmission and their main parameter features are summarized in Table 1.

Table 1. Features of *E. multilocularis* transmission dynamics models

Model feature	1	2	3	4	5	6
DH population = constant	•		•	•	•	•
DH population = f(season)		•				
DH dispersion			•	•		
Predator/prey interaction = f(season, location)		•		•		
Intermediate hosts infection	•	•	•	•		•
Environmental contamination			•			•
IH population = constant	•		•	•	•	•
IH population = f(season)		•				
IH population = f(habitat)				•		
Egg longevity = constant	•		•		•	•
Egg longevity = f(season, location)		•				
Egg longevity = f(habitat)				•		
infection pressure = constant	•				•	
infection pressure = f(season)		•				
infection pressure = f(space)			•	•		
Acquired immunity			•		•	
Control effort	•		•	•		•

* Spatially explicit models, *DH* – definitive host, *IH* – intermediate host. 1 – Roberts and Aubert 1995; 2 – Ishikawa et al. 2003; 3 – Hansen et al. 2003; 4 Milner-Gulland et al. 2004; 5 – Budke et al. 2005a, b; 6 – Takumi and Van der Giessen 2005.

The earliest models of *Echinococcus* transmission were developed for *E. granulosus*. The transmission cycle of that parasite is of the same form as *E. multilocularis* but the key intermediate and definitive hosts are generally domestic ungulates and dogs respectively (Eckert et al. 2001). The models of Roberts et al. (1986) parameterised gain and loss rates of both infection and immunity within a given host species. From a set of four ordinary differential equations they derived expressions for prevalence and abundance for both dogs and sheep. The models have proved useful for answering control related questions in closed (i.e. farm) systems where homogeneity and equilibrium assumptions may often be reasonable (Torgerson 2003; Morgan et al. 2004; Budke et al. 2005a).The abundance and prevalence equations have also been fitted to both *E. granulosus* and *E. multilocularis* infection data in dogs of Tibetan pastoral communities in north west Sichuan, China (Budke et al. 2005b). That study indicated differences in the immunological response of dogs to the two parasites. In contrast to the host specific models of Roberts et al (1986), Takumi and Van Der Giessen (2004) present a model based on the entire transmission cycle of *E. multilocularis*. Their aim was to estimate the biomass (within the Dutch endemic area) of each of the three stages of parasite development. As for the model of Roberts et al. (1986), homogeneity in space and time of host species densities and contact rates was assumed.

The requirement of a constant infection pressure may not be reasonable when host populations are dynamic in time and space. Roberts and Aubert (1995) argued that for *E. multilocularis* in France intermediate host dynamics were unimportant which was proven not to be the case in further studies (Viel et al. 1999; Raoul et al. 2003). Temporal dynamics were explored further by Ishikawa et al. (2003). Those authors present a detailed model of seasonal fluctuation in both fox and red-backed vole (*Clethrionomys rufocanus*) densities. Contact rates between the two species were also adjusted to account for the effect of snow depth which introduces into the model an environmental variable which can vary in both space and time. The model was parameterised for two different provinces of the northern Japanese island of Hokkaido.

Two teams have attempted a fuller incorporation of spatial pattern in *E. multilocularis* transmission models using spatially explicit models including a stochastic component. Milner-Gulland et al. (2004) investigated the influence of landscape fragmentation in an illustrative metapopulation model based on the isolated patch network in a semi-arid area of Kazakhstan. In a parasite-focused approach (three stages), which consider cysts as individuals, intermediate hosts population as habitat and foxes as dispersal media for worms, they used the ecological parameters of semi-arid local areas of Kazakhstan and RAMAS GIS. This model incorporate environmental parameters (vegetation, climate) believed to influence parasite distribution and predict conditions for parasite persistence in discrete isolated focus. Secondly Hansen et al. (2003, 2004) have developed "Echi", a grid-based simulation model, which has enabled numerous simulation experiments. The "Echi" simulations have been based on simulated landscapes and low resolution geo-referencing in parasitological data sets. These experiments have enabled various transmissions to be tested in relation to fox and vole data from Northern Germany. The conclusion was that, from all simulations, only the landscape scenario was found to be robust against parameter variations, and that some properties (temperature, humidity) of the intermediate host's micro-habitat is responsible for the robustness of the parasitic cycle.

4 Transmission ecology in China

All of the models reviewed above have fitted models within areas which might be considered biogeographically homogeneous. It remains a key challenge to model, and therefore map variation in infection pressure across large areas. Here "large" is taken to mean of a sufficient size such that landscape heterogeneity may contribute to detectable trends in the local biomass of the parasite. The endemic area of central China fits this description and a number of sub-areas of this focus (Fig. 2) have been investigated and are described here.

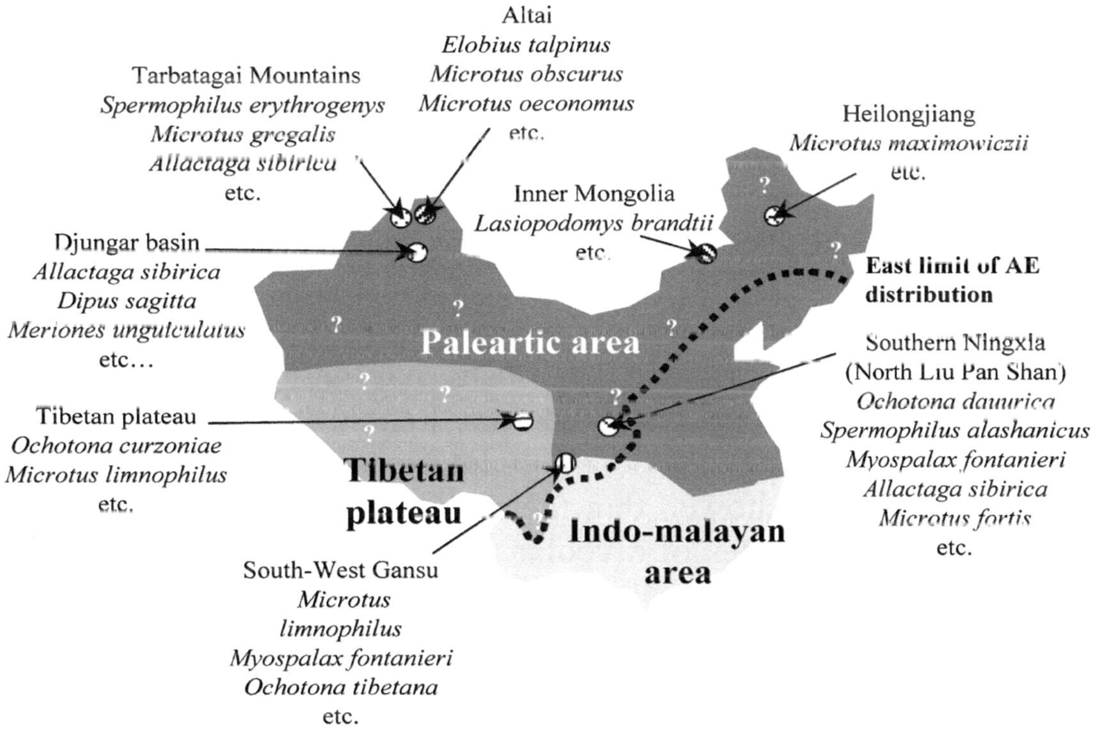

Fig. 2 China, the location of various small mammal surveys and dominant species trapped.

4.1 Eastern Tibetan plateau: western Sichuan

Serxu county, Sichuan, is situated on the Tibetan plateau at 4300 meters above sea level. The principal inhabitants of the area are Tibetan, many of whom maintain a semi-nomadic life style orientated around yak herding. The landscape is dominated by grassland and it is easy to find high densities of the black lipped pika *Ochotona curzoniae* in areas of low vegetation cover. It has long been clear that pasture degradation, overgrazing and pika populations are related. However it has been hard to ascertain causation: do high densities of *Ochotona* degrade pasture, or are the *Ochotona* symptoms of overgrazing? Recent studies suggest that *Ochotona* only enhance the soil degradation initiated by livestock (Smith and Foggin 1999). However, the dynamics of the degradation processes are still unclear and factors other than grazing, such as climate, might be involved. A recent departure from traditional practice had been the fencing of pastures motivated by landowners wishing to separate their privately owned pastures from common lands. A key motivation is to preserve areas of high quality grassland specifically for young and weak animals during the winter and spring months. This fencing has provided a semi-natural experiment in which the effects of yak grazing on the small mammal population can be observed directly.

A small mammal community survey has suggests a differential response in the relative densities of the small mammal intermediate hosts to the level of grazing pressure (Raoul et al., unpublished data). *O. curzoniae* and *Ochotona cansu* are never found in the tall grasses of protected winter pasture and, more generally, their densities are lower in areas where grazing pressure is low. Higher densities of these species are observed in the common lands where yak density is higher. Conversely, higher densities of *Microtus limnophilus* (the most abundant microtine species in the area) were observed in fenced grassland and bushy habitats, where vegetation structure offers safe cover against predators. Fencing therefore appears to augment potential intermediate host populations on both sides of the fence and Wang et al. (2004) showed that the area of fenced pasture within a township was related to rates of alveolar echinococcosis in the human population.

The Eastern Tibetan plateau has the highest rates of human alveolar echinococcosis ever recorded in the world (Li et al. 2005). Transmission there remains active and it has been proposed that the Eastern part of the Tibetan plateau might act as a large meta-stable reservoir for the parasite (Giraudoux et al. 2006). The two species of fox present on the plateau (the red fox *V. vulpes*, and the Tibetan fox *Vulpes ferrilata*) are known to be definitive hosts of *E. multilocularis*, with high prevalence rates recorded in the late 90's e.g. 44.4% (76/171) for *V. ferrilata*, (Qiu et al. 1999). The Buddhist religion on the Tibetan plateau encourages people to feed and protect dogs since they are thought to be the last reincarnation before humans. Therefore domestic dogs are numerous in and around villages and between 10% and 20% of them have been found infected by the adult worm (Budke et al. 2005a). This presents a significant risk factor for the human disease (Wang et al. 2006). Two species of arvicoline (*M. limnophilus* and *Microtus irene*), two species of lagomorph (*Lepus oiostolus* and *O. curzoniae*), one species of cricetine (*Cricetulus kamensis*) have been found to harbour larval infections (Qiu et al. 1999; Raoul et al., unpublished data). In a transect of 600km across the Eastern plateau numerous large *Ochotona* colonies were observed in areas over 3500 meters above sea level and these colonies were especially evident in the presence of degraded soils (Giraudoux et al. 2006). Moreover, multi-annual population fluctuations of *O. curzoniae* are known to occur with peak densities reaching almost 300 individuals/ha (Fan et al. 1999). However, the spatial synchrony of these outbreaks over large areas, which may be critical for transmission processes, is not documented. A new species, *Echonococcus shiquicus*, has recently been found in both *V. ferrilata* and *O. curzoniae* (Xiao et al. 2005). The relative epidemiological importance of this species has yet to be identified.

The Eastern part of the Tibetan plateau is currently a large active focus of transmission, and may feed (via fox dispersal and/or dog trading) newly favourable habitats in the eastern spurs (Ningxia and Gansu) (Giraudoux et al. 2006).

4.2 Tibetan plateau spurs: southern Gansu

The early 1990s saw the first reports of high prevalence rates of human AE in rural Han communities of southern Gansu, China (Craig et al. 1992). Trapping surveys were conducted to identify the key intermediate host reservoirs and 15 species were identified with susceptible intermediate hosts found in every habitat present. However, it was in the shrub land and scrubland areas where the vole *M. limnophilus* and the hamster *Cricetulus longicaudatus* were trapped in greatest abundance (Giraudoux et al. 1998, 2003). These species are both excellent *E. multilocularis* hosts and were highly notorious among local people for population outbreaks leading to agricultural damage (Chen et al. 1982). The end of the commune system in the late seventies coincided with increased rates of deforestation in the area. Successional growth in deforested areas combined with continued anthropogenic pressure (grazing and shrub cutting) drastically increased the proportion of shrub and scrub cover in some areas and thus increased the proportional cover of optimal *Microtus* habitat (Giraudoux et al. 1998). An index of ROMPA (Ratio of Optimal to Marginal Patch Area, see below) calculated from local land use maps was found to explain differences in the prevalence rates of the worst and least affected villages in the area (Giraudoux et al. 1996; Craig et al. 2000; Giraudoux et al. 2003). In further analyses of the epidemiology of the area landscape composition was quantified using remotely sensed data from the Thematic Mapper (TM) and Multispectral Scanner (MSS) sensors mounted on NASA's Landsat series satellites. These analyses have indicated that the proportions of grass and shrubland surrounding villages indeed help explain differences in the observed prevalence rates (Danson et al. 2003, 2004). The probability of infection was also observed to be related to dog ownership indicating a key role of the domestic dog for bringing the parasite into the human environment (Craig et al. 2000; Danson et al. 2003). A dramatic crash in the domestic dog population occurred in the early 1990's (Craig et al. 2000). More recently it has transpired that poisoning for control of agricultural rodent pests started in the mid nineties. The effects of that control program on wildlife definitive host reservoir populations can be inferred from a visibly abrupt crash in the domestic dog populations. At present then it can be said that *E. multilocularis* in the study area of southern Gansu is virtually extinct.

4.3 Tibetan plateau spurs: southern Ningxia

The Liupanshan mountains in the south of The Ningxia Hui Autonomous Region became a focus of work on rodent assemblages and *E. multilocularis* transmission in the 1980. Two early surveys in the mountain forests of Haiyuan and Guyuan Counties identified 13 small mammal species among which 2 where found infected with mature protoscolecies (*Spermophilus dauricus* and *Myospalax fontanierii*) (Li et al. 1985; Hong and Lin 1987).

A recent three year epidemiological screening survey in Xiji county identified a foci of human infection among the villages which lay in the intermittent area between the northern stretches of the Liupanshan and the subsequent Yueliangshan (Yang et al. 2006).

Current land use in Xiji contrasts to that in Zhang. Residual areas of forest are still threatened in Zhang despite protection bills whereas in Xiji remnants of forest are restricted to mountain areas, grazing pressure is now regulated and reforestation is backed by local government. Large areas of Southern Ningxia are planned to be forested and a large quantity of both ploughed and grazing land has been set-aside and transformed to plantation. Where economically feasible these plantations are being populated with young conifers and poplars, otherwise shrubs are planted instead. A recent rodent survey has indicated 16 species of small mammals in the area with both temperate zone (e.g. *Apodemus agrarius*) and desert species (e.g. *Dipus sagitta*) being present within some tens of kilometres of each other (Raoul et al., unpublished data). The trapping success of potential intermediate hosts was higher in recently planted set aside fields and abandoned grasslands (e.g. *C. longicaudatus* and *Ochotona daurica*), and in the young plantations (e.g. *Spermophilus alashanicus/dauricus*).

The observed degree of habitat change (transition from fields to forest) can be expected to disturb current rodent assemblages and therefore the transmission system of *E. multilocularis*. As in Zhang this projected increase in intermediate hosts habitat could increase the probability of a re-emergence, although methods of rodent control are clearly an over-riding factor at present. As in neighbouring Gansu, the local dog population is no longer visible save for a small number of permanently chained animals. Recent epidemiological surveys found no AE cases in children younger than 15 years of age. For now transmission appears to be controlled.

4.4 Northern Xinjiang

The landscape of Xinjiang is quite different to that of the eastern Tibetan plateau. Three mountain ranges (Altai in the north, Tianshan in the centre and Kunlun in the south) border the two natural semi-desertic basins (Junggar in the north and Tarim in the south). General ecological conditions (climate, vegetation and fauna) therefore vary sharply among these ecological units. *E. multilocularis* has been detected in red fox (11/36) and in wolf (1/2) at the border with Kazakhstan (Watihan 1987; Wang et al. 1989), but has never been formally found in dogs. Considering intermediate hosts, little information is available: *Mus musculus* and *Spermophilus erythrogenys* have been found infected but with few protoscolecies in the latter species (Zhou et al. 1998) and their importance in the transmission ecology remains questionable. An epidemiological survey has revealed that the human disease is relatively common in the Altai, western Junggar and Tianshan mountains, i.e. in the northern part of Xinjiang, and semi-nomadic lifestyle (typical of the Kazakh and Mongol ethnic groups) has been identified as a risk-factor for the disease (Zhou et al. 2000). The semi-nomadic groups move seasonally in search of suitable pastures for livestocks. Most summer pastures are located in the middle belts of the Altai and Tianshan mountain ranges. Winter pastures are generally shared with sedentary communities in the Altai foothills or in the plains of the Junggar cold semi-desert. People therefore move seasonally between very different ecological conditions with very different small mammal communities occurring within just some tens of kilometres of each other. For example communities of alpine meadows and forests, and communities of cold semi-deserts (Giraudoux et al., unpublished data). Nothing is known about the relative roles played by the various small mammal communities in *E. multilocularis* transmission in Xinjiang, and especially how each community contributes to stability in transmission. The low temperatures and high rainfall typical of the summer pasture areas make conditions favourable to the survival of *E. multilocularis* eggs. Moreover, several small mammal species, e.g. *Microtus arvalis*, *Spermophilus erythrogenys*, have been recorded as agricultural pests (Wang and Yang 1983), displaying regular high densities in grasslands of northern Xinjiang mountains. This suggests that human contamination may be more likely to occur in the summer pastures than in the winter settlements.

This north-western Chinese foci of transmission seems to be contiguous with the endemic area of Central Asia (Russia, Kazakhstan), but whether it is linked to the large but distant Tibetan foci (separated by the desertic Tarim basin) is still an open question.

5 Quantifying landscape effects

The ROMPA hypothesis is an attempt to describe the effects of both primary production and landscape composition on small mammal population dynamics (Lidicker 1995, 2000). The hypothesis suggests that the availability of optimal habitat in the landscape influences the growth rate of small mammal populations, but for grassland rodents high ROMPA also implies little suitable habitat for generalist predators and thus different cyclic patterns may emerge depending on available habitat. Percentage cover is perhaps the simplest and most popular metric by which *E. multilocularis* infection data have been compared to the environment. Motivated by the ROMPA hypothesis this metric has been used to study local variations in respect to landscape composition in eastern France (Pesson and Carbiener 1989; Giraudoux et al. 2003; Pleydell et al. 2004), northern Germany (Staubach et al. 2001) and central China (Danson et al. 2003, 2004). In computing ROMPA these studies have been confronted with a scale issue, namely which radius to use when creating buffer-zones within which ROMPA is to be obtained. Staubach et al (2001) fixed a buffer of 2, 5 km, a distance that was assumed to represent a fox home range. Danson et al.

(2004) explored a range of buffer sizes using stepwise model selection, choosing at each step the buffer which maximized the likelihood of the data given the parameters. Both approaches assume an equal weighting of pixels within the buffer regardless of their distance from the village or fox location. This simplistic parameterization of spatial effect is unlikely to represent the true nature of spatial dependence between a point observation of parasitological infection and the environmental conditions which contributed to the infection, a relationship which could reasonably be expected to decay with distance.

Potential solutions include cokriging although this is typically hampered by both non-linearity in discrete data sets and by the large number of pixels to be manipulated. A manageable alternative could lie in model averaging and Burnham and Anderson (2002) outline a simple derivation of weights for model averaging using the Akaike Information Criterion. Under that paradigm a range of models may be fitted using various buffer definitions and the final prediction map is based on a likelihood based weighting of the predictions from the various models.

6 Pattern based approaches

Attempts to derive global parameters based on landscape and the Ratio of Optimal to Marginal Patch Area (ROMPA) appear to fail when extrapolated from one endemic area to another (EchinoRisk network). For example, large areas of grassland in eastern France have been associated with *Arvicola terrestris* outbreaks and elevated transmission rates (Giraudoux et al. 2003). However, once extrapolated from eastern France an index of grassland ratio derived from the CORINE land-cover map fails to describe trends in fox prevalence data from southern Germany (Pleydell et al., unpublished data). This is likely to represent a scale mismatch between the geographical data at hand and the phenomenon giving rise to transmission: namely that a network of small microtine patches undetectable in the CORINE data set appears to sustain transmission in southern Germany (Romig, personal communication). The inability to extrapolate indices of grassland ratio from one endemic area to another is perhaps best illustrated by the example of Zürich, an urban landscape in which heightened transmission is sustained through a network of parks and fox prevalence rates have been observed to reach 60% (Hofer et al. 2000). This example suggests that optimized indices of ROMPA are specific to the transmission-systems within which they were formulated and that a new approach is required to understand the diversity of ecological systems which can give rise to transmission.

The apparent inability of ROMPA to provide a single unifying global parameter opens the question how can transmission and propagation be quantified and predicted across large areas characterised by biogeographical diversity? For example, in the endemic area of central China, which extends from the high altitude grasslands of the Tibetan plateau to the southern limits of the loess plateau, transmission is supported by a number of different small mammal assemblages and the relative roles of various definitive host species is also likely to vary. It is likely that many local differences in transmission ecology currently remain undocumented. These uncertainties prohibit fully deterministic modelling. To address this problem and to explore the local particularities in favourable environmental conditions transmission models need to be coupled with more flexible approaches capable to detecting unknown spatial patterns and non-linearities.

7 Flexible regression techniques

Modern regression techniques provide flexible approaches to the exploration of non-linearity and spatial pattern. Generalised additive models (GAM) extend generalized linear models (GLM) to include flexible functions for identifying non-linearity in the effects of continuous covariates. Traditionally interpretation of these models had been restricted by sensitivity to the modellers choice of knots, however modern methods of penalized likelihood now automate knot selection by penalizing over parameterization whilst maintaining sufficient numbers of knots to identify the key non-linearities in a data set (Wood and Augustin 2002).

Spatially explicit regression techniques are now established in disease mapping where they are used to overcome the small numbers problem. This problem essentially arises when insufficient epidemiological data exists to provide meaningful confidence intervals, thereby very high and very low prevalence rates might be observed in neighbouring areas where the true prevalence is equal simply due to large uncertainties arising from small sample sizes. In disease mapping this problem is addressed using conditional auto-regression (CAR), that is to assume *a priori* that the epidemiological situation in neighbouring areas is likely to be similar based on proximity in space and environmental conditions. This has been done in fox prevalence studies in Lower Saxony (Berke 2001) and in Austria (Duscher et al., unpublished data) where the analyses suggested emergence during the 1990s. The basis of CAR methods is an underlying random field, a smooth surface representing spatial variation. Random fields can also be used to detect spatial heterogeneity in the strength of environmental explanatory variables, a technique known as geographically weighted regression. This was done in an *E. multilocularis* survey across eastern France where a random field was fitted as a spatially-varying parameter for grassland ratio enabling the geographical limits of a grassland ratio index to be identified (Tolle et al. 2005).

Great flexibility is offered by modern regression techniques for detecting non-linearity and spatially-varying-parameters. Incorporation of these techniques into traditional transmission models provides a means of pattern detection nested within a process based system of equations which offers to overcome the limitations of homogeneity assumptions while proving an objective exploration of heterogeneity in environmental influence. This pattern based approach is particularly useful in situations where transmission processes are poorly understood such as across the central Chinese endemic area.

8 Concluding remarks

A key element in disease emergence/re-emergence is ecosystem disruption as a result of anthropogenic effects which may be as rapid as in forestry and agricultural changes. There is however difficulty in developing suitable models to study ecology of infectious diseases, wherein spatial determinants that meaningfully characterize wildlife reservoir habitat, can be linked in turn to host ecology and to dynamics of pathogen/parasite transmission. Spatial variables in the form of landscape and socio-economic characteristics should be linked to parasite transmission dynamics using an integrated modeling approach that takes into account multi-level heterogeneity at habitat, host and parasite domains and deterministic transmission parameters. The diversity of small mammals host communities and landscape worldwide offer a number of systems that sustain transmission of *E. multilocularis* at various time-space scales. It is expected that further advances will come from methods combining quantification of host communities from field surveys, landscape *via* remote sensing and parasite transmission *via* population screenings conducted on definitive hosts (e.g. dogs in villages in China or foxes in Europe) and humans, in a spatially explicit context. The combination of multi-level field approaches with modern regression techniques coupled with traditional transmission models provide a unique opportunity of investigating how a diversity of small mammal communities and anthropogenic landscapes can regulate parasite transmission.

References

Ables ED (1965) An exceptional fox movement. J Mammal 46:132

Allen SH, Sargeant AB (1993) Dispersal patterns of red foxes relative to population density. J Wildl Manag 57:526–533

Artois M (1989) Le Renard roux (*Vulpes vulpes* Linnaeus, 1758). In: Artois M, Delatre P (eds) Encyclopédie des carnivores de France. SFEPM, Nort/Erdre, pp 1–90

Berke O (2001) Choropleth mapping of regional count data of *Echinococcus multilocularis* among red foxes in Lower Saxony, Germany. Prev Vet Med 52:119–131

Budke CM, Campos-Ponce M, Wang Q, Torgerson PR (2005a) A canine purgation study and risk factor analysis for echinococcosis in a high endemic region of the Tibetan plateau. Vet Parasitol 127:49–55

Budke CM, Qiu J, Craig PS, Torgerson PR (2005b) Modelling the transmission of *Echinococcus granulosus* and *Echinococcus multilocularis* in dogs for a high endemic region of the Tibetan plateau. Int J Parasitol 35:163–70

Burnham KP, Anderson DR (2002) Model selection and multimodel inference: A practical information-theoretic approach, 2nd edn. SpringerVerlag, New York

Chen JC, Yao ZY, Liao CH (1982) Report on the damage caused by rodent pest in the grassland of South Gansu, China. Chinese J Zool 3:25

Craig PS, Liu D, Macpherson CNL, Shi D, Reynolds D, Barnish G, Gottstein B, Wang Z (1992) A large focus of alveolar echinococcosis in central China. Lancet 340:826–831

Craig PS, Giraudoux P, Shi D, Bartholomot B, Barnish G, Delattre P, Quéré JP, Harraga S, Bao G, Wang Y, Lu F, Vuitton DA (2000) An epidemiological and ecological study of human alveolar echinococcosis transmission in south Gansu, China. Acta Tropica 77:167–177

Danson FM, Graham AJ, Pleydell DRJ, Campos-Ponce M, Giraudoux P, Craig PS (2003) Multi-scale spatial analysis of human AE risk in China. Parasitology 127:133–141

Danson FM, Craig PS, Man W, Shi DZ, Giraudoux P (2004) Landscape dynamics and risk modelling of human alveolar echinococcosis. Photogram Eng Remote Sens 70:359–366

Deplazes P, Hegglin D, Gloor S, Romig T (2004) Wilderness in the city: The urbanization of *Echinococcus multilocularis*. Trends Parasitol 20:77–84

Eckert J, Schantz PM, Gasser RB, Torgerson PR, Bessonov AS, Movsessian SO, Thakur A, Grimm F, Nikogossian MA (2001) Geographic distribution and prevalence. In: Eckert J, Gemmell MA, Meslin FX, Pawlowski ZS (eds) WHO/OIE manual on echinococcosis in humans and animals: A public health problem of global concern. World Organisation for Animal Health, Paris, pp 100–142

Fan N, Zhou W, Wei W, Wang Q, Jiang Y (1999) Rodent pest management in the Qinghai-Tibet alpine meadow ecosystem. In: Singleton GR, Hinds LA, Leirs H, Zhang Z (eds) Ecologically-based management of rodent pests. Australian Center for International Agriculture Research, Canberra, pp 285–304

Giraudoux P, Vuitton D, Bresson-Hadni S, Craig P, Bartholomot B, Barnish G, Laplante JJ, Zhong SD, Wang YH, Lenys D (1996) Mass screening and epidemiology of alveolar echinococcosis in France, Western Europe, and in Gansu, Central China: From epidemiology towards transmission ecology. In: Ito J, Sato N (eds) Alveolar echinococcosis: Strategy for eradication of alveolar echinococcosis of the liver. Fuji Shoin, Sapporo, pp 197–211

Giraudoux P, Quéré JP, Delattre P, Bao GS,Wang XM, Shi DZ, Vuitton, DA, Craig PS (1998) Distribution of small mammals along a deforestation gradient in southern Gansu, central China. Acta Theriol 43:349–362

Giraudoux P, Delattre P, Takahashi K, Raoul F, Quéré JP, Craig P, Vuitton D (2002) Transmission ecology of *Echinococcus multilocularis* in wildlife: What can be learned from comparative studies and multi-scale approaches? In: Cestode zoonoses: An emergent and global problem. NATO Sciences Series, IOS press, Amsterdam, pp 251–262

Giraudoux P, Craig PS, Delattre P, Bartholomot B, Bao G, Barnish G, Harraga S, Quéré JP, Raoul F, Wang YH, Shi D, Vuitton D (2003) Interactions between landscape changes and host communities can regulate *Echinococcus multilocularis* transmission. Parasitology 127:121–131

Giraudoux P, Pleydell DRJ, Raoul F, Quéré JP, Wang Q, Yang YR, Vuitton DA, Qiu JM, Yang W, Craig PS (2006) Transmission ecology of *Echinococcus multilocularis*: What are the ranges of parasite stability among various host communities in China? Parasitol Int 55:S237–S246

Hansen F, Tackmann K, Jeltsch F, Wissel C, Thulke HH (2003) Controlling *Echinococcus multilocularis*: Ecological implications of field trials. Prev Vet Med 60:91–105

Hansen F, Jeltsch F, Tackmann K, Staubach C, Thulke HH (2004) Processes leading to a spatial aggregation of *Echinococcus multilocularis* in its natural intermediate host *Microtus arvalis*. Int J Parasitol 34:37–44

Hofer S, Gloor S, Müller U, Mathis A, Hegglin D, Deplazes P (2000) High prevalence of *Echinococcus multilocularis* in urban red foxes (*Vulpes vulpes*) and voles (*Arvicola terrestris*) in the city of Zürich, Switzerland. Parasitology 120:135–142

Hong LX, Lin YG (1987) Studies on the development and histopathology of alveolar cestode of *Echinococcus multilocularis* in human and animal hosts. Endemic Dis Bull 2:51–59

Ishikawa H, Ohga Y, Doi R (2003) A model for the transmission of *Echinococcus multilocularis* in Hokkaido, Japan. Parasitol Res 91:444–451

Kapel CMO, Torgerson PR, Thompson RCA, Deplaze P (2006) Reproductive potential of *Echinococcus multilocularis* in experimentally infected foxes, dogs, raccoon dogs and cats. Int J Parasitol 36:79–86

Kepron C, Schoen J, Novak M, Blackburn BJ (1999) NMR study of lipid changes in organs of jirds infected with *Echinococcus multilocularis*. Comp Biochem Physiol B 124:347–353

Li T, Qiu J, Yang W, Craig PS, Chen X, Xiao N, Ito A, Giraudoux P, Mumuti W, Yu W, Schantz PM (2005) Echinococcosis in Tibetan populations, Western Sichuan Province, China: A unique and unusually severe public health problem? Emerg Infect Dis 11:1866–1873

Li WX, Zhang GC, Lin YG, Hong LX (1985) The occurrence of *Echinococcus multilocularis* Leuckart, 1863, the natural animal hosts in China and its morphological study. Acta Zool Sinica 31:365–371

Lidicker WZ (1995) Landscape approaches in mammalian ecology and conservation. Univ Minnesota Press, Minneapolis

Lidicker WZ (2000) A food web/landscape interaction model for microtine rodent density cycles. Oikos 91:435–445

Milner-Gulland EJ, Torgerson PR, Shaikenov B, Morgan E (2004) The tapeworm *Echinococcus multilocularis* in Kazakhstan: Transmission dynamics in a patchy environment, In: Akçakaya HR, Burgman MA, Kindvall O, Wood CC, Sjögren-Gulve P, Hatfield JS, McCarthy MA (eds) Species conservation and management. Case studies. Oxford Univ Press, Oxford, pp 179–89

Morgan ER, Milner-Gulland EJ, Torgerson PR, Medley GF (2004) Ruminating on complexity: Macroparasites of wildlife and livestock. Trends Ecol Evol 19:181–188

Novak M, Modha A, Blackburn BJ (1993) Metabolic alterations in organs of *Meriones unguiculatus* infected with *Echinococcus multilocularis*. Comp Biochem Physiol B 105:517–521

Pawlowski ZS, Eckert J, Vuitton DA, Ammann RW, Kern P, Craig PS, Dar KF, De Rosa F, Filice C, Gottstein B, Grimm F, Macpherson CNL, Sato N, Todorov T, Uchino J, von Sinner W, Wen H (2001) Echinococcosis in humans: clinical aspects, diagnosis and treatment. In: Eckert J, Gemmell MA, Meslin FX, Pawlowski ZS (eds) WHO/OIE manual on echinococcosis in humans and animals: A public health problem of global concern. World Organisation for Animal Health, Paris, pp 20–66

Pesson B, Carbiener R (1989) Ecologie de l'échinococcose alvéolaire en Alsace: Le parasitisme du Renard roux (*Vulpes vulpes L.*). Bull d'Écol 20:295–301

Pleydell DRJ, Raoul F, Danson FM, Graham A, Craig PS, Tourneux F, Giraudoux P (2004) Modelling the spatial distribution of *Echinococcus multilocularis* infection in foxes. Acta Tropica 91:253–265

Qiu JM, Liu FJ, Schantz P, Ito A, Carol D, He JG (1999) Epidemiological survey of hydatidosis in Tibetan areas of Western Sichuan Province. Archivos Internacionales de la Hidatidosis 23:84

Raoul F, Michelat D, Ordinaire M, Decote Y, Aubert M, Delattre P, Delazes P, Giraudoux P (2003) *Echinococcus multilocularis*: secondary poisoning of fox population during a vole outbreak reduces environmental contamination in a high endemicity area. Int J Parasitol 33:945–954

Rausch RL (1995) Life cycle patterns and geographic distribution of *Echinococcus* species. In: Thompson RCA, Lymbery AJ (eds) *Echinococcus* and hydatid disease, CABI, Wallingford, pp 51–134

Roberts MG, Aubert MFA (1995) A model for the control of *Echinocccus multilocularis* in France. Vet Parasitol 56:67–74

Roberts MG, Lewis JR, Gemmel MA (1986) Population dynamics in echinococcus and cystercosis: Mathematical model of the life-cycle of *Echinococcus granulosus*. Parasitology 92:621–641

Romig T (2002) Spread of *Echinococcus multilocularis* in Europe? In: Craig P, Pawloski (eds) Cestode zoonoses: Echinococcosis and cysticercosis. IOS Press, Amsterdam Berlin Oxford Tokyo Washington, pp 65–80

Rosatte RC (2002) Long distance movement by a coyote, *Canis latrans*, and red fox, *Vulpes vulpes*, in Ontario: Implications for disease-spread. Can Field Nat 116:129–131

Saitoh T, Takahashi K (1998) The role of vole populations in prevalence of the parasite (*Echinococcus multilocularis*) in foxes. Res Pop Ecol 40:97–105

Smith AT, Foggin JM (1999) The plateau pika (*Ochotona curzoniae*) is keystone species for biodiversity on the Tibetan plateau. Anim Cons 2:235–240

Staubach C, Thulke HH, Tackmann K, Huges-Jones M, Conraths FJ (2001) Geographic information system-aided analysis of factors associated with the spatial distributions of foxes. Am J Trop Med Hyg 65:943–948

Takumi K, Van Der Giessen J (2005) Transmission dynamics of *Echinococcus multilocularis*: Its reproduction number, persistence in an area of low rodent prevalence, and effectiveness of control. Parasitology 131:133–140

Thompson RCA, McManus DP (2001) Aetiology: Parasite and life-cycles. In: Eckert J, Gemmell MA, Meslin FX, Pawlowski ZS (eds) WHO/OIE manual on echinococcosis in humans and animals: A public health problem of global concern. World Organisation for Animal Health, Paris, pp 1–19

Tolle F, Pleydell D, Combes B, Cliquet F, Piarroux M, Giraudoux P, Tourneux FP (2005) Identification of environmental risk factors for the presence of *Echinococcus multilocularis*. In: Tendorio JA, Juliao RP (eds) 14th European colloquium on theoretical and quantitative geography, September 9-13 2005, Tomar, Portugal (DVD-ROM)

Torgerson PR (2003) The use of mathematical models to simulate control options for echinococcosis. Acta Tropica 85:211–221

Veit P, Bilger B, Schad V, Schäfer J, Frank W, Lucius R (1995) Influence of environmental factors on the infectivity of *Echinococcus multilocularis* eggs. Parasitology 110:79–86

Viel JF, Giraudoux P, Abrial V, Bresson-Hadni S (1999) Water vole *(Arvicola terrestris scherman)* density as risk factor for human alveolar echinococcosis. Am J Trop Med Hyg 61:559–565

Vuitton D, Godot V, Zhang S, Harraga S, Penfornis A, Beurton I, Gottstein B, Kern P (2002) Host immunogenetics and role in epidemiology of larval cestodoses. In: Craig P, Pawloski (eds) Cestode zoonoses: Echinococcosis and cysticercosis. IOS Press, Amsterdam Berlin Oxford Tokyo Washington, pp 183–193

Vuitton DA, Zhou H, Bresson-Hadni S, Qiang W, Raoul F, Giraudoux P (2003) Epidemiology of alveolar echinococcosis in humans: An update of the situation in Europe and China and new epidemiological trends. Parasitology 127:87–107

Wang SB, Yang GY (1983) Rodent fauna of Xinjiang. Xinjiang's People Publ House, Urumqi

Wang Q, Vuitton DA, Qiu JM, Giraudoux P, Xiao YF, Schantz PM, Raoul F, Li TY, Yang W, Craig PS (2004) Fenced pasture: A possible risk factor for human alveolar echinococcosis in Tibetan pastoralist communities of Sichuan, China. Acta Tropica 90:285–293

Wang Q, Qiu J, Yang W, Schantz PM, Raoul F, Craig PS, Giraudoux P, Vuitton DA (2006) Socioeconomic and behavior risk factors of human alveolar echinococcosis in Tibetan communities in Sichuan, People's Republic of China. Am J Trop Med Hyg 74:856–862

Wang W, Wu Y, Wu JG, Ding ZX (1989) Investigation of *Echinococcus multilocularis* in Tachang, Xinjiang – new record of wolf infected with *E. multilocularis*. Endemic Dis Bull 4:8–10

Watihan (1987) The discovering of *Echinococcus multilocularis* in red fox in Xinjiang. Xinjiang Animal Industry 2:36–38

Weber JM, Aubry S (1993) Predation by foxes, *Vulpes vulpes*, on the fossorial form of the water vole, *Arvicola terrestris scherman*, in western Switzerland. J Zool Lond 229:553–559

Wood SN, Augustin NH (2002) GAMs with integrated model selection using penalized regression splines and applications to environmental modelling. Ecol Model 157:157–177

Xiao N, Qiu J, Nakao M, Li T, Yang W, Chen X, Schantz PM, Craig PS, Ito A (2005) *Echinococcus shiquicus n. sp.*, a taeniid cestode from Tibetan fox and plateau pika in China. Int J Parasitol 35:693–701

Yang YR, Ellis M, Sun T, Li Z, Liu XZ, Vuitton DA, Bartholomot B, Giraudoux P, Craig PS, Boufana B, Wang Y, Feng XH, Wen H, Ito A, McManus DP (2006) A unique family clustering of human echinococcosis cases in a Chinese community. Am J Trop Med Hyg 74:487–494

Zhou HX, Wen H, Wang YH, Delattre P, Quere JP, Vuitton DA, Craig P, Giraudoux P (1998) Experimental susceptibility of *Spermophilus erythrogenys* to *Echinococcus multilocularis*. Ann Trop Med Parasitol 93:335–337

Zhou HX, Chai SX, Craig PS, Delattre P, Raoul F, Quéré JP, Vuitton DA, Wen H, Giraudoux P (2000) Epidemiology of alveolar echinococcosis in Xinjiang Uygur Autonomous Region, China: A preliminary analysis. Ann Trop Med Parasitol 94:715–729

26 Parasites and pest population management[1]

Herwig Leirs and Grant R. Singleton

1 Introductory remarks: Small mammals, parasites and pest control

A number of small mammals are considered pests because they cause damage to humans either directly (e.g., rat bites, vampire bat blood meals) or more often indirectly by transmitting pathogens to humans and livestock, causing losses in field crops or stored harvest produce and processed food, or damaging infrastructure and natural and cultural assets that are deemed valuable by humans. Rodents, the most abundant group of small mammals, with some species thriving very well in the human environment, are common among these pests. Fruit bats can cause considerable damage in tropical fruit orchards, whereas insectivorous bats act as reservoirs for several viruses. The European rabbit is the major vertebrate pest in Australia and high on the list in New Zealand as well, only surpassed by the possum. Some small mammals are usually not considered pests but this may depend on local conditions, whereas some others are a nuisance or carry pathogens.

Pest management, micromammals and macroparasites are linked in various ways. First, in many cases, it is simply its role as a host for zoonotic macroparasites that gives a mammal species pest status. The ultimate goal of pest population management is then to control the macroparasite rather than the host per se. Second, there have also been attempts to use macroparasites as a form of biological control of the host species. Third, pest population management usually affects the density or population turnover of the small mammal species, and this may have significant interacting effects on the parasite dynamics. Fourth, changes in landscape use by humans may lead to changes in rodent population densities and consequent

[1] We thank Stephen Davis, Ron Verhagen and Dave Spratt for their valuable comments on the manuscript for this chapter.

increase of risk in zoonoses. Also, landscape changes may simplify habitats and lead to zoonoses though the accidental ingestion of invertebrates that are the intermediate hosts of many macroparasites of small mammal species that previously were not considered a pest by humans (e.g., Spratt 2005). Fifth, invasive mammal species in some instances have led to an extension of the host spectrum of macroparasites. Such events have led to conservation (e.g., black-footed ferret) and human health concerns (e.g., sylvatic plague in USA).

In this chapter, we will present these different relationships, consider their biological background, and discuss how they affect pest management approaches. We start with a brief introduction of the pest problems and management strategies.

2 Parasites and the pest status of small mammals

2.1 Parasites make small mammals a pest

A number of small mammals are considered pests, exactly because they are carrying parasites that can infect human beings or livestock. *Rattus norvegicus* is a notorious pest in and around agricultural buildings because of the damage it can cause to stored food and infrastructure. On slaughter pig farms it has an especially high pest status because it can carry *Trichinella spiralis* nematodes (Kapel 2000; Meerburg et al. 2004). When pigs consume infected rats that they find on the farm, they may become infected themselves, which constitutes a health risk for meat consumers and may cause a huge economical problem for the farm owner or even a whole country's pork meat export. *Arvicola terrestris* is considered a pest in some parts of Europe in grasslands or orchards, but is also of concern since it is an intermediate host for *Echinococcus multilocularis* (Saucy 1994; Schmitt et al. 1997). The table at the end of this chapter (see Appendix) provides an overview of the most important macroparasites that are pathogenic for humans or livestock and for which small mammals are reservoirs. In addition, small mammals are used as reservoirs by numerous microparasites. The diversity of these microparasites is huge, including rabies-like lyssa-viruses, SARS-like coronaviruses and henipaviruses in bats (Hoar et al. 1998; Fooks et al. 2003; Mackenzie and Field 2004; Li et al. 2005); Ebola and Marburg filoviruses in hitherto unknown small mammals (bats seem to be likely candidates, Leroy et al. 2005); hantaviruses (McCaughey and Hart 2000), bacteria (e.g., causing leptospirosis, rat-bite fever, plague), protozoans (e.g., *Toxoplasma gondii*) in rodents. A full treatment of all of these microparasitic zoonoses falls beyond the scope of this book, but we will nevertheless return to some of them since they are carried by arthropod vectors that feed on small mammals as well as on humans or livestock. Bubonic plague, caused by the bacterium *Yersinia pestis*, is basically a flea-vectored infection in rodents, but humans can also become infected through bites from infected rodent fleas (Gage and Kosoy 2005). The discovery of the bacterium-rat-flea relation in the late 19th century became one of the most important driving forces behind organised and enforced rodent control in urban settings and ship transport throughout the world. Even today, plague is still epithetical for any rodent borne disease. Moreover, fleas, ticks and mites also transmit between small mammals and humans a variety of other microorganisms that may cause serious disease such as the bacterium *Borrelia burgdorferi* (Lyme disease), several rickettsias (causing typhus fevers, spotted fevers, scrub fever, ehrlichiosis) or viruses (e.g., several tick-borne encephalites) (Gratz 1988; Spratt 2005).

2.2 Parasite release makes small mammals a pest

In theory, parasites can limit populations of small mammals to a certain degree (see also below and other chapters in this book). Populations of invasive species usually have only part of their native parasite (or predator) fauna, and this sometimes results in the host population reaching higher abundance or biomass. Thus, this release from the normal parasite pressure may lead such populations to become pests, while in their native area they do not have such characteristics. This has been well documented for a number of invasive plants or invertebrates (Torchin et al. 2003; Torchin and Mitchell 2004). In small mammals, invasive populations, especially on islands, are known to have a poor parasite fauna in terms of both abundance and diversity (Göuy de Bellocq et al. 2002; Milazzo et al. 2003; Pisanu et al. 2001). However, to the best of our knowledge, no examples of this being the main reason for the development of pest populations of mammals that elsewhere are harmless, are reported in the literature. Nevertheless, it is worth to consider always whether the parasite release contributes to the pest status of such populations. In fact, it has been suggested several times that neutralising the parasite release would help significantly to control invasive species.

2.3 Invasive species and landscape changes – increasing the management complexity

The introduction of new mammal species into an ecosystem either through successful dispersal and colonization (e.g., the global spread of commensal rodents such as *Rattus norvegicus, Rattus rattus, Mus musculus*) or through intentional releases (e.g., *Oryctolagus cuniculus* and *Vulpes vulpes* in Australia) has led to the range expansion of macro- and micro-parasites. A classic example is the establishment of sylvatic plague in small mammal assemblages in North America in 1908 following the plague pandemic at the turn of the 20th century. The geographic distribution of the disease has been fairly stable over the past 50 years. However, a recent study confirmed that it is distributed more widely than previously thought (Cully et al. 2000). Sylvatic plague in North America causes epizootics in prairie dog populations that lead to extirpation of colonies and is the only disease known to cause high mortality in this species (Barnes 1993). Plague also has important conservation impacts because the prairie dog is an important food resource for the highly endangered black-footed ferret (see Christe et al., this volume, for discussion of conservation impacts of macroparasites).

In Australia, native rodent species have acquired an impressive array of macroparasites from the introduced rodents (e.g., *Angiostrongylus cantonensis*, *Trichuris muris, Syphacia muris, Hepatojarakus pycnofasciatus, Heterakis spumosa, Mastophorus muris*, *Nippostrongylus brasiliensis, Strongyloides ratti, Calodium hepaticum, Eucoleus gastricus, Hymenolepis diminuta, Raillietina celebensis, Moniliformis moniliformis)* and domestic animals (e.g., *Fasciola hepatica, Spirometra erinacei* – larval stage, *Taenia taeniaeformis* – larval stage, *Linguatula serrata* – larval stage) (D.M. Spratt, personal communication). However, there is little evidence of native rodent helminths (which represent a rather rich fauna) going in the opposite direction into the introduced *R. rattus, R. norvegicus* and *M. musculus* (except for *Angiostrongylus mackerrasae*).

Landscape changes can have major consequences on the distribution and abundance of rodents and their macroparasites. In the USA, forest destruction and fragmentation reduces mammalian species diversity, with some species such as *Peromyscus leucopus* adapting well to these landscape changes. For example, they show higher population densities in small forest fragments. These population increases of an important host species for Lyme disease, have led to a dramatic increase in the density of infected tick nymphs in small (<2 ha) forest patches (Allan et al. 2003). Therefore, human induced habitat fragmentation can increase the risk of humans contracting a tick-borne disease.

An emerging issue is human-induced changes in the landscape as a result of global warming. The changes in distribution of the reservoirs or intermediate hosts of parasites, may lead to the range extension of zoonotic diseases and an increased prevalence in epizootics. In the past few decades, the expansions of ranges of rodent-borne diseases such as trypanosomiasis, Lyme disease, tick-borne encephalitis and plague have been reported (Lindgren et al. 2000; Harvell et al. 2002). Although the evidence is still being collated, if these trends continue, then there will be important management implications for the small mammal wildlife hosts of these diseases.

3 Ecologically-based rodent management

Pest management in general and small mammal management in particular have for a long time been understood to be equal to the killing of individual animals belonging to the pest species and, if possible, to the extermination of the entire pest populations. This turned out to be generally an unsustainable and, in the medium-to-long term, ineffective approach, especially in open settings such as natural, agricultural or urban environments where the entire extermination is not feasible. In the agricultural insect pest control, this led to the development of Integrated Pest Management (IPM) strategies that attempted to combine a number of different control methods. Rodent management took a longer time to change but, approximately 10 years ago, the concept of the Ecologically Based Rodent Management (EBRM) approach has been developed (Singleton 1997; Singleton et al. 1999). This approach starts from a thorough understanding of the pest rodents' biology (ecology, physiology, taxonomy) and the damage they cause, and tries to identify ecological factors to which this damage may be most sensitive. Obviously, in many cases, albeit not always, these factors are to be related to rodent abundance, and EBRM will then investigate how to best affect rodent abundance at the moment when it matters to damage. For many species, this will best be done by increasing mortality (e.g., with rodenticides). However, in other species or under other circumstances it may be more effective to affect reproduction or dispersal. For example, population ecological models suggest that the dynamics of *Phyllotis darwini* in Chili is more sensitive to changes in survival, while *Mastomys natalensis* in Tanzania is, at least in some seasons, more affected by changes in reproduction (Lima et al. 2003). A fine example of EBRM is the Trap Barrier System (TBS) used for the control of *Rattus argentiventer* in Indonesia, where the large scale foraging movements are put to use by luring rats into traps around an early planted crop. This reduces rodent densities over a much larger area than just that single field (Singleton and Sudarmaji 1998). Still, even the TBS is just one from the package of management options developed for *R. argentiventer* based on a solid understanding of its ecology (Singleton et al. 2005).

EBRM does not *a priori* prefer or exclude any technique but it does take into account the more indirect effects on the environment, particularly the whole ecological community in which the rodent population lives. When parasites are involved in some way, such community aspects are important *a fortiori* and an ecologically-based approach is especially appropriate. This requires a very sound understanding of the pest animals' as well as the parasites' population ecology, and this information unfortunately is often still lacking.

4 Parasite control through pest control

4.1 Why control parasite reservoirs?

There are basically two reasons why small mammal population management has its place in the struggle to reduce the burden of parasitic disease (in humans or any other domesticated or wild species that humans would like to protect). The first argument is the intuitive assumption that there is a positive relation between the abundance of reservoirs and the force of infection to humans, i.e. the risk for humans of becoming infected is higher when there are more individuals of the reservoir species, hence reservoir numbers should be lowered. A second argument comes from the theory on the ecology of infections, where it is derived that there is a threshold of host density below which an infection cannot persist. Again, this argues for a reduction in reservoir population numbers. Both arguments are less straightforward than often thought, so they need more consideration.

4.2 Force of infection to humans

The probability that an individual human becomes infected with a parasite is dependent on the probability that he or she comes in contact with an infective stage of the parasite, and that probability is again proportional to the availability of infective stages of the parasite in the environment (and, of course, the human's behaviour). The relationships between the availability of infective stages and the abundance of small mammal reservoir, however, are not necessarily straightforward. First, it will depend on whether the parasite is transmitted from one host to another via direct contact, a free-living infective stage, an intermediate host or a vector. When intermediate hosts and vectors are involved, their abundance (or even relative abundance in relation to humans) may contribute more to the force of infection than the abundance of the reservoir species. However, even in the case of a directly transmitted infection, the link between reservoir abundance and risk to humans can take different forms. Davis et al. (2005) explored this relation, starting from a simple model for the force of infection to humans λ, $\lambda = \beta Np$, where β is the transmission coefficient (the number of contacts per time a human has with the reservoir, times the probability that contact with an infected reservoir results in transmission of the infection), N is the population size of the reservoir and p is the prevalence of the infection in the reservoir population. At first glance, this relation is simple, but, since prevalence can be related to host abundance, the outcome is more complex (Fig. 1). If prevalence increases with host density, as in most infections with direct horizontal transmission (Grenfell and Dobson 1995; Mills et al. 1999; Hudson et al. 2001), then the force of infection to humans increases proportionally to N^2 (Fig. 1a). Thus, if the reservoir abundance is rising twofold, then the risk to humans rises fourfold. In a number of other infections, like those with frequency-dependent sexual transmission, there is no overt relationship between host abundance and prevalence. In this scenario, the force of infection to humans increases linearly with increasing abundance (Fig. 1b). In both cases, there is a positive relationship between the reservoir abundance and the risk for humans. Therefore, reducing the number of reservoir individuals should decrease the risk for humans.

If there is a threshold host abundance, below which the infection does not persist in the reservoir (e.g., Davis et al. 2004, see also below), then obviously there is also a non-linearity in the relation between abundance and risk to humans (Fig. 1c). In such cases, there is a host population density below which further reduction of this density does not have any effect on the risk to humans. Consequently, controlling the small mammal host below this threshold does not have direct beneficial effects but only makes sense if it contributes to keeping the host abundance low enough.

The last and, at first glance, somewhat counterintuitive scenario is that there is a negative relationship of some sort between population density and prevalence (Fig. 1d). The most obvious example, as also reviewed by Davis et al. (2005) is the "juvenile dilution effect". This effect is due to the recruitment of numerous uninfected young animals that increase the population abundance much faster than they can become infected.

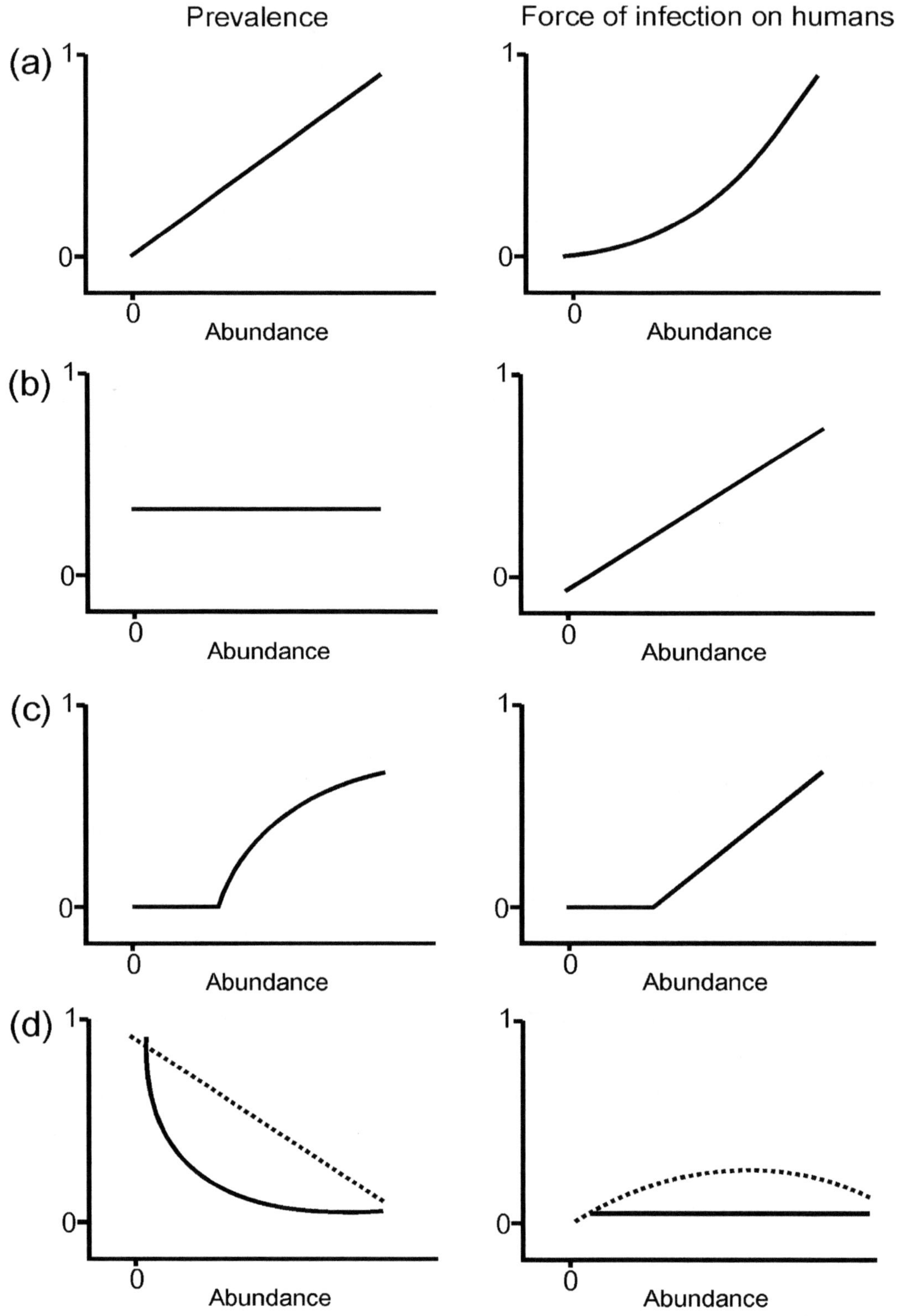

Fig. 1. Relationships between abundance and the force of infection on humans, depending on the relationship between host abundance and prevalence of disease in the host population; (a) a positive relationship, (b) no relationship, (c) a threshold relationship and (d) a negative relationship (modified after Davis et al. 2005)

There is a temporal scaling issue involved here, because while the juvenile dilution effect may work at a seasonal scale, there can still be a positive relationship between abundance and prevalence at a multiannual scale. Nevertheless, the juvenile dilution effect has its importance for population management, because it means that, within a year, an increase of the reservoir population (with uninfected juveniles) is not a reason to immediately start a control action. In fact, depending on the form of the negative relation between abundance and prevalence (e.g., the speed with which uninfected juveniles are recruited), intermediate host densities may even constitute a higher risk to humans than higher or lower densities (Fig. 1d). Similar effects can be obtained when arthropod vectors are involved in transmission of infection. The basic reproductive rate R_0, and from there p, is then proportional not to the host abundance N_h but rather to N_v/N_h, i.e. the number of available vectors per host individual. Reducing N_h may then have adverse effects on the risk to humans. That risk is further increased when such pest control action would result in the vectors looking for alternative hosts and ending up on humans. A classical example is the control of bubonic plague where a rodent control action without simultaneous action against fleas may force the latter to feed on humans, which would then effectively increase the risk to humans of becoming infected (Gratz 1999).

In conclusion, while it generally will be an appropriate disease control strategy to target small mammal species that are reservoirs of parasites, it is necessary to evaluate the relation between reservoir abundance and force of infection to humans. This requires a good understanding of the infection's ecology, but also of the reservoir host's population ecology.

4.3 Threshold population densities and herd immunity

Theories of infection ecology predict that there is a host population density below which the infection cannot persist (see references in Rosa et al., this volume). Intuitively, this is understood as a threshold density below which, on average, the chances for every infectious individual to meet an uninfected individual (to pass the infection to) are too small. More formally, it is the host density of a naive population below which the infection's basic reproductive rate R_0 becomes less than 1. There are a variety of ways to formulate R_0, but many include the number of sensitive hosts in the population. From there, the invasion threshold density S_T can be calculated. Obviously, R_0 and, thus, also S_T depend on the properties of the host (e.g., the contact rate between hosts, or between hosts and free-living parasite stages or intermediate hosts or vectors) as well as on the properties of the parasite (length of the period during which a parasite stage remains infectious, rate of success for a parasite to colonise a host given that a host comes into contact with an infectious stage). The concept of the invasion threshold densities has a very strong theoretical support, and has been linked to observed patterns in human childhood diseases (reviewed in Hastings 1997), but observational data to support it in wildlife are very scarce (Begon et al. 2003; Lloyd-Smith et al. 2005). Study of measles epidemics in humans led to the recognition of the "critical community size", which is the size of a community that produces enough susceptible individuals to maintain transmission (Bartlett 1960). For example, there are longitudinal data suggesting that threshold numbers of susceptible individuals are needed for reinvasion of phocine distemper virus in harbour seals (Swinton et al. 1998) and corona- and parvoviruses in lions (Packer et al. 1999). A clear host density threshold was observed for infection with plague bacteria *Yersinia pestis* in *Rhombomys opimus* in Kazakhstan (Davis et al. 2004).

The existence of a threshold density opens several possibilities for disease control. In short-lived small mammals, the critical community size is probably close to the actual invasion threshold. If one manages to keep a small mammal host density below the threshold value, then an existing infection will fade out, and, even when it is reintroduced, it will not be able to establish. The effect is, thus, much more profound than a reduction of the force of infection as discussed above. The effect also can be much more sustainable, because if it is technically feasible to prevent reintroduction of the pathogen, the small mammal population can even be allowed to rise again to the original abundance level, so that fewer side-effects on the natural community are expected and no continued application of environmentally or ethically undesirable control techniques (pesticides, trapping) is needed.

The existence of a threshold can also be used as a management tool for the planning of control actions. As long as the densities remain below the threshold, no control is needed, whereas if the densities are above the threshold, authorities should be alert. In the case of plague in gerbils in Kazakhstan, there was a time lag between reaching the threshold and records of the infection (Davis et al. 2004). This allows the development of the early warning systems that predict future epizootics among gerbils and, from there, the risk to humans. It should be pointed out, however, that the threshold is not an absolute boundary for the presence of an infection. Firstly, high enough densities may be a necessary but not sufficient condition for the infection to spread, and other factors (e.g., climate) may play a role above the threshold. In addition, there is the stochastic aspect of whether a parasite arrives in a population where it is absent. If vectors or intermediate hosts are involved in the transmission of a parasite, then there may be several thresholds for each component that need to be reached, or thresholds for relative vector/host proportions. Secondly, densities below the threshold mean that the establishment of the infection is impossible, which is not equivalent to saying that the infection never occurs. Infection can still arrive in such low-density populations and even be transmitted among individuals for some time before it fades out.

As mentioned above, it is the density of *susceptible* individuals, not the whole population, that is relevant in the threshold theory. If an infection elicits in a host an immune response with lasting effects, then animals that have been infected before are no longer susceptible to a new infection with the same pathogen. Vaccination allows a reduction of the number of susceptibles without a reduction of the total abundance of the host population. This is crucial in the concept of herd immunity, where a whole population is protected against establishment of a parasite by the immunisation of just a proportion of the population (Begon et al. 1996). The concept also applies to wildlife and should be considered in cases where reduction of abundance is undesirable or simply not feasible. A splendid example is the eradication of rabies from a large area of Western Europe. Foxes, the main wildlife host of this infection, have been hunted and poisoned for many decades in an attempt to eradicate rabies or, at least, reduce the force of infection to humans (Aubert 1999). This proved ineffective, mainly due to the high reproductive capacity of the foxes and their dispersal behaviour through which vacated places were quickly recolonised by the immigrant foxes. In fact, the increased dispersal movements contributed to a better spread of the infection. In the 1980's, a new approach was instigated. Fox populations were immunised with oral rabies vaccines hidden in baits that were distributed over large areas (Pastoret and Brochier 1998). Currently, rabies has disappeared from most of Western Europe despite increasing fox populations and the termination of vaccination campaigns (Selhorst et al. 2005). Of course, the fox population is now susceptible which means that much attention must be paid to avoiding reinvasion of rabies, either along the borders through natural dispersal of foxes or through human transport of infected dogs (Bugnon et al. 2004; Thulke et al. 2000).

A wildlife vaccination approach as described above could be promising in several other cases where it is not possible or desirable to reduce host population densities to a low enough level. It could also be considered as an alternative for vaccination of humans in cases where the latter is commercially not viable for the vaccine industry. Indeed, the development of the vaccine may not be so difficult or costly in itself, but tests of efficacy and safety in humans may be too costly or ethically difficult and, especially for diseases that occur in poor countries, investments in vaccine development may not pay off enough for commercial companies. Requirements for a vaccine for wildlife are much more relaxed, but there are a number of other issues that need to be considered. Of course, a good quality vaccine must be available, but an equally large problem is to find adequate delivery methods for it. An oral bait with vaccine included is the most obvious approach, but one could also consider other alternatives such as a self-disseminating agent, like a genetically modified virus that carries similar antigens as the targeted parasite. The latter approach is not without ethical and environmental controversies. The needed efficiency of bait delivery is linked to the transmission properties of the pathogen in the host population: more "infectious" pathogens have a lower S_T and, thus, a higher proportion of the population will need to be vaccinated in order to achieve the herd immunity. Also, the ecology of the targeted wildlife host species must be taken into account: absolute numbers of hosts to be vaccinated, host dispersal patterns and population turnover rates are all important. Finally, one should not overlook possible resistance from human society where it may not be easily accepted that vaccination programs are targeting wild animals rather than humans or where people traditionally kill the host, rather than to protect it against infection.

4.4 If pest control equals harvesting…

If populations of pest species show strong compensatory mechanisms to cope with the increased mortality, then the classical lethal pest control strategies will not lead to a reduction in the population size. They will result, however, in an increased turn-over in the population with the recruitment of new individuals. When the recruitment is by reproduction, then the young animals entering the population will be susceptible to the infections of concern. This process will actually increase the number of susceptible individuals in the population, even when the total population size remains constant. The larger number of susceptibles leads to more intensive transmission, and since animals are constantly removed from the population, they do not have the chance to acquire a recovered immune status. If recruitment is mainly by dispersal, then the attracted individuals can be susceptible as well, but they could also be infectious. Thus, if the pest control acts as an unintended form of sustained harvesting, a parasite's prevalence may actually increase rather than decrease.

5 Pest population management using parasites

5.1 The basis for biological control

Parasites have a negative effect on host fitness and, therefore, they have long been considered as a potential pest control method (see review as early as in Elton 1942). Host-parasite models predict that parasites play a regulatory role in the population dynamics of a host (Anderson 1978; Anderson and May 1978). This conclusion has also been supported empirically by a number of experimental studies on model systems for small mammal-helminth interactions, such as the nematode *Heligmosomoides polygyrus* in *Apodemus flavicollis* (see also Rosa et al., this volume). In laboratory populations of these mice, introduction of *H. polygyrus* reduced equilibrium density levels by 10% in comparison to control populations kept under the same conditions (Scott 1987). Soon after, it was shown that this effect was caused by the decreased survival of the infected mice rather than reduced natality, and occurred in laboratory as well as in wild populations (Scott 1990; Gregory 1991; Quinnell 1992).

The evidence for a regulatory role of parasites in host population dynamics is not necessarily a good basis for biological control with parasites. Regulation is the process through which high population numbers are depressed down to an equilibrium level, but also increased up to this level when they reach low values (Begon et al. 1996). Many pest species cause damage already at low densities, and the objective of pest control is then to reduce population size well below the level of the natural equilibrium. It is not trivial to reach this goal using parasites. It requires that the negative impact of the parasite on the host more than exceeds the population's intrinsic growth rate and that the infection and its impact are persistent.

Biological control with parasites and parasitoids, especially in insect pest management, is well accepted and widespread (Pimentel 2002). On the contrary, in the control of small mammals, this management approach has been clearly successful in two cases only, both of which involved microparasites (viruses). There may actually be a biological basis for this difference as can be demonstrated by the shape of density-dependence, expressed as the per capita rate of population change as a function of density. Åstrøm (1997) noted that if the shape of this curve is downward convex (i.e. a stronger density dependence at lower densities, as suggested for insects) then even when these populations are controlled to a very low level, they are likely to be stable. If this curve is downward concave, (i.e. a stronger density dependence at higher densities, as suggested for mammals), then strongly depressed populations are likely to fluctuate violently, suggesting that it is simply more complicated to biologically control mammalian than insect pests.

5.2 Biological control using parasites

Spratt (1990) described eight characteristics of the ideal candidate for biological control of mammals with helminths. These are as follows:

- there needs to be strong and persistent effect on mortality and/or fecundity;
- the impact of the bio-control agent would be more pronounced if there were density-dependent effects;
- success is likely to be higher if the helminth has a direct life cycle, otherwise time lags could diminish its effect on host population dynamics;
- persistence and diffusion of the parasite requires transmission via aerosol, long environmental persistence (e.g., resistance to desiccation) or a highly mobile and common vector;
- ideally, the bio-control agent should have high host specificity;
- the parasite needs to be cheap to maintain and able to cycle in the laboratory;
- little genetic resistance should be expected to develop rapidly in the host;
- if it is an exotic helminth, then it should need to pass the quarantine regulations of the target country.

These eight characteristics provide a very useful audit of the likely chances of success for a helminth that is proposed as a candidate biological control agent. If the candidate does not meet one of these prerequisites, then it is important to be aware of the likely implications on the parasite-host interactions or of the likely societal acceptance of the bio-control agent. Singleton (1994) also reviewed the prospects of macro-parasites as biological control agents, but specifically for rodent pest species.

Earlier, Singleton and colleagues explored the possibilities of using the nematode *Calodium hepaticum* (=*Capillaria hepatica)* as a biological control method to manage populations of house mice in Australia. House mice are an introduced mammal pest in Australia and display irregular population explosions of massive magnitude (>800 mice per ha), causing huge losses to farmers (Singleton 1997). *C. hepaticum* is a liver parasite that produces eggs that remain in the liver of the host. When the host is eaten by another animal, the eggs pass through this animal's gut, are excreted with faeces, and then mature in the soil until they are picked up by a new host. Alternatively, the eggs become embryonated in the environment after the death and subsequent decomposition of the infected host. Infection with this nematode has a significant negative impact on the survival of house mice (Spratt and Singleton 1986). Due to this lethal effect, the transmission without an intermediate host, the possibility to formulate infectious eggs into a bait and the capacity of the eggs to remain infectious in the environment made this worm a promising agent for potential biological control, and model simulations indicated that it would indeed have a regulatory effect on house mouse populations (McCallum and Singleton 1989). Experiments in field enclosures showed that the infection successfully persisted for at least 1.5 years after release in a mouse population, but they also indicated compensations due to density-dependent mortality so that there was no difference in abundance between control and experimental treatments (Barker et al. 1991). A number of impressively extensive replicated field experiments were then set up in Eastern and Southern Australia. In the first experiment, in four treatment and three control areas of 4 km^2 each, the parasite was released during the low density phase of the population dynamics; there was successful transmission of the parasite, and after four months *C. hepaticum* had a considerable impact on survival in the mouse population; however, since survival was poor also in the control sites, there was no treatment effect on abundance (Singleton et al. 1995). Also, *C. hepaticum* did not have a noticeable effect on the population once breeding had commenced (Singleton et al. 1995). In a second experiment, started during the increase phase of a mouse population outbreak, 10% of the mice were infected in four treatment sites of 16 km^2 and again compared with three control sites. After two months, prevalence had risen to 30% but then started decreasing again and the mouse population increased to outbreak densities (Singleton and Chambers 1996). The authors attributed the failure to reach effects in both experiments to low population densities or drought conditions which limited effective and persistent transmission.

Microparasites have been more successfully applied as biological control agents. Two examples that stand out are the myxoma virus and the Rabbit Hemorrhagic Disease (RHD) virus that were used in the management of *Oryctolagus cuniculus* populations in Australia. The myxoma virus was successfully released in 1950 with a mortality often around 99% in the rabbit populations where the virus established (Cooke and Fenner 2002). The success of the myxoma virus was highest in the temperate and semi-arid zones where there was good survival of the vectors of the virus – rabbit fleas and mosquitoes. Rabbit populations bounced back within 5 years as they became immune to the various strains of the virus. The myxoma virus also changed genetically with selection favouring a virus that was more persistent, but which had lower mortality rates. Nevertheless, the myxoma virus is still an effective biological control agent more than 50 years after it became established in field populations.

The RHD virus escaped from quarantine facilities on an island off the Australian coast in 1995 and then spread quickly through the mainland. The drastic effects on the wild rabbit populations then incited also an unauthorised release of the virus in New Zealand. The recorded reductions in the damage by rabbits and costs to control them are huge (Saunders et al. 2002). Also in the case of the RHD, the effects of the introduced infections are strongly affected by other factors, such as aridity of the environment (Cooke and Fenner 2002; Mutze et al. 2002), virulence of the RHD strains or general condition of the rabbits (Bruce et al. 2004; Bruce and Twigg 2005; Story et al. 2004), and interactions with predator effects (Reddiex et al. 2002) or myxomatosis (Mutze et al. 2002).

It is no coincidence that the use of parasites as a biological control method for small mammals has received considerable attention in Australia. The targeted pests are introduced species that colonised an environment where there was only a poor community of competitors, predators and parasites and that were, thus, regulated at much higher levels, or not regulated at all, in comparison to populations in the original areas. Moreover, due to the phylogenetic unrelatedness of the local fauna to lagomorphs, the risks for infection of non-target species was minimal.

5.3 Parasites as bio-rodenticides

The biological control described above aspires to establish an infection in a pest population so that the population stays at a low level for an extended period and the development of chronic and acute impacts are prevented. Parasites also have been thought of as biological rodenticides, where the objective was to kill individual rodents, much as a poison would do. This is not genuine biological control since the complex ecological host-parasite interactions discussed above do not apply here. The main issue under scrutiny here is the pathogenicity of the parasite to individual rodents, and how to infect as many rodents as possible and as quickly as possible. Transmission between rodents may be a useful characteristic, but it is not a strict necessity. Nevertheless, such an approach could still have its place in ecologically-based rodent management, as an alternative to "poison", as long as the bio-rodenticide is more benign environmentally and/or has a lower impact on the non-target species.

A protozoan parasite that received some attention in this respect is the apicomplexan *Sarcocystis singaporensis*. This parasite has snakes of the genus *Python* as its definitive hosts; infectious stages are excreted with snake's faeces and when these are eaten by rodents, they develop into muscle cysts in the rodent. When the rodent is eaten by a snake, the cycle is completed (Jäkel et al. 1997). The infection causes serious and often lethal disease in rodents. In an experiment in different agricultural habitats in Thailand, laying out bait pellets with a high concentration of the infectious stage of the parasite resulted in high (58-92%) parasite-induced mortality in rodents of three species (*R. norvegicus, Rattus tiomanicus, Bandicota indica*) (Jäkel et al. 1999). Subsequent field trials in rice fields in Thailand indicated that the bio-rodenticide was as effective in controlling rodents as conventional control (Jäkel et al. 2006). It is worth pointing out that these were direct effects of the infectious bait, without involvement of the definitive hosts. A commercial product containing the parasite has been developed and has been distributed in some Southeast Asian countries under the trade name "PRORODENT". There are production issues common to this and other bio-rodenticides. These include the need for very high quantities of the infectious parasite stages and the need for biological baits to have a reasonable shelf life if they are going to be available for tactical or chronic use. Also, a bio-rodenticide has to compete with the efficacy and costs of conventional chemical rodenticides. Given that these generic challenges often discourage investment in bio-rodenticides, it is encouraging that a macroparasite based bio-rodenticide is now available commercially.

Much more widespread as a bio-rodenticide are preparations of the bacterium *Salmonella enterica*. Such formulations have been produced since the first half of the 20th century and are currently still available in a number of countries (Painter et al. 2004). The most widespread formulation is "Biorat", a preparation produced in Cuba or under Cuban license and used in a number of Central American, South American and Asian countries. Reports about the efficacy of this formulation as a rodenticide are often enthusiastic, but the major concern is the health risk that it may pose to other vertebrates than the targeted rodents, including humans (Friedman et al. 1996). The producers claim that the used strain is not pathogenic to humans but there is no substantial documentation to support this. The bacteria that are isolated from Biorat are *Salmonella enterica* serotype Enteritidis phage type 6a (Painter et al. 2004). The same type of bacteria was also used in "Ratin", a Salmonella-based rodenticide that was commercially available until 1960 in Europe and that caused several outbreaks of human illness. The dissemination of salmonellosis to other species is clearly undesirable.

5.3 Parasites as vehicles for immunocontraception

The aim of controlling the reproductive potential of a mammal pest population as a means to manage its negative impacts has led to much interest in approaches to sterilize the pest species (see Bomford 1990; Gao and Short 1993; Dell'Omo and Palmery 2002 for reviews). Thus far, there has been no successful sterility management of wild populations of small mammals. One approach that has promising potential is the delivery of a reproductive protein in a form that generates a strong and persistent immune response to the protein, thereby blocking fertilisation of the animal that receives the protein. This approach is described as immunocontraception and potentially either males or females could be sterilized (see Tyndale-Biscoe 1994). Over the past decade most of the research on immunocontraception of rodents has focused on female sterility and through using either a non-infectious agent in oral baits or an infectious species-specific virus as a carrier of an infertility agent. A system that works well under laboratory conditions has been developed for viral-vectored immunocontraception (see Singleton et al. 2002 for review).

Based on the concept developed for rodents in using viruses as vectors for a sterility protein, a program of research has been developed to try to control the brush-tail possum *Trichosurus vulpecula* using the nematode *Parastrongyloides trichosuri*, as the vehicle for the delivery and subsequent spread of a sterilizing protein (Cowan et al. 2006).

6 Concluding remarks

The presence of macroparasites can affect the pest status of small mammals and the damage they cause. Pest management of small mammal populations can also affect the macroparasite populations, in a positive as well as a negative way. Despite the effects of macroparasites on small mammal fitness, there is little hope for the near future that they can be used for biological control of small mammals, except perhaps for some bio-pesticides. Small mammals and macroparasites interact in complex ways, and the implications for pest management are equally complex.

References

Allan BF, Keesing F, Ostfeld RS (2003) Effect of forest fragmentation on Lyme disease risk. Cons Biol 17:267–272

Anderson RM (1978) The regulation of host population growth by parasitic species. Parasitology 76:119–157

Anderson RM, May RM (1978) Regulation and stability of host-parasite population interactions. I. Regulatory processes. J Anim Ecol 47:219–247

Åstrøm M (1997) The paradox of biological control revisited: Per capita nonlinearities. Oikos 78:618–621

Aubert MFA (1999) Costs and benefits of rabies control in wildlife in France. Revue Scientifique et Technique de l'Office International des Epizooties 18:533–543

Barker SC, Singleton GR, Spratt DM (1991) Can the nematode *Capillaria hepatica* regulate abundance in wild house mice: Results of enclosure experiments in Southeastern Australia. Parasitology 103:439–449

Barnes AM (1993) A review of plague and its relevance to prairie dog populations and the black-footed ferret. In: Oldemeyer JL, Biggins DE, Miller BJ (eds) Management of prairie dog complexes for the reintroduction of the black-footed ferret. U.S. Fish and Wildlife Service, Washington, pp 28-37

Bartlett MS (1960) The critical community size for measles in the United States. J R Stat Soc A 123:37–44

Begon M, Harper JL, Townsend CR (1996) Ecology: Individuals, populations and communities. Blackwell Sci, Oxford

Begon M, Hazel SM, Telfer S, Bown K, Carslake D, Cavanagh R, Chantrey J, Jones T, Bennett M (2003) Rodents, cowpox virus and islands: Densities, numbers and thresholds. J Anim Ecol 72:343–355

Bomford M (1990) A role of fertility control in wildlife management? Australian Government Publishing Service, Canberra

Bruce JS, Twigg LE (2005) The reintroduction, and subsequent impact, of rabbit haemorrhagic disease virus in a population of wild rabbits in south-western Australia. Wildl Res 32:139–150

Bruce JS, Twigg LE, Gray GS (2004) The epidemiology of rabbit haemorrhagic disease, and its impact on rabbit populations, in south-western Australia. Wildl Res 31:31–49

Bugnon P, Breitenmoser U, Peterhans E, Zanoni R (2004) Efficacy of oral vaccination in the final stage of fox rabies elimination in Switzerland. J Vet Med B 51:433–437

Cooke BD, Fenner F (2002) Rabbit haemorrhagic disease and the biological control of wild rabbits, *Oryctolagus cuniculus*, in Australia and New Zealand. Wildl Res 29:689–706

Cowan PE, Ralston MJ, Heath DD, Grant WN (2006) Infection of naïve, free-living brushtail possums (*Trichosurus vulpecula*) with the nematode parasite *Parastrongyloides trichosuri* and its subsequent spread. Int J Parasitol 36:287–293

Cully JF, Carter LG, Gage KL (2000) New records of sylvatic plague in Kansas. J Wildl Dis 36:389–392

Davis S, Begon M, De Bruyn L, Ageyev VS, Klassovskiy NL, Pole SB, Viljugrein H, Stenseth NC, Leirs H (2004) Predictive thresholds for plague in Kazakhstan. Science 304:736–738

Davis S, Calvet E, Leirs H (2005) Fluctuating rodent populations and risk to humans from rodent-borne zoonoses. Vector-Borne Zoonotic Dis 5:305–314

Dell'Omo G, Palmery M (2002) Fertility control in vertebrate pest species. Contraception 65:273–275

Elton C (1942) Voles, mice and lemmings. Oxford Univ Press, London

Fooks AR, Brookes SM, Johnson N, McElhinney LM, Hutson AM (2003) European bat lyssaviruses: An emerging zoonosis. Epidemiol Infect 131:1029–1039

Friedman CR, Malcolm G, RigauPerez JG, Arambulo P, Tauxe RV (1996) Public health risk from *Salmonella*-based rodenticides. Lancet 347:1705–1706

Gage KL, Kosoy MY (2005) Natural history of plague: Perspectives from more than a century of research. Ann Rev Entomol 50:505–528

Gao Y, Short RV (1993) The control of rodent populations. Oxford Rev Reproduct Biol 15:265–310

Göuy de Bellocq J, Morand S, Feliu C (2002) Patterns of parasite species richness of Western Palaeartic micro-mammals: Island effects. Ecography 25:173–183

Gratz NG (1999) Control of plague transmission. In: Dennis DT, Gage KL, Gratz N, Poland JD, Tikhonov I (eds) Plague manual: Epidemiology, distribution, surveillance and control, vol WHO/CDS/CSR/EDC/99.2. WHO, Geneva, pp 97–134

Gratz NN (1988) Rodents and human disease: A global appreciation. In: Prakash I (ed) Rodent pest management. CRC Press, Boca Raton, pp 101–169

Gregory RD (1991) Parasite epidemiology and host population growth: *Heligmosomoides polygyrus* (Nematoda) in enclosed wood mouse populations. J Anim Ecol 60:805–821

Grenfell BT, Dobson AP (eds) (1995) Ecology of infectious diseases in natural populations. Cambridge Univ Press, Cambridge

Harvell CD, Mitchell CE, Ward JR, Altizer S, Dobson AP, Ostfeld RS, Samuel MD (2002) Ecology: Climate warming and disease risks for terrestrial and marine biota. Science 296:2158–2162

Hastings A (1997) Population biology: Models and Concepts. SpringerVerlag, New York

Hoar BR, Chomel BB, Rodriguez FDA, Colley PA (1998) Zoonoses and potential zoonoses transmitted by bats. J Am Vet Med Assoc 212:1714–1720

Hudson PJ, Rizzoli A, Grenfell BT, Heesterbeek H, Dobson AP (2001) The ecology of wildlife diseases. Oxford Univ Press, Oxford

Jäkel T, Khoprasert Y, Sorger I, Kliemt D, Seehabutr V, Suasa-Ard K, Hongnark S (1997) Sarcosporidiasis in rodents from Thailand. J Wildl Dis 33:860–867

Jäkel T, Khoprasert Y, Endepols S, Archer-Baumann C, Suasa-Ard K, Promkerd P, Kliemt D, Boonsong P, Hongnark S (1999) Biological control of rodents using *Sarcocystis singaporensis*. Int J Parasitol 29:1321–1330

Jäkel T, Khoprasert Y, Promkerd P, Hongnark S (2006) An experimental field study to assess the effectiveness of bait containing the parasitic protozoan *Sarcocystis singaporensis* for protecting rice crops against rodent damage. Crop Prot (in press)

Kapel CMO (2000) Host diversity and biological characteristics of the *Trichinella* genotypes and their effect on transmission. Vet Parasitol 93:263–278

Leroy EM, Kumulungui B, Pourrut X, Rouquet P, Hassanin A, Yaba P, Delicat A, Paweska JT, Gonzalez JP, Swanepoel R (2005) Fruit bats as reservoirs of Ebola virus. Nature 438:575–576

Li W, Shi Z, Yu M, Ren W, Smith C, Epstein JH, Wang H, Crameri G, Hu Z, Zhang H, Zhang J, McEachern J, Field H, Daszak P, Eaton BT, Zhang S, Wang LF (2005) Bats are natural reservoirs of SARS-like coronaviruses. Science 310:676–679

Lima M, Stenseth NC, Leirs H, Jaksic FM (2003) Population dynamics of small mammals in semi-arid regions: A comparative study of demographic variability in two rodent species. Proc R Soc Lond B 270:1997–2007

Lindgren E, Talleklint L, Polfeldt T (2000) Impact of climatic change on the northern latitude limit and population density of the disease-transmitting European tick *Ixodes ricinus*. Env Health Persp 108:119–123

Lloyd-Smith JO, Cross PC, Briggs CJ, Daugherty M, Getz WM, Latto J, Sanchez MS, Smith AB, Swei A (2005) Should we expect population thresholds for wildlife disease? Trends Ecol Evol 20:511–519

Mackenzie JS, Field HE (2004) Emerging encephalitogenic viruses: Lyssaviruses and henipaviruses transmitted by frugivorous bats. Arch Virol Suppl 18:97–111

McCallum HI, Singleton GR (1989) Models to assess the potential of *Capillaria hepatica* to control population outbreaks of house mice. Parasitology 98:425–437

McCaughey C, Hart CA (2000) Hantaviruses. J Med Microbiol 49:587–599

Meerburg BG, Bonde M, Brom FWA, Endepols S, Jensen AN, Leirs H, Lodal J, Singleton GR, Pelz HJ, Rodenburg TB, Kijlstra A (2004) Towards sustainable management of rodents in organic animal husbandry. Njas-Wageningen J Life Sci 52:195–205

Milazzo C, Goüy de Bellocq J, Cagnin M, Casanova J-C, Di Bella C, Feliu C, Fons R, Morand S, Santalla F (2003) Helminths and ectoparasites of *Rattus rattus* and *Mus musculus* from Sicily, Italy. Comp Parasitol 70:199–204

Mills JN, Ksiazek TG, Peters CJ, Childs JE (1999) Long-term studies of hantavirus reservoir populations in the southwestern United States: A synthesis. Emerg Infect Dis 5:135–142

Mutze G, Bird P, Kovaliski J, Peacock D, Jennings S, Cooke B (2002) Emerging epidemiological patterns in rabbit haemorrhagic disease, its interaction with myxomatosis, and their effects on rabbit populations in South Australia. Wildl Res 29:577–590

Packer C, Altizer S, Appel M, Brown E, Martenson J, O'Brien SJ, Roelke-Parker M, Hofmann-Lehmann R, Lutz H (1999) Viruses of the Serengeti: Patterns of infection and mortality in African lions. J Anim Ecol 68:1161–1178

Painter JA, Molbak K, Sonne-Hansen J, Barrett T, Wells JA, Tauxe RV (2004) *Salmonella*-based rodenticides and public health. Emerg Infect Dis 10:985–987

Pastoret PP, Brochier B (1998) Epidemiology and elimination of rabies in Western Europe. Vet J 156:83–90

Pimentel D (2002) Encyclopedia of pest management. Marcel Dekker, New York,

Pisanu B, Chapuis JL, Durette-Desset MC (2001) Helminths from introduced small mammals on Kerguelen, Crozet, and Amsterdam Islands (Southern Indian Ocean). J Parasitol 87:1205–1208

Quinnell RJ (1992) The population dynamics of *Heligmosomoides polygyrus* in an enclosure population of wood mice. J Anim Ecol 61:669–679

Reddiex B, Hickling GJ, Norbury GL, Frampton CM (2002) Effects of predation and rabbit haemorrhagic disease on population dynamics of rabbits (*Oryctolagus cuniculus*) in North Canterbury, New Zealand. Wildl Res 29:627–633

Saucy F (1994) Density dependence in time series of the fossorial form of the water vole, *Arvicola terrestris*. Oikos 71:381–392

Saunders G, Kay B, Mutze G, Choquenot D (2002) Observations on the impacts of rabbit haemorrhagic disease on agricultural production values in Australia. Wildl Res 29:605–613

Schmitt M, Saucy F, Wyborn S, Gottstein B (1997) Infestation of water voles (*Arvicola terrestris*) with metacestodes of *Echinococcus multilocularis* in the canton of Freiburg (Switzerland). Schweizer Z Obst Weinbau 139:84–93

Scott ME (1987) Regulation of mouse colony abundance by *Heligmosomoides polygyrus*. Parasitology 95:111–124

Scott ME (1990) An experimental and theoretical study of the dynamics of a mouse - nematode (*Heligmosomoides polygyrus*) interaction. Parasitology 101:75–92

Selhorst T, Muller T, Schwermer H, Ziller M, Schluter H, Breitenmoser U, Muller U, Brochier B, Pastoret PP, Mutinelli F (2005) Use of an area index to retrospectively analyze the elimination of fox rabies in European countries. Environ Manag 35:292–302

Singleton GR (1994) The prospects and associated challenges for the biological control of rodents. In: Anonymous (ed) Proceedings of the 16th Vertebrate Pest Conference, Santa Clara, California, USA, 1-3 March 1994, pp 301-307

Singleton GR (1997) Integrated management of rodents: a Southeast Asian and Australian perspective. Belgian J Zool 127:157–169

Singleton GR, Chambers LK (1996) A manipulative field experiment to examine the effect of *Capillaria hepatica* (Nematoda) on wild mouse populations in southern Australia. Int J Parasitol 26:383–398

Singleton GR, Sudarmaji S (1998) An experimental field study to evaluate a trap-barrier system and fumigation for controlling the rice field rat, *Rattus argentiventer*, in rice crops in West Java. Crop Prot 17:55–64

Singleton GR, Chambers LK, Spratt DM (1995) An experimental field study to examine whether *Capillaria hepatica* (Nematoda) can limit house mouse populations in Eastern Australia. Wildl Res 22:31–53

Singleton GR, Hinds L, Leirs H, Zhang Z (1999) Ecologically-based management of rodent pests. ACIAR, Canberra

Singleton GR, Farroway LN, Chambers LK, Lawson MA, Smith AL, Hinds LA (2002) Ecological basis for fertility control in the house mouse (*Mus domesticus*) using immunocontraceptive vaccines. Reproduction Suppl 60:31–39

Singleton GR, Sudarmaji, Jacob J, Krebs CJ (2005) Integrated management to reduce rodent damage to lowland rice crops in Indonesia. Agricult Ecosyst Environ 107:75–82

Spratt DM (1990) The role of helminths in the biological control of mammals. Int J Parasitol 20:543–550

Spratt DM (2005) Australian ecosystems, capricious food chains and parasitic consequences for people. Int J Parasitol 35:717–724

Spratt DM, Singleton GR (1986) Studies on the life-cycle, infectivity and clinical effects of *Capillaria hepatica* (Bancroft) (Nematoda) in mice, *Mus musculus*. Austral J Zool 34:663–675

Story G, Berman D, Palmer R, Scanlan J (2004) The impact of rabbit haemorrhagic disease on wild rabbit (*Oryctolagus cuniculus*) populations in Queensland. Wildl Res 31:183–193

Swinton J, Harwood J, Grenfell BT, Gilligan CA (1998) Persistence tresholds for phocine distemper virus infection in harbour seal *Phoca vitulina* metapopulations. J Anim Ecol 67:54–68

Thulke HH, Tischendorf L, Staubach C, Selhorst T, Jeltsch F, Muller T, Schluter H, Wissel C (2000) The spatio-temporal dynamics of a post-vaccination resurgence of rabies in foxes and emergency vaccination planning. Prev Vet Med 47:1–21

Torchin ME, Mitchell CE (2004) Parasites, pathogens, and invasions by plants and animals. Front Ecol Environ 2:183–190

Torchin ME, Lafferty KD, Dobson AP, McKenzie VJ, Kuris AM (2003) Introduced species and their missing parasites. Nature 421:628–630

Tyndale-Biscoe CH (1994) Virus-vectored immunocontraception of feral mammals. Reproduct Fertil Develop 6:281–287

Appendix. Macroparasites of small mammals that are relevant for humans or livestock

Macroparasites	Micromammals		Relevance for humans or livestock
	definitive hosts	intermediate hosts	
Trematoda			
Alaria spp.	mustelids, canids	rodents as paratenic hosts of metacercariae	occasional infection in domestic carnivores; human rare paratenic host of metacercariae
Brachylaima cribbi	house mouse		infection in humans
Dicrocoelium dendriticum	hares, rabbits, large rodents		parasitosis in sheep, goats
Euryhelmis squamula	shrews, mustelids		infection in mink
Opisthorchis viverrini	viverrids, felids		rare infection of cats and man
Clonorchis sinensis	mustelids		infection in humans
Nanophyetus salmincola	mustelids, canids, felids		*N. salmincola* is a vector for ricketsial fever agents to dogs and man
Fasciola hepatica	hares, rabbits, rodents		fascioliasis in sheep, cattle, humans
Echinostoma spp.	Norway rat, mustelids		infection in humans
Isthmiophora melis	mustelids, hedgehog		infection in mink
Paragonimus spp.	mustelids, viverrids, large rodents		infection in humans
Schistosoma japonicum, S. mattheei, S. mansoni	rodents		schistosomiasis
Schistosomatium sp.	small rodents, shrews, mustelids		cercaria cause dermatitis in humans

Cestoda			
Rodentolepis (= *Hymenolepis*) *microstoma*, *R. nana*, *R. diminuta*	rodents		infections in humans
Taenia pisiformis, *T. taeniaeformis*, *T. brauni*, *T. serialis*, *T. multiceps*	carnivores	lagomorphs, rodents	infection in dogs, cats, humans
Echinococcus multilocularis	canids	microtine rodents	human alveolar echinococcosis
Echinococcus vogeli	bush dogs	rodents	infection in humans
Dipylidium caninum	canids		infection in humans
Spirometra erinacei	canids, felids		infection in humans
Inermicapsifer arvicanthidis	rodents		infection in humans
Nematoda			
Toxascaris leonina	canids, felids	rodents, lagomorphs	ascariasis in dogs and cats rare in humans
Toxocara canis, *T. felis*	canids, felids	rodents as paratenic hosts	infection in dogs and cats; visceral and ocular "larva migrans" in humans
Baylisascaris procyonis	racoons	rodents	infection in humans
Ancylostoma duodenale	carnivores but only rarely		human hookworm; other *Ancylostoma* spp.may cause cutaneous "larva migrans" in humans
Uncinaria spp.	canids, felids		hookworm infection in dogs
Aelurostrongylus abstrusus	felids	rodents may act as transport hosts after eating infected snails	infection in cats
Angiostrongylus cantonensis, *Angiostrongylus* spp.	rodents, shrews, mustelids		meningitis in humans, paralysis in dogs
Gnathostoma spinigerum	canids, felids, mustelids,		erratic parasite (larval) under the skin of man - "Creeping eruption" and meningitis

Brugia sp.	viverrids, pangolins, tupaias, small primates		infection in humans
Trichinella spp.	rats		trichinosis in pigs or humans
Calodium hepaticum (=*Capillaria hepatica*)	rodents		rarely, heavy infection in humans or domestic animals
Haycocknema perplexum	dasyurids?		rare but can be fatal in adult humans
Acari			
Ornithonyssus bacoti	rodents		vector for several viruses and bacteria
Allodermanyssus sanguineus	rodents		vector for several viruses and bacteria
Echinolaelaps echidnus	rodents		vector for several viruses and bacteria
Argas sp.	some species on bats		vector for several viruses and bacteria
Ixodidae (various ticks of the genera *Ixodes*, *Amblyomma*, *Dermacentor*, *Haemaphysalis*, *Hyalomma*, *Rhipicephalus*)		larvae feed on various small mammals	adults feed on domestic carnivores, cattle, humans; vectors for several viruses and bacteria
Trombicula spp.		larvae feed on various small mammals	adults feed on domestic carnivores, cattle, humans; vectors for several pathogens
Sarcoptes scabiei	various small mammals		mange
Leptotrombidium deliensec	rodents, bandicoots		scrub typhus (*Orientia tsutsugamushi*) - humans
Insecta			
Triatoma spp.	armadillos		vectors for *Trypanosoma cruzi*
Ceratophyllus (*Nosopsyllus*) spp.	rodents		vectors for *Yersinia pestis*
Xenopsylla spp.	rodents		vectors for *Yersinia pestis*, *Rickettsia typhi*
Ctenocephalides felis	felids, opossums		vector for *Rickettsia*

			felis, *Bartonella henselae*
other Siphonaptera	rodents		vectors for *Yersinia pestis*
Mosquitoes	rodents		vector for *Plasmodium* sp., filariids
Glossina sp.	rodents		vectors for *Trypanosoma* sp.

27 Biological conservation and parasitism

Philippe Christe, Serge Morand and Johan Michaux

1 Introductory remarks

1.1 Threats to biodiversity

Expansion of the human population has inexorably led to the destruction and degradation of ecosystem diversity with the consequence of a biological diversity crisis on the Earth. The example of the thylacine (*Thylacinus cynocephalus*) in Tasmania is among the best known cases of recent mammal extinctions. What is less recognized is the fact that in addition to habitat deterioration and hunting, a disease could have played a central role in this extinction (Guiler 1961). Another example is the white-tailed deer (*Odocoileus virginianus*). During the last century, the range of the white-tailed deer has expanded greatly in North America as a result of human forestry activities. During the colonisation of new territories, white-tailed deer was a carrier for meningeal worm (*Parelophostrongylus tenuis*). Whereas meningeal worm was not virulent for white-tailed deer, moose (*Alces americana*) and woodland caribou (*Rangifer tarandus*) were highly susceptible to the nematode and succumbed to neurologic diseases (Anderson 1972). This demonstrates that habitat deterioration may promote range expansion of one host species with a parasite that may function like a biological weapon against its potential competitors. These two cases, among others, highlight the importance of taking into account diseases and parasites when studying the causes of the decline of threatened populations.

This chapter investigates how parasites are involved and interact with the main causes of population declines and also emphasizes the positive roles that parasites may play in the maintenance of biodiversity.

1.2 Do we need to conserve parasite species?

Pandas, tigers, right whales and gorillas are emblematic and charismatic species worldwide and a consensus exists for the need of their conservation. The same is true for economically important species such as salmonids and sturgeons. But do we really need the conservation of parasitic species? Not really! During our childhood, we all have favourite stuffed animals (representing one of these emblematic species) we take to sleep. However, do you ever see a child squeezing a stuffed worm, flea or tick? The lack of affect for cryptic species and the perception of parasites as disgusting creatures among the public certainly leads to a disinterest among governmental and conservation agencies to preserve them. Fortunately, in 1990, Donald A. Windsor expressed his concern for this matter in the famous slogan: "Equal Rights for Parasites!" (Windsor 1990). Five years later he pleaded once more for parasite conservation in an editorial that appeared in "Conservation Biology" (Windsor 1995). During the same time, the ominous term "co-extinction" was proposed to characterize the dual extinction of a host and its specialized parasite (Stork and Lyal 1993). Despite the passing of 15 years since these passionate declarations and the exponential increase of an interest in conservation biology, we can point out that parasitic species are far from being in a leading position among current conservation priorities. Very few parasites are listed on the IUCN Red List of Threatened Species (IUCN, 2003; Whiteman and Parker 2005). Some parasite extinctions have been even intentionally provoked as revealed by the will to remove parasites from hosts in captive breeding programs (Stork and Lyal 1993).

To convince resource managers that parasites are an important component of all ecosystems, the following arguments, which mainly rely on their potential utilitarian effects, are advocated by the parasites' defenders. First, parasites are living organisms and are *de facto* part of biodiversity. They shape community structure by reducing competitive abilities and vulnerability to predation of their hosts and have strong impact on ecosystem functioning (Hudson 2005). Moreover, parasites could maintain biodiversity by mediating competitive interactions between different members of an ecosystem. Because the rate of molecular evolution is usually faster in parasite DNA than that within the homologous loci of their hosts (Moran et al. 1995; Nieberding et al. 2004), the study of the evolution of parasite DNA sequences could provide valuable information on past population dynamics, evolutionary history and current demographic processes of endangered hosts (Whiteman and Parker 2005). Parasites could thus be used as a biological "magnifying glass" (Nieberding et al. 2004). Another utilitarian effect of parasites is their potential use as indicators of environmental quality and ecosystem health (Marcogliese 2005). Indeed, parasites may be used as accumulation indicators of heavy metal contamination, particularly in aquatic ecosystems (Sures et al. 1999). In addition, parasite species and composition revealed perturbations in ecosystem structure and function (Marcogliese 2005). Furthermore, the use of parasite in human medicine is a new promising field of investigation, as illustrated by the use of helminths as therapeutic agents for inflammatory disease (Hunter and McKay 2004).

2 Parasite resistance and stress

2.2 Environmental stress and parasite susceptibility

Wild animals in their natural habitat have to cope not only with predictable environmental changes such as the cycles of seasons and their associated modifications in resource availability and temperature but also with unpredictable events such as catastrophes, spread of new diseases and human disturbances. Whereas animals react adaptively by behavioural and physiological modifications to predictable changes, unpredictable disturbances may have negative effects on population dynamics of living organisms. Increasingly rapid disappearing and fragmentation of habitats, which may be considered as unpredictable environmental changes, translates in a cascade of negative effects and can result in physiological stress on animals (Suorsa et al. 2003, 2004). The first physiological responses of an animal to stressful stimuli include cardiovascular effects and a hormonal response involving synthesis and secretion of glucocorticosteroids (Romero 2004). Consequently, a corticosteroid response might be a good indicator of a stress response (Hofer and East 1998). It is, therefore, not surprising that corticosteroid level is measured in many studies in ecology and conservation biology that have evaluated the effect of different environmental and human perturbations on the stress level of wild animals (Creel et al. 1997; Creel et al. 2002; Mostl and Palme 2002; Romero 2004; Palme et al. 2005). The consequences of a stress response on parasite resistance are complex and alter host immunocompetence in different ways (Apanius 1998). The immune system appears to be down regulated under stressful environmental conditions (von Holst 1998), particularly under severe chronic stress with prolonged periods of high cortisol concentrations (Mostl and Palme 2002). Stress stimuli may arise due to different factors in a perturbed environment. Habitat fragmentation may be related to chronic food shortage (Zanette et al. 2000). Thus degradation of environmental conditions may decrease resource availability that in turn affects body condition and immune defences (Chandra and Newberne 1977; Klasing 1998; Christe et al. 2003). As body condition is usually positively correlated with immune defences (Møller et al. 1998), individuals with poor body condition will be especially vulnerable to attacks of parasites (Christe et al. 1998; Christe et al. 2000). Edge effect due to fragmentation may also be a source of stress because predators may have easy access to dense forest patches which were previously inaccessible. It has been experimentally demonstrated that exposure to predators reduced the ability of hosts to cope with parasitism mediated through effects on immune function (Navarro et al. 2004). Consequently parasitism may be favoured in fragmented habitat through the effect of predators. Thus, parasites, which can also be considered as an environmental stressor, may reinforce the effect of habitat degradation and participate in the reduction of a population.

In addition to habitat degradation and fragmentation, anthropogenic factors such as environmental pollution, hunting, tourism and leisure activities exert a negative pressure on wildlife and are thought to cause stress (Fowler 1999; Mullner et al. 2004). Clearly, more studies are needed to investigate the relationship between anthropogenic factors, level of stress and parasitism in endangered populations.

3 Parasitism in isolated and declining population

3.1 Review of theory on parasitism and extinction risks

Conservation biology deals with two major paradigms: population invasion and population decline. Both are related to each other (i.e. decline may be a result of an invasion) and both emphasize the potential roles of parasites and/or pathogens (Prenter et al. 2004).

Theoretical, experimental and empirical studies have established clearly that parasites play important roles in regulating population dynamics (Scott 1987; Scott and Dobson 1989; Albon et al. 2002; Rosa et al. in this volume) and structuring free-living communities (Minchella and Scott 1991; Morand and Arias Gonzalez 1997; Hudson and Greenman 1998; Tompkins et al. 2001). Parasites then have a large impact on biological conservation (Dobson and May 1988; McCallum and Dobson 1995; Sasal et al. 2000), as parasites and pathogens may compromise reintroduction or translocation programs (Viggers et al. 1993). They may have a higher impact on threatened species generally characterized by a lower level of genetic variability, particularly on genes associated to immune system (Hedrick, 2003).

3.1.1 Threshold in host-parasite population dynamics

Small-sized host populations may be prone to extinction due to stochastic events. Several processes, including the Allee effect (Stephens and Sutherland 1999; see below), may increase the probability of extinction of small populations. These processes operate when host population size decreases to under a critical or threshold level, below which populations are almost doomed. Population viability analysis is one approach that has been developed for management purposes of small-sized endangered populations.

Threshold size has been also extensively studied in the case of host-parasite dynamics (Dobson 1989). The basic reproductive number, R_0, is the major concept in host-parasite population dynamics. This quantity is defined as the number of new infections occurring after introduction of one parasite, or one infected host, into a naïve and susceptible host population. R_0 is positively linked to host density in the case of direct-transmitted parasites (see Rosa in this book). Parasites, or infection, can spread in the population when $R_0>1$ and as R_0 depends on host density, the condition of parasite invasion corresponds to a case when the density of host population exceeds a threshold density. Obviously, host-parasite dynamics are viewed in terms of parasite invasion or parasite invisibility. The task of disease management is then to decrease R_0 below one, i.e., below the threshold density.

The interplay between host and parasite thresholds has not been considered adequately. Deredec and Courchamp (2003) emphasized the importance of the relative position of the host and parasite thresholds: when the parasite threshold is higher than that of the host, the parasite is driven to extinction and the host population is relieved of its parasite; when the host population threshold is higher than that of the parasite, the host is driven to extinction while the parasite continues to exert strong pressure on the host until it reaches its own threshold. Hence, mathematical epidemiology and population dynamics are important tools for investigating thresholds and persistence of both hosts and parasites. They may help in determining the conditions to maintain a high level of parasite threshold in comparison to the host threshold.

3.2 Which diseases are important for conservation?

Microparasites are generally considered as an important threat in conservation biology (Daszak et al. 2000; Cleaveland et al. 2001). All conservation textbooks refer to the canine distemper virus, rinderpest and the avian malaria as examples of pathogen-driven extinction. Introduced diseases have been implicated in the local extinction of a number of species (McCallum and Dobson 1995; Vitousek et al. 1997) and the global species extinction of Hawaiian birds (VanRiper et al. 1986) and the thylacine (Guiler 1961) among others. Daszak et al (2000), in their review, mentioned 19 microparasites and no macroparasites as important threats for conservation and as zoonotic threats for human health through spill-over. The lack of reference to macroparasites may suggest that they are indeed less important, and that their survey is not of major interest, with the notable exception of ectoparasites (ticks, fleas) because of their roles as vectors of numerous virus, bacteria and protozoans.

Moreover, results based on comparative analyses in carnivores show that host species that live in low density within a restricted geographic area experience low parasitic pressure in terms of parasite species diversity, suggesting that parasites may not represent a particularly important risk for these host species (Torres et al. 2006). In contrast, widespread host species that live in high density are exposed to a wide range of parasite species that may affect drastically the population dynamics of these carnivores, suggesting that macroparasites may regulate them at least locally. These results lead to the paradox that parasite infection, and particularly that of macroparasites, is less crucial for small and isolated populations than for large populations. This paradox is apparent and resolved by considering the investment in immune defences, which is directly related to the prevalence and/or diversity of parasites as a mean to control infection (Martin et al. 2001). Evidence comes again from comparative studies, which suggest that hosts allocate their investment in immune function as a function of their probability of exposure to parasites (Møller and Legendre 2001; Møller et al. 2005). Large populations are composed of highly immunocompetent individuals and small populations of low immunocompetent ones. Hence, parasites and pathogens are threats to small and isolated populations because of poor performance of their immune system against pathogen introduction, but parasites (and parasite diversity) are probably necessary to maintain high levels of immune defence, which helps against new pathogens.

3.3 Allee effect

The Allee effect may be defined as "a positive relationship between any component of individual fitness and either numbers or density of conspecifics" (Stephens et al. 1999). The beneficial effects of conspecifics not only include antipredator vigilance, predator dilution, social thermoregulation, reduction of inbreeding but also social facilitation of reproduction through helpers (Stephens and Sutherland 1999; Courchamp et al. 2000). When population size reaches a low density, animal species that are subject to an Allee effect will suffer from a reduction in some aspects of their fitness that in turn will affect negatively growth rate of populations. Because of their potential role in extinctions of declining species, the Allee effects have thus become much studied in conservation biology (Stephens and Sutherland 1999; Lafferty and Gerber 2002). Interestingly, Allee effects and parasitism have several features in common that are of interest when studying population dynamics in conservation biology (Deredec 2005). For example, theoretical models demonstrated the importance of host density in the probability for a parasite to become established in a host population (see above) and empirical studies have shown a positive relationship between host sociality or density and parasite prevalence and intensities (Anderson and May 1978; Brown and Brown 1986; Møller et al. 1993; Stanko et al. 2002; Altizer et al. 2003). Thus, animal species that aggregated as a behavioural response to the strong Allee effects, would be more prone to suffer the negative effects of parasites. Parasite species may also be subject to the Allee effects that influence the occurrence and the severity of epidemics as illustrated by patchy distributions of worms in hosts as a result of the necessity for female worms to find a mate in order to reproduce (Cornell et al. 2004).

It has been suggested that sexual selection, in particular female mate preferences, could lead to an Allee effect (Møller and Legendre 2001). If only males of poor quality are available for mating in a small population, females may refrain from reproduction or reproduce at a low rate. As a consequence of mating with a male of a non-preferred phenotype, females could decrease their parental investment resulting in poor reproductive success (Møller and Legendre 2001). Parasite-mediated sexual selection has been the focus of numerous studies since the influential hypothesis of Hamilton and Zuk (1982). A meta-analysis of the available studies related to this topic has revealed a negative relationship between parasite load, immunocompetence and the expression of male secondary sexual characters (Møller et al. 1999). Thus, parasites, by decreasing the expression of male secondary sexual characters, may contribute and reinforce the potential Allee effects created by sexual selection.

4. Invasive species and parasites

4.1 Parasite mediated competition

Mediation of competition by parasites is one mechanism of parasite interference (Anderson 1972; Hudson and Greenman 1998; Poulin 1999). Parasite-mediated competition is inferred when two different host species have different susceptibilities to the same non-specific parasite species. The presence of a given host species may decrease the fitness of the other host species simply by transmitting a pathogen to the more vulnerable host species (Hudson and Greenman 1998). The abundance of the more vulnerable host to the parasite is then depleted, potentially under the host threshold. Moreover, as the parasite infects two host species, the parasite threshold is obviously low. This "apparent" competition, mediated via a shared pathogen, differs from the classical competition for limited resources. Strong evidence of this competition was obtained not only from experiments but also from the field (Tompkins et al. 2000), e.g. red and grey squirrels in England (Tompkins et al. 2002b) and pheasant and grey partridge in England (Tompkins et al. 2002a). Parasite-mediated competition may operate for introduced host species, as they can be best competitors simply by introducing and transmitting a new parasite to the naïve native species. This can lead to a non-fit combination that can be more pathogenic (Hudson and Greenman 1998; Prenter et al. 2004). In this case, the invader uses parasites as biological weapons. Immune-naïve residents will be weakened or even killed by the new pathogens.

The most famous example of such a process comes from the history of the expansion of European humans through America where million of native people were killed by the influenza and other pathogens that accompanied conquistadors.

4.2 Parasite release hypothesis

Parasite mediated competition is not the only way by which parasites may interfere in competition processes. Recently, it was shown that many introduced species lost most of their parasites from their native habitats when introduced to new ones (Torchin et al. 2002; Torchin et al. 2003). This could be responsible for the demographic explosion of some introduced species, formulated as the "parasite release hypothesis".

The parasite release hypothesis was proposed as an ecological mechanism to explain the success of introduced species. As the introduced species lose their parasites when invading new habitats, they have a competitive advantage over local species. Mitchell and Power (2003) and Torchin et al. (2003) found that parasitism is significantly reduced in organisms in their introduced range, supporting the "parasite release hypothesis". One cause to explain that invaders may leave behind their parasites is that many parasites have complex life cycle stages with more than one host. If one of those hosts is absent in the new colonized area, the life-cycle of the parasite would be disrupted.

4.3 Immunity

In the invasion process, invasive host species harbouring a high diversity of parasites in their native habitat have advantages as they lose a large number of parasites and pathogens (see above). Invasive host species have another advantage if they have invested in strong immune defences in their natural range, which may then subsequently confer a better capacity to control parasites that they may acquire in the introduced habitat. Hosts having evolved strong immune defences are prime candidates for successful invasion (and also more resistant towards invaders). This hypothesis was proposed in the case of introduced plants and recently for the case of introduced animals (Lee and Klasing 2004; Møller and Cassey 2004). In contrast, hosts that are exposed to a low diversity of parasites may invest less in immune defences. Maintaining a strong immune system for threat ened host species, or for individual hosts maintained in captivity in the view of reintroduction, is a new task for conservation biologists.

5 Conservation genetics and parasites

5.1 Genetic diversity and pathogen resistance

Habitat fragmentation and its degradation is probably one of the main factors leading to the disappearance of many species. Indeed, it often leads to a decrease in population sizes as well as to the appearance of barriers to gene flow between isolated populations. The small populations that result from this fragmentation often suffer from reduction of genetic diversity associated with genetic drift and inbreeding effects. This loss of genetic variation can result in a rapid reduction of fitness (lower possibility to adapt to long term changes in environment, poor reproductive ability associated with a lower sperm quality, higher juvenile mortality, lower general survival, etc) (O'Brien 1994). Several recent studies (Cassinello et al. 2001; Keller and Waller 2002; Spielman et al. 2004) also showed that populations with a low genetic variability are generally more susceptible to infectious viruses, bacteria and other pathogens. The case of the cheetah (*Acinonyx jubatus*) is probably one of the best known concerning this phenomenon., The two major subspecies of cheetah (*A. jubatus jubatus* from southern Africa and *A. jubatus raineyi* from eastern Africa) display markedly reduced levels of genetic variability compared to other mammal species (O'Brien 1994). This would result in intensive inbreeding. When a breeding colony of this species was contaminated by feline infectious peritonitis (FIP) in Oregon state (USA), 100% of the captive animals showed morbidity symptoms and 60% of them died (O'Brien 1994). In contrast, in domestic cats, the mortality incidence of this virus is very rare (around 1%). According to O'Brien (1994), the high sensitivity of this cheetah colony to the FIP would be directly linked to the very low (almost monomorphic) level of variation of the Major Histocompatibility (MHC) genes characterising the cheetah.

A wide variety of gene classes (where the MHC is the most notable but see also the eosinophil-associated RNase (EARs) genes, the tumor necrosis factor gene promoter, the interleukine receptor or the γ-interferon receptors; Hill 1998; Zhang et al. 2000) are normally variable in natural populations and could contribute to disease resistance. MHC genes encode cell-surface glycoproteins, binding antigens derived from pathogens and parasites and constitute the most polymorphic genes in vertebrates (Parham 1999; Charbonnel et al. in this volume). They present antigens to T-lymphocytes which develop the appropriate immune responses. Two major groups of MHC genes are recognised: the MHC class I genes are specific to the immune defence against intracellular pathogens by binding peptides mainly derived from viral proteins or cancer infected cells. The MHC class II genes present with T-lymphocytes, peptides essentially derived from extra-cellular parasites (bacteria, nematodes, cestodes, etc.). The variability of MHC genes is correlated with the diversity of the T-lymphocyte receptors, which, in turn, determine the resistance of an organism to pathogens and parasites (Parham 1999).

Therefore, the cheetah, with its very low variability of MHC genes, is not well protected against the FIP and probably against many other pathogens and therefore is at a high risk of extinction. However, according to several recent studies, several processes would help to maintain high levels of MHC genes diversity. Indeed, these studies demonstrated that the antigen binding sites (ABS) display more non-synonymous than synonymous substitutions compared to what would be observed under neutral theory (in this condition, the rate of synonymous substitution is predicted to be larger than the rate of non-synonymous substitution as the latter change the amino acid composition and would be likely deleterious) (Sommer 2005). This phenomenon cannot be explained by higher mutation rates in this region (Hughes and Yeager 1998) and the hypothesis accepted at present is that this particular nucleotide diversity in MHC genes would be the result of balancing selection. This would allow the maintenance of large numbers of alleles in populations and also the persistence of allelic diversity over long periods of time. Following this strategy, the binding of a large set of antigens would be possible.

Two main types of balancing selection have been proposed to explain high levels of genetic diversity in MHC genes of vertebrates:

- "Overdominance" strategy (Hedrick 1998; Richman 2000), where the heterozygotes are expected to have higher fitness than parental homozygotes as the latter will carry less divergent allelic sequences and, therefore, will have less chance to resist a large panel of antigens and/or multiple types of pathogens and parasites.
- "Frequency dependent selection" strategy (Hedrick 1998). This occurs when an allele or genotype is favoured at one frequency, but disadvantaged at another frequency. This hypothesis is based on the fact that host-parasite dynamics is considered as a co-evolutionary race. Pathogens adapt to infect the most common genotype, leaving rare genotypes least infected. If alleles are favoured when they are rare, but selected against when they are common, this will result in a balanced polymorphism (Sommer 2005)

Different studies confirmed the effect of balancing selection on the high MHC diversity. One of the best examples concerns the Nicolas Island fox (*Urocyon littoralis dickeyi*) (Aguilar et al. 2004). On the basis of different neutral markers (microsatellites, minisatellites and allozymes), this species is considered as one of the most monomorphic among sexually reproducing species. Regarding the low variability of these markers, this species would have many problems of fitness as well as low resistance to pathogens. However, it is characterised by a surprising high level of MHC diversity which makes it much more resistant to what could be expected. This observation is interpreted as being the result of intense periodic balancing selection at the MHC which may have allowed the persistence of variation within this species despite strong genetic drift.

5.2. Inbreeding, MHC and risk of extinction

Under some circumstances (for example, particular historical events such as bottlenecks or founder effects), strength of selection acting on MHC genes can be insufficient to maintain variation in small or fragmented populations over a long period of time (Sommer 2005). In these cases, the power of genetic drifts can be stronger than the power of selection. This can lead to a loss of genetic diversity not only on the neutral markers but also on the MHC genes. This would explain the very low genetic variability in highly threatened species such as the cheetah (*Acinonyx jubatus*) (see above), the Asian lion (*Pantera leo persica*) (O'Brien, 1994), the common hamster (*Cricetus cricetus*) in the Netherlands (Smulders et al. 2003), the Scandinavian beaver (*Castor fiber*) (Ellegren et al. 1993), the Northern elephant seal (*Mirounga angustirostris*) (Hoelzel et al. 1999) and the Scandinavian moose (*Alces alces*) (Ellegren et al. 1996). Under these circumstances, threatened species present a high risk of extinction as they can be very sensitive to new diseases and changes in environment.

However, other studies demonstrated that endangered species such as the Przewalski's horse (*Equus przewalski*) (Hedrick et al. 1999), the Arabian oryx (*Oryx leucoryx*) (Hedrick et al. 2000) and the Malagasy giant jumping rat (*Hypogeomys antimena*) (Sommer 2003) are characterised by a low number of MHC alleles but which are separated by a high level of nucleotide and amino acid divergence. Analysis at the ABS showed that non-synonymous substitutions were higher than synonymous ones, suggesting selection leading to an increase of amino acids changes in the ABS region and thus to higher divergences between MHC alleles (Sommer 2003). These studies indicated that other selection processes are able to maintain some MHC polymorphism (not on the number of alleles but rather on the genetic difference between the existing alleles) even in species surviving bottlenecks. This would be sufficient to prevent immediate pathogen-induced declines. However, such kind of adaptive processes to changing conditions is probably limited and does not predict the outcome effects of introduced pathogens, which differ from commonly encountered diseases. Probably, the maintenance or even renewal of variation in functional important regions of the MHC, either from mutation, recombination or immigration from other populations, would be an important genetic component to allow an appropriate immune response (Sommer 2005). However, too strong genetic bottlenecks, leading to important inbreeding depressions, do not permit such kind of processes to operate and this explains why some species like the cheetah or the Asian lion are so sensitive nowadays to diseases.

6. Management

6.1 Breeding program and risk of parasite transmission

As mentioned by McCallum and Dobson (1995), "diseases and parasites pose particularly severe problems in captive populations, in which animals are held at high density, may be stressed and may be exposed to cross-species transmission". During the last 20 years, a great amount of zoos worldwide have participated in the management of endangered species., Many threatened species have captive populations that act as insurance against extinction in the wild and, indeed, captive breeding programs have saved some endangered species from extinction (e.g. Père David's deer, European bison, etc) (Frankham et al. 2002). Because parasites may have negative effects on their host, veterinarians in zoos take great care to reduce or even to remove entirely parasite loads on captive animals. As the ultimate goal of breeding programs in zoos is to increase threatened populations or to reintroduce individuals into the wild, parasites play an important role. What could be the consequences of maintaining hosts during many generations in a parasite-free environment? The potential risk is to releasc into the wild individuals that have lost their defences against pathogens and diseases. Once in the wild, they will be in contact with a vast array of parasite species and may be unable to resist to their detrimental effects. Maintaining some parasites on individual hosts kept in captivity could be a way to solve part of this problem.

6.2 Beneficial effects of parasites and parasite conservation in captive breeding program

Macroparasites, because of chronic infections, have evolvcd several kinds of immune evasion strategies (Charbonnel et al. in this volume). Some strategies of immunomodulation displayed by many macroparasites may have some beneficial effects on their hosts by regulating Th1/Th2 cytokine responses (Weinstock et al. 2004). Th1 responses induce inflammatory cell activity to control intracellular infections while Th2 responses drive humoral immune responses to control extra-cellular parasites (see Weil et al. in this volume).

Mice with helminths have blunted Th1 responses while helminths promote Th2 responses associated with production of interleukin 4 (IL-4), which helps impede Th1 cell differentiation. Thus, induction of IL-4 could underlie the alterations seen in host immunity (i.e. high inflammatory activities). Helminths also appear to protect the host from aberrant Th2 diseases such as asthma and food allergy (Weinstock et al. 2004), and there is now an immunological basis for protection by helminths. Human epidemiological data and several animal studies support the notion that helminths protect the host from immunological disease (Elliott et al. 2005), particularly those caused by the activation of the Th1 response by microparasites. For example, helminths protect mice and rats from experimental autoimmune encephalomyelitis, as welll as other diseases of immunity. Thus, natural exposure to helminths may guard animals from developing severe immunological diseases, suggesting that helminths should be useful in conserving both endangered and captive species.

Gompper and Williams (1998) proposed a series of measures to maintain endangered parasite species originating from threatened hosts in captive breeding program. However, they pointed out that because most of the public disapprove of protecting parasite species, attempts to conserve unique species of parasites could result in a hostile public response against efforts to preserve hosts. Therefore they proposed a series of measures aimed to save parasite species without damaging attempts to conserve hosts. One of those measures was to find alternative hosts to maintain parasite populations for potential reintroduction once the host population was restored. However, the problem of parasite conservation concerns mainly highly host-specific parasites. To find alternative hosts on which specialist parasite populations would be viable may be a difficult task because experiments on cross-species infection have shown a strong decrease on both parasite survival and reproductive success on the foreign host, even if this new host species belongs to the same host genus (Giorgi et al. 2004).

7 Concluding remarks

The consequence of human population growth is closer contact between human and reservoir hosts of numerous diseases. The spread of disease to endangered wildlife species due to contact with humans and domestic animals, and vice versa, increases as humans and their domestic animals get in more contact with these species due to habitat fragmentation. Emergence of new diseases and particularly those from small mammals such as rodents or bats are of great public health concern (Leroy et al. 2005). Conservation medicine, a new theme within the field of conservation biology, has been viewed as the application of medicine to improve the conservation of wildlife and ecosystems (Aguirre et al. 2002). Conservation medicine, according to Otsfeld et al. (2002) is "devoted to understanding the interactions among human-induced and natural changes in (1) climate, habitat and land use; (2) pathogens, parasites, and pollutants; (3) biodiversity and health within animal communities; (4) health of humans" (Ostfeld et al. 2002). The 2005 "*Anus horribilis*" for bats worldwide illustrates the importance of this new field of investigations. While it was discovered in China that bats are the reservoir of SARS virus (Lau et al. 2005; Li et al. 2005), it was found in Africa that they are probably the reservoir for Ebola virus (Leroy et al. 2005).

We strongly hope that this chapter will convince ecologists and conservation biologists that pathogens and parasites, mostly investigated by veterinarians and physicians, should not be ignored or eradicated because of their crucial importance to wild and domestic animals and humans.

References

Aguilar A, Roemer G, Debenham S, Binns M, Garcelon D, Wayne RK (2004) High MHC diversity maintained by balancing selection in an otherwise genetically monomorphic mammal. Proc Natl Acad Sci USA 101:3490–3494

Aguirre AA, Ostfeld RS, Tabor GM, House C, Pearl MC (2002) Conservation medicine. Ecological health in practice. Oxford Univ Press, Oxford

Albon S, Stien A, Irvine RJ, Ropstad R, Halvorsen O (2002) The roles of parasites in the dynamics of a reindeer population. Proc R Soc Lond B 269:1625–1632

Altizer S, Nunn CL, Thrall PH, Gittleman JL, Antonovics J, Cunningham AA, Dobson AP, Ezenwa V, Jones KE, Pedersen AB, Poss M, Pulliam JRC (2003) Social organization and parasite risk in mammals: Integrating theory and empirical studies. Ann Rev Ecol Evol Syst 34:517–547

Anderson RC (1972) The ecological relationships of meningeal worm and native cervids in North America. J Wildl Dis 8:304–310

Anderson RM, May RM (1978) Regulation and stability of host-parasite population interactions. 1. Regulatory processes. J Anim Ecol 47:219–247

Apanius V (1998) Stress and immune defense. Adv Study Behav 27:133–153

Brown CR, Brown MB (1986) Ectoparasitism as a cost of coloniality in cliff swallows (*Hirundo pyrrhonota*). Ecology 67:1206–1218

Cassinello J, Gomendio M, Roldan ER (2001) Relationship between coefficient of inbreeding and parasite burden in endangered gazelles. Cons Biol 15:1171–1174

Chandra RK, Newberne PM (1977) Nutrition, immunity, and infection. Plenum, New York

Christe P, Møller AP, DeLope F (1998) Immunocompetence and nestling survival in the house martin – the tasty chick hypothesis. Oikos 83:175–179

Christe P, Arlettaz R, Vogel P (2000) Variation in intensity of a parasitic mite (*Spinturnix myoti*) in relation to the reproductive cycle and immunocompetence of its bat host (*Myotis myotis*). Ecol Lett 3:207–212

Christe P, Giorgi MS, Vogel P, Arlettaz R (2003) Differential species-specific ectoparasitic mite intensities in two intimately coexisting sibling bat species: Resource-mediated host attractiveness or parasite specialization? J Anim Ecol 72:866–872

Cleaveland S, Hess GR, Dobson AP, Laurenson MK, McCallum HI, Roberts MG, Woodroffe R (2001) The role of pathogens in biological conservation. In: Hudson PJ, Rizzoli A, Grenfell BT, Heesterbeek H, Dobson AP (eds) The ecology of wildlife diseases. Oxford Univ Press, New York, pp 139-150

Cornell SJ, Isham VS, Grenfell BT (2004) Stochastic and spatial dynamics of nematode parasites in farmed ruminants. Proc R Soc Lond B 271:1243–1250

Courchamp F, Clutton-Brock T, Grenfell B (2000) Multipack dynamics and the Allee effect in the African wild dog, *Lycaon pictus*. Anim Cons 3:277–285

Creel S, Creel NM, Monfort SL (1997) Radiocollaring and stress hormones in African wild dogs. Cons Biol 11:544–548

Creel S, Fox JE, Hardy A, Sands J, Garrott B, Peterson RO (2002) Snowmobile activity and glucocorticoid stress responses in wolves and elk. Cons Biol 16:809–814

Daszak P, Cunningham AA, Hyatt AD (2000) Wildlife ecology – Emerging infectious diseases of wildlife – Threats to biodiversity and human health. Science 287:443–449

Deredec A (2005) De la menace à l'outil: Modélisation de l'impact du parasitisme en biologie de la conservation. Ph. D. Thesis, Paris-Sud University, Orsay

Deredec A, Courchamp F (2003) Extinction thresholds in host-parasite dynamics. Ann Zool Fenn 40:115–130

Dobson AP (1989) The population biology of parasitic helminths in animal populations. In: Levin SA, Hallam TG, Gross LJ (eds) Applied mathematical ecology. Biomethematic 18, SpringerVerlag, Berlin, pp 145–175

Dobson AP, May M (1988) Restoring island ecosystems: The potential of parasites to control introduced mammals. Cons Biol 2:31–39

Ellegren H, Hartman G, Johansson M, Andersson L (1993) Major histocompatibility complex monomorphism and low levels of DNA fingerprinting variability in a reintroduced and rapidly expanding population of beavers. Proc Natl Acad Sci USA 90:8150–8153

Ellegren H, Mikko S, Wallin K, Andersson L (1996) Limited polymorphism at major histocompatibility complex (MHC) loci in the Swedish moose *A. alces*. Mol Ecol 5:3–9

Elliott DE, Blum A, Metwali A, Ince NM, Setiawan T, Winkler S, Wang Y, Urban JF, Summers RW, Weinstock JV (2005) Helminths modulate mucosal immunity and inflammation. Immunology 116:20–20

Fowler GS (1999) Behavioral and hormonal responses of Magellanic penguins (*Spheniscus magellanicus*) to tourism and nest site visitation. Biol Cons 90:143–149

Frankham R, Ballou JD, Briscoe DA (2002) Introduction to conservation genetics. Cambridge Univ Press, Cambridge
Giorgi MS, Arlettaz R, Guillaume F, Nussle S, Ossola C, Vogel P, Christe P (2004) Causal mechanisms underlying host specificity in bat ectoparasites. Oecologia 138:648–654
Gompper ME, Williams ES (1998) Parasite conservation and the black-footed ferret recovery program. Cons Biol 12:730-732
Guiler ER (1961) The former distribution and decline of the thylacine. Austral J Sci 23:207-210
Hamilton WD, Zuk M (1982) Heritable true fitness and bright birds: A role for parasites? Science 218:384–387
Hedrick PW (1998) Balancing selection and MHC. Genetica 104:207–214
Hedrick PW, Parker KM, Miller EL, Miller PS (1999) Major histocompatibility complex variation in the endangered Przewalski's horse. Genetics 152:1701–1710
Hedrick PW, Parker KM, Gutierrez-Espeleta GA, Rattink A, Lievers K (2000) Major histocompatibility complex variation in the Arabian oryx. Evolution 54:2145–2151
Hill AVS (1998) The immunogenetics of human infectious diseases. Ann Rev Immunol 16:593–617
Hoelzel AR, Stephens JC, O'Brien SJ (1999) Molecular genetic diversity and evolution at the MHC DQB locus in four species of pinnipeds. Mol Biol Evol 16:611–618
Hofer H, East ML (1998) Biological conservation and stress. Adv Study Behav 27:405–525
Hudson P (2005) Introduction – Parasites, diversity, and the ecosystem. In: Thomas F, Renaud F, Guégan J-F (eds) Parasitism and ecosystems. Oxford Univ Press, Oxford
Hudson P, Greenman J (1998) Competition mediated by parasites: Biological and theoretical progress. Trends Ecology Evol 13:387–390
Hughes AL, Yeager M (1998) Natural selection at major histocompatibility complex loci of vertebrates. Ann Rev Genetics 32:415–435
Hunter MM, McKay DM (2004) Review article: Helminths as therapeutic agents for inflammatory bowel disease. Aliment Pharmacol Therapeut 19:167-177
Keller L, Waller D (2002) Inbreeding effects in wild populations. Trends Ecol Evol 17:230–241
Klasing KC (1998) Nutritional modulation of resistance to infectious diseases. Poultry Sci 77:1119–1125
Lafferty KD, Gerber LR (2002) Good medicine for conservation biology: The intersection of epidemiology and conservation theory. Cons Biol 16:593–604
Lau SKP, Woo PCY, Li KSM, Huang Y, Tsoi HW, Wong BHL, Wong SSY, Leung SY, Chan KH, Yuen KY (2005) Severe acute respiratory syndrome coronavirus-like virus in Chinese horseshoe bats. Proc Natl Acad Sci USA 102:14040–14045
Lee KA, Klasing KC (2004) A role for immunology in invasion biology. Trends Ecol Evol 19:523–529

Leroy EM, Kumulungui B, Pourrut X, Rouquet P, Hassanin A, Yaba P, Delicat A, Paweska JT, Gonzalez JP, Swanepoel R (2005) Fruit bats as reservoirs of Ebola virus. Nature 438:575–576

Li WD, Shi ZL, Yu M, Ren WZ, Smith C, Epstein JH, Wang HZ, Crameri G, Hu ZH, Zhang HJ, Zhang JH, McEachern J, Field H, Daszak P, Eaton BT, Zhang SY, Wang LF (2005) Bats are natural reservoirs of SARS-like coronaviruses. Science 310:676–679

Marcogliese DJ (2005) Parasites of the superorganism: Are they indicators of ecosystem health? Int J Parasitol 35:705–716

Martin TE, Møller AP, Merino S, Clobert J (2001) Does clutch size evolve in response to parasites and immunocompetence? Proc Natl Acad Sci USA 98:2071–2076

McCallum H, Dobson AP (1995) Detecting disease and parasite threats to endangered species and ecosystems. Trends Ecol Evol 10:190–194

Minchella DJ, Scott ME (1991) Parasitism – a cryptic determinant of animal community structure. Trends Ecol Evol 6:250–254

Mitchell CE, Power AG (2003) Release of invasive plants from fungal and viral pathogens. Nature 421:625–627

Møller AP, Cassey P (2004) On the relationship between T-cell mediated immunity in bird species and the establishment success of introduced populations. J Anim Ecol 73:1035–1042

Møller AP, Legendre S (2001) Allee effect, sexual selection and demographic stochasticity. Oikos 92:27–34

Møller AP, Dufva R, Allander K (1993) Parasites and the evolution of host social behavior. Adv Study Behav 22:65–102

Møller AP, Christe P, Erritzøe J, Mavarez J (1998) Condition, disease and immune defence. Oikos 83:301–306

Møller AP, Christe P, Lux E (1999) Parasitism, host immune function, and sexual selection. Quart Rev Biol 74:3–20

Møller AP, Christe P, Garamszegi LZ (2005) Coevolutionary arms races: Increased host immune defense promotes specialization by avian fleas. J Evol Biol 18:46–59

Moran NA, vonDohlen CD, Baumann P (1995) Faster evolutionary rates in endosymbiotic bacteria than in cospeciating insect hosts. J Mol Evol 41:727–731

Morand S, Arias-Gonzalez E (1997) Is parasitism a missing ingredient in model ecosystems? Ecol Model 95:61–74

Mostl E, Palme R (2002) Hormones as indicators of stress. Domestic Anim Endocrinol 23:67–74

Mullner A, Linsenmair KE, Wikelski M (2004) Exposure to ecotourism reduces survival and affects stress response in hoatzin chicks (*Opisthocomus hoazin*). Biol Cons 118:549–558

Navarro C, DeLope F, Marzal A, Møller AP (2004) Predation risk, host immune response, and parasitism. Behav Ecol 15:629–635

Nieberding C, Morand S, Libois R, Michaux JR (2004) A parasite reveals cryptic phylogeographic history of its host. Proc R Soc Lond B 271:2559–2568

O'Brien SJ (1994) A role for molecular genetics in biological conservation. Proc Natl Acad Sci USA 91:5748–5755

Ostfeld RS, Meffe GK, Pearl MC (2002) Conservation medicine. The birth of another crisis discipline. In: Aguirre AA, Ostfeld RS, House CA, Tabor GM (ed) Conservation medicine. Ecological health in practice. Oxford Univ Press, Oxford, pp 17-26

Palme R, Rettenbacher S, Touma C, El-Bahr SM, Mostl E (2005) Stress hormones in mammals and birds. Comparative aspects regarding metabolism, excretion, and noninvasive measurement in fecal samples. Trends Comp Endocrinol Neurobiol 1040:162–171

Parham P (1999) Virtual reality in the MHC. Immunol Rev 167:5–15

Poulin R (1999) The functional importance of parasites in animal communities: Many roles at many levels? Int J Parasitol 29:903–914

Prenter J, MacNeil C, Dick JTA, Dunn AM (2004) Roles of parasites in animal invasions. Trends Ecol Evol 19:385–390

Richman A (2000) Evolution of balanced genetic polymorphism. Mol Ecol 9:1953–1963

Romero LM (2004) Physiological stress in ecology: Lessons from biomedical research. Trends Ecol Evol 19:249–255

Sasal S, Durand P, Faliex E, Morand S (2000) Experimental approach to the importance of parasitism in biological conservation. Marine Ecol Progr Ser 198:293–302

Scott ME (1987) Regulation of mouse colony abundance by *Heligmosomoides polygyrus* (Nematoda). Parasitology 95:111–129

Scott ME, Dobson AP (1989) The role of parasites in regulating host abundance. Parasitol Today 5:176–183

Smulders MJ, Snoek LB, Booy G, Vosman B (2003) Complete loss of MHC genetic diversity in the common hamster (*Cricetus cricetus*) population in the Netherlands. Consequences of conservation strategies. Cons Genetics 4:441–451

Sommer S (2003) Effects of habitat fragmentation and changes of dispersal behaviour after a recent population decline on the genetic variability of noncoding and coding DNA of a monogamous Malagasy rodent. Mol Ecol 12:2845–2851

Sommer S (2005) The importance of immune gene variability (MHC) in evolutionary ecology and conservation. Front Zool 2:1–34

Spielman D, Brook BW, Briscoe DA, Frankham R (2004) Does inbreeding and loss of genetic diversity decrease disease resistance? Cons Genetics 5:439–448

Stanko M, Miklisova D, Goüy de Bellocq J, Morand S (2002) Mammal density and patterns of ectoparasite species richness and abundance. Oecologia 131:289–295

Stephens PA, Sutherland WJ (1999) Consequences of the Allee effect for behaviour, ecology and conservation. Trends Ecol Evol 14:401–405

Stephens PA, Sutherland WJ, Freckleton RP (1999) What is the Allee effect? Oikos 87:185–190

Stork NE, Lyal CHC (1993) Extinction or co-extinction rates. Nature 366:307–307
Suorsa P, Huhta E, Nikula A, Nikinmaa M, Jantti A, Helle H, Hakkarainen H (2003) Forest management is associated with physiological stress in an old-growth forest passerine. Proc R Soc Lond B 270:963–969
Suorsa P, Helle H, Koivunen V, Huhta E, Nikula A, Hakkarainen H (2004) Effects of forest patch size on physiological stress and immunocompetence in an area-sensitive passerine, the Eurasian treecreeper (*Certhia familiaris*): An experiment. Proc R Soc Lond B 271:435–440
Sures B, Siddall R, Taraschewski H (1999) Parasites as accumulation indicators of heavy metal pollution. Parasitol Today 15:16–21
Tompkins DM, Draycott RAH, Hudson PJ (2000) Field evidence for apparent competition mediated via the shared parasites of two gamebird species. Ecol Lett 3:10–14
Tompkins DM, Greenman JV, Hudson PJ (2001) Differential impact of a shared nematode parasite on two gamebird hosts: Implications for apparent competition. Parasitology 122:187-193
Tompkins DM, Parish DMB, Hudson PJ (2002a) Parasite-mediated competition among red-legged partridges and other lowland gamebirds. J Wildl Manag 66:445–450
Tompkins DM, Sainsbury AW, Nettleton P, Buxton D, Gurnell J (2002b) Parapoxvirus causes a deleterious disease in red squirrels associated with UK population declines. Proc R Soc Lond B 269:529–533
Torchin ME, Lafferty KD, Kuris AM (2002) Parasites and marine invasions. Parasitology 124:S137–S151
Torchin ME, Lafferty KD, Dobson AP, McKenzie VJ, Kuris AM (2003) Introduced species and their missing parasites. Nature 421:628–630
Torres J, Miquel J, Casanova J-C, Ribas A, Feliu C, Morand S (2006) Parasite species richness of Iberian carnivores: Influences of host density and range distribution. Biodivers Cons (in press)
VanRiper C, VanRiper SG, Goff ML, Laird M (1986) The epizootiology and ecological significance of malaria in Hawaiian land birds. Ecol Monogr 56:327–344
Viggers KL, Lindenmayer DB, Spratt DM (1993) The importance of diseases in reintroduction program. Widl Res 20:687–698
Vitousek PM, Dantonio CM, Loope LL, Rejmanek M, Westbrooks R (1997) Introduced species: A significant component of human-caused global change. New Zealand J Ecol 21:1–16
von Holst D (1998) The concept of stress and its relevance for animal behavior. In: Møller AP, Milinski M, Slater PJB (eds) Stress and behavior. Acad Press, San Diego, pp 1-132
Weinstock JV, Summers R, Elliott DE (2004) Helminths and harmony. Gut 53:7–9
Whiteman NK, Parker PG (2005) Using parasites to infer host population history: a new rationale for parasite conservation. Anim Cons 8:175–181
Windsor DA (1990) Heavenly hosts. Nature 348:104–104

Windsor DA (1995) Equal rights for parasites. Cons Biol 9:1–2
Zanette L, Doyle P, Tremont SM (2000) Food shortage in small fragments: Evidence from an area-sensitive passerine. Ecology 81:1654–1666
Zhang JZ, Dyer KD, Rosenberg HF (2000) Evolution of the rodent eosinophil-associated RNase gene family by rapid gene sorting and positive selection. Proc Natl Acad Sci USA 97:4701–4706

Part VI. Conclusion

28 Global changes and the future of micromammal-macroparasite interactions

Serge Morand, Robert Poulin and Boris R. Krasnov

1 Introductory remarks

Recent changes and scenarios of future changes pose serious threats for living resources, conservation of biodiversity and health. Global changes are the results of increases in all factors of anthropogenetic origins: atmospheric conditions, land uses, over-exploitation of resources, biotic invasions and pollutants. Global changes affect biodiversity by increasing the rate of extinction, by modifying the functioning of ecosystems and then by affecting the health of plants, animals and humans. The causes and consequences of global changes and global warming are the matter of numerous scientific papers, reviews, books and reports (Vitousek et al. 1997; Ottersen et al. 2001; Stenseth et al. 2002; Walther et al. 2002; Cury and Morand 2004; Lovejoy and Hannah 2005). Several reviews have dealt with the consequences of global changes for host -parasite and pathogen interactions (Shope 1991; Dobson and Carter 1992; Colwell 1996; Daily and Erhlich 1996; Patz et al. 1996, 2000; Sutherst 1998, 2001, 2004; Sutherst et al. 1998; Rogers and Randolph 2000; Wilson 2000; Harvell et al. 1999, 2002; Marcogliese 2001; Kovats et al. 2002; Mouritsen and Poulin 2002; Poulin 2006). This chapter aims to explore how host-parasite interactions, and in particular small mammals and their endo- and ectoparasites, are affected by the rapid and pronounced changes that are affecting our planet.

Daily and Erhlich (1996) used the epidemiological environment concept in an attempt to embed the conditions and processes, both biophysical and social, that may influence the interaction between animals, humans and their activities, and pathogen agents. Here, we follow this approach, as did Randolph (2003), who exemplified this concept with tick-borne zoonoses in the context of global changes. In order to study how changes affect the epidemiology of host-parasite relationships, one must use epidemiological models with the derivation of the basic reproductive number (see Rosà et al. in this volume).

How co-adaptation will be affected by global changes also needs to be questioned. The fundamental asymmetry in host-parasite relationships, i.e. a host faces several parasites whereas a parasite interacts with only one or few hosts, may become exacerbated following global changes with more frequent parasite spill-overs and emerging pathogens.

2 Global changes and parasite-mammal outcomes

2.1 Climate change

Anthropogenic climatic change is now well established, although the magnitude and the regional and local consequences are still difficult to predict (see IPCC reports, Intergovernmental Panel on Climate Change at http://www.ipcc.ch/; IPCC 2001). Temperatures but also precipitation patterns and soil humidity will be affected globally, regionally and locally with both increases in average values and more frequent extreme events (storms, hot events…). Such changes will affect the geographical distributions of animals, plants, pathogens, reservoirs, and vectors, as well as the competitive ability among predators and their prey.

Global climate change is altering the ecology of pathogens and parasites, and the interactions among hosts and their parasites. Climate change is driving the emergence of disease in humans, domestic animals, and wildlife. Indeed, temperature is one of the most important abiotic parameters that may affect parasites at all their life-cycle stages. Temperature affects the release of eggs or larvae by adult worms and adult ectoparasites, embryonic development and hatching rates, longevity of free-living stages, infectivity to intermediate hosts, development in these hosts, infectivity to definitive hosts, time to maturation, and the longevity and mortality of adults (Marcogliese 2001; Poulin 2006). Temperature also plays a key role in host feeding and behavior, host range size and general ecology, and host resistance to infection.

Kutz et al. (2005) have recently shown how a host-parasite system may respond to climate change. They presented an empirical and predictive model to elucidate the impact of climate warming on development rates of a parasitic nematode of muskoxen in the Canadian Arctic, a region that is particularly vulnerable to climate change. Kutz et al. (2005) showed that warming in the Arctic may have already radically altered the transmission dynamics of this parasite with the number of parasite generation increasing due to the lessened detrimental effect of the outside environment. The infection pressure is expected to continue to escalate for muskoxen.

A second example is given by two tick-borne diseases for which the tick vector is affected by climate change, tick-borne encephalitis (TBE) and Lyme disease. The ticks may live for several years and their survival, reproduction rate and activity are affected by seasonal climate, which indirectly influences the risk of disease. Warmer temperatures increase vector and pathogen reproduction and blood feeding activity (Patz et al. 2002). Changes in rainfall, humidity but also large-scale meteorological phenomena such as ENSO (El Niño-Southern Oscillation) may affect the number and the quality of vector breeding sites (Liang et al. 2002). These influences may operate in synergy with other changes in the habitat and in biodiversity. Modifications of the tick's habitat and of the occurrence of animals (including small mammals) that are carriers of the different pathogens result in changes in the spread of tick-borne pathogens. Several studies have shown that in recent decades the tick *Ixodes ricinus*, transmitting Lyme borreliosis and TBE, has spread into higher latitudes (e.g., Sweden) and altitudes (e.g. Czech Republic), and has become more abundant in many places (Daniel et al. 2003; Materna et al. 2005). Climate change in Europe seems likely to facilitate the spread of Lyme borreliosis and TBE into higher latitudes and altitudes, and to contribute to the extension and transmission of these diseases in some new areas. For example, modeling suggests that the disease will no longer persists along the southern edge of its present range but may have new foci in northern parts of Europe (Randolph 2001).

Global warming has a significant impact on host-parasite interactions. However, scenarios of the future of these interactions are not easy to produce without the help of modeling because of the non-linear interactions between temperature and various aspects of transmission.

2.2 Ozone layer depletion and UV

Ozone depletion, which affects mostly high latitude and high altitude ecosystems, will certainly affect host-parasite interactions as it is known that ozone may directly suppress immune responses in various animals (Jones and Wigley 1989).

Penetration of UV radiation in fresh water is expected to increase as a result of warming, acidification, and consequent reduction of dissolved organic carbon (Schindler et al. 1996). This increasing UV penetration consequently may affect intermediate hosts and fresh water stages of trematodes and cestodes. While studies have been done on the harmful effects of UV radiation on free living organisms (Bothwell et al. 1994), little work has been done on the potential effects of UV radiation on parasites and on their hosts exposed to increased UV-B radiation, although UV-B has been shown to cause immunosuppression (Patz et al. 1996). Kiesecker and Blaustein (1995) showed a synergistic interaction between UV-B radiation and pathogens that increases amphibian mortality in nature. How changes in UV radiation will impact the interactions between micromammals and their macroparasites remains to be seen.

2.3 Land use and over-exploitation of living resources

Fragmentation of ecosystems, mostly due to the conversion of forests and wildlife habitats into agriculture, results in human-made island ecosystems. Knowing that true islands are places of high extinction rates, the fragmentation processes are likely to drive many isolated populations and species to extinction.

One other consequence of fragmentation is the increasing contacts among domestic animals, wild animals and humans. Increasing the proximity among these results in the increased transmission of both zoonotic or anthropozoonotic diseases (Blouin et al. 1984; Wallis and Lee 1999; Daszak et al. 2000). For example, some management practices in national parks and natural reserves allowing or favoring multi-usages, in which domestic animals graze the same lands as wild animals, facilitate the spread of diseases among them (Deem et al. 2001). Increasing encroachment of farms on wildlife habitats has also increased the overlap between livestock and wild animals, with the consequence that the vast majority of emerging diseases of livestock have been acquired from wild animals (Cleaveland et al. 2001).

The epidemiological environment is greatly altered by the recent intensification of agriculture, with massive uses of nutrients and pesticides. This may favor rodents as they are important consumers of agricultural crops. Rodents are hosts of many parasites that may cause diseases to humans, including bacteria, viral hemorrhagic fevers, tick-borne encephalites, but also macroparasites such as trematodoses and nematodoses (see Duplantier and Sene; Leirs and Singleton; Casanova et al. in this volume). Agricultural intensification favours rodent outbreaks through the removal of predators and other natural enemies, while supplementing their food supply, with subsequent disease outbreaks in rural human populations.

In many Western countries the epidemiological environment is significantly altered by reforestation and suburbanization. These changes in land uses may have consequences for humans and their domestic animals by increasing the contact between wild animals, and in particular rodents. A good example of this is given by Lyme disease, which has spread over the last decades due to increases in deer populations, and also rodent and tick populations, human recreational activities and contacts with ticks (see Daily and Ehrlich 1996).

2.4 Biotic invasions

Biotic invaders are species that establish in a new area in which they proliferate to the detriment of local species and the environment. They are the biggest ecological outcomes of the global alterations and distributions of the earth's biota due to human transport and commerce (Williamson 1996). As emphasized by Mack et al. (2000), biotic invasions can be compared with epidemics because many important factors in disease epidemiology are common to invasions. These factors are the minimum population size necessary for successful establishment, population growth and the fate of interacting species in the new range.

The movement of parasites, potential vectors, or disease reservoirs is greatly facilitated by intense modern transport. A recent case of transfer of a dangerous vector was the introduction from Asia to the United States of a mosquito (*Aedes albopictus*) capable of transmitting dengue.

Identifying future parasite or vector invaders and vulnerable communities, and taking effective measures to prevent their establishment and their dispersal, are great challenges for conservation biology and ecological health. The features associated with bio-invasions are:

1. Attributes of parasite invaders. Parasite invaders are generally those parasites that are found in high local abundances or prevalences, with direct life-cycles and low host specificity. Local abundances of a parasite species are positively correlated with geographical distribution, with highly abundant parasites found in a high number of host populations (Nieberding et al. 2006). The distribution of parasites with indirect life-cycles is dependent on those of all hosts in the life cycle. Any modification of host distribution will determine if parasites can persist and where they colonize (Dobson and Carper 1992; Magnanou and Morand in this volume). Nevertheless, parasite invaders can have indirect life-cycles such as *Schistosoma mansoni*, *Fasciola hepatica* or *Angiostrongylus* spp. Low host specificity may not be a crucial characteristic of parasite invaders, because a more important characteristic is the ability of the parasite to host switch.

2. Attributes of invading hosts, vectors and reservoirs. Many morphological and ecological characteristics may favour invasion success. The escape from native parasites and predators is one these. The introduced species may have a competitive advantage over local species because in the new area they are released from control by their natural enemies. Mitchell and Power (2003) and Torchin et al. (2003) found that parasitism is significantly reduced in organisms in their introduced range compared to their native range, supporting the "parasite release hypothesis".

Invasive hosts are also at an advantage during invasion when they have evolved strong immune defences in their original range. High immune investment confers a better capacity to control parasites that they may acquire or be in contact with in the introduced range (Lee and Klasing 2004; Møller and Cassey 2004; see Magnanou and Morand in this volume).

3. Community vulnerability. Vacant niches within communities, i.e. low species richness, both in hosts and in their parasites is widely assumed to provide opportunities for the settlement and spread of biotic invaders (Elton 1958). The vacant niches hypothesis suggests that species-poor communities cannot offer biological resistance to invasion (see below, biodiversity loss).

2.5 Pollutants, endocrine distorters, immuno-suppressors, insecticide and drug resistance

Animals are affected by several pollutant stressors that may potentially cause endocrine perturbation (reproductive or stress hormones) and/or immune suppression. Long-term exposures to low-level molecules due to pollution are known to be detrimental. The negative effects of estrogenic agents that disrupt endocrine function and the negative effects of PCBs (Polychlorinated biphenyls) on immunocompetence are now well documented (Yamamoto et al. 1996; O'Hara and Rice 1996).

The massive use of insecticide in agriculture, as well as rodenticide, nematicide, acaricide and other drugs in veterinarian medicine, favors the emergence and spread of resistance in ectoparasites and intestinal helminths. This renders the control of parasites and vector-borne diseases even more difficult.

Animal stress affects the risk of infection and the induced pathogenicity of parasites (Esch et al. 1975). Stress induced by climate change, exacerbated by several anthropogenic environmental effects, may result in immunosuppression and increased susceptibility to parasites (Holmes 1996). This, in turn, may enhance adult parasite growth, fecundity, and survival, thus promoting transmission, pathogenicity, and potential evolutionary responses from the parasites (Holmes 1996; Lafferty and Holt 2003).

2.6 Biodiversity loss, the diversity-disease hypothesis and a dilution effect

The massive ongoing extinction of populations and species is mostly the consequence of land conversion for agricultural purposes (Wilson 1992). Biodiversity loss is usually quantified in terms of the rate of loss of species diversity. A conservative estimate of the global rate of eukaryote species loss is one extinction per hour (Wilson 1992), which exceeds by at least four orders of magnitude the rate of evolution of novel species (Lawton and May 1995). However, biodiversity loss also concerns the reduction of genetic variation among populations and the modification of food-webs linked to the loss of interactions within communities (symbiosis, host-parasite interactions), which may greatly affect the functioning of ecosystems and the goods and services they provide.

Recent evidence indicates that high species diversity may reduce exposure to parasites and pathogens (Keesing et al. 2006). Initial evidence came from plant-pathogens interactions. Decreased plant species diversity has been hypothesized to increase the severity of diseases caused by specialist pathogens, i.e. the diversity-disease hypothesis (Elton 1958). The mechanism hypothesized is that both parasite intensity and spread are inversely related to host species abundance (Mitchell et al. 2002). Decreased diversity allows the remaining species to achieve higher abundance, which facilitates the spread of pathogens specific to these hosts. In an in-situ experiment, Mitchell et al. (2003) showed that three components of global change, i.e. elevated CO_2, nitrogen addition and decreased plant species richness ("diversity"), increased both the abundance and the impact of plant pathogens. The diversity-disease hypothesis in its original formulation has never been tested, to our knowledge, in the case of macroparasites and their hosts.

In the case of vector-borne diseases, it has been suggested that high biodiversity may reduce the risk of disease by a mechanism of "dilution effect". The dilution effect predicts that infection rates among vectors will be lower in highly diverse host communities, where transmission to the competent host reservoir is diluted due to the presence of unsuitable hosts. The dilution effect has been demonstrated in several vector-borne diseases (Ostfeld and Keesing 2000; Schmidt and Ostfeld 2001; Holt et al. 2003; Telfer et al. 2005), and recently in the case of West Nile (Ezenwa et al. 2005).

LoGuidice et al. (2003) tested the dilution effect hypothesis by showing that high host species diversity reduces the infection prevalence of the spirochete *Borrelia burgdorferi*, the agent of Lyme disease, in its tick vector. Higher numbers of host species dilute the effects of the most competent disease reservoir, the white-footed mouse *Peromyscus leucopus*. Species-poor host communities tend to have much more mice, and less other hosts, whereas species-rich communities have mice but many other potential hosts. LoGuidice et al. (2003) demonstrated that some host species are poor reservoirs of *B. burgdorferi*, and thus reduce the prevalence of the disease by providing food for, but rarely infecting, ticks. Important dilution host species include small mammals like squirrels, which are characterized by high tick burdens, low reservoir competence and high population density. When such host species are added to a depauperate community, the prevalence of infection in ticks declines. Some other small mammals, like shrews, are "rescue hosts", which are capable of maintaining a high disease risk when mouse density is low.

3 The use of epidemiological modelling

3.1 Models and the basic reproduction number R_0

The determinants of the reproductive success of a parasite include environmental conditions such as temperature and moisture, the life history traits of the parasite such as its fecundity, the density of the hosts, the level of parasite virulence, the response of host immunity and the resistance of the parasites. All these parameters should be incorporated and quantified in mathematical epidemiological models (see Rosà et al. in this volume).

Mouritsen et al. (2005) have explored the potential consequences of climate change by building a simulation model, and more precisely by studying the consequence of increasing temperature on coastal host-parasite system. They parameterized the model with experimental and field data, and used a scenario in which temperature rises by up to 5 °C, which corresponds to the range predicted by the year 2075 for the area where the data were obtained. They showed that an increase of 3.8 °C in ambient temperature would result in a parasite-induced collapse of the second intermediate host population. Due to the importance of the second intermediate host, their population decline is expected to impact the coastal ecosystem as a whole.

Mathematical models allow us to derive the basic reproduction number (R_0). A parasite must achieve a basic reproduction greater than one in order to establish and spread in a host population. R_0 is by definition the number of secondary cases produced in a population of naïve hosts after the introduction of one primary infected host. There is a critical host population size below which the disease or the parasite cannot establish and spread. This host threshold population size is necessary for the perpetuation of most epidemic diseases. Formulae for R_0 can be derived for macroparasites, and for microparasites transmitted by vectors such as ticks (tick-borne diseases) or fleas, which allow us to theoretically investigate how parasite epidemiology is affected by climate change or other global changes.

The use of R_0 can be illustrated with the case of arthropod-borne diseases such as TBE, Lyme disease or plague. The simplest R_0 formula can be written as follows (Randolph and Rogers 2000):

$$R_0 = \frac{a^2 \beta M p^n}{Nr(-\ln p)} \tag{1}$$

where a is the vector biting rate, M is the vector population size, N is the host population size, r is the infectivity period length in the host, p is the daily survival rate of the vector, and β is the transmission rate.

All these parameters and variables are affected by climate factors, like temperature, as shown in Fig. 1.

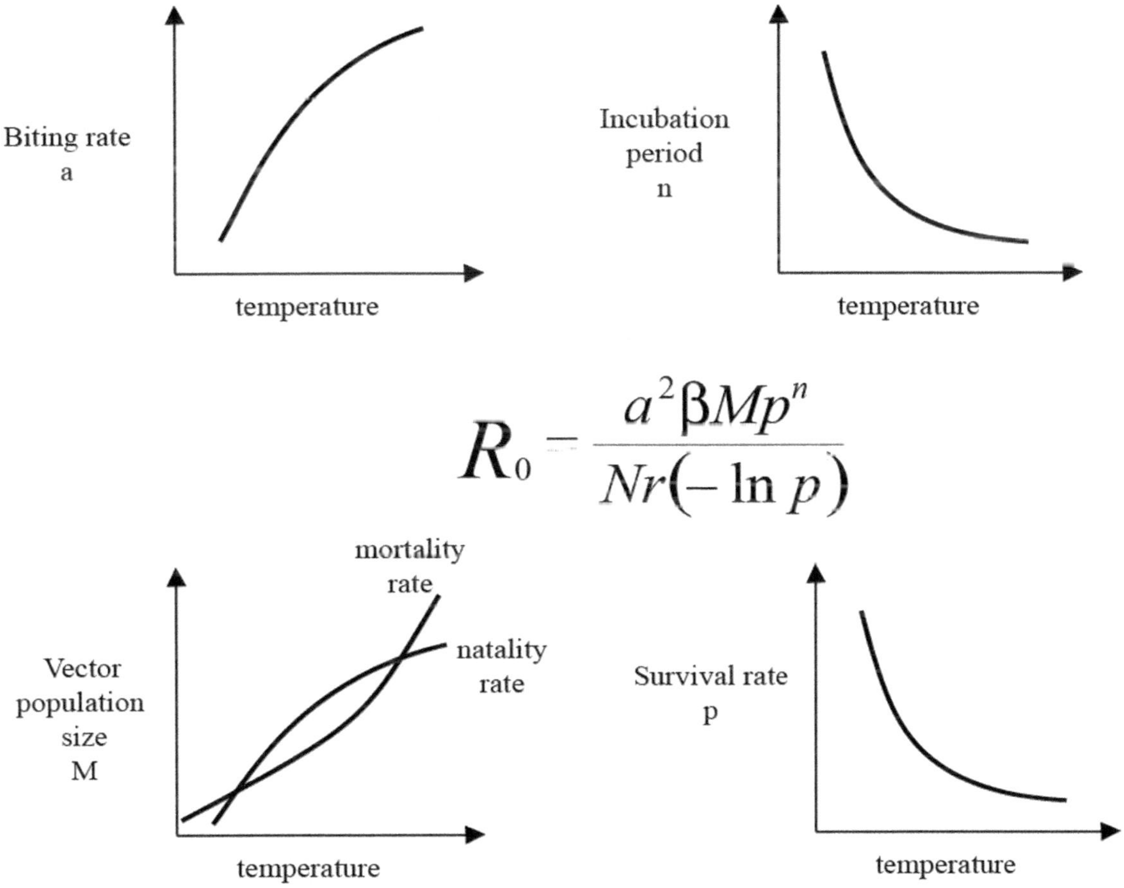

Fig. 1. Effects of climate factors on parameters and variables of the basic reproduction number (R_0) (modified after Randolph 2003)

For example, a shortening of the generation time at higher temperature has been reported for flea species (Krasnov et al. 2001) and for ticks (Randolph 2003). The incubation time, *n*, of the agent within the vector often decreases with increasing temperature. Vector biting rate is positively related to temperature, perhaps in relation to the increase of the metabolic rate. For example, the metabolic rate of fleas is known to increase with increasing temperature (Fielden et al. 2004).However, we should note the highly non-linear interactions among abiotic factors such as temperature and parameters related to the population biology of both hosts and vectors. Moreover, any environmental changes that may affect the population densities of both hosts and vectors are likely affecting R_0.

3.2 Risks of outbreak

The possible magnitude of an infectious disease outbreak is related to both the basic reproduction number R_0 and the numbers of introduced infected hosts (or the number of introduced parasites) (Fig. 2). For pathogens that are minimally transmissible (R_0 close to 0), outbreak size depends largely on the number of introductions of infected hosts or parasites. For highly transmissible pathogens (R_0>>1), outbreak size is determined largely by the size of the susceptible host population. For pathogens that are moderately transmissible (corresponding to $R_0 \approx 1$), notable outbreaks are possible (especially if multiple introductions occur), but the scale of these outbreaks is very sensitive to small changes in R_0 (Woolhouse and Gowtage-Sequeria 2005).

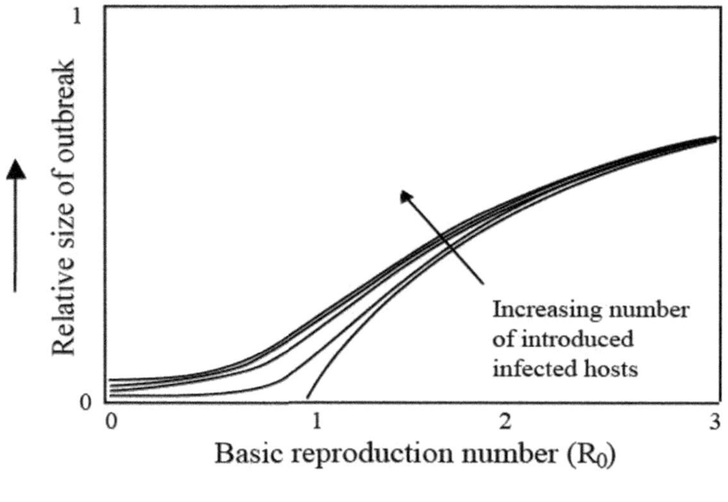

Fig. 2. Relationship between outbreak size (in percent of infected hosts) and two epidemiological parameters, the number of primary cases (or introduced infected hosts) and the basic reproduction number R_0 (modified after Woolhouse and Gowtage-Sequeria 2005)

Any change affecting the nature of the epidemiological environment is likely to modify the host-parasite interaction through (a) the increase of R_0 or (b) the increase of the number of introduced hosts.

3.3 Risks of emergence

The epidemiological environment is affected in a way that humans increase their potential contacts with parasites and pathogens of wildlife (Dobson and Foufopoulos 2001; Woolhouse et al. 2005). Woolhouse and Gowtage-Sequeria (2005) surveyed human pathogen diversity and produced a count of 1,407 human pathogen species. Among all pathogen species, 177 (13%) species were considered as emerging or reemerging. Microparasites such as viruses and bacteria are the most numerous. Only 287 helminth species were recorded with 10 emerging or reemerging (3%). Most emerging or re-emerging helminth parasites originated from rodents (see Casanova et al. in this volume).

Dobson and Foufopoulos (2001) developed matrix models for quantifying R_0 for a variety of potential types of emergent pathogens that may cause outbreaks. In the models, R_0 was highly sensitive to heterogeneities that are created by the spatial structure of the host population and the pathogen's ability to use multiple host species.

4 Coadaptation and maladaptation

With global climate change come alterations in the distribution and abundance of species, and by logical extension, their parasites. The habitat, including the potential host range, of parasites will be modified. Parasites are experiencing range contractions and extensions, not only geographically, but in terms of the number of host species they exploit.

The assemblages of parasites at local scales are likely the results of adaptation to regional environmental conditions. Krasnov et al. (2005) showed that the diversification of flea assemblages on mammals is associated with climatic variables. In warm regions, a greater number of congeneric species per flea assemblage is observed then in colder regions, which may be the result of intrahost speciation. When regional temperature increases, intrahost speciation becomes a relatively more important mode of diversification than acquisition of fleas via host switching.

There should be intense selection pressure on parasites to adapt to new conditions and new hosts or new host populations. Environmental changes imply that parasites will be faced with a new epidemiological environment, with new habitats, and will experience contact with new host species.

The ongoing changing geographic structure of species and worldwide redistribution of hosts and parasites will change the global dynamics of co-evolutionary interactions (Thompson 2005). Local adaptations of parasites to their hosts are the results of evolutionary lags between parasite genotypes and host genotypes (Morand et al. 1996; Lajeunesse and Forbes 2002), that depend on both host and parasite migration (Gandon et al. 1996). Local adaptation then depends of the geographical structure of interactive species, and any alteration of this structure may disrupt adaptive processes, eventually leading to maladaptation.

These alterations may have consequences for different interactions, and are likely to shift the coevolutionary dynamics back toward earlier stages of non-equilibrium dynamics. However, the co-evolutionary dynamics can be changed in a way that makes it very difficult to predict the consequences and the outcomes of new interaction networks (Thompson 2005).

Global changes may also affect the evolution of parasite virulence. Models have demonstrated that the evolution of parasite life-history traits is driven by the age specific mortality rate of the parasite (Morand and Poulin 2000), among other factors. The evolution of virulence may depend on many intrinsic, i.e. the force of the host immune response, and extrinsic factors, such as the survival of free-living infective stages. For example, acceleration in parasite development time in the growing season or lower mortality in certain parasite stages may lead to an increase in virulence. In such global context, it remains extraordinarily difficult to predict the evolutionary direction of virulence, which depends on so many factors that act synergistically and non-linearly.

5 Co-extinction

Global changes will extend the geographical ranges of many species along with their parasites, potentially overlapping with endangered species and driving them to extinction (Dobson and Carper 1992). The seasonal and spatial distributions of parasites and their hosts are often temperature dependent and the synchronicity of their population dynamics is threatened by climatic changes. Many parasites have adapted their life cycles, reproduction, and transmission to overlap with definitive and intermediate hosts (Poulin 2006). Any increase in temporal asynchrony between infective parasite stages and hosts has significant effects on the persistence of the interaction. Some parasites will adapt to these new conditions, whereas others may not. We may expect parasite extinctions to occur.

The global extinction of parasites may be more pronounced than that of free-living animals. According to Poulin and Morand (2004), the 11% of threatened mammalian species imply that a total of 409 helminth species are at risk of extinction. A similar picture emerges in the case of mammalian ectoparasites (Fig. 3), though in this case the orders of mammals at risk are not those harboring the highest ectoparasite diversity. This suggests that there is an additional asymmetry in the host-parasite relationship, one concerning extinction risk.

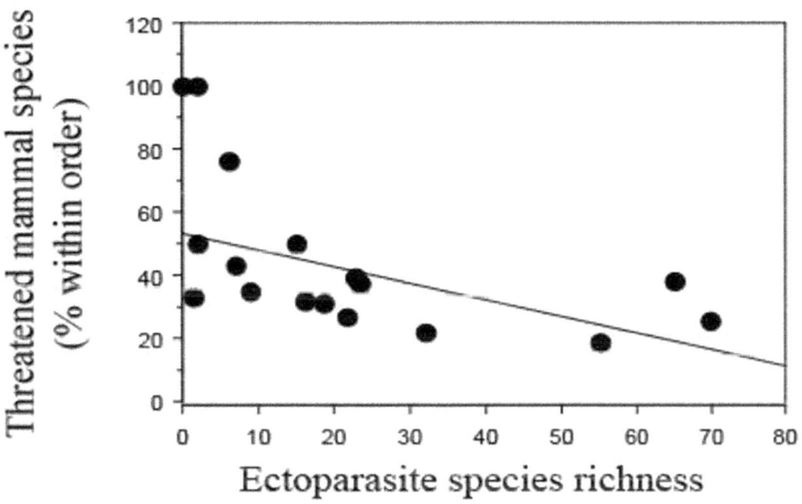

Fig. 3. Relationship between percentage of mammals at risk of extinction and their ectoparasite species richness (data for mammals are from the ICUN Red List, whereas those for ectoparasites are from Kim 1985)

6 Concluding remarks

Escalating human activities, habitat fragmentation and degradation, and increasing the proximity between wildlife and humans have greatly impaired the health of both humans and wildlife (Daszak et al. 2000). These global phenomena are occurring at an unprecedented rate and speed, which creates opportunities for parasites to negatively affect their hosts through emergence, outbreak and higher virulence. Diseases and pathogens are now considered as important factors in the conservation of biodiversity (Meffe 1999; Daszak et al. 2000; Christe et al. in this volume). Several reviews have emphasized the potential importance of parasites and pathogens as mediators of host population dynamics under changing climatic conditions in general, and global warming in particular. Several studies have now emphasized the need to preserve vertebrate biodiversity and community composition in order to significantly reduce the risk of emergence (LoGiudice et al. 2003).

For any host extinction, at least one parasite may go with it. This may not be seen as a negative consequence. Parasites should not only be thought of as harmful, however. Parasites have their own intrinsic value (Sprent 1992; Poulin and Morand 2004). They contribute to maintain high diversity as ecological engineers (Thomas et al. 1999), they may control biotic invaders, and they stabilize ecosystems (Marcogliese and Cone 1997; Mouritsen et al. 2005; Sukhdeo and Hernandez 2005; Arias-Gonzalez and Morand 2006). At the individual host level, parasites may help to equilibrate immune responses and prevent the spread of other parasites. There is then a balance between the need to control parasites and the benefits of their presence.

Many researchers now recognize the need to preserve biodiversity in order to maintain high ecological health (Aguirre et al. 2002). This task needs to be given a higher priority and increase its rate of discovery, in particular by collecting data, and organizing databases on parasites and their hosts (Brooks and Hoberg 2000).

References

Aguirre AA, Ostfeld RS, House CA, Pearl MA, Tabor GM (2002) Conservation medicine: Ecological health in practice. Oxford Univ Press, Oxford

Arias-Gonzalez E, Morand S (2006) Trophic functioning with parasites: A new insight for ecosystem analysis. Mar Ecol Prog Ser (in press)

Blouin EF, Kocan AA, Glenn BL, Kocan KM, Hair JA (1984) Transmission of *Cytauxzoon felis* Kier, 1979 from bobcats, *Felis rufus* (Schreber), to domestic cats by *Dermacentor variabilis* (Say). J Wildl Dis 20:241–242

Bothwell ML, Sherbot DMJ, Pollock CM (1994) Ecosystem response to solar ultraviolet-B radiation: Influence of trophic level interactions. Science 265:97–100

Brooks DR, Hoberg EP (2000) Triage for the biosphere: The need and rationale for taxonomic inventories and phylogenetic studies of parasites. Comp Parasitol 67: 1–25

Cleaveland S, Laurenson MK, Taylor LH (2001) Diseases of humans and their domestic mammals: Pathogen characteristics, host range and the risk of emergence. Phil Trans R Soc Lond B 356:991–999

Colwell RR (1996) Global climate and infectious disease: The cholera paradigm. Science 274:2025–2031

Cury P, Morand S (2004) Marine biodiversity and global change: A dynamic interaction with a human component. In: Barbault R, Chevassus-au-Louis B (eds) Biodiversity and global change. Apdf, Paris, pp 50–79
Daily GC, Ehrlich PR (1996) Global change and human susceptibility to disease. Ann Rev Energy Environ 21:125–144
Daniel M, Danielova V, Kriz B, Jirsa A, Nozicka J (2003) Shift of the tick *Ixodes ricinus* and tick-borne encephalitis to higher altitudes in central Europe. Eur J Clin Microbiol Infect Dis 22:327–328
Daszak P, Cunningham AA, Hyatt AD (2000) Emerging infectious diseases of wildlife: threats to biodiversity and human health. Science 287:433–449
Deem SL, Karesh WB, Weisman W (2001) Putting theory into practice: Wildlife health in conservation. Cons Biol 15:1224–1233
Dobson AP, Carper R (1992) Global warming and potential changes in host-parasite and disease-vector relationships. In: Peters RL, Lovejoy TE (eds) Global warming and biological diversity. Yale Univ Press, New Haven, pp 201–217
Dobson AP, Foufopoulos J (2001) Emerging infectious pathogens of wildlife. Phil Trans R Soc Lond B 356:1001–1012
Elton CS (1958) The ecology of invasions by animals and plants. Methuen, London
Esch GW, Gibbons JW, Bourque JE (1975) An analysis of the relationship between stress and parasitism. Am Midl Nat 93:339–353
Ezenwa VO, Godsey MS, King RJ, Guptill SC (2005) Avian diversity and West Nile virus: Testing associations between biodiversity and infectious disease risk. Proc R Soc Lond B 273:109–117
Fielden LJ, Krasnov BR, Khokhlova IS, Arakelyan MS (2004) Respiratory gas exchange in the desert flea *Xenopsylla ramesis* (Siphonaptera: Pulicidae): Response to temperature and blood-feeding. Comp Biochem Physiol A 137:557–565
Gandon S, Copowietz Y, Dubois Y, Michalakis Y, Olivieri I (1996) Local adaptation and gene-for-gene coevolution in a metapopulation model. Proc R Soc Lond B 263:1003–1009
Harvell CD, Kim K, Burkholder JM, Colwell RR, Epstein PR, Grimes DJ, Hofmann EE, Lipp EK, Osterhaus ADME, Overstreet RM, Porter JW, Smith GW, Vasta GR (1999) Emerging marine diseases-climate links and anthropogenic factors. Science 285:1505–1510
Harvell CD, Mitchell CE, Ward JR, Altizer S, Dobson AP, Ostfeld, RS, Samuel MD (2002) Climate warming and disease risks for terrestrial and marine biota. Science 296:2158–2162
Holmes JC (1996) Parasites as threats to biodiversity in shrinking ecosystems. Biodivers Cons 5:975–983
Holt RD, Dobson AP, Begon M, Bowers RG, Schauber EM (2003) Parasite establishment in host communities. Ecol Lett 6:837–842
IPCC (Intergovernmental Panel on Climate Change) (2001) Climate change 2001. The scientific basis. Cambridge Univ Press, Cambridge

Jones R, Wigley T (1989) Ozone depletion: Health and environmental consequences. John Wiley, New York

Keesing F, Holt RD, Ostfeld RS (2006) Effects of species diversity on disease risk. Ecol Lett 9:485–498

Kiesecker JM, Blaustein AR (1995) Synergism between UV-B radiation and a pathogen magnify amphibian embryo mortality in nature. Proc Natl Acad Sci 92:11049–11052

Kim KC (ed) (1985) Coevolution of parasitic arthropods and mammals. John Wiley, New York

Kovats RS, Campbell-Lendrum DH, McMichael AJ, Woodward A, Cox JSH (2001) Early effects of climate change: Do they include changes in vector-borne disease? Proc R Soc Lond B 356:1057–1068

Krasnov BR, Khokhlova IS, Fielden LJ, Burdelova NV (2001) Development rates of two *Xenopsylla* flea species in relation to air temperature and humidity. Med Vet Entomol 15:249–258

Krasnov BR, Shenbrot GI, Khokhlova IS, Poulin R (2005) Diversification of ectoparasite assemblages and climate: An example with fleas parasitic on small mammals Global Ecol Biogeogr 14:167–175

Kutz SJ, Hoberg EP, Polley L, Jenkins EJ (2005) Global warming is changing the dynamics of Arctic host-parasite systems. Proc R Soc Lond B 272:2571–2954

Lafferty KD, Holt RD (2003) How should environmental stress affect the population dynamics of disease? Ecol Lett 6:654–664

Lajeunesse MJ, Forbes MK (2002) Host range and local parasite adaptation. Proc R Soc Lond B 269:703–710

Lawton JH, May RM (1995) Extinction rates. Oxford Univ Press, Oxford

Lee KA, Clasing KC (2004) A role for immunology in invasion biology. Trends Ecol Evol 19:523–529

Liang SY, Linthicum KJ, Gaydos JC (2002) Climate change and the monitoring of vector-borne disease. J Am Med Assoc 287:2286

LoGiudice K, Ostfeld RS, Schmidt KA, Keesing F (2003) The ecology of infectious disease: Effects of host diversity and community composition on Lyme disease risk. Proc Natl Acad Sci USA 100:567–571

Lovejoy TE, Hannah L (2005) Climate change and biodiversity. Yale Univ Press, New Heaven

Mack RN, Simberloff D, Lonsdale WM, Evans H, Clout M, Bazzaz FA (2000) Biotic invasions: Causes, epidemiology, global consequences, and control. Ecol Appl 10:689–710

Marcogliese DJ (2001) Implications of climate change for parasitism of animals in the aquatic environment. Can J Zool 79:1331–1352

Marcogliese DJ, Cone DK (1997) Food webs: A plea for parasites. Trends Ecol Evol 12:320–325

Materna J, Daniel M, Danielova V (2005) Altitudinal distribution limit of the tick *Ixodes ricinus* shifted considerably towards higher altitudes in central Europe: results of three years monitoring in the Krkonose Mts. (Czech Republic). Cent Eur J Public Health 13:24–28

Meffe GK (1999) Conservation medicine. Cons Biol 13:953

Mitchell CE, Power AG (2003) Release of invasive plants from fungal and viral pathogens. Nature 421:625–627
Mitchell CE, Tilman D, Groth JV (2002) Effects of grassland species diversity, abundance, and composition on foliar fungal disease. Ecology 83:1713–1726
Mitchell CE, Reich PB, Tilman D, Groth JV (2003) Effects of elevated CO_2, nitrogen deposition, and decreased species diversity on foliar fungal plant disease. Global Change Ecol 9:438–451
Møller AP, Cassey P (2004) On the relationship between T-cell mediated immunity in bird species and the establishment success of introduced populations. J Anim Ecol 73:1035–1042
Morand S, Poulin R (2000) Optimal time to patency in parasitic nematodes: Host mortality matters. Ecol Lett 3:186–190
Morand S, Manning SD, Woolhouse MEJ (1996) Parasite-host coevolution and geographic patterns of parasite infectivity and host susceptibility. Proc R Soc Lond B 263:119–128
Mouritsen KN, Poulin R (2002) Parasitism, climate oscillations and the structure of natural communities. Oikos 97:462–468
Mouritsen KN, Tompkins DE, Poulin R (2005) Climate warming may cause a parasite-induced collapse in coastal amphipod populations. Oecologia 146:476–483
Nieberding C, Goüy De Bellocq J, Morand S (2005) The biogeography of host-parasite interactions: From nested assemblages to comparative phylogeography. Mammal Study 30:S87–S93
O'Hara TM, Rice CD (1996) Polychlorinated biphenyls. In: Fairbrother A, Locke LN, Hoff GL (eds) Noninfectious diseases of wildlife. Iowa State Univ Press, Ames, pp 71–86
Ostfeld RS, Keesing F (2000) Biodiversity and disease risk: The case of Lyme disease. Conser Biol 14:722–728
Ottersen G, Planque B, Belgrano A, Post E, Reid PC, Stenseth NC (2001) Ecological effects of the North Atlantic oscillation. Oecologia 128:1–14
Patz JA, Epstein PR, Burke TA, Balbus JM (1996) Global climate change and emerging infectious diseases. J Am Med Assoc 275:217–223
Patz JA, Graczyk TK, Geller N, Vittor AY (2000) Effects of environmental change on emerging parasitic diseases. Int J Parasitol 30:1395–1405
Patz JA, Hulme M, Rosenzweig C, Mitchell TD, Goldberg RA, Githeko AK, Lele S, McMichael AJ, Le Sueur D (2002) Climate change: Regional warming and malaria resurgence. Nature 420:627–628
Poulin R (2006) Global warming and temperature-mediated increases in cercarial emergence in trematode parasites. Parasitology 132:143–151
Poulin R, Morand S (2004) Parasite biodiversity. Smithsonian Inst Press, Washington
Randolph SE (2001) The shifting landscape of tick-borne zoonoses: Tick-borne encephalitis and Lyme borreliosis in Europe. Proc R Soc Lond B 356:1045–1056
Randolph SE (2003) Wildlife, climate and politics: Possible causes of the recent increases in tick-borne zoonoses. J Parasitol 89:S69–S71

Randolph SE, Rogers DJ (2000) Fragile transmission cycles of tick-borne encephalitis virus may be disrupted by predicted climate change. Proc R Soc Lond B 267:1741–174

Rogers DJ, Randolph SE (2000) The global spread of malaria in a future, warmer world. Science 289:1763–1766

Schindler DW, Curtis PJ, Parker BR, Stainton MP (1996) Consequences of climate warming and lake acidification for UVB penetration in North American boreal lakes. Nature 379:705–708

Schmidt KA, Ostfeld RS (2001) Biodiversity and the dilution effect in disease ecology. Ecology 82:609–619

Shope R (1991) Global climate change and infectious diseases. Environ Health Perspect 96:171–174

Sprent JFA (1992) Parasites lost? Int J Parasitol 22:139–151

Stenseth NC, Mysterud A, Ottersen G, Hurrell JW, Chan K-S, Lima M (2002) Ecological effects of climate fluctuations. Science 297:1292–1296

Sukhdeo MVK, Hernandez AD (2005) Food webs patterns and the parasite perspective. In: Thomas F, Renaud F, Guégan JF (eds) Parasitism and ecosystems. Oxford Univ Press, Oxford, pp 54–67

Sutherst RW (1998) Implications of global change and climate variability for vector-borne diseases: Generic approaches to impact assessments. Int J Parasitol 28:935–945

Sutherst RW (2001) The vulnerability of animal and human health to parasites under global change. Int J Parasitol 31:933–948

Sutherst RW (2004) Global change and human vulnerability to vector-borne diseases. Clin Microb Rev 17:136–173

Sutherst RW, Ingram JSI, Scherm H (1998) Global change and vector-borne diseases. Parasitol Today 14:297–299

Telfer S, Brown KJ, Sekules R, Begon M, Haydon T, Birtles R (2005) Disruption of a host-parasite system following the introduction of an exotic species. Parasitology 130:661–668

Thomas F, Poulin R, De Meeüs T, Guégan JF, Renaud F (1999) Parasites and ecosystem engineering: What roles could they play? Oikos 84:167–171

Thompson JN (2005) The geographic mosaic of coevolution. Univ Chicago Press, Chicago

Torchin ME, Lafferty KD, Dobson AP, McKenzie VJ, Kuris AM (2003) Introduced species and their missing parasites. Nature 421:628–630

Vitousek PM, Mooney, HA, Lubchenco J, Melillo JM (1997) Human domination of Earth's ecosystems. Science 277:494–499

Wallis J, Lee DR (1999) Primate conservation: The prevention of disease transmission. Int J Primatol 20:803–826

Walther G-R, Post E, Convey P, Menzel A, Parmesan C, Beebee TJC, Fromentin J-M, Hoegh-Guldberg O, Bairlein F (2002) Ecological responses to recent climate change. Nature 416:389–395

Williamson M (1996) Biological invasions. Chapman and Hall, London

Wilson ME (2000) Environmental change and infectious diseases. Ecosyst Health 6:8–12

Wilson EO (1992) The diversity of life. Harvard Univ Press, Cambridge
Woolhouse MEJ, Gowtage-Sequeria S (2005) Host range and emerging and re-emerging pathogens. Emerg Infect Dis 11:1842–1847
Woolhouse MEJ, Haydon DT, Antia R (2005) Emerging pathogens: The epidemiology and evolution of species jumps. Trends Ecol Evol 20:238–44
Yamamoto JT, Donohoe RM, Fry RM, Golub MS, Donal JM (1996) Environmental estrogens: Implications for reproduction in wildlife. In: Fairbrother A, Locke LN, Hoff GL (eds) Noninfectious diseases of wildlife. Iowa State Univ Press, Ames, pp 31-51

Index

GPSR Compliance

The European Union's (EU) General Product Safety Regulation (GPSR) is a set of rules that requires consumer products to be safe and our obligations to ensure this.

If you have any concerns about our products, you can contact us on ProductSafety@springernature.com

In case Publisher is established outside the EU, the EU authorized representative is:

Springer Nature Customer Service Center GmbH
Europaplatz 3
69115 Heidelberg, Germany

Batch number: 10762625

Printed by Printforce, the Netherlands